7th International Conference on Cooling and Heating Technologies (ICCHT 2014)

IOP Conference Series: Materials Science and Engineering Volume 88

Selangor, Malaysia
4 – 6 November 2014

ISBN: 978-1-5108-1311-3
ISSN: 1757-8981

Printed from e-media with permission by:

Curran Associates, Inc.
57 Morehouse Lane
Red Hook, NY 12571

Some format issues inherent in the e-media version may also appear in this print version.

Printed by Curran Associates, Inc. (2015)

Institute of Physics
Dirac House, Temple Back
Bristol BS1 6BE UK

Phone: 44 1 17 929 7481
Fax: 44 1 17 920 0979

techtracking@iop.org

Additional copies of this publication are available from:

Curran Associates, Inc.
57 Morehouse Lane
Red Hook, NY 12571 USA
Phone: 845-758-0400
Fax: 845-758-2634
Email: curran@proceedings.com
Web: www.proceedings.com

TABLE OF CONTENTS

IOP Conf. Series: Materials Science and Engineering **88** (2015) 011001 doi:10.1088/1757-899X/88/1/011001

7th INTERNATIONAL CONFERENCE ON COOLING & HEATING TECHNOLOGIES 2014 (ICCHT 2014)

Sustainable And Innovation in Heating & Cooling Technologies

4th – 6th November 2014

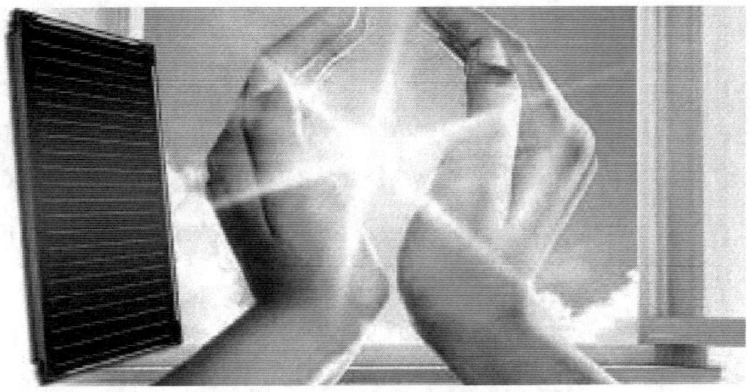

GRAND DORSETT SUBANG HOTEL
Jalan SS 12/1
47500 Subang Jaya
Selangor Darul Ehsan

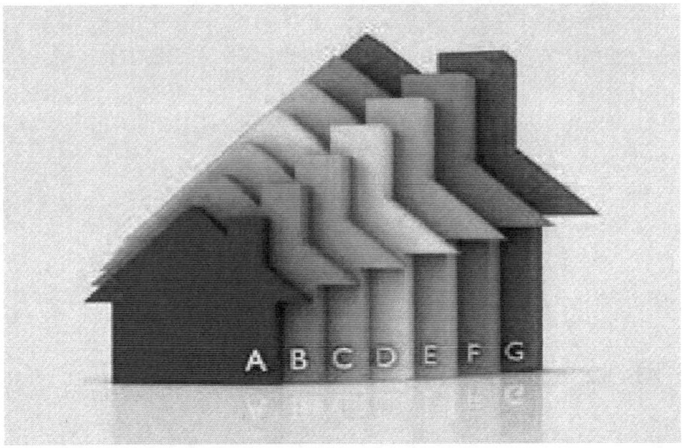

Organised by:

MECHANICAL ENGINEERING TECHNICAL DIVISION
THE INSTITUTION OF ENGINEERS, MALAYSIA

Endorsing Organizations:

Board of Engineers, Malaysia

Jabatan Kerja Raya Malaysia

Construction Industry Development Board

IOP Conf. Series: Materials Science and Engineering **88** (2015) 011001 doi:10.1088/1757-899X/88/1/011001

INTRODUCTION

The Kyoto protocol has initiated a pledge from almost all developing and developed countries to be committed to reducing CO_2 emissions. Development of new renewable energy technologies are also of interest in this conference. Greenhouse gases have contributed to global warming and other man-made disasters. Cooling and Heating communities also have responsibilities towards the commitment of reducing the greenhouse gas emissions. In addition, depleting natural resources also act as a threat to the Cooling and Heating industries, causing them to develop highly efficient equipment and innovative technologies.

The 1st International Conference on Cooling & Heating Technologies was held in Hanoi Vietnam (Jan. 2005). Whereas the 2nd, 3rd, 4th and 5th ICCHT conferences were held in Dalian, China (Jul. 2006), Tokyo, Japan (Jul. 2007), Jinhae, Korea (Oct. 2008) and Bandung, Indonesia (Dec. 2010) respectively. The 6th International Conference on Cooling & Heating Technologies (ICCTH2012) was held in Xi'an in China on November 9-12, 2012.

It is our pleasure to welcome you to the 7th International Conference on Cooling & Heating Technologies (ICCTH2014) on 4th – 6th November 2014 at the Grand Dorsett Subang Hotel, Subang Jaya, Selangor Darul Ehsan, Malaysia

The Theme of the Conference is “Sustainability and Innovation in Heating & Cooling Technologies". The sub-themes are:-

- CO2 Reduction and Low Carbon Technologies
- HVAC System and Natural Ventilation
- Energy & Alternative Energy
- Computational Fluid Dynamics
- Low Temperature & Refrigeration Engineering

EXHIBITION

In conjunction with the Conference, an Exhibition will be organized as an integral part of the Conference. Project experiences, product solutions, new applications and state-of-the-art information will be highlighted.

IOP Conf. Series: Materials Science and Engineering **88** (2015) 011001 doi:10.1088/1757-899X/88/1/011001

ICCHT 2014 – ORGANISING COMMITTEE

Chairman: Ir. Fam Yew Hin

Advisor: Y.Bhg Datuk Ir. Prof. Dr Ow Chee Sheng

Members:
Ir. Noor Hisham Yahaya
Ir. Dr. Cheong Thiam Fook
Ir. Gopal Narian Kutty
Ir. Luk Chau Beng
Ir. Dr. Kannan M. Munisamy
Ir. Dr. Tan Chee Fai
Dr. Chan Hoy Yen

Co-chairmen:

Prof. Maogang He, Xi'an Jiaotong University, China
Prof. S.H. Winoto, National University of Singapore, Singapore
Prof. Aryadi Suwono, Institut Teknologi Bandung, Indonesia
Prof. Hanshik Chung, Gyeongsang National University, Korea

Co-organising Committee Members:

Maogang He, Xi'an Jiaotong University, China
Yongping Yang, North China Electric Power University, China
Hao Zhou, Zhejiang University, China
Y.C.Ahn, Pusan National University, Korea
H. Abdurrachim, Institut Teknologi Bandung, Indonesia
E. Erham, Politeknik Negeri Bandung, Indonesia

IOP Conf. Series: Materials Science and Engineering **88** (2015) 011002 doi:10.1088/1757-899X/88/1/011002

Peer review statement

All papers published in this volume of *IOP Conference Series: Materials Science and Engineering* have been peer reviewed through processes administered by the proceedings Editors. Reviews were conducted by expert referees to the professional and scientific standards expected of a proceedings journal published by IOP Publishing.

On generating counter-rotating streamwise vortices

S H Winoto[1], H Mitsudharmadi[2], A C Budiman[1], S M Hasheminejad[1], T Nadesan[1], Tandiono[3], H T Low[1], and T S Lee[1]

[1*]formerly at National University of Singapore, Singapore 119260, now retired.
[1]National University of Singapore, Singapore 119260.
[2]formerly at Temasek Laboratory at National University of Singapore, Singapore 117411, now at Clean Combustion Research Center, KAUST, Thuwal 23955, Saudi Arabia.
[3]Institute of High Performance Computing, Singapore 138632.

E-mail: jacky.mucklow@iop.org

Abstract. Counter-rotating streamwise vortices are known to enhance the heat transfer rate from a surface and also to improve the aerodynamic performance of an aerofoil. In this paper, some methods to generate such counter-rotating vortices using different methods or physical conditions will be briefly considered and discussed.

1. Introduction

A well known method to enhance the heat transfer rate from a surface or to improve the aerodynamic performance of an aerofoil is by generating counter-rotating streamwise vortices in the surface or aerofoil boundary layer. Such method has recently attracted much interest due to the potential applications in fluid and thermal engineering. In this paper, some types of counter-rotating streamwise vortices which are generated using different methods or conditions will be briefly considered and discussed

2. Caused by centrifugal instability

The well known Görtler vortices which occur in concave surface boundary layer flows, as sketched in Fig.1, are in fact a system of counter-rotating streamwise vortices caused by centrifugal instability [1]. These vortices will appear if the Görtler number G_θ, as defined by [2]:

$$G_\theta = (U_\infty \theta/\nu)\sqrt{\theta/R} \tag{1}$$

is greater than a critical value $G_{\theta cr}$, where ν is the fluid kinematic viscosity, θ the momentum thickness based on Blasius flat plate boundary-layer solution, U_∞ the free-stream velocity, and R the concave surface radius of curvature. The vortices will grow downstream resulting in a three-dimensional boundary-layer flow due to streamwise momentum distribution which causes spanwise variation in the boundary-layer thickness, due to the formation of the so-called "upwash" region, where low momentum fluid moves away from the surface, and the "downwash" region, where high speed outer fluid moves towards the surface (Fig. 1). At "upwash" regions, the boundary-layer is thicker and the shear stress is lower than those at "downwash" regions.

[1] To whom any correspondence should be addressed.

The effects of Görtler vortices on boundary layer development, heat transfer and deposition (for the case of turbine blades) cannot be ignored since concave surfaces exist in many fluid engineering applications.

Since the analytical work of Görtler [1], the early experimental works were mainly focused on visualizing these vortices to confirm their existence, as reviewed by Winoto *et al.* [3].

The heat transfer effect of Görtler vortices was first investigated by McCormack *et al.* [4] who reported 100 to 150% increase in Nusselt number on a concave surface in the presence of such vortices, compared with a reference flat plate. An empirical relationship for laminar heat transfer enhancement by the vortices was also proposed based on measurements on a turbine cascade [5]. The more recent workson the thermal effects of Görtler vortices were done by Crane & Sabzvari [6] , Crane & Umur [7], and Momayez *et al.* [8, 9].

Numerical works on Görtler vortex instability were initially focused to find a unique stability curve. Floryan and Saric [10] found that the neutral curve appears to asymptotically level off at $G_{\theta,cr}$ = 0.4638 which can be considered as a critical value. Finnis & Brown [11] found that the minimum point of the unstable region occurs at $G_\theta = 1.38$ for the dimensionless wave number $\alpha\theta = 0.28$. Different values of $G_{\theta,cr}$ were also obtained from experimental works. Moreover, Kottke & Mpourdis [12] did not detect any sign of instability when the screens that act as a source of disturbance were placed sufficiently far upstream. The above results show that the concept of a unique stability curve is not tenable in Görtler problem. The growth, as well as the wavelength selection mechanism of Görtler vortices, is fully governed by the receptivity process, as discussed by Denier *et al.* [13] and Bassom & Hall [14].

Assuming the most amplified vortices will occur in an experiment, a method based on the Görtler vortex stability diagram of Smith [2], for example, can be used to predict the experimental wavelength of Görtler vortices, for which the non-dimensional wavelength parameter Λ is defined as:

$$\Lambda = (U_\infty \lambda_m/\nu)\,(\lambda_m /R)^{0.5} \tag{2}$$

where λ_m is the most amplified Görtler vortex wavelength, and Λ represents a family of straight lines which cross the Görtler vortex stability diagram of $G_\theta$versus $\alpha\theta$ where $\alpha\theta$ is called dimensionless wave number and α (= $2\pi/\lambda$) is called wave number. Luchini & Bottaro [15] found that the most amplified wavelengths are for Λ = 220 to 270, while Floryan [16], Smith [2], and Meksyn [17] respectively proposed Λ = 210, 272, and 227.

Since the wavelengths of naturally developed Görtler vortices are not uniform, experimental studies were biased due to the choice of "good" pairs of such vortices. Hence, Peerhossaini & Bahri [18], Ajakh *et al.* [19], Toe *et al.* [20] and Mitsudharmadi *et al.* [21, 22] used a series of thin wires placed upstream and perpendicular to the concave surface leading edge to pre-set or "force" the wavelength of Görtler vortices to be uniform and equal to the spanwise distance between the wires.

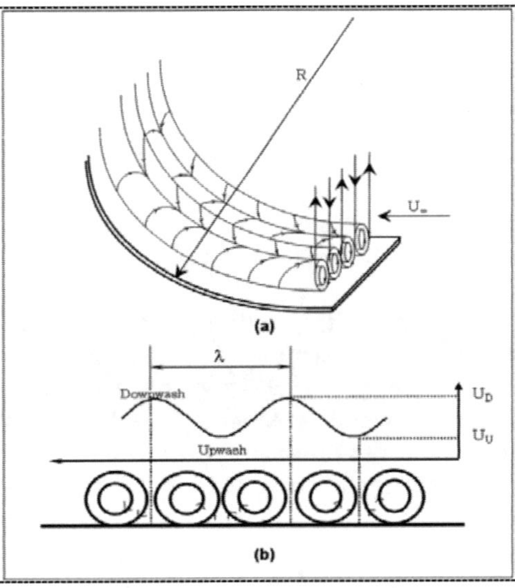

Figure 1. Sketch of Görtler vortices on a concave surface: (a) Flow configuration, and (b) Spanwise distribution of mean streamwise velocity U showing the definitions of vortex wavelength λ, downwash, and upwash regions.

Using hot-wire anemometer measurement, Mitsudharmadi *et al.* [23] were probably the first who experimentally showed the splitting and merging of Görtler vortices. The splitting of vortices occurred when the pre-set vortex wavelength was larger than the most amplified vortex wavelength λ_m and the merging occurred when the pre-set vortex wavelength was smaller than λ_m. Mitsudharmadi *et al.* [24] also experimentally confirmed the existence of its secondary instability.

Tandiono *et al.* [25] also used hot-wire anemometry to study the linear and non-linear development of Görtler vortices and later investigated the development of wall shear stress in concave surface boundary layer in the presence of Görtler vortices [26]. The wall shear stress τ_w was "measured" or estimated by near-wall hot-wire measurements using the near-wall velocity gradient technique [27] on concave surface of 1.0 m radius of curvature in the presence of controlled Görtler vortices for three different. It was found that τ_w at downwash decreases at a slightly lower rate than the Blasius curve, and increases after its minimum point. In contrast, τ_w at upwash decreases at a higher rate than the Blasius curve. The minimum τ_w is found to be 59% of the Blasius value at that position. After its minimum point, τ_w increases slightly due to the secondary instability as the onset of secondary instability is just slightly before the location of the minimum τ_w at the upwash [28]. The spanwise-averaged wall shear stress coefficient $\overline{C_f}$, which initially follows the Blasius curve, increases well above the local turbulent boundary layer value farther downstream due to the nonlinear effects of Görtler vortex instability and the secondary instability modes.

Görtler type vortices can also appear on flow over corrugated plates [29]. Flow in a channel with one of the simplest corrugation geometry of a sinusoidal plate is studied for which a corrugated plate with amplitude of 7.5 mm and wavelength of 78 mm, is mounted into a small wind tunnel. A flat plate is also mounted with the gap between these two plates is fixed at H = 50 mm, as shown in Fig. 2. Layer of smoke is generated using smoke-wire flow visualization technique for which droplets of paraffin oil through the electrically heated thin wire and green laser sheet is mounted perpendicular to the channel. A digital camera is placed at certain distance after the exit of the wind tunnel to capture the visualization images.

IOP Conf. Series: Materials Science and Engineering **88** (2015) 012001 doi:10.1088/1757-899X/88/1/012001

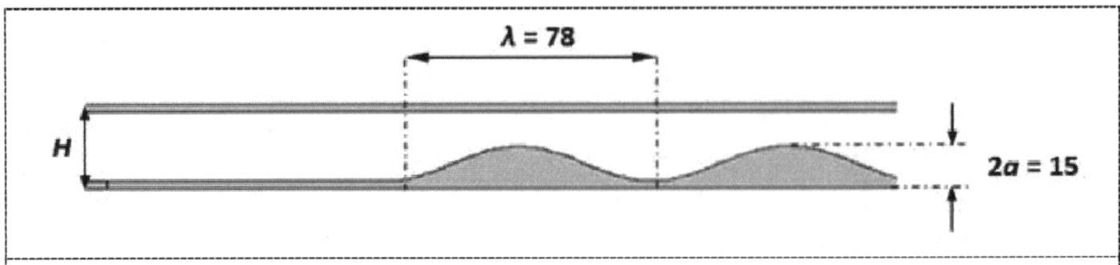

Figure 2. Side view of the channel used in the experiment [30].

From the smoke-wire flow visualization, the vortices can be found at the first peak of corrugation. In Fig. 3, for Re (= U_∞.½H/ ν) = 3350, at least two vortex structures are shown in the midspanwise plane. However, these naturally-developed vortex structures are not really similar since they have different spanwise wavelengths. Nevertheless, both structures show the stem and head of mushroom-like structures. The stem is related to the low momentum fluid that pulled out from the wall and returned back toward the wall with maximum shear and formed the mushroom hat [21]. Similar to the concave surface case, these two structures refer to the most amplified counter-rotating vortices.

Non-uniformity in size and in spanwise wavelength of such vortices will cause experimental bias in studying the phenomenon [31]. Thus, to obtain a more uniform size and wavelength for such vortices, for example as shown in Fig. 4, it is necessary to pre-set the vortex wavelength by some geometric patterns like triangles on the flat plate leading edge. This patterned leading edge is used as a substitute for perturbation wires used in some earlier works, such as by Peerhossaini & Bahri [18] and Ajakh *et al.* [19].

From the visualization, the downstream development of the vortices is also examined. After the first peak, separation bubble formed at the valley of corrugation and the vortices were floating above it. These vortices remained visible until the second valley, where the mixing of the flow was improved and the structures broke down prior to turbulence. As the Reynolds number increased, the onset of turbulence is moved upstream. While for Re = 3350 the vortices remain visible up to the second peak, for Re = 5300 the vortex structures can be clearly observed only at the first peak.

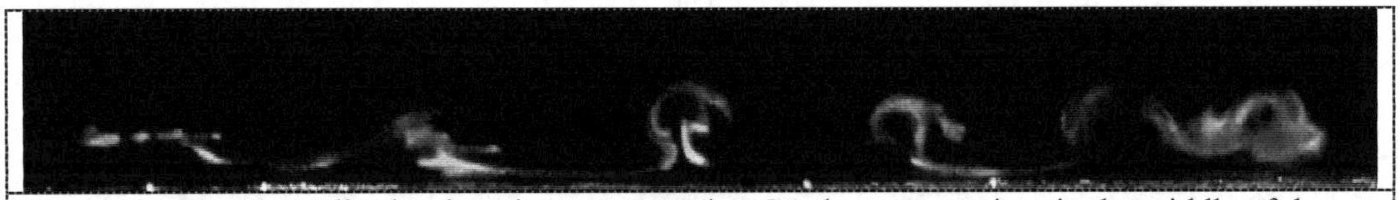

Figure 3. Two naturally-developed counter-rotating Görtler type vortices in the middle of the spanwise plane at the first peak of corrugation, caused by the centrifugal effect from sinusoidal plate. In this case, Re = 3350.

Figure 4. Counter-rotating Görtler type vortices with nearly uniform spanwise wavelength at the first peak of sinusoidal plate with the modified leading edge, Re = 3350 (taken from [30]).

3. Caused by leading edge patterns

Some simple geometric patterns on the leading edge of a flat plate or an airfoil, patterns, such as, saw tooth shape, can generate counter-rotating streamwise vortices in the surface boundary layer flow which can improve momentum exchange which results ina better mixing zone and increased heat transfer.

The use of leading edge pattern was actually inspired by nature, for example, the owl wings generate mini vortices by comb like feathers, so to retard the boundary layer separation, and consequently, give this predator the ability of quietly approaching its prey by high stall angle of attack [32].These vortices were interpreted like vortex lift over the delta wings, which create low pressure region over the surface. As the angle of attack increases, the vortex lift increases until it breaks down.

Studies by Fish and Battle [33] on humpback whales' pectoral flipper demonstrated that the flipper has high aspect ratio, symmetric profile with large sinusoidal tubercles (protuberances on the leading edge) which can cause a considerable increase in the stall angle of attack and the maximum lift coefficient. The combination of tubercles and the high aspect ratio of the flippers allow humpback whale to perform sharp banking turns lead to a rise in proportion of lift to drag coefficients [34] despite of its huge and inflexible body.

These examples help to better understand the actual mechanism of the leading edge shape on generating vortices. Soderman [35] examined the effect of attached saw-tooth to the leading edge of a 2-D airfoil and found that this type of leading edge can reduce the separation and drag. Johari *et al.* [36] experimentally studied the role of sinusoidal protuberances on the leading edge of a 2-D airfoil. They determined that the amplitude of the protuberances plays more significant role on the performance than the wavelength and also found from visualization that the flow separation forms chiefly in the region between two ridges. Cranston *et al.* [37] recently showed the size of the serrations on flat plates at low Reynolds numbers significantly affects the aerodynamic characteristics and separation bubble.

Some practical applications have been considered, for example, Weber *et al.* [38] used tubercles on the leading edge of rudder and found the tubercles enhance lift for high angles of attack. It causes more favourable control during turning.

An industrial fan with leading edge protuberances on blades was reported to produce higher flow rates at lower speeds and lower noise level and hence improved its efficiency and electricity consumption [39]. The use of tubercles technology has also been successfully applied in renewable power generation. A 35 kW wind turbine utilizing tubercles technology on its blades was recently introduced. This new design produced more electricity at more moderate wind speeds [40].

Since the mechanism of the leading edge pattern on the flow structures and the effective parameters are still not well understood yet, Hasheminejad*et al.* [41] studied the effect of different leading edge patterns on the vortex structures generated using smoke visualizations on a flat plate with 14 mm wavelength of saw-teeth, semi-circulars and slots. The experiments were performed in a small open-circuit low speed wind tunnel with its test section of 160mm x 160mm and free-stream turbulent intensity of about 0.25%. The smoke-wire flow visualization was used to visualize the cross-sections of the vortex structures generated in the flat plate boundary layers at some streamwise positions from the flat plate leading edge for different Reynolds numbers (based on wavelength of leading edge patterns) ranging from 1250 to 1800.

IOP Conf. Series: Materials Science and Engineering **88** (2015) 012001 doi:10.1088/1757-899X/88/1/012001

For all the leading edge patterns considered, the visualization results show that pairs of counter-rotating vortices are formed in the laminar part of boundary layer and grow in the streamwise direction until they break down to turbulence. However, for the case of the leading edge with slots, there are additional counter-rotating vortices which occur in the spaces between the slots which may be a secondary flow caused by the intense upwash flow made by the abrupt change of the flow direction in a narrow span. The cross-sections of the vortex structures due to different leading edge patterns are visualized in Fig. 5. Hot-wire anemometer measurements within these vortex structures will be conducted to obtain quantitative data.

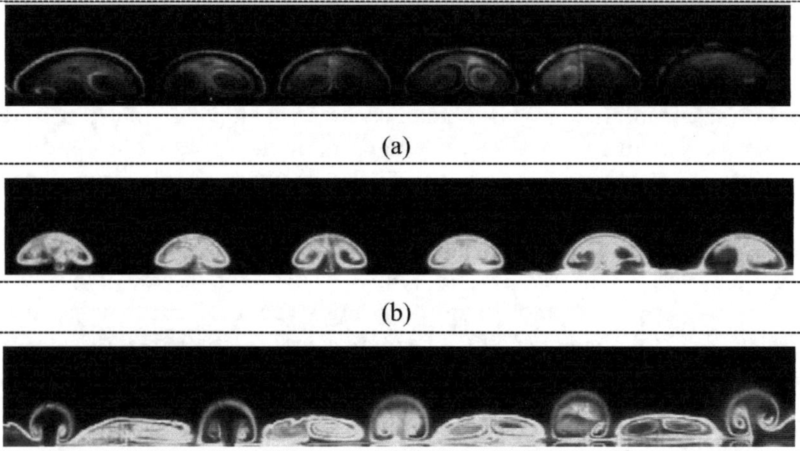

(a)

(b)

(c)

Figure 5. Cross sectional visualizations of flow at x=10 mm and Re= 1250 for different leading edge patterns: (a) saw tooth, (b) semicircle and (c) slot.

4. Due to vortex generator

Mushroom like counter-rotating vortex pairs are predominant structures that occur in the transitional boundary layers [42]. To study the dynamics of these vortices in the transitional boundary layer, evolving spatially and temporally, is cumbersome [43]. Hence, a more convenient environment to characterize these embedded vortices is necessary. Görtler vortex system is a good example to characterize such vortices since they appear as steady, spatially evolving structures in concave surface laminar boundary layer flow. Initiation or simulation of such counter-rotating streamwise vortex pair in a zero pressure gradient flat plate boundary layer is possible by passive means, such as vortex generators which have been effectively employed in aerodynamics for flow control, as well as in heat transfer enhancement. There are many different types of vortex generators and a brief overview is given by Lin [44].

Besides the vortex generators, scales of fast swimming sharks were reported to have directional surface roughness elements known as convergent and divergent riblets [45,46]. Recently, the application of convergent and divergent riblets in manipulating turbulent boundary layer, to reduce the skin friction drag and to control very large scale motion, was initiated [47]. The highly inflexional velocity profiles created around the center of convergence, by these roughness elements are similar to those of Görtler vortices [45,48]. The convergent and divergent riblets used in these studies consist of multiple riblets arranged in tandem with a constant pitch. The aim of this work is to incorporate a single set of convergent riblets (Fig. 6) in a flat plate boundary layer to study the growth and formation of these vortices using smoke-wire visualization technique.

IOP Conf. Series: Materials Science and Engineering **88** (2015) 012001 doi:10.1088/1757-899X/88/1/012001

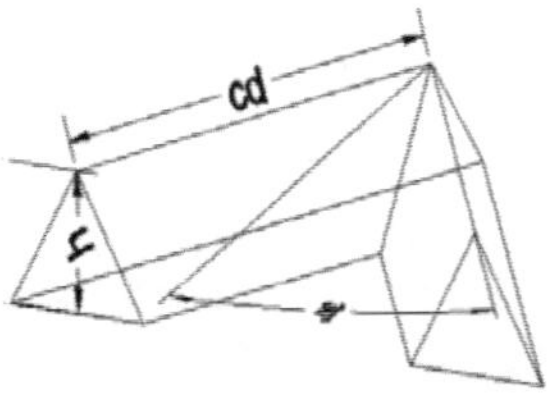

Figure 6. A single set of convergent riblets

The experiment was carried out on a flat plate where a single set of convergent riblets are mounted. The riblet height is 1mm and the aspect ratio of the riblet is defined as the ratio of cd to h, which is set to 2 in this study, where cd is the riblet length. The riblet set is placed at a distance from the leading edge such that the ratio of Blasius boundary layer thickness to the height of the riblet (δ/h) is 0.5. The angle of the riblet to the flow is 30°as shown in Fig. 6. The m odel is mounted inside a low speed wind tunnel with free stream turbulent intensity of about 0.25% and with test section of 160 x 160mm cross-section. Smoke sheet is created by applying a constant current through a wire coated by paraffin oil dripped along through gravity. The co-ordinates has its origin at the trailing end of the riblets. Any distance mentioned henceforth is with respect to the origin.

As shown in Fig. 7, the vortices are initiated by the convergent riblets grows along the downstream direction. The mushroom like structure comprises the stem and the counter-rotating vortex pair distributed on its either side. As the flow encounters the convergence, the fluid stream in the boundary layer, encountering the riblets, is lifted up due to continuity thus initiating a transverse velocity. The low momentum fluid from the boundary layer is lifted up to a distance until it overcomes the normal pressure gradient. As the low momentum fluid is ejected away from the wall, the outer high momentum fluid tries to penetrate the boundary layer to satisfy continuity. The high momentum fluid thus envelopes to form the mushroom hat exhibiting maximum shear stress, similar to Görtler vortices.

Along the downstream direction, the vortices gets tilted which might indicate the onset of transition as shown in Fig. 7 (c) and far downstream, the structures broke down. As the ratio δ/h decreased, the observed phenomenon occurred earlier upstream.

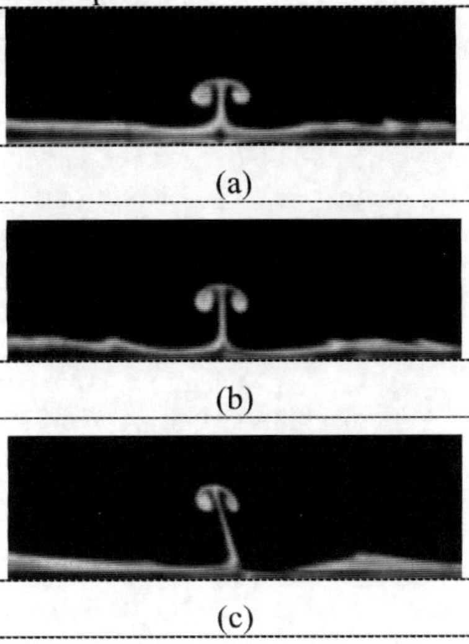

Figure 7. Streamwise counter-rotating vortex structures at different cross stream planes for $\delta/h=2$, (a)at x=30mm, (b) at

x=60mm, and (c) at x=150mm.

5. Conclusions

Some types of counter-rotating streamwise vortices generated by different methods or physical conditions have been briefly considered and visualized using smoke-wire flow visualization technique. The experimental work on these vortices are still on going to obtain quantitative data from the vortex structures by means of hot-wire anemometry and/or particle image velocimetry.

6. Acknowledgement

The authors would like to thank the Temasek Laboratories at the National University of Singapore (NUS) for conductingsome of the experiments at its Aeroscience Laboratory.

The third, fourth and fifth authors are currently recipients of the NUS Research Scholarship, while the second and sixth authors are former recipients of the NUS Research Scholarship.

7. References

[1] Görtler, H., Über eine dreidimensionale Instabilität laminarer Grenzschichten an konkaven Wänden, *Ges. D. Wiss. Göttingen, Nachr. a. d., Math.*, 2 (1) (1940); translated as "On the three-dimensional instability of laminar boundary layers on concave walls", NACA TM 1375, 1954.

[2] Smith, A.M.O., On the growth of Taylor-Görtler vortices along highly concave wall,. Quart. Appl. Mech., 13, pp 233-262, 1955.

[3] Winoto, S.H., Mitsudharmadi, H. and Shah, D.A., Visualizing Görtler vortices, Journal of Visualization, 8, no. 4, pp 315-322, 2005.

[4] McCormack, P.D., Welker, H. and Kelleher, M., Taylor- Görtler vortices and their effect on heat transfer, ASME Journal of Heat Transfer, 92, pp. 101-112, 1970.

[5] Kan, S., Miwa, K., Morishita, T., Munakata, Y. and Nomura, M., Heat transfer of a turbine blade, Proceedings of Joint International Gas Turbine Conference, Tokyo, Japan, October 4-7, pp 219-226, 1971.

[6] Crane, R.I. and Sabzvari, J., Heat transfer visualization and measurement in unstable concave-wall boundary layers, ASME Journal of Turbomachinery, 111, pp 51-56, 1989.

[7] Crane, R.I. and Umur, H., Concave-wall laminar heat transfer and Görtler vortex structure: Effects of pre-curvature boundary layer and favourable pressure gradients, ASME Paper No. 90-GT-94, 1990.

[8] Momayez, L., Dupont, P. and Peerhossaini, H., Some unexpected effects of wavelength and perturbation strength on heat transfer enhancement by Görtler instability, International Journal of Heat and Mass Transfer, 47, pp 3783-3795, 2004a.

[9] Momayez, L., Dupont, P. and Peerhossaini, H., Effects of vortex organization on heat transfer enhancement by Görtler instability, International Journal of Thermal Sciences, 43, pp 753-760, 2004b.

[10] Floryan, J.M. and Saric, W.S., Stability of Gortler vortices in boundary layers, AIAA Journal, 20, no. 3, pp 316-324, 1982.

[11] Finnis, M.V. and Brown, A., Stability of a laminar boundary-layer flowing along a concave surface, ASME Journal of Turbomachinery, 111, pp 376-386, 1989.

[12] Kottke, V and Mpouradis, B, On the existence of Taylor-Görtler vortices on concave walls, Proceedings of the 4th International Symposium on Flow Visualization, Paris, 1986, pp. 475-480, Hemisphere Publishing Corp., Washington, 1987.

[13] Denier, J.P., Hall, P., and Seddougui, S.O., On the receptivity problem for Görtler vortices: vortex motions induced by wall roughness, Philosophical Transactions: Physical Sciences and Engineering, 335, no. 1636, pp 51-85, 1991

[14] Bassom, A.P. and Hall, P., The receptivity problem for O(1) wavelength Görtler vortices, Proceedings of Mathematical and Physical Sciences, 446, no. 1928, pp 499-516, 1994

[15] Luchini, P. and Bottaro, A, Görtler vortices: a backward-in-time approach to the receptivity problem, Journal of Fluid Mechanics, 363, pp 1-23, 1998
[16] Floryan, J.M., On the Görtler instability of boundary-layer, Progress in Aerospace Science, 28, pp 235-271, 1991.
[17] Meksyn, D., Stability of viscous flow over concave cylindrical surfaces, Proceedings .Roy. Soc., 203, no. 1073, pp 253-265, 1950.
[18] Peerhossaini, H. and Bahri, F., On the spectral distribution of the modes in non-linear Görtler instability, Experimental Thermal and Fluid Science, 16, no. 3, pp 195-208, 1998.
[19] Ajakh, A., Kestoras, M.D., Toe, R., and Peerhossaini, H., Influence of Forced Perturbation in the Stagnation Region on Görtler Instability, AIAA Journal, 37, no. 12, pp 1572-1577, 1999.
[20] Toe, R., Ajakh, A., and Peerhossaini, H., Heat transfer enhancement by Görtler instability, International Journal of Heat and Fluid Flow, 23, no. 2, pp 194-204, 2002.
[21] Mitsudharmadi, H., Winoto, S. H. and Shah, D. A., Development of boundary layer flow in the presence of forced wavelength Görtler vortices, Physics of Fluids, Vol. 16, no. 11, pp 3983-3996, 2004.
[22] Mitsudharmadi, H., Winoto, S.H. and Shah, D.A., Development of most amplified wavelength Görtler vortices, Physics of Fluids, no. 18, 014101, 1 – 12, 2006.
[23] Mitsudharmadi, H., Winoto, S.H., and Shah, D.A., Splitting and merging of Görtler vortices, Physics of Fluids, no. 17, 124102, 1 – 12, 2005.
[24] Mitsudharmadi, H., Winoto, S.H., and Shah, D.A., Secondary instability in forced wavelength Görtler vortices, Physics of Fluids, no. 17, 074104, 1 – 8, 2005.
[25] Tandiono, Winoto, S.H. and Shah, D.A.,On the linear and nonlinear development of Görtler vortices, Physics of Fluids, 20, no. 8, 094103, 1 – 15, 2008.
[26] Tandiono, Winoto, S.H. and Shah, D.A.,Wall shear stress inGörtler vortex boundary layer flow, Physics of Fluids, 21, no. 8, 084106, 1 – 9, 2009.
[27] Hutchins, N. & Choi, K.S., Accurate measurements of local skin friction coefficient using hot-wire anemometry, Progress in Aerospace Sciences 38 (4-5), 421, 2002.
[28] Bottaro, A.&Klingmann,B. G. B., On the linear breakdown of Görtler vortices, European Journal of Mechanics, B/Fluids,15 (3), 301, 1996
[29] Nishimura, T., Yano, K., Yoshino, T., & Kawamura, Y. 1990. Occurrence and structure of Taylor-Goertler vortices induced in two-dimensional wavy channels for steady flow. Journal of Chemical Engineering of Japan, no. 23 (6), pp. 697-703
[30] Budiman, A.C., Mitsudharmadi, H., Low, H.T. & Winoto, S.H. Visualization of counter-rotating streamwise vortices in a rectangular channel with one-sided wavy surface. Proceeding of the 31st AIAA Applied Aerodynamics Conference. San Diego, CA., 2013.
[31] Mitsudharmadi, H., Jamaludin, M.N.A. & Winoto, S.H. Streamwise vortices in channel flow with a corrugated surface. Proceedings of the 10th WSEAS International Conference on Fluid Mechanics & Aerodynamics (FMA'12). Istanbul, Turkey, 2012.
[32] Anderson, G., W. An Experimental Investigation of a High-Lift Device on the Owl Wing. Master Dissertation, Air Force Institute of Technology, Wright-Patterson Air Force Base, Ohio, 1973
[33] Fish, F.E & Battle, J. M. Hydrodynamic Design of the Humpback Whale Flipper. Journal of Morphology, 225, pp. 51-60, 1995.
[34] Fish, F.E., Weber, P.W., Murray, M., M. &Howle, E.. The Tubercles on Humpback Whales' Flipper: Application of Bio- Inspired Technology, Vol. 51, number 1, pp 203-213, 2011.
[35] Soderman, P, T. Aeordynamic Effects of Leading Edge Serrations on a Two-Dimensional Airfoil. National Aeronautics and Space Administration. Ames Research Center, Moffett Field, CA, 1972.
[36] Johari, H., Henoch, C., Custodio, D., and Levshin, A. 2007. Effects of Leading-Edge Protuberances on Airfoil Performance. AIAA Journal, Vol. 45, No. 11, Nov. pp 2634–2642.

[37] Cranston, B., Laux, C., Altman, A. 2012. Leading Edge Serrations on Flat Plates at Low Reynolds Number. 50th AIAA Aerospace Sciences Meeting including the New Horizons Forum and Aerospace Exposition, Nashville, Tennessee, 0053.

[38] Weber, P., W., Howle, L., E., Murray, M., M. 2010. Lift, drag and cavitation onset on rudders with leading edge tubercles. Mar Tech 47:27–36.

[39] Ontario Power Authority. 2010. Energy efficient fans take their cue from the humpback whale.http://archive.powerauthority.on.ca/Storage/122/16957_AgNews_July231.pdf.

[40] Howle, L.E. WhalePower Wenvor blade. A report in the efficiency of a WhalePower Corp. 5 meter prototype wind turbine blade. BelleQuant Engineering, PLLC, 2009.

[41] Hasheminejad, S., M., Mitsudharmadi, H. & Winoto S., H. Visualization of Boundary Layer on Flat Plate with Leading Edge Patterns. 2013. 12thAsian Symposium on Visualization, Tainan, Taiwan.

[42] Bernard, P.S. The hairpin vortex illusion. Journal of Physics. 318, 2011.

[43] Swearingen J.D. &Blackwelder R.F., The growth and breakdown of streamwise vortices in the presence of a wall. Journal of Fluid Mechanics. Vol.182, pp. 255-290, 1987

[44] Lin J.C., Review of low profile vortex generators to control boundary layer separation, "Progress in Aerospace Sciences. 38, pp.389-420, 2002.

[45] Koeltzsch, K,Dinkelacker, A &Grundmann, R. Flow over convergent and divergent wall riblets, Experiments in Fluids. Vol. 33, pp. 346 - 350, 2002.

[46] Nugroho, B, Kulaindaivelu, V, Harun, Z, Hutchins, N &Monty, J. P. An investigation into the effects of highly directional surface roughness on turbulent boundary layers, Proceedings of the 17th Australasian Fluid Mechanics Conference, Auckland, New Zealand, 2010.

[47] Nugroho, B, Hutchins, N, &Monty, J. P. Effects of diverging and converging roughness on turbulent boundary layers. Proceedings of the 18th Australasian Fluid Mechanics Conference, Launceton, Australia, 2012.

[48] Nadesan, T, Mitsudharmadi, H, Lee, T.S. &Winoto, S.H.. Effects of converging riblets in flat plate boundary layer, The 12th International Symposium of Fluid Control Measurement and Visualization, Nara, Japan, 2013.

Mechanical Engineering Education and Its Challenges

C S Ow[1], M M Kanan[2]
[1]Professor, Mechanical Engineering Faculty, Universiti Teknologi MARA, Selangor, Malaysia.
[2]Senior Lecturer, Universiti Tenaga Nasional(UNITEN), Selangor Malaysia

E-mail: iodised@gmail.com
E-mail: kannan@uniten.edu.my

Abstract. The paper addresses historical development in Engineering Education in the country, its evolution till present day efforts toward the formation of Professional Engineers (PE). Of particular interest is the proposed recognition of more than one pathways towards PE Certification amongst member countries of the International Engineering Alliance (IEA). However, Engineering Education of Gen Y poses challenges at maintaining relevant benchmarks at the basic degree level. The widespread use of sophisticated software for simulation work in any undergraduate programme has its sacrifices with respect to emphasis on depth of knowledge. A definite mismatch between what is perceived by the educators and the actual performance of graduates had been identified in a forum and an employment survey conducted by the Institution of Engineers Malaysia (IEM). Suggestions as to how this can be addressed include the setting up of a Board of Educators to regulate the education industry.

[1] To whom any correspondence should be addressed.

1. Introduction

Earning an Accredited Engineering degree is one of the assured pathways towards professional registration as a Professional Engineer [1] in Malaysia. Whilst the option rests with the individual engineering graduate as to pursue this pathway or otherwise, the subsequent postgraduate work- experience defines whether one is up to the mark to be certified or not. Redmill F [2] presented a paper on strategic perspectives on engineering education focussing mainly on aims of curriculum in targeting the kind of graduates and hence professionals for the professing. The paper had its weaknesses in that the strategy should also focus on the educators and the relevant skills they need to have so that the graduates have a chance to start with.

At the recent IEA meeting in New Zealand, Katy Turff [3] from the Engineering Council of United Kingdom (UK) shared with those attending the meeting the good practice in competence-based registration of professional engineer. Her presentations included both academic pathways as well as non-academic pathways in the development of professional engineers. Development of engineering education in UK is quite innovative to the extent that one is sometimes at a complete loss as to which is the best pathway to follow. Recognition of the diversity of students some of whom are not academically inclined, dreamers, late bloomers to the academically brilliant ones is the way forward in strategising any curriculum. The knowledge profile for Graduate entry to the UK engineers register may follow one of two discernible pathways. These are via formal programme or via a series of informal on the job learning.

Knight J et al.[4] highlighted a professional pathway for UK students who would like an accelerated entry into the workplace for allied professions. Rather than attending full time study the computing students are permitted to sit for the same exam but attending classes one day a week having successfully completed the first year of the course. This philosophy is well reflected amongst university entry requirement in New Zealand where for certain courses a good command of English (IELTS score exceeding 6.5) is sufficient.[5] The graduate should be able to undertake a journey of discovery albeit under guidance and close monitoring of the mentor.

The historical development of Engineering Education in United States of America is well documented by Mark and Buchanan [6] highlighting the central role played by American Society for Engineering Education over the past 120 years. It started with the Morrill Act 1862 also known as the Land Grant College Act and the subsequent numerous engineering schools set up are broadly classified into one of two camps namely the school culture which stressed on theoretical education and the shop culture which stressed on hands on learning in the laboratory. Further reforms saw the setting up of a body known as the Engineering Council for Professional Development which evolved to become the present Accreditation Board for Engineering and Technology (ABET).

This year the Institution of Professional Engineers New Zealand (IPENZ) celebrated its Centenary (1914-2014). Formal engineering education has similarly evolved from technical education along similar route as the USA with early Learned Engineering societies similarly evolving into the present form of an IPENZ. [7]. The Engineering Council of South Africa (ECSA) is a statutory body set up under the auspicious of the country's Engineering Profession Act (2000). Its core functions are the accreditation of engineering programmes, registration of persons as professionals in specified categories and the regulation of the practice of registered persons. [8]

In Malaysia perhaps one of the more comprehensive summary of development in Engineering Education is by TT Chiam [9] who traced the foundation for education in Malaysia began with the founding of three trust- english-medium schools namely Penang Free School in 1816, Singapore Free School (renamed Raffles Institution) in 1823 and the Malacca Free School (renamed Malacca

High School) in 1836. After the Second World War, the establishment of Technical College in 1949 provided an opportunity for Malaysians to obtain sub-professional technical education locally. The original University of Malaya was established in Singapore in 1949 and with it the Department of Engineering formed in 1955 within the Faculty of Science. A Faculty of Engineering was established in Kuala Lumpur in 1958 and UM was divided into two divisions, one in Singapore and the other in Kuala Lumpur. The UM in Singapore evolved to be present day National University of Singapore whilst that in Kuala Lumpur remained as University of Malaya.

Megat et al [10] traced the beginning of formal 4 year engineering education in the country offered by major Public Universities and the subsequent reduction to 3 years instead of 4 in 1966 since universities in the United Kingdom were doing just that with the same entry qualifications. The system at UM then was still a 3 term system in a full academic year although Institute Teknologi MARA (presently Universiti Teknologi MARA) conducted a three year Diploma programme with those attaining good grades proceed to the Advanced Diploma programme for an additional two and a half years. The entry qualification to this programme was at the SPM level and was accredited initially by The Institution of Engineers on a regular 5 yearly basis. The desire to establish a suitable basis for 3 or 4 years education led to the Institution of Engineers Malaysia hosting of a forum during the Presidency of Dr Ting Wen Hui [11] to decide on the appropriate equivalent credit hours of engineering and non engineering studies needed for an engineering graduation. The main reason for this development was that whilst the oldest university still runs a 3 term per academic year programme, most of the other government sponsored universities have migrated to the American Semester based system.

Although Institut Teknologi MARA (ITM) was amongst the earliest institutions to introduce the semester system in the country in 1976 by dividing the 8 subjects per academic year into 4 each for each of the semester (half year) programme, the ongoing discussions then did not affect ITM as it run a two-tier system and maintained as such until the institution changed its status to that of a University at the turn of the century. The IEM forum resulted in general agreement that at least 120 credit hours of study with at least 80 hours for engineering subjects for graduating engineering students would be sufficient to be equated to the 3 term one academic year system but no timeline was set for the duration for the study which is normally completed comfortably within 4 years.

In the 90s numerous private colleges offered twinning-type engineering degree programmes with established overseas universities. The initial setup of the National Accreditation Board (LAN) in the 90s and its subsequent migration to Malaysian Qualifications Agency (MQA) presently led to the widespread upgrade of private colleges and the promulgation of private universities beginning with the three industry led giants namely Universiti Tenaga Nasional (UNITEN), Multimedia University (MMU) and Universiti Teknologi PETRONAS (UTP). Accreditation of programmes were conducted by LAN for private Institution of Higher Learnings (IHLs) only but the then accreditation conducted jointly by BEM and IEM continued with appropriate representation from LAN and the Public Services Department (JPA). The joint committee is the start of the Engineering Accreditation Council (EAC) which subsequently saw the inclusion of stakeholders such Malaysian Council of Engineering Deans (MCED) and additional engineers from the industry. Numerous innovations were introduced by the Engineering Education industry and this lead to widespread development in engineering education.

2. Educational preparation

Under the present educational system, varsity entrants are supposed to be equipped with the depth and breadth of physical and mathematical sciences. Any inadequacies at the preparatory level shall have to be addressed at the university level in order to maintain benchmarks. The semester system of education permits such corrections by allowing candidates to repeat and re-sit the subjects for which they have yet to achieve the required benchmark.

Where does the education process begin? At Home? At the Kindergarten? During Primary Education? Lower Secondary?, Upper Secondary? Finally, at the Universities and Colleges? Continual Education till the day we drop dead?

Let us begin with education at home. I would like to share an experience which I had with a granddaughter. Her mum decided to raise her with lots of Positives rather than Negatives. Words like "Cannot", "Don't", "No" etc. are banned from our dictionary when replying the little girl at her formative ages less than 5 years. It takes quite a transformation within the family to cope with such a change. Adults have to innovate for appropriate responses to questions and along with the innovative responses, the child is observed to be equally innovative in her subsequent actions. When this pathway is reinforced at the Kindergarten school and subsequent Primary, Secondary and Tertiary levels, one certainly makes learning interesting and educational activity fill with innovative responses. Unfortunately somewhere along this journey more than one word of "NO" shall contribute towards derailing the entire spirit and perhaps put off any further interest in whatever the subject may be.

We have witnessed the traditional methods of teaching science and mathematics shifting gradually from prescriptive to student-centred learning at schools and universities with evaluation focussing on use of tools rather than hard memories of multipliers etc. to the extent that one uses a calculator to do simple multiplications for an answer. What is the significance of this development? A lack of mental arithmetic or lack of depth of fundamentals and how it works? Naturally there are lots of arguments for this "natural" development because we now use IT tools in a digital world. Even the English language as we know it is transformed by Short Message Services (SMSes) to a "sound" version such as C U B4 6 ☺ etc. There is a clear difference in the SMSes between a Traditionist (born between 1922-45), the Baby Boomers generation (40s-50s), X-generation (60s-70s) and Y-Generation (80s-90s) as we know it. A consequence of Y-Gen is that more information can be stored and made available to the user at any instant but the negative side of this is the lack of basic understanding of fundamentals at all levels for the average student.

3. Mechanical engineering crossroads

Peck Cho [12] commented that the arrival of the information age at the turn of the century saw a paradigm shift which changed all aspects of education. These changes were with respect to Access, Certification, Content and Organisation. The same time, same age and same place educational window can shift to any time, anyone, any place manifested in the form of Continuing Education, Distance Education and a Learning Society. With regards to Certification, rather than treating an accumulation of sufficient credits and grade points, the shift has been towards assessment of acquisition of specific skill sets such as those in an OBE system. A corresponding shift in acquiring (static) knowledge which can be cumulative, permanent, transmitted and cognitive towards information (dynamic) which is transient , transitive , transacted and affective. A manifestation of these development is the student-centred learning (SCLs), problem-based (PBLs) teaching and lifelong learning which we hear today.

Traditional curriculum which addresses 5 out of 11 Programme Outcomes (POs) of the original ABET Engineering Criteria 2000 had left out the remaining six which focussed on "soft" skills etc. Present day driving forces from the industry require emphasis of soft skills amongst graduating engineering students under a false assumption that the core skills are not compromised. Together with proliferation in specialisation within each subject matter, the task of identifying what is really needed and what is irrelevant has become more formidable. It is also worthwhile to note that a Mechatronic programme is neither Mechanical nor Electronic but an integration of certain elements from each added to it the Assembly level software programming skills.

Recent development with Washington Accord graduate attributes speak volumes about what skills an engineering graduate should have. When it comes to registration of engineers with the Board of Engineers Malaysia (BEM), therein lies the frequent debate of what e.g. an Electrical Engineer does and must acquire in the formative years, etc. Whether the industrial development in this country is ready to employ e.g. Mechatronic graduates is another subject for debate. How about Mechanical Precision Engineering Programme and the list of options is never ending. The main purpose for accreditation of engineering programme within Washington Accord or IEA Signatories is to maintain some sense of Substantial Equivalence (yet to be defined since its inception). As an example, Engineers Canada defined "Substantial Equivalency" to mean comparable in program content and educational experience. It therefore implies reasonable confidence that the graduates possess the academic competencies needed to begin a professional practice at the entry level but such programmes may not be absolutely identical. Taking it one step further, is there a need to BEM to register all disciplines of engineering? Perhaps Singapore PE board has got it right by registering those disciplines whose work has direct impact upon safety of the Public.

Presently the EAC 2012 Manual [13] spells out clearly the core subjects required for a programme to be classified under the Mechanical Discipline. These are reproduced here in Appendix. The question is how do we establish/ measure or have an inkling of the amount of emphasis given to understanding of fundamentals underpinning the formation of a mechanical engineering graduate even with the form of accreditation exercise we now conduct. I do feel that the examples of the types of courses listed in the Appendix should be reanalysed only for core requirement for the registration of Mechanical Engineers whose work affected public safety. Stifling further evolution and innovation in development of any programme shall severely handicap graduates in facing future technological challenges

The accreditation process conducted is largely a peer review process dependent upon the calibre of peer reviewers as well as the time available to do a simple audit. The proof of employability is with the market which the next section shall address. The specifications for relevant must-have-ingredients are listed in an accreditation manual hosted by the BEM. It is done not only for the Mechanical discipline but also Electrical, Civil, Chemical and Electronic Engineering disciplines and implies the need to define an overall basic core knowledge for those disciplines whose subsequent work has an impact upon public safety. Are these "requirement" restrictive towards meeting future market needs? Can we afford to move forward without such a mechanism? The answer may be both yes and no.

4. Technological challenges

Do the core competencies spelt out in Appendix for Mechanical engineering programme adequately prepare graduates to meet current challenges? At a recent meeting of HVAC engineers in Seattle, USA, a Technical Plenary presentation by Dennis Hayes [14] which entitled a building which lives like a tree indicated the possible options presently day technology could provide. Short of producing the much desired oxygen for the atmosphere, this building has most of the functions of a tree producing net positive energy overall. Do we have an energy crisis if all buildings are designed and built to such functions? Certainly the role of learned societies such as ASHRAE and the development of its Codes and Standards which focuses towards Net Zero Energy building would soon be replaced by towards net positive outcomes.

Another development which has yet to make a future impact is the magnetically levitated shaft using advanced magnetic bearing in a 800-TonsR HFC 134A compressor as highlighted by M.S Jeng [15]. The only shortcoming reported is its extra cost of 30% more than conventional bearings. Self-centering within a 200 microns air gap at all speeds and loads, it is truly a bearing of the future with minimal friction and hence better energy efficiency. One cannot help speculating further when such bearings are the norm for all road vehicles in the near future.

There is price to be paid in the supply of such products in that the Total Equivalent Energy consumed (TEE) in producing it from scratch must be tagged on at some stage. Whilst Global Warming issues have placed the blame on some green house gases such as carbon-dioxide, there are however contradictory arguments by R I Nigmatulin [16] towards this in that historically the planet has undergone cycles of global warming and cooling with or without the corresponding increases of CO2. Perhaps a greater source of basic heat, greenhouse gas, moisture and flux for the atmosphere is the ocean which has a heat capacity 1000 times that of the atmosphere and a mass 270 times as great.

Bill Coad, [17] even predicted the use of a Seasonal Ferris-wheel to overcome the cooling/heating needs globally across the tropics. The then available technology was simply too expensive to be implemented. Nevertheless use of Solar Panels and other Green Technologies (GT) should also consider the incorporation of TEEs and carbon footprints into their computation. A possible future direction for GT is to utilise energy supply at source rather than to first having to convert to electricity the process of which does not exceed 33% efficiency, burning coal or oil at source. New materials such as development of Metal Organic Framework (MOF) sponge [15] for hydrogen gas at low pressure when successful shall see more widespread use of hydrogen-fuel-PEM fuel cell powered vehicles and the brand new Hydrogen Economy albeit that in certain European countries H2 refilling stations have already been set up. With the limited resources of platinum and other metals needed for such an economy to be sustainable, a system of recycling and reuse has to be in place at the onset for it to be successful. There is a need to define adequate fundamentals in the education of Mechanical Engineering graduates for them to address these and other future technological challenges.

5. Role of learned societies

The Institution of Engineers Malaysia (IEM) hosted a National Forum on Benchmarking the Quality of Engineers on 16 January 2012 in which various stakeholders such as academics, industry practitioners, employers of engineers as well as regulatory bodies attended the forum. Several examples were cited with regards to the low standard or quality of graduate engineers entering the industry such as:

IOP Conf. Series: Materials Science and Engineering **88** (2015) 012002 doi:10.1088/1757-899X/88/1/012002

(i) Graduates with CGPA 3.5 and above were severely handicapped at drawing bending moment diagrams and shear force diagrams (in the Civil Discipline)
(ii) Electrical engineers cannot explain why birds standing on live wires are not electrocuted.
(iii) Poor understanding of design fundamentals
(iv) Mistakes in design calculations
(v) Lack of engineering knowledge
(vi) Unfamiliar with the basic codes of design
(vii) Not able to apply engineering knowledge

The consensus reached at the end of the forum was that:
(i) There is a need to raise the standard of engineering graduates.
(ii) Schools need to improve on the education delivery system to ensure quality students enter universities (some term it as "unfit" for industry and "untrainable").
(iii) Institutions of Higher Learning (IHLs) to engage more practicing engineers to teach while lecturers should gain more industrial experience.

Arising from the forum, IEM established a Task Force on 3 December 2012 to prepare a Position Paper on this matter. The first task undertaken was to commission an Employer's Survey targeted at engineers who are most likely in positions capable of judging the quality of fresh graduates, i.e. those of 35 years of age or older. Results from the survey shall be duly reported in the Position Document to be issued in the near future. The IEM survey points out clearly to the lack of DESIRED OUTCOMES of recent graduates. Private communications with IEM volunteers indicated that young graduates are not able to approach the design of taller buildings as he/she was trained to do a 4 storey only and vice versa. This brings up the question of whether are we conducting Educational or merely Training activities at those IHLs?

The IEM survey results further reinforced the fact that a kind of decoupling exist between education providers on measured and perceived outcomes which do not meet the desired outcomes of employers. Perhaps a fresh survey on desired outcomes of graduates should be undertaken so as to close the loop as discussed by certain accreditation panels during the exit meeting.

As an educator from yesteryear, I do perceive one commonality in that we do not produce graduates to meet the needs of any specific industry but to equip him/her with adequate academic competencies and interpersonal skills to face the "real" world. This is a tall order to achieve when the students who entered varsity also lacks the depth in analytical skills needed to solve science and mathematics problems. At the school level the over indulgence of "exam-knowledge" has led to a culture of "how to answer" examination questions rather than the inquisitive "why this is so" as commented by D Rahman [18]. Much has been written about the country's educational blueprints, nevertheless, the important factor is the quality of an education delivered and acquired as perceived in the plans.

6. Changing landscapes and complex engineering problems

Does accreditation of engineering programmes guarantee quality? Similarly does the presence of an ISO Quality System assure confidence? How are our future lecturers tenured? The requirement of the 2012 Engineering Accreditation Manual for greater emphasis on Complex Engineering Problems as defined in Appendix has been debated and continually improved upon at International Engineering Alliance (IEA) meetings. However they do not guarantee that graduates meet the desired results from the above survey. Perhaps the inherent weakness

IOP Conf. Series: Materials Science and Engineering **88** (2015) 012002 doi:10.1088/1757-899X/88/1/012002

within the process has not been truly addressed. At a recent IEA meeting in Wellington, complex engineering problems involving current research issues faced difficulties in defining boundary conditions for the students to solve. The definition of complex engineering problems is clearly spelt out in Appendix. However to come up with such problems in undergraduate studies requires much thought and may also be a function of the amount of industrial exposure one has with real problems which are non routine in nature and job specific. This reminds me of a conversation with a past principal of Ngee Ann Polytechnic in Singapore in which he insisted that a guarantee for employment be made by each of his staff for the graduates of any new programme being proposed by them. This may be out of the norm but that is the crux of the issue, are we producing unemployment with some mandatory survey conducted as a matter of routine to establish needs for a new program?

The Peer Review Process carried out during Accreditation Exercises is dependent upon the Peer Benchmark which is predominantly transient. The IEM survey results clearly pointed out to a probable FAILed (First Attempt In Learning) system as it stands. Civil Engineers knows well the need to accurately determine platform levels based on the Ordinance Datum Level (ODLs). On a similar note, how can we benchmark graduate outcomes, based on the multitude of PBLs, Student-Centred Learning etc, in a system of marks which are given in-house albeit objectively and subjectively? Can the commercialisation of education such as the setting up of private IHLs help to sustain benchmarks? Where are the conflict of interests? Can we afford to let market forces of demand and supply determine the quality of Mechanical Engineering Graduates?

I am impressed by the present day practices of Continuous Quality Improvement (CQI) embedded within an Outcome Based Education (OBE) in which we defined clearly the Learning Outcomes and Course Outcomes which we plan to deliver with all the necessary measurements in place. Reflecting upon the IEM Employers feedback result, one wonders whether we have missed something important in the entire exercise. Perhaps an independent method of assessment is needed across all educational providers so as to maintain benchmark. Is this too difficult a call to make?

Is there a need for a Board of Educators to be set up along the lines of BEM, ensuring ethics and accountability amongst the educators/teachers? An established scheme for the development of educators so that they are professionally qualified to provide the right type of education is equally important in addition to the OBE system. Most of all is the availability of an avenue for the public to raise appropriate concerns/issues with education- providers if there is proof of negligence anywhere. At the school level is it ethical for a teacher to provide tuition classes to the same students after school?

7. Crisis and choices

Perhaps the country could look beyond Washington Accord Signatory in setting suitable benchmarks. A scheme for the tenure and continuous professional development of engineering educators is needed. Engineers Canada may have got it right in that all of the academics who teach an accredited engineering programme are Professional Engineers by registration. In such a structure, a platform is available to define and maintain benchmarks within a profession of ethical practices. Are good researchers also great educators and vice versa? How can we guarantee quality of deliverer as this is the most basic requirement for any OBE to be successful albeit with the current system well establish. Staring at the past, whatever the type of education system proposed, we need to educate graduates who are able to contribute to society via moving with the current technology and well equipped with all the relevant soft and hard (fundaments) skills.

Simply stated, it should never be a difficult task to achieve and shall not be too in the future. Use of software in design and simulations at the undergraduate level demands a good understanding of the underlying principles which govern the use of such IT tools, their limitations and mathematical limitations.

8. Conclusions

More questions than answers have been raised in this paper. A kind of decoupling seems to be present between perceived graduate attributes achieved and those desired by the industry. Even with and OBE system with CQI in-placed as required by the Accrediting Body, engineering education in the country has yet to satisfy industrial needs with respect to graduate competencies and skills. A Board for regulating Educators is proposed to assure ethical standards in place for the continual development of engineering educators. The survey results from amongst employers conducted by IEM is an eye opener to many education providers in that a mismatch do exist between what an employer sought and what is produced. Whilst there is a need to provide emphasis on appropriate fundamentals the regulatory framework in place must not stifle innovations to produce graduates who are able to face future technological challenges.

9. References

[1] Registration of Engineers Act 138 (Malaysia)
[2] Redmill F, Strategic Perspectives on Engineering Education, Engineering Education 2002: Professional Engineering Scenarios (Ref. No 2002/056), IEE (Vol2)
[3] Turff, Katy, Sharing good practice in competence-based registration, International Engineering Alliance Meeting, Wellington New Zealand, June 2014
[4] Knight J,Ballie L, Williams V, Osmon P E, Industrial Placement and Sponsorship: The Professional Pathway, Engineering Education 2002: Professional Engineering Scenarios (Ref No. 2002/056), IEE (Vol 1)
[5] Hwei Ling Ow, (private communications), 2008.
[6] M Clark and W Buchanan, The American Society for Engineering Education and the Morrilll Act of 1862, Proc 2012 JSEE Annual Conference and AEESEAP Workshop, Tokyo, Japan August 2012.
[7] Institution of Professional Engineers New Zealand webpage (www.ipenz.org.nz) 2014.
[8] Engineering Council of South Africa webpage (www.ecsa.co.za), 2014
[9] T T Chiam, Personal Experience and Observations of the Development of Engineering Education and Profeesional Practices in Malaysia for the Past 50 Years, "Jurutera", Pub. The Institution of Engineers, Malaysia, pp 5-9, 2009
[10] M M N Megat Johari, A A Abang Abdullah, M R Osman, M S Sapuan, N Mariun, M S Jaafar, A H Ghazali, H Omar and M Y Rosnah, A New Engineering Education Model for Malaysai, Int. J Engng Edu. Vol. 18, No. 1, pp8, 2002.
[11] W H Ting, Past President, The Institution of Engineers Malaysia
[12] Peck Cho, Future of Mechanical Engineering Education – Crisis, Crossroads and Choices, Proc ASME Forum on Mechanical Engineering Education, Korea 2000.
[13] Engineering Programme Accreditation Manual, Pub. Engineering Accreditation Council, Malaysia, ppB, 2012.
[14] D Hayes, A Net Positive Building that Functions Like a Tree: Bullitt Center, Technical Plenary, ASHRAE ANNUAL MEETING, Seattle, June 2014
[15] M S Jeng, ITRI Green Campus Programme, Technical Seminar, ASHRAE 17th Region XIII Chapter Regional Conference, Taiwan Aug 2014.

[16] Robert I Nigmatulin, Multiphase and Multiscale in Oceanology, Climatology and Economics, Proc. 14th Asian Congress of Fluid Mechanics- 14th ACFM, Vietnam, Vol 1, pp1, 2013
[17] Bill Coad, (private communications), Past President ASHRAE
[18] Danial Rahman, A Change of Emphasis, The Star Online, 18th Sept. 2014

APPENDIX

ENGINEERING CONTENT FOR SELECTED ENGINEERING DISCIPLINES AND INNOVATIVE PROGRAMMES

(a) Engineering Sciences, Principles, and Applications

An accredited programme is expected to cover the broad areas of the respective disciplines at an appropriate level. The following are examples of underpinning courses that may be introduced for the respective disciplines:

CIVIL	MECHANICAL	ELECTRICAL	CHEMICAL	ELECTRONICS
Strength of Materials	Materials	Circuits and Signals	Chemical Thermodynamics	Circuits and Signals
Structural Analysis and Design	Statics and Dynamics	Electro-magnetic Fields and Waves	Material and Energy Balance	Electro-magnetic Fields and Waves
Fluid Mechanics/ Hydraulics	Fluid Mechanics	Instrumentation and Control	Chemical Kinetics & Reactor Design	Instrumentation and Control
Soil Mechanics/ Geotechnical Engineering	Thermodynamics and Heat Transfer	Digital and Analog Electronics	Momentum Transfer	Digital and Analog Electronics
Civil Engineering Materials	Mechanical Design	Machines and Drives	Heat Transfer	Microprocessor System
Statics and Dynamics	Instrumentation and Control	Power Electronics	Mass Transfer	Programming Techniques

Construction Engineering	Vibrations	Electrical Power Generation and High Voltage Engineering	Separation Process	Introduction to Electrical Power System
Surveying	Solid Mechanics	Communications System	Process Design	Computer Architecture
Water Resources and Hydrology	Manufacturing/ Production	Power System Analysis	Process Control & Instrumentation	Communications System
Highway and Transportation	Electrical Power and Machines	Electronic Drives and Applications	Safety and Environmental Protection	Electronic System Analysis and Design
Environmental Studies	Electronics and Micro-processors	Electrical Energy Utilisation	Environmental Studies	Multimedia Technology and Applications
	Computer Aided Engineering		Plant and Equipment Design	
			Viability – Legal Framework, Economics, Operability and Reliability	

(b) Mathematics, Statistics and Computing
These courses should be studied to a level necessary to underpin the engineering courses of the programme and with a bias towards application. The use of numerical methods of solution is encouraged, with an appreciation of the power and limitations of the computer for modelling engineering situations. Wherever practicable, it is preferred that mathematics, statistics and computing are taught in the context of their application to engineering problems and it follows that some mathematical techniques may be learnt within other subjects of the course. In addition to the use of computers as tools for calculation, analysis and data processing, programme should introduce their application in such area as:-

CIVIL	MECHANICAL	ELECTRICAL	CHEMICAL	ELECTRONICS
Computer Aided Analysis and Design	Computer Aided Design and Manufacture	Mathematical Applications	Computer Analysis and Design	Mathematical Applications
Economics Analysis for Decision Making	Economics analysis for decision making	Statistical and Numerical techniques	Economics Analysis for Decision Making	Statistical and Numerical techniques
Data bases and Information Systems	Data bases and Information Systems	Computer Applications	Numerical Methods and Optimisation	Computer Applications
Operational Research	Operational Research		Operational Research	
Business and Management	On-line Control of Operations and		Databases and	

Systems	Processes		Information Systems	
Statistical and Numerical Techniques				

(c) Engineering Applications

Emphasis on Engineering Applications in degree programmes aims to ensure that all engineering graduates have a sound understanding of up-to-date industrial practice, in particular:

Civil Engineering:

- to appreciate the characteristic and structural behaviour of materials in a variety of user environments
- to be able to analyse and design structural components from these materials
- to appreciate the range of construction technology currently available and the skills which they require in people
- to appreciate the cost aspects of material selection, construction methods, operation and maintenance in their interaction with design and the delivery of civil engineering facilities and services
- to understand the whole process of industrial decision-making in design, manufacturing and use and how it must be influenced not only by technical ideas but also by the practical constraints of financial and human resources and by the business and social environment of engineering.

Mechanical Engineering:

- to appreciate the characteristic behaviour of materials in a variety of user environments
- to appreciate the range of manufacturing methods currently available and the skills which they require in people
- to appreciate the cost aspects of material selection, manufacturing methods, operation and maintenance in their interaction with design and product marketing
- to understand the whole process of industrial decision-making in design, manufacturing and use and how it must be influenced not only by technical ideas but also by the practical constraints of financial and human resources and by the business and social environment of engineering.

Electrical and Electronic Engineering:

- to appreciate the characteristic behaviour of materials in electrical and electronic systems
- to be able to analyse and design electrical and electronic systems from devices/components made of various materials
- to appreciate cost effectiveness of component/device equipment selection, manufacturing process and integration process
- to appreciate the range of manufacturing methods currently available and the skills which they require in people
- to understand the whole process of industrial decision making in design, manufacturing and use and how it must be influenced not only by technical ideas but also by the practical constraints of financial and human resources and by the business and social environment of engineering.

Chemical Engineering

- to appreciate the characteristic and structural behaviour of materials in a variety of user environments
- to be able to adopt these materials in process design and analysis
- to understand the general sequence of processing steps for any given type of chemical process
- to calculate and analyse the material and energy flows for a given chemical process
- to understand the selection or estimation of process operating conditions, selection of process equipment, maintenance and process troubleshooting
- to analyse the various types of unit operations and processing steps and to decide their relative advantages or disadvantages on the basis of environment, economics, safety and operability
- to understand the various process control schemes for the purpose of maintaining production quality, ensuring process safety and preventing waste.

(d) Evaluating an Innovative Programme
It is a challenge for an accreditation process to promote innovation, experimentation and dissemination of good practice, while maintaining standards that can be objectively certified nationally and internationally. Innovation by its nature challenges existing wisdom, but not every programme that departs from existing norms can be said to be innovative or desirable. All fundamentals required in the programme must be maintained.

The EAC accreditation system encourages innovation by minimising prescriptiveness in how the required outcomes are attained. Programme evaluation will always focus on the intent of the criteria and on the demonstrated capability of graduates to enter engineering practice at a professional level. Clearly however, a programme which departs radically from the methods normally thought necessary – for example, by employing only a fraction of the normal complement of staff – may expect a searching examination of method as well as outcomes. The EAC and the Evaluation Panel are expected to be receptive to new approaches, and to use the best judgement available to evaluate the substance and merit of the programme.

Continuing innovation and development can be expected to lead to restatement of the criteria and policy of accreditation.

(e) Definition of Complex Problem Solving
The range of complex problem solving as required by the Programme Outcomes in Section 4.0 is defined as follows:

Range of Problem Solving	Attribute	Complex Problems
1	Preamble	Engineering problems which cannot be resolved without in-depth engineering knowledge, much of which is at, or informed by, the forefront of the professional discipline, and have some or all of the following characteristics:
2	Range of conflicting requirements	Involve wide-ranging or conflicting technical, engineering and other issues
3	Depth of analysis required	Have no obvious solution and require abstract thinking, originality in analysis to formulate suitable models
4	Depth of knowledge required	Requires research-based knowledge much of which is at, or informed by, the forefront of the professional discipline and which allows a fundamentals-based, first principles analytical approach
5	Familiarity of issues	Involve infrequently encountered issues

6	Extent of applicable codes	Are outside problems encompassed by standards and codes of practice for professional engineering
7	Extent of stakeholder involvement and level of conflicting requirements	Involve diverse groups of stakeholders with widely varying needs
8	Consequences	Have significant consequences in a range of contexts
9	Interdependence	Are high level problems including many component parts or sub-problems

IOP Conf. Series: Materials Science and Engineering **88** (2015) 012003 doi:10.1088/1757-899X/88/1/012003

Thermodynamic Losses in Multi-effect Distillation Process

S Shen[1]
Key Laboratory for Desalination of Lioning Province, School of Energy and Power Engineering, Dalian University of Technology, Dalian, Lioning, China

Abstract.. The multi-effect distillation (MED) is one of desalination technologies. It is also applied in chemical engineering and other industries for evaporation, distillation, crystallization, etc. In a large multi-effect distillation plant, some tiny thermodynamic losses might have a great influence to the performance and design parameters. For the detailed analysis and design of a MED desalination plant, a series of experiments were carried out. The thermodynamic losses in a MED desalination plant is analyzed as an example to show its effect on the performance and structure parameters. The thermodynamics losses have a cumulative effect. With the increase of effect number and the concentration ratio, the thermodynamic losses shall be a dominant factor for the operation performance of a MED plant.

1. Introduction

Desalination is becoming one of the important measures for fresh water supplement in offshore area. It developed very fast in the past decades. The global desalination production is 7.48×107m3•d-1 now, and it will reach 108m3•d-1 in 2015[1]. China is one of the countries in the world facing severe water shortage. Up to now the desalination capacity in China is 7×105m3•d-1 and it will be increased to 2.6×106m3•d-1[2]. Nowadays the dominant commercial techniques of desalination are multiple stage flash (MSF), reverse osmosis (RO) and multiple effect evaporation (MEE) [3-5]. In the history the development of MEE desalination technique was restrained by the problems of scaling and capital cost. With the design of low temperature multiple effect evaporation technology, the application of thermal vapor compressor (TVC) and the cheap materials in the plant, it gets a fast development [6]. It seems that the market of MSF desalination technique is partially displaced by the MEE desalination technique. Compared with the RO sea water desalination technique, the MEE desalination shows advantages for high salinity, sea water with pollution and low sea water temperature [7]. The low temperature multiple effect evaporation technique has the characteristics of low scaling and corrosion, simplicity of sea water pretreatment, high quality

[1] To whom any correspondence should be addressed.

of distilled water, possible use of low temperature exhaust heat and suitable for low temperature sea water, etc. Its capacity reached 8% of the global desalination market [8].

MEE desalination device belongs to heat equipment. Different from the normal heat transfer process in common heat exchange equipment, when we analyze and design MEE desalination device, it is crucial to accurately understand the thermodynamic losses. There are some analyses on thermodynamic losses in MEE desalination system, but little attention and special discussion is given on this. The methods to treat the thermodynamic losses are different. For example, Darwish [9,10] and Kamali [11,12] built the simplified model of systematic analysis, where the boiling point elevation of seawater (BPE) is set to be a fixed value, and other resistance losses are ignored; El-Dessouky's [13,14] model is classic and comprehensive including thermodynamic losses caused by the BPE, flow resistance of demister, flow resistance in tubes and the condensation resistance in tubes, but the loss caused by pressure drop when steam flowing through heat transfer tube bundle. Not all the correlative formulas are from the experiments in the running conditions of MEE.

The apparent heat transfer temperature difference of LT-MEE desalination device is only between 2°C and 4°C and the device is running in the vacuum condition, when the saturation temperature is sensitive to pressure difference, so the effect of thermodynamic loss on the system performance should not be ignored. The concentration shortage of thermodynamic loss is the main reason of engineering calculation deviation. In this paper, the correlative formula between the LT-MEE desalination device resistance and the physical property are derived from the experimental results. The thermodynamic loss in the desalination device and their distribution are accurately calculated. The impacts of evaporator/condenser effects on the various thermodynamic losses in the system are analyzed. The theory on the operation characteristics of the LEE desalination system is concluded.

2. Desalination processes and analysis on thermodynamic losses

LT-MEE desalination device consists of horizontal tube falling film evaporator and condenser, the rear condenser, flash tank and non-condensable gas drawn out system, etc., as shown in figure 1.

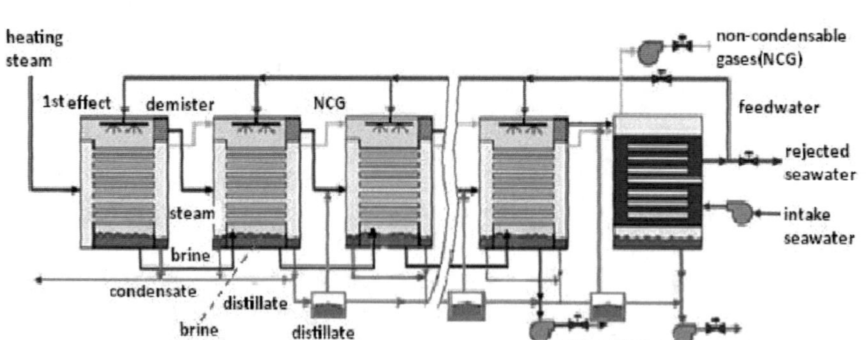

Figure 1. Schematic of LT-MEE desalination system

The intake seawater is divided into two parts after coming out of the rear condenser, where it is used to cool the steam. One part is fed into the evaporators as feed water, and the other part is rejected back into the sea as rejected seawater. The feed water flows through a distributor and is equably distributed on the top row tube in the evaporators. Then under the action of gravity, it flows downward along the tube bundle in the form of horizontal tube falling film. The seawater is heated by the steam in the tube and some turns into steam, which is called secondary steam and flows into the steam tank in the evaporator. Then it flows through a demister, where the seawater droplets carried by the steam are gotten rid of, and goes into the tubes of next effect evaporator as the heating steam. The rest of the concentrated brine flows into the brine space of next effect evaporator, where some strong brine flashes because of the pressure change. The strong brine is finally flow into the last effect evaporator, where it is discharged out of the system.

The heating steam, provided by external heat source, flows into the steam-inlet tube header of first effect evaporator, and releases heat and condenses inside the horizontal tubes to heat the seawater outside of the tubes. The condensed water flows into the fresh water tube chamber, then either returns to the boiler feed water system or into the flash tank according to the different quality of heating steam. The secondary steam produced in the first effect flows into the tubes of the second effect evaporator, releasing heat and condensing. The condensed water goes into the flash tank, where some flash into steam because of the reduced pressure. The flashed steam flows into the steam-inlet tube header of next effect evaporator, and the condensed water flows into the fresh water tube chamber of next effect evaporator. By that analogy, the steam produced in the last effect evaporator condenses into water in the rear condenser. All of the condensed water is discharged out of the unit as product water.

The average apparent temperature difference is the ratio of total temperature difference, which is the temperature difference between the temperature of heating steam in the first effect evaporator and the saturation temperature of secondary steam produced in the last effect evaporator, to the number of effect. Compared with fresh water system under the same pressure, in desalination system the condensing temperature of the steam in the tubes decreases because of boiling point elevation of the seawater and the flow resistance of the steam. That leads to an obvious reduction of the effective heat transfer temperature difference in evaporators. The difference between the apparent heat transfer temperature difference and effective heat transfer temperature difference is known as thermodynamic loss of the heat transfer process.

In multi-effect evaporation desalination process, the flow direction of the steam produced in evaporators is shown by the arrows in figure 2. When the steam flows through the tube bundle 5 to steam collection chamber, the shadow part in the right figure in figure 2, there exists the pressure drop Δp_t, so the saturation temperature of steam close to the steam collection chamber will be lower than the evaporation temperature in center of tube bundle. The steam flows through the demister 6 and passes by the steam channel 3 and goes into the steam-inlet tube header of next effect evaporator. In the process, another pressure drops Δp_d, Δp_f, Δp_b form and pressure drops down further. The corresponding saturation temperature also further reduces. The steam flow in the tubes of next effect evaporator will produce pressure drops Δp_c. Therefore, along the tube , the saturation temperature of steam condensation in the tube will gradually reduce. All of the temperatures reductions lead to condensation temperature falling, which makes the heat transfer temperature difference in the evaporator reduce. This is the thermodynamic loss caused by the steam flow resistance in the device.

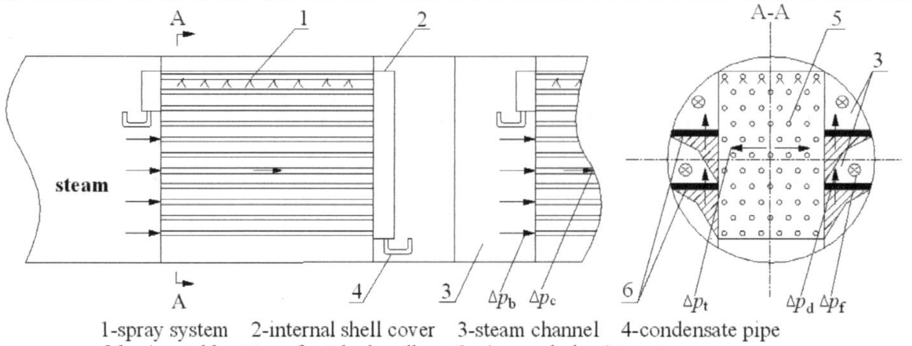

Figure 2. Schematic of resistance loss in LT-MEE desalination system

In addition, the boiling point of the falling film evaporation seawater outside the tubes is higher than fresh water under the same pressure, which means that the heat transfer temperature difference further reduces. While the steam evaporated from the seawater is superheated corresponding to the pressure inside the evaporator, and its condensation temperature will reduce a little. That is the heat loss caused by seawater properties. The seawater boiling point elevation (BPE) and temperature loss (Δt_p) caused by steam flow resistance can be reflected qualitatively in figure 3.

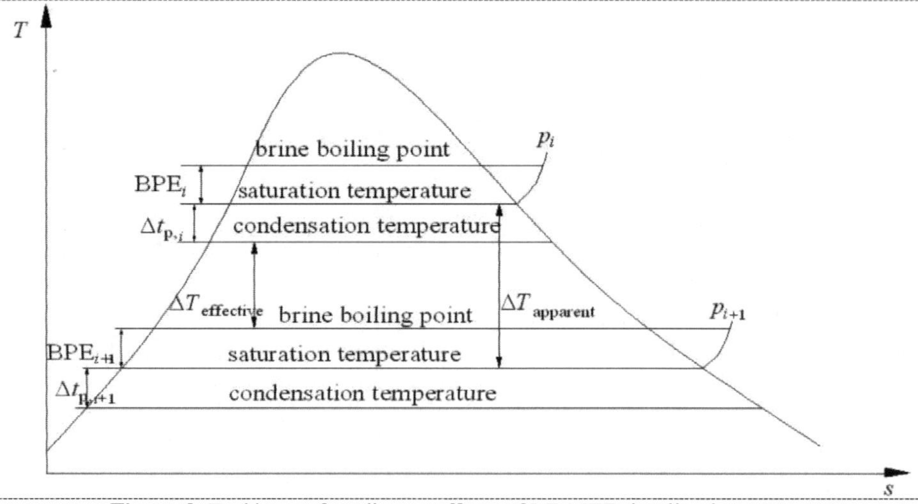

Figure 3. Tephigram for adjacent effects of LT-MEE desalination system

3. Mathematical model of thermodynamic loss

The thermodynamic mathematical model of LT-MEE desalination system is similar to that in literature [15], the formulas to calculate the thermodynamic losses are as follows.

The formula to calculate seawater boiling point is given in literature [16] :

$$\text{BPE} = AX + BX^2 + CX^3 \quad (1)$$

Where,

$$A = \left(8.325\times10^{-2}+1.883\times10^{-4}T+4.02\times10^{-6}T^2\right)$$

$$B = \left(-7.625\times10^{-4}+9.02\times10^{-5}T-5.2\times10^{-7}T^2\right)$$

$$C = \left(1.522'\ 10^{-4} - 3'\ 10^{-6}T - 3'\ 10^{-8}T^2\right)$$

Where , X is salinity, g/kg. T is water temperature, °C.

The flow resistance through the mesh demister is expressed as [17],

$$\Delta p_{\rm d}=3.88178\left(\rho_{\rm d}\right)^{0.375798}\left(v\right)^{0.81317}\left(d_{\rm w}\right)^{-1.56114147}L_{\rm d} \quad (2)$$

where, $\rho_{\rm d}$ is the density of the mesh, kg/m^3. v is the velocity of vapor phase , m/s. d_w is the diameter of the metal mesh wire, mm. L_d is the thickness of the mesh, m.

The flow resistance in the steam channel, that is the pressure drop when the steam flow after the mesh demister and go into the steam tube inlet, including local resistance $\Delta p_{\rm b}$ and flow resistance along the passage $\Delta p_{\rm f}$, could be calculated from[18]:

$$\Delta p_{\rm b} = \rho\alpha\left(\frac{v^2}{2}\right) \quad (3)$$

$$\Delta p_{\rm f}=\frac{1.306\times10^{-4}M^2L\left(1+3.6/\delta_i\right)}{\rho\delta^5} \quad (4)$$

where, M is the steam flow through the passage, kg/s. v is the flow velocity when the saturated steam flows through the minimum cross section in the tube bundle, m/s. R is the radius of the bend , m. β is the bend angle, °. L is the length of the steam passage, m. ρ is the density of the steam in the passage, kg/m^3. δ is the equivalent diameter of the passage, m. α is the local loss coefficient, calculated by:

$$\alpha=\left(0.00146+0.00181\times\left(\frac{\delta_i}{R}\right)^{3.5}\right)\beta \quad (5)$$

The flow resistance through the tube bundle could be calculated from the formula in the literature[19] , which is obtained from the experimental results that steam sweeps across the horizontal falling film tube bundle.

$$\Delta p_{\rm t}=\frac{fNG^2}{2\rho} \quad (6)$$

where, N is the number of columns. G is the steam mass flux per unit cross section area, kg/(m^2·s). f is the resistance correction coefficient. When the equilateral triangle arrangement of tubes is adopted, the outer diameter of the pipe is 25.4mm, is the spray density, kg/(m·s). D is the outer diameter of the tube, m. μ is the dynamic viscosity of saturated spray seawater, Pa/s. η is the kinematic viscosity of saturated steam, m^2/s.

The flow resistance in the tube condensation could be calculated according to the literature[20]:

$$\Delta p_{\rm c} = L_{\rm t}\cdot c\rho v^2 Re_{\rm v}{}^a \overline{x}^b \quad (8)$$

IOP Conf. Series: Materials Science and Engineering **88** (2015) 012003 doi:10.1088/1757-899X/88/1/012003

where, L_t is the length of the pipe, m. *c*, *a* and *b* are empirical constants. *v* is the steam inlet velocity, m/s. $\bar{x}$ is the integral average dryness of steam. The inner diameter of the pipe is24mm, Re_v<3000 , a=−0.234, b=5/3, c=4.6.

4. Analysis of examples

Based on above models, through the calculation and analysis on the examples, we obtain the thermodynamic loss in large LT-MEE desalination device and the distribution in every evaporator. It is shown that the total heat transfer difference loss, caused by the resistance loss and seawater boiling point elevation, changing with the effect number of evaporators. The calculation parameters and the configuration parameters of evaporator/condenser in LT-MEE desalination plants with various effects are listed in table 1 and table 2 respectively.

Table1. Parameter range for calculation

Parameter	Value	Parameter	Value
Distillate production/(t·d^{-1})	15000	Concentration ratio	1.5
Heating steam temperature/℃	68	Salinity/‰	32.0
Evaporating temperature in the last effect/℃	46	Feed seawater temperature/°C	41
Length of steam channel/m	2	Diameter of steam channel/m	3

Table 2. Evaporator/condenser configuration parameters of desalination plants with various effects

Parameter	Value				
Arrangement mode of tube bundle			Regular triangle		
Tube spacing-outer diameter ratio			1.3		
Row-column ratio of tube bundle			2.2		
Outer diameter of tube/mm			25.4		
Thickness of tube/mm			0.7		
Length of tube/m			7		
Wire diameter of demister/mm			0.3		
Packing density of demister/kg·m^{-3}			150		
Pad thickness of demister/m			0.05		
Number of evaporators/ condensers *n*	6	8	10	12	14
Area of demister/m^2	7.6	5.8	4.8	4.2	3.7
Heat transfer area of each evaporator /m^2	7828	8115	8504	9087	9799
Column number of tube bundle	80	81	83	86	89
Row number of tube bundle	177	179	184	190	197

Take the example of 6 effects LT-MEE desalination plant for example. Proportions of various resistance losses are shown in figure 4. It can be seen that the thermodynamic loss caused by flow resistance occupy 10.8%, where the proportion of thermodynamic loss caused by the steam passage flow resistance is minimal, only 0.1%, that can be neglected.

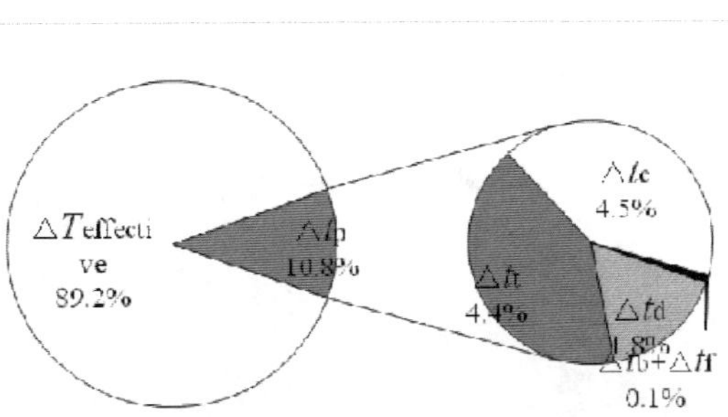

Figure 4. Proportions of various resistance losses

The Variation of thermodynamic loss, due to the demister flow resistance, the tube bundle flow resistance and the condensation flow resistance in the tubes, with the number of effects are shown in figure 5 to figure 7 respectively. With the fixed value of effect number, the thermodynamic losses caused by all the flow resistances increase accelerating with the rise of the effect number. With the rise of the effect number, the evaporation temperature reduces gradually, and the latent heat of vaporization increases, while the amount of condensing water and the concentrated seawater gathered effect by effect increase, and the flash steam from the condensing water and the concentrated seawater in each evaporator (m_{df}, m_{bf}) increases effect by effect. Both these factors lead to the amount of the steam vapor decreases first then increases in each of evaporators, as shown in figure 8. The density of saturated vapor in the evaporators decreases gradually, it would affect the steam velocity together with the amount of the steam, and then further affect various flowing resistance. According to steam characters, the lower of evaporating temperature, the more saturated temperature reduction caused by unit pressure drop, so the heat loss in low temperature evaporator is more obvious.

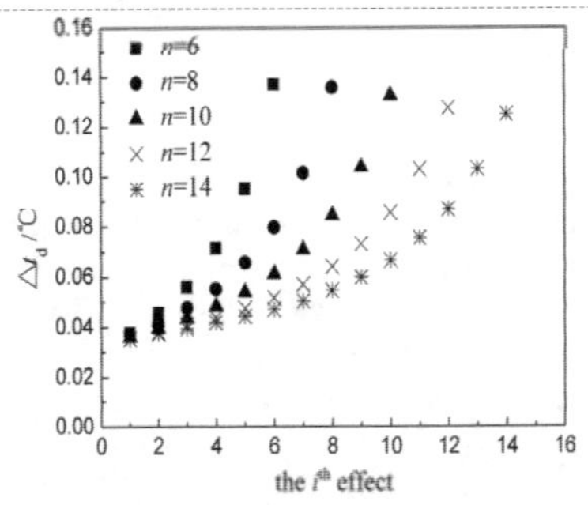

Figure 5. Variation of thermodynamic loss due to the demister with the number of effects

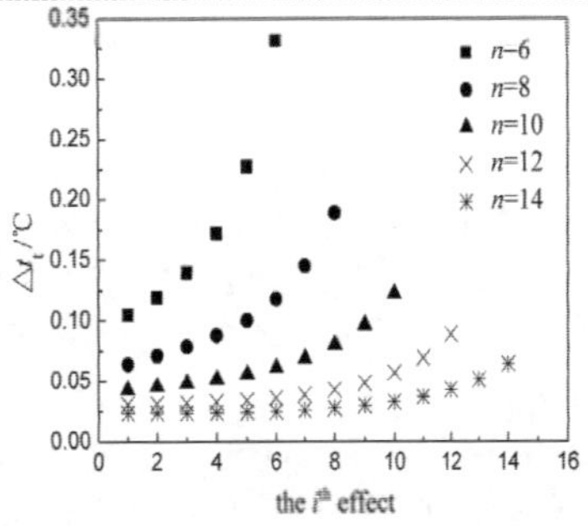

Figure 6. Variation of thermodynamic loss due to the tube bundle with the number of effects

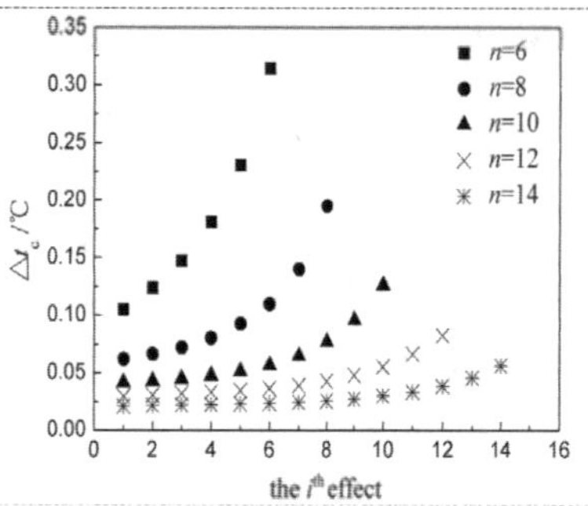

Figure 7. Variation of thermodynamic loss due to the condensation process with the number of effects

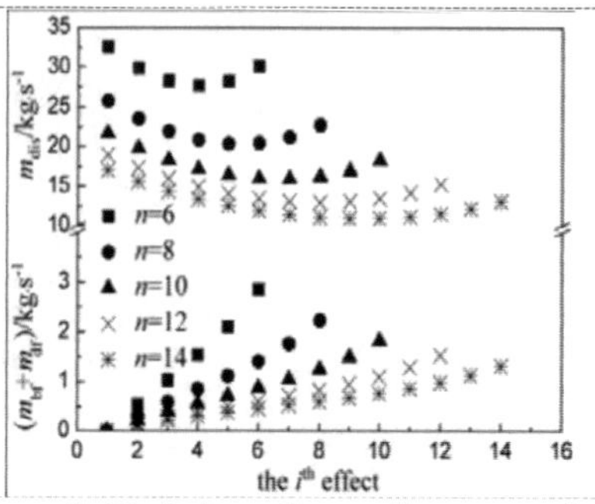

Figure 8. Variation of mass flow rate of the generated vapor with the number of effects

The thermodynamic loss caused by various resistances in the same sequence evaporator decreases with rise of evaporator number. Because of the fixed total amount of fresh water, with the increase of the evaporator number, the load of every effect decreases. At the same time, the saturated temperature in the same sequence evaporator increases and the latent heat of vaporization decreases, which makes the steam in the evaporator slowly decline, as shown in figure 8. When the saturation temperature increases, saturated steam density increases, skimmer flowing resistance in the same sequence evaporator and tube condensing flow resistance decrease with increasing number of evaporator. Keep the steam velocity through the demister in the first effect evaporator fixed, then the demister area is determined. so with the rise of evaporator number, the flowing resistance of the first effect skimmer keeps unchangeable. The tube bundle flowing resistance are not only affected by the above factors, but also affected by the falling film flow spray density [15]. Calculation example in this paper is based on the condition of consistent total amount of feed water, with the increase of evaporator number, the seawater spray density in the same sequence evaporator decreases, as well as the vapor production decreases. The co-action of steam flow, steam density and the water spray density makes the tube flow resistance in same sequence evaporator decreasing with the increasing of evaporator number. Meanwhile, with the increase of the evaporating temperature, the saturation temperature reduction caused by the unit pressure drop turns smaller. Therefore, various thermodynamic losses in the same sequence evaporator decreasing with the increasing number of evaporator/condenser.

Table 9 shows the proportion of the thermodynamic loss caused by each resistance and the total thermodynamic loss caused by all the resistances changing with evaporator effect numbers. It is shown that with the increase of evaporator number, the total heat loss caused by flow resistance decreases. The proportion of thermodynamic loss caused by skimmer flow resistance continuously increases. When the effect number increases from 6 to 14, the proportion increases from 16.8% to 49.9%, and the proportion of tube bundle flow resistance and tube condensation flow resistance decrease from 41.4% and 41.7% to 26% and 24% respectively.

In this paper the total resistance ranges from 1260Pa to 1830Pa, so the average flow resistance per effect ranges from 90Pa-305Pa, and the maximum flow resistance in single evaporator is about 395Pa. therefore, the characteristic of "low flow resistance" could be concluded in low temperature multi-effect evaporation desalination device. The apparent heat transfer temperature difference is only between 2°C and 4°C, which shows typical "little temperature difference" heat transfer. And the heat transfer process is operated under the vacuum and "in saturation state". The pressure has a close relation to temperature. The change of saturation temperature can affect the effective heat transfer temperature difference, and it is "highly sensitive" to pressure change. Take the evaporating temperature of 60°C for example, resistance loss 395Pa corresponds to 0.43°C, which make the effective heat transfer temperature difference decrease by 10.7%-21.5%. It is seen that if the low flow resistance that less than 400Pa, is not considered carefully, it is enough to make the disruptive results of water desalination device performance. The structure design of the desalination device should reduce the vapor flow resistance as possible, as well as be forecast accurately in thermodynamic calculation and consider the thermodynamic loss caused by flow resistance, and the "low flow resistance" is required in the device performance calculation and design.

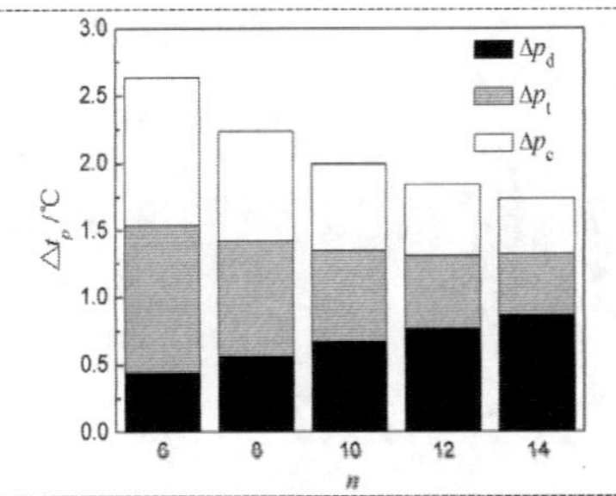

Figure 9. Variation of thermodynamic loss due to the condensation process with the number of effects

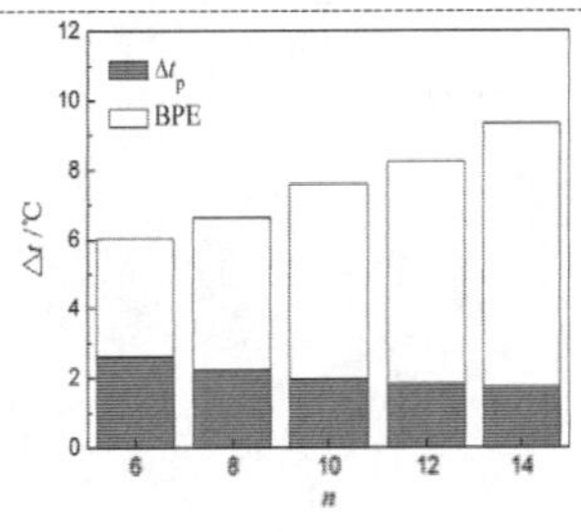

Figure 10. Variations of total thermodynamic losses with the number of effect

We have discussed thermodynamic losses caused by the flow resistance device, seawater boiling point elevation further exacerbates the loss. Because seawater boiling point is higher than that of fresh water, on the premise of fixing the pressure inside and outside of the evaporator tubes, that makes heat transfer temperature difference reduced between the inside and outside, which could be calculated from formula (1). The loss of total heat transfer temperature difference for the device caused by the flow resistance loss and seawater boiling point elevation loss is shown in Figure 10. With rise of evaporator effect number, the total thermodynamic loss increases, where the proportion of the loss caused by seawater boiling point elevation increases significantly. When the effect number is 6-14, the loss caused by seawater boiling point elevation accounts for 56.3%-81.4%.

As shown in formula (1), BPE is the function of salinity and temperature, and the salinity affects BPE more greatly. Therefore, when the other parameters are unchanged, BPE in every effect evaporator does not change a lot. In fact, it is the increased evaporator number makes rise of the total temperature drop. With the evaporator number increasing, temperature drop caused by the resistance loss decreases, while that caused by BPE increases, under the co-action of both the factors, the total thermodynamic losses rises.

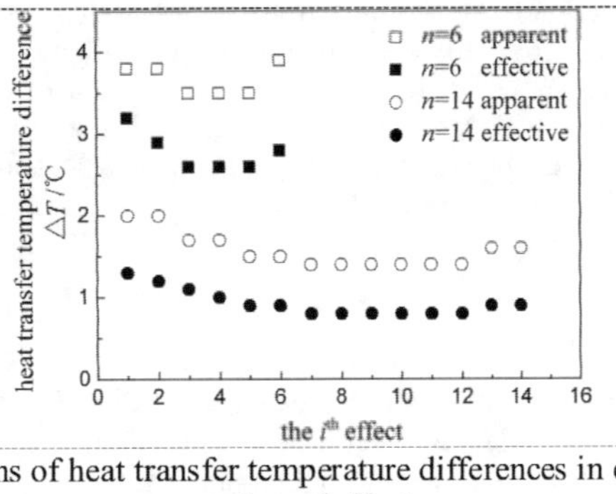

Figure 11. Variations of heat transfer temperature differences in each effect with the number of effects

Figure 11 shows the apparent and effective temperature difference in each effect changing with the effect numbers. It is shown that when the effect number is 6, the apparent temperature difference in each effect is 3.5-3.9°C, and the effective temperature difference is 2.6-3.2 °C. The proportion of the corresponding thermodynamic loss is 15.8% -28.2%. When the effect number is 14, the apparent temperature difference in each effect is 1.4-2 °C, the effective temperature difference 0.8-1.3 °C. The proportion of the corresponding thermodynamic loss reaches up to 35% -43.8%. Obviously desalination plant is "high sensitive" to various types of losses.

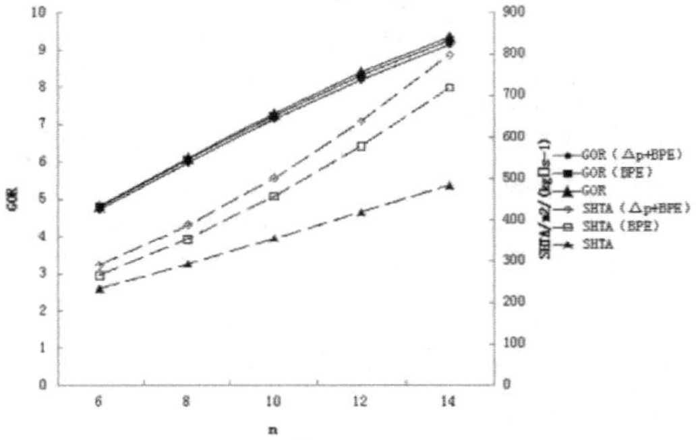

Figure 12. Effect of thermodynamic loss on performance of LT-MEE desalination plant

The performances of LT-MEE desalination plant, including the gained output ratio (GOR) and the specific heat transfer area (SHTA), are compared in figure 12 when the thermodynamic losses are considered and ignored and only resistance loss is ignored. It can be seen that when the thermodynamic losses are considered the values of GOR are same for the same effect number plants, but there is bigger difference in SHTA. That is the heat transfer area of the evaporator need to be increased, and with the increase of effect number, the difference is becoming more and more obvious. Within the scope of this paper, the relative deviation of SHTA reaches 39.4% if the thermodynamic losses are ignored compared with when the losses are considered. the relative deviation of SHTA also could reach 10% when only resistance loss is ignored. Therefore, a large design error will occur if the thermodynamic losses are ignored.

Increasing the number of evaporator can improve GOR and decrease the resistance loss, but intensify the thermodynamic loss caused by seawater boiling point and increase the total thermodynamic loss, and then increase the heat transfer area. Therefore, the two factors should be weighed when analyzing the economy to determine the optimal effect number.

The theory of "small temperature difference, low flow resistance, saturated state, high sensitive" on the operation characteristics of LT-MEE desalination plant is obtained according to the analysis on the thermodynamic losses in the plant. It is the basis to scientifically understand the theory on characteristics and fully analyze every thermodynamic loss in the plant for building integrated theory and method to design LT-MEE desalination plant.

5. Conclusion

According to the operation characteristics of large LT-MEE desalination plant, calculation and analysis on the distribution of thermodynamic losses caused by various factors in each evaporator are carried on. And the influences of effect number on every thermodynamic loss is investigated as well. In the scope of calculation in this paper, the following conclusions can be obtained:

(1) The theory of "small temperature difference, low flow resistance, saturated state, high sensitive" on the operation characteristics of LT-MEE desalination plant is obtained according to the analysis on the thermodynamic losses in the plant.

(2) When the effect number is fixed, the thermodynamic losses cause by demister flow resistance, tube bundle flow resistance and condensation flow resistance in the tube increase with the increase of evaporator sequence, and the increase rate gradually becomes larger.
(3) With the increase of the effect number, the thermodynamic loss cause by steam flow resistance in the same effect sequence evaporator slowly declines. The gross thermodynamic losses decreases caused by flow resistance. The proportion of thermodynamic loss cause by demister flow resistance rises. The tube bundle flow resistance and condensation flow resistance in the tube decrease. The thermodynamic loss caused by seawater boiling point elevation increases, and its proportion increases significantly.
(4) The total thermodynamic losses increases with the increase of the effect number, and the proportion is fairly considerable. It must be payed full attention in designing the plant.

6. References

[1] Global Water Intelligence. IDA desalination yearbook 2012-2013: Market profile[R]. 2012.
[2] The General Office of the State Council. Opinions of the General Office of the State Council on accelerating the development of the seawater desalination industry[J]. *State Council Gazette*, 2012, 13: 57-59.
[3] Ruan Guoling, Feng Houjun. Technical process in seawater desalination technology at home and abroad [J]. *China Water & Wastewater*, 2008, 24 (20): 86-90.
[4] Al-Sahali M, Ettouney H. Developments in thermal desalination processes: Design, energy, and costing aspects[J]. *Desalination*, 2007, 214 (1–3): 227-240.
[5] Deng R Y, Xie L X, Lin H, Liu J, Han W. Integration of thermal energy and seawater desalination[J]. *Energy*, 2010, 35 (11): 4368-4374.
[6] Mezher T, Fath H, Abbas Z, Khaled A. Techno-economic assessment and environmental impacts of desalination technologies.[J]. *Desalination*, 2011, 266: 263-273.
[7] Ghaffour N, Missimer T M, Amy G L. Technical review and evaluation of the economics of water desalination: Current and future challenges for better water supply sustainability [J]. *Desalination*, 2013, 309: 197-207.
[8] ESCWA (Economic and Social Commission for Western Asia). Role of desalination in addressing water scarcity 2009 [R]. 2009: 9.
[9] Darwish M A, Al-Juwayhel F, Abdulraheim H K. Multi-effect boiling systems from an energy viewpoint[J]. *Desalination*, 2006, 194(1-3): 22-39.
[10] Darwish M A, Abdulrahim H K. Feed water arrangements in a multi-effect desalting system [J]. *Desalination*, 2008, 228 (1-3): 30-54.
[11] Kamali R K, Abbassi A, Sadough Vanini S A, Saffar Avval M. Thermodynamic design and parametric study of MED-TVC [J]. *Desalination*, 2008, 222(1-3): 596-604.
[12] Kamali R K, Abbassia A, Sadough Vanini S A. A simulation model and parametric study of MED–TVC process [J]. *Desalination*, 2009, 235(1–3): 340-351.
[13] El-Dessouky H T, Ettouney H M. Multiple-effect evaporation desalination systems. thermal analysis[J]. *Desalination*, 1999, 125 (1–3): 259-276.
[14] El-Dessouky H T, Ettouney H M, Mandani F. Performance of parallel feed multiple effect evaporation system for seawater desalination[J]. *Applied Thermal Engineering*, 2000, 20(17): 1679-1706.
[15] Shen S Q, Zhou S H, Yang Y, Yang L P, Liu X H. Study of steam parameters on the performance of a TVC-MED desalination plant[J]. *Desalination and Water Treatment*, 2011, 33(1-3): 300-308.
[16] El-Dessouky H T, Ettouney H M. Fundamentals of Salt Water Desalination [M]. 2002: 587.
[17] El-Dessoukey H T, Alatiqi I M, Ettouney H M, Al-Deffeeri N S. Performance of wire mesh mist eliminator [J]. *Chemical Engineering and Processing*, 2000: 129-139.
[18] Oak Ridge National Laboratory (ORNL). Research and Development Progress [R]. USA: United States Department of the Interior, 1967.

[19] Shen Shengqiang, Liu Hua, Feng Yin, Chen Shi，Gong Luyuan，Liu Rui，Chen Xue. Steam flow resistance across horizontal tube bundle under vacuum condition [J]. *CIESC Journal*, 2013, 64(3): 886-890.
[20] Shen S Q, Liu R, Yang Y, Liu X H, Chen J X. Condensation character of a stratified flow inside a horizontal tube [J]. *Desalination and Water Treatment*, 2011, 33(1-3): 218-223.

Sensitivity study of Bubble diameter for prediction of flow pattern in homogeneous bubble column regime

M.Pourtousi[1] , P Ganesan[1*], J N Sahu[2], Ghufran Redzwan[3]
[1]Department of Mechanical Engineering, Faculty of Engineering, University of Malaya, 50603, Kuala Lumpur, Malaysia
[2]Department of Petroleum and Chemical Engineering, Faculty of Engineering, Institut Teknologi Brunei, Tungku Gadong, P.O. Box 2909, Brunei Darussalam
[3] Institute of Biological Sciences, University of Malaya, 50603 Kuala Lumpur, Malaysia

E-mail: poo_ganesan@um.edu.my, mo_poortoosi@yahoo.com, sahujaya@gmail.com, ghufran@um.edu.my

Abstract. Determining the bubble diameter size in a bubble column rector plays an important role to accurately predict flow pattern in a bubble column reactor. This paper employs the Eulerian-Eulerian method to numerically investigate the sensitivity study of bubble diameter size in a cylindrical bubble column reactor. Existing experimental results in the literature are used to validate the implementation of the proposed numerical method. In our simulation various bubble diameter size (i.e., 3-5.5mm) are used to find an appropriate bubble size inside the bubble column when the regime is homogeneous (superficial gas velocity = 0.005m/s). The result shows that bubble diameter 4mm is a reasonable size for flow pattern prediction inside the column.

1. Introduction

Bubble column reactors are used as multiphase contactors and reactors in chemical, petrochemical, biochemical and metallurgical industries such as fermentation and biological wastewater treatment [1-4]. The bubble column reactors are often cylindrical vessels equipped with a gas distributor at the bottom. The gas distributer spargers fluids bubbles into vessel containing either a liquid phase or a liquid–solid suspension. Bubble column reactors have many advantages over other types of reactors such as high heat and mass transfer coefficients, low maintenance and operating costs and high durability of the catalyst or other packing material [4-6].

To date, the numerical approaches have been used in many studies to predict the bubble column reactors because of the difficulties that are still found in designing and scaling-up the bubble columns [7, 8]. Computational fluid dynamics (CFD) has improved our partial knowledge of the complex hydrodynamic processes occurring inside the bubble column [9].

[1] To whom any correspondence should be addressed.

Inside the bubble column, an interaction exist between the dispersed gas and the liquid that affects the interphase forces (e.g. drag force, lift force, etc.,), turbulence and bubble diameter in the column. Therefore, a proper solution for the bubble columns is dependent on the correct modeling of interphase forces, turbulence models and appropriate bubble diameter inside the column [7, 10]. Several models for interphase forces and turbulence have been reported in the literature, [7, 10, 11]. For instance the drag force, lift force, turbulent dispersion and added mass model have been studied by [7, 10]. In addition, various turbulence models (e.g. standard *k–ε* model, Reynolds Stress Model (RSM), Large Eddy Simulation (LES)) have been used to predict flow pattern inside the bubble column [8, 10, 12, 13]. Since many investigations studied interface forces and turbulence models on prediction of bubble column, this study specifically focuses on effect of bubble diameter size on prediction of flow pattern using Eulerian–Eulerian framework in a homogeneous flow regime.

2. Methodologies

2.1 Governing equations

The two-phase model based on the Eulerian–Eulerian approach is used to simulate dispersed gas-liquid flows in bubble columns. Each fluid (or phase) is treated as a continuum in the domain under consideration in this approach. The phases share this domain and interpenetrate as they move within it. The Eulerian modelling framework is described as follows:

Continuity equation:

$$\frac{\partial}{\partial t}(\rho_k \epsilon_k) + \nabla(\rho_k \epsilon_k u_k) = 0 \quad (1)$$

Momentum transfer equation:

$$\frac{\partial}{\partial t}(\rho_k \epsilon_k u_k) + \nabla(\rho_k \epsilon_k u_k u_k) = -\nabla(\epsilon_k \tau_k) - \epsilon_k \nabla p + \epsilon_k \rho_k g + M_{I,k} \quad (2)$$

The terms on the right hand side of Eq. (2) represents the stress, the pressure gradient, the gravity and the ensemble-averaged momentum exchange between the phases due to interface forces. The pressure is shared by the both phases. The stress term of phase *k* is described as follows:

$$\tau_k = -\mu_{eff,k}\left(\nabla u_k + (\nabla u_k)^T - \frac{2}{3} I(\nabla u_k)\right) \quad (3)$$

where $\mu_{eff,k}$ is the effective viscosity. The effective viscosity of the liquid phase is composed of three contributions: the molecular viscosity, the turbulent viscosity and an extra term due to bubble induced turbulence as given below;

$$\mu_{eff,L} = \mu_L + \mu_{T,L} + \mu_{BI,L} \quad (4)$$

The model proposed by Sato and Sekoguchi [14] was used to take account the turbulence due to the movement of bubbles, which can be written as

$$\mu_{BI,L} = \rho_L C_{\mu,BI} \varepsilon_G d_B |u_G - u_L| \quad (6)$$

with the model constant of $C_{\mu,BI}$ is set to 0.6 as recommended in ref [14]. The calculation of the effective gas viscosity is based on the effective liquid viscosity [23] which is:

$$\mu_{eff,G} = \frac{\rho_G}{\rho_L} \mu_{eff,L} \quad (5)$$

The total interfacial force acting between the two phases are based on the interphase drag force and turbulent dispersion force, which can be written as follows:

$$M_{I,L} = -M_{I,G} = M_{D,L} + M_{TD,L} \quad (7)$$

The interphase momentum transfer between the gas and liquid due to the drag force is given by:

$$M_{D,L} = -\frac{3}{4} \epsilon_G \rho_L \frac{C_D}{d_B} |u_G - u_L|(u_G - u_L) \quad (8)$$

In our numerical study a constant drag coefficient of 0.44 is used which is suggested in references [12, 15]. The turbulent dispersion force derived by Lopez de Bertodano [16] is used, which is

$$M_{TD,L} = -M_{TD,G} = -C_{TD} \rho_L k \nabla \epsilon_L \quad (9)$$

where k and C_{TD} are the liquid turbulent kinetic energy per unit of mass and turbulent dispersion coefficient respectively. In our simulations, values of 0.3 is used for the turbulent dispersion coefficient as suggested in References [10, 16]. Turbulence has been modelled using a 'standard' k-ε model.

2.2 Geometrical structure and Simulation cases

A three-dimensional (3D) domain of a bubble column with a height and diameter of 2.6 m and 0.288 m, respectively is used. Air bubbles are sparged in the quiescent water in the column using a ring sparger. The gas has a superficial velocity of 0.005 m/s at the ambient condition.

2.3 Grid

A structured grid based on hexahedral grid as shown in Figure.1 is used throughout the domain. The type of grid adopted herein is almost identical to that used in Boutet et al. [17]. The axial length of the domain is divided into 60 grid elements. The grid namely Grid 1, typically has about 40500 elements.

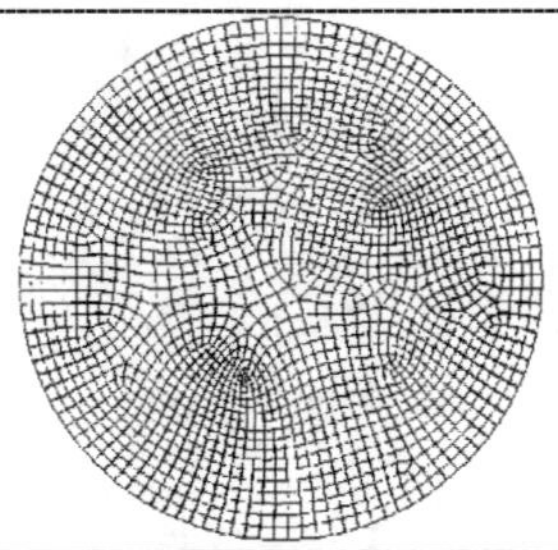

Figure 1. Grid intensity of the computational model consisting of 40500 structural elements.

2.4 Boundary conditions and numerical methods

The gas velocity from each nozzle of the ring spargers is calculated based on a superficial gas velocity of 0.005 m/s. The top surface of the bubble column is treated as the 'degassing' boundary condition. On the side walls, a no-slip boundary condition is used for the liquid phase and a free-slip condition for the gas phase. Simulations were carried out using the commercial CFD software package of ANSYS-CFX 13. The time step of 0.1s is used in our simulations and the sensitivity of a much lower time step, i.e., 0.01s are investigated.

3. Results and discussion

In industrial bubble column, bubble usually sparges trough the sparger. As bubbles travel to the top surface of the column, they merges and break into number of small bubbles. These interaction results in bubbles with different diameters inside the bubble column. In homogeneous regime, bubbles have almost same diameter and velocity and the rate of interaction between bubbles are small. Therefore, we can select one bubble diameter instead of range of bubble size and this way requires less computational time in comparison with other methods (e.g., population balance model). We specify only one bubble diameter for one case study and the sensitivity of the use of different diameters of gas bubble in predictions of axial liquid velocity and gas hold-up in the bubble column have been also investigated. The radial distribution of the time-averaged axial liquid velocity and the gas hold-up at column height 1.6 m is shown in Figure. 2 and Figure. 3, respectively based on bubble diameters of 4 mm to 5.5mm. Figure. 4 shows the planar averaged gas hold-up at various column heights. Such results from the experimental investigation of Pfleger and Becker [12], which both used a bubble diameter of 4mm, are included in the Figures for a comparison. The use of a bubble diameter of 4.5mm, an increase about 0.5 mm from that of bubble with 4mm diameter, only results in small differences in the radial of the axial liquid velocity (Figure. 2), the gas hold-up (Figure. 3) and the averaged gas hold-up (Figure. 4) in the column. However, this is not the case for the bubble diameter of 5 mm (or 5.5 mm) with an increase of 1mm (or 1.5mm) from that of bubble with 4mm size. The differences are significant. For example, the radial profile of the axial liquid is asymmetric with more positive velocity at right side of the column. Consequently, the radial profile of the gas rises asymmetrically in the column (Figure.3) and has a two-peak profile at the central region of the column.

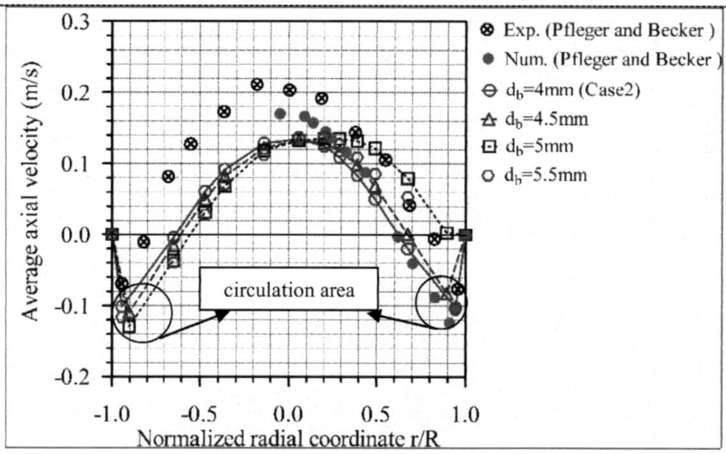

Figure 2. Comparison between the average of axial liquid velocity from various bubble diameter sizes and that from experimental and numerical from Pfleger and Becker [10] at height 1.6m.

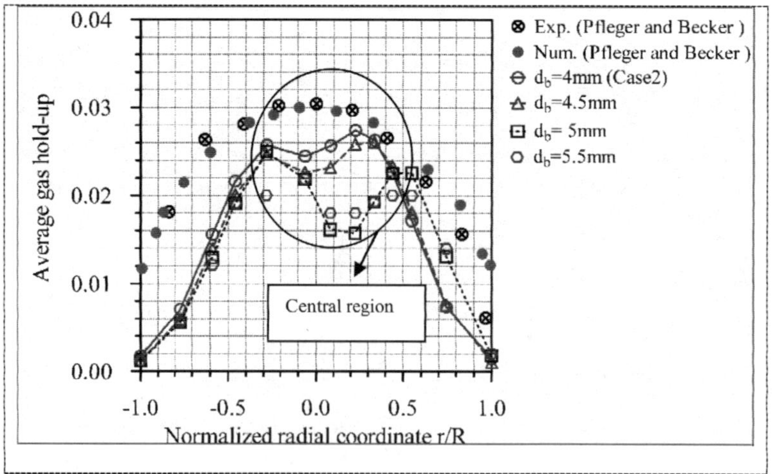

Figure 3. Comparison between the average of gas hold-up from various bubble diameter sizes and that from experimental and numerical from Pfleger and Becker [10] at height 1.6m.

The amount of gas for case of bubble diameter 5 and 5.5mm at different heights of bubble column is lesser than smaller bubble (i.e., 4 and 4.5mm). For example, at 2m height the planner average gas hold-up for 4 and 5mm are 0.0156 and 0.0138 respectively (see Figure.5). The increase of gas hold-up may attribute to the fact that, as bubble size decreases, bubble interfacial area increases. This finding is consistent with previous studies [18].

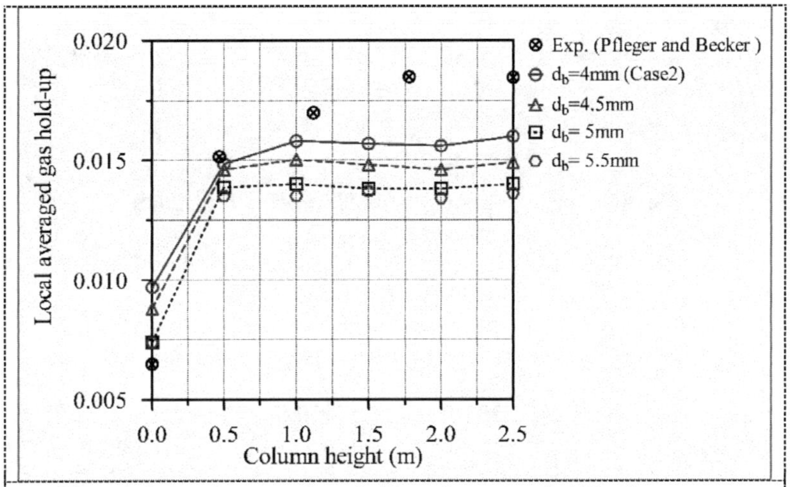

Figure 4. Local averaged gas hold-up for various bubble diameters at different column heights.

Figure.5 shows that the amount of gas tends to move to the center specially near the sparger region. As bubbles travel to the surface, the fraction of the gas decreases. It is interesting to say that merging bubble near the sparger may be more than other parts of the bubble column and this is a good reason that near this region bigger bubble have a good agreement with Pfleger' experimental study.

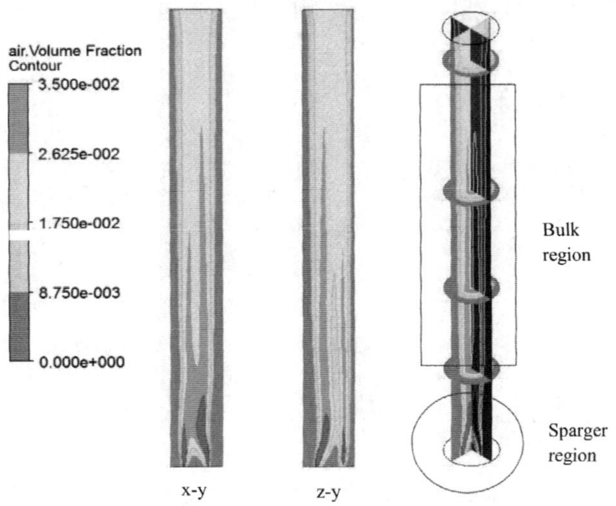

Figure 5. Contour plot for air fraction inside the bubble column.

4. Conclusions

In this investigation we studied various bubble diameter sizes to show the appropriate prediction of the flow pattern inside the bubble column. In view of all bubble diameter sizes, the bubble diameter 4mm is in reasonable agreement with the experimental data for the overall and local gas hold-up. In addition, the results show that the bubble diameters near the sparger region are 5mm and 5.5mm.

5. Acknowledgments

This research is financially supported by University of Malaya, Ministry of Higher Education High Impact Research (UM.C/HIR/MOHE/ENG/13) and University Malaya Research Grant (UMRG: RG121/11AET).

6. References

[1] Shah Y et al. Design parameters estimations for bubble column reactors AIChE Journal 28 no. 3 pp 353-379 1982
[2] Kantarci N Borak F and Ulgen K O Bubble column reactors Process Biochemistry 40 pp 2263-2283 2005
[3] Sahu J N et al 2009 Performance of a modified multi-stage bubble column reactor for lead(II) and biological oxygen demand removal from wastewater using activated rice husk J Hazard Mater 161 no 1 pp 317-24
[4] Pourtousi M Sahu J and Ganesan P 2014 Effect of interfacial forces and turbulence models on predicting flow pattern inside the bubble column Chemical Engineering and Processing: Process Intensification 75 pp 38-47 2014.
[5] Şal S Gül Ö F and Özdemir M 2013 The effect of sparger geometry on gas holdup and regime transition points in a bubble column equipped with perforated plate spargers Chemical Engineering and Processing: Process Intensification
[6] Salierno G L et al. 2013 Discrete axial motion of a radioactive tracer reconstructed from the response of axially aligned detectors: Application to the analysis of a bubble column dynamics Chemical Engineering Science 100 no 0 pp 402-412
[7] Gupta A and Roy S 2013 Euler–Euler simulation of bubbly flow in a rectangular bubble column: Experimental validation with Radioactive Particle Tracking Chemical Engineering Journal 225 no 0 pp 818-836
[8] Deen N Solberg T and Hjertager B 2001 Large eddy simulation of the gas–liquid flow in a square cross-sectioned bubble column Chemical Engineering Science 56 pp 6341-6349
[9] Elena Díaz, M., Montes, F.J. and a Galán, M. 2009 Influence of the lift force closures on the numerical simulation of bubble plumes in a rectangular bubble column Chemical Engineering Science. 64 pp 930-944.
[10] Tabib M V a Roy S and Joshi J B 2008 CFD simulation of bubble column—An analysis of interphase forces and turbulence models Chemical Engineering Journal 139 pp 589-614
[11] Zhang D Deen N G and Kuipers J 2006 Numerical simulation of the dynamic flow behavior in a bubble column: A study of closures for turbulence and interface forces Chemical Engineering Science 61 pp 7593-7608
[12] Pfleger D and Becker S 2001 Modelling and simulation of the dynamic flow behaviour in a bubble column Chemical Engineering Science 56 pp 1737-1747
[13] Dhotre M Niceno B and Smith B 2008 Large eddy simulation of a bubble column using dynamic sub-grid scale model Chemical Engineering Journal 136 pp 337-348
[14] Sato Y and Sekoguchi K 1975 Liquid velocity distribution in two-phase bubble flow International Journal of Multiphase Flow 2 pp 79-95
[15] Dhotre M T and Smith B L 2007 CFD simulation of large-scale bubble plumes: Comparisons against experiments Chemical Engineering Science 62 no 23 pp 6615-6630
[16] Lopez de Bertodano, M., Lahey Jr., R.T. and Jones, O.C. 1994 Turbulent bubbly two-phase flow data in a triangular duct Nuclear Engineering and Design. 146 pp 43-52.

IOP Conf. Series: Materials Science and Engineering **88** (2015) 012004 doi:10.1088/1757-899X/88/1/012004

[17] Laborde-Boutet C et al 2009 CFD simulation of bubble column flows: Investigations on turbulence models in RANS approach Chemical Engineering Science 64 no 21 pp 4399-4413
[18] Bouaifi M et al. 2001 A comparative study of gas hold-up, bubble size, interfacial area and mass transfer coefficients in stirred gas–liquid reactors and bubble columns Chemical Engineering and Processing: Process Intensification 40 no 2 pp 97-111

IOP Conf. Series: Materials Science and Engineering **88** (2015) 012005 doi:10.1088/1757-899X/88/1/012005

Wavelet Transform Of Acoustic Signal From A Ranque-Hilsch Vortex Tube

Y Istihat[1a] and W Wisnoe[b]
Faculty of Mechanical Engineering
Universiti Teknologi MARA (UiTM)
40450 Shah Alam, Selangor Darul Ehsan

E-mail: [a]yusman@unisel.edu.my, [b]wira_wisnoe@salam.uitm.edu.my

Abstract. This paper presents the frequency analysis of flow in a Ranque-Hilsch Vortex Tube (RHVT) obtained from acoustic signal using microphones in an isolated formation setup. Data Acquisition System (DAS) that incorporates Analog to Digital Converter (ADC) with laptop computer has been used to acquire the wave data. Different inlet pressures (20, 30, 40, 50 and 60 psi) are supplied and temperature differences are recorded. Frequencies produced from a RHVT are experimentally measured and analyzed by means of Wavelet Transform (WT). Morlet Wavelet is used and relation between Pressure variation, Temperature and Frequency are studied. Acoustic data has been analyzed using Matlab® and time-frequency analysis (Scalogram) is presented. Results show that the Pressure is proportional with the Frequency inside the RHVT whereby two distinct working frequencies is pronounced in between 4-8 kHz.

1. Introduction

The invention of Vortex Tube by Georges J. Ranque was begun since 1933 and further developed by Hilsch 1947. Some explanation on the temperature separation has been stated by the inventors [1] but the real phenomena still a curious dispute to the others. Figure 1 shows the schematic of RHVT.

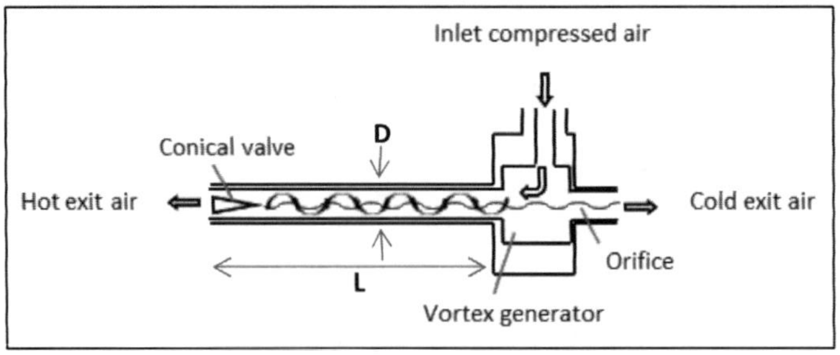

Figure 1 Schematic representation of a RHVT operating principle

[1] To whom any correspondence should be addressed.

The compressed air flows through the RHVT's inlet and went out through the hot and cold exits [2]. Swirling generator located at the inlet chamber initiate a swirl flow inside the tube. This swirling flow will creates temperature distribution along the tube [3]. The compressed air was injected tangentially in the inlet and moving air near the peripheral will release at hot exit to produce hot air. Conical valve will reflect the moving air in the core side, bounced back and pass through the orifice as cold air at the cold exit.

Increasing the supply pressure, L/D ratio and number of nozzles will increase the performance of RHVT [4]. Aydin et. al [5] introduced a new geometry which called "helical swirl flow generator" for the cold end side for the improvement of RHVT performance meanwhile Wisnoe et. al [6] determined that the conical valve shape, pressure source and swirl generator influence the cold air at cold outlet, the mass cold fraction and the isentropic efficiency of RHVT.

Flow behaviour in RHVT consists of energy separation due to viscous friction as result of interaction between layers. Besides generating heat, the phenomena produce some noise can be heard clearly. Kurosaka et al. [7] conclude about the mechanism of energy separation in RHVT is due acoustic streaming induced by the vortex whistle. Yunpeng et al. [8] study the flow structure in a counter flow vortex tube by measuring velocity distribution and has come with velocity fluctuation by time domains. Chang-Soo & Chang-Hyun [9] concluded that vortex tube has two distinct kinds of frequency, low and high frequency periodic fluctuations.

The main objective of this paper is to do a preliminary analysis of acoustic signal produced from flow inside RHVT using Wavelet Transform (WT). The aim is to obtain the relation between pressure, frequency, temperature and time of the flow.

2. Experimental Setup

The schematic diagram of experimental setup is shown in Figure 2. The intention of this study is to record acoustic signal from RHVT using two microphones located at hot and cold exits of RHVT. The RHVT and microphones were placed in a special box which is appropriate to diminish surrounding noise.

Data Acquisition System (DAS), Multi-Channel Spectrum Analyser (PULSE System Type 3560D) has been used to acquire wave data. Five inlet pressures 20, 30, 40, 50 and 60 psi were supplied to the system so that temperature, flowrate and acoustic signal can be recorded. Sampling frequencies of 51200 Hz and 4096 wave data was captured for each microphone within 80 miliseconds. Method of renaming and processing raw data was referring to Istihat Y. et. al. [10]. The experiment was held in the thermodynamics lab at 1006 mbar atmospheric pressure, 23.7°C of room temperature and 72% humidity.

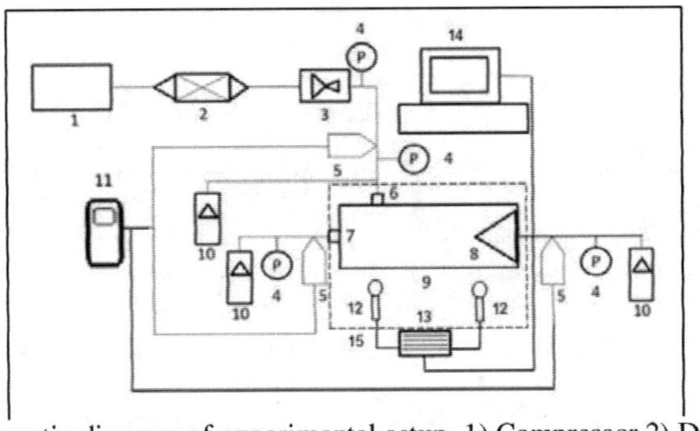

Figure 2 Schematic diagram of experimental setup. 1) Compressor 2) Dehumidifier 3) Pressure regulator (source) 4) Pressure gage 5) Thermocouple 6) Inlet 7) Orifice (cold end) 8) Conical valve (hot end) 9) Vortex tube 10) Flowmeter 11) Thermometer 12) Microphone probe 13) Data Acquisition System (DAS) 14) Computer 15) Box

3. Wavelet Transform (WT) and Matlab Setup

The Wavelet Transform (WT) is a method for time-frequency analysis. Richard (2007) [11] found that the continuous wavelet transform (CWT) of acoustic signals is a promising method to obtain the time-frequency energy distribution of a signal. The time-frequency analysis was originated from further study of frequency domain. Frequency domain is a useful technique to identify frequency at a given time domain using Fourier Transform (FT) but it has poor localization in both time and frequency [12].-The WT of a signal *x(t)* is a time-scale decomposition obtained by dilating and translating along the time axis of a chosen analysing function (wavelet). The CWT is defined as:

$$1_1\ 11,11 = \frac{1}{\sqrt{1}} 1_{11}^{1}\ 1111\ {}^{*}1\frac{111}{1}1\ 1\ 1 \quad (1)$$

where b is the parameter localising the wavelet function in the time domain, a is the dilation parameter defining the analysing window stretching ψ^* is the complex conjugate of the basic wavelet function. In other words, $x(t)$ is the acoustic signal given in the time domain and $W_\psi(a,b)$ is the WT result as a function of a and b. The wavelet function used in this study is Morlet Wavelet which is defined as:

$$1\ 111 = \frac{1}{\sqrt[1]{1}} 1^{11}{}_{1}{}^{1}1^{11^1}/1 \quad (2)$$

where ω_o is the wavelet frequency.

The computation is executed using Matlab® command tfrscalo [13].

4. Results and Analysis

4.1 Experiment Results

Data for experiments was tabulated in Table 1. Temperature different between hot and cold exits was calculated as Delta T (°C) and graph of temperature profiles within working inlet pressure was draw in Figure 3.

Table 1 Experimental data

Pressure (psi)			**Temperature (°C)**			**Delta T (°C)**	**Flowrate (lpm)**		
inlet	out-hot	out-cold	inlet	out-hot	out-cold	Th - Tc	inlet	out-hot	out-cold
20	0	0	23.6	36.9	7.1	29.8	70	70	50
30	0	0	23.3	42.0	0.5	41.5	80	100	68
40	0	1	23.3	43.9	-2.0	45.9	105	135	70
50	3	1	23.4	38.4	-1.5	39.9	125	172	80
60	7	10	23.6	32.4	-0.1	32.5	100	235	45

From the graph, pressure of 40 psi gives more significant temperature different (45.9°C). Inlet temperature has a constant value around 23°C and no changes with the increment of pressure. For the out-hot (hot exit) temperature, the line's profile is same with the Delta T (temperature different). The value is increase gradually and meet a peak value at 45.9°C during 40 psi reading. Lines profile turn to negative slope until last reading of pressure at 60 psi. For the out-cold (cold exit) region, temperature was detected decrease from 7.1°C to -2.0°C at 40 psi and the value is incrase to -1.5 and -0.1°C at 50 and 60 psi but in small ammount.

Figure 4-6 designate signal analysis of wave data for 20, 40 and 60 psi of hot and cold exits of RHVT. There are 2 rows of frequency at around 5 and 8 kHz (except for P = 60 psi out-cold, there are 3 area of frequency). Table 2 show the correlation of parameter studied with frequency. From the scalogram observation, the higher frequency is more visible in out-hot (hot exit) of the tube. This frequency does not have constant amplitude but it oscillates at several periodic times. For 20 psi out-hot, a peak reading was detected at 62 ms. Pressure 40 and 60 psi gives peak values at 36 and 10 ms.

IOP Conf. Series: Materials Science and Engineering **88** (2015) 012005 doi:10.1088/1757-899X/88/1/012005

The lower frequency is more visible in cold side. From 20 to 60 psi reading, the peaks for each oscillation was charted at 18, 12 and 40 ms. All readings were not constant and this is due to unsteady flow inside the tube.

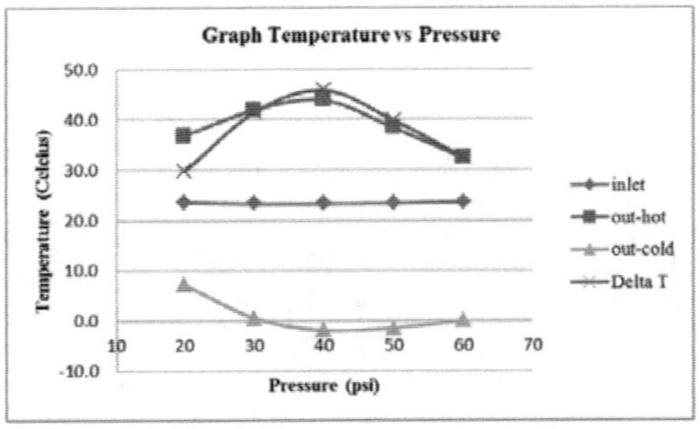

Figure 3 Graph Temperature versus Pressure

4.2 Scalogram of acoustic signal

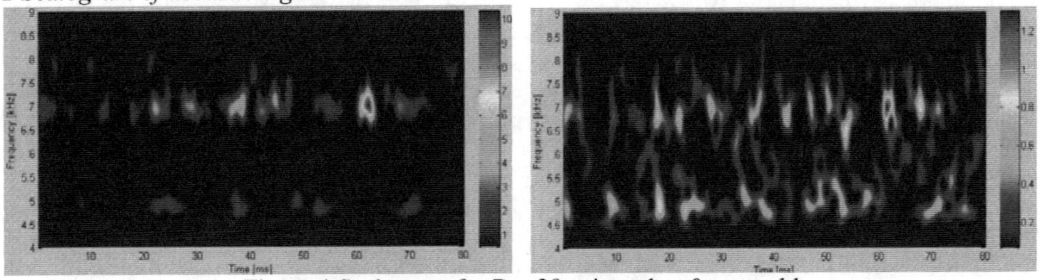

Figure 4 Scalogram for P = 20 psi out-hot & out-cold

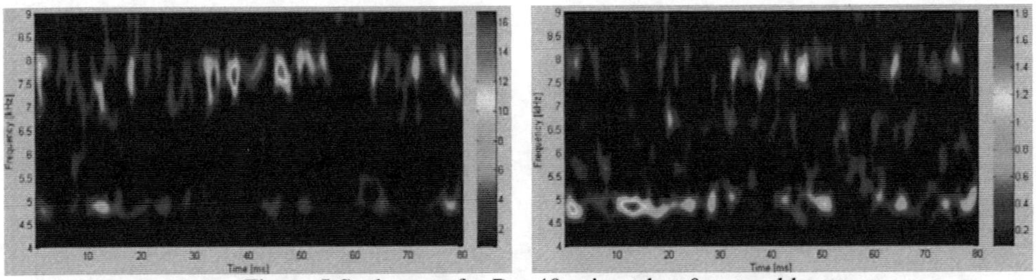

Figure 5 Scalogram for P = 40 psi out-hot & out-cold

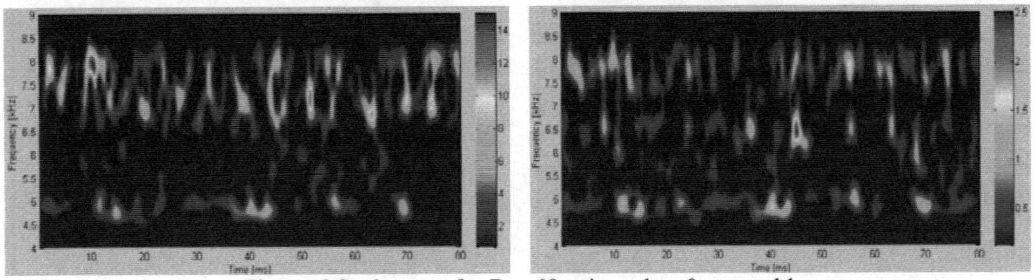

Figure 6 Scalogram for P = 60 psi out-hot & out-cold

Table 2 Correlation Data of Pressure, Temperature, Temperature Difference and Frequency

P(psi)	**Temperature (C)**		**Delta T (C)**	***f,hi (Hz)***	***f,lo (Hz)***
inlet	out-hot	out-cold	hot - cold	freq-high	freq-low
20	36.9	7.1	29.8	7.0	4.8
30	42.0	0.5	41.5	7.5	4.8
40	43.9	-2.0	45.9	7.7	4.8
50	38.4	-1.5	39.9	7.8	4.7
60	32.4	-0.1	32.5	8.0	4.8

The higher frequency is related to the swirling flow in the hot tube and has a close agreement with Chang-Soo & Chang-Hyun (2006). They found that a RHVT has two distinct kinds of frequency, low and high frequency periodic fluctuations. Low frequencies are actually consistent with the calculated Helmholtz frequency and high frequencies are due to precession oscillation by the offset of swirl centre with geometric centre.

Figure 7 shows the curve of frequency versus pressure. It can be observed that pressure is proportional with frequency. Pressures influence the high frequency (freq-high) in a sense of the value is increased to 1 kHz (from 7-8 kHz). On the other hand, pressure does not influence the low frequency (freq-low) and can be classified as negligible effect.\

Figure 8 shows the relation of temperature with frequency-high among out-hot, Delta T and out-cold. The optimum value of pressure for this study is 40 psi. Frequency-high for out-hot and Delta T increase with the temperature. The values were decrease after this point. But for the out-cold, it gives a vice versa effects.

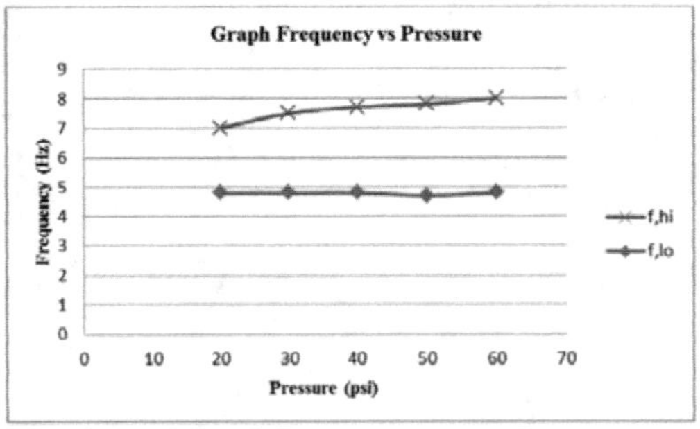

Figure 7 Graph Frequency versus Pressure

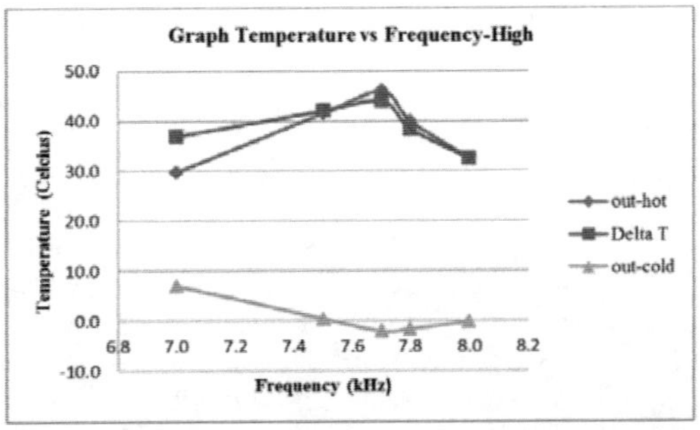

Figure 8 Graph Temperature versus Frequency-high

5. Conclusion and Recommendation

This paper shows swirling frequency inside the RHVT was studied and analysed by using WT. From the observation it can be concluded that the swirling frequency oriented were laid in between 4-8 kHz. Pressure of 40 psi was an optimum working pressure to generate high temperature different. Scalograms of all analysis signals were presented and it characterizes the frequency for each hot and cold exits.

For the recommendation, a box divider between hot and cold section should be arranged. The idea is to get pure frequency tone from each section. Attached microphones should be able to capture two different tones from hot and cold region. Further investigation is required to observe the relation between the frequencies observed with the flow in hot and cold region.

6. Acknowledgment

Authors would like to thank the Malaysian Ministry of Higher Education (MOHE), the Research Management Institute (RMI) and the Faculty of Mechanical Engineering, Universiti Teknologi MARA (UiTM) for providing fund and support for this research under Fundamental Research Grant Scheme (FRGS) 600-RMI/ST/FRGS 5/3 (74/2014).

7. References

[1] Xue, Y., Arjomandi, M., and Kelso, R. (2010). A Critical Review of Temperature Separation in a Vortex Tube. *Experimental Thermal and Fluid Science*, *34*(8), 1367–1374.

[2] Ismail, N., Wisnoe, W., & Remeli, M. F. (2014). Experimental Investigation on the Effect of Orifice Diameter and Inlet Pressure to the Ranque-Hilsch Vortex Tube Performance. *Applied Mechanics and Materials Vols. 465-466*, 515-519

[3] Wisnoe, W., Shukri, E. S., Zailani, R., Che Mi, M.H. and Zakaria, M. F. (2013). Numerical Investigation of Temperature Distribution in a Diffuser Equipped with Helical Tape. *Applied Machanics and Materials,* vol 393. Pp. 793-798.

[4] Dincer, K., Tasdemir,S., Baskaya, S., and Uysal., B. Z. (2008). Modeling of The Effects of Length to Diameter Ratio and Nozzle Number On The Performance of Counterflow Ranque-Hilsch vortex Tubes Using Artificial Neural Networks. *Applied Thermal Engineering,* vol.28, pp. 2380-2390.

[5] Aydin, O., Markal, B. and Avci., M. (2010). A New Vortex Generator For a Counter-Flow Ranque-Hilsch Vortex Tube. *Applied Thermal Engineering,* vol. 30, pp. 2505-2511.

[6] Wisnoe, W., Ismail, N., Remeli, M. F. and Zakaria, M. F.. (2013). Experimental Investigation On The Effect of Conical Valve Shape and Swirl Generator to The Performance of Ranque-Hilsch Vortex Tube . *2013 IEEE Business Engineering and Industrial Applications Colloquium,* pp. 815-819.

[7] Kurosaka, M., Chu, J. Q. and Goodmann, J. R.b(1982). Ranque Hilsch effect revisited: Temperature Separation Effect Traced to Orderly Spinning Waves or Vortex Wistle. *In: AIAA 82-0952, AIAA/ASME third joint thermo-physics of fluid plasma and heat transfer conference.*

[8] Xue,Y., Arjomandi, M. and Kelso, R. (2012). Experimental Study of The Flow Structure in a Counter Flow Ranque–Hilsch Vortex Tube. *International Journal of Heat and Mass Transfer* 55 (2012) 5853-5860

[9] Chang-Soo, K. and Chang-Hyun, S. (2006). Dynamic Characteristrics of an Unsteady Flow Through a Vortex Tube. *Journal of Mechanical Science and Technology (KSME Int. J.).* Vol. 20, No. 12, pp. 2209-2217.

[10] Istihat,Y., Nuawi, M. Z., Bahari, A. R., and Mahmood, W. M. F. W. (2012). Development of Ultrasonic Test Rig System for Fuel Injector. *International Review of Mechanical Engineering (I.RE.M.E.), Vol. 6, N. 3 ISSN 1970 - 8734 March 2012.* pp. 625-629

[11] Richard, B. (2008). An Algorithm for the Continuous Morlet Wavelet Transform.

[12] Liner, C. (2010). An Overview of Wavelet Transform Concepts And Applications, 1–17.

[13] Time-Frequency Toolbox Reference Guide, October 26, 2005.

IOP Conf. Series: Materials Science and Engineering **88** (2015) 012006 doi:10.1088/1757-899X/88/1/012006

A Study on the Air flow outside Ambient Vaporizer Fin

G Oh[1], T Lee[2], H Jeong[3], H Chung[4]
[12]Department of mechanical and Energy Engineering, Gyeongsang National University, TongYoung, Republic of Korea
[34]Department of Mechanical and Energy Engineering, Gyeongsang National University·Institute of Marine Industry, TongYoung, Republic of Korea

E-mail: goldstar89@nate.com

Abstract. In this study, we interpreted Fog's Fluid that appear in the Ambient Vaporizer and predict the point of change Air to Fog. We interpreted using Analysis working fluid was applied to LNG and Air. We predict air flow when there is chill of LNG in the air Temperature and that makes fog. Also, we interpreted based on Summer and Winter criteria in the air temperature respectively. Finally, we can check the speed of the fog when fog excreted.

1. Introduction

Natural gas that used in Korea comes from oil-producing countries after liquefaction and compression process. Vaporization process is a step for using the imported LNG. Usually, the vaporization process is divided into two categories. One is the way to steal the cold from the LNG by using sea water vaporizer in the supply bases. The other way is to transport LNG from the bases of supply to small satellite terminals that built near the market and vaporize when it is needed.

For producing NG in small satellite terminals, the satellite centers use ambient vaporizer that works with air. It, however, causes fog because of temperature difference between the atmosphere and the chilly air from the LNG. Since the fog influence surrounding environment, it is necessary to be removed. IN this study, we compared the flow of air in the vaporizer with the fan and without the fan by using FLUENT.

2. Preparing the Calculation

The analysis was carried out so as to meet the installation criteria of the ambient vaporizer that is used in Korea Gas Corporation and we used GAMBIT and FLUENT.

[1 234] 38, Cheondaegukchi-gil, Tongyeong-si, Gyeongsangnam-do, Korea.

IOP Conf. Series: Materials Science and Engineering **88** (2015) 012006 doi:10.1088/1757-899X/88/1/012006

3. Calculation Condition

Table 1. Fan uninstallation

Velocity	0 m/s
Air	300 k
LNG early entry temperature	110 k
Air Entry	pressure outlet
Air Exit	pressure outlet

Table 2. Fan installation

Velocity	10 m/s
Air	300 k
LNG early entry temperature	110 k
Air Entry	velocity inlet
Air Exit	outflow

4. Calculation Results

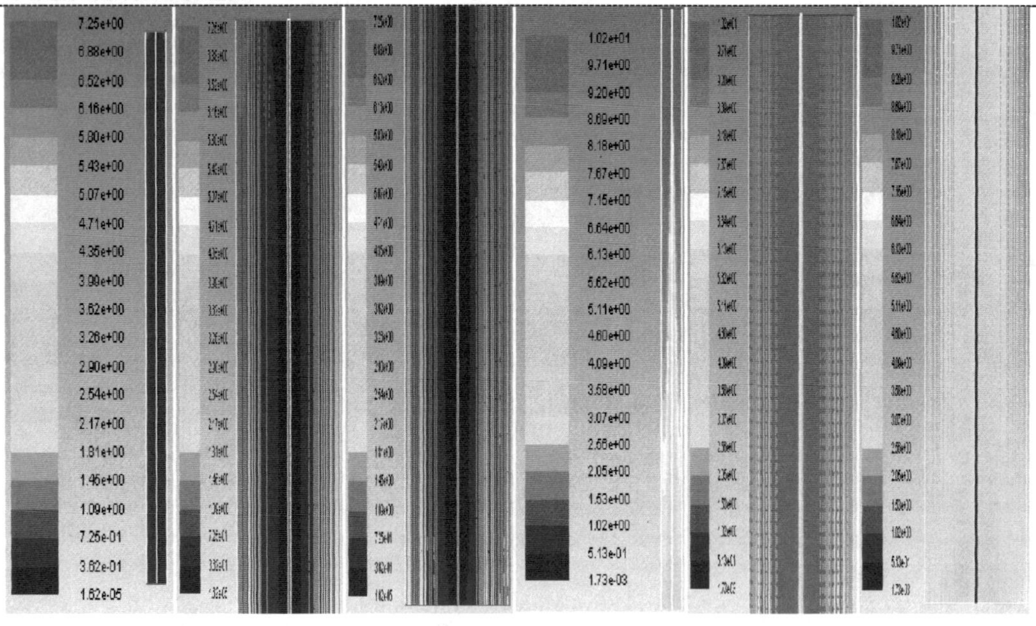

Figure 1. Velocity of whole, top, bottom (Velocity=0m/s)
Velocity of whole, top, bottom(Velocity=10m/s)

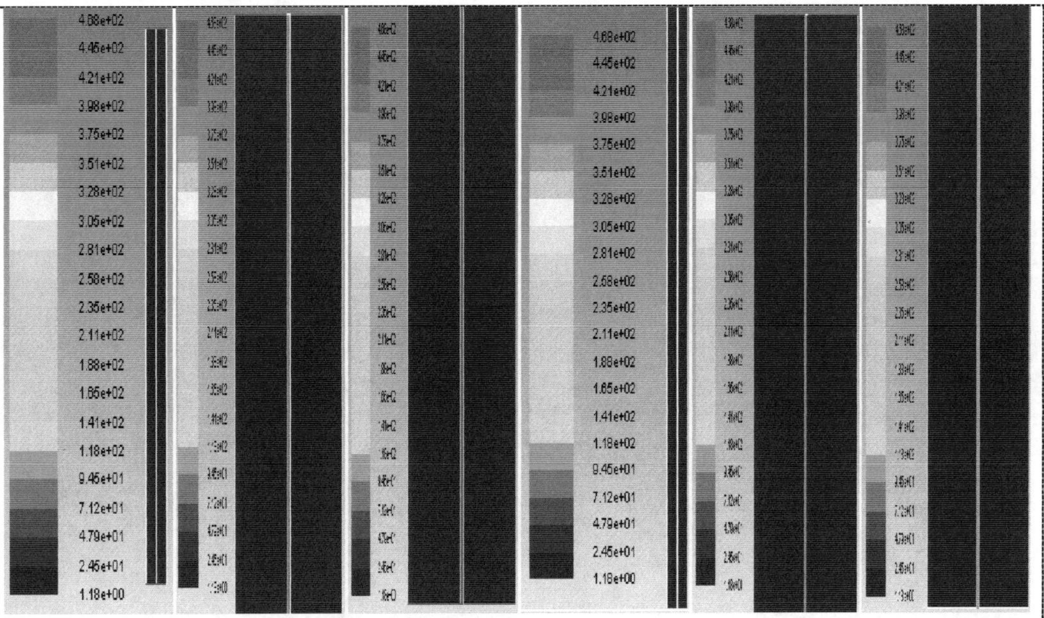

Figure 2. Density of whole, top, bottom(Velocity=0m/s).

Density of whole, top, bottom(Velocity=10m/s).

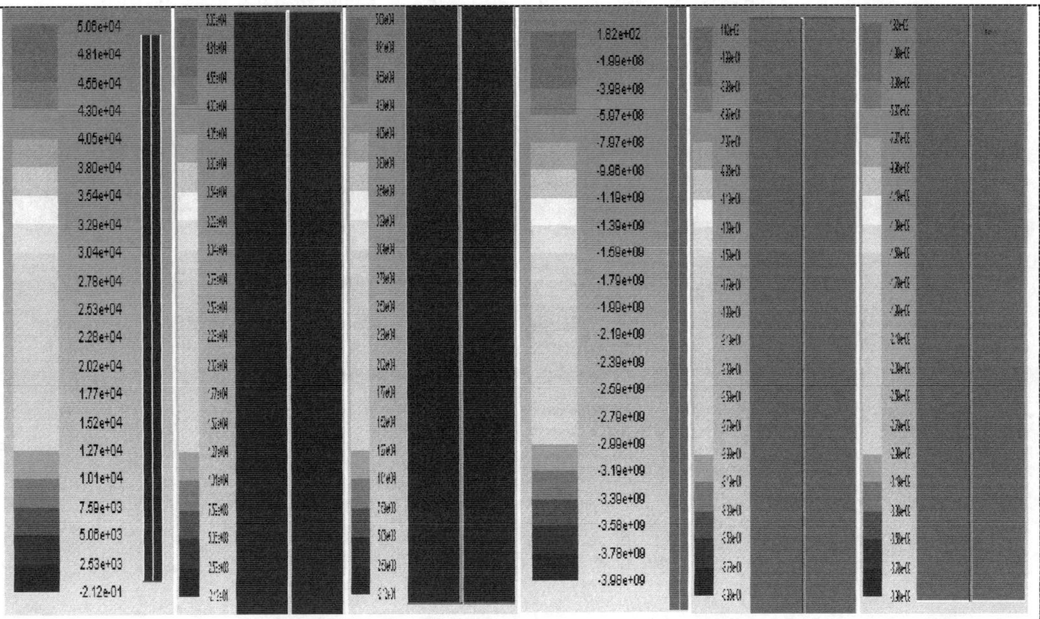

Figure 3. Pressure of whole, top, bottom(Velocity=0m/s)

Pressure of whole, top, bottom(Velocity=10m/s).

IOP Conf. Series: Materials Science and Engineering **88** (2015) 012006 doi:10.1088/1757-899X/88/1/012006

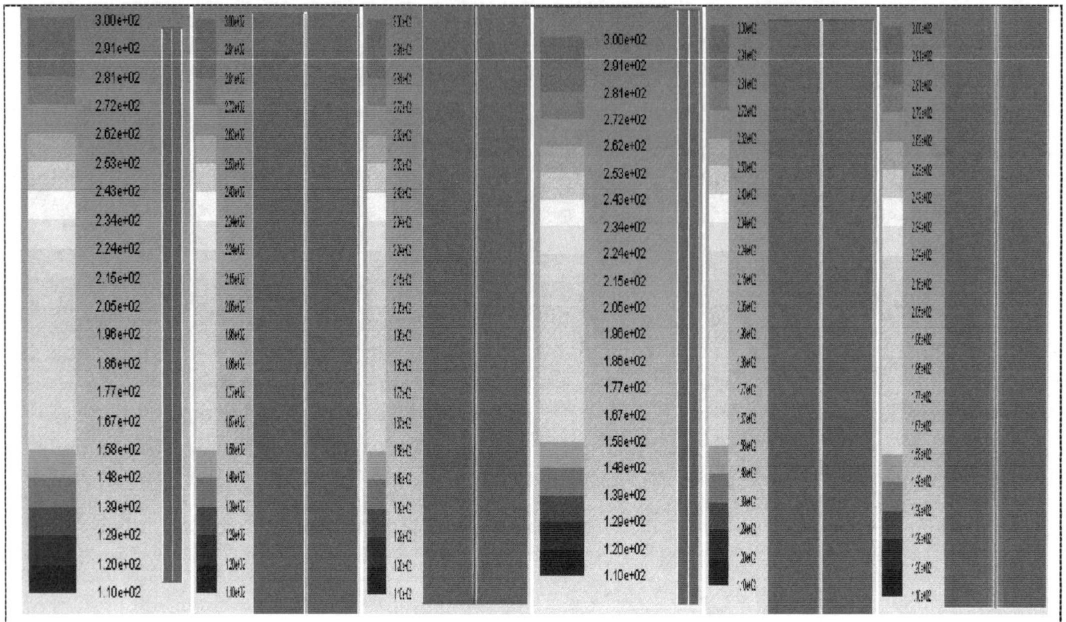

Figure 4. Temperature of whole, top, bottom(Velocity=0m/s).

Temperature of whole, top, bottom(Velocity=10m/s).

Fig.1 to Fig.4 show the result values of velocity, pressure, density and temperature after giving a boundary condition of being uninstalled to the Fan and being installed to the Fan.

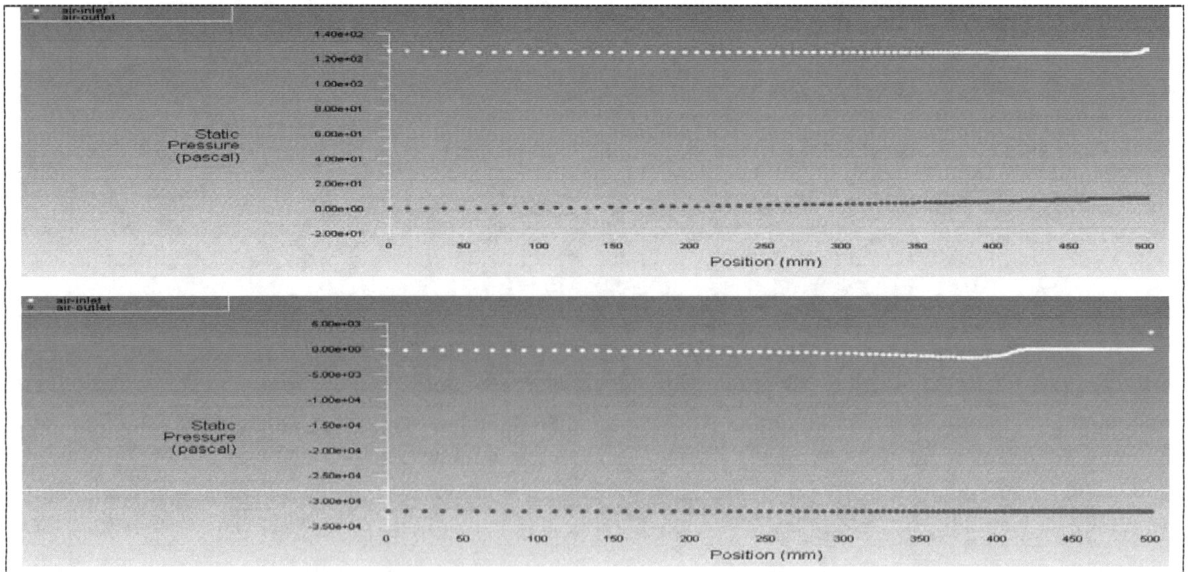

Figure 5. Pressure Graph When the air velocity is 10 m/s and 0 m/s

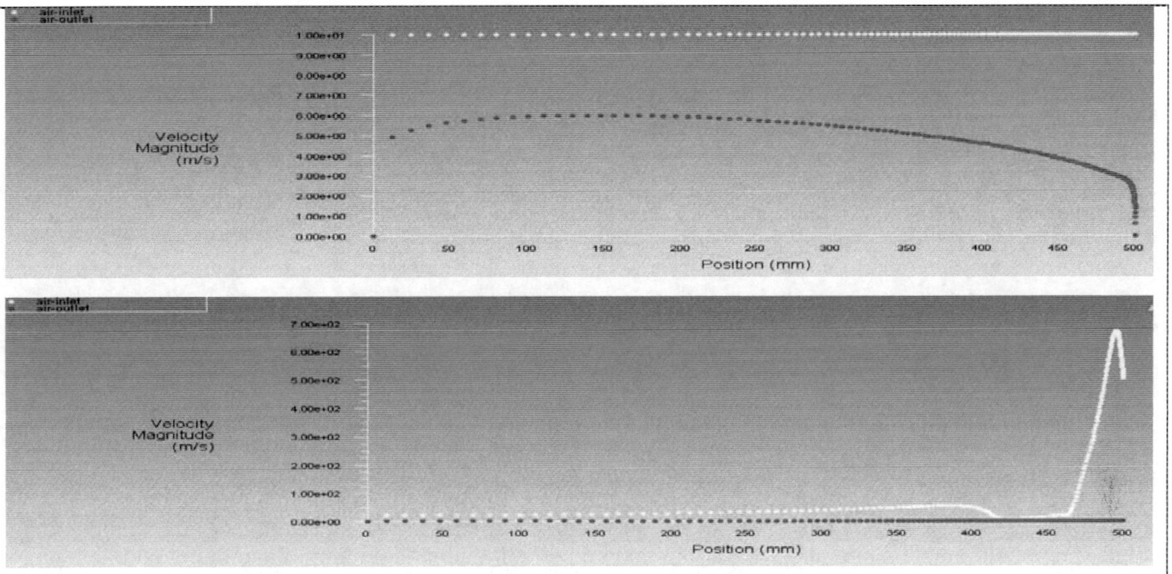

Figure 6. Velocity Graph When the air velocity is 10 m/s and 0 m/s

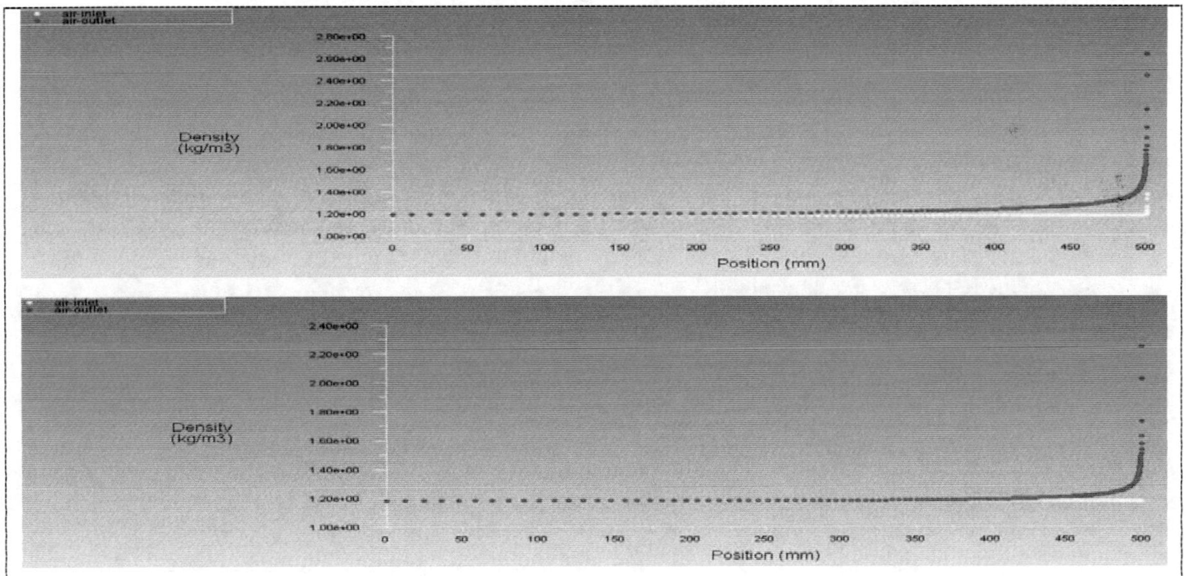

Figure 7. Density Graph When the air velocity is 10 m/s and 0 m/s

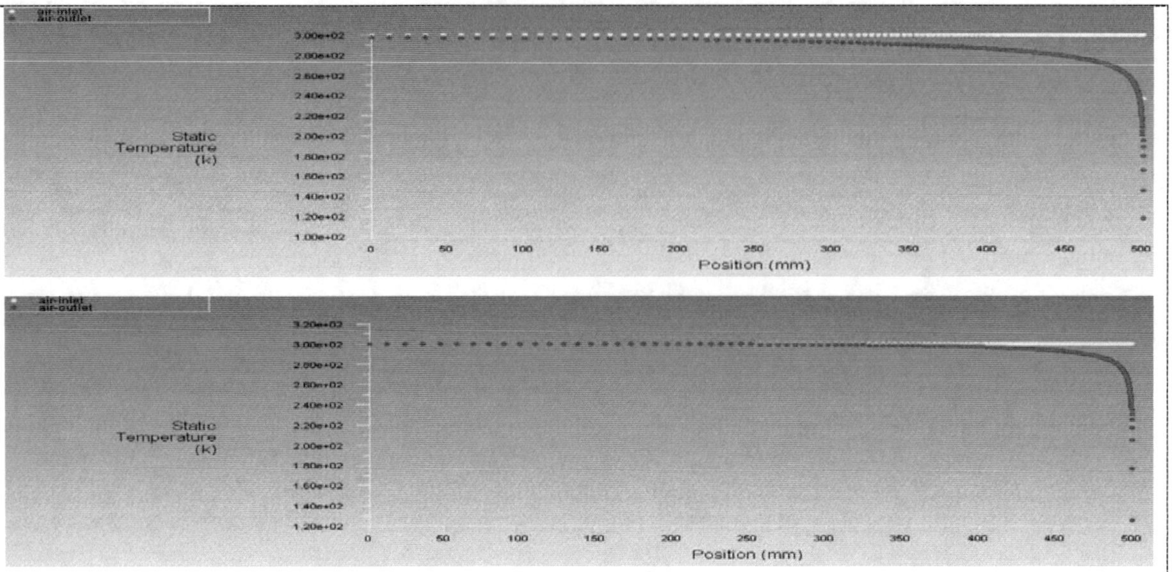

Figure 8. Temperature Graph When the air velocity is 10 m/s and 0 m/s

Fig.5 to Fig.8 show the result graph of velocity, pressure, density and temperature After giving a boundary condition of being uninstalled to the Fan and being installed to the Fan

5. Conclusion

In Fig.1 and Fig.5, It could be seen that the vortex phenomenon is occurring from the vaporizer upper side when the speed is 0m/s. When the air speed is 10m/s, the disappearance of the vortex phenomenon was confirmed. Further, It is able to know that the LNG is vaporized more quickly by comparing the temperature change and density change

6. Postscript

This research was supported by Basic Science Research Program through the National Reseach Foundation of Korea(NRF) funded by Ministry of Education, Science and Technology(NO. NRF-2011-0021376) and the Future Leading Project through the Small and Medium Business Administration(No.S2044441)

7. References

[1] Ansys 14.0 Fluent user guide
[2] Kye-Bock Lee, Prediction of the frost layer growth in LNG air vaporizer, Journal of Industrial Science and Technology Institute Vol.26, No.1, pp.67-72, June 2012

IOP Conf. Series: Materials Science and Engineering **88** (2015) 012007 doi:10.1088/1757-899X/88/1/012007

Adaptive flux-based nodeless variable finite element formulation with error estimation for thermal-structural analysis

S Traivivatana[1], **S Phongthanapanich**[2] **and P Dechaumphai** [3,4]

[1] Energy Research Institute, Chulalongkorn University, Bangkok 10330, Thailand

[2] Department of Mechanical Engineering Technology, College of Industrial Technology, King Mongkut's University of Technology North Bangkok, Bangkok 10800, Thailand

[3] Department of Mechanical Engineering, Chulalongkorn University, Bangkok 10330, Thailand

Abstract. A posteriori error estimation for the nodeless variable finite element method is presented. A nodeless variable finite element method using flux-based formulation is developed and combined with an adaptive meshing technique to analyze two-dimensional thermal-structural problems. The continuous flux and stresses are determined by using the flux-based formulation while the standard linear element interpolation functions are used to determine the discontinuous flux and stresses. To measure the global error, the L_2 norm error is selected to find the root-mean-square error over the entire domain. The finite element formulation and its detailed finite element matrices are presented. Accuracy of the estimated error is measured by the percentage relative error. An adaptive meshing technique, that can generate meshes corresponding to solution behaviors automatically, is implemented to further improve the solution accuracy. Several examples are presented to evaluate the performance and accuracy of the combined method.

1. Introduction

The finite element method has been widely used to analyze many engineering problems [1,2]. Both triangular and tetrahedral elements are frequently used due to the simplicity for constructing meshes for complex geometry problems but the solution accuracy is normally low. Since the thermal stresses are sensitive to the predicted thermal gradients, a more accurate thermal analysis is required to predict temperature distribution. The solution accuracy may be improved using the *h*-method of adaptation where the mesh is globally or locally refined or coarsened [3,4].

The objective of this paper is to develop a nodeless variable finite element method to improve the predicted solution of the two-dimensional Poisson's equation. The nodeless variable finite elements use the quadratic interpolation functions to express the solution distribution over the element without requiring additional actual nodes. In addition, the paper introduces and applies the flux-based formulation to derive the finite element matrices for such nodeless variable element.

An adaptive unstructured meshing technique [4,5] is implemented to improve the solution accuracy. The efficiency of the combined nodeless variable finite element method using flux-based formulation and the adaptive meshing technique is evaluated by several examples with exact solutions.

2. Nodeless variable finite element method

[1] To whom any correspondence should be addressed.

IOP Conf. Series: Materials Science and Engineering **88** (2015) 012007 doi:10.1088/1757-899X/88/1/012007

2.1. Governing equations and boundary conditions

The Poisson's equation for two-dimensional domain in *x-y* coordinate system as shown in figure 1 can be written in the conservation form as,

$$\frac{\partial\{E\}}{\partial x}+\frac{\partial\{F\}}{\partial y}=f(x,y) \tag{1}$$

where $f(x,y)$ denotes the source function. The flux vector components $\{E\}$ and $\{F\}$ contain the heat flux components for thermal analysis or the stress components for structural analysis given by,

$$\{E\}=-c\frac{\partial\{U\}}{\partial x} \quad \text{and} \quad \{F\}=-c\frac{\partial\{U\}}{\partial y} \tag{2}$$

where U is the primary variable and c is the material property that depends on types of problem. The Poisson's equation shown in equation (1) is to be solved together with appropriate boundary conditions that may consist of,

$$U_1(x,y) \quad \text{on } S_1 \tag{3a}$$

$$c\frac{\partial\{U\}}{\partial n}+d\left(\{U\}-U_\infty\right)=q \quad \text{on } S_2 \tag{3b}$$

where d and U_∞ are constants and q is the secondary variable.

2.2. Nodeless variable flux-based finite element formulation

The flux-based formulation [6] is implemented to derive the corresponding finite element equations for a nodeless variable element. For the triangular nodeless variable element, the distribution of the primary variable over the element is assumed in the form,

$$U(x,y)=\sum_{i=1}^{6} N_i(x,y)U_i=\lfloor N(x,y)\rfloor\{U\} \tag{4}$$

where $\lfloor N(x,y)\rfloor$ consists of the element interpolation functions, and $\{U\}$ is the vector of the unknown primary variables and the nodeless variables. The nodal primary variables are U_1 through U_3, while U_4 through U_6 are the nodeless variables. The element interpolation functions, N_1, N_2, N_3 are identical to the element interpolation functions L_1, L_2, L_3 used for the standard three-node triangular element. The nodeless variable interpolation functions implemented in this paper are,

$$N_4=L_2L_3 \;;\; N_5=L_1L_3 \;;\; N_6=L_1L_2 \tag{5}$$

Each nodeless variable interpolation function varies quadratically along one edge and vanishes along the other edges. To derive the finite element matrices by means of the flux-based formulation, the method of weighted residuals is first applied to equation (1). The Gauss's theorem is then applied to the flux derivative terms to yield,

$$\int_\Omega \frac{\partial N_i}{\partial x}\{E\}d\Omega+\int_\Omega \frac{\partial N_i}{\partial y}\{F\}d\Omega=\int_S N_i\{E\}n_x\,d\Gamma+\int_S N_i\{F\}n_y\,d\Gamma-\int_\Omega N_i\,f(x,y)\,d\Omega \tag{6}$$

where Ω is the element domain and S is the element boundary.

In the flux-based formulation, the element flux distributions are computed from the actual nodal fluxes as,

$$\{E\}=\sum_{i=1}^{3}\overline{N}_i\{E_i\}=\lfloor\overline{N}\rfloor\{E^n\} \quad \text{and} \quad \{F\}=\sum_{i=1}^{3}\overline{N}_i\{F_i\}=\lfloor\overline{N}\rfloor\{F^n\} \tag{7}$$

where $\lfloor\overline{N}\rfloor$ are the standard linear element interpolation functions, i.e., $\lfloor L_1 \quad L_2 \quad L_3\rfloor$. The $\{E^n\}$ and $\{F^n\}$ are the vectors of the actual nodal fluxes.

Determination of nodal fluxes depends on the types of problem considered. For structural analysis problem, the stress components σ_x, σ_y, τ_{xy} are related to the strain components by Hook's law. The compact form of equation (6) for structural analysis can be written as,

$$\int_{\Omega} [B]^T \{\sigma\} d\Omega = \int_{S} [N]^T \{T_S\} d\Gamma \tag{8}$$

where

$$[B] = \begin{bmatrix} \frac{\partial N_1}{\partial x} & 0 & \cdots & \frac{\partial N_6}{\partial x} & 0 \\ 0 & \frac{\partial N_1}{\partial y} & \cdots & 0 & \frac{\partial N_6}{\partial y} \\ \frac{\partial N_1}{\partial y} & \frac{\partial N_1}{\partial x} & \cdots & \frac{\partial N_6}{\partial y} & \frac{\partial N_6}{\partial x} \end{bmatrix}^T \quad \text{and} \quad \begin{array}{l} [N] = \begin{bmatrix} N_1 & 0 & \cdots & N_6 & 0 \\ 0 & N_1 & \cdots & 0 & N_6 \end{bmatrix} \\ \{\sigma\} = \lfloor \sigma_x \quad \sigma_y \quad \tau_{xy} \rfloor^T \\ \{T_s\} = \lfloor T_{Sx} \quad T_{Sy} \rfloor^T \end{array} \tag{9}$$

and $\{T_s\}$ is the boundary tractions vector which can be written as,

$$\{T_s\} = [\overline{N}]\{\overline{T}_s\} \tag{10}$$

where $[\overline{N}]$ and $\{\overline{T}_s\}$ are interpolation matrix and boundary tractions vector, respectively. The interpolation functions in equation (10) needed for integration along a typical element side S are,

$$N_1 = 1 - \frac{x}{L};\ N_2 = \frac{x}{L};\ N_3 = \frac{x}{L}\left(1 - \frac{x}{L}\right) \tag{11}$$

where L is the length of element edge and x is the local coordinate along the edge starting from node 1 as shown in figure 2. Appropriate boundary conditions of the structural analysis problem are then applied. Two types of the boundary conditions considered herein are the specified displacements and boundary tractions, as given by equations (3a) and (3b), respectively.

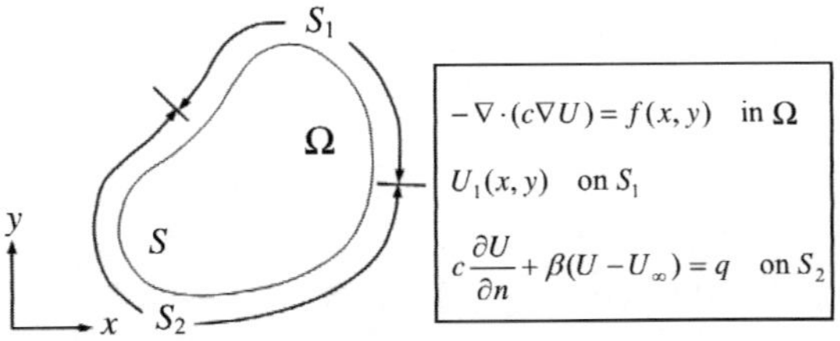

Figure 1. Two-dimensional domain and boundary conditions for Poisson's problem.

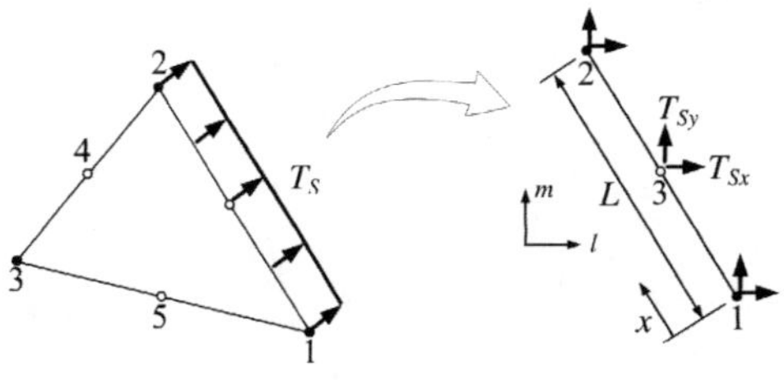

Figure 2. Discretization of boundary tractions into nodal quantities.

3. A posteriori error estimation

3.1. Flux-based formulation for nodal fluxes

Figure 3(a) shows the typical flux (heat flux or stress) distributions computed by using the conventional finite element formulation for standard triangles. Figure 3(b) shows the continuous flux distributions for the same triangular mesh by using the flux-based formulation.

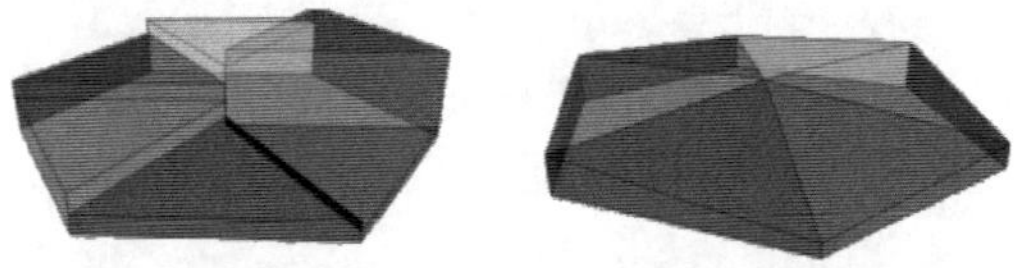

a) Discontinuous b) Continuous

Figure 3. Flux distributions.

Errors of fluxes are determined from the difference between the computed continuous and discontinuous fluxes. The errors of flux gradients are defined by

$$e_g = g^{FB} - g^{ST} \tag{12}$$

where g^{FB} are the flux-based element gradients and g^{ST} are the standard element gradients.

The method of weighted residuals is applied to equation (12) and substituting the element flux distributions from equation (7) to yield,

$$[M]\{\bar{g}^{FB}\}=[G]\{U\} \tag{13}$$

where $[M]$ is the mass matrix and $\{\bar{g}^{FB}\}$ is the vector of the flux-based nodal fluxes.

3.2. Error estimator

To measure the global error, the L_2 norm error is selected to find the root-mean-square error over the entire domain [7]. From equation (12), the square of the L_2 norm error for the gradients is

$$\left\| e_g^{FE} \right\|_{L_2}^2 = \int_\Omega (g^{FB} - g^{ST})^2 \, d\Omega \tag{14}$$

By applying the same concept onto the errors of gradients, the finite element equations for the finite element error square are,

$$\left\| e_g^{FE} \right\|_{L_2}^2 = \lfloor \bar{g}^{FB} \rfloor [M] [\bar{g}^{FB}] + \lfloor \bar{g}^{FB} \rfloor [G]\{U\} + \lfloor U \rfloor [K]\{U\} \tag{15}$$

The finite element global error for gradients is computed from

$$\left\| e_g^{FE} \right\|_g = \left[\sum_{element} \left\| e_g^{FE} \right\|_{L_2}^2 \Big/ A_d \right]^{1/2} \tag{16}$$

where A_d is the total domain area. The square of exact L_2 norm error for gradients is

$$\left\| e_g^{FB} \right\|_{L_2}^2 = \int_\Omega (g^{EX} - g^{FB})^2 \, d\Omega \quad \text{and} \quad \left\| e_g^{ST} \right\|_{L_2}^2 = \int_\Omega (g^{EX} - g^{ST})^2 \, d\Omega \tag{17}$$

where g^{EX} are exact element gradients, $\left\| e_g^{FB} \right\|_{L_2}^2$ and $\left\| e_g^{ST} \right\|_{L_2}^2$ are the square of error for flux-based element and standard element, respectively. The exact global error for both flux-based element and standard element are applied with the same procedure. From the concept of measuring discretization error in the elastic problem, the Z^2-error estimate [8], the percentage relative error is defined by,

$$\eta = \frac{\|\mathbf{e}\|_{L_2}}{\|\mathbf{u}\|_{L_2}} \times 100 \tag{18}$$

where $\|\mathbf{e}\|_{L_2}$ is the error in the L_2 norm and $\|\mathbf{u}\|_{L_2}$ is L_2 norm of the solution. The mesh adaptation is terminated when the value of η is less than 5% which is reasonable for engineering applications [9,10].

4. Numerical examples

To evaluate the performance and accuracy of the combined adaptive nodeless variable finite element method together with the flux-based formulation and the proposed error estimation method, two example problems are studied and presented. These problems are: (1) a square domain with source function and (2) an L-shaped plate under uniform loading.

4.1. Square domain with source function

A unit square domain with the specified boundary conditions is shown in figure 4. The specified source function over the plate [11] is given by,

$$f(x,y) = -14x(1-x)(1-2y) - 4y(1-y)(1-2x) + 2(1+2x+7y)\left[x(1-x)+y(1-y)\right] \tag{19}$$

The exact solution for the primary variable distribution is,

$$U(x,y) = x(1-x)\,y(1-y)(1+2x+7y) \tag{20}$$

Figure 5 shows the three structure finite element mesh models with 32, 128 and 512 nodeless variable finite elements, respectively. The first T3NL–M1, the second T3NL–M2 and the third

IOP Conf. Series: Materials Science and Engineering **88** (2015) 012007 doi:10.1088/1757-899X/88/1/012007

T3NL–M3 meshes consist of 25, 81 and 289 nodes, respectively. The figure also shows the predicted solution contours obtained from these mesh models.

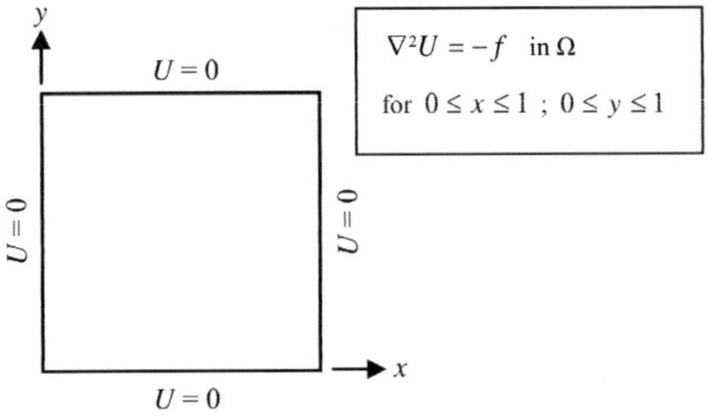

Figure 4. Problem statement of a unit square domain with specified source function.

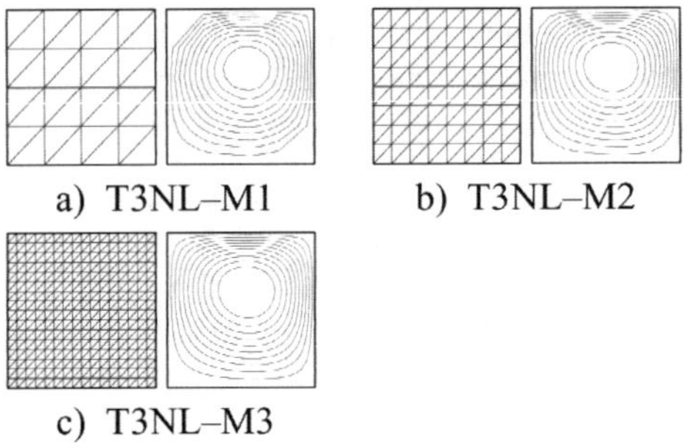

Figure 5. Three structured mesh models and their solution contours.

The comparisons between the exact and three nodeless variable flux-based finite element solutions along the edge x = 0.5 are shown in figure 6. The L_2 norm and percentage relative errors of the exact global error using the flux-based and standard elements together with the approximate global error are presented in Table 1. The table shows that the computed global errors for the three finite element mesh models are in the same trend as the exact errors. These global errors can be used for measuring the solution accuracy obtained from the finite element mesh models with different discretization.

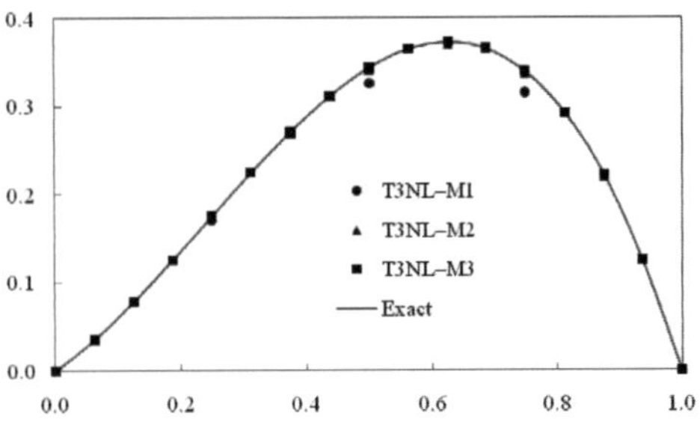

Figure 6. Comparison between the exact and three finite element solutions along the edge x = 0.5.

Table 1. Quantitative comparisons of global errors for a unit square domain with specified source function.

Element	$\|e_g^{FB}\|_g$, (η_{FB})	$\|e_g^{ST}\|_g$, (η_{ST})	$\|e_g^{FE}\|_g$, (η_{FE})
T3NL–M1	0.210 (19.98%)	0.283 (25.20%)	0.190 (19.22%)
T3NL–M2	0.065 (6.96%)	0.089 (9.29%)	0.061 (6.52%)
T3NL–M3	0.018 (1.99%)	0.025 (2.72%)	0.017 (1.88%)

4.2. L-shaped plate under uniform loading

This test case is a L-shaped domain [8] subjected to a tensile load of $\sigma = 1$ as shown in figure 7. The plane stress condition is assumed with the Poisson's ratio of 0.3 and Young's modulus E = 100,000.

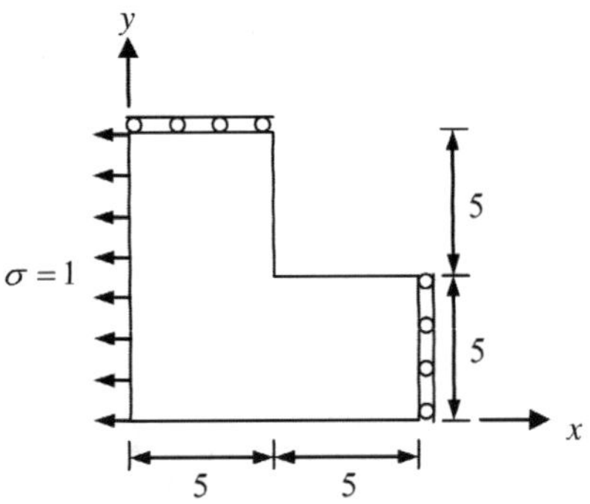

Figure 7. Problem statement of an L-shaped plate under uniform loading.

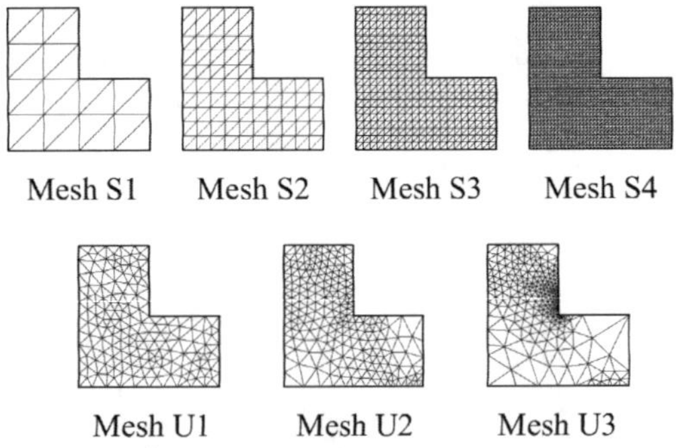

Figure 8. Meshes used in analysis of L-shaped plate under uniform loading.

Figure 8 shows the four uniform meshes (S1 to S4), an initial unstructured mesh (U1) and two adaptive meshes (U2 and U3) used in the analysis. The number of nodes in different meshes and the computed percentage errors of their solutions are tabulated in Table 2. The table also highlights performance of the adaptive meshing technique that the number of nodes are almost constant (228 and 257 nodes in mesh U2 and U3, respectively) during the adaptive process, whereas the final uniform mesh contains five times greater of the number of nodes (1,281 nodes in mesh S4) at the same error level. The comparison for the solution convergence rates between the uniform refinement meshes and the adaptive meshes are presented in figure 9 where N denotes the number of nodes. The solution convergence rates of the uniformly refine meshes is 0.36 as compared to 2.59 for the adaptive meshes. The convergence rates of the uniform refinement meshes are relatively low due to the solution singularity at the center corner.

Table 2. Quantitative comparisons of global errors for analysis of L-shaped plate under uniform loading

Element	Number of nodes	η_{FE}
Mesh S1	21	12.08%
Mesh S2	96	6.35%
Mesh S3	341	4.07%
Mesh S4	1,281	2.67%
Mesh U1	166	6.21%
Mesh U2	228	3.53%
Mesh U3	257	2.35%

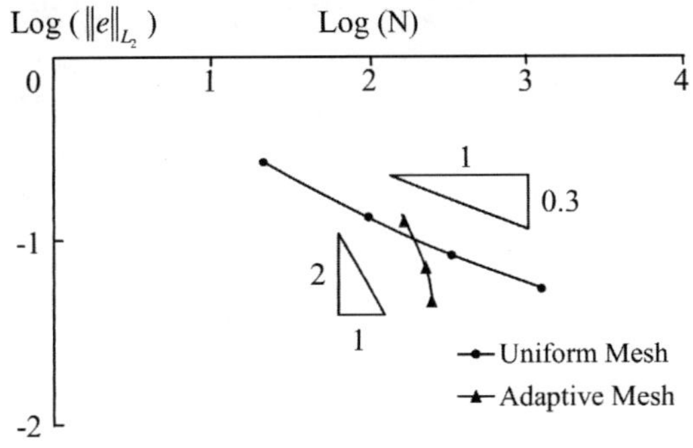

Figure 9. Comparison of solution convergences between the uniform refinement and adaptive meshes.

5. Conclusions

The nodeless variable finite element method using flux-based formulation was developed to analyze two-dimensional thermal-structural problems. The flux-based formulation was developed and applied to the nodeless variable finite element in order to reduce the computational complexity as compared to the conventional finite element method. An adaptive meshing technique was implemented to further improve the solution accuracy. The performance of the combined procedure was evaluated by using two problems. These problems demonstrate that the combined nodeless variable finite element method using the flux-based formulation with the adaptive meshing technique helps increasing the solution accuracy while reducing the unknowns as compared to the conventional method.

References

[1] Phongthanapanich S and Dechaumphai P 2006 *Adv. Eng. Software* **27** 797-804
[2] Cheng Z and Paraschivoiu M 2003 *Finite Elem. Anal. Des.* **39(7)** 581-97
[3] Eibner T and Melenk J M 2007 *Comput. Mech.* **39(5)** 575-95
[4] Phongthanapanich S and Dechaumphai P 2004 *Trans. of the CSME* **27** 319-39
[5] Phongthanapanich S and Dechaumphai P 2004 *Trans. of the CSME* **28** 531-50
[6] Phongthanapanich S, Traivivatana S, Boonmaruth P and Dechaumphai P 2006 *Acta Mech. Sinica* **22(2)** 138-47
[7] Babuska I, Banerjee U and Osborn J E 2007 *Numer. Math.* **107(3)** 353–95
[8] Zienkiewicz O C and Zhu J Z 1992 *Int. J. Numer. Methods Engrg.* **33** 1365-82
[9] Becker R, Hansbo P and Larson M G 2003 *Comput. Meth. Appl. Mech. Eng.* **192(5-6)** 723-33
[10] Huang C-S, Leeb C-F and Cheng A H-D 2007 *Eng. Anal. Bound. Elem.* **31(7)** 614-23
[11] Lesaint P and Zlamal M 1979 *RAIRO Anal. Numer.* **13** 139-66

IOP Conf. Series: Materials Science and Engineering **88** (2015) 012008 doi:10.1088/1757-899X/88/1/012008

Indoor Airflow Simulation inside Lecture Room: A CFD Approach

S Lin, B T Tee[1] and C F Tan
Centre for Advanced Research on Energy (CARe)
Faculty of Mechanical Engineering, Universiti Teknikal Malaysia Melaka,
Hang Tuah Jaya, 76100 Durian Tunggal, Melaka, Malaysia.

E-mail: tee@utem.edu.my

Abstract. Indoor air flow distribution is important as it will affect the productivity of the occupants. Poor air flow distribution not only cause discomfort to the occupants but also influence their ability to conduct their activities. The main purpose of this study is to investigate the indoor air flow inside lecture rooms through CFD simulation approach. Two types of air-conditioning configuration system in lecture rooms have been selected for this study which includes the split unit and centralized system. The air flow distribution between these two systems are analyzed and compared. Physical measurement is conducted using a velocity meter for validation purpose. CFD simulation is developed by using ANSYS Fluent software. The results specifically the air velocity and temperature data are compared and validated. Based on the findings, design recommendation is proposed with the aim to improve on the current air flow distribution in the lecture rooms.

1. Introduction

Most of the students spend majority of their time in lecture rooms during their lecture or tutorial session. In this case, air flow distribution from the air-conditioning system becomes important as it will determine whether students perceive the air velocity and air temperature in the acceptable comfort range [1]. This study emerged based on a need to further investigate the current air flow distribution system in lecture rooms of UTeM. The air flow in a typical lecture room in UTeM is not uniform for the current arrangement of air-conditioning units in the room. It was also discovered that the position of air-conditioning units in the lecture rooms will affect the air flow distribution in the room, which then causes changes in thermal sensation among the occupants. Therefore, a Computational Fluid Dynamics (CFD) study is necessary in order to investigate the air flow pattern and air flow distribution, which will help to identify the locations where air distribution is inadequate. This has been supported by previous studies (Lim et.al [1], Lim et.al [2], Cheong et.al[3], Ruponen & Thinker [4] and Gon et.al [5]).

[1] To whom any correspondence should be addressed.

2. Methodology

2.1 Room Selection Procedure
Two lecture rooms (Figure 1 & Figure 2) with different type of air conditioning configuration have been selected for this study. The details of the rooms are given in Table 1 below:

Table 1. The details of the rooms.

Items	Lecture Room 1 (LR1)	Lecture Room 2 (LR2)
Dimension (m)	12 (L) x 9 (W) x 3 (H)	14.85 (L) x 7.35 (W) x 3.5 (H)
Air-conditioning Configuration	Central Unit	Split Unit

Figure 1. Lecture Room 1 (LR1)

Figure 2. Lecture Room 2 (LR2)

2.2 Indoor Air Parameter Measurement
The indoor air parameter which to be measured in this study are the air velocity and air temperature. The air velocity meter was used for data measurement and the height of the sensor was adjusted to 1.1 m above ground level. The range of air temperature for this device is from 17.8°C to 93.0°C while the range of air velocity is from 0 m/s to 30m/s. The selected rooms were divided into 20 zones to obtain the desired measurements as suggested by previous studies [1] [2].

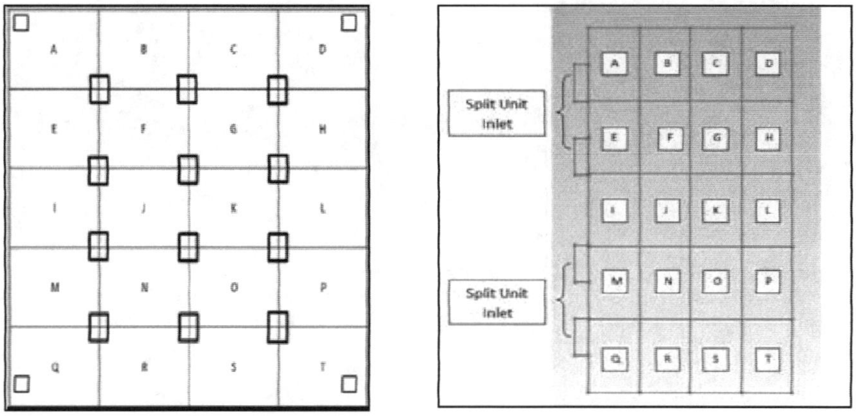

(a) LR1 (b) LR2

Figure 3. Zonal division of LR1 (left) and LR2 (right)

2.3 Simulation Procedures

The simulation is conducted by using the ANSYS Fluent CFD software. From this study, it is learned that the general room modeling procedures including defining geometry of the model, defining fluid domains, boundary conditions, initial guess, meshing and solver control. The assumptions made for simulation are given as follows:

(a) The air conditioning system in selected lecture rooms is fully functioning and running well.
(b) The rooms are fully sealed and enclosed without any holes or gaps (excluding doors, windows and exhaust vent).
(c) The outside temperature on the surface of the room is constant
(d) Internal heat source emitted from the digital devices and lights will be neglected due to minimal effect on the temperature.
(e) The furniture (chairs and tables) are included in the simulation.

The simulation parameter is summarized in Table.

Table 2. Summary of simulation parameter.

Items	LR1	LR2
Domain		
Viscous model	k-epsilon (2 equation)	k-epsilon (2 equation)
k-∈ Model	RNG	RNG
Near-wall treatment	Standard Wall Functions	Standard Wall Functions
Meshing		
Number of Nodes	30779	27886
Number of Elements	155986	141285
Solution Method		
Turbulent Kinetic Energy	Second Order Upwind	Second Order Upwind
Turbulent Dissipation Rate	Second Order Upwind	Second Order Upwind

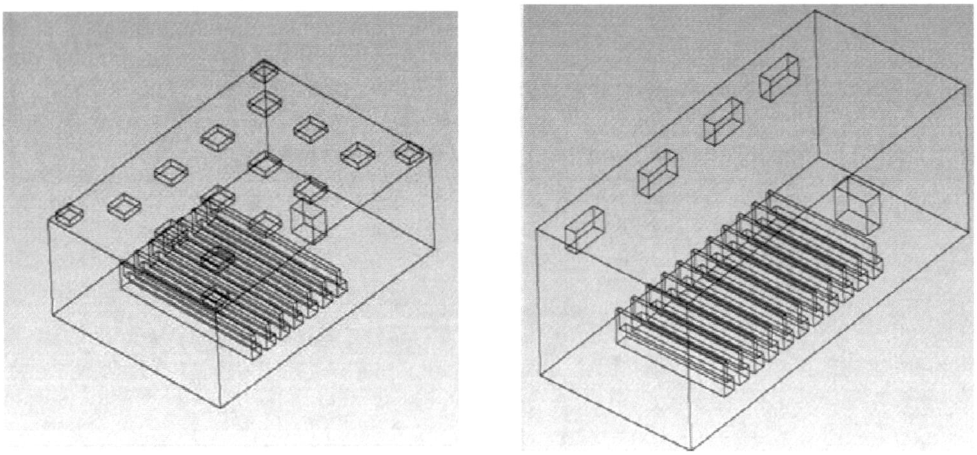

(a) LR1 (b) LR2

Figure 4. Geometry sketch of LR1 (left) and LR2 (right).

3. Results and Discussions

Based on the experimental results as shown in Table 3, the average air temperature for LR1 was 27.3 °C and the average temperature for LR2 was 21.8 °C. The average air velocity for LR1 was 0.12 m/s and the average air velocity for LR2 was 0.15 m/s. Referring to this data, the results for LR1 does not conform to the standard MS 1525:2007 [6] and further consideration can be taken to improve the overall air-conditioning system. The results for LR2 however are reasonably good with better air flow in the lecture room.

According to CFD approach, the analysis was successfully performed to obtain the air parameters for both lecture rooms. The simulation modeling was influence with the standard procedures which are defining the geometry of the lecture room, meshing, fluid properties, boundary conditions, solver settings, solution methods and the results are obtained from CFD post. The average air temperature for LR1 was 26.9°C and the average temperature for LR2 was 18.7°C. The average air velocity for LR1 was 0.10 m/s and the average air velocity for LR2 was 0.18 m/s. Similar with the results from the physical measurement, the data obtained for LR1 and LR2 does not conform to the standard of MS 1525:2007.

Figure 5 and 6 shows the CFD stream plot of the air velocity inside both rooms. Based on the figures, it can be seen that the air distribution pattern for centralized air conditioning unit performed fairly better than the split unit system even though the air velocity present in LR1 was lower.

Table 3. Summary of results of Lecture Room 1 and Lecture Room 2

Items	Recommended Standard MS 1525:2007	Physical Measurement of LR1	CFD Simulation of LR1	Physical Measurement of LR2	CFD Simulation of LR2
Average Air Velocity (m/s)	0.15-0.5	0.12	0.10	0.15	0.18
Average Temperature (°C)	22-26	27.3	26.9	21.8	18.7

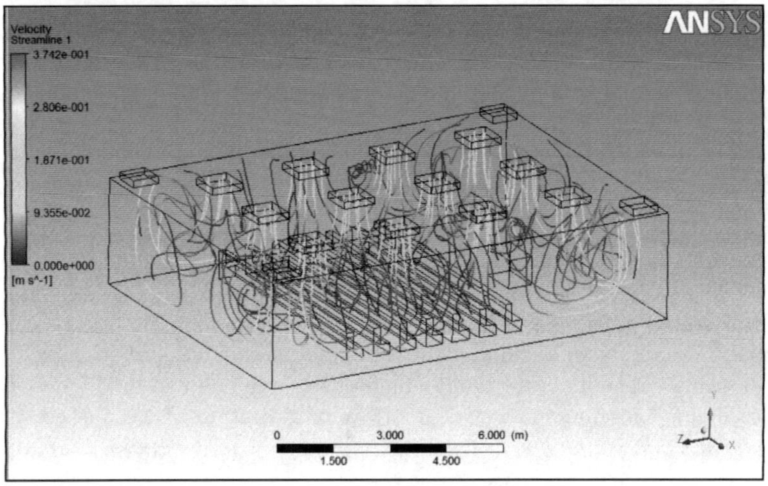

Figure 5. Stream plot for LR1.

IOP Conf. Series: Materials Science and Engineering **88** (2015) 012008 doi:10.1088/1757-899X/88/1/012008

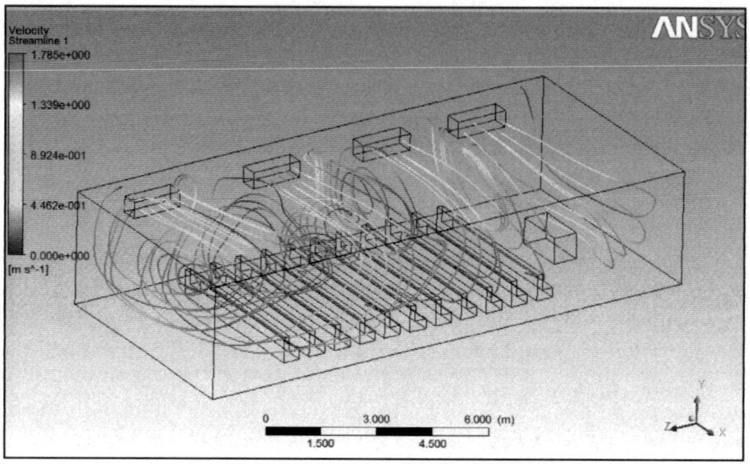

Figure 6. Stream plot for LR2.

The orientation and positioning of the split unit air conditioning unit in LR2 could be altered to obtain a better air flow pattern. Figure 7 below shows that a different orientation of the air conditioning unit could give a better air flow pattern in the lecture room.

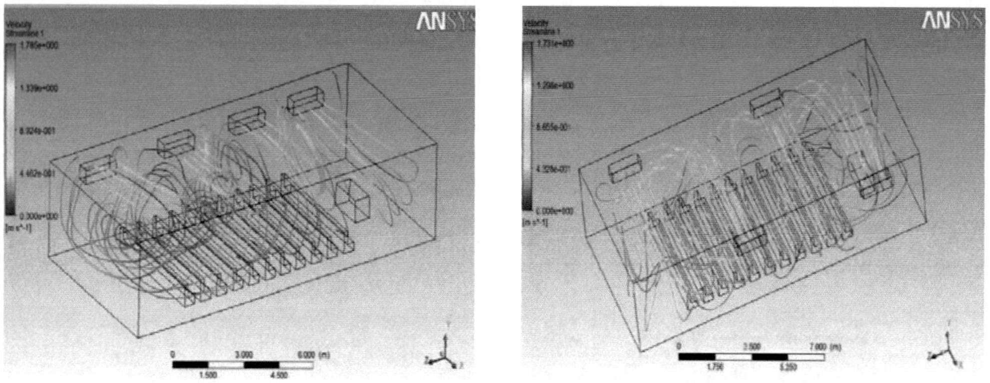

(a) (b)

Figure 7. Original position of the split units (a) and new orientation of the split units (b) with the air flow pattern.

4. Conclusions

The comparison results for LR1 between physical measurement and CFD simulation gives 16.7% difference for air velocity and 1.4% for the temperature. The results for LR2 give percentage difference of 20.0% for air velocity and 14.22% for temperature. The errors that exist are mainly due to the uncontrolled infiltration in actual room which difficult to be quantified for CFD simulation. In term of air-conditioning configuration, the air flow distribution pattern in LR1 has a better streamline pattern than LR2. Even though the air velocity in LR1 is below than the recommended standard, the air distribution tends to be steady and well distributed across every part of the room.

Acknowledgment
The authors would like to thank Universiti Teknikal Malaysia Melaka for the financial support through Short Term Research Grant award (PJP/2012/FKM(6A) S01077.

References

[1] Lim J, Tee B T & Husain A 2005 Simulation of Airflow in Lecture Room *The AEESEAP International Conference 2005 (Kuala Lumpur, Malaysia 7th – 8th June 2005)* (Kuala Lumpur, Malaysia:Universiti Malaya)

[2] Lim C M, Tee B T, & Lim J 2006 Airflow Simulation Studies Inside KUTKM's Lecture Rooms *National Seminar on Science and Its Applications in Industry (Malacca, Malaysia 14th – 15th February 2006)* (Malacca:KUTKM) p 27 – 39.

[3] Cheong K, Djunaedy E, Tham C K, Sekhar S, Wong N & Ullah M 2003 *Building and Environment* **38(1)** 63–73

[4] Ruponen M & Tinker J A 2007 CFD Validation of the thermal comfort in a room using draft rates. *Proceedings of the Clima 2007 Wellbeing Indoor Congress (Helsinki, Finland 10th - 14th, 2007)* (Helsinki:Finland) ISBN 978-952-998-3-6

[5] Kim G, Schaefer L, Lim T S & Kim J T 2013 Energy and Buildings **64** 323–331

[6] Malaysian Standard 2007 *Code and Practice on Energy Efficiency and Use of Renewable Energy for Non-Residential Buildings* (Putrajaya:Department of Standards Malaysia)

Large eddy simulation of natural ventilation for idealized terrace houses due to the effect of setback distance

L Tuan[1,3], A Abd Razak[1,4], S A Zaki[2], A F Mohammad[1], and M K Hassan[1]

[1]Faculty of Mechanical Engineering, Universiti Teknologi MARA (UiTM) Shah Alam, Malaysia

[2]Malaysia-Japan International Institute of Technology, Universiti Teknologi Malaysia, Kuala Lumpur, Malaysia

Abstract. Similar to most tropical countries, Malaysia have low wind speed and airflow characteristics to provide an effective natural ventilation system for comfortable living especially in terrace houses. Even so, by designing them with suitable threshold height/width, *H/W*, ratio may help reduce heat sink, or even the accumulation of contaminants within the setback distance. Through this study, the downstream building of these terrace houses will be investigated due to the effects from an upstream building. With the use of Large-Eddy Simulation (LES) method, the formation of the vortex within the threshold *H/W* ratio will be clearly simulated and allow the observation of flow regimes developed by each model. With increasing threshold *H/W* ratios the models will exhibit some wake interference flow and skimming flow which will determine the negative or positive effect of ventilation from the upstream building towards the downstream building. The airflow characteristics of the downstream house will also be analysed and the most effective layout in providing a better air circulation may be determined. Improving the natural ventilation of such houses could significantly reduce these negative effects such as the accumulation of dust, smoke or bacteria. In turn, with the alarming rate of depletion in natural resources and its effects to the environment, this study can significantly reduce energy usage for ventilation and space cooling.

1. Introduction

Natural ventilation has been long used for space cooling and the removal of contaminants in buildings for comfortable living. By permitting natural ventilation, the indoor air will be cooler, cleaner and fresher without the use of any mechanical forces. Since, there have been a high rate of depletion in natural resources in the recent years, the use of mechanical ventilations need to be greatly reduced. This has led to the reintroduction of natural ventilation for heat removal and ventilation. Hence, ways to promote natural ventilation inside a building need to be further studied especially in the tropical climates.

Malaysia is a tropical country that experiences hot and humid climate throughout the year. The climate within this region is generally known to have high ambient temperatures at an average minimum of 23°C and an average maximum of 33°C with small diurnal and annual ranges [1]. Also, in this region, the wind speed varies between the range of 1 m/s to 3 m/s (light breeze) in major towns [2]. The high temperature and low wind speed experienced by this region contributes to the lower comfort level indoors, as the weather is unable to provide good ventilation for buildings. In many cases, this has caused the overheating of outdoor environment in urban areas.

In many countries like Malaysia mechanical cooling and ventilation techniques are highly used for comfortable living. In Malaysia, the percentage of the yearly energy consumption for domestic use is 29% [3]. From that 29% energy consumption, 7.105 GJ/year is yielded for the cooling and ventilation of

[3] Author to whom any correspondence should be addressed.
[4] Author to whom any details of research funding should be addressed.

buildings. Based on the Building and Climate Change 2007 report it is targeted for Malaysia to reduce the use of electricity by 5.6% and fuel demand by 26.7% annually [4]. This is one of the many challenges in building design of homes in the tropics to achieve comfortable living. During the past few decades, there have been many experiments and research carried out on naturally ventilated buildings in order find solutions for achieving energy saving, green energy, energy efficiency and sustainable energy goals due to the depletion of natural resources. In the area of cross ventilation, one of the recent researches conducted particularly on the improvement of natural ventilation of the terrace housing community in Malaysia is the work by Mohamad et al [5].

Mohamad et al. [5] stated that there are many challenging factors that influence the analysis of natural ventilation such as wind velocity, wind direction, temperature difference, opening design, building shape and adjacent building and much effort are being put in to hurdle over these challenges. One of the many studies in simulating the effects of natural ventilation using the LES method is by Yi Jiang et al. [6]. In justification to its capabilities, the LES method is very outstanding in accounting unsteady flow and generating detailed information on the flow structure, including turbulence statistics, has been proven in many studies to date [7].

2. Simulation Setting

2.1. Building Models

The building model of this study adopted a basic intermediate single-storey terrace house which is similar in layout and dimension to a previously conducted study by Mohamad et al. [5]. In comparison to his study, this current study is modelled with an opposing downstream building set at a separation distance, *d*, of *0.5H*, *H* and *1.5H*. The cases evaluated are as summarized in Table 1. The target building for this study is the downstream structure (*H2*) and will be analysed at each point *P1*, *P2*, *P3*, *P4*, and *P5* of each consecutive room. Point *P1* is the Living room, point *P2*, *P3* and *P4* are the bedrooms, and point *P5* is the utility room (*see*, Figure 1). Abiding the Uniform Building by-laws [8] the openings of the rooms are the doors which are made greater than 10% of the floor area with a minimum dimension of 2m in height and 0.75m in width [8]. Apart from that, the building height is modelled with a standard height of 3m [8].

2.2. Computational Domain and Boundary Conditions

The computational domain that was modelled is as in Figure 2. The limiting boundary conditions are set up similar to Mohamad et al [5]. The inlet, outlet and lateral boundaries of the domain of the building are treated as periodic condition, and the top boundary of the domain is treated as free-slip condition. The bottom boundary and surface of the building model are treated as non-slip conditions that will guide the computations throughout the simulation. For all the models that will be carried out, the initial velocity of the wind will be directed at a zero incidence angle perpendicular to the front façade of the building (in the horizontal x-direction), at 5m/s. The respective axis orientation and domain sizes are as labelled. A summary of the steps carried out in OpenFOAM for this particular simulation is as in Figure 3.

Table 1. Evaluated cases

Simulation Case	Setback distance
Case 0.5H	1.5m
Case H	3m
Case 1.5H	4.5m

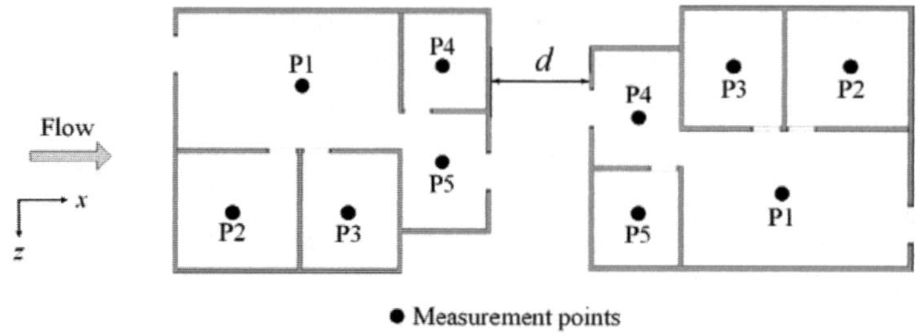

Figure 1. Model layout of the buildings

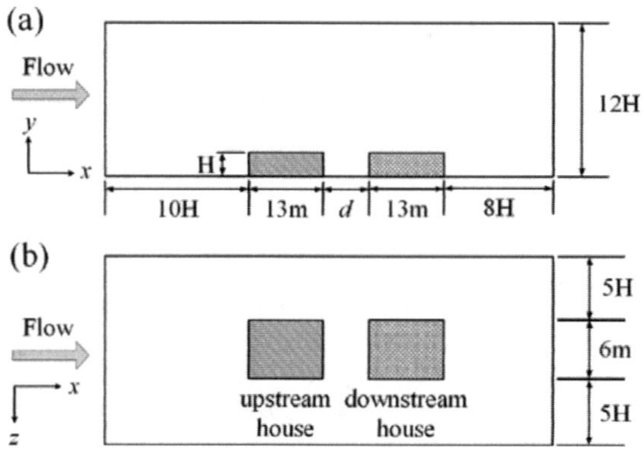

Figure 2. Computational domain size (a) Sectional side view (b) Plan view

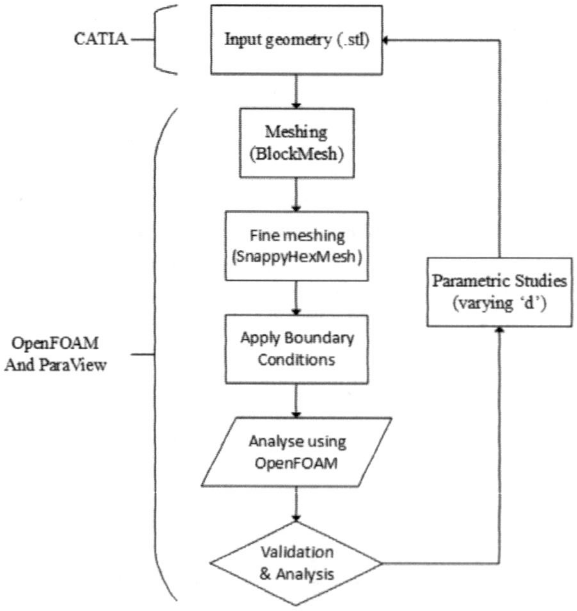

Figure 3. Summary of simulation steps

3. Result and Discussion

The validations of the cases were compared with the experimental data obtained from Yi Jiang et al. [6] as in Figure 4. The mean velocity distribution obtained from this study generally agreed to the experimental results of his study. However, there were some discrepancies found especially at the front façade of the building (y/H <1) and near the roof height (y/H =1). This may be due to the coarseness of the grid as stated by Yi Jiang et al. [6].

Focusing on the target building, the velocity distribution at each consecutive room of the downstream house starting from the point of entry at, *P5H2* until the exit route, *P1H2* is as in Figure 5. Generally, the flow across the downstream target building will be affected by the shear stress of the ceiling. This can be seen in the range of y/H >0.9 where the profile seem to "stick" to the ceiling or backflow occurs. This backflow may also occur due to short ceiling height where a higher grid discretization level is needed to better simulate the profile near the ceiling. Despite the discrepancies from the effect of the ceiling some similarities between the cases can also be seen at a height ratio between 0.6 < y/H > 0.9 (between the door height and the ceiling). Here, the velocity tends to increase at an almost constant value along that range of height ratio. Apart from that it can be seen that the profile generated by Case *1.5H* over-estimated the other two cases, Case *0.5H* and Case *H*. This shows that the velocity across the rooms are better in each of the consecutives rooms for Case *1.5H* which shows that the target building is better ventilated with a higher setback distance.

Looking closely at point of measurement, *P1H2*, it was found that with a higher setback distance there is a higher chance for the air from the downstream house to be replenished outdoors. As we can see from Figure 5, the flow generated is more drawn to the exit or door (y/H <0.6) for Case *1.5H*. However, for Case *H* and Case *0.5H*, the flow generated are less drawn to the exit than Case *1.5H* but rather, it is more drawn to the ceiling. Thus, this clearly shows that by increasing the setback distance can possibly improve the ventilation rate of the downstream house.

Even so, from this induced wind flow there may be some other impacts of the setback distance that results in carrying negative effects from the upstream house such as "dirt and dust". This may be observed as in Figure 6. This figure shows the streamtracer and velocity profile across the upstream and downstream house. From this figure we can clearly see the wind flow across the cascading buildings and it can be seen that a higher setback distance will induce a stronger wind flow within the upstream house. When this happens there are more chances of the wind to "pick up" any negative effects from the upstream house and carry it to the downstream house especially from isolated rooms i.e. bedroom 3 (*P1H4*). However, with the formation of wake interference flow (highlighted by yellow box in Figure 6) within the threshold *H/W* further downstream of the wind distribution the accumulated particles may be dispersed from the area.

4. Conclusion

The simulation focuses on the flow fields inside and around a basic single-storey intermediate terrace house in Malaysia. The simulated model was varied with three different setback distance configurations of *0.5H*, *H* and *1.5H*. For the three cases, the velocity distributions inside the consecutive rooms were determined. It was found in the present study that the higher the setback distance, *d*, the higher the chances of ventilation to happen. However, with a higher setback distance, there is a higher chance of some negative effects from the upstream house to be carried into the downstream house such as dust and dirt. On the other hand, after further analysis of the results, the formation of wake interference flow within the threshold *H/W* will help to disperse these negative effects from the area. Depending on the setback distance, the higher the *H/W* ratio, the slower the rate for the carrying winds to disperse.

5. Acknowledgement
This research is financially supported by grant-in aid for scientific research (600-RMI/FRGS 5/3 (73/2013)) from the Ministry of Education, Malaysia.

6. References
[1] Rajalingam J K 2005 *Kolej Universiti Teknikal Kebangsaan Malaysia*
[2] Kubota T and Ahmad S 2006 *J of Asian Arch and Building Eng* **5** 199-206
[3] Kubota T, Jeong S, Toe D H C, Ossen D R 2011 *J of Int Dev and Coop* **17** 61-9
[4] Initiative C 2007 *Buildings and Climate Change*
[5] Mohamad M F, Hagishima A, Tanimoto J, Ikegaya N and Omar AR 2013 *13th Conf of Int Building Performance Simulation Assoc Chambery France*
[6] Jiang Y, Alexander D, Jenkins H, Arthur R and Chen Q 2003 *J of Wind Eng and Ind Aerodynamics* **91** 331-53
[7] Walton A, Cheng A and Yeung W 2002 *Atmospheric Environment* **36** 3601-13
[8] Uniform Building By-Law (UBBL) 1984 Malaysia

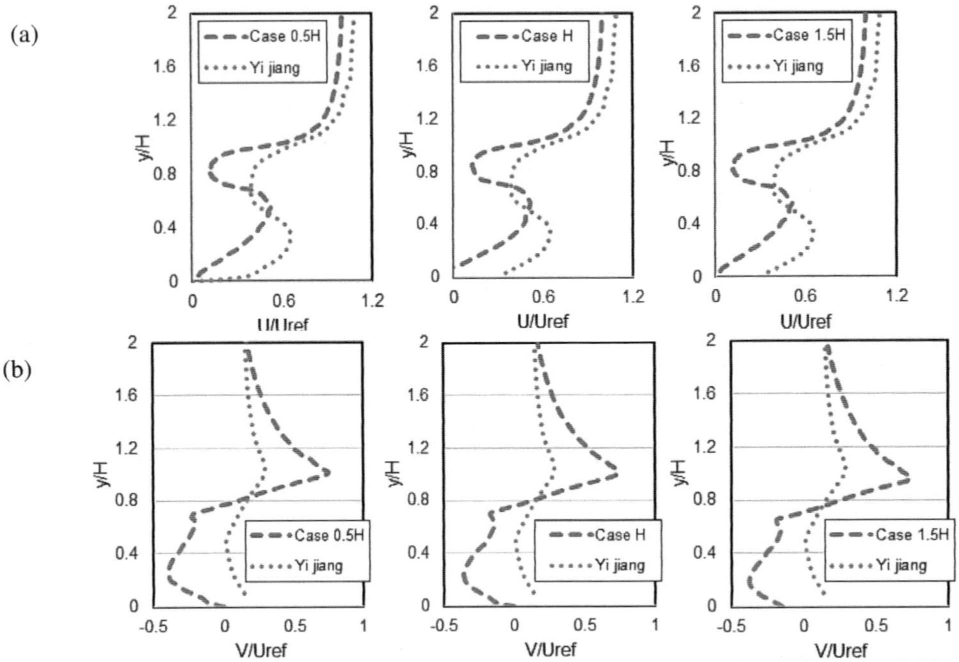

Figure 4. Mean velocity distribution for cross ventilation (a) streamwise (U/Uref) and (b) vertical (V/Uref)

IOP Conf. Series: Materials Science and Engineering **88** (2015) 012009 doi:10.1088/1757-899X/88/1/012009

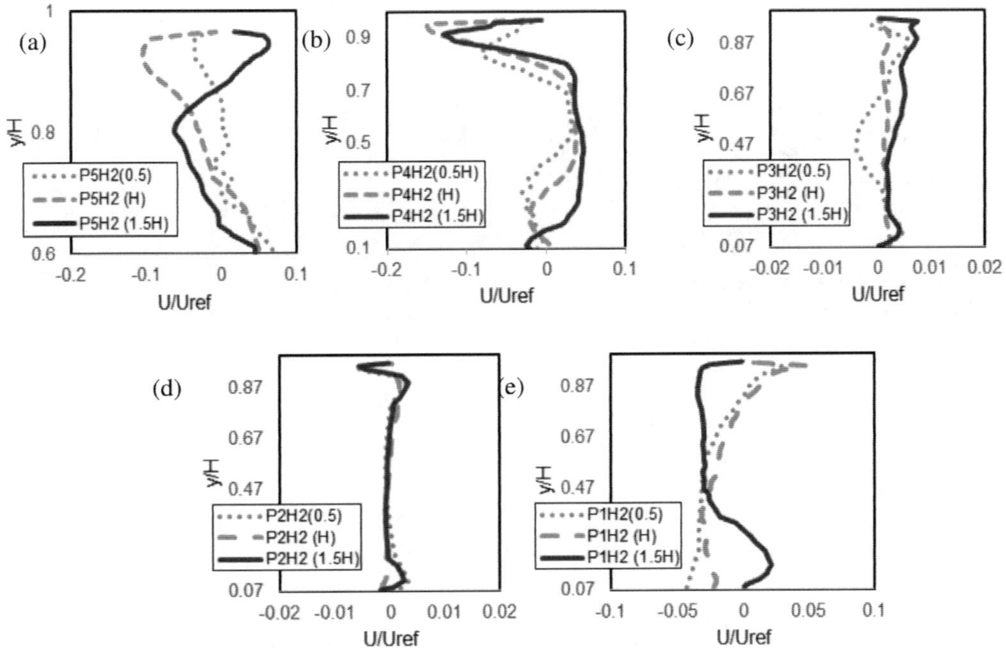

Figure 5. Streamwise mean velocity distribution at each consecutive rooms (a) *P5H2*, point of entry (b) *P4H2* (c) *P3H2* (d) *P2H2* (e) and *P1H2*, exit route

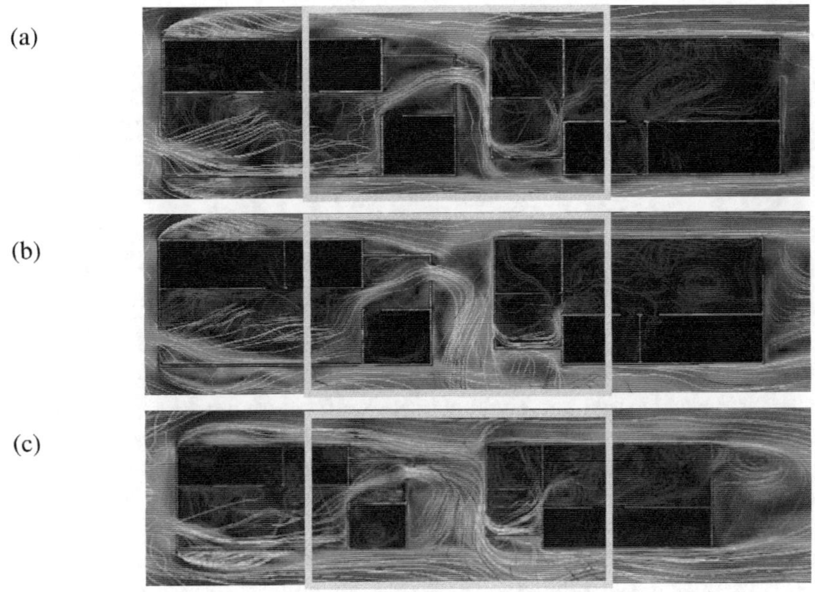

Figure 6. Flow Regime Generated at Threshold H/W (a) Case *0.5H* (b) Case *H* (c) Case *1.5H*

Natural convection heat transfer in vertical triangular sub-channel in Zirconia-water nanofluid

N P Tandian[1], **A A K Alkharboushi**[2] and **K Kamajaya**[3]
[1] Faculty of Mechanical and Aerospace Engineering, Institut Teknologi Bandung, Bandung 40132, Indonesia
[2] Al Zawiya University, Motred – Zawiya, Libya
[3] Center for Applied Nuclear Science and Technology, National Nuclear Energy Agency of Indonesia, Bandung 40132, Indonesia

E-mail: n4th4n.t4nd14n@gmail.com

Abstract. Natural convection heat transfer in vertical triangular sub-channel has important role in cooling mechanism of the APWR and the PHWR nuclear reactors. Unfortunately, natural convection correlation equations for such geometry are scarcely available. Recent studies showed that ZrO_2-water nanofluid has a good prospect to be used in the nuclear reactor technology due to its low neutron absorption cross section. Although several papers have reported transport properties of ZrO_2-water nanofluids, practically there is no correlation equation for predicting natural convection heat transfer in a vertical triangular sub-channel in ZrO_2-water nanofluid. Therefore, a study for finding such heat transfer correlation equation has been done by utilizing Computational Fluid Dynamics software and reported in this paper. In the study, natural convection heat transfer in a vertical triangular sub-channel has been simulated at several values of heat transfer flux within 9.1 to 30.9 kW/m^2 range and ZrO_2concentrations of 0 (pure water), 0.27, and 3 volume-% of ZrO_2. The study shows that the ZrO_2 concentration has no significant influence to the natural convection heat transfer at those concentration levels. The obtained theoretical heat transfer correlation equations were verified through experiment, and they showed very similar results. The correlation equations are reported in this paper.

1. Introduction

Heat transfer fluids have many industrial and civil applications, including processes in transportation, energy supply, air-conditioning and electronic cooling, etc. Traditional base fluids, such as water, oils, glycols and fluorocarbons, have inherently relatively poor heat transfer performance due to their low values of thermal conductivities and heat capacities. Research and development activities are being carried out to improve the heat transfer properties of fluids. There are innovative ideas trying to enhance the thermal conductivity by adding solid particles of heat transport improvement agent into base fluids since Maxwell initiated it in 1881 [1]. At the very beginning, solid particles of micrometer, even millimeter magnitudes were blended into the base fluids to make suspensions. However, large

[1] To whom any correspondence should be addressed.

solid particles cause troublesome problems, such as abrasion on the surface, clogging the micro channels, eroding the pipeline and increasing the pressure drop, which substantially limits the practical applications. Then, the development of nanofluid gives a new chance to overcome the problems found in conventional suspensions. Several nanoparticles that can be applied in nanofluids, such as Al_2O_3, Cu, CuO, and TiO_2 have been widely investigated. But so far nanofluid of ZrO_2-water is rarely investigated although the ZrO_2 nanofluid has a better possibility to be used in the nuclear reactor technology due to its low neutron absorption cross section [2]. Therefore, the ZrO_2-water nanofluid has been chosen as the nanofluid studied in this paper.

Meanwhile, natural convection from a bundle of hot cylinders to a cooling fluid flowing around the cylinders becomes more and more important in modern nuclear reactor applications. The heat transfer geometric configuration consisting of a bundle of cylinders and fluid among the cylinders are commonly known as a sub-channel. Unfortunately, correlation equations for predicting natural convection heat transfer in sub-channel configuration are scarcely available. Although a couple of studies have been done in seeking for such correlation equations, but the studies were limited to conventional coolants, mainly with pure water.

This paper deals with highlights of a recent research that seeks for convective heat transfer correlation equations of ZrO_2-water nanofluid as working fluid in sub-channels formed among vertical cylinders was done based on the CFD analysis. Several simulations of heat convection within a vertical sub-channel were done in the research by using commercial CFD software. The simulations were done with a various concentration of ZrO_2 in heating power levels. The results obtained form the simulation were compared to various experimental data done in the National Nuclear Energy Agency and other available data in the literature. The heat transfer correlation equations obtained in this research can be used for predicting heat transfer coefficients in sub-channels formed among vertical cylinders that are commonly utilized in nuclear reactors or heat exchangers.

2. Numerical Modeling

2.1. Geometrical model

As was mentioned in the previous section, this paper covers highlights of the theoretical study (CFD simulations) of the research that currently is being done. The main objective of the research is finding a new correlation equation for calculating natural convection heat transfer coefficient within a vertical triangular sub-channel formed among vertical cylinders to a zirconia-water nanofluid. The vertical cylinders simulate a bundle of fuel rods of a nuclear reactor or tubes of a heat exchanger. The size and geometry of vertical cylinders used in the current research specifically replicate the size of fuel rods in the Bandung TRIGA 2000 reactor of the National Nuclear Energy Agency, Indonesia.

A sketch of important parts of the test section used in the experimental part of this research is shown in figure 1.a. The experiment test section consists of a test section outer box made of glass sheets so that they are transparent, a main test section that confines a test cylinder assembly, and a distributor plate. The test cylinder assembly consists of three main (heated) cylinders and three dummy (unheated) cylinders that arranged to form triangular sub-channels. The main cylinders are equipped with electric heaters, while the dummy cylinders do not have heaters. The dummy cylinders are installed to generate similar flow pattern as that of sub-channel surrounded by a very large number of cylinders. Each main cylinder equipped with four thermocouple junctions uniformly installed along heated part of the test cylinder and one thermocouple junction installed in the unheated part of the cylinder to measure wall temperature distribution along the main cylinder length.

During the experiment, fresh water or ZrO_2-water nanofluid enters the space between the test section outer box and the main test section through a test section main inlet that is available at the bottom of the test section outer box. Most of the cooling water or ZrO_2-water nanofluid flows upward through the test section annular between the test section outer box and the main test section and leaves the test section at the main outlet. Meanwhile, a part of cooling fluid leaves the test section annular and enters the lower part of the main test section through entrance holes that are available at the lower

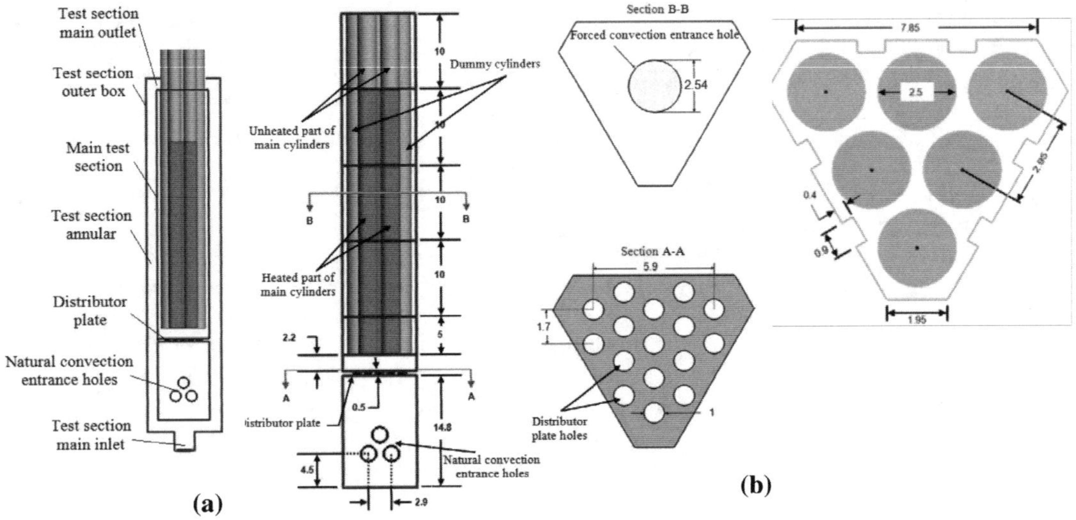

Figure 1. (a) Important parts of the experimental test section, and (b) sketch of geometrical model for the CFD simulation. All dimensions are in cm.

part of main test section walls. The cooling fluid then flows through the distributor plate that distributes the flow at the upstream end of the cylinder assembly.

A geometrical model utilized in the CFD simulation in this study is shown in figure 1.b. The calculation domain for the CFD analysis only covers the volume inside the main test section, excluding volumes of the main and dummy cylinders. As simplification, the domain around the cylinder assembly is divided into several sub domains as shown in figure 1.b, so that average temperature inside the sub-channel and average temperature on the main cylinder wall inside each sub-domain can be compared to their associated measured temperatures.

2.2. Operating condition, assumptions and governing equations

The operating condition considered in this study follows the operating condition of the experimental test of the research. During the experiment, the heater power on each cylinder varies between 250 and 850 W, which are associated to heat fluxes of 9.1 and 30.9 kW/m^2, respectively. It is assumed that the heat flux is constant and uniform over the cylinder surfaces. To observe the influence of nanoparticle concentration, there are three values of ZrO_2 volumetric concentration are simulated in the research, i.e. 0% (pure water), 0.027%, and 3%. Unfortunately, due to a limitation in the amount of available ZrO_2 nanoparticles the experiment with ZrO_2 concentration of 3% has not been done.

Several important assumptions are considered in this study and some of them are also used as boundary conditions for the CFD analysis. These assumptions are:

- The nanoparticles can be easily fluidized in the base fluid, thus, the nanofluid mixture is considered as a single phase.
- The nanoparticles and the base fluid are in thermal equilibrium with each other.
- The calculation assumes a steady operating condition.
- Pressure based computational scheme is used for the whole fluid domain.
- Since the outer box is excluded from the calculation domain, therefore the pressure at the outlet water surface is constant at the atmospheric pressure, while pressures at domain inlet are at their respective hydrostatic pressures.
- The effective thermo-physical properties of the ZrO_2-water nanofluid depend on the temperature and volume concentration of the nanoparticles and their values are obtained from literature [3]. Since the data in the literature [3] are obtained from experiments, therefore

influences of important phenomena exist in the nanofluids, e.g.: thermophoresis and the Brownian motion effects, have been reflected in the thermo-physical properties data.

- The nanofluid flow is expected to be relatively slow and the gaps between wall and cylinder surfaces are small, therefore it is assumed that the flow is laminar.
- Since the water velocity is low, therefore the viscous dissipation effect is negligible.
- No-slip condition occurs at boundaries between the cooling fluid and solid surfaces, i.e.: cylinder walls, test section walls, distributor plate, etc.
- All boundaries between the cooling fluid and solid surfaces are adiabatic, except on the heated area of the main cylinders where there are constant and uniform heat transfer fluxes.

Basically the governing equations consist of continuity equation, momentum equation, energy equation implemented in the Ansys Fluent 15.0. These equations can be expressed as following [4].

Continuity equation. The continuity equation, or equation for conservation of mass, can be written in the tensor notation as follows:

$$\nabla \cdot (\rho \vec{v}) = 0 \qquad (1)$$

where ρ is the fluid mass density and $\vec{v}$ fluid velocity vector.

Momentum equation. Conservation of momentum in an inertial (non-accelerating) reference frame can be described by the following equation:

$$\nabla \cdot (\rho \vec{v}\vec{v}) = -\nabla p + \nabla \cdot \vec{\vec{\tau}} + \rho \vec{g} \qquad (2)$$

where p is the static pressure, $\vec{\vec{\tau}}$ is the stress tensor (described below), and $\rho \vec{g}$ is the gravitational body force. The stress tensor can be expressed as following:

$$\vec{\vec{\tau}} = -\mu\left[\left(\nabla \vec{v} + \nabla \vec{v}^{T}\right) - \tfrac{2}{3}\nabla \cdot \vec{v}\vec{\vec{I}}\right] \qquad (3)$$

where μ is the molecular viscosity, $\vec{\vec{I}}$ is the unit tensor, and the second term on the right hand side of equation (3) represents the effect of volume dilation.

Energy equation. The energy conservation equation can be expressed in the following form:

$$\nabla \cdot \left[\vec{v}\rho\left(h + \tfrac{1}{2}v^{2}\right)\right] = \nabla \cdot (k \nabla T) \qquad (4)$$

where k is the thermal conductivity of the fluid, and T is temperature of the fluid.

3. Physical properties of nanofluid

Due to the idea of dispersing solid particles of zirconia in water base fluid is newly discovered, there is no significant study about physical properties of ZrO_2-water nanofluid found in literature. The physical properties of nanofluid used in this study were based on the properties suggested by Williams et al. [5] and modified by Rea et al. [3]. Basically physical properties of the ZrO_2-water nanofluid, such as its mass density, heat capacity, viscosity, and thermal conductivity are influenced by the ZrO_2 nanoparticle volumetric concentration and the nanofluid temperature. Detailed information related to these physical properties is beyond the coverage of this paper.

4. Results and discussion

4.1. Temperature distribution

Typical distribution of fluid and cylinder surface temperature along the main cylinder length at various heat flux values are shown in figure 2. The data in figure 2 are obtained from the CFD simulations with ZrO_2-water nanofluid at solid volumetric concentration of 0.027%. CFD simulations at other nanofluid concentrations also gave very similar results as those are shown in figure 2. The position shown in figure 2 is measured in vertical direction from the lower end of the main cylinder.

Figure 2.a shows that the fluid temperature distribution curves have relatively larger slopes near the upstream end of the cylinder compared to those at more downstream position, although theoretically slope of the fluid temperature distribution curve has to be linear if the heat flux is constant. This relatively larger fluid temperature distribution slope suggests that there is *entrainment* or lateral flow near the upstream end of the main sub-channel, i.e. sub-channel between three main cylinders. At further downstream locations the entrainment becomes less significant so that the fluid temperature slope practically is constant. As what is generally expected, the fluid temperature distribution curves also show that the fluid temperature curve is steeper at higher heat flux.

Figure 2.b shows the cylinder surface temperature distribution along the cylinder axial position at several heat flux values. The surface temperature distribution shows similar tendency as that of the fluid temperature distribution due to previously discussed reasoning. The variation of surface temperature slope also related to the developing flow condition, i.e. the flow has not been fully developed, when its flow boundary layer thickness increases with increasing elevation.

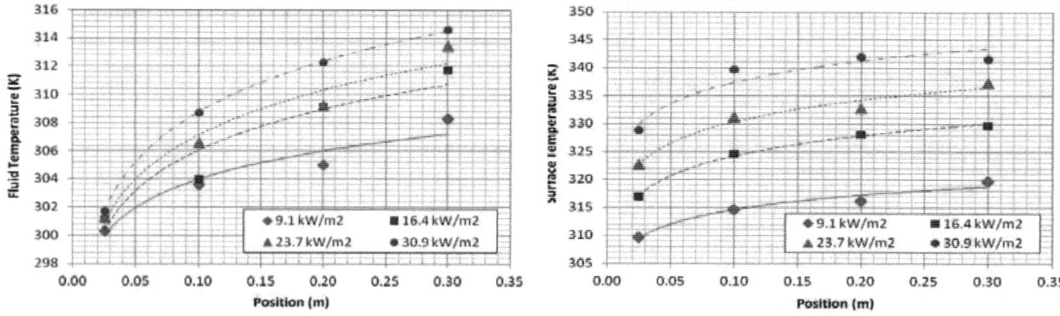

Figure 2. Temperature distribution along the cylinder length at various heat flux values. (a) Fluid temperature. (b) Cylinder surface temperature.

4.2. Natural convective correlation equations

The postprocessor of the CFD software used in this study can evaluate area-weighted average heat transfer coefficient and mass-weighted average temperature. The obtained area-weighted average heat transfer coefficient is needed for evaluating Nusselt number *Nu*, while the mass-weighted average temperature is used for evaluating the cooling fluid physical properties needed for calculating other dimensionless parameters commonly used in natural convective correlation equations, such as Grashof number *Gr* and Rayleigh number *Ra*.

The CFD simulation results can be presented in terms of dimensionless parameters and plotted as shown in figure 3. The figure compares the natural convective correlation equations at various values of ZrO_2 volumetric concentrations obtained from the CFD simulation and their corresponding experimental data. The figure shows that the convective characteristics of these ZrO_2-water nanofluids are not significantly influenced by ZrO_2 concentration. All CFD simulation regression lines are practically the same, while the discrepancies between experimental regression lines are insignificant since they are in the same order as the data spread. The regression lines can be expressed as general equation (5) with a, b, and c values are given in table 1. The experimental Nusselt numbers in figure 3 are significantly larger than the CFD simulation Nusselt number, since the calculation domain used in the CFD simulation excludes the fluid outside the main test section so that influences of the pressure drop inside the test section annular between the inlet and outlet of the main test section was neglected (i.e. the chimney effect along the main test section was neglected in the CFD simulation), which leads to smaller pressure difference between the inlet and outlet of the test section and slower simulated fluid flow inside the main test section.

$$\lg Nu = a \lg\left(Ra\frac{D_h}{Y}\right) + b \quad \text{or} \quad Nu = c\left(Ra\frac{D_h}{Y}\right)^a \tag{5}$$

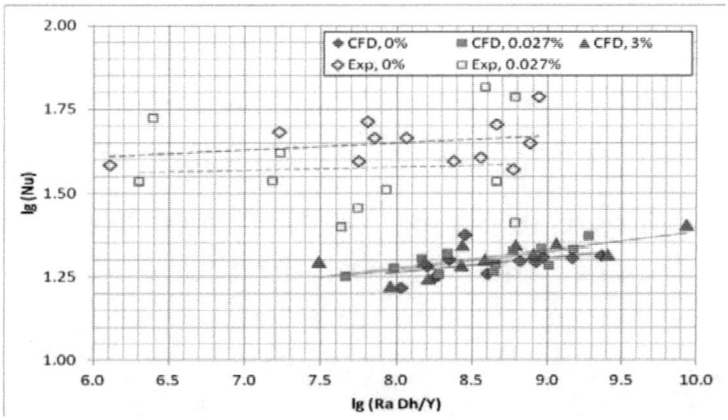

Figure 3. Natural convective correlation curves.

Table 1. Coefficients for the natural convective correlation equation.

No.	Natural Convection Case	*a*	*b*	*c*
1	CFD simulation, ZrO_2-H_2O 0% (pure H_2O)	0.042	0.927	8.445
2	CFD simulation, ZrO_2-H_2O 0.027%	0.053	0.850	7.075
3	CFD simulation, ZrO_2-H_2O 3%	0.054	0.844	6.981
4	Experiment, ZrO_2-H_2O 0% (pure water)	0.022	1.477	29.992
5	Experiment, ZrO_2-H_2O 0.027%	0.009	1.505	31.982

5. Closing Remarks

The theoretical study covered in this paper is associated to a more detailed experimental study that is still being conducted. Several natural convective correlation equations for various concentration of ZrO_2 have been obtained from the study. Both the theoretical and experimental data show that the concentration of ZrO_2 has insignificant influence to natural convection Nusselt number, for the ZrO_2 concentration is less than 3%. The correlation equations can be used for predicting the natural convection heat transfer in sub channel in ZrO_2-water nanofluids with similar geometry.

6. Acknowledgement

The authors would like to thank the Center for Applied Nuclear Science and Technology, National Nuclear Energy Agent of Indonesia for the courtesy in providing the experimental data.

7. References

[1] Yu W, France D M, Choi S U S and Routbort J L 2007 *Review and assessment of nanofluid technology for transportation and other application* (Argonnne: Energy System Division - Argonne National Laboratory)

[2] Hidayanti D, Tandian N P, Suwono A and Umar E 2014 Investigation on modeling the thermal conductivity of zirconia-water based nanofluids *Proc. 3rd Appl. Sci. for Tech. Applications - ASTECHNOVA 2014* (Yogyakarta, 13-14 August 2014)

[3] Rea U, Tom M, Hu L W and Buongiorno J 2009 Laminar convective heat transfer and viscous pressure loss of alumina-water and zirconia-water nanofluids *Int. J. Heat and Mass Transf.* **52** 2042-8

[4] Anonym, 2013, *Ansys Fluent user's guide* (Canonsburg, PA: Fluent Inc.)

[5] Williams W, Buongiorno J and Hu L W 2008 Experimental investigation of turbulent convective heat transfer and pressure loss of alumina/water and zirconia/water nanoparticle colloids (nanofluids) in horizontal tubes *J. Heat Transf.*

IOP Conf. Series: Materials Science and Engineering **88** (2015) 012011 doi:10.1088/1757-899X/88/1/012011

Numerical modelling of multi-pass solar dryer filled with granite pebbles for thermal storage enhancement

M W Kareem[1,3], K Habib[1] and M H Ruslan[2]
[1]Department of Mechanical Engineering, Universiti Teknologi PETRONAS, 31750 Tronoh, Perak, Malaysia
[2]Solar Energy Research Institute (SERI), Universiti Kebangsaan Malaysia, 43600 UKM Bangi, Selangor Darul Ehsan, Malaysia

E-mail: kareem_g01949@utp.edu.my

Abstract. In this paper, a theoretical modelling of a cheap solar thermal dryer for small and medium scale farmers with multi-pass approach has been investigated. Comsol Multiphysics modelling tool was employed using numerical technique. The rock particles were used to enhance the thermal storage of the drying system. The local weather data were used during the simulation while parameters and coefficients were sourced from literature. An improvement on efficiency of up to 7% was recorded with error of 10^{-5} when compared with the reported double pass solar collector. A fair distribution of hot air within the cabinets was also achieved. Though the modelling tool used was robust but the characterization of the system materials need to be done to improve the system accuracy and better prediction.

1. Introduction

The United Nations has constantly monitoring its targets set for renewable energy sector under renewable energy for all by the year 2050 [1]. Therefore effort to improve the efficiency of the renewable energy systems is a major concern for researchers and this has made the use of latest modelling tools a must in the research on solar energy, in order to accomplish the noble goal of reducing the carbon content of our atmosphere due to the frequent usage of energy sourced from fossil fuels.

Carbonell et al. [2] examined both modified correlation model and an improved Duffie-Beckman model under a transient condition. Unsteady parameters such as solar irradiance, mass flow rate and inlet temperature of fluid were employed to get better system performance. These models were validated by virtual test approach that was based on numerical technique. However, it was difficult to determine the physical behaviour of the system through this approach.

Wu and Wang [3] developed a model using numerical technique to study the dynamic behaviour of volumetric solar thermal air absorber when subjected to different working conditions. A two dimensional discretization of solar absorber made of ceramic material was generated and analysed using CFD software. This dynamic model was confirmed to be accurate in predicting the volumetric solar hot air absorber. It was also affirmed that the ceramic solar absorber has smooth reaction to variation in thermal flux during the cause of operation. Ceramic porous absorber was investigated

[3] To whom any correspondence should be addressed.

IOP Conf. Series: Materials Science and Engineering **88** (2015) 012011 doi:10.1088/1757-899X/88/1/012011

numerically by Wu et al. [4] to determine the pressure drop of the air and the flow field pattern inside the porous material. A pressure drop of up to 70% at air velocity of $5ms^{-1}$ was achieved. A model based on non-equilibrium local temperature of both fluid and porous matrix was studied by Wu et al. [5] to determine the characteristics of thermal transfer that exist between both the air and the ceramic porous material. They developed a thermal transfer coefficient correlation as a function of porosity, velocity and the cell geometry by using CFD modelling tool.

Significant increase in rate of drying was reported with solar dryer that has pebble bed as energy storing material with energy charging at the early hours of operation and discharge the heat energy at the later period. This extends the drying period after sunset, though the geometry of the porous matrix, air mass flow rate, porosity, physical and thermal characteristics are functions of thermal storage [6]. Pebble bed material was introduced in 1970s to address energy shortage that was frequent. Recent report shows that porous matrixes has improved thermal application technique, cost effective and reliable means of storing thermal energy [7].

Interest in solar drying is gaining momentum as a result of shifting the energy source form hydrocarbon based to renewable energy sources. Hence, solar energy source remains the most economical and eco-friendly option suitable for food preservation methods such as drying technique. This was emphasised by Fudhoil et al [8] that solar energy source remains the best option to ensure all year round food security for fast growing global population. They remarked that solar drying is gaining more attention of researchers than ever based on its present and future relevance in global food security. Kareem et al. [9] reported the advantages of drying applications and emphasised the solar energy as the clean source of drying agricultural crops.

Many solar dryers have been reported in literature but a cost effective multi-pass dryer has been left undone. Hence, a robust tool that captures all relevant physics of the solar drying system was utilised to model a cost effective dryer for small and medium scale farmers.

2. Theoretical Analysis

The system model was based on the fundamental principle of thermodynamics as shown in equation (1), while equation (2) and equation (3) are the extension of equation (1). The velocity field (U) at a point (x,y) and other thermo physical variables were declared during simulation based on local weather data using heat transfer platform of Comsol Multiphysics modelling tool.

$$C_p\rho\frac{\partial T}{\partial t}+C_p\rho U.\nabla T=\nabla.[k\nabla T]+Q \tag{1}$$

$$C_p\rho\left[\frac{\partial T}{\partial t}+U.\nabla T\right]=k\nabla^2 T+Q \tag{2}$$

$$\nabla^2 T=\frac{\partial^2 T}{\partial x^2}+\frac{\partial^2 T}{\partial y^2}+\frac{\partial^2 T}{\partial z^2} \tag{3}$$

The performance efficiency of the drying system can be computed by equation (4) [10].

$$\eta_d=\frac{\dot{m}\rho C_p dT}{IA_c} \tag{4}$$

The wet crop temperature or material to be dried can be modelled by expression proposed by Pabis et al. [11] as shown in equation (5).

$$\frac{\partial T}{\partial t} = \frac{dM}{dt}\left(\frac{h_v \rho_d}{C\rho_w}\right) - [T - T_a]A\left(\frac{h}{VC_p\rho_w}\right) \tag{5}$$

Oliveira and Haghighi [12] presented a detailed expression that models the drying mechanisms of porous media such as transportation of water capillarity, diffusion and convection as depicted in equation (6)

$$\lambda \frac{\partial \Omega_i}{\partial t} = \sum_{n=1}^{m} \nabla.(P_{in} \nabla \Omega_n) + Q_i \qquad i = 1, m \tag{6}$$

3. Geometrical System Model and Discretization

A two dimensional approach was employed in this investigation. Figure 1 shows the combinations of fourteen basic plane figures used to define the geometry of the drying system on Comsol geographical user interface. The model boundary conditions were implemented through the selected physics module that has the capability of analysing and numerically solved the problem.

The composition was made of seven different materials such as air as the thermodynamic fluid, aluminium as solar collector, granite as sensible energy storage, water properties of material library was conditioned to model mass transfer of moisture, while the rest materials were used to replicate the system casing.

Surface to surface and porous thermal transfers were considered during the simulation. The outcome of dryer finite element discretization is shown in figure 2 in which 77379 degree of freedom was achieved. This was obtained after various boundary conditions were defined and meshing was performed under a time dependent situation.

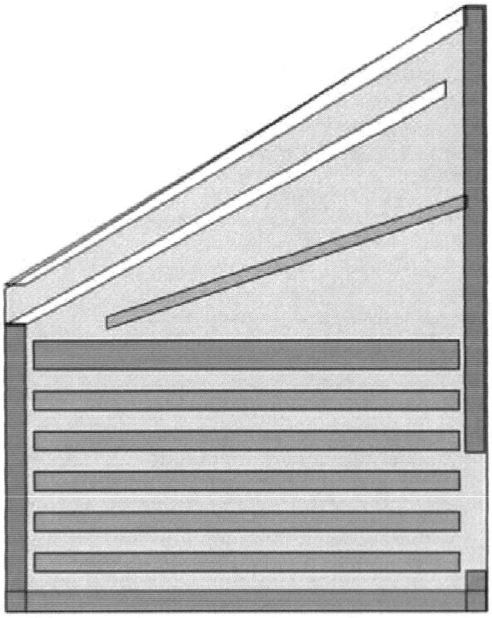

Figure 1. Drying system geometry

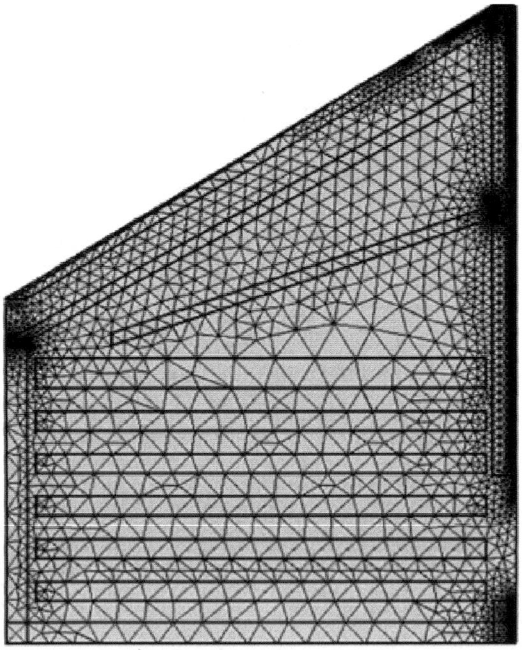

Figure 2. Discretization of the system geometry

4. Results and Discussion

The modelling was simulated for a period of 32 s and the convergence of the numerical iteration was achieved within this period. The drying chamber mean temperature of about 348.50K was recorded which is suitable for drying of common crops. The surface temperature in figure 3 shows the distribution of temperature within the modelled system while the isothermal graph in figure 4 shows the areas with equal temperature value.

The force convective source imposed at the base of system influenced the rate of drying which reached the peak value when the air mass flow rate was at 0.042 kgs^{-1}. The granite material used as alternative heat source has prolonged and enhanced the thermal transfer within the drying system, especially during the discharge period. The economic value of all materials used for modelling was considered for cost optimization of the dying system. This made it affordable for the small and medium scale farmers.

It was also noticed that the system operating conditions were suitable for drying of medicinal herbs which its demand is currently on the increase since Beijing Declaration of World Health Organization in 2008.

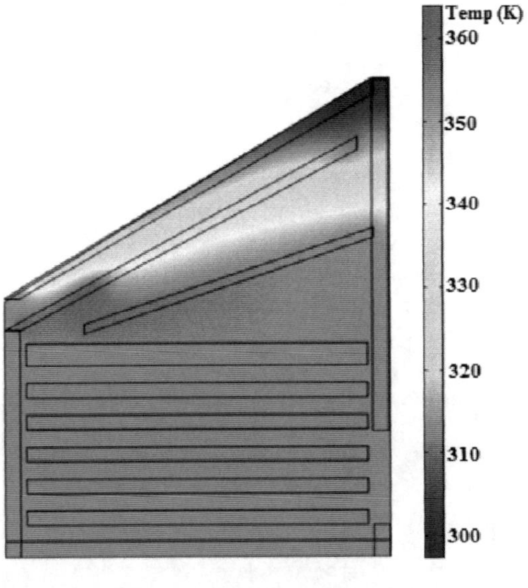

Figure 3. System surface temperature

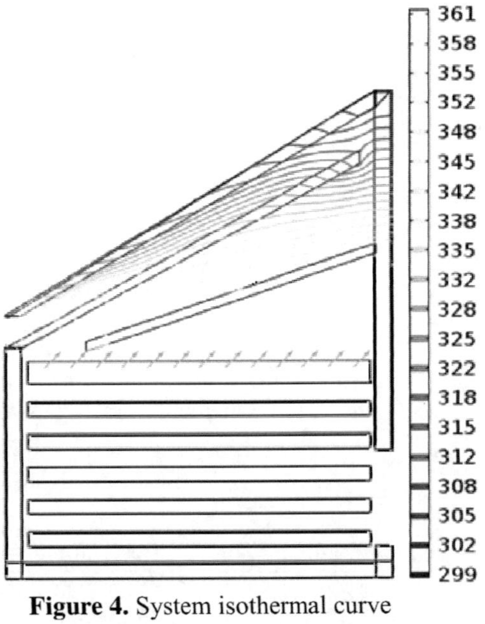

Figure 4. System isothermal curve

The trend of simulation convergence is depicted in figure 5. The error value was plotted against the number of iterations for the nonlinear solver which reflects the relevance of the modelling tool for solar thermal drying system. The numerical solver was able to give solution after the seventh iterative procedure with moderate computer memory usage.

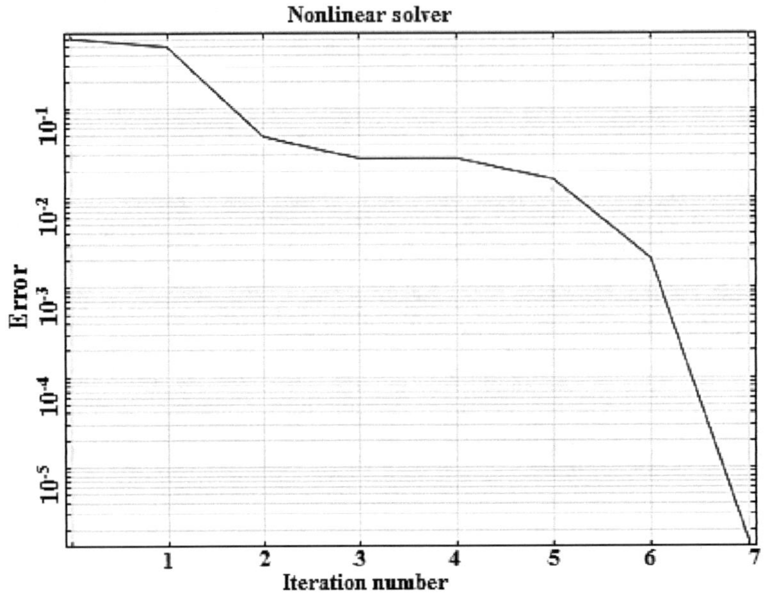

Figure 5. System error-iteration profile

5. Conclusion

The modelling of solar drying system has been done with heat transfer module of Comsol Multiphysics. The results indicate the relevance of the tool employed in solar energy system modelling with inclusion of porous media. The error recorded at the end of iterations was 10^{-5}. Granite was imposed as a sensible thermal storage matrix which acted as heat source during discharge period. This has positive contribution on system performance efficiency which was improved by 5-7 % when compared with a reported study on double pass solar dryer [13]. Despite the result obtained, there was still need to do specific thermal characterization for each material that constitutes the system for better accuracy.

Nomenclature

Symbol	Description	Unit
C	Specific heat capacity	$kJkg^{-1o}C^{-1}$
ρ	Fluid density	Kgm^{-3}
T	Temperature	^oC
U	Velocity field (x,y)	ms^{-1}
t	Time	S
k	Thermal conductivity	$W(m.^oC)^{-1}$
Q	Heat flux	W
A_c	Collector aperture area	m^2
I	Solar Irradiance	Wm^{-2}
V	Volume	m^3
A	Crop surface area	m^2
h	Heat transfer coefficient	$Wm^{-2}.^oC^{-1}$
$\dot{m}$	Mass flow rate	Kgs^{-1}
M	Moisture content	Kg

λ	Capacity coefficient of drying material
P	Kinetic coefficient of drying material
Subscripts	
p	Constant pressure
c	Collector
d	Dry, drying
w	Wet
a	Ambient

6. Acknowledgement

The authors appreciate the URIF grant awarded by Universiti Teknologi PETRONAS to accomplish this study.

7. References

[1] Renné D, "Path to 100% renewable energy conference," in *SunBurst* ed. Germany: ISES, 2013.

[2] Carbonell D, Cadafalch J, and Consul R, "Dynamic modelling of flat plate solar collectors. Analysis and validation under thermosyphon conditions," *Solar Energy,* vol. **89**, pp. 100-112, 2013.

[3] Wu Z and Wang Z, "Fully coupled transient modeling of ceramic foam volumetric solar air receiver," *Solar Energy,* vol. **89**, pp. 122-133, 2013.

[4] Wu Z, Caliot C, Bai F, Flamant G, Wang G, Zhang J, *et al.*, "Experimental and numerical studies of the pressure drop in ceramic foams for volumetric solar receiver applications," *Applied Energy,* vol. **87**, pp. 504-513, 2010.

[5] Wu Z, Caliot C, Flamant G, and Wang Z, "Numerical simulation of convective heat transfer between air flow and ceramic foams to optimise volumetric solar air receiver performances," *International Journal of Heat and Mass Transfer,* vol. **54**, pp. 1527-1537, 2011.

[6] Alkilani M M, Sopian K, Alghoul M A, Sohif M, and Ruslan M H, "Review of solar air collectors with thermal storage units," *Renewable and Sustainable Energy Reviews,* vol. **15**, pp. 1476-1490, 2011.

[7] Xu J, Wang R Z, and Li Y, "A review of available technologies for seasonal thermal energy storage," *Solar Energy,* vol. **103**, pp. 610-638, 2014.

[8] Fudholi A, Sopian K, Othman M Y, and Ruslan M H, "Energy and exergy analyses of solar drying system of red seaweed," *Energy and Buildings,* vol. **68**, pp. 121-129, 2014.

[9] Kareem M W, Habib K, and Gilani S I, "A review of solar air heater for drying of agricultural products," *Advance Material Research,* vol. **903**, pp. 239-244, 2014.

[10] Duffie J A and Beckman W A, *Solar engineering of thermal processes*, 3rd ed.: John Wiley and Son Inc. USA, 2006.

[11] Pabis S, Jayas D S, and Cenkowski S, *Grain drying*. New York: John Wiley and Sons, Inc., 1998.

[12] Turner I and Mujumdar A S, Eds., *Mathematical modeling and numerical techniques in drying technology* (A synopsis of the strategies and efficient resolution techniques used for modelling and numerically simulating the drying process. New York: Marcel Dekker, Inc., 1997, p.^pp. Pages.

[13] Sopian K, Alghoul M A, Alfegi E M, Sulaiman M Y, and Musa E A, "Evaluation of thermal efficiency of double-pass solar collector with porous–nonporous media," *Renewable Energy,* vol. **34**, pp. 640-645, 2009.

IOP Conf. Series: Materials Science and Engineering **88** (2015) 012012 doi:10.1088/1757-899X/88/1/012012

Geometric Optimization of Thermo-electric Coolers Using Simulated Annealing

D V K Khanh[1], P M Vasant[2], I Elamvazuthi[3] and V N Dieu[4]
[1, 2]Department of Fundamental & Applied Sciences, Universiti Teknologi PETRONAS, Malaysia.
[2]Department of Electrical & Electronic Engineering, Universiti Teknologi PETRONAS, Malaysia.
[3]Department of Power System, HCMC University of Technology, Vietnam.

E-mail: [1]kimkhanh2906@gmail.com, [2]pvasant@gmail.com, [3]irraivan_elamvazuthi@petronas.com.my, [4]vndieu@gmail.com

Abstract. The field of thermo-electric coolers (TECs) has grown drastically in recent years. In an extreme environment as thermal energy and gas drilling operations, TEC is an effective cooling mechanism for instrument. However, limitations such as the relatively low energy conversion efficiency and ability to dissipate only a limited amount of heat flux may seriously damage the lifetime and performance of the instrument. Until now, many researches were conducted to expand the efficiency of TECs. The material parameters are the most significant, but they are restricted by currently available materials and module fabricating technologies. Therefore, the main objective of finding the optimal TECs design is to define a set of design parameters. In this paper, a new method of optimizing the dimension of TECs using simulated annealing (SA), to maximize the rate of refrigeration (ROR) was proposed. Equality constraint and inequality constraint were taken into consideration. This work reveals that SA shows better performance than Cheng's work.

Nomenclature

Symbol	Description	Unit
STEC	Single-stage thermo-electric cooler	
L	Height of the confined volume	*mm*
A	Cross-sectional area of STEC legs	mm^2
N	Number of thermo-electric couple	-
ROR	Rate of refrigeration	*W*
COP	Coefficient of performance	-
T_h, T_c	Hot side and cold side temperature	^oK
T_{ave}	Average of cold and hot side temperature	^oK

[1] To whom any correspondence should be addressed.

$\alpha_n = \alpha_p$	Seebeck coefficient of n-type and p-type thermo-electric element	V/K
$\rho_n = \rho_p$	Electrical resistivity of n-type and p-type of thermo-electric element	Ωm
$\kappa_n = \kappa_p$	Thermal conductivity of n-type and p-type of thermo-electric element	$W / m^o K$
r_c	Electrical contact resistance	Ωm^2

1. Introduction

Mud-Pulse High-Temperature Measurement While Drilling (MWD) is a system developed to perform drilling related measurements down-hole and transmit information to the surface while drilling a well [1]. In a typical down-hole environment, the temperature can reach 230°C. Therefore, maintaining optimal payload temperatures is very important and requires a cooling system which is capable of pumping a significant load and requires a low thermal resistance path on the heat rejection (hot side). It has been identified that this can be accomplished using thin-film thermo-electric cooling devices.

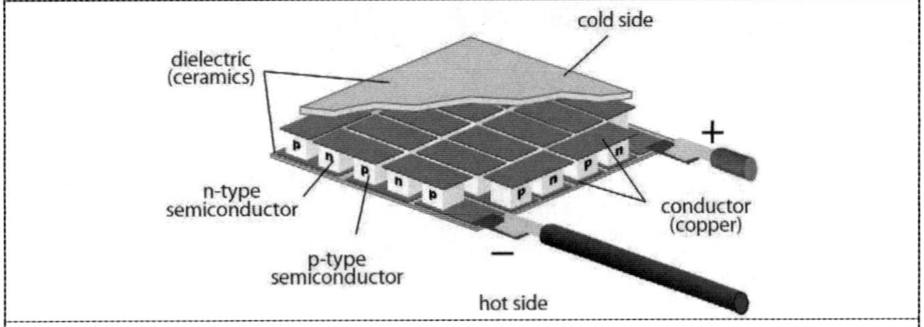

Figure 1. Single-stage thermo-electric cooler

Thermo-electric coolers (TECs) (figure 1) are solid state cooling devices that use the Peltier effect through p-type and n-type semiconductor elements. These types of coolers are used to convert electrical energy into a temperature gradient. TECs use no refrigerant and have no dynamic parts which make these devices highly reliable and require low maintenance. However, for this application the most attractive feature of TECs is that they have the capacity for cooling instruments such as MWDs under extreme physical conditions [2]. Two parameters play crucial role in characterization of TECs are the *ROR* and *COP*. The main drawback of TECs is the poor *COP* and *ROR* [10]. Improvement on the performance of TECs can be achieved by means of developing new materials with a higher figure of merit (*Z*), optimizing module design and fabrication, improvement of the heat exchange efficiency and employment of multi-stages system [2]. Several intelligent techniques that can be used for engineering design optimization are discussed in Deb [3]. However, one of the most effective and non-traditional method used as an optimization technique for TECs is Non-dominated Sorting Genetic Algorithm [4]. Similar sophisticated techniques such as Simulated Annealing (SA), other evolutionary algorithms (Differential Evolution and Particle Swarm Optimization) can be used in their pure and hybrid form to enhance the effectiveness of the optimization of TECs [5, 6]. SA is a point-based meta-heuristic which has the ability to escape from local optima [7], flexibility and ability to approach global optimality [8]. SA is easy to code, even for complex system and can deal with highly nonlinear models with many constraints [9]. This study focuses on optimizing the design parameters of STEC using SA optimization technique, to create the maximum *ROR* under some defined constraints. The *L, A, N* of thermo-electric module were optimized.

2. Mathematical modelling of TECs

Operation of TECs is based on the Peltier effect. When a DC current passes through a pair of p- and n-type semiconductor materials, one side of the junctions is cooled and the other side is heated. The energy balance equations at the hot junction and the cold junction for TECs can be described as in equations (1) and (2) [13]. The input electrical power and coefficient of performance can be calculated using equations (3), (4). α, ρ_r, κ represent for thermo-electric material properties. *A*, *L*, *N* are geometric properties. *COP* is a common metric used to quantify the effectiveness of a heat engine [10, 11]. The material properties are considered to be dependent on the average temperature of the cold side and hot side temperatures of each stage and their values can be calculated using equations (5), (6), (7) [12].

$$ROR = N\left[\alpha I T_c - \frac{1}{2}I^2(\rho_r\frac{L}{A} + \frac{2r_c}{A}) - \frac{\kappa A(T_h - T_c)}{L}\right] \quad (1)$$

$$Q_h = N\left[aIT_h + \frac{1}{2}I^2(\rho_r\frac{L}{A} + \frac{2r_c}{A}) - \frac{\kappa A(T_h - T_c)}{L}\right] \quad (2)$$

$$P = Q_h - ROR \quad (3)$$

$$COP = \frac{ROR}{Q_h - ROR} \quad (4)$$

$$\alpha = \alpha_p = -\alpha_n = (-263.38 + 2.78\mathrm{T}_{ave} - 0.00406T_{ave}^2)10^{-6} \quad (5)$$

$$\rho = \rho_p = \rho_n = (22.39 - 0.13T_{ave} + 0.00030625T_{ave}^2)10^{-6} \quad (6)$$

$$\kappa = \kappa_p = \kappa_n = 3.95 - 0.014T_{ave} + 0.00001875T_{ave}^2 \quad (7)$$

3. Optimization process

This work used SA in optimizing the design parameters of STEC. The objective of the optimization calculation is to determine the optimal *L, A* and *N*. One objective, namely maximizing *ROR* of STEC is considered for single objective optimization. The design variables *A, L, N* of STECs were put in an inequality constraint which was bound by upper and lower limits of the design variables and the total area *S* of STEC. Additionally, STEC is put in some requirements which are the confined volume of STEC (*S*), the minimum requirement of *COP* and the maximum cost of material [2]. Because optimization of STEC geometry may cause the reduction in the *COP*, the *COP* is used as a constraint condition during the optimization in order to guarantee that the TECs with the optimal geometry has a relatively high *COP* [10]. Referring to equations (1), (2), (3) and (4), the parameters T_c, T_h, *I* are defined in the beginning of the calculation. The unknown term is material properties of TECs will be determined based on the equations (5), (6) and (7) with values of T_h and T_c.

SA is a method for solving constrained and bound-constrained optimization problems [11]. The method models the physical process of heating a material and then slowly lowering the temperature to decrease defects, thus minimizing the system energy.

4. Simulation results

Preliminary tests for finding the optimal value of geometric properties of STEC were conducted by using SA and run the system in MATLAB. Parameters selection of STEC were referred from [13] and was listed in Table 1. The purpose of this test is to understand the operation and benefits of SA in the optimization of STEC design, especially the stability and reliability of SA.

Table 1. Parameters setting of STEC.

Group	Parameters setting	Specific values
1	Objective function	Maximize *ROR*
2	Variables	$0.03mm < L < 1mm$ $0.09mm^2 < A < 100mm^2$

		$1 < N < 1000$
3	Fixed parameters	$S = 100mm^2$ $T_h = T_c = 323\,^oK$ $r_c = 10^{-8}\ \Omega m^2$
4	Constraints	- $A.N < 100mm^2$ - Maximum cost USD 385 - Minimum requirement $COP = 1$

Table 2. Parameters setting of simulated annealing algorithm.

Parameters setting	Specific values
Initial temperature	100
Final temperature	10^{-10}
Temperature reduction	0.95
Boltzmann annealing	1
Stopping criteria	- Final stopping temperature - Function tolerance 10^{-6}

Table 3. Parameters setting of SA.

Maximum fitness	Minimum fitness	Mean fitness	Standard deviation
9.975	8.663	9.589	0.386

SA code for programming was used and adjusted its parameters so that it can be used for finding the optimal design of TECs [14]. Choosing the good algorithm parameters is very important because it affects the whole optimization process [15]. Parameters setting of SA are listed in Table 2. Using this strategy, SA was run in 30 trials randomly on STEC system. The first case is for nonlinear inequality constraint $A.N < 100\text{mm}^2$, the second case is for equality constraint $COP = 1$. The best, average value and the lowest value of 30 trials were noted down to evaluate the performance of the algorithm such as reliability.

After testing, SA converged quickly in several seconds. As shown in figure 2, when running under nonlinear inequality constraint ($A.N < 100mm^2$), SA converged quickly in several seconds and satisfied the constraint. The obtained results of 30 trials runs are nearly same. Maximum and minimum values of best fitness are collected from figure 2 and calculate the standard deviation (Table 3). A low standard deviation 0.386 indicates that data points tend to be very close to the mean. Therefore, the optimal results of SA are stable and reliable. SA is robust with nonlinear inequality constraint. Problem happened when running SA under nonlinear equality constraint ($COP = 1$). As shown in figure 3, the maximum value of *ROR* is very small when compare with Cheng's work [13]. The optimal dimensions of STEC were unable to satisfy the constraint, SA gets stuck at local optima. Finding optimal dimension of STEC when running SA under nonlinear equality constraint is not reliable. In the following step, STEC system was run with SA under the two nonlinear inequality constraints of total area $A.N < 100mm^2$, *COP* requirement was neglected. I and T_c of STEC were varied independently. Table 4 presents the optimal design variables for various input currents from $0.1A$ to $2A$. The maximum values of *ROR* seem unchanged when the input current is larger than $0.5A$. Table 4 demonstrates that TECs can reach its maximum *ROR* even under various inputs current. Table 5 presents the optimal dimension of STEC for various cold side temperatures from $293\,^oK$ to $330\,^oK$. The maximum *ROR* increases when T_c is increased. The increasing of *ROR* is more obvious when cold side temperature exceeds hot side temperature. And as the cold side temperature is increased, the optimal leg area decreases and the number of thermo-element couple increases. In figures 4 and 5, the comparison with Cheng's work [13] in same conditions are shown, SA obviously demonstrate better performance when SA can find good dimensions of STEC with bigger values of *ROR*.

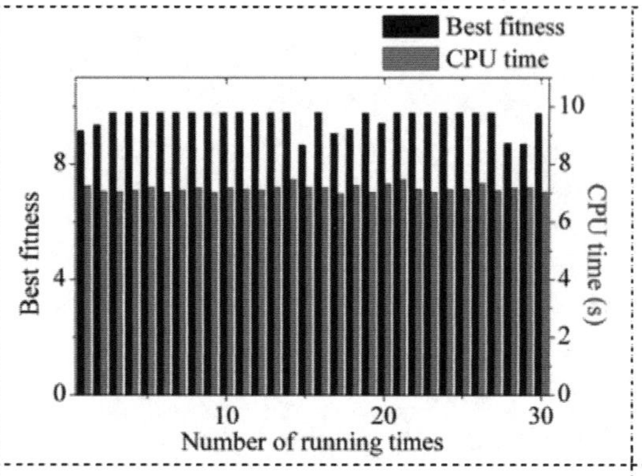

Figure 2. Run STEC using SA under nonlinear inequality constraint (A.N<100mm^2)

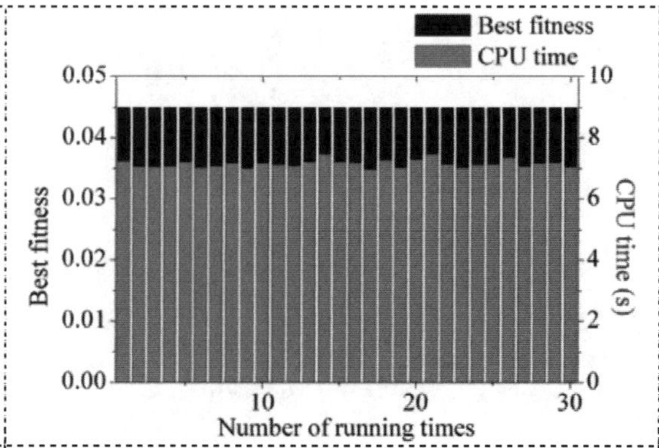

Figure 3. Run STEC using SA under nonlinear equality constraint (*COP* = 1).

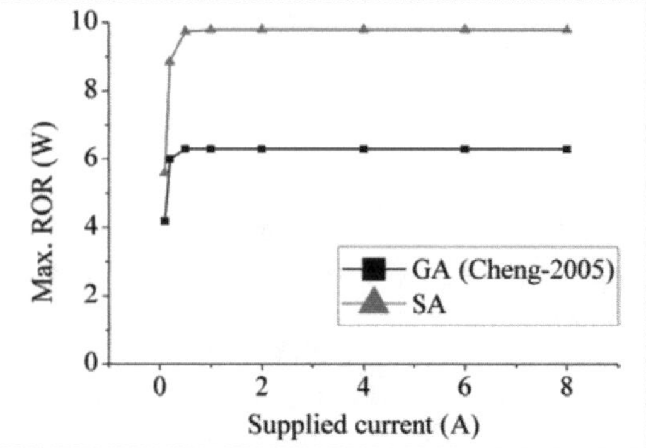

Figure 4. Comparison between SA and GA [13] on maximum *ROR* under various input currents

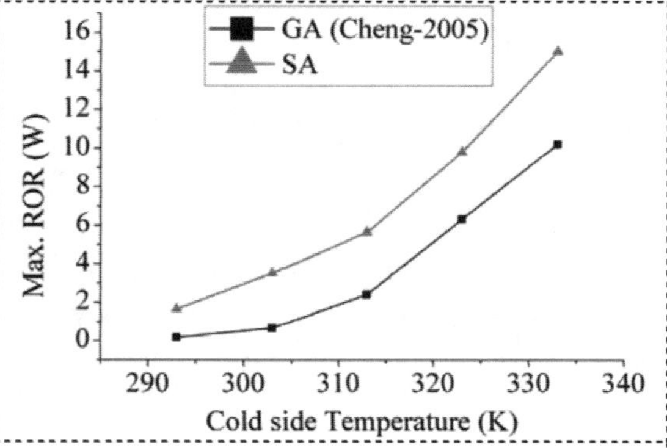

Figure 5. Comparison between SA and GA [13] on the maximum *ROR* under various cold side temperature.

Table 4. Optimal dimension of STECs under various input currents.

I (A)	*[A, L, N]*	Max. *ROR*
0.1	[0.099, 0.300, 1000]	5.6
0.2	[0.099, 0.300, 1000]	8.86
0.5	[0.183, 0.306, 546]	9.74
1	[0.354, 0.300, 283]	9.79
2	[0.700, 0.300, 143]	9.79
4	[1.388, 0.300, 72]	9.79
6	[2.081, 0.300, 48]	9.79
8	[2.786, 0.300, 36]	9.79

Table 5. Optimal dimension of STECs under various input cold side temperature.

T_c (^oK)	*[A, L, N]*	Max. *ROR*
283	-	-
293	[0.517, 1.000, 194]	1.65

303	[0.504, 1.000, 198]	3.52
313	[0.412, 0.618, 242]	5.63
323	[0.346, 0.300, 289]	9.79
333	[0.342, 0.300, 293]	15.01

5. Conclusion
The obtained results showed that simulated annealing performed very well when running system with nonlinear inequality constraint, but simulated annealing got stuck at local optimum when run the system with nonlinear equality constraint. Simulated annealing gave better optimal dimensions of STECs with bigger value of rate of refrigeration compared with genetic algorithm which was used in Cheng's work.

6. Acknowledgments
This research work is supported by University Teknologi Petronas under the Exploratory Research Grant Scheme-PCS-No. 0153AB-121 (ERGS) of Ministry of Higher Education Malaysia (MOHE).

7. References
[1] Scherbatskoy, S.A., *Systems, apparatus and methods for measuring while drilling*. 1982, Google Patents.
[2] Goldsmid, H.J., *The Thermoelectric and Related Effects*, in *Introduction to Thermoelectricity*. 2009, Springer. p. 1-6.
[3] Deb, K., *Multi-objective optimization.* Multi-objective optimization using evolutionary algorithms, 2001: p. 13-46.
[4] Deb, K., et al., *A fast and elitist multiobjective genetic algorithm: NSGA-II.* Evolutionary Computation, IEEE Transactions on, 2002. **6**(2): p. 182-197.
[5] Blum, C., A. Roli, and M. Sampels, *Hybrid metaheuristics: An emerging approach to optimization*. Vol. 114. 2008: Springer.
[6] Vasant, P., et al. *Solving deterministic non-linear programming problem using Hopfield artificial neural network and genetic programming techniques.* in *AIP Conference Proceedings-American Institute of Physics*. 2012.
[7] Van Laarhoven, P.J. and E.H. Aarts, *Simulated annealing*. 1987: Springer.
[8] Kalyanmoy, D., *Optimization for engineering design: Algorithms and examples*. 2004: PHI Learning Pvt. Ltd.
[9] Sohrabi, B., *A Comparison Between Genetic Algorithm and Simulated Annealing Performance in Preventive Part Replacement.* Management, 2006(72): p. 120-112.
[10] Huang, Y.-X., et al., *Geometry optimization of thermoelectric coolers using simplified conjugate-gradient method.* Energy, 2013. **59**(0): p. 689-697.
[11] Boussaïd, I., J. Lepagnot, and P. Siarry, *A survey on optimization metaheuristics.* Information Sciences, 2013.
[12] Nain, P.K.S., S. Sharma, and J.M. Giri, *Non-dimensional Multi-objective Performance Optimization of Single Stage Thermoelectric Cooler*, in *Simulated Evolution and Learning*, K. Deb, et al., Editors. 2010, Springer Berlin Heidelberg. p. 404-413.
[13] Cheng, Y.-H. and W.-K. Lin, *Geometric optimization of thermoelectric coolers in a confined volume using genetic algorithms.* Applied Thermal Engineering, 2005. **25**(17–18): p. 2983-2997.
[14] Vasant, P. and N. Barsoum. *Hybrid Simulated Annealing and Genetic Algorithms for industrial production management problems*. 2009.
[15] Vasant, P., *Hybrid LS–SA–PS methods for solving fuzzy non-linear programming problems.* Mathematical and Computer Modelling, 2013. **57**(1): p. 180-188.
[16] Production of thermo-electric modules, URL http://kryothermtec.com/technology.html

Solidification of Cu-Water nanofluid in a trapezoidal cavity: A CFD study

R K Sharma[1*], P Ganesan[1], I H Metselaar[1]
[1*]Department of Mechanical Engineering, Faculty of Engineering, University of Malaya, 50603, Kuala Lumpur, Malaysia

E-mail: [1*]ravipvb@gmail.com

Abstract. A numerical study has been carried out to investigate the solidification of a binary mixture of water and Cu nanoparticles inside a horizontal trapezoidal cavity of different aspect ratio under specific given boundary conditions for temperature and concentration gradients. The vertical side walls of the cavity are insulated while the top wall temperature is kept lower than that of the bottom wall. The effect of parameters such as the ratio of the cavity length to height (aspect ratio), the cold wall temperature (-5 to -30 °C) and the initial temperature of the nanofluid (0 °C to16 °C) on solidification time is investigated. The moving solid-liquid interface is obtained using Enthalpy-porosity technique in the model. We found that the solidification time decreases with the increase of the aspect ratio (i.e., a longer trapezoidal cavity) and the decrease of cold wall temperature. Aspect ratio is found to give a prominent effect. However, the initial temperature of fluid does not affect the solidification time much.

1. Introduction

In order to bridge the gap between energy availability and its uses, the latent heat thermal energy storage (LHTES) system is an effective way to store the energy for later use by melting/freezing of Phase Change Materials (PCMs). These devices have applicability in wide range of applications such as buildings, electronic cooling, material processing and thermal management of air and space-craft. PCMs with nano size are preferred than the micro size particles to avoid clogging during flow. In the current study a PCM (water + Cu nanoparticles) is used to investigate the solidification process. A mixture of a base fluid and solid particles of nano size is termed as nanofluid [1]. The solidification process of water-Cu nanofluid of different volume concentration of Cu nano particles was numerically investigated by Khodadadi and Hosseinizadeh [2] and they found that the nanofluid freezes faster than the base fluid. PCMs are encapsulated for LHTES because it provides the larger heat transfer area, reduces PCMs interaction with external environment, and controls the variation in volume change during phase change process. In a pioneer work done by Gau and Viskanta [3], the effect of buoyancy driven flow in gallium filled in a rectangular cavity is experimentally investigated. They found that as the time progresses, convection takes over the melting process over conduction. Later this work was numerically validated by Brent et al. [4] using the Enthalpy-porosity technique. A numerical study carried out by Assis et al. [5] for melting of RT27 (Rubitherm GmbH) filled in spherical enclosure shows that the melting process incorporate the convection phenomenon in liquid phase. Recently an

[1] To whom any correspondence should be addressed.

IOP Conf. Series: Materials Science and Engineering **88** (2015) 012013 doi:10.1088/1757-899X/88/1/012013

experimental and numerical study [6] has been carried out for constrained melting of PCMs in spherical shell and observed that with time convection gets strengthened. There has been number of studies considering the three regular geometries namely square/rectangular, cylindrical and spherical. The selection of geometrical shape of cavity mainly depends upon the application [7]. Because of the large surface area normal to the direction of heat transfer, trapezoidal cavity has received significant attention of researchers for encapsulation. This geometry has wide range of application such as solar heater, casting and mold design etc. An experimental study [8] was carried out to investigate the effect of concentration of NH_4Cl (0-19.8%) and wall temperatures (-30 to 0 °C) on the solidification of binary alloy (NH_4Cl+H_2O) filled in trapezoidal cavity and they found that the rate of solidification diminishes with increase of NH_4Cl concentration.

Despite of much experimental and numerical investigation carried out using regular geometries such as square/rectangular, spherical and cylindrical, very little work has been done to investigate the solidification/melting process by the conduction or convection heat transfer of PCM in a trapezoidal cavity. Therefore, the purpose of the present work is to analyze the solidification of water-Cu nanofluid filled in a horizontally placed isosceles trapezoidal cavity using the Ansys-Fluent 13.0 CFD commercial package. We found that the horizontal placement of the cavity saves the solidification time when compared with vertically placed one.

3. Research Methodology

3.1 Mathematical formulation

Fig. 1 shows the two dimensional (2D) geometry considered in the current study. Isosceles trapezoidal cavity placed in horizontal direction with 10 mm^2 internal area is used. The effect of the ratio of trapezoidal length to height, namely aspect ratio (AR = L/H) is investigated. Different AR, i.e., 1.0, 1.2, 1.4, 1.6, and 1.8, are tested, but the internal area is kept same for all the cavities. The nanofluid within the cavity is Newtonian, laminar, and incompressible. Thermo-physical properties (Table 1) of the nanofluid are assumed to be constant, whereas the density variation in the buoyancy force is based on the Boussinesq approximation. The nanoparticles are assumed to have a uniform shape and size (10 nm diameter). The left lower corner of the cavity was the origin of the coordinate system. Gravity acts in the negative y-coordinate direction, $\imath_1 = 0$, and? $_1 = -\imath$.

The initial and boundary conditions for the present investigation are follows:

at the top inclined wall $T=T_c$

at the bottom inclined wall $T=T_c$

at left and right vertical surfaces $\frac{\partial T}{\partial y}=0$

at all solid boundaries $u=v=0$

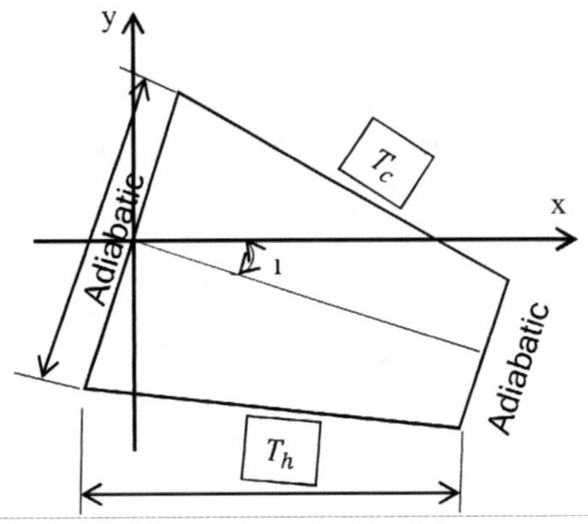

Figure 1. Sketch of the two dimensional trapezoidal.

Table 1. Thermo-physical properties of the base fluid (water) and the Cu nanoparticles.

Property	Copper nanoparticles	Base fluid (water)
ı [[ı ı = ı ı]	8954	997.1
ı [[ı ı ı]	-	8.9 × 10ı ı
ı, [ı/ı ı ı]	383	4179
ı [[ı /ı ı]	400	0.6
ı [[1/ı]	1.67 × 10ı ı	2.1 × 10ı ı
ı [[ı/ı ı]	-	3.35 × 10ı
ı ı	-	6.2
ı ıı	-	0.125
ı, [ı]	10ı ı	-

The mathematical governing equations are as follows:
The continuity, momentum considering the Boussinesq approximation and energy equation for the above mentioned assumptions can be written in the following form:
Continuity

$$\frac{\partial u}{\partial x}+\frac{\partial v}{\partial y}=0, \tag{1}$$

X-momentum

$$\frac{\partial u}{\partial t}+u\frac{\partial u}{\partial x}+v\frac{\partial u}{\partial y}=\frac{1}{\rho_{nf}}\left(-\frac{\partial p}{\partial x}+\mu_{nf}\nabla^2 u+(\rho\beta)_{nf}\,g_x\left(T-T_C\right)\right), \tag{2}$$

Y-momentum

$$\frac{\partial v}{\partial t}+u\frac{\partial v}{\partial x}+v\frac{\partial v}{\partial y}=\frac{1}{\rho_{nf}}\left(-\frac{\partial p}{\partial y}+\mu_{nf}\nabla^2 v+(\rho\beta)_{nf}\,g_y\left(T-T_C\right)\right), \tag{3}$$

Energy equation

$$\frac{\partial T}{\partial t}+u\frac{\partial T}{\partial x}+v\frac{\partial T}{\partial y}=\frac{\partial}{\partial x}\left[\frac{(k_{nfo}+k_d)}{(\rho c_p)_{nf}}\frac{\partial T}{\partial x}\right]+\frac{\partial}{\partial y}\left[\frac{(k_{nfo}+k_d)}{(\rho c_p)_{nf}}\frac{\partial T}{\partial y}\right], \tag{4}$$

The density of the nanofluid is given by:

$$\rho_{nf} = (1-\phi)\rho_f + \phi\rho_s, \quad (5)$$

The viscosity of nanofluid is given by Brinkman [9]:

$$\mu_{nf} = \frac{\mu_f}{(1-\phi)^{2.5}}, \quad (6)$$

The heat capacitance of the nanofluid and part of the Boussinesq are:

$$(\rho c_p)_{nf} = (1-\phi)(\rho c_p)_f + \phi(\rho c_p)_s \quad (7)$$

$$(\rho\beta)_{nf} = (1-\phi)(\rho\beta)_f + \phi(\rho\beta)_s \quad (8)$$

The latent heat of the nanofluid is evaluated using

$$(\rho L)_{nf} = (1-\phi)(\rho L)_f \quad (9)$$

Where L is the latent heat and ϕ is the nanoparticle volume fraction.

Thermal conductivity of the quiescent (subscript 0) nanofluid is:

$$\frac{k_{nfo}}{k_f} = \frac{k_s + 2k_f - 2\phi(k_f - k_s)}{k_s + 2k_f + 2\phi(k_f - k_s)} \quad (10)$$

Whereas the effective thermal conductivity of the nanofluid is:

$$k_{eff} = k_{nf0} + k_d, \quad (11)$$

And the thermal conductivity enhancement term due to thermal dispersion is given by:

$$k_d = C(\rho c_p)_{nf}\sqrt{u^2 + v^2}\,\phi d_p. \quad (12)$$

The empirically-determined constant C is evaluated following the work of Wakao and Kaguei (1982).

3.2 Numerical method

The numerical solution of the problem uses the enthalpy-porosity approach. In this technique, the porosity in each cell is set equal to the liquid fraction in that cell and this fraction is computed at each iteration. The SIMPLE method of the commercial code ANSYS FLUENT is used for solving the governing Eqs. 1-4. Orthogonal and uniform grid size of 6500 elements is used. The time step size for all the simulations in this study is 0.5 s and number of iterations for each time step are 800. The under-relaxation factor all the components, such as velocity components, pressure correction, thermal energy etc. is kept at 0.3. Convergence criteria are set at 10^{-7} for continuity and momentum and 10^{-9} for thermal energy. The QUICK differencing scheme was used for solving the momentum and energy equations, whereas the Pressure Staggering Option (PRESTO) scheme was used for pressure correction equations. In enthalpy method, the solution is based on a fixed grid and governing equations are modified such that they are valid for both phases. Also the mushy zone constant is set to 10^5 [kg/m^{3}s].

4.0 Results

4.1 Validation of the model

Experimental results of Gau and Viskanta (1984), numerical predictions by Brent et al. (1988) and Khodadadi and Hosseinizadeh (2007) for melting of solid gallium in a rectangular cavity are compared with the current numerical data in Fig 2. A two dimensional rectangular cavity of 8.89 cm × 6.35 cm size completely filled with solid gallium was considered and qualitatively, the trends are agreeable among all four approaches, whereas the present computational results are more close to the experimental data and previous numerical predictions.

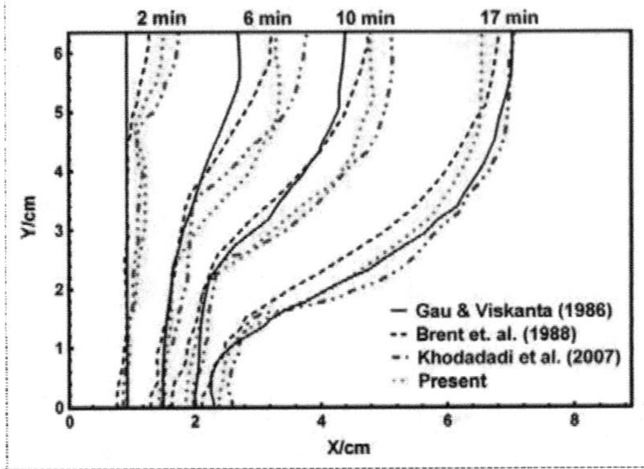

Figure 2. Comparison of the progress of the melting phase front with time.

3.2 Effect of aspect ratio on solidification time

Fig. 3 shows the variation in the heat flux in the nanofluid of ϕ = 20% filled in square cavity (AR = 1.0) and trapezoidal cavity (AR = 1.2-1.8) until 1600 sec (or complete solidification). Note that the tilt angle is kept 0° for these cases. The increasing heat flux with increasing aspect ratio is seen in the figure. As the solidification progresses, layer of solidified nanofluid increases which enhances the thermal resistance, resulting, decrease of heat flux from cold wall. As the aspect ratio increases, the surface area of the cold wall normal to the direction of heat flow increases which increases the heat flux at this wall. Increasing heat flux indicates the rapid solidification of nanofluid in the cavity.

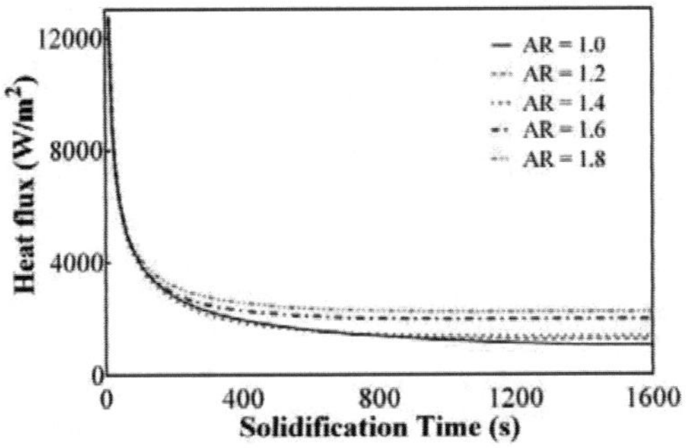

Figure 3. Heat flux variation with time from cold wall for different aspect ratio and ΔT = 10 °C.

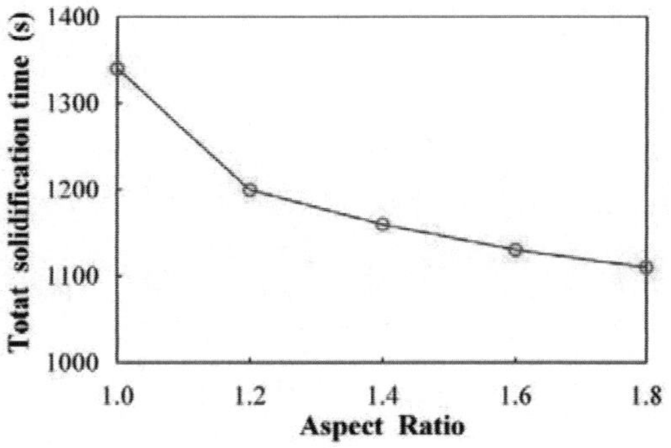

Figure 4. Effect of aspect ratio (AR) on the total solidification time of nanofluid of 20 % nanoparticle volume fraction and $\theta = 0°$.

Fig. 4 shows the total solidification time of nanofluid in trapezoidal cavity for different aspect ratios and placed at 0° tilt angle. The figure shows that the solidification time significantly reduces with the increase of the aspect ratio (a longer trapezoidal cavity). This implies that for a specific internal area cavity, the solidification process can be expedited by changing the geometrical structure. Since the internal area (10 mm^2) remains constant for all aspect ratios, the amount of PCM filled in the cavity is also equal, so, by changing the aspect ratio, we change the surface area of cold wall normal to the direction of heat transfer, which increases the heat transfer rate. Almost 8.5% increment was seen in the cold wall surface area when the aspect ratio was increased from 1.0 to 1.8 and because of this increment the solidification time of PCM was decreased by more than 17%. Therefore, the aspect ratio of this cavity can be utilized to control the solidification time.

3.3 Effect of wall and fluid temperatures

Fig 5 shows the effect of cold surface temperature on the total solidification time of nanofluid (ϕ = 20%) filled in trapezoidal cavity of aspect ratio 1.4. The effect of five different cold surface temperature, T_h = -5°, -10°, -15°, -20°, -25°, and -30° is investigated in simulation Cases 15-20. It was found, as expected, that decreasing the cold surface temperature (T_c) increases the heat transfer rate, resulting, decreases the solidification time. It was found that when T_c was decreased from -5 to -10 °C, the reduction in solidification time was more than half (from 2540s to 1160s) while further reducing the wall temperature from -10 to -15 °C, and so on does not decrease the solidification time this much. This is because, in the first step the temperature was reduced by exact half (-5 to -10 °C) but from -10 to -15 °C and onwards the reduction is less than half so as the saving in solidification time.

Fig 6 shows the effect of cavity aspect ratio on the nanofluid's solidification time for initial temperature T_i = 0°, 4°, 8°, and 16°. Simulation Cases 26-45 are used to investigate the effect of these parameters on solidification time. This shows that aspect ratio plays a significant role in controlling the solidification time but initial temperature does not affect the total solidification time and it is nearly same for all aspect ratio.

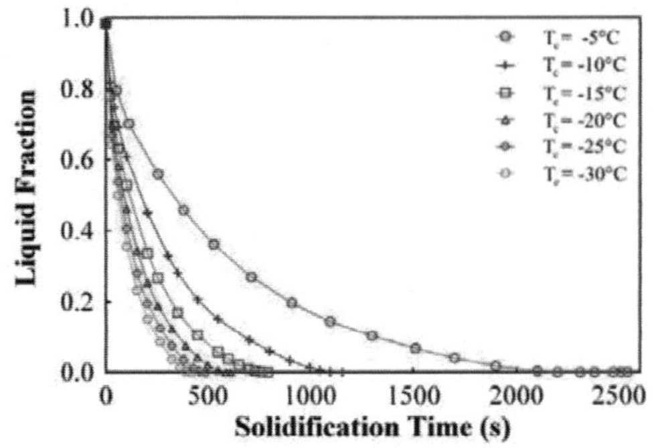

Figure 5. Effect of cold surface temperature on total solidification time of nanofluid (ϕ = 20%, AR = 1.4 and θ = 0°).

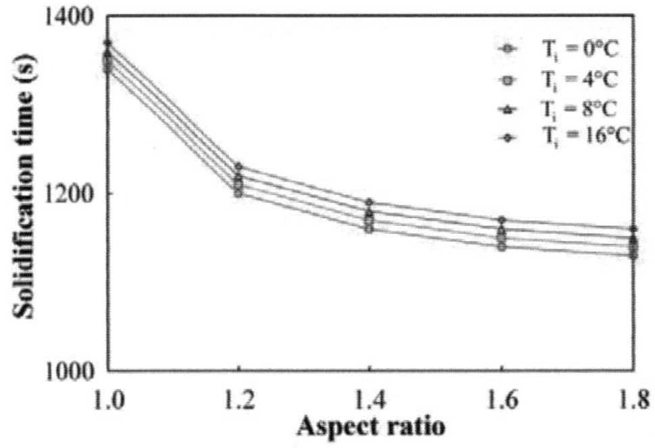

Figure 6. Effect of initial temperature (T_i) of PCM on total solidification time for various aspect ratio and θ = 0°.

4. Conclusions

The solidification phenomenon of PCM in the horizontal trapezoidal cavity is studied using CFD. The effect of various aspect ratios, cold wall temperature, PCM's initial temperature, cavity tilt angle, and Grashof number on heat transfer rate in PCM is investigated. The results of this numerical study lead to the following conclusions:

1. Solidification time is significantly decreased with the increase of the aspect ratio; e.g., the trapezoidal cavity with an AR=1.8 requires almost 17% lesser solidification time than that of the square cavity (AR = 1.0) having the same internal area. Therefore cavity aspect ratio can be used as a controlling parameter to improve the heat transfer rate.
2. The decrease of the cold surface temperature (T_c) decreases the solidification time significantly (nearly 50%) for the case where T_c was reduced from -5 to -10 °C but relatively less significant for other cases in the square cavity as well as trapezoidal cavity.

The theoretical prediction in this paper is hoped to be a useful guide for experiments dealing with the study of effectiveness of aspect ratio of trapezoidal cavity filled with PCM in improving the heat transfer rate.

5. Acknowledgement

This research is financially supported by University of Malaya, Ministry of Higher Education High Impact Research (UM.C/HIR/MOHE/ENG/21), and Exploratory Research Grant Scheme (ERGS: ER013-2013A), Kuala Lumpur, Malaysia

6. References

[1] Choi S U S Eastman J A Eastman Enhancing thermal conductivity of fluids with nanoparticles 1995.

[2] Khodadadi J M Hosseinizadeh S F Nanoparticle-enhanced phase change materials (NEPCM) with great potential for improved thermal energy storage, International Communications in Heat and Mass Transfer 34 pp 534-543 2007

[3] Gau C Viskanta R Melting and Solidification of a Pure Metal on a Vertical Wall, Journal of Heat Transfer 108 pp 174-181 1986

[4] Brent A D Voller V R Voller K J Reid Enthalpy-Porosity Technique For Modeling Convection-Diffusion Phase Change: Application To The Melting Of A Pure Metal Numerical Heat Transfer 13 pp 297-318 1988

[5] Assis E Katsman L Ziskind G Letan R Numerical and experimental study of melting in a spherical shell, International Journal of Heat and Mass Transfer 50 pp 1790-1804 2007

[6] Tan F L Hosseinizadeh S F Khodadadi J M Fan J Experimental and computational study of constrained melting of phase change materials (PCM) inside a spherical capsule, International Journal of Heat and Mass Transfer 52 pp 3464-3472 2009

[7] Al-abidi A A Bin Mat S Sopian K Sulaiman M Y Mohammed A T CFD applications for latent heat thermal energy storage: a review, Renewable and Sustainable Energy Reviews pp 353-363 2013

[8] Duggirala R K Lin C K Ghenai C Investigation of double-diffusive convection during the solidification of a binary mixture (NH4Cl–H2O) in a trapezoidal cavity, Experiments in Fluids 40 pp 918-927 2006

[9] Brinkman H C The Viscosity of concentrated suspensions and solutions the viscosity of concentrated suspensions and solutions J Chem Phys 20 pp 571-581 1952

Thermal comfort investigation on a naturally ventilated two-storey residential house in Malaysia

N A Malek[1], **M H Khairuddin** [1], **M F Rosli** [2]
Department of Mechanical Engineering,
Universiti Tenaga Nasional, Jalan IKRAM-UNITEN, 43000 Kajang, Selangor, MALAYSIA

E-mail: norhazwani@uniten.edu.my

Abstract. This paper presents a case study to investigate the human thermal comfort on a naturally ventilated two-storey residential house in Malaysia. Three parameters were investigated in this study, namely the air temperature, air velocity and air humidity. These parameters were measured using the appropriate measuring device to obtain the actual data and compared with simulation results. The level of thermal comfort in the house was found to be poor as the parameters measured are over the limits specified by ASHRAE standards. Simulation on the model of the house was performed using the Computational Fluid Dynamics (CFD) commercial code, FLUENT to visualize the temperature distribution and air flow pattern and velocity in the house. The error between these two sets of data was acceptable and thus the simulation used in this study was validated. Comparison was also made in the CFD simulation to see the effects of using a ceiling fan installed in the house and without ceiling fan. The level of thermal comfort was poor in both cases as it did not satisfy the standards set by ASHRAE but more uniform temperature distribution inside the house was found when the ceiling fan was turned on. The thermal comfort level became critical in the afternoon as during this period, the house receives direct sunlight which causes the temperature inside the house to increase. Although the mechanical ventilation devices did not help to improve the thermal comfort in the house being studied, the CFD simulation results can be used by building designers to further improve the level of thermal comfort in naturally ventilated residential houses.

1. Introduction

Due to Malaysian climate that is hot and humid throughout the year, the weather is not consistent and changing. High temperature with high humidity causes discomfort to the human and it affects the thermal comfort of the occupants in a house. Most residential buildings in Malaysia are not equipped with air-conditioning systems, but relies on natural ventilation, passive cooling system and mechanical ventilation devices such as ceiling and extractor fans to achieve certain level of thermal comfort. To optimize the thermal comfort effects, special attention must be given to the design and installation of these devices. Mechanical ventilation is more convenient compared to natural ventilation, but there is

[1] Universiti Tenaga Nasional, Jalan IKRAM-UNITEN, 43000 Kajang, Selangor, MALAYSIA
[2] Nichicon (M) Sdn Bhd, No.4 Jln P/10, Kawasan Perusahaan Bangi, 43650, Selangor, MALAYSIA

IOP Conf. Series: Materials Science and Engineering **88** (2015) 012014 doi:10.1088/1757-899X/88/1/012014

a downside for it. As the mechanical ventilation consumes electricity, it is costly while natural ventilation is free.

This study was conducted to investigate the thermal comfort of a naturally ventilated double-storey terrace house in Bangi, Selangor, Malaysia. The aim of this study is to investigate the house ventilation with the thermal comfort of the occupants in the house. Data measurements were taken and some parameters such as air temperature, air velocity and air humidity were studied. Simulation of the house using a CFD software, FLUENT was also done. When the results from each method were obtained and recorded, analysis of data was made by comparing these results together. The simulation was also done to compare the thermal comfort of the house with the fan turned on and off. All the results obtained were then compared with the standard set by the American Society of Heating, Refrigerating and Air-Conditioning Engineers (ASHRAE) for human thermal comfort. On top of that, the objective of this study is also to find out the best improvement method to enhance the human thermal comfort.

2. Literature Review

To assess the thermal comfort, it is necessary to determine the air velocity, temperature distribution, and relative humidity of the air in the residential building. Numerical methods such as computational fluid dynamics (CFD) can be utilized to assist in the analysis of thermal comfort. FLUENT is a useful tool that can be used to create a virtual model of the interior of the residential building and simulate air flow and temperature distribution which are directly related to thermal comfort. Many researchers have used CFD to analyze thermal comfort in building spaces and investigated the effects of natural and stratified ventilations on the thermal comfort. This includes Bastide et.al. [1], who has utilized CFD to predict building energy efficiency, Serra & Semiao [2] who evaluated and compared between two different ventilation strategies. Stravrakakis et.al. [3] also evaluated the effect of window sizes to thermal comfort and hygiene in buildings using CFD. Modifications to any existing residential building can also be simulated using CFD prior to any physical renovations.

3. Methodology

3.1 Actual Data

Data collection and analysis were performed for the parameters investigated which are the air temperature, air velocity and air humidity. Standard values set by ASHRAE were compared with the data measured to determine the thermal comfort of the house occupants. The measurement of the parameters investigated was carried out when the house is ventilated naturally. The source of natural ventilation is from windows and doors which are open to allow the air from outside the house to enter the house and vice versa. The readings were taken periodically at 12am, 6am, 8am, 12pm, 2pm, 6pm, 8pm and 10pm for each day in two weeks. The measuring devices were placed in the hall, dining room, kitchen, master bedroom, room 1 and room 2. The air temperature was measured using anemometer and hygrometer.

3.2 CFD Simulation

The simulation was performed using CFD software, FLUENT for house with natural ventilation and fan turned off and when the fan is turned on. The house was drawn and meshed using Gambit software and analyzed using FLUENT to investigate the behaviour of the air temperature and air velocity inside the house. The software was unable to simulate the air humidity. Thus, only air temperature and air velocity were simulated and investigated in both simulations. The boundary conditions were set the same with the value measured during the data collection. Other physical values were also measured such as the wall temperature, glass temperature, velocity inlet (wind speed), size of openings (window and door) and other. Some parts of the house were neglected as it does not affect the parameters investigated. Figures 1 (a) and (b) show the floor plan with dimension for ground floor and first floor.

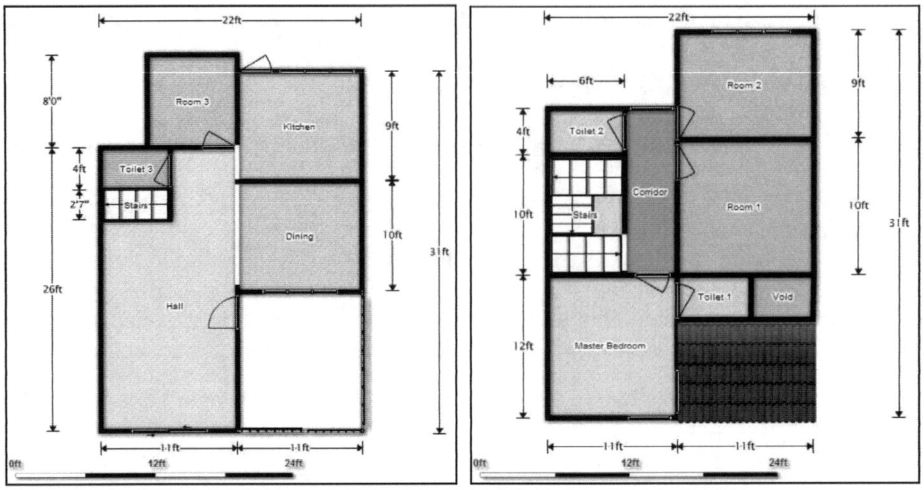

Figure 1. (a) Ground Floor Plan (b) First Floor Plan

The measured physical value used in the simulation is the opening and fan dimensions. The openings are windows and doors, which is the source of natural ventilation. The dimensions of doors, windows and fans were considered to be standard in size for all residential houses in Malaysia. The dimension of the fan is described as span. Span means the diameter of the fan or the distance between one ends of the fan to another end. 52″ span of fan was chosen as it is in standard size that applicable for all room dimensions. This size is usually used for standard bedroom and dining room [4]. Table 1 shows the boundary conditions set in the simulation. The velocity inlet is the front door which is the source of the natural ventilation and the back door is the pressure outlet for the air to flow out which is shown in Figure 2. Wall will be set by default automatically for faces that are not set with boundary condition.

Table 1. Boundary Conditions

Boundary Condition/ Characteristic	Wall Thickness (t), m	Wall Temperature (T_w), K	Air Velocity (v), m/s	Air Temperature (T_a), K
Outer Wall	0.125	303	N/A	N/A
Inner Wall	0.125	300	N/A	N/A
Glass Sliding Door	0.005	303	N/A	N/A
Velocity Inlet	N/A	N/A	0.2	300
Fan	N/A	N/A	12.7	300
Pressure Outlet	N/A	N/A	N/A	N/A

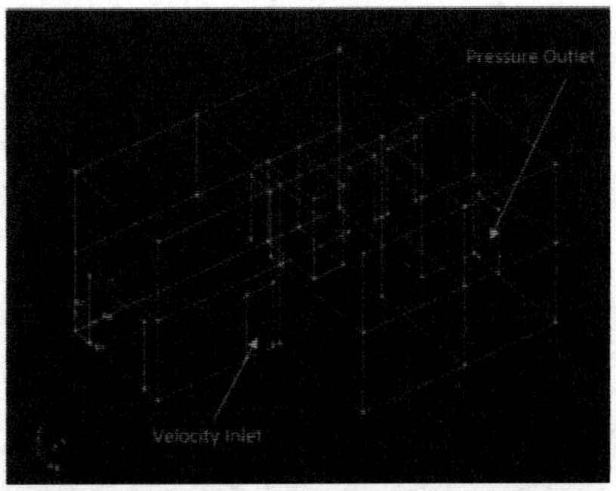

Figure 2. Gambit House Drawing with Boundary Condition

4. Results, Analysis and Discussions

After the simulation data was obtained, validation of the data was performed with the actual data. Comparisons were made in both cases and represented in Figure 3 (a) and (b). The percentage error for air temperature and air velocity is 0.24% and 13.11% respectively. These errors are acceptable and thus the simulation results are valid and acceptable.

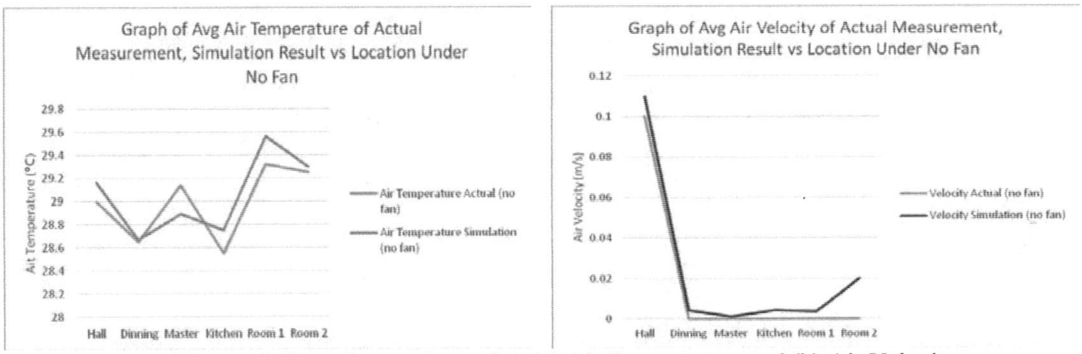

Figure 3. Data Comparisons for: (a) Air Temperature and (b) Air Velocity

The velocity vector of the internal house was obtained to investigate the pattern of air flow inside the house with the fan turned on and off (see Figure 4 (a) and (b)). When there is only natural ventilation, the air flow is only existed on the ground floor because at the ground floor, there is a source of natural ventilation which is the front door. The back door which acts as the outlet for the wind flow also helps in encouraging the air flow at the ground floor. Small or no air flow at the first floor cause the temperature distribution at the first floor for the case of fan off is higher compared to temperature distribution for case of fan turned on. From the results obtained, it shows that the natural ventilation only exists on the ground floor because the source of the wind is located at the hall and the outlet is at the kitchen. The air velocity at the first floor is approximately zero. The results for both actual and simulations were then compared in a graph together with the ASHRAE set of standards in Figure 5. As the velocity for the actual and simulation when the fan turned off are almost the same, both values are overlapped in the graph. Due to the very small value, the graph plotted for the velocity when fan turned off shows straight line near to zero. The velocity is below the limit set by the ASHRAE. When the fan is turned on, the magnitude of velocity is very large and exceeds the limit set by ASHRAE. With the presence of mechanical ventilation, the temperature distribution of the house is able to be regulated and maintained.

IOP Conf. Series: Materials Science and Engineering **88** (2015) 012014 doi:10.1088/1757-899X/88/1/012014

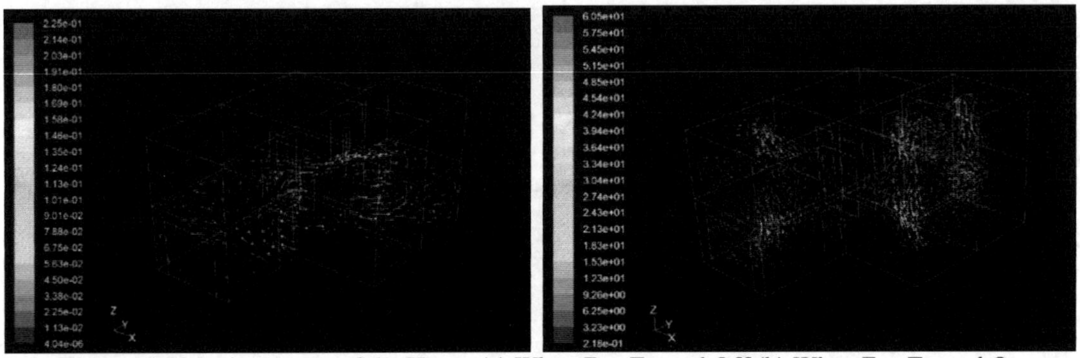

Figure 4. Velocity Vector of the House (a) When Fan Turned Off (b) When Fan Turned On

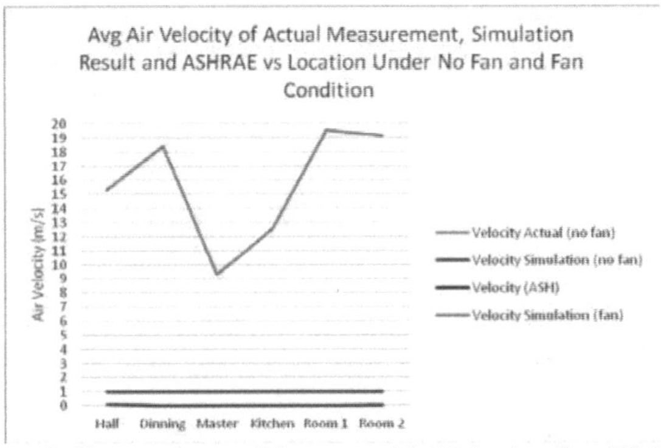

Figure 5. Comparison of Average Air Velocity for All Cases

The heat absorbed by the wall is one of the factors that cause the temperature inside the house to increase. With natural ventilation available at the ground floor, the temperature distribution inside the house is maintained to 302 K compared to the temperature of the wall which is 303 K (see Figures 6 (a) and (b)).

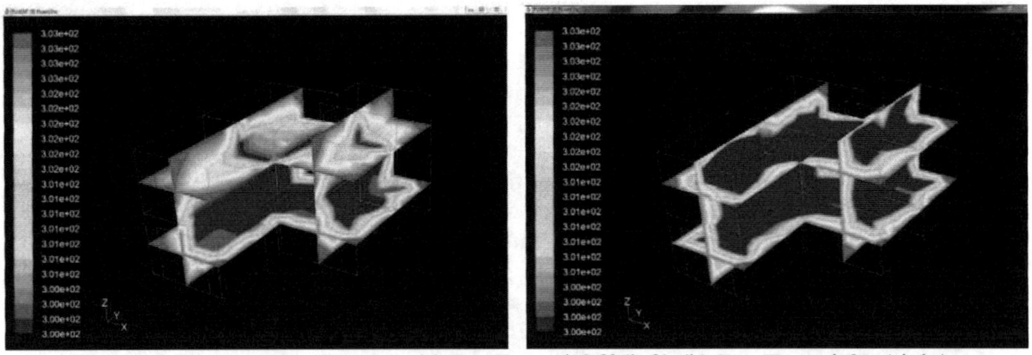

Figure 6. Temperature Contour (a) Fan Turned Off (left) (b) Fan Turned On (right)

Some part of the house at the first floor has temperature of 303 K, which is almost the same with the wall temperature. This causes discomfort to the occupant due to the high temperature. The results obtained show that mechanical ventilation which is the fan, helps a lot in maintaining the temperature distribution inside the house. The results for both actual and simulations were then compared in a graph together with the ASHRAE set of standards in Figure 7. It is shown in the figure that the air temperature for case when fan is turned on is close to the ASHRAE standard, while the other results are farther away from the ASHRAE standard. Natural ventilation is not enough to maintain the temperature as it is changing throughout the time and not consistence.

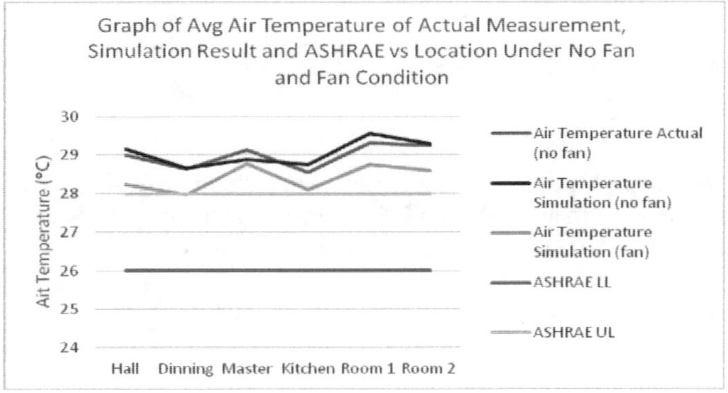

Figure 7. Comparison of Average Air Temperature for All Cases

5. Conclusion

As conclusions, the thermal comfort of the house is generally poor and does not satisfy the standards set by ASHRAE. However, the level of thermal comfort for the house with fan turned on is considered good compared to that of with fan turned off. Both cases show that the thermal comfort level becomes critical during afternoon as during this period, the house receives direct sunlight which causes the temperature inside the house to increase. With the fan turned on, the temperature inside the house is able to be maintained and regulated. Improvement is needed in order to achieve the thermal comfort. For air velocity, mechanical ventilation such as fan can be used to encourage the air flow inside the house. As the house design is fixed and house modification is restricted, other improvement should be considered for example, installing the latest mechanical ventilation system that is more effective and energy saving. Glass doors and windows are tinted with protective film to block direct sunlight from penetrating the house. Planting shade trees is also another method in reducing the direct sunlight and reduce the temperature distribution around the house. Others opening available such as small windows can be opened to encourage natural ventilation flowing from outside to inside house.

6. References

[1] P. L. F. G. a. H. B. A. Bastide, "Building energy efficiency and thermal comfort in tropical climates," Presentation of a numerical approach for predicting the percentage of well-ventilated living spaces in buildings using natural ventilation, vol. 38, pp. 1093-1103, 2006.

[2] A. V. S. N. Serra, "Comparing displacement ventilation and mixing ventilation as HVAC strategies through CFD," Engineering Computation: International Journal for Computer-Aided Engineering and Software, vol. 26, no. 8, pp. 950-971, 2009.

[3] D. K. P. Z. H. S. N. M. G.M. Stavrakakis, "Selection of window sizes for optimizing occupational comfort and hygiene based on computational fluid dynamics and neural networks," Building and Environment, vol. 46, pp. 298-314, 2011.

[4] D. Mar, "What Size Ceiling Fan Do I Need?," 2013. [Online]. Available: http://www.delmarfans.com/educate/basics/what-size-ceiling-fan-do-i-need/ . [Accessed 8 November 2013].

CFD Analysis of the Anti-Surge Effects by Water Hammering

Tae-oh Kim[1], Hyo-min Jeong[2], Han-shik Chung[3], Sin-il Lee[4], Kwang-sung Lee[5]
Department of Energy Mechanical Engineering, Institute of Marine Industry, Gyeongsang University, Tongyeong, Republic of Korea

Email: Taeo_k@naver.com[1], hmjeong@gnu.ac.kr[2], hschung@gnu.ac.kr[3], lsi4493@naver.com[4], Lks815@naver.com[5]

Abstract. Water hammering occurs due to the surge effect that comes from operating the pump, sudden stop during the operating due to a blackout and rapid open and close of the valve. By the water hammering of the pipeline and the pump, the valve are damaged. In this paper, transient analysis is conducted by CFD (Computational Fluid Dynamics). The purpose of this paper is to provide the research data about the change of the pressure and flow in the pipe that caused by the water hammering.

1. Introduction

Pressure rise and down occur due to the surge effect that comes from operating the pump, sudden stop during the operating due to a blackout and rapid open and close of the valve. Water hammering is a pressure surge or wave caused when a fluid in motion is forced to stop or change direction suddenly. A water hammer commonly occurs when a valve closes suddenly at an end of a pipeline system, and a pressure wave propagates in the pipe. This pressure wave can cause major problems, Vapor Cavity and Column Separation that draw pipe collapse. It is possible to reduce the effects of the water hammer pulses with Air Chamber, Surge Tanks, Surge Relief Valve, and other features. Among them, Surge Relief Valve is more useful than others to reduce the water hammering.

In this study, we did numerical analysis of water hammering in the pipeline to improve the surge relief valve by using CFD.

2. Methods and numerical analysis

2.1 *Analysis model*

Fig. 1 is the basic 2D pipe model with length and height that were used in the CFD analysis. In addition, the given area of the pipe is X * Y = 30m * 0.3m. **Fig. 2** the boundary conditions were the valve inlet minus the remaining three equal parts at regular intervals which showed that the inlet and the outlet valve serves.

Numerical methods

We used unsteady flow calculation for water hammering analysis. And we make the water hammering by locking the valve of the pipe that instantly obstructs the steady flow in the pipe.

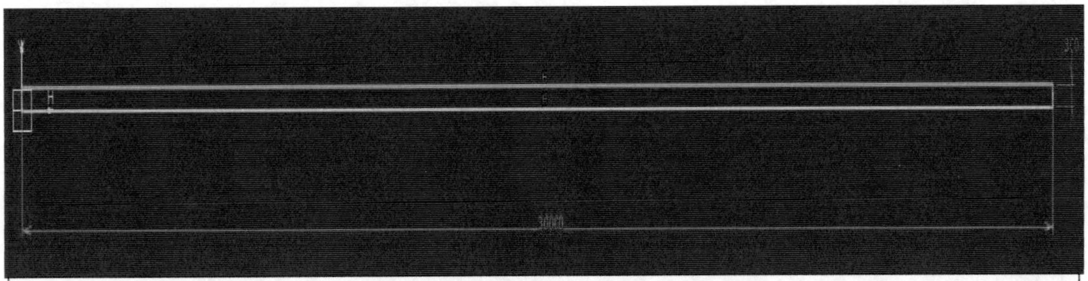

Fig. 1 Geometry model

Fig. 2 boundary conditions

3. numerical value analysis result

fig. 3-5 and **fig. 6-7** were analyzed in respective inlet pressure 6bar and 30bar. Time step for the transient state analysis was conducted 0.0001 seconds. Interprets the pipe valve to 0.0006 seconds is opened, analyzing while the valve is closed is 0.0007 seconds, and the maximum pressure. Pressure drops more and more from 0.0009 seconds, the pressure of 6bar pressure of 30bar can be seen to become constant at the initial pressure of 0.0015 sec 0.0022 seconds.

Fig. 3 Pressure changes in the pipe for water hammer (0.0006(sec))

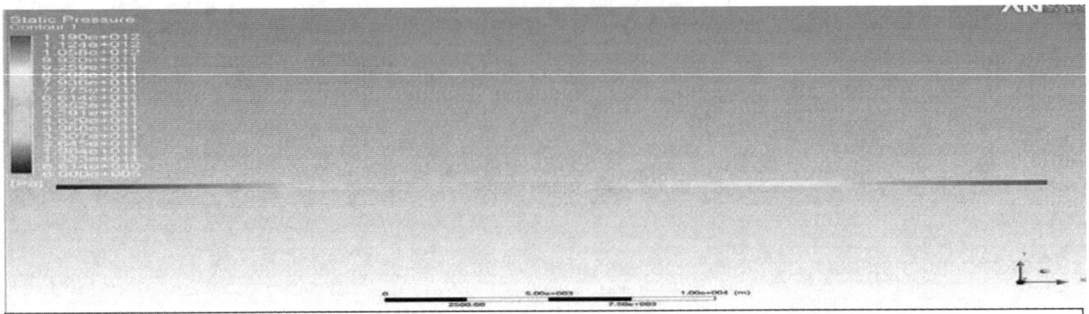

Fig. 4 Pressure changes in the pipe for water hammer (0.0007(sec))

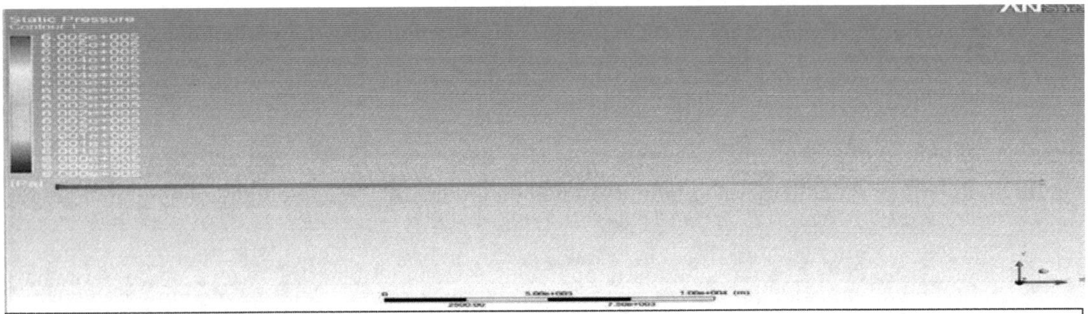

Fig. 5 Pressure changes in the pipe for water hammer(0.0022(sec))

Fig. 6 Pressure changes in the pipe for water hammer (0.0007(sec))

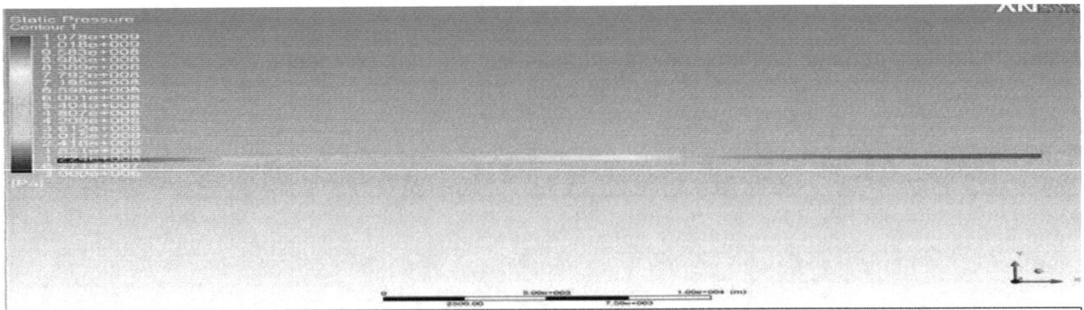

Fig. 7 Pressure changes in the pipe for water hammer (0.0014(sec))

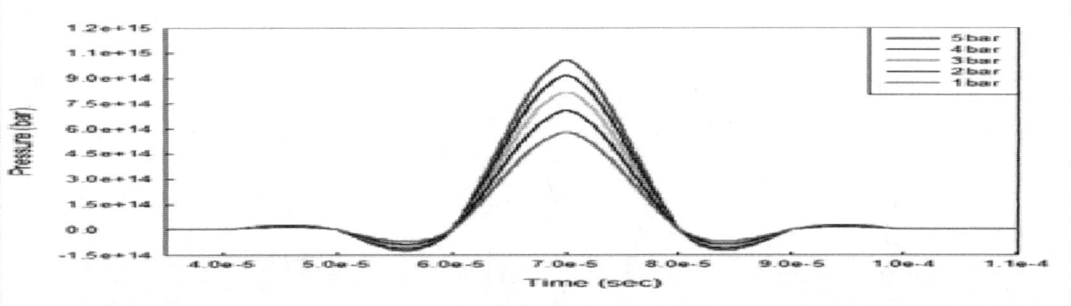

Fig. 8 has represented the pressure curve for the 10 meter long pipe where pressure range is 1-5bar.

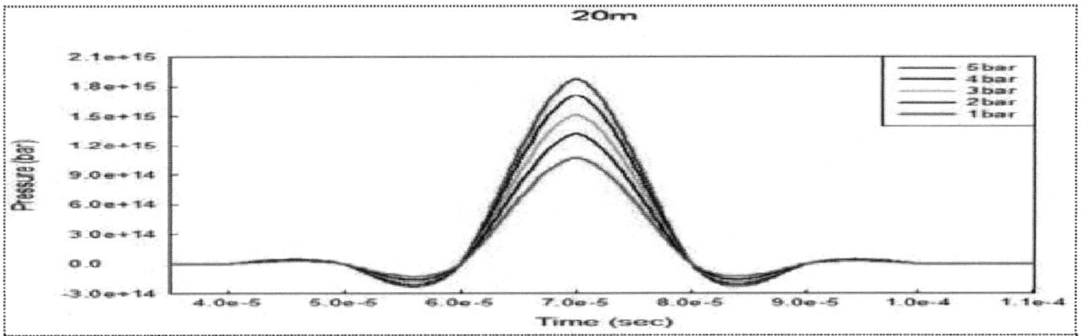

Fig. 9 has represented the pressure curve for the 20 meter long pipe where pressure range is 1-5bar.

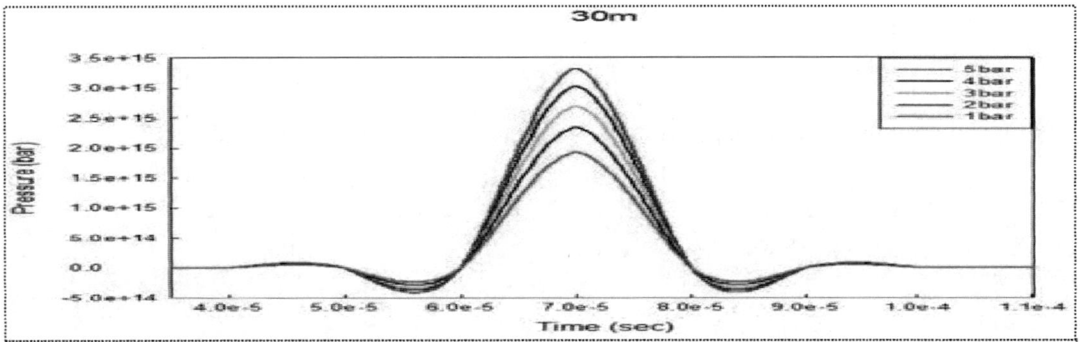

Fig. 10 has represented the pressure curve for the 30 meter long pipe where pressure range is 1-5bar.

Table. 1 Measured and Results total Pressure

	10m	20m	30m
1bar	5.82693E+14	1.08143E+15	1.93275E+15
2bar	7.11921E+14	1.32231E+15	2.34669E+15
3bar	8.20482E+14	1.51627E+15	2.70599E+15
4bar	9.18226E+14	1.71175E+15	3.03296E+15
5bar	1.00848E+15	1.87913E+15	3.32602E+15

Fig. 8, **Fig. 9** and **Fig. 10** have represented the pressure profiles of outlets for the three different sizes of pipes where applied pressure range is 1-5bar. **Table. 1** has shown that when same pressure is applied to the three different sizes of pipes (10m, 20m and 30m), high pressure has developed in the smaller pipe i.e. more pressure in 10 m long pipe and this value of pressure has decreased with the increase the length of pipe.

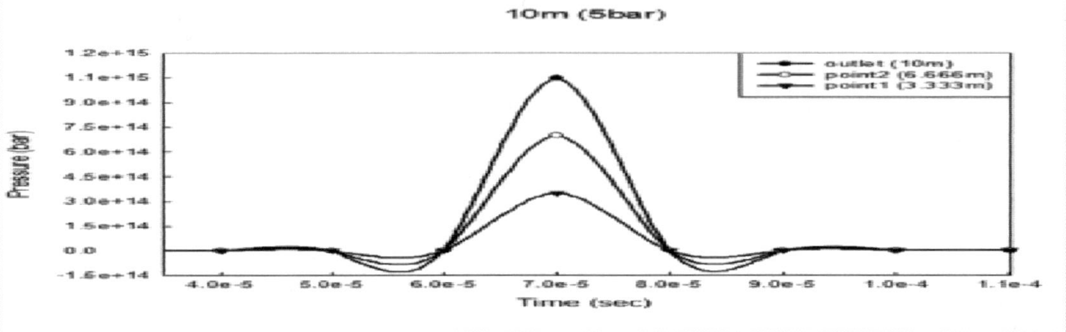

Fig. 11 Calculating the measured pressure curve using three valves (10m)

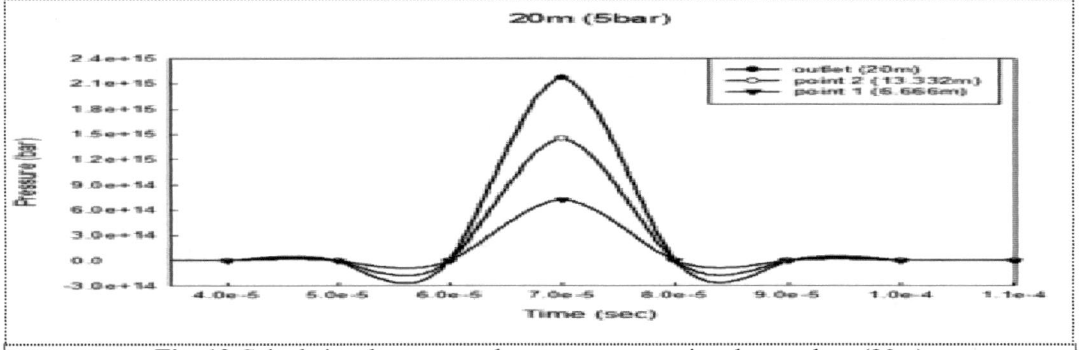

Fig. 12 Calculating the measured pressure curve using three valves (20m)

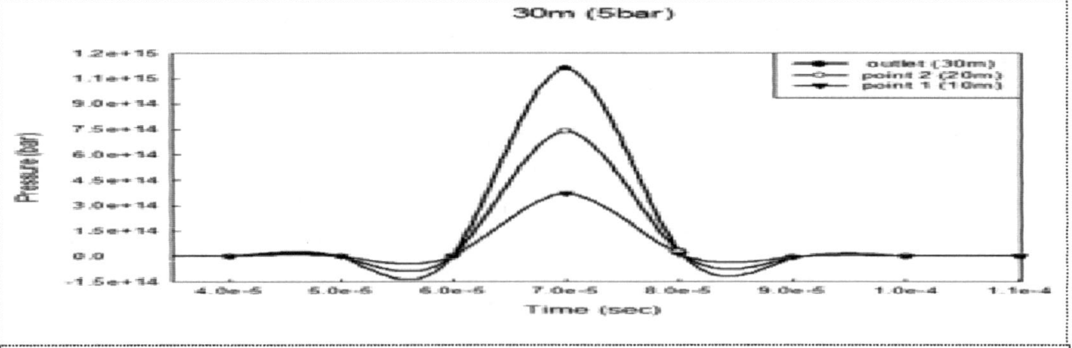

Fig. 13 Calculating the measured pressure curve using three valves (30m)

Fig. 11-Fig. 13 The interpretation of the three valves in the single pipe. Moment comes the pressure is higher in the three valves near the inlet. Even if the same pressure at the inlet to the pipe 10m, the 20m, 30m and interpreted as 1-5bar pressure, high pressure valves hitting the short piping. Moment comes the pressure is higher in the three valves near the inlet. The higher the pressure is coming out will force greater the damage will be greater in water hammer.

4. Conclusion

In this paper, the pressure condition and the variation of it inside the pipe caused by the water hammering (surge) was analyzed with the help of CFD (computational fluid dynamics). The study was conducted for the improvement of the surge relief valve. The respective pressures were given for each pipe and the values of pressure and force were obtained at every 0.0001 seconds.

- High pressure appears in the smaller pipe and this value of pressure decreases when the length of pipe is increased.
- There is an existence of high pressure near the inlet and the value of the pressure decreases with the increase the distance from the inlet.
- The maximum pressure of 6bar and 30bar are 3.59E +14 and 7.56E +15 respectively and the maximum power of 6bar and 30bar are 1.08E +14 and 2.269E +15 respectively.

5. Reference

1. 2013. Performance Evaluation of 3D CFD Code for Water Hammer Phenomenon in Pipeline. The Korean Society of Mechanical Engineers. Vol 12. No 1. pp 4098-4102. Republic of Korea.

Double droplets simultaneous impact on liquid film

Y Guo[1], G Chen, S Shen, J Zhang
Key Lab for Desalination of Lioning Province, School of Energy and Power Engineering, Dalian University of Technology, Dalian, China

E-mail: zzbshen@dlut.edu.cn

Abstract. The evolution of double droplets simultaneously impinging on flat liquid film are obtained with CLSVOF method(Combined Level Set and VOF). the impinging velocity, liquid film thickness, and the horizontal distance between the two droplets were investigated to analyze the factors that affect the evolution.

NOMENCLATURE:
$\vec{u}$ velocity vector
ρ density
p pressure
δ surface tension
κ total curvature of the interface
T temperature
S source term
ϕ distance function
g gravitational constant
t time
$\vec{n}$ interface normal vector
μ dynamic viscosity
t^* non-dimensional time
h liquid film thickness
h^* dimensionless film thickness
d initial droplet diameter
v droplet velocity before impact
We Weber number
Oh Ohnesorge number
Re Reynolds number
r droplet radius

1. Introduction

Drop impact phenomenon is an important and common process in industry, such as ink jet printing, liquid drops impact on turbine blades, oil drops impact on walls of the combustion chamber in diesel engines, spray coating and cooling, plasma spraying, drops impact on outer surfaces of heat transfer tubes in falling film evaporators, as well as liquid atomization and cleaning. In nature, erosion of soil, dispersal of spores and microorganisms, rain drops falling onto ground and wings of aircrafts also

[1] To whom any correspondence should be addressed.

involve drop impact. For most applications, a thin liquid film can be formed on the solid surface after drop impact, then the subsequent impact target becomes a liquid film rather than a solid surface. Thus, the phenomenon of drop impact on a liquid film should be paid more attention.

Some experimental results on droplet impact on the liquid film have been obtained by using high-speed camera. Hu Hai Bao[1] did experiments to investigate the process of droplet impact on micro groove, and found the maximum diameter changing characteristics. Guo[2] experimentally investigated the phenomena of spray formation, splashing and especially the bell spry after a droplet impacted onto a liquid film. Rioboo[3] experimentally observed three typical outcomes, depending on the impact speed, after the impingement: deposition, crown formation without splashing and splashing. Cossali[4] conducted experiments of the droplet impact on films, and the results showed that the upper external diameter was consistently smaller than the lower external diameter, and the evolution of the crown height depended on *We*, whereas its growing velocity and the crown thickness evolution were almost independent of *We*. Motzkus[5] carried out an experimental investigation on the influence of different parameters such as liquid properties, impact droplet properties and the thickness of liquid film on the emission of airborne particles produced by the impact of millimeter-size droplets onto a liquid film. Computational simulation on this thesis emerged later than the experimental observation, but it developed rapidly with the development of the computer technology. Guo[6] simulated the processes of horizontally lined two droplets impact on liquid film simultaneously and vertically lined two droplets impact on liquid film one after the other using lattice Boltzmann method. They analyzed the effects of droplets speeds and the center distance of the two droplets on the fluid flow. Shi[7]utilized lattice Boltzmann method to simulate the flow of a drop impacting on a liquid film. Three typical outcomes, i.e., deposition, crown and splash, are obtained. Lee[8] proposed the two-dimensional level set (LS) method for the interface tracking of the two-phase flow to simulate the impingement process numerically. The results showed that, the crown spread more strongly as the drop impact velocity and film thickness increased. Guo[2] concluded that the kinetic energy of the droplet was the main reason that led the spray and splash. When the thickness of the liquid film increased, the splash is weakened. However, if only the Weber number and the Reynolds number were close, the fluid flow caused by the droplet impact was similar. Shen[9] simulated droplet impact on curved surfaces with lattice Boltzmann method, and they analyzed the effects of impact speeds on the droplet impacting dynamics.Cossali[10] and Vander Wal[11] associated splashing with the production of satellite drops separating from the crown liquid sheet after the impact, which were named as secondary drops. Pasandideh-Fard[12] established the numerical model for heat transfer process after single droplet impact on solid surface. H.Fujimoto [13] experimentally investigated double droplets continuously impact on solid surface with different temperatures. T.Minamikawa[14] numerically and experimentally studied double droplets continuously impact on heated surface, they found the results are in agreement. Sivakumar[15]experimentally investigated the droplets impact on liquid films during spraying, compared the results with single droplet and discussed the interaction of multi-droplets.

Though there are many studies on the drop impact phenomenon, from the information above mentioned, it is found that there is few researches particularly focused on the evolution characteristics during double droplets simultaneously impact on very thin films. Therefore, in the present study, we mainly demonstrate the evolution characteristics after double droplets simultaneously impinging on flat liquid film. And the effects that affect the evolutions are discussed.

2. Numerical Method

The evolution after droplet impact on liquid film is a complex two phase flow process with deformation of the interface between liquid and gas, resulting in great difficulty for the simulation of the two-phase flow. The key problem is to capture the interface effectively. CLSVOF combines VOF method and Level Set method, where Level Set function is used to accurately compute the curvature and the normal vector to the interface while the VOF function is used to reconstruct the interface and conservation is guaranteed.

In the VOF method, the volume fraction α is introduced throughout the whole computational domain. α is defined as:

$$\alpha = \frac{\text{volume of the liquid}}{\text{total volume of the control volume}} \tag{1}$$

When the control volume is full of liquid, α = 1, and α = 0 as the control volume is full of gas. The interface lies within control volume with a volume fraction of 0 < α < 1. Governing equation of the volume fraction can be written as:

$$\frac{\partial \alpha}{\partial t} + \vec{u} \cdot \nabla \alpha = 0 \tag{2}$$

In the LS method, in order to track the interface between the gas and liquid, a signed distance ϕ, is used which is defined as the distance from the interface, and the interface is described as the regions with ϕ= 0. Governing equation of ϕ is described as follows:

$$\frac{\partial \phi}{\partial t} + \vec{u} \cdot \nabla \phi = 0 \tag{3}$$

The continuity and Navier-Stokes equation for the incompressible flow and the energy conservation equation can be given in Eq.(4) to Eq.(6).

$$\nabla \vec{u} = 0 \tag{4}$$

$$\frac{\rho \partial \vec{u}}{\partial t} + \rho \nabla \cdot (\vec{u}\vec{u}) = -\nabla p + \nabla \cdot \mu \left[\nabla \vec{u} + (\nabla \vec{u})^T \right] - \sigma \kappa \delta(\phi) \nabla \phi + \rho g \tag{5}$$

$$\frac{\rho(\phi) \partial T}{\partial t} + \rho(\phi) \nabla \cdot (UT) = \nabla \cdot \left(\frac{\lambda}{c_p} \nabla T\right) + S \tag{6}$$

where, κ is the interface curvature, which is calculated by using the following equation:

$$\kappa = \nabla \cdot \frac{\nabla \phi}{|\nabla \phi|} \tag{7}$$

and $\delta(\phi)$ is distance function and defined as:

$$\delta(\phi) = \begin{cases} \dfrac{1 + \cos(\pi \phi / a)}{2a} & |\phi| < a \\ 0 & |\phi| \geq a \end{cases} \tag{8}$$

where a = 1.5w and w is the minimum size of the cell.

The interface normal vector $\vec{n}$ can be calculated by Eq.(9).

$$\vec{n} = \frac{\nabla \phi}{|\nabla \phi|} \tag{9}$$

When the moving interface is tracked, the interface profile always becomes vague or generates oscillations due to numerical dissipation, so it is necessary to reconstruct the interface for each time step. In the CLSVOF method, values of both the VOF function and the LS function are used. Namely, the VOF function provides the size of portion that the interface may pass through, and the interface normal vector in Eq.(9) calculated by LS function determines the direction of the interface.

3.0 Simulation Results and Discussion

3.1 Physical model

The double droplets with same size, whose center distance is *Sh*, simultaneous impact the liquid film with the same velocity, and the time when the bottoms of the droplets initially contact the film is taken as initial time t=0, as shown in figure 1.The velocity at initial time is V, and the thickness of the liquid film is h. Both the droplets and the film are water, and the viscosity and surface tension are constants. The simulation region is 50mm×20mm, and the mesh number is1000×400.

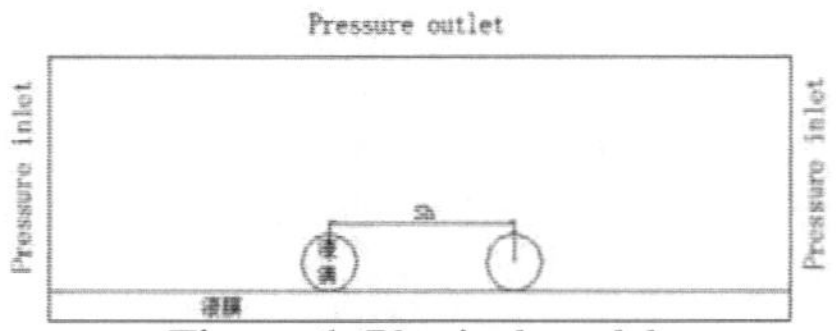

Figure 1. Physical model

3.2 Results and discussion

Figure 2 is the evolution after 2mm double droplets with the 4mm horizontal spacing simultaneous impact 1.2 mm thickness liquid film with the velocity of 0.5m/s,1.0 m/s and 2.0 m/s respectively. It can be seen from 2(a) that when the impacting velocity is smaller, the area and height of spray are smaller. Both the sprays produced by the two droplets impact expand outward and meet at the middle of the impacting point. Then laminar jet is formed along the intersection line. The laminar jet is higher than the sprays around it. There are several wave crests for the around sprays (3ms), which are lower than the central laminar jet. That means slight fluctuation is caused by the collision. The height of the central laminar jet is about 2.42 times to the height of the surrounding sprays. When the velocity is 1m/s, compared to figure 2(a), figure 2(b) shows with the rise of impact velocity, the spray diameter and height increase, the time when the sprays produced by the two droplets meet at the middle of the double droplets turns earlier, and the laminar jet becomes higher. For the bigger impacting velocity contributes to the faster convergence of the droplet and the film, so a set of sprays with obvious crests are formed at 3 ms. The height of laminar jet along the intersection line is 2.63 times to the height of its surrounding sprays. From figure 2(c) it can be seen that when the impacting velocity is up to 2m/s, after double droplets impact, two crown sprays are formed going around the two impact points. The top of sprays quickly breaks up into many small droplets. Laminar jet is formed along the spray intersection line after the two crown sprays impact, and time when this jet forms is much earlier than that of the smaller velocity, just as shown in figure 2. It is obvious that for velocity 2m/s the laminar jet can be found at 0.8ms, while for velocity 1m/s, it is at 2ms that the jet is obvious. For velocity 0.5m/s, even at 3msthe height of the jet is far smaller than other two conditions. In figure 2(c), the height of laminar jet along the intersection line is about 1.68 times to that of surrounding sprays, and finally the top of the laminar jet also breaks up into many small droplets.

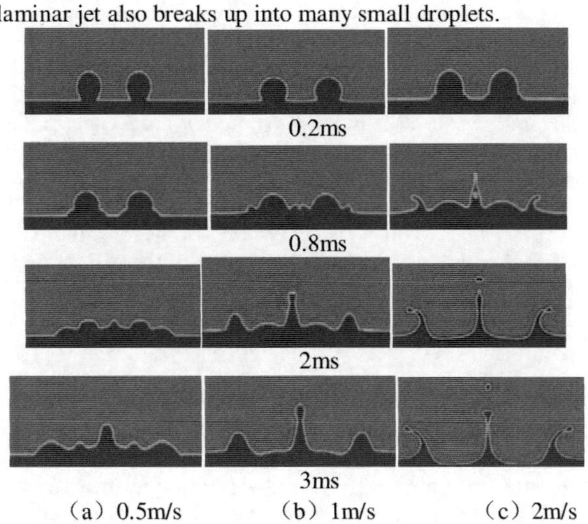

Figure 2. The evolution after double droplets simultaneous impact liquid film at different impact speed （*Sh*=2D）

Figure 3 is the pressure distribution after double droplets simultaneous impact on liquid film with the velocity of 2m/s. It indicates at the neck regions there are four pressure stagnation fields. The pressure within the stagnation area is much higher than the atmosphere. So great pressure gradient is created from inner to free surface, which causes the jets upward and outward at the neck. With the outward movements of both the droplets, the pressure gradient decreases gradually at both sides, while in the middle area, the two droplets meet and collide, and pressure stagnation point is produced at the impact point (0.8ms) where the pressure is higher. Because of the bigger pressure gradient to the free surface, the upward jets are produced at the interaction of the two droplets. The surrounding fluid goes into jet area constantly, and the laminar jet goes higher and higher. The kinetic energy of the liquid gradually transforms into potential energy of the jet. In the area of impact point, that is non-neck region, because of the smaller pressure difference between the inner fluid and the free surface, no jet appears, and this part of fluid enters into the jet area as supplement, so the film layer of this area is becoming thinner and thinner.

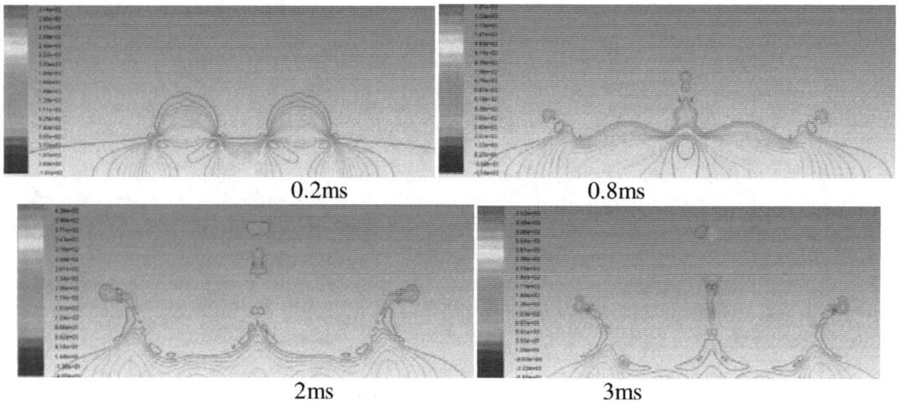

Figure 3. Pressure distribution after droplets simultaneously impact liquid film

Figure 4 is the evolution after 2mm double droplets with the 6mm horizontal spacing simultaneous impact 1.2 mm thickness liquid film with the velocity of 0.5m/s,1.0 m/s and 2.0 m/s respectively. Compare figure 4 with figure 2, we can see that for the big horizontal spacing, the time sprays needed to meet is longer after two droplets impacting on the liquid film. The evolution difference is more obvious when the impacting velocity is bigger. Just as shown in the figure 4 (c), there exists hollow between the top and the bottom of the converged crown sprays in the middle of two droplets. In figure 4(a), when the velocity is as small as 0.5m/s, the sprays are just meeting at 2ms, while in figure 4(b), at that time the central jet has formed. In figure 4(a) the height of the central jet is equivalent to that of surrounding sprays. In figure 4(b) when the impacting velocity is increased to 1.0m/s, the time when sprays form is earlier than that in figure 4 (a). The sprays meet, collide and converge with each other and move upward. As shown in figure 4(b), the height of spray is 1.76 times to that of surrounding sprays at 3.0ms. When the impacting velocity is increase to 2.0m/s, splashing occurs quickly after two droplets impacting on the film simultaneously. The splashing sprays collide earlier at the higher position, furthermore, the spray on the bottom also collide, just as shown in figure 4(c) at1.0ms. As the result, an upward velocity normal to the wall is produced, which makes the fluid continues to move upward. In addition, the collision at higher position also brings upward movement, that makes part of liquid break away from surface tension and secondary droplets are produced. The height of the jet at the intersection is equivalent to that of surrounding sprays.

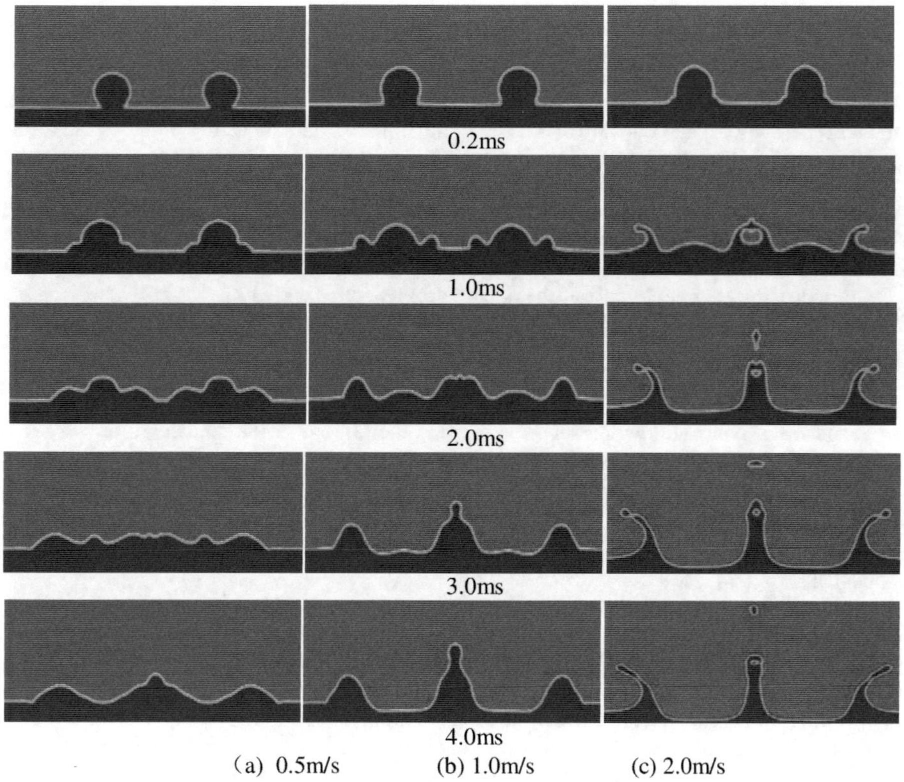

Figure 4. The change of the spray after double droplets simultaneously impact liquid film at different impact speed（Sh=3D） 4. Conclusions

The solidification phenomenon of PCM in the horizontal trapezoidal cavity is studied using CFD. The effect of various aspect ratios, cold wall temperature, PCM's initial temperature, cavity tilt angle, and Grashof number on heat transfer rate in PCM is investigated. The results of this numerical study lead to the following conclusions:

1. Solidification time is significantly decreased with the increase of the aspect ratio; e.g., the trapezoidal cavity with an AR=1.8 requires almost 17% lesser solidification time than that of the square cavity (AR = 1.0) having the same internal area. Therefore cavity aspect ratio can be used as a controlling parameter to improve the heat transfer rate.
2. The decrease of the cold surface temperature (T_c) decreases the solidification time significantly (nearly 50%) for the case where T_c was reduced from -5 to -10 °C but relatively less significant for other cases in the square cavity as well as trapezoidal cavity.

The theoretical prediction in this paper is hoped to be a useful guide for experiments dealing with the study of effectiveness of aspect ratio of trapezoidal cavity filled with PCM in improving the heat transfer rate.

5. Conclusions

CLSVOF method is applied to simulate the evolution after double droplets simultaneously impinging on flat liquid film. The conclusions are as follows:

1) After double droplets simultaneously impinging on flat liquid film, there are sprays produced by both droplets. They expand outward and meet at the middle of the impacting point, as the result, the laminar jet is formed along the intersection line, which is higer than the surrounding sprays.

2) The impact velocity affects the evolution a lot after double droplets simultaneously impinging on flat liquid film. The spray diameter and height increase with the rise of impact velocity and the laminar jet becomes higher. The times that the height of the laminar jet to the height of surrounding sprays denpend on the impact velocity. At higher impact velocity, the time when the sprays produced by the two droplets meet at the middle of the double droplets turns earlier.

3) The horizontal distance between double droplets affects the evolution a lot after double droplets simultaneously impinging on flat liquid film. When the horizontal distance is big, the time when the two sprays meet caused by double droplets impact is later than that when the distance is small. Furthermore, for big impact velocity both the tops and bottoms of the two sprays meet and converge, and there exists hollow between the top and the bottom of the converged crown sprays in the middle of two droplets. However, for the small impact velocity, there is not this phenomenon.

5. Acknowledgement

Supports from National Natural Science Foundation of China (No. 51376037) for this research are acknowledged with thanks.

6. References

[1] Hu H B, Huang S H, Chen L B 2013 Chin. Phys. B 22 084702
[2] Guo Jiahong, Dai Shiqiang, Dai Qin. Experimental research on the droplet impacting on the liquid film (in Chinese), Acta Physica Sinica, 2010, 59(4): 2601-2609.
[3] R. Rioboo, C. Bauthier, et al. Experimental investigation of splash and crown formation during single drop impact on wetted surfaces, Experiments in Fluids, 2003, 35 (6): 648-652.
[4] G. E. Cossali, M. Marengo, et al. The role of time in single drop splash on thin film, Experiments in Fluids, 2004, 36 (6): 888-900.
[5] C.Motzkus, F.Gensdarmes, E.Gehin.Parameter study of microdroplet formation by impact of millimeter-size droplets onto a liquid film, Aerosol science, 2009, 40: 680-692.
[6] Guo Jiahong, Dai Shiqiang, Numerical study on the mechanism of the normal impact of two droplets onto a thin film, J. Shanghai University, 2007, 11(3): 210-213.
[7] Shi Zi-yuan, YanYong-hua, Yang Fan, Qian Yue-hong , Hu Guo-hui. A lattice Boltzmann method for simulation of a three dimensional drop impact on a liquid film, Journal of hydrodynamics, 2008, 20(3): 267-272.
[8] S. Lee, N. Hur, et al. A numerical analysis of drop impact on liquid film by using a level set method, Journal of Mechanical Science and Technology, 2011, 25 (10): 2567-2572.
[9] Shengqiang Shen, Feifei Bi, Yali Guo. Simulation of droplets impact on curved surfaces with lattice Boltzmann method, International Journal of Heat and Mass Transfer, 2012, 55 : 6938–6943.
[10] G.E. Cossali, A. Coghe, M. Marengo, The impact of a single drop on a wetted solid surface, Exp. Fluids, 1997, 22 (6): 463-472.
[11] R. Vander Wal, G. Berger, S. Mozes, Droplets splashing upon films of the same fluid of various depths, Exp. Fluids, 2006, 40 (1): 33-52.
[12] Pasandideh-Fard M, Aziz S D, Chandra S, Mostaghimi J Int. J. Heat Fluid Fl. 2001, 22: 201
[13] Fujimoto H, Ito S, Takezaki I, Exp. Fluids 2002, 33: 500
[14] Minamikawa T, Fujimoto H, Hama T, Takuda H. Numerical Simulation of Two Droplets Impinging Successively on a Hot Solid in the Film Boiling Regime. ISIJ International, 2008, 48: 611-615.

[15] Sivakumar D, Tropea C. Splashing impact of a spray onto a liquid film. Phys. Fluids, 2002, 14(12): L85-L88.

[16] Yarin, A. L. and Weiss, D. A. Impact of drops on solid surfaces: self-similar capillary waves, and splashing as a new type of kinematic discontinuity, Journal of Fluid Mechanics, 1995, 283: 141-173.

[17] N. Nikolopoulos, A. Theodorakakos and G. Bergeles. Three-dimensional numerical investigation of a droplet impinging normally onto a wall film, Journal of computational physics, 2007, 225 (1): 322-341.

[18] H. N. Oguz and A. Prosperetti. Surface-tension effects in the contact of liquid surfaces, Journal of Fluid Mechanics, 1989, 203: 149-171.

Study of heat transfer due to turbulent flow of nanofluids through rib-groove channel

A N Al-Shamani[1,2*], K Sopian[1], A M. Abed[1], M A Alghoul[1], M H Ruslan[1], S Mat[1].
[1]Solar Energy Research Institute (SERI), Universiti Kebangsaan Malaysia, 43600 Bangi, Selangor, Malaysia
[2]Technical College Al-Musaib, Al-Furat Al-Awsat Technical University, Iraq.

E-mail: ali.alshamani@yahoo.com

Abstract. Nanofluids for improve characteristics flow in a rib-groove channel are investigate. The continuity, momentum and energy equations were solved by FLUENT program. The bottom wall of channel is heated while the upper wall is symmetry, the left side velocity inlet, and the right side is outlet (pressure out). Four different rib-groove shapes are used. Four different types of nanoparticles, Al_2O_3, CuO, SiO_2, and ZnO with different volumes fractions in the range of 1 % to 4 % and different nanoparticle diameter in the range of 25 nm to 70 nm, are dispersed in the base fluid water are used. In this paper, several parameters such as different Reynolds numbers in the range of $10000 < Re < 40000$ are investigated. The numerical results indicate that the trapezoidal with increasing height in the flow direction rib- trapezoidal groove has the best heat transfer and high Nusselt number; the nanofluids with SiO_2 have the best behavior. The Nusselt number increases as the volume fraction increases and it decreases as the nanoparticle diameter increases.

1. Introduction

They are many researchers studied the forced convection heat transfer in a rib-groove channel. Recently, researchers have shown an increased in forced convection heat transfer. Attempts to enhance heat transfer in the heat exchangers by using roughen surfaces such as rib, groove and helical rib in disturbing the flow and in providing transverse/longitudinal vortices or three dimensional mixing. There many geometric shapes of the rib-groove channel had been studied in the past. The common geometric shape include rectangular, triangular, square, and circle. All these types of rib-groove channel were used in many engineering application such as cross-flow heat exchanger, gas turbine airfoil cooling design, solar air heater blade cooling system, and gas cooled nuclear reactor. One of the ways to enhance heat transfer in the separated regions is to employ nanofluids. Nanofluids are fluids that contain suspended nanoparticles such as metals and oxides. These nano-scale particles keep suspended in the base fluid. Thus, it does not cause an increase in pressure drop in the flow field. Past studies showed that nanofluids exhibit enhanced thermal properties, such as higher thermal conductivity and convective heat transfer coefficients compared to the base fluid.

Kamali and Binesh [1] studied turbulent heat transfer and friction characteristics in a square duct with various-shaped ribs mounted on one wall. The simulations were performed for four rib shapes, i.e., square, triangular, trapezoidal with decreasing height in the flow direction, and trapezoidal with increasing height in the flow direction. Bilen et al. [2] presented an experimental study of surface heat transfer and friction characteristics of a fully developed turbulent air flow in different grooved tubes. Tests were performed for Reynolds number range 10,000 – 38,000 and for different geometric groove shapes (circular, trapezoidal and rectangular). Eiamsa-ard and Promvonge [3] performed the combined effects of rib-grooved turbulators on the turbulent forced convection heat transfer and friction characteristics in a rectangular duct under a uniform heat flux boundary condition (UHF). Promvonge and Thianpong assessed turbulent forced convection heat transfer and friction loss

[1] To whom any correspondence should be addressed.

behaviors for air flow through a constant heat flux channel (CHF) fitted with different shaped ribs. The rib cross-sections used were triangular (isosceles), wedge (right-triangular) and rectangular shapes. Jaurker et al [4] presented the heat transfer and friction characteristics of rib-grooved artificial roughness on one broad heated wall of a large aspect ratio duct shows that Nusselt number can be further enhanced beyond that of ribbed duct while keeping the friction factor enhancement low. . Manca et al. [5] presented a numerical investigation on air forced convection in a rectangular channel with constant heat flux applied on the bottom and upper external walls.

Several investigations have been reported in the literature on the convective heat transfer in nanofluids; see, for example, Mohammed et al. [6] investigated nine different rib–groove shapes which were three different rib shapes with three different groove shapes including rectangular, triangular and trapezoidal and they were interchanged with each other. Four different types of nanoparticles Al_2O_3, CuO, SiO_2, and ZnO with different volume fractions in the range of 1% to 4% and different nanoparticle diameters in the range of 25 nm to 80 nm were used. They found that the rectangular rib–triangular groove has the highest Nusselt number among other rib–groove shapes. The SiO_2 nanofluid has the highest Nusselt number compared with other nanofluid types. Nguyen et al. [7] investigated experimentally the behavior and heat transfer enhancement of Al_2O_3/water nanofluid flowing inside a closed system that is used for cooling of micro-electronic components. For a 6.8 vol.% concentration, the heat transfer coefficient was found to increase as much as 40 % compared to that of the base fluid. Liu et al. [8] investigated the thermal conductivity of copper–water nanofluids produced by chemical reduction method. Results showed 23.8% improvement at 0.1% volume fraction of copper particles. Higher thermal conductivity and larger surface area of copper nanoparticles were attributed to this improvement. It was noted that thermal conductivity increases with particles volume fraction but it decreases with the elapsed time.

The forced convection heat transfer in rib-groove channel occurred in many industrial devices with the base heat transfer fluids such as air, water, or ethylene glycol. However, the low thermal conductivity has always been the primary limitation in the development of energy-efficient heat transfer fluids, performance and compactness of many engineering equipment such as electronic devices and heat exchanger. To overcome this limitation, there is a strong motivation to improve advanced heat transfer fluids with substantially higher thermal conductivity. Hence, in recent years, nanofluids have attracted more attention for cooling in various industrial applications. This new generation of heat transfer fluids consists of suspended nanoparticles, which have a better suspension stability compared to millimeter or micrometer size ones.

2. Numerical Model

2.1. Physical Model

Schematic diagrams of rib-groove channel for geometrical model and four cases (five rib-five groove) are shown in **Fig.1** (a) Geometrical model, (b) Trapezoidal with increasing height in the flow direction Rib-Isosceles Trapezoidal Groove. (c) Isosceles Trapezoidal Rib- Trapezoidal with increasing height in the flow direction Groove. (d) Trapezoidal with increasing height in the flow direction Rib-Groove. (e) Trapezoidal with decreasing height in the flow direction Rib- Isosceles Trapezoidal Groove. The general case is Isosceles Trapezoidal Rib-Groove, Trapezoidal with increasing height in the flow direction Rib-Groove and Trapezoidal with decreasing height in the flow direction Rib-Groove. Both the height and length of the rib-groove channel are fixed and the shapes of rib and groove were changed. The channel height is set to H=60 mm while the channel rib width w=20 mm, rib height e=10 mm, and groove width w=20 mm, and the groove height e=10 mm. To ensure a fully developed flow, the entrance section smooth channel L1= 200 mm, and the exit L3= 90 mm. As shown in **Fig.1** the left side is velocity inlet and the right side is pressure outlet, the down wall is uniform heat flux q=10000 w/m^2, and the up wall is symmetry.

2.2. Boundary Conditions

The boundary conditions for this study are specified for the computational domain as shown in **Figs 1 b-e.** These figures show the general rib-groove channel model selected from this study, whose top wall Symmetry and bottom wall is subject to heat flux while the left side is velocity inlet and the right side is pressure outlet. The turbulence model is very important to accommodate the flow behavior of each application. To attain accurate prediction in the rib-groove channel, the standard k-ε turbulence model, the Renormalized Group (RNG) k-ε turbulence model were selected. To evaluate the pressure field, the pressure-velocity coupling algorithm SIMPLE was selected. At the inlet, uniform velocity profile has been imposed. The turbulence intensity was kept at 1% at the inlet. The solutions are considered to be converged when the normalized residual values reach 10^{-5} for all variables.

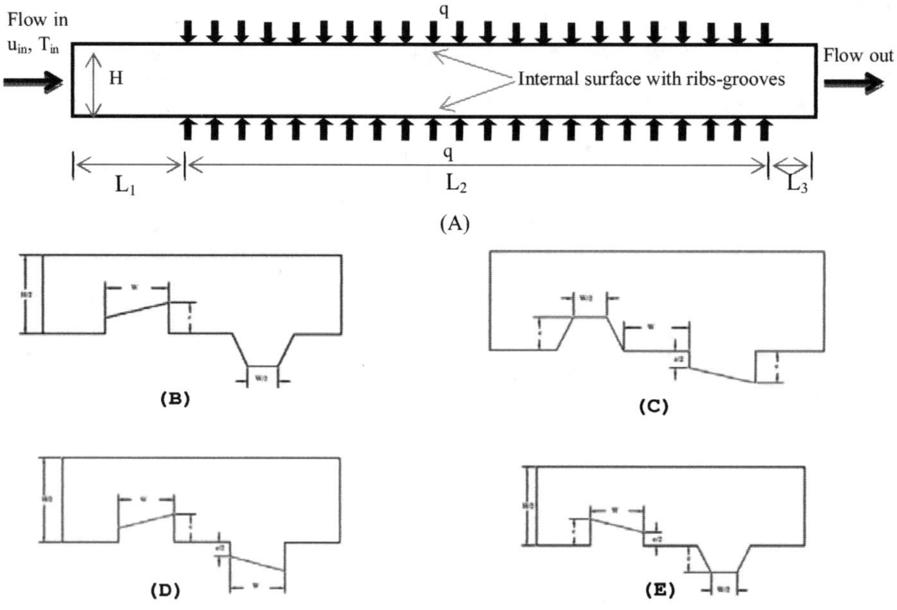

Figure 1. Schematic diagram of Rib-Groove Channel.

2.3. Thermophysical Properties of Nanofluids

In order to carry out simulations for nanofluids, the thermal properties of nanofluids must be calculated first. The nanoparticles being used are Al_2O_3, CuO, SiO_2, and ZnO. The effective properties of mass density, specific heat, thermal conductivity, and viscosity are actually calculated according to the mixing theory.

The density of nanofluid, ρ_{nf} can be obtained from the following equation [9]:

$$\rho_{nf} = (1-\phi)\rho_f + \phi\rho_{np} \tag{8}$$

The effective heat capacity at constant pressure of nanofluid, $(\rho\, C_p)_{nf}$ can be calculated from the following equation [9]:

$$(\rho\, C_p)_{nf} = (1-\phi)(\rho\, C_p)_f + \phi\,(\rho\, C_p)_{np} \tag{9}$$

By using Brownian motion of nanoparticles in rib-groove channel, the effective thermal conductivity can be obtained by using the following mean empirical correlation [10]:

$$k_{eff} = k_{static} + k_{Brwnian} \tag{10}$$

$$k_{static} = k_f \left[\frac{(k_{np}+2k_f) - 2\phi(k_f - k_{np})}{(k_{np}+2k_f) + \phi(k_f + k_{np})}\right] \tag{10.1}$$

$$k_{brwnian} = 5 * 10^4\, \beta\, \phi\, \rho_f\, C_{p,f} \sqrt{\frac{KT}{2\rho_{np} R_{np}}}\, f(T,\phi) \tag{10.2}$$

Where: Boltzmann constant: $k = 1.3807 * 10^{-23}\, J/K$

Modeling, $f(T,\phi)$

$$f(T,\phi) = (2.8217 * 10^{-2}\phi + 3.917 * 10^{-3})\left(\frac{T}{T_0}\right) + (-3.0669 * 10^{-2}\,\phi - 3.3.91123 * 10^{-3})$$

For 1% $\le \phi \le$ 4% and $300K < T > 325K$.

The effective viscosity can be obtained by using the following mean empirical correlation [10]:

$$\mu_{eff} = \mu_f * \frac{1}{(1 - 34.87\,(d_p/d_f)^{-0.3}\,\phi^{1.03})} \tag{11}$$

$$d_f = \left[\frac{6M}{N\pi\rho_{fo}}\right] \tag{11.1}$$

Where: M is the molecular weight of base fluid, N is the Avogadro number = $6.022*10^{23}$ mol^{-1}, ρ_{f0} is the mass density of the based fluid calculated at temperature T_0=293K.

3. Results and Discussion

3.1. The Effect of Different Rib-Groove Shapes

In this study, four different Rib-Groove shapes which (Trapezoidal with increasing height in the flow direction Rib-Isosceles Trapezoidal Groove, Isosceles Trapezoidal Rib-Trapezoidal with increasing height in the flow direction Groove, Trapezoidal with increasing height in the flow direction Rib-Groove, and Trapezoidal with decreasing height in the flow direction Rib- Isosceles Trapezoidal Groove) on thermal and fluid fields. The variation of heat transfer Nusselt number with Reynolds number is presented in **Fig.2**. As shown in **Fig.2** the heat transfer rate increase with the increase of Reynolds number. The Trapezoidal with increasing height in the flow direction Rib-Isosceles Trapezoidal Groove (Trap+R-TrapG) channel provided the mean heat transfer enhancement over the smooth channel. This is because of a strong mixing of the fluid induced form turbulent flow, leading to higher temperature gradients.

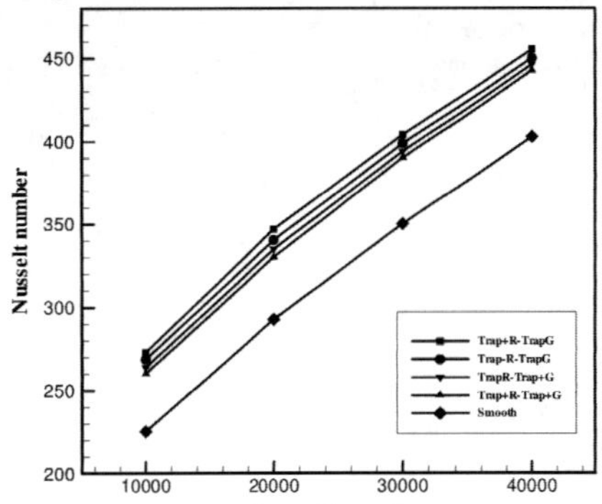

Figure 2. Normalized Nusselt number for channel with different Rib-Groove shapes at different Reynolds number.

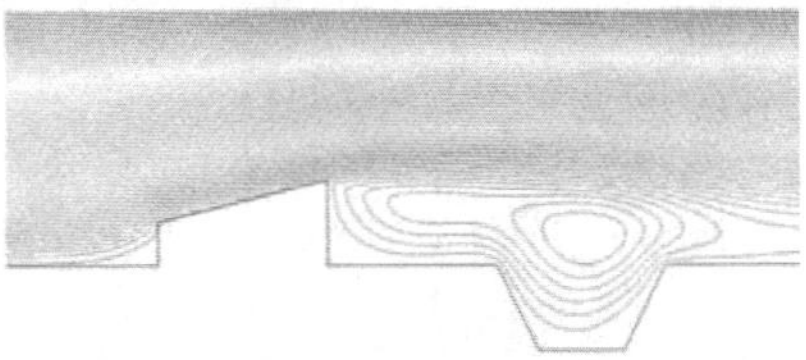

(A) Trap.+ Rib-Trap. Groove

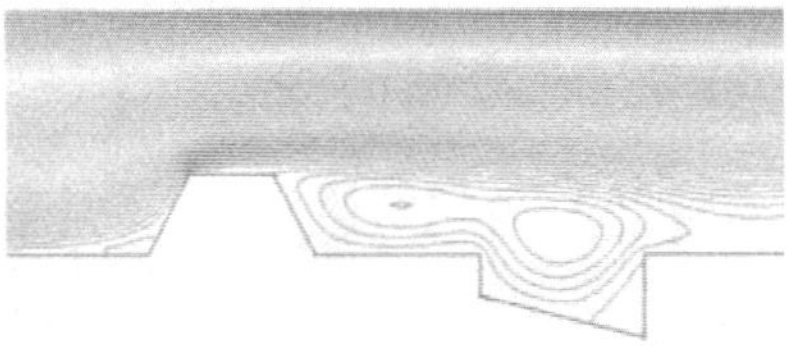

(B) Trap. Rib-Trap.+ Groove

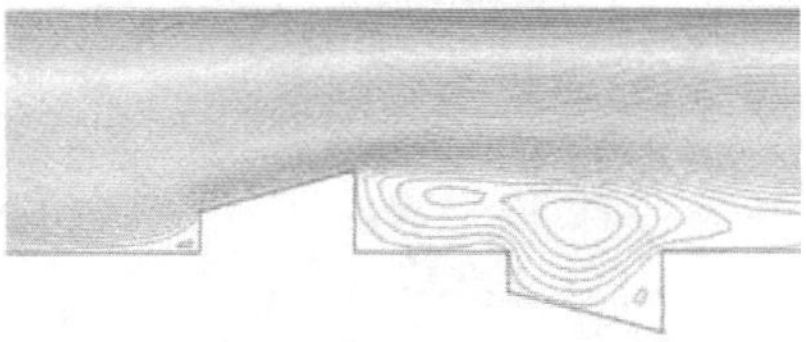

(C) Trap.+ Rib-Trap.+ Groove

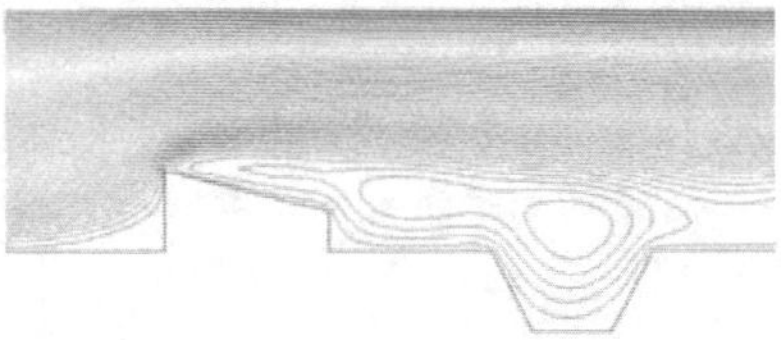

(D) Trap.- Rib-Trap. Groove

Figure 3. Velocity streamlines contours at Reynolds number = 40000

As shown in **Fig.3** the velocity streamline for four different rib-groove channel, at Re=40000. It is clear to show the recirculation zones (vortex) behind ribs and grooves.

3.2. The Effect of Different Types of Nanoparticles

Four different types of nanoparticles Al_2O_3, CuO, SiO_2, and ZnO and pure water as a base fluid are used. The Nusselt number for different nanofluids and different values of Reynolds number are shown in **Fig.4**. This figure show for the different range of Reynolds number, it is clear that nanofluid with SiO_2 has the best heat transfer rate, followed by Al_2O_3, ZnO, CuO and pure water respectively. This phenomenon can be attributed to the fact

IOP Conf. Series: Materials Science and Engineering **88** (2015) 012017 doi:10.1088/1757-899X/88/1/012017

of the high thermal properties of SiO_2-water compared with the lower thermal properties of other nanofluids types or with pure water.

3.3. The Effect of Different Volume Fractions, φ of Nanoparticles

The volume fraction of nanoparticles is actually referred to the volume of nanoparticles constituent divided by the volume of the all constituents of the mixture prior to mixing. In fact, pure water has zero volume fractions of nanoparticles. The volume fraction range of 1-4% with different values of Reynolds number is investigated. It is clear shown in **Fig.5**, increasing nanoparticles volume fraction enhances the Nusselt number. By increasing the volume fraction of nanofluid the thermal conductivity of the fluid increases which enhanced the heat transfer. It is clear that the volume friction, φ=0.01, has the best heat transfer rate followed by 0.03, 0.02, 0.01, 0.00 respectively.

3.4. The Effect of Different Nanoparticles Diameters, d_p

In this study used SiO_2-water as a working fluid with fixed other parameters such as volume fraction 4% except Reynolds number was in the range of 1000 – 40000. The range of nanoparticles diameter is 25-70 nm. As shown in **Fig.6**, the results of Nusselt number increase with decreasing the nanoparticles diameter. This can be attributed to in increment of the thermal conductivity due to nanoparticles diameter decreases. It is concluded that by using smaller diameter of nanoparticles will lead to get better heat transfer enhancement. In all cases, it can be observed that Nusselt number increase with increasing Reynolds number for all nanoparticles diameter, d_p leading to better heat transfer.

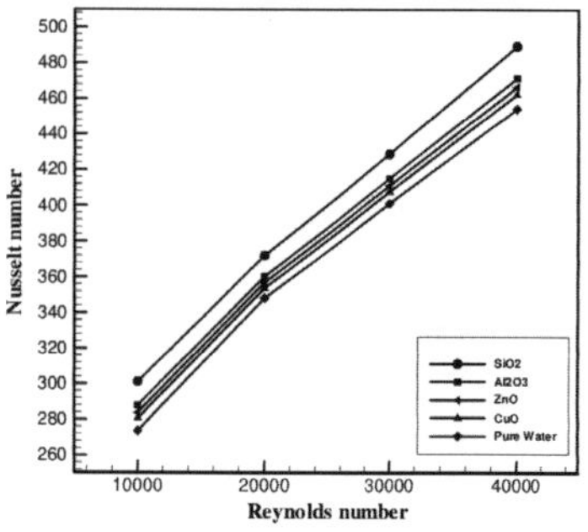

Figure 4. Normalized Nusselt number for channel with different nanofluids types at different Reynolds number.

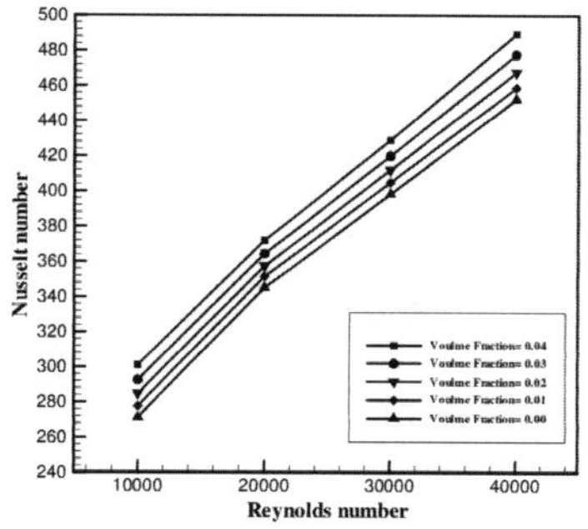

Figure 5. Normalized Nusselt number for channel with different volume fraction, φ of nanoparticles at different Re.

3.5. The Effect of Different Reynolds number

This study was done at Reynolds number in the range of 10000 – 40000. From **Fig.7**, it can observe that as the Re increases the average Nusselt number also increases. As shown in **Fig.7** the Reynolds number increases the convective current becomes more and more strong and the maximum value of isotherms reduces. The variation of heat transfer Nusselt number with x-position at different Reynolds number is presented in **Fig.7**. The Reynolds number Re=40000 provided the mean heat transfer enhancement over other Reynolds number. This is because of a strong mixing of the fluid induced form turbulent flow and appearance of reverse flow between the adjacent rib-groove elements.

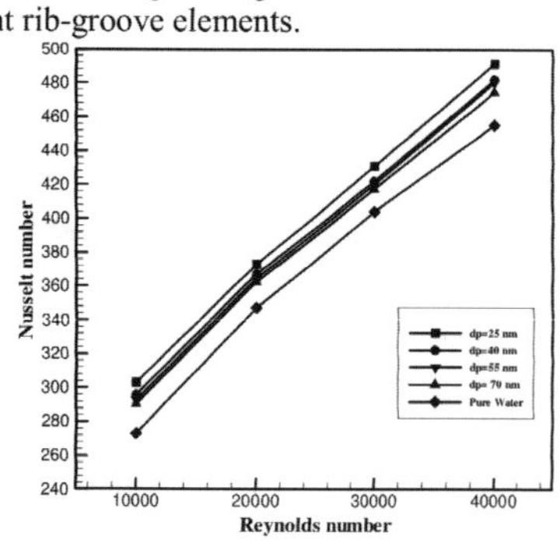

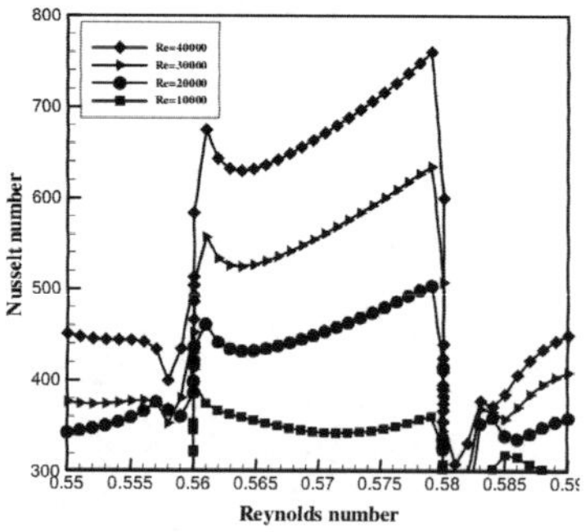

Figure 6. Normalized Nusselt number for channel with different nanoparticles diameters, d_p at different Re.

Figure 7. Normalized Nusselt number with x-position for different Reynolds number.

4. Conclusion

Numerical simulation of turbulent forced convection heat transfer in a rib-groove channel was carried out. The heat transfer enhancement resulting from various parameters, which include the different shapes of rib-groove channel, the type of nanofluids (Al_2O_3, CuO, SiO_2, and ZnO with water), volume fraction of nanoparticles in the range of $1\leq\varphi\leq4$, nanoparticles diameter in the range of $25\leq dp\leq70$, and the Reynolds number in the range of $10000\leq Re\leq40000$. The governing equations were solved utilizing finite volume method with certain assumptions and appropriate boundary conditions to provide a clear understanding of the modelling aims and conditions for the present study. CFD software (GAMBIT and FLUENT) are employed in this study to simulate the current study.

A number of conclusions can be drawn from the current work as follows:

1- By changing the rib-groove shapes on thermal and fluid fields, the results show that Trapezoidal with increasing height in the flow direction Rib-Isosceles Trapezoidal Groove (Trap+R-TrapG) gives the highest Nusselt number and the best heat transfer from thermal enhancement factor test.
2- By changing the types of nanoparticles, (Al_2O_3, CuO, SiO_2, and ZnO), the results show that SiO_2 gives the highest Nusselt number followed by Al_2O_3, ZnO, and CuO respectively while pure water gives the lowest Nusselt number.
3- The Nusselt number increased with increasing the volume fraction of nanoparticles.
4- The Nusselt number increase gradually when decreasing the nanoparticles diameter.
5- The Nusselt number increased gradually by increasing the Reynolds number, Re in the range of 10000 – 40000.

5. References

[1] R. Kamali and A. Binesh, "The importance of rib shape effects on the local heat transfer and flow friction characteristics of square ducts with ribbed internal surfaces," *International Communications in Heat and Mass Transfer,* vol. 35, pp. 1032-1040, 2008.
[2] K. Bilen, M. Cetin, H. Gul, and T. Balta, "The investigation of groove geometry effect on heat transfer for internally grooved tubes," *Applied Thermal Engineering,* vol. 29, pp. 753-761, 2009.
[3] S. Eiamsa-Ard and P. Promvonge, "Thermal characteristics of turbulent rib-grooved channel flows," *International Communications in Heat and Mass Transfer,* vol. 36, pp. 705-711, 2009.
[4] A. R. Jaurker, J. S. Saini, and B. K. Gandhi, "Heat Transfer and Friction Characteristics of Rectangular Solar Air Heater Duct Using Rib-Grooved Artificial Roughness," *Solar Energy,* vol. 80, pp. 895–907, 2006.
[5] O. Manca, S. Nardini, and D. Ricci, "Numerical Study of Air Forced Convection in A Rectangular Channel Provided With Ribs," *Frontiers in Heat and Mass Transfer,* 2011.
[6] H. Mohammed, A. Al-Shamani, and J. Sheriff, "Thermal and hydraulic characteristics of turbulent nanofluids flow in a rib–groove channel," *International Communications in Heat and Mass Transfer,* vol. 39, pp. 1584-1594, 2012.
[7] C. T. Nguyen, G. Roy, C. Gauthier, and N. Galanis, "Heat Transfer Enhancement Using Al_2 O_3–Water Nanofluid for an Electronic Liquid Cooling System," *Applied Thermal Engineering,* vol. 27, pp. 1501–1506, 2007.
[8] M.-S. Liu, M. C.-C. Lin, C. Y. Tsai, and C.-C. Wang, "Enhancement of Thermal Conductivity with Cu for Nanofluids Using Chemical Reduction Method " *International Journal of Heat and Mass Transfer* vol. 49, pp. 3028–3033, 2006.
[9] O. Manca, S. Nardini, and D. Ricci, "A numerical Study of Nanofluid Forced Convection in Ribbed Channel," *Applied Thermal Engineering* vol. 37, pp. 280-292, 2012.
[10] R. S. Vajjha and D. K. Das, "Experimental Determination of Thermal Conductivity of Three Nanofluids and Development of New Correlations," International Journal of Heat and Mass Transfer, vol. 52, pp. 4675–4682, 2009.

IOP Conf. Series: Materials Science and Engineering **88** (2015) 012018 doi:10.1088/1757-899X/88/1/012018

The theoretical analysis of the Fog removal in the LNG Ambient Vaporizer

T Lee[1], D Lee[2], H Jeong[3], H Chung[4]
[12]Department of Mechanical and Energy Engineering, Gyeongsang National University, TongYoung, Republic of Korea

[34]Department of Mechanical and Energy Engineering, Gyeongsang National, University·Institute of Marine Industry, TongYoung, Republic of Korea

E-mail: hansumguy1@nate.com

Abstract. The fog removal process is one of the important process in LNG Ambient Vaporizer. In this study we carried out theoretical study of the fog removal process in LNG Ambient Vaporizer. The LNG Ambient Vaporizer in Incheon area was used in our study. The fog temperature and the required energy produced from air fan to remove fog in LNG Ambient Vaporizer were calculated using average temperature of Incheon area in 2012 by Psychometruc Chart method. As a result we can be remove fog in LNG Ambient Vaporizer using Enthalpy[kW] energy in summer season and Enthalpy[kW] in winter season respectively.

1. Introduction

In generally, Liquefied and compressed natural gas of production center NG(Natural Gas) if transported, Korea is used via the re-initialization process. LNG(Liquefied Natural Gas) is re-liquefied natural gas transport in order to import like this step for using LNG is a vaporization process. The process of evaporation can be classified into two. First, the case where vaporized by robbing from the LNG cold mostly introduced seawater to the vaporizer to be used in the supply base and LNG. Second, it is a case to be vaporized when you need to land transport of LNG liquefaction state the main supply base, it is allowed to save the construction of a small LNG satellite base in the center of each source of demand. Therefore, in order to produce on this NG satellite base is obtained by using the atmospheric air. When Ambient Vaporizer is working, the fog is formed by the temperature difference of the ambient temperature and LNG cool air. Lots of fog around the Ambient Vaporizer have to be removed because it gives the bad influence to a surrounding. In this study we did the enthalpy calculation by using Psychrometric Chart for the effectively fog removal.

2. Preparing the Calculation

4 seasons of Korea is clear. Therefore we selected the summer and winter as the representative season. We used the Psychrometric Chart program(Psy-Chart-SAREK) provided by SAREK(The Society of Air-conditioning and Refrigerating Engineers of Korea). Also, we selected the climate condition of the

[1 234] 38, Cheondaegukchi-gil, Tongyeong-si, Gyeongsangnam-do, Korea.

Incheon city in the Korea, because the LNG conveyance wit is in that place. The climate condition data did 2012 with a standard, and we got that data from KMA(Korea Meteorological Administration).

3. The Climate Condition Standard

We selected a ambient temperature, land temperature, dew point temperature and the cool air happening in LNG as the standard of climate condition.

3.1. Summer Season

Table 1. The Vaporizer bottom condition standard of the summer.

Season	Vaporizer bottom cool air		Ambient Temperature	Land Temperature	Dew Point Temperature
	Temperature	Velocity			
Summer	-8℃	1.39m/s	27℃	32℃	22.2℃(ambient temperature 27℃, relative humidity 75%)

3.2. Winter Season

Table 2. The Vaporizer bottom condition standard of the winter.

Season	Vaporizer bottom cool air		Ambient Temperature	Land Temperature	Dew Point Temperature
	Temperature	Velocity			
Winter	-33℃	1.48m/s	2℃	4℃	-6.24℃(ambient temperature 2℃, relative humidity 55%)

We had researched for 1year based on the database which provided by KMA(Korea Meteorological Administration). Data is Temperature, Humidity, Land Temperature, Wind Velocity.

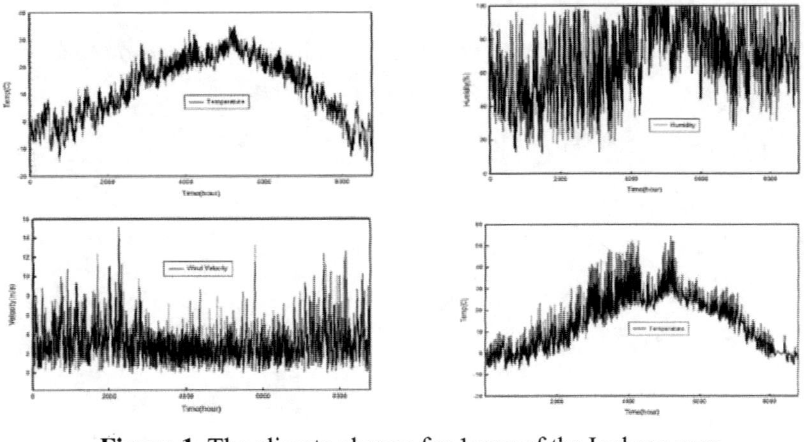

Figure 1. The climate change for 1year of the Incheon area

IOP Conf. Series: Materials Science and Engineering **88** (2015) 012018 doi:10.1088/1757-899X/88/1/012018

Fig 2.~Fig 3. Show the climate change of June(summer) and December(winter)

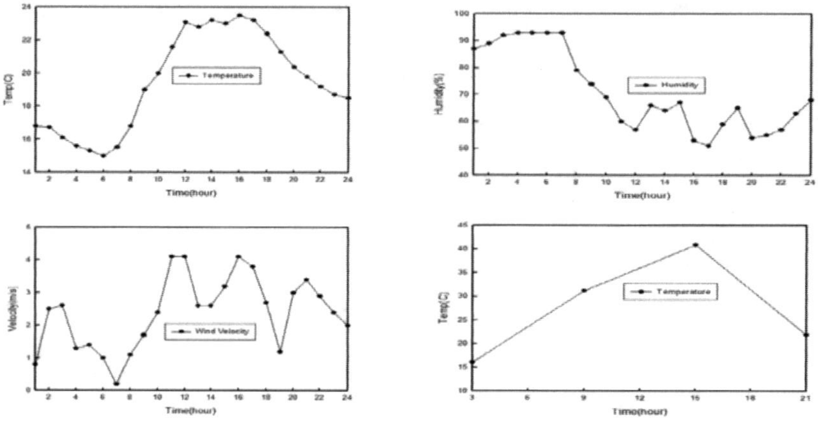

Figure 2. The climate change for 1st June of the Incheon area

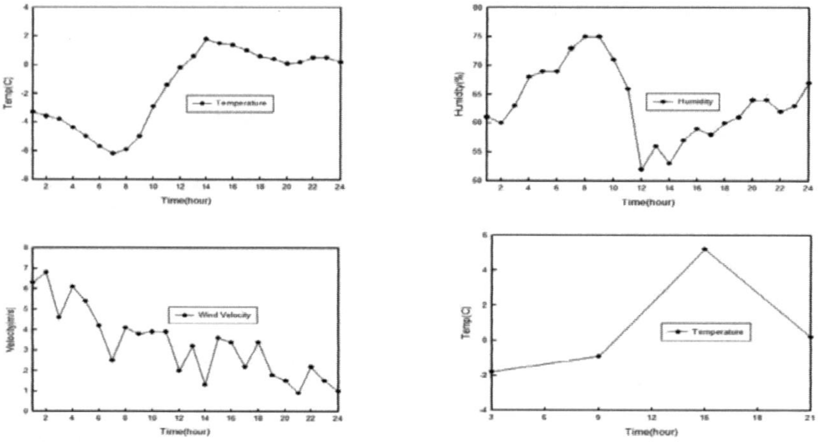

Figure 3. The climate change for 1st December of the Incheon area

4. Calculation Results

4.1 Summer Season Total Enthalpy calculation

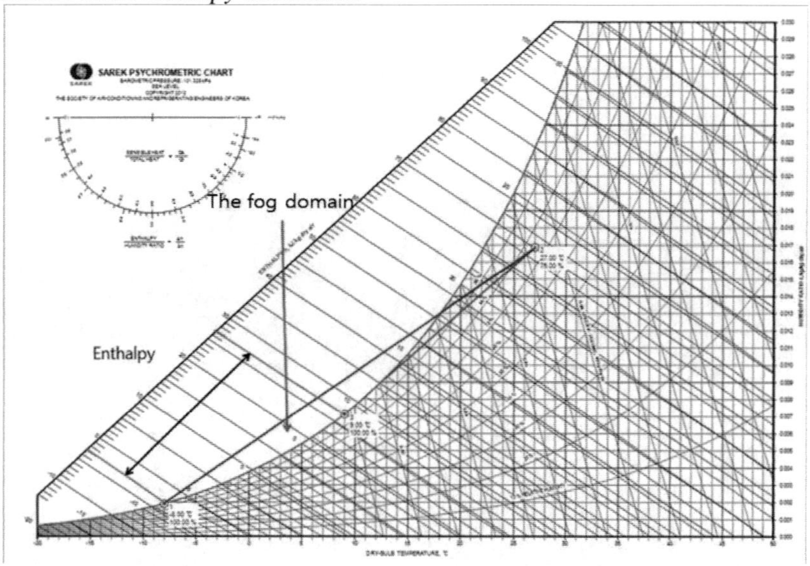

Figure 4. Summer season Psychrometric Chart

Figure 4 is provided in the table using the data of a drawing of Psychrometric Chart. Point 1 is frost the bottom of the vaporizer and Point 2 is the ambient temperature. When we use psychrometric chart, it is shown fog generation point. Fog generation point temperature is 9℃. We find Enthalpy is a picture of -3.31 to 27. Therefore E=30.31 [KJ/kg]. If cool air velocity is assume in Table 3, Total Enthalpy is shown.

Table 3. Summer Season Total Enthalpy calculation results.

Season	Vaporizer bottom cool air		Mass flow rate(m=ρAU[kg/s], ρ=1.2, A=3.35*3.35m^2)	Total Enthalpy E·m[kW]
	Temperature	Velocity		
Summer	-8℃	0.5m/s	6.7	203.1
		1.0m/s	13.5	409.2
		1.5m/s	20.2	612.3
		2.0m/s	26.9	815.3
		2.5m/s	33.7	1011.4

4.1 Winter Season Total Enthalpy calculation

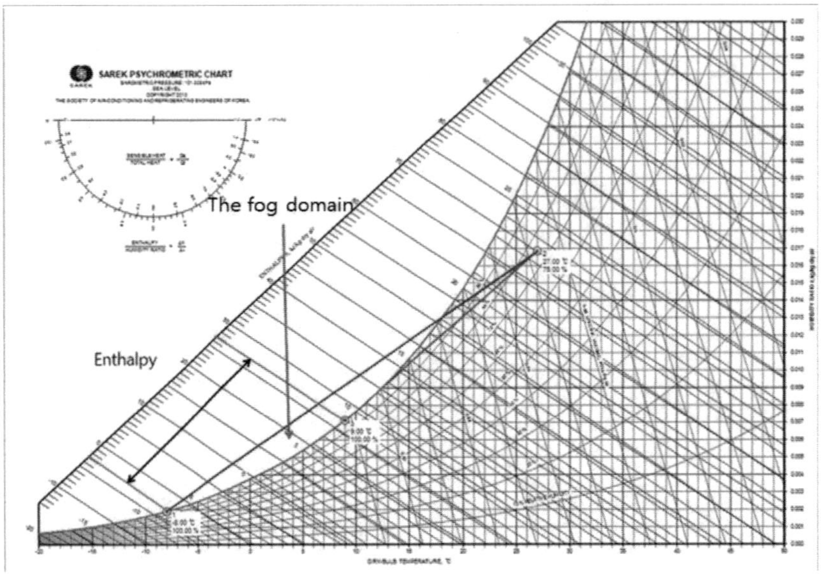

Figure 5. Winter season Psychrometric Chart

Figure 5 is provided in the table using the data of a drawing of Psychrometric Chart. Point 1 is frost the bottom of the vaporizer and Point 2 is the ambient temperature. When we use psychrometric chart, it is shown fog generation point. Fog generation point temperature is -18℃. We find Enthalpy is a picture of -32.78 to -16.22. Therefore E=17 [KJ/kg]. If cool air velocity is assume in Table 4, Total Enthalpy is shown

.

Table 3. Winter Season Total Enthalpy calculation results.

Season	Vaporizer bottom cool air		Mass flow rate(m=ρAU[kg/s], ρ=1.2, A=3.35*3.35m^2)	Total Enthalpy E·m[kW]
	Temperature	Velocity		
Winter	-33℃	0.5m/s	6.7	113.9
		1.0m/s	13.5	229.5
		1.5m/s	20.2	343.4
		2.0m/s	26.9	457.3
		2.5m/s	33.7	572.9

5. Conclusion

As the Table 3 and Table 4 showed, compared summer and winter supply for fog in the summer E=30.31[KJ/kg] and winter E=17[KJ/kg]. Therefore, supply enthalpy for fog removal was found to be 56% lower than the summer in winter.

6. Postscript

This research was supported by Basic Science Research Program through the National Reseach Foundation of Korea(NRF) funded by Ministry of Education, Science and Technology(NO. NRF-2011-0021376) and the Future Leading Project through the Small and Medium Business Administration(No.S2044441)

7. References

[1] KMA(Korea Meteorological Administration). 2012. Research on the climate of the year of Incheon city.
[2] SAREK(The Society of Air-conditioning and Refrigerating Engineers of Korea) provided by PSY-Chart.

Theoretical model of static semi-ellipsoidal droplet on a horizontal surface

S Chen, D Li, S Shen[1], Y Tao
School of Energy and Power Engineering, Dalian University of Technology, Dalian 116024, China)

Abstract: The objective of this work is to establish a theoretical static force equilibrium equation of the semi-ellipsoidal droplet on a horizontal wall by analyzing the forces acting on the differential element and deducing the expressions of surface tension as well as the internal pressure acting on the droplet. The relationship between the static spreading radius, the density, the volume, the surface tension coefficient and the contact angle of the droplet is given. The results from the equation are compared with the experimental data and show a good agreement. The data from the numerical simulation using the volume of fluid method also agree well with those from the theoretical equation.

The collision processes of droplet on wall are widespread in industrial and agricultural production as well as in nature [1-4], such as the collision of raindrops on the surface of airplanes and trains, the collision of droplets in the fire-fighting [5] and in the spray cooling, the fuel spray impingement in the cylinder of internal combustion engine [6] and the hitting of droplet on the blades in the turbo-machinery [7] etc. The process of droplet colliding with wall is an important research content of the two-phase flow. It is also a basic content of heat and mass transfer studies.

Predecessors have done a lot of researches in the process of droplet impacting on wall. In the aspect of theoretical studies, Chandra [8] theoretically analyzed the maximum diameter of droplet hitting on wall according to the energy conservation, and got an expression of the maximum spreading diameter of droplet. Pasandideh et al. [9] improved the existing model by taking the viscous boundary layer into account and ignoring the gas-liquid surface energy of droplet. Vafaei et al. [10] thought that the axial symmetric droplet did not meet the initial Young equation. [11] They gave an equation of contact angle depending on the mass of droplet and properties of the three phases by analyzing the surface tension and the gravity. Borisov et al. [12] analyzed the adhesive attraction between droplet and wall. They got a theory that the viscous force does not have to be considered for droplets in minor diameter. They also concluded that the *We* number and the wetting angle have majority influences on the motion state of droplet after the collision on wall. In the experiments, Sikalo et al. [13-14] got the droplet deformation at different collision speed and the viscosity after collision on the paraffin and

1 Correspondence author: zzbshen@dlut.edu.cn

glass panel by using high speed camera。 They also got the effect of collision speed and viscosity on the droplet spreading. Bi et al. [15] did an experiment about the spreading characteristics of droplet impacting on solid surface. They got a result that the viscosity has an important influence on spreading of droplet. They also found that the surface tension plays a crucial role for the retraction process. In the meanwhile, the collision speed influences the largest spreading factor. In the numerical simulation. Li Chunxi et al. [16] established the film thickness and concentration evolution model by using lubrication theory. They got the spreading characteristics of droplet and the influence law of the substrate structural parameters. Li Yanpeng et al. [17] simulated the flow processes of droplet impacting on a curved surface with low impact energy. To sum up, the researches on droplet impingement tends to use experimental method and simulation method. Because of the complexity of droplet flow, there are huge difficulties in analyzing the stresses acting on droplet theoretically. Thus, few relative studies had been done on this area.

This article analyzes the forces on the differential element of the static semi-ellipsoidal droplet on flat wall, the mathematical expression deduction of surface tension and internal pressure on the element, the force equilibrium equation of static droplet, the static spreading radius and the height of the semi-ellipsoidal droplet on horizontal wall by solving the equation numerically, and the comparison of the radius and height from the equation with experimental data and numerical simulation results.

1. Theoretical model of static spreading radius

In certain conditions, the droplet oscillates once it collides with wall under the action of the internal pressure, surface tension and inertial force. The droplet spread out along the wall at the beginning of collision. When the droplet reaches its maximum spreading radius, the spreading speed reduces to zero and the droplet begins to retract. Similarly, once it retracts to the minimum spreading radius, the retraction speed becomes zero the spreading restarts. Under the action of viscous damping dissipation, the amplitude of spreading and retraction decreases gradually. In the last, the droplet reaches the stationary state. At this time, the spreading radius of the droplet is the static spreading radius.

1.1 Droplet geometric model and stress analysis

There are three types of contact angle of the droplet with the wall, as shown in Fig.1. Fig.1a, Fig.1b and Fig.1c show the contact angles less than, equal to and larger than 90° respectively. The direction of the contact force is same as the direction of the contact angle.

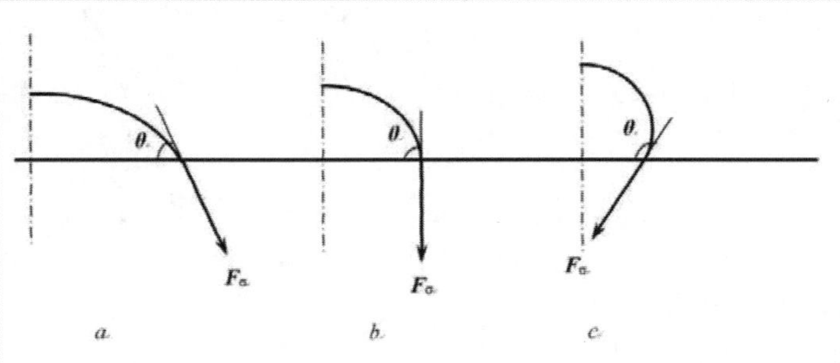

Figure 1. Contact angle of actual droplet.

Under certain conditions, the shape of a real droplet is close to semi-ellipsoidal. [18] Therefore, this article assumes the droplet having the shape of semi-ellipsoid. This article also assumes that the magnitude and the direction of contact force acting on semi-ellipsoidal droplet are same as that acting on real droplet.

In static state, the semi-ellipsoidal geometry model of a droplet is shown in Fig. 2a. The radius of droplet in the horizontal direction is r. The height is z. A differential fluid element is taken in the semi-ellipsoid. The differential angle of the element is dα. The angle between the tangent line at point (r', z ') and the axis r is β. The angle between the tangent line at point *(r '+dr ', z '+dz ')* and the axis r is *β+dβ*.

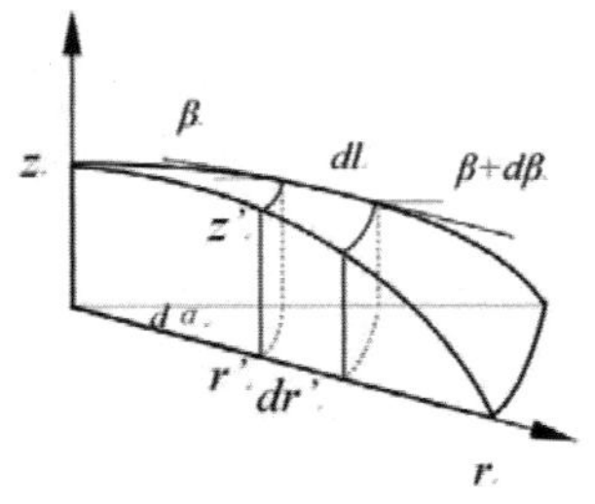

(a) Geometric model

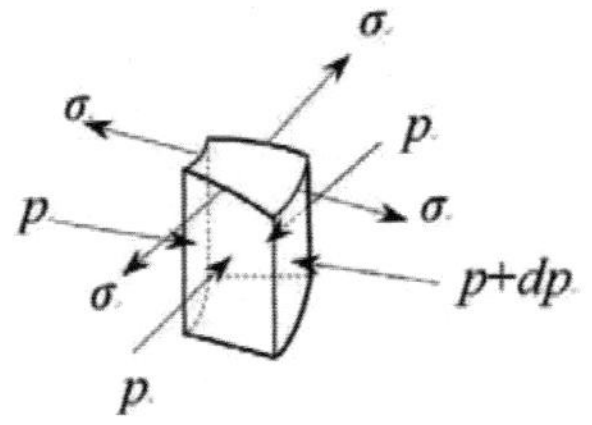

(b) Force acting on element

Figure 2. Semi-ellipsoidal droplet model.

The shape equation of semi-ellipsoidal droplet is:

$$\frac{r'^2}{r^2}+\frac{z'^2}{z^2}=1 \ \text{or} \ \begin{cases} r' = r\cos(\varphi) \\ z' = z\sin(\varphi) \end{cases} \quad 且\ r>0,\ z>0,\ 0<\varphi<\frac{\pi}{2} \tag{1}$$

The relationship between the slope and the angle β is:

$$\frac{dz'}{dr'}=-\tan\beta \tag{2}$$

The stresses acting on the droplet element are shown in Fig. 2 b. F_σ is the surface tension acting on the curved liquid surface of element and F_p is the internal pressure acting on the side face of element. Their expressions are as follows:

The magnitude of surface tension:

The left edge of surface: $F_{\sigma1} = \sigma r' d\alpha$ (3)

The right edge of surface: $F_{\sigma1}' = \sigma(r' + dr')d\alpha$ (4)

The front edge of surface: $F_{\sigma2} = \sigma dl$ (5)

The rear edge of surface: $F_{\sigma 2}' = \sigma dl$ (6)

Where, l is the 1/2 circumference of a semi-ellipsoidal droplet, $l = \frac{\pi z}{2} + r - z$

The magnitude of internal pressure:

Internal average pressure applied on the side face of element is:

$$p = 2\sigma k + \frac{1}{2}\left[\rho g z + \rho g (z - z')\right] \quad (7)$$

Where:

$2\sigma k$ — the Laplace pressure, i.e., the pressure difference between the inside and the outside of the curved liquid surface of element

k — the curvature at point (0, z), $k = \frac{z}{r^2}$

$\frac{1}{2}\left[\rho g z + \rho g (z - z')\right]$ — The average static pressure, i.e., the average gravitational pressure between point (r', z') and point (r', 0).

The left side of element: $F_{p1} = pr'z'd\alpha$ (8)

The right side of element: $F_{p1}' = (p + dp)(r' + dr')(z' + dz')d\alpha$ (9)

The front side of element: $F_{p2} = pz'dr'$ (10)

The rear side of element: $F_{p2}' = pz'dr'$ (11)

Adding up the projections of above forces (equation (3) to (6), and (8) to (11)) in the r-axis direction and omitting the higher order infinitesimal, we can obtain the static force equilibrium equation of element:

$$\begin{aligned}&(-\sigma r'\sin\beta d\alpha d\beta + \sigma\cos\beta d\alpha dr') - 2\sigma dl\sin\frac{d\alpha}{2} + \\ &(-pr'd\alpha dz' - pz'd\alpha dr' - r'z'd\alpha dp) + 2pz'dr'\sin\frac{d\alpha}{2} = 0\end{aligned} \quad (12)$$

The contact force between droplet and wall:

$$F_\sigma = \int_0^{2\pi} \sigma r\cos\theta d\alpha = 2\pi\sigma r\cos\theta \quad (13)$$

Integrating Eq. (12) in the entire droplet and adding Eq. (13), we can get:

$$\begin{aligned}&\iint(-\sigma r'\sin\beta d\alpha d\beta + \sigma\cos\beta d\alpha dr') - \iint 2\sigma dl\sin\frac{d\alpha}{2} \\ &+\iint(-pr'd\alpha dz' - pz'd\alpha dr' - r'z'd\alpha dp) + \iint 2pz'dr'\sin\frac{d\alpha}{2} \\ &+2\pi\sigma r\cos\theta = 0\end{aligned} \quad (14)$$

Solving Eq. (1) and (2) simultaneously, we can get the relationship between the geometric parameters of droplet:

$$d\beta = \frac{rz(\cos\beta)^2}{\left(\sqrt{r^2 - r'^2}\right)^3} dr' \tag{15}$$

$$\begin{cases} \sin\beta = \dfrac{r'z}{\sqrt{r^2(r^2 - r'^2) + (r'z)^2}} \\ \cos\beta = \dfrac{r\sqrt{(r^2 - r'^2)}}{\sqrt{r^2(r^2 - r'^2) + (r'z)^2}} \end{cases} \tag{16}$$

$$dp = -\frac{1}{2}\rho g dz' = \frac{1}{2}\frac{\rho g r' z}{r\sqrt{r^2 - r'^2}} dr' \tag{17}$$

Substituting Eq. (15) to (17) into Eq. (14) and simplifying the result, we can obtain the integral static force equilibrium equation of droplet:

$$-2\pi r\sigma(l - r\cos\theta) + \frac{9V^2\sigma}{4r^5} + \left(\frac{3}{4} - \frac{1}{\pi}\right)\frac{3\rho V^2 g}{2r^3} = 0 \tag{18}$$

Where V is the volume of semi-ellipsoidal droplet:

$$V = \frac{2}{3}\pi r^2 z \tag{19}$$

Eq. (18) gives the theoretical relationship between the static spreading radius, the density, the volume, the surface tension coefficient and the contact angle of droplet. In Eq. (18), the first item is the resultant surface tension, and its direction is towards the center of droplet, which plays a decisive role in droplet retraction; and other items are the resultant internal pressure, and its direction is away from the center, which plays a leading role in droplet spreading. When droplet is in static, the resultant surface tension is equal to the resultant internal pressure, but their direction is opposite. In this paper, Eq. (18) is solved by using the method of dichotomy to obtain the static spreading radius.

2. Numerical simulation of static spreading radius

The oscillation process of droplet on wall is simulated by VOF method of commercial software Fluent[19]. When the oscillation stops under the action of viscosity damping, the droplet reaches its stationary state, and the static spreading radius can be obtained. The droplet may have different initial velocity, i.e., the oscillation process is different, but its final static spreading radius is the same.

Fig. 3 shows the axial symmetric region for numerical calculation. The radius and the height of the region are 3mm and 2 mm respectively. The uniform quadrilateral grid has 0.015 mm side length which meets the requirements of the precision of numerical calculation. The pressure boundary is considered as an atmospheric pressure. The wall condition is considered as no relative slipping between the droplet and the wall. The coupling algorithm of droplet velocity and pressure is the PISO method. The iteration converges when the residual is less than 10^{-6} with the time step is 4×10^{-6}s.

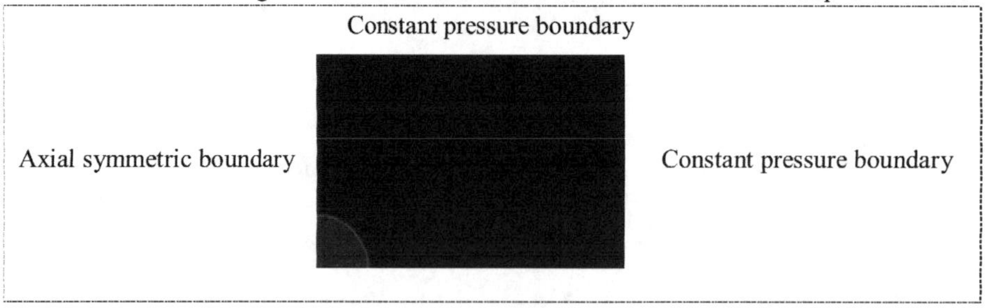

Wall

Figure 3. Simulation domain and boundary conditions.

The droplet is water and its initial shape is hemispherical. The ambient medium is air. The gas and the liquid are both assumed incompressible in the calculation. The physical parameters and initial conditions are shown in Table 1. Where μ is the coefficient of viscosity, or the dynamic viscosity, σ is the coefficient of surface tension, and v_0 is the initial spreading speed.

Table 1. Physical parameters and initial conditions of the droplet

Droplet	ρ	μ	σ	θ	V	v_0
	$kg\cdot m^{-3}$	$Pa\cdot s$	$N\cdot m^{-1}$	°	m^3	$m\cdot s^{-1}$
Water	1000	0.001	0.073	64	2.62×10^{-10}	0.5

3. Model Validation

3.1. Experimental verification of the theoretical model

In order to verify the reliability of the theoretical model of static spreading radius, the comparison between the shape of the theoretical droplet and that of the experiment is shown in Fig. 4. The experimental result comes from the literature [20]. The droplet volume is $8.18\times10^{-9}m^3$, the measured equilibrium contact angle is 95 °, the static spreading radius is 1.56mm, and the rest of the parameters are in reference [20]. The static spreading radius from Eq. (18) is 1.588 mm, the relative error between theory and experiment is 1.79%. The static height of the theoretical models and the experimental droplets are 1.550mm and 1.59mm respectively, and the relative error is -2.52%. The theoretical calculations and experimental results are fairly close, though the static spreading radius of theoretical model is slightly larger than that of the experimental data. It can be seen easily from Fig.4 that the theoretical shape of droplet is in good agreement with that of the experiment.

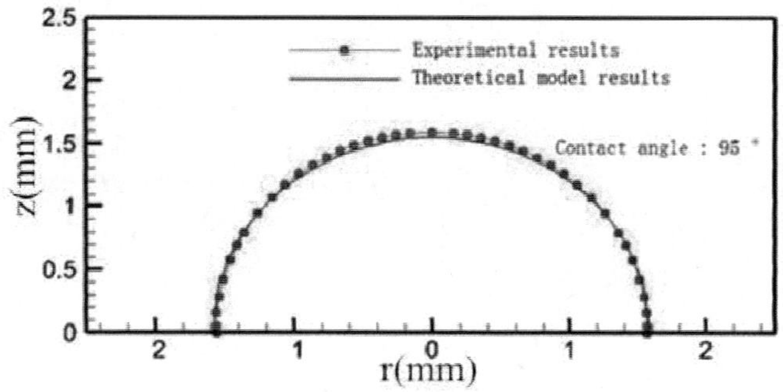

Figure 4. Comparison of the static spreading form.

3.2. Comparisons between numerical simulation and theoretical model

IOP Conf. Series: Materials Science and Engineering **88** (2015) 012019 doi:10.1088/1757-899X/88/1/012019

The droplet static shapes of theoretical model and of numerical simulation are shown in Fig. 5. The droplet volume is $2.62\times10^{-10}\,m^3$. The contact angle of Fig. 5(a), Fig.5 (b) and Fig.5 (c) are 85 °, 90 ° and 95° respectively. It can be seen easily that when the contact Angle is 85 °, the static spreading radius of theoretical model is slightly less than that of the numerical simulation; when the contact Angle is 90 °, their results are in good agreement; when the contact Angle is 95 °, the static spreading radius of theoretical model is slightly greater than that of the numerical simulation.

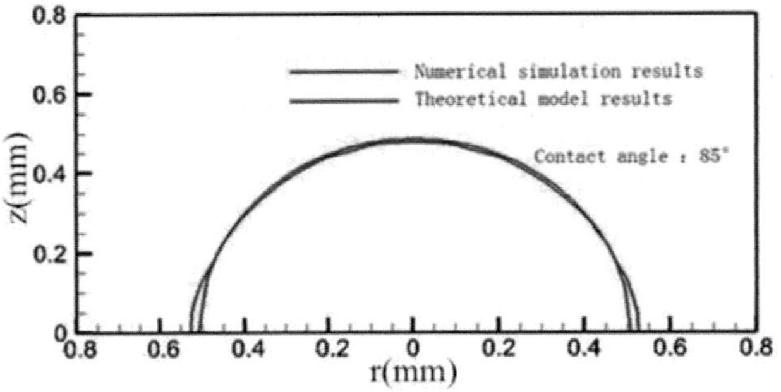

Fig.5a

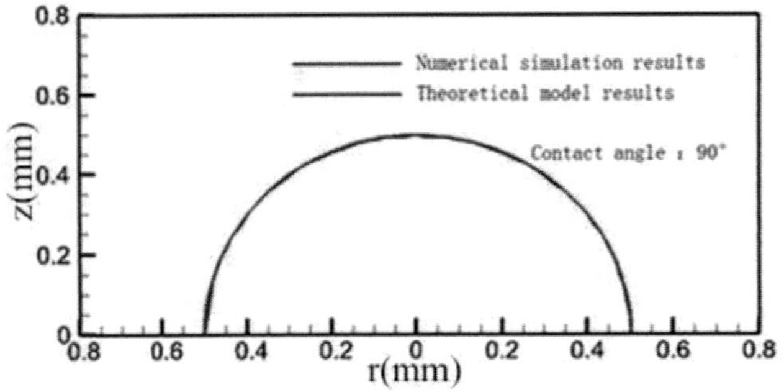

Fig.5b

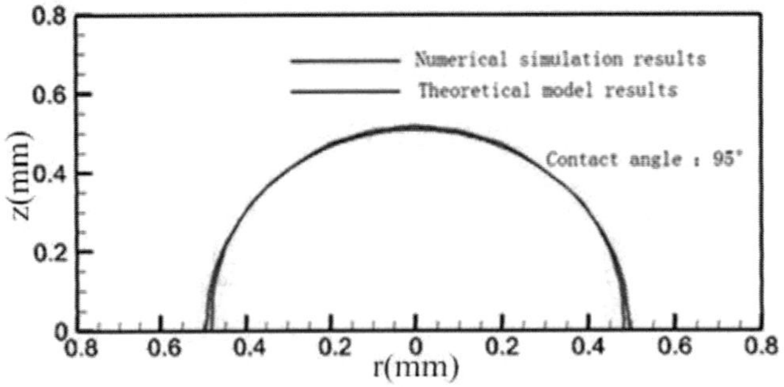

Fig.5c

Figure 5. Comparison of the static spreading form.

The static spreading radii of the theoretical model and of the numerical simulation are shown in Table 2. It can be seen from table 2 that when $64° < \theta < 90°$, the radius will become smaller as the contact angle becomes larger, and the absolute value of the error between the theoretical model and the numerical simulation becomes smaller, but the error is negative; when the droplet contact angle is 90 °, the error is -0.4%, the absolute value of the error is smallest; when $90° < \theta < 105°$, the radius will become smaller as the contact angle becomes larger, the error becomes positive and larger. The maximum absolute value of the error of the static spreading radius is within 13%.

Table 2. Static spreading radii of simulation and of theory.

Program	Contact angel (°)	Static spreading radii (mm)		Error (%)
		Simulation solutions	Theoretical solutions	
1	64	0.617	0.538	-12.80%
2	70	0.591	0.528	-10.66%
3	85	0.525	0.507	-3.43%
4	90	0.503	0.501	-0.40%
5	95	0.482	0.496	2.90%
6	105	0.440	0.486	10.45%

The static height of theoretical model can be calculated from Eq. (19). The results of the static height of theoretical model and of numerical simulation are listed in Table 3. It can be seen from the results that when $64° < \theta < 90°$, the height will become larger as the contact angle becomes larger, and the error between theoretical model and numerical simulation becomes smaller and maintains positive; when the droplet contact angle is 90 °, the error is 0.2% and minimum; when $90° < \theta < 105°$, the height will become larger as the contact angle becomes larger, the error is negative and the absolute value of the error becomes larger. The maximum absolute error of the static height is also within 13%.

Table 3. Static height of simulation and of theory.

Program	Contact angel (°)	Static height (mm)		Error (%)
		Simulation solutions	Theoretical solutions	
1	64	0.388	0.432	11.34%
2	70	0.412	0.448	8.74%
3	85	0.476	0.486	2.10%
4	90	0.497	0.498	0.20%

5	95	0.518	0.508	-1.93%
6	105	0.564	0.529	-6.21%

It can be seen from Table 2 and Table 3 that the relative errors of the radius and of the static height have different sign, but their absolute values are within 13%. This fact shows that the theoretical solutions and the numerical results are in good agreement. There are differences in shape between the semi-ellipsoidal droplet and the real droplet. The edge of semi-ellipsoid droplet is vertical with the wall, and a contact force is applied along the direction of contact angle of real droplet to simulate the forces acting on the droplet by the wall. When the contact angle of the real droplet is 90 °, the semi-elliptical model is closest to the actual droplet, so the errors between theoretical formula and numerical simulation are minimal. When the contact angle of real droplet is close to 90 °, the semi-elliptical model is also close to the real droplet. And the absolute values of error between theoretical model and numerical simulation are smaller, So that the theoretical model and the numerical simulation have a higher fit.

Comparing the results of theoretical model with that of experiment and of numerical simulation, we can find that there is a good agreement between the three. This shows that the model of force acting on the static semi-ellipsoidal droplet is correct and can be used farther to study the motion state of the droplet.

Solving Eq. (18), we can find that the static spreading radius has nothing to do with the initial velocity of droplet, and that the static spreading radius increases with the increasing of the density and the volume of droplet and decreases with the increasing of the surface tension coefficient and the contact angle.

4. Conclusion

Based on the purely theoretical stress analysis of the stationary state of semi-ellipsoidal droplet, we get the relationship between the radius and the density, the volume, the surface tension coefficient and the contact angle. Comparing the theoretical model results with the experimental and the numerical results, we get the following conclusions:

1. The theoretical relationship of static spreading radius can be used to analyze the spreading radius and the static height of droplet on the horizontal wall, and to calculate accurately the size of area covered by droplet;

2. The error of static spreading radius between theoretical model and experiment result is 1.79%, and the error of static height is 2.52%. In the range of the contact angle which has been simulated, the error absolute values of the radius and of the height between theoretical model and numerical simulation have the same changing rule and are less than 13%. When the contact angle is 90 °, the errors of the radius and of the height are minimum values, and the values are 0.4% and 0.4% respectively;

3. The analysis on the theoretical model of semi-ellipsoidal static droplet shows that the static spreading radius has nothing to do with the initial speed of droplet impacting on the wall; the static spreading radius increases with the increasing of droplet density and volume and decreases with the increasing of surface tension coefficient and contact angle.

Acknowledgement
The authors are grateful for the support of the Key Project of the National Natural Science Foundation of China (No.51336001).

References

[1] SANADA T, WATANABE M, SHIROTA M, YAMASE M, SATIO T. Impact of high-speed steam-droplet spray on solid surface[J]. Fluid dynamics research, 2008, 40(7): 627-636
[2] CHIU H R, MA H K. Solidification Models of an Impinging Metal Droplet[J]. Particulate Science and Technology, 2009, 27(1): 57-67
[3] MEHDIZADEH N Z, CHANDRA S. Boiling during high-velocity impact of water droplets on a hot stainless steel surface[J]. Proceedings of the Royal Society A: Mathematical, Physical and Engineering Science, 2006, 462(2074): 3115-3131
[4] BOLLEDDULA D A, BERCHIELLI A, ALISEDA A. Impact of a heterogeneous liquid droplet on a dry surface: Application to the pharmaceutical industry[J]. Advances in colloid and interface science, 2010, 159(2): 144-159
[5] LAN MEIJUAN, LI YUAN, WANG XISHI. Drop impact onto charring surface of heated Betula costata wood[J]. CIESC Journal, 2013, 64(8): 2807-2812
[6] CHENG XIAOBEI, HUANG RONGHUA, WANG ZHI, et.al. Study on Spray Impingement and Fuel Film Formation [J]. TRANSACTIONS OF CSICE, 2002, 20(3): 205-210
[7] ZHANG BIN, HAN QIANG, YUAN XIAOFANG, et al. Simulation of Impact Force of Drop Impacting on Horizontal Plate[J]. 2013, 47(9): 23-27.
[8] CHANDRA S, AVEDISIAN C T. On the collision of a droplet with a solid surface[J]. Proceedings: Mathematical and Physical Sciences, 1991, 13-41
[9] PASANDIDEH M, QIAO Y M, CHANDRA S, et al. Capillary effects during droplet impact on a solid surface[J]. Physics of fluids, 1996, 8(3): 650-659
[10] VAFAEI S, PODOWSKI M Z. Theoretical analysis on the effect of liquid droplet geometry on contact angle[J]. Nuclear engineering and design, 2005, 235(10): 1293-1301
[11] YOUNG T. Philosophical Transactions of the Royal Society of London, 1805, 95: 65-87
[12] BORISOV V T, CHEREPANOV A N, PREDTECHENSKII M R, et al. Effect of wettability on the behavior of a liquid drop after its collision with a solid substrate[J]. Journal of applied mechanics and technical physics, 2003, 44(6): 803-808
[13] ŠIKALO Š, TROPEA C, GANIĆ E N. Dynamic wetting angle of a spreading droplet[J]. Exp. Therm. Fluid Sci., 2005, 29(7): 795-802
[14] ŠIKALO Š, GANIĆ E N. Phenomena of droplet–surface interactions[J]. Exp. Therm. Fluid Sci., 2006, 31(2): 97-110
[15] BI FEIFEI, GUO YALI, SHEN SHENGQIANG, et al. Experimental study of spread characteristics of droplet impacting solid surface[J]. Acta Phys. Sin., 2012, 61(18): 295-300
[16] LI CHUNXI, PEI JIANJUN, YE XUEMIN. Spreading of liquid droplet containing insoluble surfactant over corrugated topography surface[J]. CIESC Journal, 2013, 64(9): 3308-3316
[17] LI YANPENG, WANG HUANRAN, Numerical Study on Deposition Characteristics of a Droplet Impinging onto Spherical Surface with Low Impact Energy[J]. 2009, 43(7): 21-24.
[18] CLANET C, BéGUIN C, RICHARD D, et al. Maximal deformation of an impacting drop[J]. Journal of Fluid Mechanics, 2004, 517: 199-208.

[19] FLUENT 6.3 User's Guide. Fluent Inc. 2006.9

[20] ROISMAN I V, OPFER L, TROPEA C, et al. Drop impact onto a dry surface: role of the dynamic contact angle[J]. Colloids and Surfaces A: Physicochemical and Engineering Aspects, 2008, 322(1): 183-191.

IOP Conf. Series: Materials Science and Engineering **88** (2015) 012020 doi:10.1088/1757-899X/88/1/012020

New vehicle bumper design for pedestrian protection during Impact

H Samaka[1*], A Manap[1], F Tarlochan[2],
[1]Faculty of Mechanical Engineering, Universiti Tenaga Nasional, Malaysia
[2]Department of Mechanical and Industrial Engineering, Qatar University, Al Tarfa Doha 2713, Qatar

E-mail: samaka65@yahoo.com

Abstract. This study discusses the influence of the active bumper design on the performance of the selected current bumper in terms of pedestrian protection by using finite element method. The legform impactor is used to test the bumper models created according to EEVC/WG regulations (European Enhanced Vehicle-Safety Committee). The simulation was performed using LS-DYNA. The lower leg injury risk was discussed based on the performance of the bumper. Results of the study show a significant improvement in the bumper performance to mitigate the impact injury of the pedestrian's lower leg.

1. Introduction

Pedestrian accidents are increasing in the entire world annually. World Health Organization 2013 (WHO) reported more than 1.24 million people die on the roads annually while 50 million are injured. 75% of the total accident deaths are pedestrian [1]. According to the NHTSA 2012 (National Highway Traffic Safety Administration) statistics, 4,280 pedestrians were killed in United States in 2010 and an estimated 70,000 were injured in traffic crashes [2]. Globally pedestrian fatalities percentage is about 50% to the total accident in the world (22% pedestrian, 23% Motorized 2-3 wheelers and 5% Cyclists). About 75% of the pedestrian accident occurs with sedans (passenger vehicle) type, and 54% at the front of these cars [3]. The "pedestrian friendly cars" designs were proposed as a solution to reduce pedestrian fatalities and mitigate injuries. These designs start from the car hood and bumper in compliance to EEVC/WG regulations for pedestrian protection. About 35% of all pedestrian body non fatal injuries occurs in the lower leg [4]. In this study, the performance of the bumper was investigated by introducing the active bumper function and compared to that of the original bumper without the active function. The bumper model was developed in Solidworks, simulated in LS-DYNA and anlyzed using finite element (FE) method. The bumper performance was evaluated in terms of upper tibial acceleration, lower leg injury risk, lower legform bending angle and knee shear displacement.

2. Numerical Models

2.1 FE Current Bumper Model

The current bumper design was modelled using finite element method as shown in Figure 1. The fascia material is plastic PVC fixed on the carbon steel material beam without absorber. All material

[1] To whom any correspondence should be addressed.

properties and parameters used in the FE model were determined with reference to material databases and related studies [5, 6]. Material specifications of the bumper is given in Table 1. The total weight of the fascia is 5.6 kg.

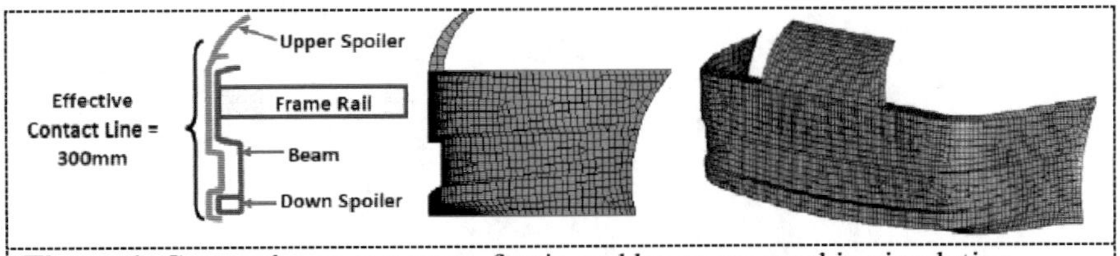

Figure 1. Current bumper system, fascia and beam are used in simulation

Table 1. Fascia and beam material specifications

Item	PVC Fascia	Steel Beam
Density ρ	1.4×10^{-6} kg/mm^3	7.8×10^{-6} kg/mm^3
Yield stress σy	46 Mpa	215 MPa
Young's modulus E	2800 MPa	210×10^3 MPa
Poisson's ratio	0.45	0.3

2.2 FE Active Bumper Model

The active bumper design was modelled using finite element method as shown in Figure 2. It was proposed as a potential solution to improve the performance of the bumper for pedestrian protection. This model allows the bumper to absorb more impact energy by separating the beam for a specific distance by creating a gap between the beam and the bumper. This will prevent direct collision of the lower leg impactor with beam, as shown in Figure 3 [7]. This active bumper moves by means of an electric motor drive that is linked to a sensor system to push the bumper using two arms. Figure 3 shows the cross sections of the active bumper with beam and the gap between them.

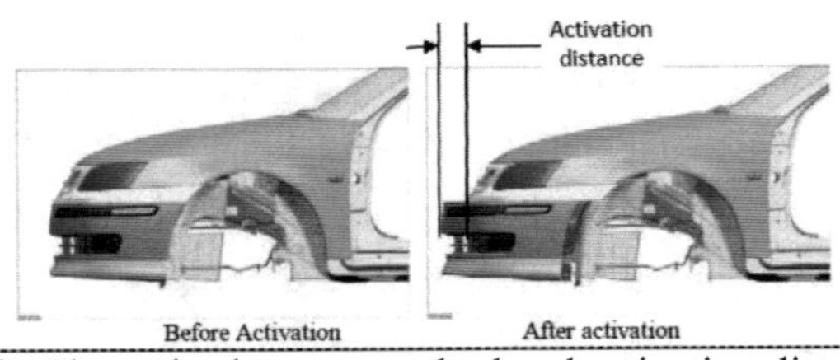

Figure 2. The active bumper method and activation distance [7]

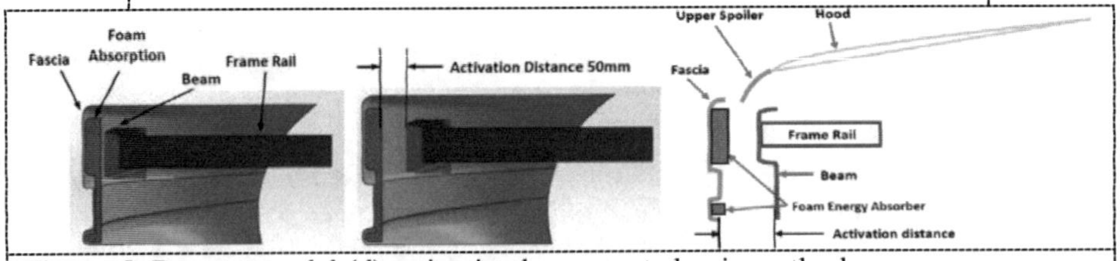

Figure 3. Bumper model (d) activation by current classic method

2.3 Impact Model

The impact test conditions according to EEVC/WG regulation [8] are; the legform impact velocity is 40 km/h = 11.1 m/s, parallel to the longitudinal axis of the vehicle. The impact point between the legform and the current bumper is 50 mm under the knee center at the tibia. The impact occurs at the centre of the bumper as shown in Figure 4.

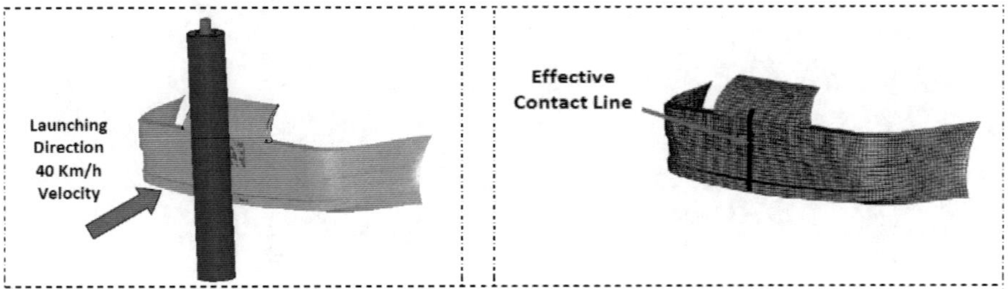

Figure 4. Launching direction of legform impactor and effective contact line

3. Results and Discussion

Figure 5 shows the impact simulation at different time period. Upon impact, the upper tibial acceleration, lower leg injury risk, lower legform bending angle and knee shear displacement are analysed and plotted.

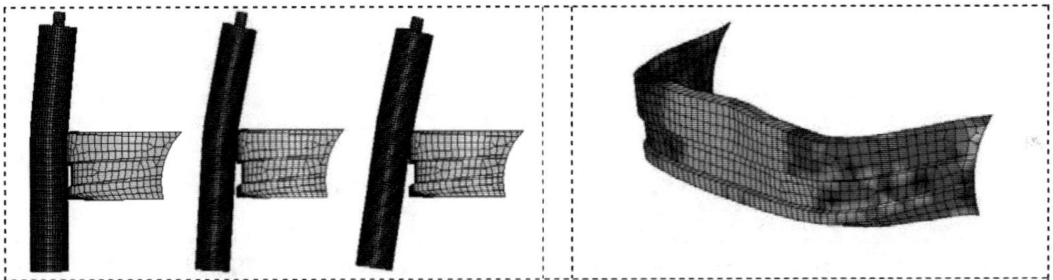

Figure 5. Impact simulation at 4, 7.5 and 9ms respectively and bumper stress

Figure 6 shows the upper tibial acceleration for current and active bumper. The higher the upper tibial acceleration is, the higher the risk of tibia and fibula bones fractures. For current bumper the maximum upper tibial acceleration is 190 G, whereas the maximum upper tibial acceleration for active bumper is greatly reduced to 150 G. The acceleration changes from negative to positive during the test period.

The lower leg injury level is described in the Injury Severity Score (ISS) as a moderate injury AIS2 for fibula fracture and series injury AIS3 for tibia fractures [9]. At 190 G, the current bumper results in a 20% injury risk for fibula and tibia fractures as shown in Figure 8(a). This percentage is considered high which indicates that the current bumper design is unsuccessful in mitigating pedestrian leg injury during accidents. On the other hand, at 150 G, the active bumper results in a 10% injury risk for fibula and tibia fractures as shown in Figure 8(b). This percentage is much lower compared to that of the current bumper which indicates a significant improvement in the bumper performance. The activation distance is limited through the deformation that occurs in the fascia resulting from the collision, which depends on the design of the bumper and material specification of fascia.

Figure 9 and 10 are shows the legform maximum bending angle and knee shear displacement respectively. The knee bending angle and the knee shear displacement indicates the severity of rupture of connective tissue surrounding the knee cartilage damage. Increased values of angle and displacement indicates a more severe rupture. Although, displacement is increased, angel is however reduced for active bumper compared to the conventional bumper.

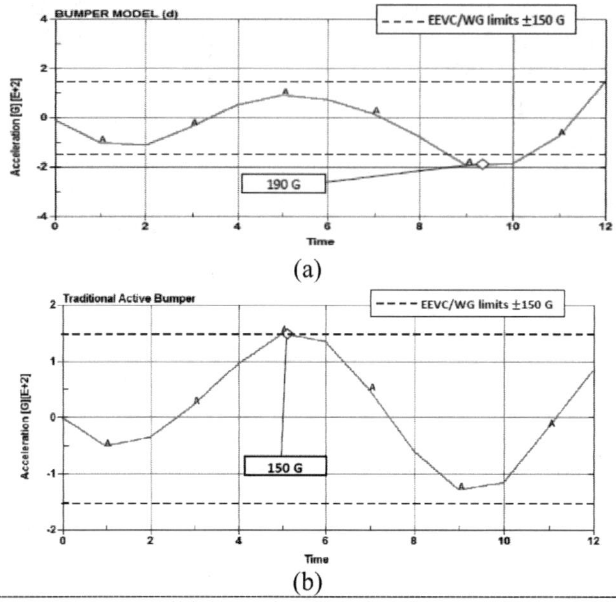

Figure 6. Upper tibial acceleration for (a) current bumper (b) active bumper

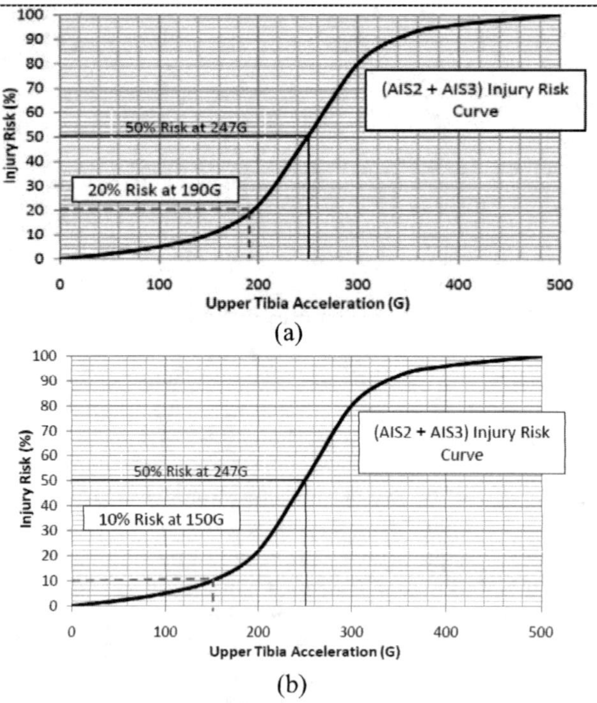

Figure 7. "Atsuhiro" Lower leg injury risk curve (AIS2+AIS3), Fibula and

Tibia fractures for (a) current bumper (b) active bumper

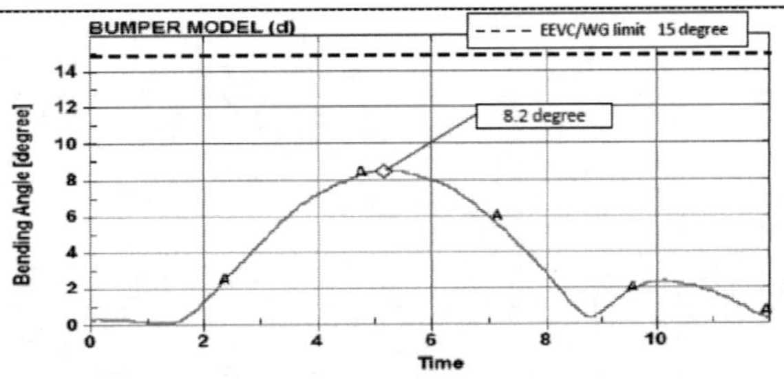

(a)

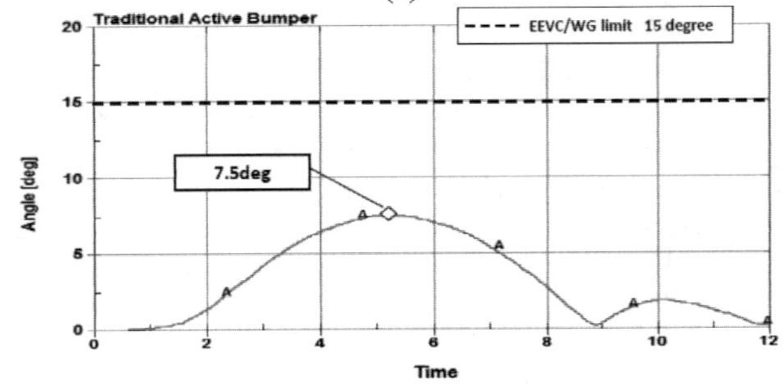

(b)

Figure 8. Lower legform bending angle for current bumper model for (a) current bumper (b) active bumper

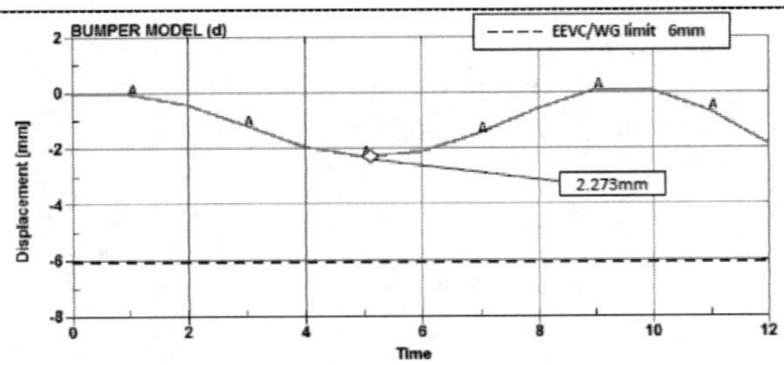

(a)

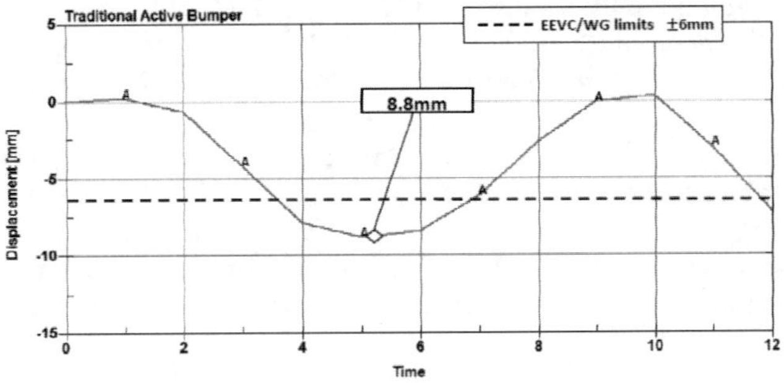

(b)

Figure 9. Knee shear displacement for (a) current bumper (b) active bumper

Table 2 shows the comparison of current and active bumpers simulation test results with EEVC/WG regulation required limits.

Table 2. Comparison of current bumper simulation test results with EEVC/WG limits

Item	EEVC/WG Max. Limits	Simulation Test Results: Current Bumper	Active Bumper
Upper tibial acceleration (G)	150	190	150
Knee shear displacement (mm)	6	2.3	8.8
Knee bending angle (deg)	15	8.2	7.5

For the active bumper design, the upper tibial acceleration is reduced to a value that falls within the limits of the EEVC/WG regulations. This reduction also leads to a decrease in the injury risk to 10% which is half of the risk without the active bumper. The analyses of these tests suggest that leg injury is caused by the hardness of the beam. The effectiveness of the impact energy distribution and absorption depends on the bumper stiffness which is directly proportional to the amount of energy absorption. The active bumper controls the amount of stiffness by creating a gap between the bumper and the beam, thus reducing the severity of the impact on the leg. These findings indicate that the active bumper design is effective and has greatly improved the bumper performance for pedestrian safety.

4. Conclusions

The current bumper design and active bumper design were studied to evaluate its performance in terms of pedestrian protection during impact. The findings can be summarized as follows:

- The maximum upper tibial acceleration for active bumper was greatly reduced to 150 G which falls within the limits of the EEVC/WG regulations
- The "Atsuhiro" lower leg injury risk for active bumper was reduced to 10% which is half of that of the current bumper
- The active bumper design reduces the bumper stiffness thus reducing the severity of the impact on the leg and greatly improved the bumper performance for pedestrian safety.

5. Acknowledgements

This work was supported by Fundamental Research Grant Scheme (FRGS20130108) funded by Ministry of Higher Education.

6. References

[1] World health organization 2013.
[2] National Highway Traffic Safety Administration, August 2012,
[3] German In-Depth Accident Study
[4] J. Hu, K.D. Klinich, *Toward Designing Pedestrian-Friendly Vehicles*. The University of Michigan, 2012
[5] Material Data Book 2003 Edition, Cambridge University Engineering Department
[6] http://textilelearner.blogspot.com/2012/09/glass-fiber-composites-properties of.html#ixzz326Jf4Ltn
[7] Ingemar Söderlund. *Reversible Active Pedestrian Safety System*, IVSS Project Report. 2009.
[8] EEVC Working Group 17 Report. *Improved Test Methods to Evaluate Pedestrian Protection Afforded by Passenger Cars*, 2002.
[9] A. Konosu, H. Ishikawa, *Reconsideration of Injury Criteria for Pedestrian Subsystem Legform Test- Problems of Rigid Legform Impactor*. Paper Number 01-S8-O-263

Burner tilting angle effect on velocity profile in 700 MW Utility Boiler

K M Munisamy [1], M Z Yusoff[1], S K Thangaraju[1] H Hassan[2], A Ahmad[2]
[1]Universiti Tenaga Nasional, Jalan IKRAM-UNITEN, 43009, Kajang
[2]Tenaga Nasional Research, Jalan Ayer Hitam, 43000 Kajang.

E-mail: kannan@uniten.edu.my

Abstract. 700 MW of utility boiler is investigated with manipulation of inlet burner angle. Manipulation of burner titling angle is an operational methodology in controlling rear pass temperature in utility boilers. The rear pass temperature unbalance between right and left side is a problem caused by fouling and slagging of the ash from the coal fired boilers. This paper presents the CFD investigation on the 0° and -30° of the burner angle of the utility boiler. The results focusing on the velocity profile. The design condition of 0° burner firing angle is compared with the off-design burner angle -30° which would be the burner angle to reduce the rear pass temperature un-balance by boiler operators. It can be concluded that the -30° burner angle reduce the turbulence is fire ball mixing inside the furnace. It also shift the fire ball position in the furnace to reduce the rear pass temperature.

Introduction

A large portion of the electrical energy generated for global usages are produced from the combustion of fossil fuel resources in thermal power plants. Tangentially fired coal fired boilers are the mostly used type of boiler in the power generation. The coal combustion process in thermal power plant consists of phenomena such as turbulence, heat transfer by convection and radiation, and chemical reactions. Many factors affect the phenomenon stated above such as furnace scale thickness, air velocity, mass flow rate and temperature of the heated air and many more. By varying these factors we can actually vary the temperature distribution In the coal fired boiler furnace which has direct impact on the efficiency of the boiler. The figure below shows the parts of a coal fired boiler.

A re-heater is a component of a coal fired power plant boiler which is responsible to reheat the steam known as cold reheat which is exiting the high pressure turbine. After it is reheated it is send back to the second steam turbine. This re-heater has boiler tubes placed in panels. While this process is going on, it was identified that there are un-even temperature distribution in the re-heater. This un-even temperature distribution can cause heat to be concentrated only on a certain region at the boiler tubes causing tube to tube temperature differences. When the temperatures are different along the tubes it will cause irregular expansion processes which will lead to cracking in the boiler tubes of the re-heater. Crack in boiler tubes will reduce the efficiency of the boiler system.

The purpose of this project is to find a solution to reduce the un-even temperature distributions in the left and right-hand side of the re-heater of a 700 MW boiler. The fluid that is being studied in this project is the flue gas. Flue gas or also known as combustion exhaust gas which is exiting to the

[1] To whom any correspondence should be addressed.

atmosphere via a flue, which is a pipe or channel for conveying exhaust gases from a furnace, boiler or steam generator. It's composition usually consists of mostly nitrogen derived from the combustion of air, carbon dioxide, and water vapour as well as excess oxygen. It further contains a small percentage of a number of pollutants, such as particulate matter (like soot), carbon monoxide, nitrogen oxides, and sulphur oxides. The first CFD simulation is done for the existing plant operation in terms of the fuel and air flow configurations by using 3-dimensional CFD simulation. Once the simulation for the existing plant operation is done, then it will be followed by the designing of the improved configuration of the plant. Once the new design of the plant operation is obtained, then CFD simulation will be performed for the new design in terms of the fuel and air flow configurations. Data from both the simulations will recorded and compared.
The results gained in this investigation will be useful to solve the un-even temperature distributions in the re-heater which will save cost of repair of the boiler tubes and further increases the efficiency of a coal fired boiler system.

2. Summary of CFD simulation in Caol and Gas fired Boiler Applications

A model of a waste heat recovery boiler was set up to study the effect of temperature of char and dust within the boiler on the deposition of these particles on the boiler walls by using Computational Fluid Dynamics (CFD) software[1]. According to Manickam et al.[1] , the deposition of particles onto the walls of the boiler can be reduced if the temperature at which the particle becomes sticky is higher than the particle temperature when it hits the wall of the boiler. The deposition of these particles on the walls causes a non-uniform temperature distribution on the surface of the wall[1].

R. Vuthaluru and H.B. Vuthaluru [2] , proposed a study on the effects of different operating conditions in a wall fired furnace using Computational Fluid Dynamics (CFD) software, FLUENT. R. Vuthaluru and H.B. Vuthaluru [2], used a 3-D combustor model to determine the temperature and heat flux profiles for a 500MW utility boiler using volatile coal as their firing medium. Different operating conditions such as effect of excess air on the furnace temperature pattern and temperatures achieved in different parts of the boiler during the combustion process have been analysed. Figure...shows the specifications of the 500MW furnace geometry. Figure 2 shows the operating conditions of the boiler.

There are 4 operating conditions in this study. First is the main boiler operating condition which is the Based on the FLUENT software simulation results, R. Vuthaluru and H.B. Vuthaluru [2], concluded that the full and partial loads with different burners in operation produces a significant impact on the temperature distribution and furnace exit temperatures.

R. Ben-Mansour et al. [3] ,proposed that the temperature distribution in the combustion chamber and in the exhaust gas are effected by various operating conditions such as excess air factor with varying the air mass flow rate, combustion air temperature and the number of tripped burners. This study is done by using CFD FLUENT software.

According to R. Ben-Mansour et al. [3] , as the combustion air temperature increases, furnace temperature increases. The furnace average temperature decrease as the excess air to fuel ratio factor increases for a given air mass flow rate [3]. According to R. Ben-Mansour et al. [3], tripping one or two burners either adjacent or opposite or tripping four results in regions of high temperature gases close to the walls and the temperature is distorted when any of the burners are tripped.

Indrusiak et al. [4] , studied the effect of air leakage in the bottom of the boiler which was simulated using CFD code CFX and the result obtained was compared with the situation where the leakage does not exist. According to Indrusiak et al. [4] , three situations were taken into account which is the study of flow field, temperature field and the heat flux generated in the furnace.The tree situations are Cases A,B and C.Case A are the value of bottom air leakage evaluated by the power plant staff [4]. Case B is the value of total amount of air which is assigned to the bottom air leakage that is computed by stoichiometric balance [4]. Case C is the value assigned to the air leakage in case A which is now added to the secondary air [4].

S. Mohammad et al. [5] , presented an investigation on the flow and combustion process in an oil fired furnace by using RNG turbulence model based commercial CFD-code FULENT. Two stages of simulation were performed, one with the effect of heat transfer to the furnace wall for full load and the other is for reduced load condition [5] .

Based on the simulation results, it can be observed that the temperature at the centre of the furnace is higher for the full load case compared to the reduced load case [5]. For both the cases the temperature distribution are high and non-uniform at the windbox region where combustion process takes place [5]. The temperature distribution is found to be more uniform at lower temperature at location beyond the windbox region due to heat loss radiation and heat transfer to the furnace wall [5]. Higher swirling and intense combustion can be observed for the full load case compared to reduced load conditions [5].

3. Model Development for 700mW Utility Boiler

The model of the 700 MW boiler is developed in 3-D CAD software then it is imported to CFD software for mesh generation. The inlet of the primary and secondary air is modelled. The inlet condition is exploited with 0° and -30° inlet condition and the velocity profile is presented. The inlet condition is determined from the actual data measured from the utility plant as figure 1.

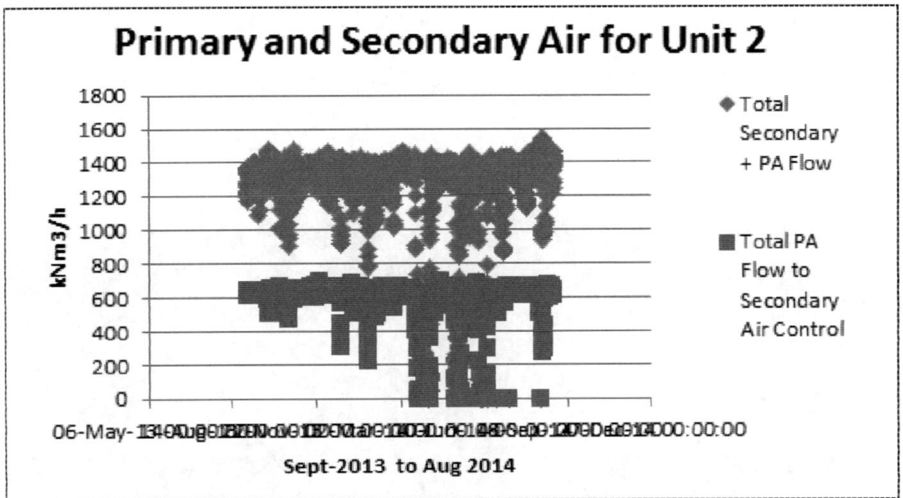

Figure 1. Primary and secondary air flow for Unit 2 – 700MW Boiler.

IOP Conf. Series: Materials Science and Engineering **88** (2015) 012021 doi:10.1088/1757-899X/88/1/012021

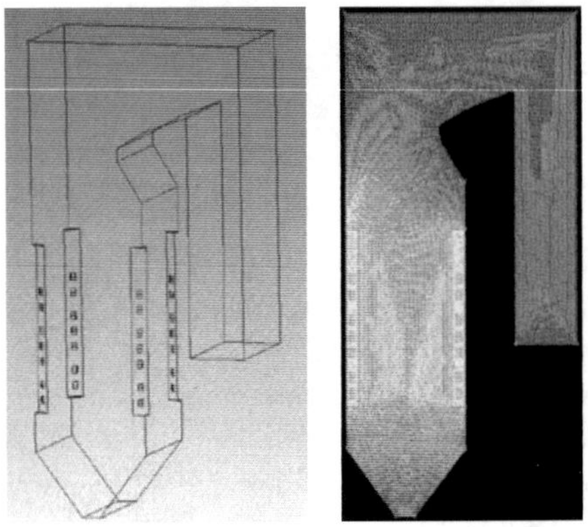

Figure 2 : 3-Dimensional Model and Mesh generated

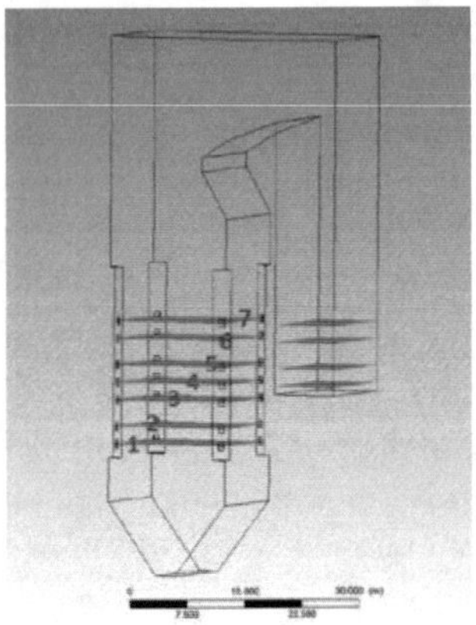

Figure 3 : 3-Dimensional Model with result planes defined

4. Result and Discussion

The results of the velocity profile from Plane 1 to Plane 6. The figure 4 shows the 0° results show the fire ball from plane 1 to 6. In comparison, figure 5 for -30° burner tilting angle shows the fire tracing is not as clear as the design condition. The fire ball is shifted below than the plane 1.

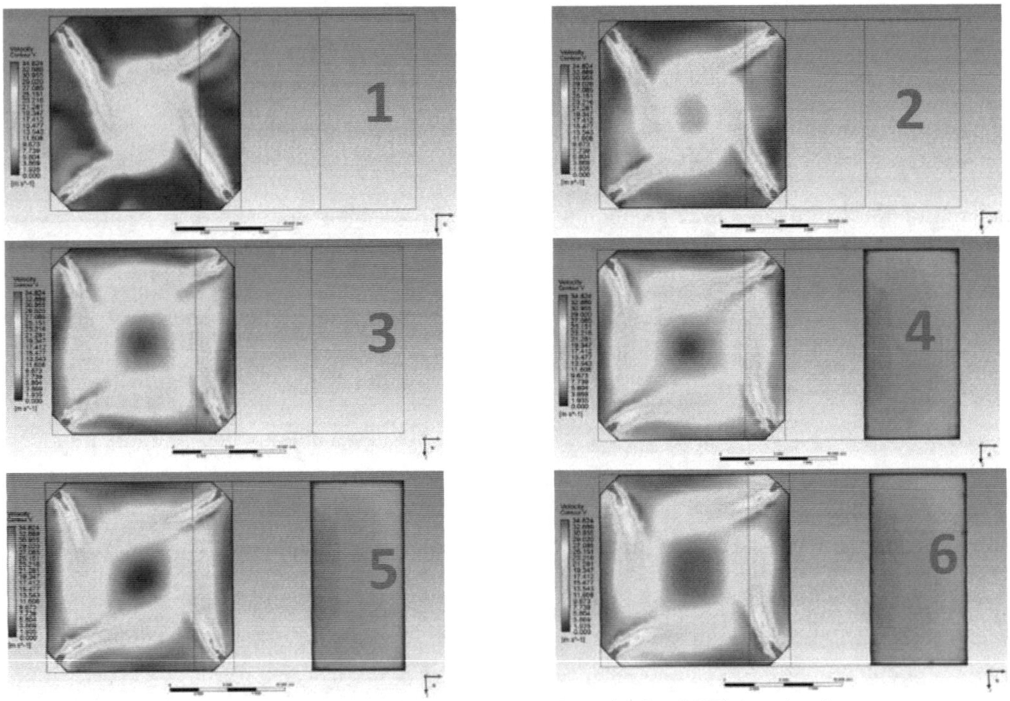

Figure 4 : Velocity Contour Results for 0° Firing Angle

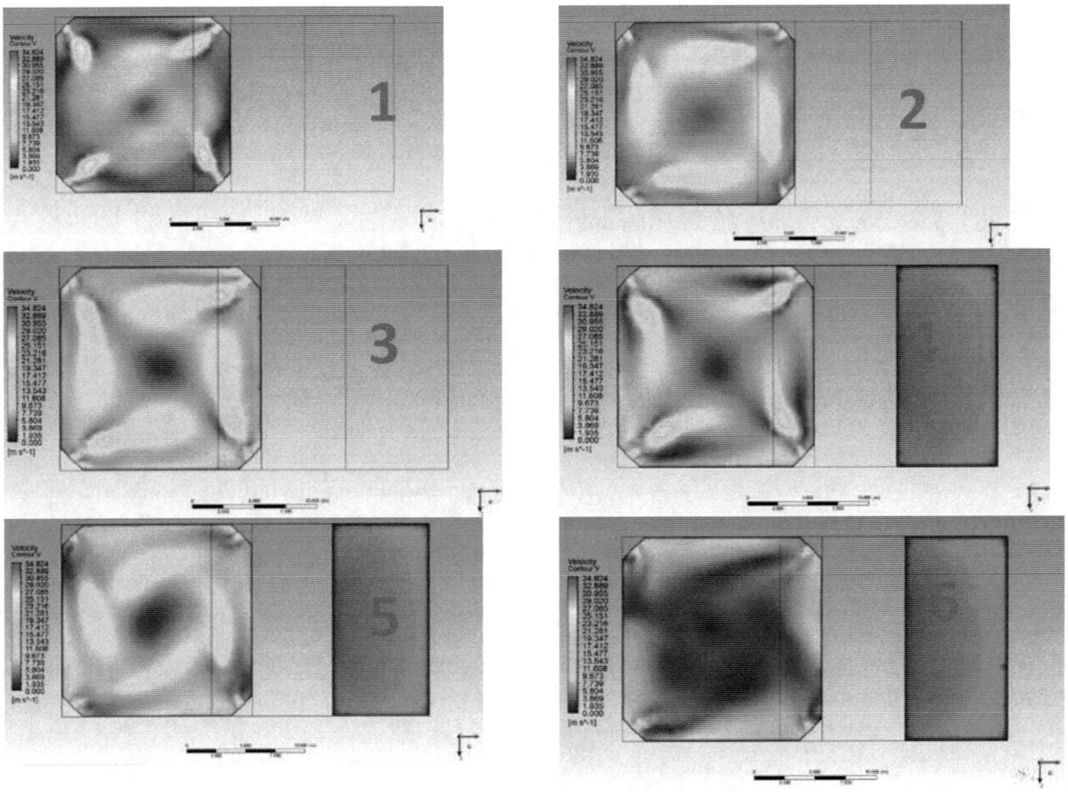

Figure 5 : Velocity Contour Results for -30° Firing Angle

5. Conclusions

It can be concluded that the -30° burner angle reduce the turbulence is fire ball mixing inside the furnace. It also shifted the fire ball position in the furnace to reduce the rear pass temperature. However, the efficiency of the furnace will be penalised by reducing the burner tilt angle.

References

[1] M. S. a. J. M.Manickam, "CFD modelling of waste heat recovery boiler," *Applied mathematical modelling,* vol. 22, p. 823 to 840, 1998.

[2] H. V. R. Vuthaluru, "Modelling of wall fired furnace for different operating conditions using FLUENT," *Fuek Processing Technology,* vol. 87, pp. 633-639, 2006.

[3] M. H. a. H. I. A. R.Ben-Mansour, "Influence of Inlet Conditions and Burner Tripping on NO and Temperature Characteristics in a Tangentially Fired Furnace," *International Journal of Environmental Application & Science,* vol. 4, no. 1, pp. 1-21, 2009.

[4] C. L. a. H.-C. Zhou, "Deduction of the two-dimensional distribution of temperature in a cross section of a boiler furnace from images of flame reduction," *Combustion and Flame,* vol. 143, pp. 97-45, 2005.

[5] E. H. a. Z. S.Mohammad, "Simulation of 120MW Industrial Boiler at Design Condition," *Journal of Advanced Science and Engineering Research,* vol. 3, pp. 286-293, 2013.

Review of Operational Challenges and Changing Conditions associated with Offshore Submerged Vertical Pumping Station

ML Samsudin[1], KM Munisamy[2] and S. K. Thangaraju[2]
[1]Malakoff Power Berhad, Kuala Lumpur, Malaysia,
mlutfi.samsudin@malakoff.com.my
[2]College of Engineering, Universiti Tenaga Nasional, Kajang, Malaysia,
Kannan@uniten.edu.my

Abstract:

Cooling water system is one of the essential auxiliary systems for a power generating plant. The decision to have an offshore pumping station, instead of onshore station, may have been driven by the scarcity in land area or for direct water supply access. However, the design has several disadvantages such as tendency for sediment intakes, limited condition monitoring activities and inaccessible sump for maintenance and modification.

The paper reviews and lists several potential operational challenges and difficulties experienced with the offshore station. The influences of changing conditions such as varying seawater tide level, clogged intake screen, intake sedimentation issues and marine fouling on the sump walls and structures, are discussed and mitigating solutions are identified. While it is impractical to conduct physical model test for studying flow pattern, the CFD method can be an attractive assessment tools. In addition to this, a review of plant operating parameters can assist in identifying flow anomalies to anticipate potential deterioration. Finally the difficulty in incorporating the solutions to vortex and sediment problems within the sump is highlighted with preference for those which can be managed by limited access or those which are moveable with the vertical pump.

1. Introduction to Cooling Water System for Power Plants

Cooling water system is one of the essential auxiliary systems for a power generating plant. One type of the employed design involves the supply of seawater to a once-through condenser system. For this, a number of pumps would draw water from a reservoir into supply piping system and later discharges back into the sea at downstream of the condenser.

The pumping station can be located onshore or offshore. At onshore, the pumps are located at the mainland with or without a suction piping stretching to the sea for water intake. Its main advantage is the ease of maintenance access. The pump sump can be isolated with stop gate and water drained off for pump overhaul as well as sump cleaning, inspection and rectification. On the other hand, the offshore station would have its structure a few stretches away from the mainland. The sump is bounded by either concrete structures or sheetpiles and submerged vertical pumps or pipes will draw the water needed for the cooling water supply. While the offshore station is normally selected due to limited land area or for direct access to water supply, it is with several disadvantages. These include

IOP Conf. Series: Materials Science and Engineering **88** (2015) 012023 doi:10.1088/1757-899X/88/1/012023

tendency for sediment intakes, limited condition monitoring activities and inaccessible sump for maintenance and modification.

2. Information on Referred Plant

The referred plant, Perai Power Plant uses an offshore pumping station for its supply of cooling water to its once-through seawater cooled condenser. Located 500m from the mainland, the wet well pumping station is a rehabilitated pump house structures for an old plant built in the 1970s. The sump is bounded by sheetpile in a double-trapezium shape with three openings for pump intakes with dedicated 20cm meshed barscreens. The 3 x 35% capacity pumps are of vertical, submerged mixed flow type, each supplying 11,700 m^3/hr of seawater to the common discharge header pipe. Refer to Figure 1 for the general assembly of the pumping station in the referred plant. It employs sodium hypochloride dosing at the sheetpile openings for marine fouling prevention. The full plant capacity is achieved with all three pumps placed in service.

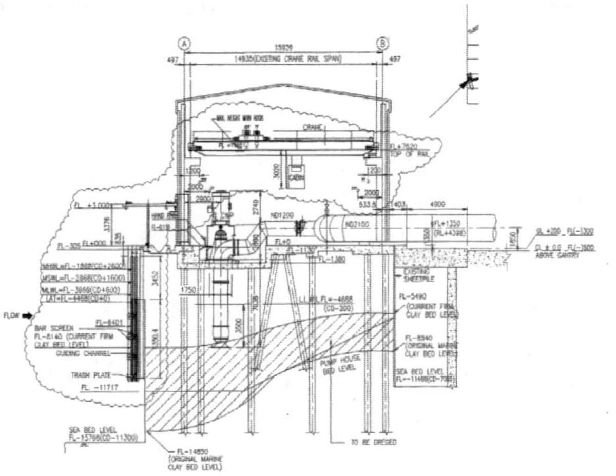

Figure 1: General assembly of the pumping station in the referred plant

This paper reviews and lists several potential operational challenges and difficulties experienced with the offshore vertical submerged pumping station, taking the referred plant as its case study. Emphasis was made to the changing conditions experienced during the operating stage which significantly deviate from the system original design. The issues faced would also apply, fully or partly, to other designs of the offshore pumping station.

3. Changing Conditions

The pumping station is normally designed with a set of parameters that would define its acceptable operating conditions. ANSI/HI 9.8 1998 listed recommended dimensions which include the minimum submergence level, pump sump length and clearances to the floor, side and rear walls [1]. For example, to avoid surface vortex occurrence, the pumping station would be designed to be higher than the critical submergence level, given by,

$$S_{Cr} = D\,(1.0 + 2.3\,F_D) \quad (1)$$

Where D is the pump intake bell outside diameter and F_D is the Froude number (dimensionless) given by;

$$F_D = \frac{V}{\sqrt{gD}} \tag{2}$$

With V as the velocity at suction inlet and g as the gravitational acceleration.

The design parameters are normally good within a few years of operations, but the subsequent years would see these parameters deviate significant to cause potential problems to the pumping system.

For the offshore pumping station such as the referred plant, several changing conditions have been observed which affect its operations.

3.1. Varying seawater tide level.

Normally the pumping station is designed with sufficient margin for the seawater tide level which can have direct impact to the pump sump water level. Low seawater tide level can influence the sump water level to reach the critical submergence. Surface vortices can start to form, leading to the development of an air core that allows continuous air supply to the pump intake. Surface vortices of ascending strength develop with decreasing sump water level and the minimum water level leading to surface vortex formation is higher with high flow velocity compared with that of low flow velocity [2]. As shown by [3], low submergence level and high flow rate can also lead to subsurface vortex occurrence accompanying the air core vortex. With the vortices, cavitation and imbalance can occur at the pump impeller leading to vibration, noise and damage to some key pump components.

For a normal pumping station, such as the referred plant, the lowest astronomical tide (LAT) level is accounted for during its design, providing sufficient margin for its operations. For the plant, the calculated Froude number is 0.41 and critical submergence at 3.10m. The pump has a minimum submergence of 3.50m against LAT level. Refer to Figure 1.

3.2. Operations with clogged intake screen

Intake screens are installed in front of the pump sump for 2 purposes which are to prevent debris from coming into the sump area and to correct the approach flow. The screens can be either passive type (e.g. meshed barscreen), or active type (e.g. screens with automatic trashraking system). As per ANSI/HI 9.8, they are recommended to be installed at a minimum distance of 6 times the pump bell diameters, 6D. Clogged screen can create severely skewed flow patterns, high velocity jets and severe instability near the pumps [1]. It also has tendency to form eddies which can result in vortex formation with non-uniform approach flow. Localised high velocity jets also have tendency to pull more debris through the screens than it can tolerate [4]. Especially closer to the water surface, small surface vortices can start accompanying the local high velocity spots but these do have significant impact to the pump downstream of the screen provided the recommended distance is adhered. Frequent inspection and cleaning of the screens are recommended to correct the approach flow.

The passive screens at the referred plant are normally clogged with marine organisms (such as jelly fish, seaweed and barnacle) and human pollutants (such as plastic bags and disposed waste). Sample photo of the clogged meshed barscreen from the referred plant is as per Figure 2 of which almost 90% of the flow areas can be clogged up.

Figure 2: Condition of the meshed barscreen in the referred plant, looking from the pump sump (left) and from the sea (right)

Another problem faced with badly clogged screen is the approach flow restriction that can lead to lower pump sump level relative to the seawater tide level. This occurs when the water flow rate through the meshed screen is less than the water being pumped out from the sump. This can lead to surface vortex formation of type 4 which is able to pull floating debris into the cooling water supply piping.

With critical submergence depth, surface vortices of type 6 can be formed with an air core connecting from the free surface to the pump bellmouth with potential to cause cavitation, rotor imbalance and discharge performance degradation. Figure 3 shows the difference in pump sump water level in the referred plant in comparison to the high and low seawater tide levels with the intake screen badly clogged.

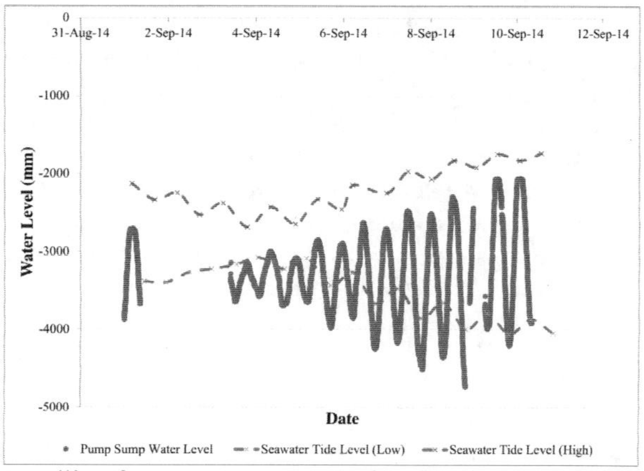

Figure 3: Comparative trendline for pump sump water level and seawater high and low tide levels. (Note: Intake screen was clogged and cleaned on 9 Sept 2014)

As preventive measure, a monitoring of the sump pump level is done on continuous basis. The measurement is compared to the seawater tide level to assess the required cleaning of the meshed screen.

An evaluation of biofouling was conducted by submerging two sets of pipe specimen in the seawater at the pumphouse in the referred plant. The growth of the marine organisms is monitored on selected days with the first set of specimen. The other set of specimen is meant for control purpose which is lifted up for observation at the end of the monitoring period. Refer to Figure 4. The observation provides insight on the local fouling rate which is very useful to plan for the cleaning frequency of the intake meshed screen.

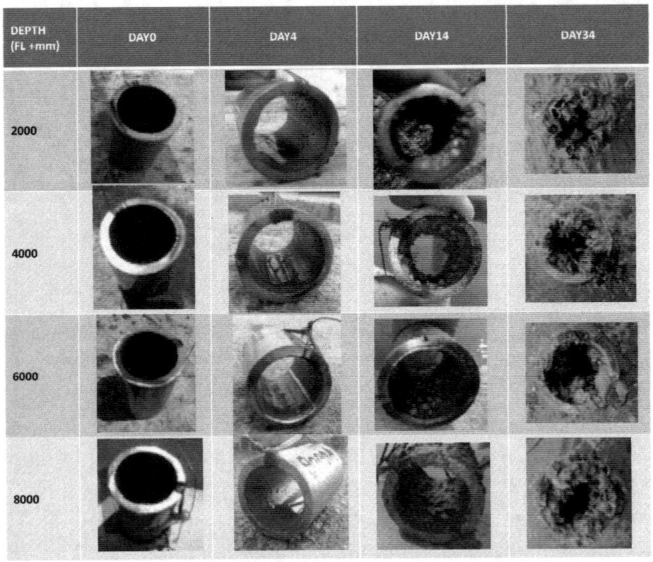

Figure 4: A review of marine fouling growth at the pumphouse in the referred plant.

For an onshore pumping station, the installation and operation of the active screens is comparatively better than the offshore station. Stop gate can be installed and crane facility can be made available to allow access for periodic thorough cleaning and maintenance for the screens. For the offshore pumping station, this can be quite challenging and expensive, thus it would not be feasible to do this on frequent basis.

3.3. Intake sedimentation issues

2 types of sedimentation are normally faced for pumping stations, typically referred to as coarse and fine sediments, distinguished by the particulate sizes. For an onshore pumping station, a settling basin can be installed prior to the pump sump to separate the sediments from the water. For an offshore station, to have this would require a relatively larger pumping station and thereby higher capital investment.

Coarse sediment is associated with particulates of size normally larger than medium sand (0.25mm – 0.125mm) [5]. Carryover of coarse sediment can lead to increased sedimentation onto the sump floor. Increased floor level reduces the sump water retention volume which can lead to fast drop in sump water level when the intake screen is badly clogged.

Over a long period, the sedimentation can also be compacted under hydrostatic pressure to produce new permanent floor profile, especially where the flow velocity is low or water is stagnant within the confined pump sump area. This new profile deviates from the earlier intended design for the sump and influences the intake flow field.

As per [6,7], the pump sump design with small clearance between the intake bellmouth and floor is desired to obtain adequate submergence. However, with very small floor clearance, the flow velocity will increase at the pump intake. Based on a study with single pump intake [3], the critical submergences for both air cored surface and sub-surface vortices are proportional to the flow velocity in the sump. As such, both vortices will occur and it would be difficult to apply anti-vortex device within the small clearance.

The new floor profile created by coarse sediment can reduce this clearance over years of operations. Reduced cross-sectional areas with converging profile can be formed, creating high velocity flow and low pressure region under the pump bellmouth. The presence of this low pressure region promotes the formation of sub-surface vortices and cavitation [2]. In addition to this, reduced floor clearance can pull in the coarse particles which can lead to impact and erosion damage of pump and piping components.

On the other hand, fine sediment is associated with fine sand normally with size 0.25mm – 0.062mm and silt of size 0.062mm – 0.004mm [5]. Fine sediment can flow with the pump water intakes and with high particle concentration, it can cause problem with the pump internal flushing and lubricating water system.

A study involving sandy clay composing of very thin particles and 15% fine sand showed that intake of high fine sediment concentration at high flow rate results in higher pump NPSH curve for silt mixture compared to that for water alone. This, being sensitive to viscous flows rather than gravitational forces, accelerates the cavitation occurrence for centrifugal pump [8].

The sediment also adds up to the density of the intake fluid mixture, thereby reducing the pump water discharge capacity and would cause problems for the equipment located downstream such as the heat exchangers and electrolyser plating at the electrochlorination plant. When any one or all pumps are not in service for extended period, the fine sediment settles down to the sump floor and leads to the similar floor issues associated with the coarse sediment.

In the referred plant, the sedimentation level and floor profile are measured and monitored on yearly basis. Diver is sent underwater to take the depth of the top layer of the sedimentation level in the pump sump. Figure 5 and 6 show the measured profile of the floor in the referred plant.

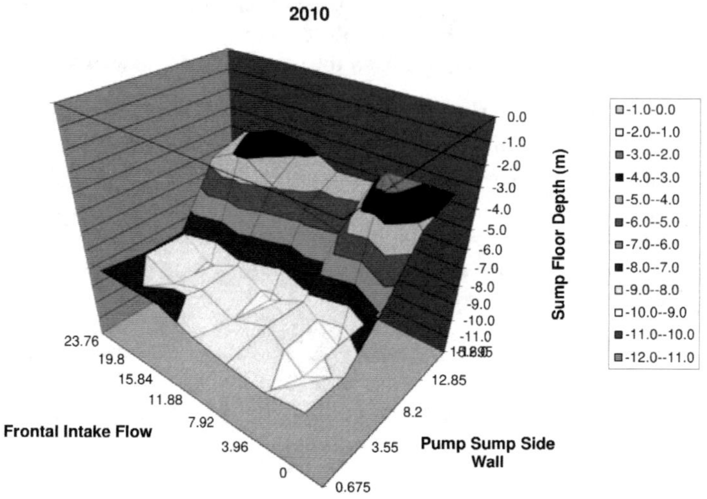

Figure 5: Sample floor profile as measured in the referred plant.

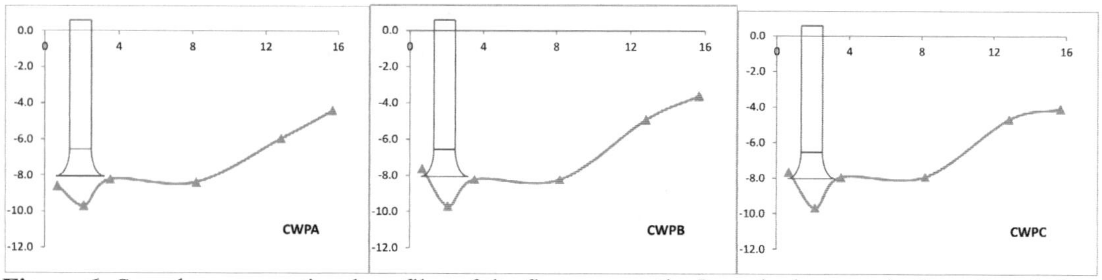

Figure 6: Sample cross sectional profiles of the floor across the 3 vertical pumps in the referred plant

A key disadvantage of an offshore pumping station when compared to an onshore station is its limited accessibility to enable inspection and cleaning of the sediments. Dredging is scheduled periodically to restore the floor profile. Since the work is very expensive, it is prudent for the sub-layers of the floor to be assessed periodically to determine the extent of the sedimentation and the best dredging method to restore the condition. Silts and soft clay layer can easily be removed with suction dredger. More extensive and expensive dredging methods, such as grab, trailer or hopper type dredger, would need to be employed for the removal of thick layers of hard and coarse sand and grades, though not all method can obtain access into the pump sump for the dredging work.

3.4. Marine fouling on pump sump walls and structures

The pumping station would have been designed in the beginning without wall roughness taken into consideration. Model test conducted for pumping station tends to be with clean and smooth walls. Over years of operations, the walls can develop roughness that would affect the flow boundary layers which can influence the flow field and the formation of vortices. For the onshore station, these can be cleaned and cleared during shutdown and overhaul period but it would not be possible for the offshore station.

For the referred plant, the pump sump is surrounded and bounded by sheetpile structures. Within the sump, there are concrete columns including those located between the pumps. Both sheetpile and columns are normally fully covered with marine growth such as barnacle and seaweed.

IOP Conf. Series: Materials Science and Engineering **88** (2015) 012023 doi:10.1088/1757-899X/88/1/012023

The effect of wall roughness on flow field inside a single pump intake was investigated through a simulation with the high Reynolds number and a $k-\omega$ model. It was found that wall roughness can reduce the intensity of both surface and sub-surface vortices. The study suggested for potential use of an artificial roughness as vortex suppressor on pump intake [9]. This would work provided the wall roughness can managed and sustained within the stipulated tolerances.

4. Model Test

ANSI/HI 9.8 1998 recommends physical hydraulic model to be done for pump intakes with specific design and operating conditions to ensure the final design produce favourable flow conditions for the pump [1]. Sump design alteration and anti-vortex devices can be incorporated and assessed in addressing any notable vortices and unwanted flow profiles. Physical model tests are mostly done during design stage of the pumping station but may not be suitable for in-service system due to changing conditions and unobtainable information such as on the marine fouling, floor profile and reduced clearance, carryover of sediments and extent of flow restriction by the intake screens which are difficult and complex to evaluate and determine. To model them would be expensive, resource- and time-consuming while the result may not be accurately applied throughout the pump operational period.

Computational fluid dynamic (CFD) methodology is an attractive alternative to study and assess the pumping station with changing conditions. The application of this technique for the study of flow profiles in pump intakes is not new and had found good agreement with the experimental methods [2,3,10,1112,13] and found good agreement between the two methods. Various topics pertaining to CFD techniques had been studies, such as two-phase modelling [7,14,15,16] and turbulence modelling [17,18,19].

5. Condition Based Assessment

It would be of high value to be able to assess the apparent symptoms of the sump flow problems such as vortex formation and sedimentation issues. There are numerous condition based assessment techniques available, such as vibration analysis, oil analysis, thermography inspection, and motor current signal analysis, to detect pump mechanical. The use of ultrasonic methods which were time of flight, Doppler and imaging techniques, had been studied to detect vortex formation in open volumes and closed ducts and were found successful [20]. However, most of these techniques can only identify the problem when the condition has already worsened.

A review the presence of flow anomalies is attempted with the available operating data such as the sump water level and pump common discharge pressure in the referred plant. Starting with a generic pumping system equation,

$$\dot{Q}_E - \dot{W} = \Delta E_n + \Delta \dot{KE} + \Delta \dot{PE} \tag{3}$$

Which is expanded into;

$$\Delta H_P = \left(\Delta H_{F1-2} + \Delta H_{EQ1-2}\right) + \frac{1}{2g}\left({v_2}^2 - {v_1}^2\right) + \left(z_2 + H_2\right) - \left(z_1 + H_1\right) \tag{4}$$

v_x, z_x, H_x are local velocity, height and pressure respectively at any given x position. Point 1 is the position on the free water surface in the pump sump where its level is measured from a reference level while point 2 is the position at the common pump discharge piping where pressure is measured. ΔH_P

IOP Conf. Series: Materials Science and Engineering **88** (2015) 012023 doi:10.1088/1757-899X/88/1/012023

is the pressure increase by the pump while $(\Delta H_{F1-2} + \Delta H_{EQ1-2})$ is the total losses within the pumping system [21].

The data plots for the relationship between the total system losses and the measured sump water level (z_1) and pump common discharge head (H_2) for the referred plant are given in Figure 7 and 8 for all the 3 pumps in service. The data includes the event of low sump water level as refered in Figure 3 above. Theoretically, energy would be loss in generating vortex flow, thus, the total system losses is expected to be register increasing values at low sump water level which would produce the trend that deviate slightly from the normal straight line. As per the plots, the total loss is inversely directly proportional to both parameters and does not indicate any apparent abnormal losses.

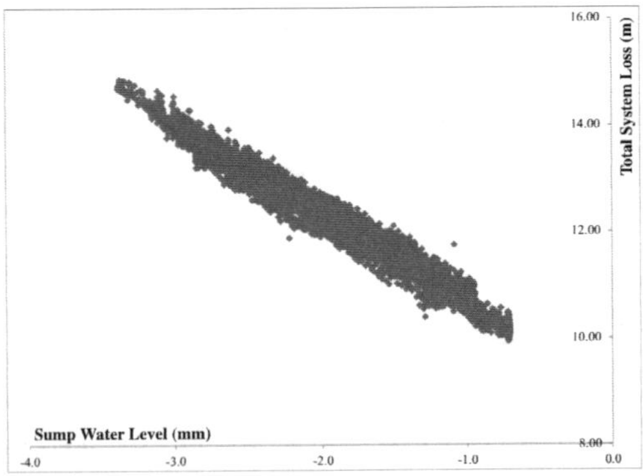

Figure 7: Relationship between calculated total system loss and measured sump water level (z_1) for the referred plant

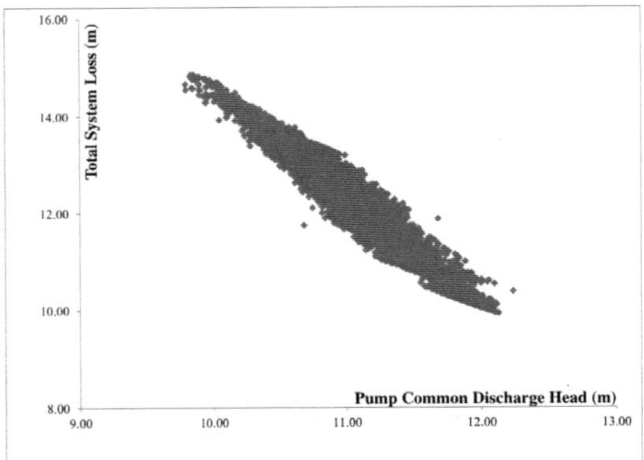

Figure 8: Relationship between calculated total system loss and measured pump common discharge head (H_2) for the referred plant

It can be of valuable future research to further refine and look at the detection methods of flow and other sump problems by analysing the apparent symptoms and reviewing the operational parameters.

6. Prevention of Vortex and Sediment Problems within the Sump

Vortex and sediment problems can be eliminated either by adhering to several published design guidelines for the proper design of the sump, or with the installation of anti-vortex and sediment control devices [1,5,6,22,23]. Solutions for vortex prevention vary based on the sump design and operating conditions. Studies using both the test models and CFD models had been done to check the effectiveness of the various employed solutions such as floor splitter [2], conical device with eight fins [24], sidewall and backwall fillets, centre floor splitters and curtain wall [25] and rectangular bar with trident shaped fixture [26].

Solutions for the vortex and sediment problems are best studied and taken during the design stage and when they occur during the operational stage, the solutions can be very difficult to be incorporated. Most of the design fixes from the guidelines, especially those for eliminating subsurface vortex, would require the access to the major parts of the pump sump and would not be economically practical for the offshore station.

For the offshore station, the preferred solutions are with features that can be managed with limited access or moveable with the vertical pump during its routine overhaul. A study on 3 vortex elimination methods involving bellmouth circle ring, backwall clapboard and approaching vertical clapboard was made using CFD. While the first 2 methods were found beneficial to correct the flow field, the vertical clapboard was not advised due to its contributing to flow pattern deterioration [27]. This is a potential future research area that would be of beneficial to the offshore station.

7. Conclusion

The paper reviews and lists several potential operational challenges and difficulties experienced with the offshore pumping station. The influences of changing conditions such as varying seawater tide level, clogged intake screen, intake sedimentation issues and marine fouling on the sump walls and structures, are discussed and mitigating solutions are identified. While it is impractical to conduct physical model test for studying flow pattern, the CFD method can be an attractive assessment tools. In addition to this, a review of plant operating parameters can assist in identifying flow anomalies to anticipate potential deterioration. Finally the difficulty in incorporating the solutions to vortex and sediment problems within the sump is highlighted with preference for those which can be managed by limited access or those which are moveable with the vertical pump.

8. References

[1] ANSI/HI 9.8 1998 *American National Standard for Pump Intake Design* American National Standards Institute Inc.

[2] Kang, W T, Yu K H and Shin B R 2011 An Investigation of Surface Vortices Behaviour in Pump Sump *11th Asian Int Conf on Fluid Machinery and 3rd Fluid Power Tech Exh (IIT Madras, Chennai, India) Nov 21-23, 2011*

[3] Okamura T, Kamemoto K and Matsui J 2007 CFD Prediction and Model Experiment on Suction Vortices in Pump Sump *9th Asian Int Conf on Fluid Machinery (Jeju, Korea) Oct 16-19, 2007*

[4] Karassik I J, Messina J P, Cooper P and Heald C C 2001 *Pump Handbook 3rd Edition* (McGraw-Hill)

[5] Zheng Y, Alsaffar M A 2004 Water Intake Sediment Problems in Thermal Power Plants *Joint Conf on Water Resources Engineering and Water Resources Planning and Management (Minnesota, USA) Jul 30 – Aug 2, 2000*

[6] Ettema R 1998 *Flow in Water-Intake Pump Bays: A Guide for Utility Engineers* EPRI, Palo Alto, CA Electric Power Research Institute TR-110948

[7] Bayeul-Laine A C, Simonet S, Bois G and Issa A 2012 Two-Phase Numerical Study of the Flow Field Formed in Water Pump Sump: Influence of Air Entrainment *26th IAHR Symp on Hydraulic Machinery and Systems; IOP Conf Series: Earth and Env Science* **15** *(2012) 022007*

[8] Ladouni A and Nemdili A 2005 Experimental Investigation of the Effect of Solid-Mixture on the Cavitation Characeristic of a Centrifugal Pump *Fluid Structure Interaction and Moving Boundary Problems, WIT Transactions on the Built Environment* **84** pp 81-87

[9] Constantinescu G S and Patel V C 2000 Role of Turbulence Model in Prediction of Pump-Bay Vortices *Journal of Hydraulic Engineering* **126** 387-391

[10] Li S, Silva J M, Lai Y, Weber L J and Patel V C 2006 Three-Dimensional Simulation of Flows in Practical Water-Pump Intakes *Journal of Hydroinformatics* **8** 111-124

[11] Mohd Remy Rozainy M A Z, Abutan I, Abdullah M Z, Mohd Ashraf M I 2008 Application of Computational Fluid Dynamics (CFD) in Physical Model of Pump Sump to Predict the Flow Characteristics *Int Conf on Construction and Bldg Tech (ICCBT-2008) (Kuala Lumpur, Malaysia) Jun 16-20, 2008 pp 79-90*

[12] Lucino C, Liscia S and Duro G 2010 Vortex Detection in Pump Sumps by Means of CFD *XXIV Latin American Congress on Hydraulics, Punta Del Este, Uruguay, Nov 2010*

[13] Desmukh T S and Gahlot V K 2010 Simulation of Flow Through a Pump Sump and Its Validation *Int J of Research and Reviews in App Sciences* ***4(1)*** 7-17

[14] Shukla N S, Kshirsagar J T 2008 Numerical Prediction of Air Entrainment in Pump Intakes *Proc of the 24th Int Pump Users Symp (Texas A&M University, USA)* pp 29-33

[15] Park Y K, Li K M, Chi Y H and Lee Y W 2011 A Study on the Generation of Vortices around a Sump Intake with Multiphase Flow *11th Asian Int Conf on Fluid Machinery and the 3rd Fluid Power Tech Exh (IIT Madras, Chennai, India) Nov 21-23, 2011*

[16] Kurger S, Bouziad Y A and Maurer W 2009 Pump Sump CFD for Vertical Pump Design *Proc ASME 2009 Fluids Eng Div Summer Mtg FEDSM2009 (Vail, Colorado, USA) Aug 2-6, 2009*

[17] Chuang W L, Hsiao S C and Hwang K S 2014 Numerical and Experimental Study of Pump Sump Flows *Math Problems in Eng* **2014** 1-20

[18] Skerlavaj A, Skerget L, Ravnik J and Lipej A 2011 Choice of a Turbulence Model for Pump Intakes *Journal of Power and Energy* **225(6)** 764-778.

[19] Skerlavaj A, Skerget L, Ravnik J and Lipej A 2014 Predicting Free-Surface Vortices with Single-Phase Simulations *Eng App of Comp Fluid Mech* **8(2)** pp 193-210

[20] Oddie G M 1996 A Note on the Use of Ultrasonic Techniques for the Detection of Vorticity and Gas-cored Vortices Upstream of the Inlets of Pumps *Ultrasonics* **34** pp 495-497

[21] Chaurette J 2003 *Pump System Analysis and Sizing 5th Edition* (Fluide Design Inc)

[22] Ettema R 1998 *Flow in Water-Intake Pump Bays: A Guide for Utility* Engineers EPRI, Palo Alto CA Electric Power Research Institute TR-110948

[23] Ettema R and Nakato T 1998 *Bed-Sediment Control at Riverside Water Intakes: A Guide for Utility Engineers* EPRI, Palo Alto CA Electric Power Research Institute TR-110551

[24] Ahmad Z, Jain B, Kumar S and Mittal M K 2011 Rational Design of a Pump-Sump and its Model Testing *J Pipeline Syst Eng Prac* **2** 53-63

[25] Kim C, Choi Y D, Choi J W and Lee Y H 2012 A Study on the Effectiveness of an Anti Vortex Device in the Sump Model by Experiment and CFD *26th IAHR Symp on Hydraulic Machinery and Systems,* IOP Conf. Ser.: Earth Environ. Sci. **15** (2012) 072004

[26] Shabayek S A 2010 Improving Approach Flow Hydraulics at Pump Intakes *Int J of Civ & Env Eng* ***10(6)*** pp 23-31

[27] Cheng L and Liu C 2012 Research on Vortex Eliminating of the Sump of Cooling Pump by CFD *Proc of the 2012 20th Int Confon Nuclear Eng collocated with ASME 2012 Power Conf (ICONE20-POWER2012) (Anaheim, California, USA) Jul 30 – Aug 3, 2012*

IOP Conf. Series: Materials Science and Engineering **88** (2015) 012024 doi:10.1088/1757-899X/88/1/012024

Application of multiphase modelling for vortex occurrence in vertical pump intake – a review

M L Samsudin[1], KM Munisamy[2] and S. K. Thangaraju[2]
[1]Malakoff Power Berhad, Kuala Lumpur, Malaysia, mlutfi.samsudin@malakoff.com.my
[2]College of Engineering, Universiti Tenaga Nasional, Kajang, Malaysia, Kannan@uniten.edu.my

Abstract. Vortex formation within pump intake is one of common problems faced for power plant cooling water system. This phenomenon, categorised as surface and sub-surface vortices, can lead to several operational problems and increased maintenance costs. Physical model study was recommended from published guidelines but proved to be time and resource consuming. Hence, the use of Computational Fluid Dynamics (CFD) is an attractive alternative in managing the problem. At the early stage, flow analysis was conducted using single phase simulation and found to find good agreement with the observation from physical model study. With the development of computers, multiphase simulation found further enhancement in obtaining accurate results for representing air entrainment and sub-surface vortices which were earlier not well predicted from the single phase simulation.

The purpose of this paper is to describe the application of multiphase modelling with CFD analysis for investigating vortex formation for a vertically inverted pump intake. In applying multiphase modelling, there ought to be a balance between the acceptable usage for computational time and resources and the degree of accuracy and realism in the results as expected from the analysis.

1. Introduction on Pump Intake Vortex Problem

In many power plants, the cooling water pump system is one of the essential auxiliaries to ensure uninterruptible power dispatch. As reported by [1], two fundamental types of problems for the water intakes are sedimentation and swirling flow problems, occurring at the entrance to or within the pump intake. Sedimentation problems can lead to pump operational difficulties due to it being either conveyed with the diverted flow or deposited in the intake and interferes with the flow. Meanwhile, swirling flow problems, which are commonly related to vortex formation, can lead to swirl and air entrainment at the pump suction bell.

The problem of vortices had been classified into (a) free surface vortices and (b) sub-surface vortices. Surface vortices originate from the free water surface and can have the strength and appearance varying from a slight dimple to a strong circulation or an open funnel or core that would connect the free surface to the pump bellmouth. Floating debris or air can be sucked into the pump column by this strong vortex.

On the other hand, sub-surface vortices are further classified into the floor-, backwall- and sidewall-attached vortices, each referring to the connectivity of the vortices at its one end to the respective boundary walls. Strong subsurface vortex has a relatively strong core and with low pressure, it can lead to the separation of dissolved air and water to form bubbles within the pump intake flow.

Several difficulties resulted from swirl and air entrainment as highlighted by [2], among others, include operational problems by the intakes of debris, reduced flow rate and head, reduced efficiency, increased vibration and noise and damages incurred on pump critical components. All of these can lead to increased maintenance costs and frequent pump downtime.

The remedies for eliminating these vortices are with proper design of the pump intake and by having adequate submergence. There are several published guidelines such as ANSI/HI 9.8 (1998) [3] and BHRA (1977) [4], to name a few, which provide recommendations for general pump intake design in reducing the vortex formation. ANSI/HI 9.8 also recommends for physical hydraulic model study to be done for pump intakes whose design deviates from the guidelines, having non-uniform or non-symmetrical approach flow, a single pump with flow greater than 2,520 litres per second or a multi-pump station with all pumps running with flow greater than 6,310 litres per second, or where the cost of a model study is 10 times less than the cumulative impact of pump inadequate performance or failure [3].

Model study is restrictive since it is very expensive, site specific, time- and resource-consuming and requires experience from the modellers. Hence, an attractive emerging technique for studying vortex formation is through CFD analysis which is relatively cheaper, quicker and readily available with the commercial codes such as CFX, FLUENT, STAR-CCM+, etc.

The purpose of this paper is to describe the application of multiphase modelling with CFD analysis for investigating vortex formation for a vertically inverted pump intake.

2. CFD Analysis and Multiphase Modelling

The governing equations applicable for the analysis of flow field of pump intake are the continuity equation and the Navier-Stokes equation of motion.

Continuity equation;

$$\frac{\partial \rho}{\partial t} + \frac{\partial \rho u_i}{\partial x_i} = 0 \tag{1}$$

Navier-Stokes equation of motion;

$$\frac{\partial(\rho u_i)}{\partial t} + \frac{\partial(u_i u_j)}{\partial x_j} = \rho f_i - \frac{\partial p}{\partial x_i} + \frac{\partial}{\partial x_j}\left[(\mu + \mu_t)(\frac{\partial u_i}{\partial x_j} + \frac{\partial u_j}{\partial x_i})\right] \tag{2}$$

To match numerical simulation with available computing resources and time, a number of assumptions and simplifications would have to be considered. One of the elements in the CFD analysis is for the fluid flow to be of either single or multiple phases, i.e. liquid water only or liquid water with gas/air/solids.

The differences between multicomponent and multiphase were distinguished. Multicomponent is when the substances are mixed at a microscopic or molecular level and would have a common field of velocity, temperature, etc, for all components in the mixture. Meanwhile, multiphase is when the substances are mixed at macroscopic level and it can be either homogeneous or inhomogeneous. Homogeneous is when common fields of variables would apply for all fractions of the mixture while inhomogeneous is when only common pressure field would apply for all fractions [5].

When a vortex is weak, the air content in water is low and its presence does not exert any significant influence to the flow field. As the vortex's strength increases, the increased air content would lead to a decrease in the pump head, capacity and power consumption together with the change of the air-water mixture density. For simulating such flow, one can apply (a) the Lagrangian approach (or Particle Transport Model), (b) the Eulerian Approach with Particle Model, or (c) the Algebraic Slip Model [5].

On the other hand, when a vortex is strong, the air content in water is high and air bubbles will merge together and form an air core that would connect from the free surface to the pump bellmouth. The air core substantially changes the flow pattern within the intake and exerts influence on the pump performance parameters. Both air and water phases are treated as continuous medium and in simulating this, one can apply (a) the free surface flows model, either homogeneous or inhomogeneous, or (b) the mixture model [5].

The study of vortex formation using multiphase modelling had been done with the homogenous model with clearly defined interface between the phases, by applying the Volume of Fluids (VOF) method. The technique is applied to a fixed Eulerian mesh with objective to track the movement of the interface of two (or more) immiscible fluids. The fluids share the same set of the continuity, momentum and energy equations and the volume fraction of each fluid in each computational cell is tracked over the domain [6].

3. Multiphase Modelling of Single Pump Intake

The modelling of single pump intake for single and two-phase analysis was done almost similarly. For a single phase modelling, 3 blocks of computational domains were constructed consisting of the approach channel, pump lower area and pump outlet/upper area. For a complete two-phase (air and water) modelling, 3 blocks of computational domains were constructed consisting of the free surface upper area (fully air), free surface lower area (fully water) and the intake pipe (fully water). Refer to Figure 1 and 2 for the single pump intake model.

Figure 1: Single pump intake model for single phase modelling.

Figure 2: Single pump intake model for two phase modelling.

From [7], the boundary conditions was taken as the total hydraulic head equalled to the assumed water elevation and the submergence level at the inlet while a constant volume flow rate inside the intake pipe at the outlet. Water is considered as monophasic and incompressible with physical models activated by gravity and turbulence. A third order approximation was used to calculate the momentum advection, while choosing the Split Lagrangian method for the advection of the VOF method.

There are generally 2 approaches used by past researchers to model the air entrainment from the surface vortex occurrence which is the conventional fixed water surface method (C-FWSM) and multiphase free water surface method (M-FWSM).C-FWSM method is economical with lower computational resources, time and cost, but it is only effective with flat and rigid water surface and the formation of vortices tends to be undermined. M-FWSM, on the other hand, is resource- and time-consuming but it is more flexible and accurate. Although C-FWSM needs twice the physical time compared to M-FWSM, to reach periodicity, it is about 5 to 7 times quicker with the same computational settings [8].

For C-FWSM, a free slip boundary is assumed at the water-air interface with particles sharing the same velocity, pressure and temperature field. As such, the same set of basic equations for momentum, energy and mass transport are used for the flow field. For M-FWSM, the simulation of the water surface disruption and distortion can be made without any assumptions on the nature of the fluid interface.

In a study to compare the two methods, the simulation using M-FWSM was able to predict both surface and sub-surface vortices as observed from the experiment whereas C-FWSM was found to be able to predict all except those originated from the back-wall [8].

Volume fraction is an important factor to cause surface vortex appear in the multiphase simulation. A study with steady single phase simulation showed that for a volume fraction of 0.5, i.e. the amount of water is the same as that of air, no vortex appeared. Using volume fraction of 0.95, the vortex becomes visible while increasing the volume fraction to 0.99, the vortex becomes strongly represented [9].

4. Steady and unsteady single phase flow simulation

Most of the earlier research applied steady single phase simulation to study pump intake vortex formation and its strength and position are comparable to the experimental results. As per [10, 11], in modelling the surface vortex formation, using this with k-ε turbulence model, the results were with good convergent, symmetric and stable that agreed well with the experimental results. Meanwhile when used with k-ω turbulence model, the results were intermittent, asymmetric and unstable and hardly converged.

Steady state simulation had been successfully used to check the effectiveness of anti-vortex device which saw no presence of vortices in the pump intake [8, 12, 13].

5. Steady and unsteady multiphase flow simulation

Steady two-phase flow simulation was applied in studying air entrainment at lowest water level which agreed well with the lab experimental study. As per [14], the simulation was also applied to confirm the effectiveness of the anti-vortex modification by raising the height of the curtain wall, of which the absence of vortex was confirmed.

As per [10, 11], as one of the employed calculation cases, the steady state condition was used as initial condition for unsteady flow simulation which gave better convergence results. The occurrence of air entrainment was found to increase with increasing mass flow rate. It was easier to be visualised by using volume of fluid of air in defined cut planes but the great difficulty was to locate these planes due to unsteady position of the air core. The strength of the predicted vortices was not equal and changes with time. It seemed to follow a specific frequency which is based on mean shear velocity and pipe diameter.

The basic flow characteristic of water within the pump intake is highly unstable, unsteady and intermittent, independent of the turbulence model used [10, 11]. In most cases, a fully unsteady state simulation is not always needed for analysis and troubleshooting as the steady state simulation can be sufficient in explaining the phenomenon for vortex formation with significantly lower computational time.

6. Multiphase Modelling of the Sub-surface Vortex

As per [15], VOF method can be applied to predict the incidence of sub-surface vortices if being used with appropriate phase-change model, e.g. cavitation.

A CFD study was made of a practical model and model built as per Turbomachinery Society of Japan (TSJ) Standard S002:2005, as per [9]. Applying single phase simulation for unsteady flow for sub-surface vortex formation of single pump intake showed that the floor-attached vortex was observed just until 7.5 seconds and then it was not represented with low vorticity. When two-phase simulation was applied, the development of the vortex always changes and not fixed, in agreement with the observation from the experimental study. It was recommended that the unsteady multiphase modelling is best applied to study the sub-surface vortices.

7. Multiphase Modelling of the Sediment

The use of multiphase, or perhaps multicomponent, modelling for pump intakes can further be extended to simulating other unexplored real-life factors influencing the vortex formation. As considered and assumed at design stage, the intake fluid flow is taken as single phase in a sense that it is free of air as well as solid particles. Although there are several design fixtures that can be taken in minimising course and fine sediment transport, as highlighted in [16], this phenomenon does occur following years of operations, especially for the case of offshore pumping stations with seawater intakes.

The sediments would have several influences to the flow field within the pump intake. Course sediment would settle to the bottom, accumulate and solidify under hydrostatic pressure over time to permanently change the floor profile. Excessively low floor clearance and converging floor profile can increase the mass flow rate towards the pump bellmouth, thereby inducing vortex formation. On the other hand, fine sediment would flow with the water intake and carried along, influencing the density and energy of the moving fluid.

As per [17], an experimental study was conducted to study the effect of silt concentration in pumping water, using samples of clay composed of very thin particles and containing approximately 15% of fine sand. It was found that the NPSH curves are not influenced by the concentration at low flow rate. At high flow rate, the curves of silt mixture appeared higher than those of water only. The cavitation accelerates with the silt concentration, being only sensitive to the viscous flow and not to the gravitational forces.

The use of Eulerian model is recommended for sedimentation analysis [6].

8. Conclusion

Multiphase modelling for flow field analysis within pump intake is becoming key consideration with CFD. It has the potential to produce accurate results in determining the formation of both surface and subsurface vortices, especially for the case of air core and entrainment phenomenon. There ought to be a balance between the high demand for computational time and resources and the degree of accuracy and realism in the results as expected from the analysis.

9. References:

[1] Melville B W, Etterna R and Nakato T 1994 *Review of Flow Problems at Water Intake*

Pump Sumps EPRI, Palo Alto CA Electric Power Research Institute TR-103474

[2] Knauss J 1987 *Swirling Flow Problems at Intakes (IAHR Hydraulic Structures Design Manual) (* New York: Taylor & Francis).

[3] ANSI/HI 9.8 1998 *American National Standard for Pump Intake Design* American National Standards Institute Inc.

[4] Prosser M 1977 *The Hydraulic Design of Pump Sumps and Intake* British Hydromechanics Research Association/Construction Industry Research and Information Association.

[5] Kochevsky A N 2005 Capabilities of Numerical Simulation of Multiphase Flows in Centrifugal Pumps using Modern CFD Software http://arxiv.org/abs/physics/0509193

[6] ANSYS FLUENT Theory Guide Rel 14.0 2011.

[7] Lucino C, Liscia S and Duro G 2010 Vortex Detection in Pump Sumps by Means of CFD *XXIV Latin American Congress on Hydraulics, Punta Del Este, Uruguay, Nov 2010*

[8] Kurger S, Bouziad Y A and Maurer W 2009 Pump Sump CFD for Vertical Pump Design *Proc of the ASME 2009 Fluids Eng Div Summer Mtg, FEDSM2009, Vail, Colorado, US Aug 2-6, 2009*

[9] Park Y K, Li K M, Chi Y H and Lee Y W 2011 A Study on the Generation of Vortices around a Sump Intake with Multiphase Flow *11th Asian Int Conf on Fluid Machinery and the 3rd Fluid Power Tech Exh, IIT Madras, Chennai, India Nov 21-23, 2011*

[10] Bayeul-Laine A C, Simonet S, Bois G and Issa A 2012 Two-Phase Numerical Study of the Flow Field Formed in Water Pump Sump: Influence of Air Entrainment *26th IAHR Symp on Hydraulic Machinery and Systems; IOP Conf Series: Earth and Env Science* **15** *(2012) 022007*

[11] Bayeul-Laine A C, Simonet S, Issa A and Bois G 2013 The Importance of the Knowledge of Flow Sream in Water Sump Pump *Congres Francais de Mechanique (21;2013; Bordeux), France, 2013-08-26*

[12] Kim C G, Choi Y D, Choi J W and Lee Y H 2012 A Study on the Effectiveness of an Anti Vortex Device in the Sump Model by Experiment and CFD *26th IAHR Symp on Hydraulic Machinery and Systems, IOP Conf Series: Earth and Env Sciences* **15** *(2012) 072004*

[13] Cheng L and Liu C 2012 Research on Vortex Eliminating of the Sump of Cooling Pump by CFD *Proc of the 2012 20th Int Confon Nuclear Eng collocated with ASME 2012 Power Conf (ICONE20-POWER2012) Anaheim, California, USA, Jul 30 – Aug 3, 2012*

[14] Shukla N S, Kshirsagar J T 2008 Numerical Prediction of Air Entrainment in Pump Intakes *Proc of the 24th Int Pump Users Symp (Texas A&M University, USA)* pp 29-33

[15] Zhao L J, Nohmi M 2012 Numerical Simulation of Free Water Surface in Pump Intake *26th IAHR Symp on Hydraulic Machinery and Systems, IOP Conf Series: Earth and Env Science* **15** *(2012) 052035*

[16] Zheng Y, Alsaffar M A 2004 Water Intake Sediment Problems in Thermal Power Plants *Joint Conf on Water Resources Engineering and Water Resources Planning and Management (Minnesota, USA) Jul 30 – Aug 2, 2000*

[17] Ladouni A and Nemdili A 2005 Experimental Investigation of the Effect of Solid-Mixture on the Cavitation Characeristic of a Centrifugal Pump *Fluid Structure Interaction and Moving Boundary Problems, WIT Transactions on the Built Environment* **84** pp 81-87

IOP Conf. Series: Materials Science and Engineering **88** (2015) 012025 doi:10.1088/1757-899X/88/1/012025

CFD Investigation on Long-Haul Passenger Bus

C F Tan[1], B T Tee[2], H C Law[3], T L Lim[4]
[1,2,3] Centre for Advanced Research on Energy (CARe)
Faculty of Mechanical Engineering, Universiti Teknikal Malaysia Melaka,
Hang Tuah Jaya, 76100 Durian Tunggal, Melaka, Malaysia

[4] TKH Manufacturing Sdn Bhd
Kota Damansara, Selangor, Malaysia

E-mail: cheefai@utem.edu.my

Abstract. Air flow distribution is one of the important factors that will influence the bus passenger comfort during long haul travel. Poor air flow distribution not only cause discomfort to the bus passenger but also influence their travel mode as well. The main purpose of this study is to investigate the air flow performance of the bus air-conditioning system through CFD simulation approach. A 3D CAD model of air ducts was drawn and hence analysed by using CFD software, namely ANSYS Fluent, to determine the airflow rate for every outlets of the air-conditioning system. The simulated result was then validated with experimental data obtained from prototype model of air duct. Based on the findings, new design concepts is proposed with the aim to meet the industry requirement as well as to improve the bus passenger comfort during long haul travel.

1. Introduction

Automotive air-conditioning (AAC) for thermal comfort in passenger cabins is now a thing of necessity rather than luxury and cooling is especially needed when travelling in summer or throughout the year in countries of hot and humid climate [1]. Application of automotive air-conditioning in the field of passenger transportation was necessary in order to ensure the comfort of passenger and driver while travelling long distance. Thus, control of airflow rate and temperature distribution of automotive air-conditioning system is essential to ensure optimal humidity and comfort inside a long-haul passenger bus. This study emerged based on a need to further investigate the current air flow distribution system in the long haul express bus that build by the Malaysian local coach manufacturer. The express bus is facing the issue of uneven airflow rate when the air is delivered from the air-conditioning system. Therefore, a Computational Fluid Dynamics (CFD) study is necessary in order to investigate the air flow pattern and air flow distribution of the express bus air conditioning ducting, which will help to identify the locations where air distribution is inadequate.

---

[1] To whom any correspondence should be addressed.

2. Methodology

2.1 Parameter Description

The current air ducts that fitted to the Hino RK-1J SLL long-haul passenger bus have been studied. The overall dimension for the front portion of air ducts at the bus driver's side was measured by using a measuring tape. A 3D CAD modelling of the air ducts with outlets was drawn by using CATIA according to the measured dimensions. The airflow rate from the air ducts outlets was measured by using integral vane anemometer. After that, analysis will be carry out by running simulation with defined parameters on air ducts CAD model with the aid of ANSYS Fluent. Figure 1 showed the outlook of Hino RK1J-SLL long-haul passenger bus.

Figure 1. Outlook of Hino RK1J-SLL long-haul passenger bus.

The overall air ducts dimension was obtained through measurement by using measuring tape at Pioneer Coachbuilders Sdn. Bhd. Furthermore, location and the size of the outlets were measured in order to construct an accurate 3D modelling of air ducts inside CATIA. Figure 2 showed the actual outlook of air ducts while Figure 3 and 4 shown the round outlets and long outlets respectively that fitted to the air ducts.

Figure 2. Air ducts outlook.

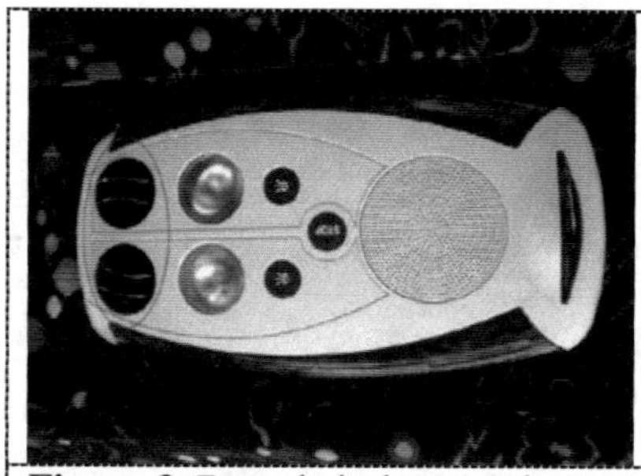

Figure 3. Round air ducts outlets.

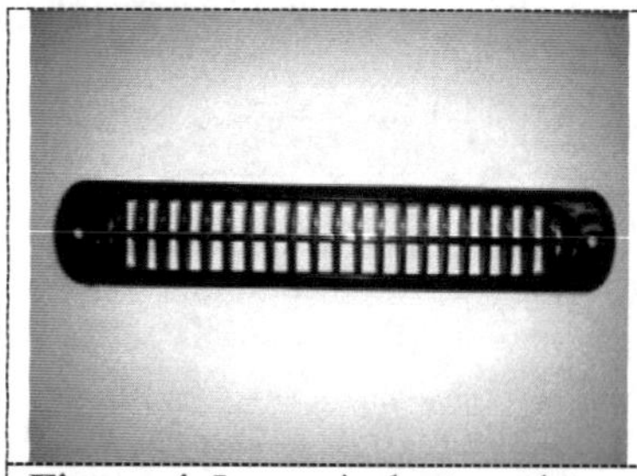

Figure 4. Long air ducts outlet.

2.2 Measurement Procedure

The rotating vane anemometer (RVA) which had been used widely in various industries has proved useful in air flow measurement. This device used a lightweight rotating vane as the sensor to measure the airflow rate [2]. Airflow rate from the air ducts outlets was measured by using an AZ Instrument 8903 integral vane anemometer. The measurement took place inside of Hino RK1J-SLL long haul passenger bus. The bus was at stationary during the measurement and the data was taken every interval of 30 minutes. The location of outlets was shown in Figure 5.

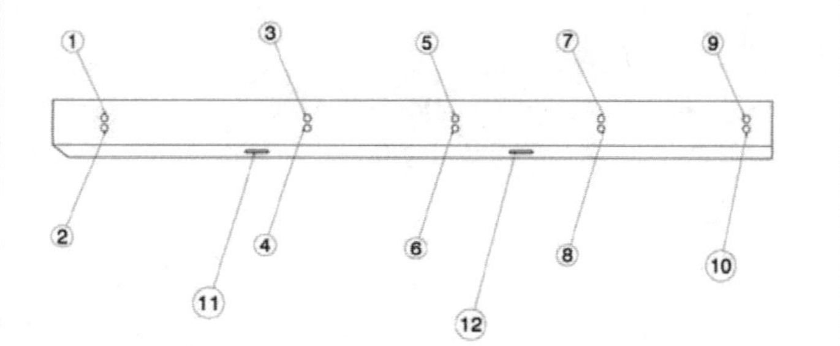

Figure 5. Location of outlets (bottom view).

2.2 Simulation Procedures

The excessive edge of the actual air ducts was eliminated and the outlets was simplified by eliminated the details of vane structure. Hence, a simplified 3D CAD model of air ducts with inlet and multiple outlets was constructed inside CATIA according to the measured dimension. The 3D CAD model of air ducts was as shown in Figure 6 below.

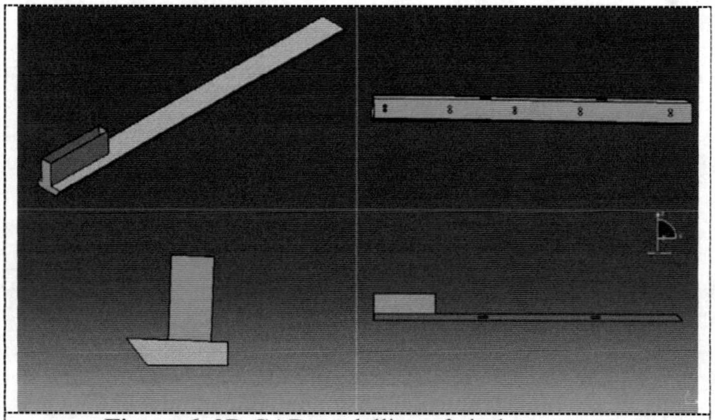

Figure 6. 3D CAD modelling of air duct.

The CFD analysis begins with meshing of the internal volume of air ducts inside ANSYS Design Modeller environment. Hence the meshed model was simulated inside ANSYS Fluent environment with defined boundary conditions. Fluent having all the physical capabilities needed to model flow, turbulence, heat transfer and reactions for industrial applications. Finite volume approach was used in Fluent to solve the equations governing the flow and is widely used in numerical simulations of different flow conditions of various complexities [3]. It is chosen for the simulation because its proven capability and validity in flows. The simulated results will then compared to the results obtained from experimental results.

The internal volume of air ducts was meshed inside ANSYS Design Modeller environment by using the automatic meshing function. The following meshing parameter was defined:

a) Relevance centre = Fine
b) Smoothing = High
c) Inflation = Program Controlled
d) Boundary Layer = 10 Layers

The other meshing parameter was remain default and the meshed model generated having 60253 of total elements.

3. Results and Discussions

The simulation results were obtained after the simulation comes to converged, that the solution no longer changes with subsequent iterations. Overall mass, momentum, energy, and scalar balances are achieved after converged. The velocity magnitude (m/s) for the internal volume of air ducts was shown in Figure 7.

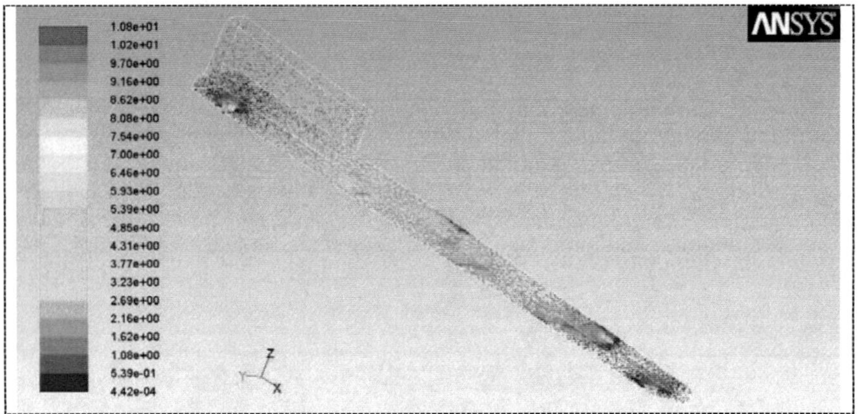

Figure 7. Velocity magnitude (m/s) for the internal volume of air ducts.

The simulation results that simulated under ideal condition usually different from the experimental results that obtained under actual operation [4]. The experimental and simulation results were compared in order to obtain the percentage of difference between experimental and simulation results. The data of experimental results and simulation results were tabulated in Table 1. Based on the table, it is shown that there is a high percentage of difference between experimental and simulation results for the outlets. The highest percentage of difference was 79.22% for outlet 2 while the lowest was 39.62% for outlet 9. This difference occurred as the simulation was run at ideal condition as the wall friction of the air ducts was not considered during the simulation.

IOP Conf. Series: Materials Science and Engineering **88** (2015) 012025 doi:10.1088/1757-899X/88/1/012025

Table 1. Comparison of experimental and simulation results.

Outlet	Experimental (m/s)	Simulation (m/s)	% of Difference
1	5.64	8.89	57.62
2	5.10	9.14	79.22
3	5.80	9.08	56.55
4	5.39	9.29	72.36
5	6.23	9.30	52.49
6	5.83	9.19	57.63
7	5.67	9.32	64.37
8	5.67	9.11	60.67
9	7.42	10.36	39.62
10	7.21	10.32	43.13
11	6.26	9.03	44.25
12	5.92	9.76	64.86

Besides that, the complex vane structure for the outlets was eliminated in the simulation, which the air will flow freely through the outlets without any resistance from the vane structure. Furthermore, the inlet flow was assumed to be constant during the simulation, that is not applicable to the actual air conditioning unit as the air conditioning unit that supply the airflow not always running at highest efficiency.

Despite the big difference between the experimental and simulation results, the simulation results showed that the outlet 1 having the lowest airflow rate of 8.89 m/s. This indicated that the outlets that located on bus driver's top having lower airflow rate compare to other outlets for both experimental and simulation results.

Design modification to the air ducts later was conducted in order to obtain a desired solution to improve the airflow rate. The internal volume of new air ducts design with deflector was constructed inside CATIA. The deflector was place at the behind the flow path of outlet 1 and outlet 2 in order to redirect more airflow to outlet 1 and outlet 2. Besides, the deflector also served the purpose to prevent the air flow to the end of air ducts while ensure the airflow directly exit through outlet 1 and outlet 2. Heuristic approach was used to determine the most effective angle of deflector, thus internal volume of air ducts with deflector at 90°, 75°, 60° and 45° was constructed. The CFD simulation of airflow rate for outlet 1 and 2 with deflector at 90° was shown in Figure 8 and 9 respectively.

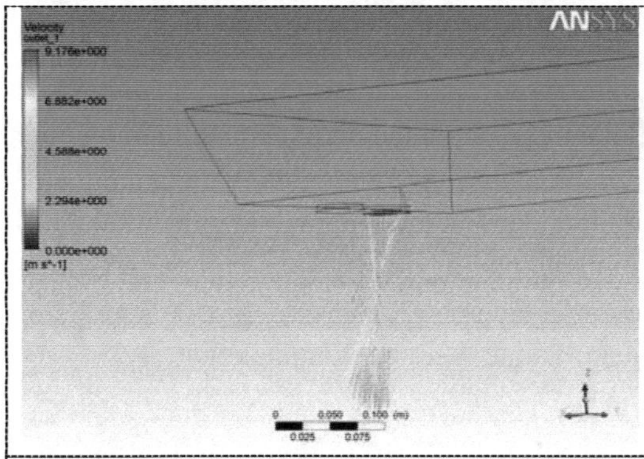

Figure 8. Airflow rate of outlet 1 for deflector 90°.

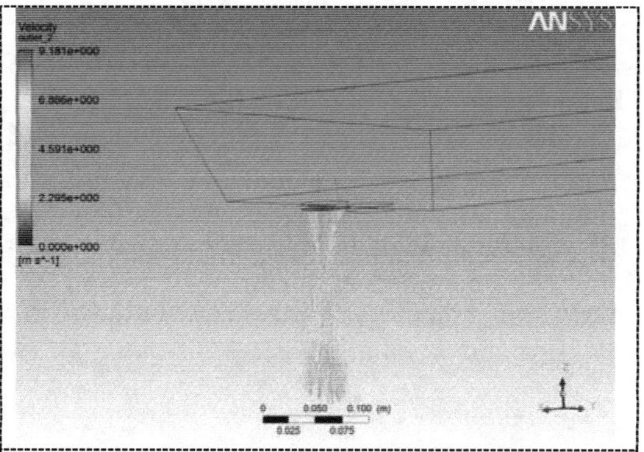

Figure 9. Airflow rate of outlet 2 for deflector at 90°.

The simulation results for original air ducts and new air ducts design with deflector at 90°, 75°, 60° and 45° was compared in Table 2 for outlet 1 and Table 3 for outlet 2. The percentage of improvement was determined by setting the simulation results of original air ducts as datum. From Table 2, it shown that the air flow rate for outlet 1 of new air ducts design with deflector at 60° having significant improvement of 6.52% over the original air ducts design. It poses higher percentage of improvement compare to 3.26% for deflector at 90, 4.50% for deflector at 75° and 4.72% for deflector at 45°.

Table 2. Percentage of improvement for outlet 1.

Degree of Deflector (°)	Datum (m/s)	Airflow Rate of Outlet 1 (m/s)	% of Improvement
90	8.89	9.18	3.26
75	8.89	9.29	4.50
60	8.89	9.47	6.52
45	8.89	9.31	4.72

Meanwhile from Table 3, it shown that the air flow rate for outlet 2 of new air ducts design with deflector at 75° showed the highest improvement of 2.63% over the original air ducts design. Its percentage of improvement followed by 2.41% for deflector at 60, 1.20% for deflector at 45° and 0.44% for deflector at 90°.

Table 3. Percentage of improvement for outlet 2.

Degree of Deflector (°)	Datum (m/s)	Airflow Rate of Outlet 2 (m/s)	% of Improvement
90	9.14	9.18	0.44
75	9.14	9.38	2.63
60	9.14	9.36	2.41
45	9.14	9.25	1.20

4. Conclusions

CFD analysis had been carried out to study the airflow rate of original air ducts and new air ducts design with deflector. The study focussed on percentage of improvement for the airflow rate at outlets along bus air-conditioning system. The objective of this study was achieved as the airflow rate delivered from outlets of air conditioning air ducts for long-haul passenger bus was improved with new design of air ducts. The new design of air ducts was fitted with deflector to redirect the airflow and the deflector was found to be effective at the angle of 60° through heuristic approach.

References

[1] Haslinda M K Mohd Y S & Nazri K 2012 Computerized Simulation of Automotive Air-Conditioning System: Development of Mathematical Model and Its Validation *International Journal of Computer Science* **9(2)** 2
[2] Burgess W A Ellenbecker M J & Treitman R D 2004 *Ventilation for Control of The Work Environment* 2nd ed Hoboken (New Jersey: John Wiley & Sons Inc)
[3] Fluent 2011 *Fluent 14.0 User's Guide* (Fluent Inc: USA)
[4] Shiu H R Ou F C & Chen S L 2002 3C Duct Design Method *The Chinese Journal of Mechanics – Series A* **18(2)**

IOP Conf. Series: Materials Science and Engineering **88** (2015) 012026 doi:10.1088/1757-899X/88/1/012026

Experimental Investigation on Design Enhancement of Axial Fan Using Fixed Guide Vane

K M Munisamy[1]*, R Govindasamy, S K Thangaraju

College of Engineering, Universiti Tenaga Nasional, Jalan IKRAM-UNITEN, 43009, Kajang, Malaysia

Email: kannan@uniten.edu.my

Abstract: Airflow passes through the rotating blade in an axial flow fan will experience a helical flow pattern. This swirling effect leads the system to experience swirl energy losses or pressure drop yet reducing the total efficiency of the fan system. A robust tool to encounter this air spin past the blade is by introducing guide vane to the system. Owing to its importance, a new approach in designing outlet guide vane design for a commercial usage 1250mm diameter axial fan with a 30° pitch angle impeller has been introduced in this paper. A single line metal of proper curvature guide vane design technique has been adopted for this study. By choosing fan total efficiency as a target variable to be improved, the total and static pressure on the design point were set to be constraints. Therefore, the guide vane design was done based on the improvement target on the static pressure in system. The research shows that, with the improvement in static pressure by 29.63% through guide vane installation, the total fan efficiency is increased by 5.12%, thus reduces the fan power by 5.32%. Good agreement were found, that when the fan total efficiency increases, the power consumption of the fan is reduced. Therefore, this new approach of guide vane design can be applied to improve axial fan performance.

1. Introduction

Fans are widely used in industrial and commercial applications such as ventilation, material handling, boilers, refrigeration, dust collection, cooling applications and others. In industrial applications fans are commonly used to supply ventilation or combustion air, to circulated air or other gases through equipment and to exhaust air or other vapors from equipment. There are two primary fans that are centrifugal and axial fan. Centrifugal fans moves the air radically outwards by centrifugal action, and then tangentially away from the blade tips. This type of fans are normally known as blowers and suitable for harsh operating condition, such as system with high temperatures, moist and material handling. Where else, the axial flow fan drives the air through the impeller in approximation of an axial direction. Axial fans utilize blades that force air to move parallel to the shaft about which the blades rotate, hence the

[1] To whom any correspondence should be addressed.

name. Owing to its cost effectiveness and multi-functional nature, the axial fan has been widely adopted in the industrial, commercial, consumer, institutional, and residential applications. [13], [15]

Many development and researches have being carried out in optimizing the design, application and efficiency improvement of the fan. The experimental investigation and numerical modeling techniques have been widely used to develop and optimize the fan [1], [2]. Initially, researches have been carried out on the design and optimization of the rotating impeller, such as investigation on the effect of installation angle of rotor blade, forward-skewed blade effects, aerofoil filled or aerodynamic effects on blade cross section and etc. In reference [1] the downstream flow resistance method has been developed for predicting the fan performance and the stalling effect in a fan system. Besides that, study on effects of installing angle setting for the rotor blade was explored in reference [2] on a axial flow fan and [4] on wells turbine , prior to improve the overall efficiency of the system. Apart from impeller design, investigation such as inlet flow distortions effect, tip clearance losses estimation, noise improvement, and swirl energy losses also have been demonstrated in references [5], [8] and [9]. Other then designs, research have been carried on the system or method development such as, an optimization technique based on the gradient method by reference [6] and [7], modification on the mining exhaust fan in reducing the 'Air Stray Losses' prior to reduce the electric consumption to overcome the situation by [10]. From the available published literatures as mentioned, there were less works have been reported on the guide vane design and application effect in improving axial fan performance. A paper by T.Setoguchi et.al. [3], presented the guide vane on the both inlet and outlet side of the rotor in overcoming the lower efficient and poor starting characteristic of the wells turbine. And in another paper by YE Zeng-Minget,al.[6] has introduced straighter vanes onto a large generator in overcoming, thus changing the swirl energy losses into useful energy in improving the efficiency of the system.

However, there is still room for improvement for guide vane design onto commercial usage axial fan to be explored. Therefore, an investigation on new approach in designing a fixed guide vane for a 1250mm diameter axial fan and its effect on improving total fan efficiency were demonstrated in this paper. An appropriate guide vane design can improve the airflow in a fan system by reducing swirling and get a higher efficiency, thus reducing the power consumption of the system. So, in this study, the fixed vane was design by choosing static pressure as a target variable to be maximized in the fan system, prior to improve the overall fan efficiency.

2. Experimental Apparatus

The experiment was conducted on a Type D test rig setup referred to AMCA 210-07 (Air Movement and Control Association, Inc.) [12] Standard with ducted inlet and outlet schematically shown in Figure 1. The inlet duct has a cross sectional area of 4.9 m^2 and a length of 5 m. An expansion connects the inlet duct and the test chamber which has a 64.2 m^2 cross sectional area and an equivalent length of 10 m. The downstream end of the test chamber on the other hand is connected to an outlet duct with a cross sectional area identical to the inlet duct with a length of 3.75 m through a gradual contraction. The fan is located at the inlet duct where at the upstream of the fan host an inlet bell and the transition allows the air to smoothly accelerate from a static condition to a high speed flow. In this study, a 12 profile rotor blade of 1250mm diameter, 30° pitch angle axial fan with a hub to tip ratio of 0.28 and 7 mm tip clearance selected (figure 2). Additionally, a 13 profile fixed vane is designed to compare and improve the fan performance.

The study was conducted using an approach termed as "downstream flow resistance" method or DFR for short to analyze the performance of the axial fan. DFR method is an approach that uses flow resistance that exist in a fan system to gain access to the fan performance where the pressure rise before the flow resistance and the pressure drop across the resistance region are used for the purpose of computing the static pressure and flow rate respectively. [1] It is possible to perform an uncertainty analysis to identify a range of values within which the true value might lie. Therefore, AMCA 210 chose an acceptable probability of 95%. According to AMCA standards, the uncertainties in the experimental results can be expressed in two parts, the measurement and combined uncertainties.

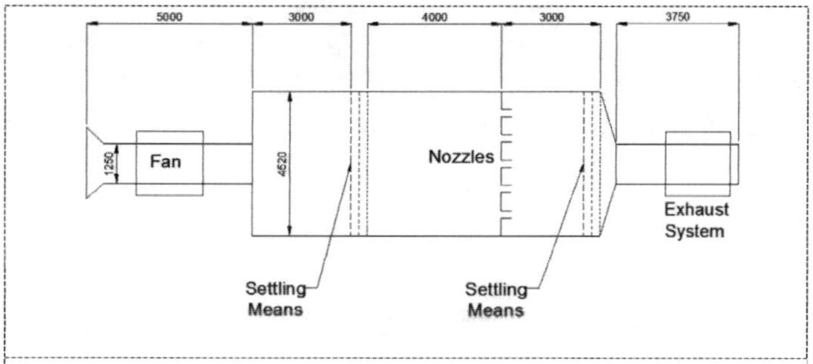

Figure 1. Type D test rig setup (dimension in mm).

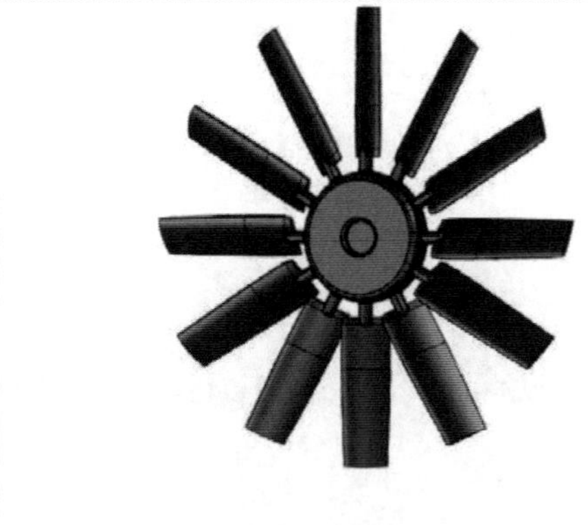

Figure 2. Rotor blade (outer diameter 1250mm).

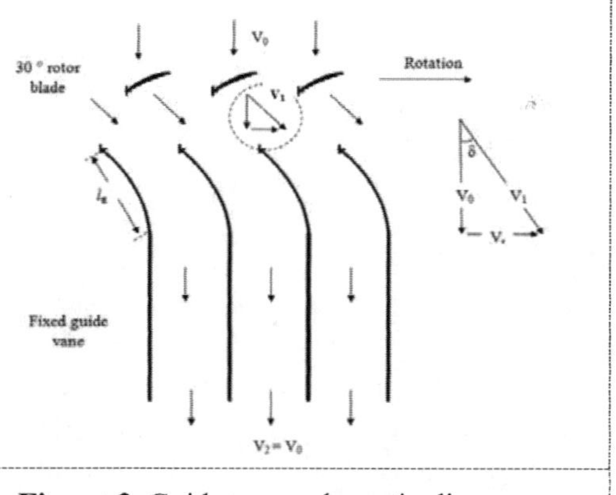

Figure 3. Guide vane schematic diagram.

2.1. Calculation

The fan total pressure was calculated from measurements of the pressures in ducts or chambers, corrected for pressure losses that occur in the measuring duct between the fan and the plane of measurement. A derivation of this calculation is Eq. 1. The fan static pressure was calculated from Eq. 2 [14].

$P_t = P_{s7} + P_v$ (1)	$P_s = P_t - P_v$ (2)
$H = \dfrac{2\pi TN}{60}$ (3)	$\eta_t = \dfrac{QP_tK_p}{H} \times 100\%$ (4)
$\eta_s = \eta_t \left(\dfrac{P_s}{P_t}\right)$ (5)	

As a torsion element is used to measure torque, the fan power input is calculated from the torque and the fan rotational speed using Eq. 3[14].The fan total efficiency is the ratio of the fan power output to fan power input, as shown in Eq. 4. With the derivation of total efficiency, the static efficiency was calculated using Eq. 5[14].

3. Design of guide vane

There were two type of guide vane can be provided to an axial fan to reduce the rotational energy losses. It can be located either on the outlet side called as outlet guide vane or on the inlet side called as inlet guide vane. The outlet guide vane produce static pressure partially to the system, where else the inlet guide vane will prepare the airflow for the blades. Since in this study, static pressure becomes the target objective to improve the fan efficiency, therefore the outlet guide vane would be the best choice. Figure 3 shows the 2D schematic diagram of the fixed vane design respective to the 30° rotating impeller.

The guide vane with a chord length of l_g = 300mm, and the end of the curvature is extended to 600mm for the purpose of effective airflow correction and to utilize the motor place as a hub for the guide vane. Good agreement found that the longer the straightener vane, the better airflow correction can be achieved. [6] The leading angle, δ of the guide vane is calculated from Eq.6 [14],

$\tan\delta = \dfrac{V_r}{V_a}$ (6)	$V_a = \dfrac{cfm}{A_a}$ (7)
$V_r = \dfrac{233 \times 10^5}{rpm} \times \dfrac{SP}{r}$ (8)	$A_a = \dfrac{(D^2 - d^2)\pi}{576}$ (8)

Where $V_{a,}$ is the axial air velocity in feet per minute through the annulus by Eq.7 and the V_r is the rotational component in feet per minute of the helical air velocity past the blades by Eq.8.And the Annular area in square feet if D and d are in inches, given by Eq.9 .[14]. It may be noted here that, determination on the static pressure on the guide vane design equation will be discuss future in section 4.0 (results and discussion). Therefore, based on this calculation the angle of fixed guide vane is set to be 40° with a constant thickness of 3 mm.

4. Results and Discussion.

In this section, results comparison has been made between without and with guide vane application for the 1250mm, 30° angle rotor blade axial fan. Figure 5 shows the efficiency and static pressure vs. volume airflow graph for the application of without guide vane. The highest efficiency for the without guide vane application is 64.96% with static pressure of 300Pa for the flow rate, Q of 15.986 m^3/s.(optimum point shown in figure 4)

From the Eq.4 (section 2), it is found that the improvement in static pressure in fan system will contribute to the overall fan efficiency to be increase, thus reduces the power consumption of the fan. Therefore, in this study the highest point of the static pressure, 351Pa in the fan operation curve region was selected to be improved with the guide vane application onto the system. Based on this selected target in static pressure, the guide vane angle was calculated in section 3. The selections point the static pressure improvement in shown in figure 5. From the application of the fixed guide vane, the static pressure were improvement approximately to 29.63% from the without guide vane application. The comparison and the improvement region in static pressure graph for with and without guide vane application is shown in figure 5.

By comparing the results at approximately at the same flow rate, Q, found that the total efficiency is increase by 5.12% at the static pressure of 304Pa for the application of with guide vane. Therefore the optimum efficiency for the guide vane application onto 1250mm axial fan would be 70.08% as shown in figure 6. Good agreement found that, the improvement in the total fan efficiency were obtained approximately at the same flow rate and static pressure value on the both cases of with and with guide vane application. This shows that, the energy losses in the system have been recovered by the guide vane installation to the fan system without any additional energy requirement. Thus, by increase in the total efficiency, the total power consumption of the fan have been reduce by 5.32% from the without guide vane application. Figure 7 shows the comparison of the power consumption for with and without guide vane application.

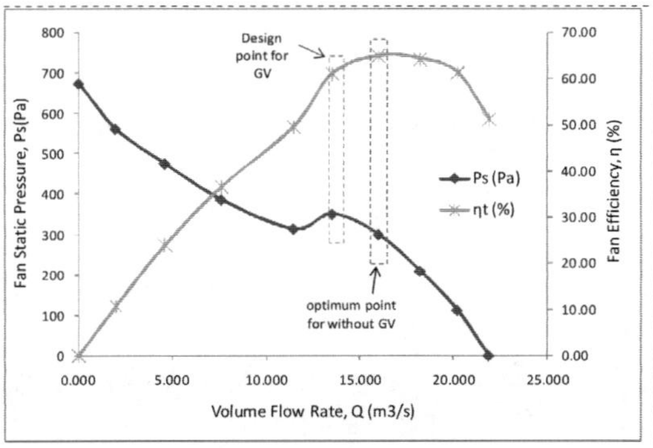

Figure 4. Efficiency and static pressure curve for without guide vane application.

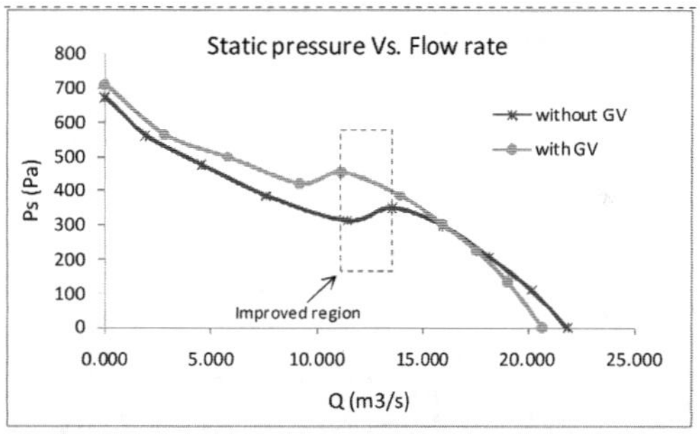

Figure 5. Static pressure vs. flowrate for with and without guide vane application.

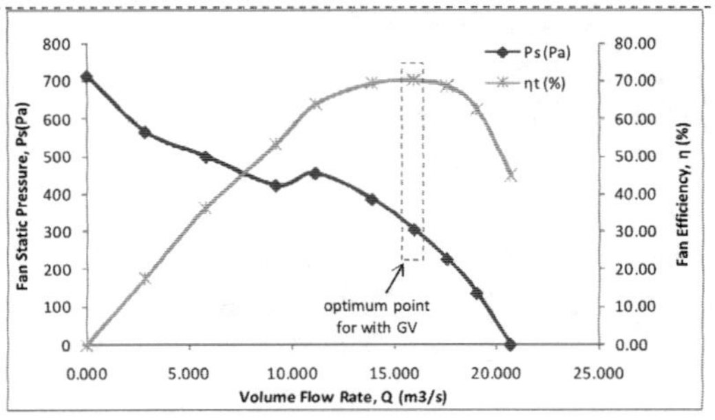

Figure 6. Efficiency and static pressure curve for with guide vane application.

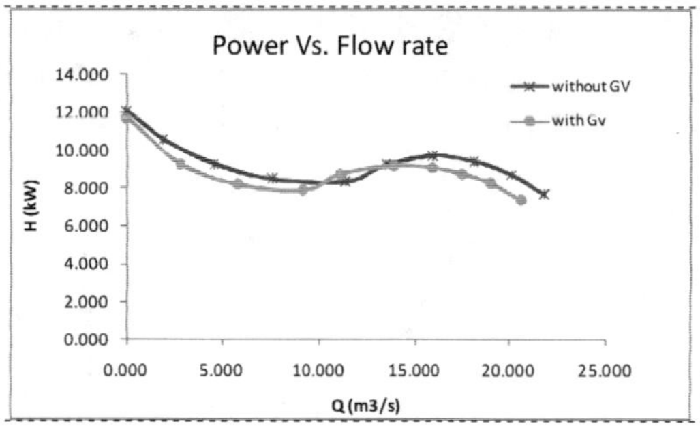

Figure 7. Power consumption comparison graph.

5.0 Conclusion

From the study, found that guide vane application onto axial fans would be one of the suitable solutions prior to encounter the air swirling or losses occur in a fan system. Reducing the swirl energy losses in a fan system can improve the total efficiency of the system. In this paper, 40° fixed guide vanes were designed onto 30° angle blade 1250mm axial fan. The single line proper curvature guide vane designing method was adopted with static pressure as a target objective in this study. From the comparison of the experimental results in section 4, found that the improvement in the static pressure for a fan system, increases the total fan efficiency of the system thus reduces the power consumption of the fan at the same time.

References

[1] Szu Hsien Liu, Rong Fung Huang, Chuang An lin, "Computational and experimental Investigations of performance curve of an axial fan using downstream flow resistance method", Experimental Thermal and Fluid Science, vol 33, pp. 827-837, 2010.

[2] Yan Xiao-kang,Wang li-jun,et.al. "Numerical and experimental investigation on effect of installation angle of rotor blade on axial flow fan", International Conference on Mechanical and Electrical Technology (ICMET), 2010.

[3] Qi Mei,Liang Huizhen,Guo Chuanjiang, " Optimized Design and Application of Axial-flow Fan", International Symposium on Intelligence Information Processing and Trusted Computing, 2010.7],[8]

[4] Setoguchi T., Santhakumar S., Takao M., Kim T.H., Kaneko K., " Effect of Guide vane shape on the performance of a Wells turbine", Renewable Energy, vol.23, pp.1-15,2001

[5] Setoguchi T., Santhakumar S., Takao M., Kim T.H., Kaneko K, " Amodified Wells turbine for wave energy conversion", Renewable Energy, vol 28,pp 79-91,2003.

[6] YE Zeng-Ming,ZHU Ting-Ting, " Research of Guide Vane Proposal for the Single-stage Axial-flow used in the Large Generator", Sustainable Power Generation and Supply, 2009.

[7] Chong-hyun CHO, Soo-yong CHO, Chaesil KIM, "Development of an axial-type fan with an optimization method", Energy Power Eng.China, 3(4), pp.414- 422,2009

[8] C-H Cho,S-Y Cho,K-Y Ahn,Y-C Kim, " Study of an axial-type fan design technique using an optimization method", Journal of Process Mechanical Engineering, vol. 223, 2009

[9] X-Sun, Z.Yang, X.Wang, " Effect of fan outlet guide vane on the acoustic treatment design in aeroengine nacelle", Journal of Sound and Vibration, vol.302,pp. 287-312, 2007

[10] C-D Pitis , " Energy Efficient Single Stage Axial Fan (ENEF), Canada Electrical Power Conference, 2007

[11] L.M.C.Ferro,L.M.C Gato,A.F.O Falcao, " Design and experimental validation of the inlet guide vane system of a mini hydraulic bulb-turbine", Renewable Energy,vol. 35,pp. 1920-28, 2010

[12] P Laboratory Methods of Testing Fans for Aerodynamic Performance Rating (AMCA 210/ASHRAE 51). American Society of Heating, Refrigerating and Air Conditioning Engineers, 2008.

[13] Ivor da Cunha , Terry Strack, and Saul Stricker, " FANS &BLOWERS " Energy Efficiency Reference Guide ,CEATI International Inc, 2008

[14] F. Bleier, Fan Handbook: Selection, Application, and Design, 1st ed. McGraw-Hil Professional, 1997.

[15] R. A. Wallis, Axial flow fans and ducts. Wiley, 1983.

IOP Conf. Series: Materials Science and Engineering **88** (2015) 012027 doi:10.1088/1757-899X/88/1/012027

A Study of the Heat Transfer Coefficient of a Mini Channel Evaporator with R-134a as Refrigerant

E B Dollera[1], E P Villanueva[2]
[1]College of Engineering, Xavier University-Ateneo de Cagayan, Cagayan de Oro City, Philippines
E-mail: edollera@xu.edu.ph

[2]College of Engineering
Iligan Institute of Technology, Mindanao State University
Iligan City, Philippines
E-mail: ep_villanueva@yahoo.com

Abstract. The present study is to evaluate the heat transfer coefficient of the mini-channel copper blocks used as evaporator with R-134a as the refrigerant. Experiments were conducted using three evaporator specimens of different channel hydraulic diameters (1.0mm, 2.0mm, 3.0mm). The total length for each channel is 640 mm. The dimension of each is 100mm.x50mm.x20mm. and the outside surfaces were machined to have fins. They were connected to a standard vapour compression refrigeration system. During each run of the experiment, the copper block evaporator was placed inside a small wind tunnel where controlled flow of air from a forced draft fan was introduced for the cooling process. The experimental set-up used data acquisition software and computer-aided simulation software was used to simulate the pressure drop and temperature profiles of the evaporator during the experimental run. The results were then compared with the Shah correlation. The Shah correlation over predicted and under predicted the values as compared with the experimental results for all of the three diameters and high variation for D_h=1.0mm. This indicates that the Shah correlation at small diameters is not the appropriate equation for predicting the heat transfer coefficient. The trend of the heat transfer coefficient is increasing as the size of the diameter increases.

1. Introduction

Vapour compression refrigeration system is the most widely used refrigeration system today. One of the main components of a vapour compression system is the evaporator. It is a type of heat exchanger which absorbs heat and rejects it through another component, the condenser. The heat rejected in the condenser consists of heat absorbed by the evaporator and the heat from the energy input in the compressor. The compressor is the heart of the vapour compression refrigeration system, discharges hot, high-pressure refrigerant gas into the condenser, which rejects heat from the gas to some cooler environment or heat sink. The high pressure, high temperature, superheated refrigerant from the compressor, condenses and passes through a capillary tube acting as an expansion valve. The refrigerant used in this study is R-134a.

The application of mini channel technology is increasingly used to achieve high heat transfer rates with compact heat exchangers. Boiling and evaporization of R-134a inside small hydraulic diameter tube and channels find many applications in compact heat exchangers for cooling electronic gadgets

[1] To whom any correspondence should be addressed.

IOP Conf. Series: Materials Science and Engineering **88** (2015) 012027 doi:10.1088/1757-899X/88/1/012027

and other refrigeration systems. The adoption of minichannel also promotes the reduction of the refrigerant charged, which is beneficial to mass and energy conservation. Minichannel phase change may differ from the conventional channels due to differences in the relative influence of gravity, shear stress and surface tension. As dimensions of electronic components reduce at mini scale, it is necessary for us to use a cooling system which can apply to the same scale. Fluid flow inside mini channel is one of the efficient candidates for small scale cooling. According to Kandlikar, mini channels are classified based on the hydraulic diameter (D_h) of 200μm to 3 mm. There has not been any comprehensive study to determine the limits of macro channel correlations to smaller data[1]. There is a need for such a study and this paper attempts to fulfill this need to some extent.

In this study, three mini channel evaporators of hydraulic diameters 1.0 mm, 2.0 mm and 3.0 mm, with each mini channel length of 640 mm. were designed, fabricated and tested. They were experimentally tested in an experimental rig. Data were recorded and stored using data acquisition system and with the use of a Lab View Software. Heat transfer coefficients were calculated using thermodynamics principles and compared with Shah Correlations. Solid Works software was used to simulate the thermodynamic parameters involved in this study.

2. Theoretical Considerations

2.1 Refrigeration Vapor Cycle

The evaporator is a heat exchanger that absorbs heat from the refrigeration system. The evaporator receives low pressure R-134a from the pressure reducing device, absorbs heat from its environment and delivers it to the compressor as a superheated gas. From the compressor, R-134a rejects heat in the condenser and enters back to the pressure reducing device. The reduced pressure in the expansion device enables the evaporator to absorb heat from the environment at a lower and cooler temperature. Evaporators are equipment that allows two-phase flow in the system which makes them difficult to calculate the exact coefficient of heat transfer. It is normal for a conventional evaporator to experience two boiling regimes, nucleate boiling and convective boiling.

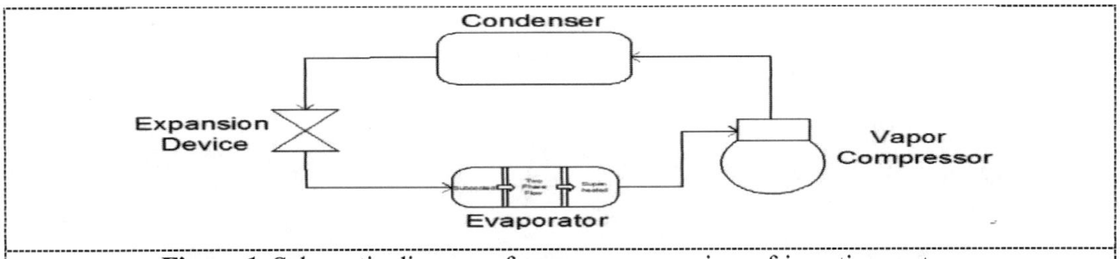

Figure 1. Schematic diagram of vapour compression refrigeration system

There are two ways in which flow boiling in small diameter channels are expected to be implemented. They are, two-phase entry after a throttle valve and sub-cooled liquid entry into the channel[2]. Because of the requirement of the liquid flow meter being used, this study belongs to the sub-cooled liquid entry into the channel. Sub-cooled liquid entry is an attractive option, because of higher heat transfer coefficients associated with sub-cooled flow boiling[2].

2.2 Two-Phase Flow Heat Transfer Coefficient, H_{TP}

From Newton's law of cooling,

$$Q = H_{TP,Exp}\,(A_i)(T_{si} - T_w) = m_r(h_{out} - h_{in}) \quad (1)$$

where Q=heat absorbed by R134a,kJ/s, $H_{TP,Exp}$ =experimental heat transfer coefficient,kW/m^2-K, A_i=inside channel surface area,m^2, T_{si} = inside saturation temperature of R134a,K, T_w=channel wall temperature,K, m_r=mass flow rate of R-134a,kg/s, h_{in}=enthalpy of R134a at mini-channel inlet,kJ/kg, h_{out}=enthalpy of R134a at mini-channel outlet,kJ/kg

From equation 1, the experimental heat transfer coefficient, $H_{TP,Exp}$, is calculated as,

$$H_{TP,Exp} = m_r\,(h_{out} - h_{in})/[\pi D_h L(x)(T_{si} - T_w)] \tag{2}$$

where D_h =hydraulic diameter,m, L = total length of the mini channel,m, and x=mini-channel length ratio.

In this study, Shah correlation [3,4] is used to determine the two-phase heat transfer coefficient. Shah correlation is the only two-phase flow correlation that approximates a decreasing experimental heat transfer coefficients as the length ratio of the mini-channel increases. The Shah correlation, $H_{TP,Shah}$, is expressed as,

$$\Psi = \frac{H_{TP,Shah}}{h_{LO}} \tag{3}$$

where Ψ =two-phase multiplier for Shah correlation, Ψ_{nb} =nucleate boiling multiplier, Ψ_{cb}=convective boiling multiplier, and $h_{LO=}$heat transfer coefficient for liquid-phase only and calculated using Dittus-Boelter[5] correlation,

$$h_{LO} = 0.023\left(\frac{k_l}{D_h}\right)\left(\frac{G(1-x)D_h}{\mu_l}\right)^{0.8}(P_r)^{0.4} \tag{4}$$

where k_l =liquid phase thermal conductivity,W/m-K, G=mass flux,kg/s-m^2, μ_l=liquid phase kinematic viscosity,kg/s-m, P_r =Prandtl number.

For nucleate boiling,

$$\Psi_{nb} = \frac{H_{TP,Shah}}{h_{LO}} \tag{5}$$

Solving for the boiling number, B_o,

$$B_o = \frac{q"}{(G)(h_{LV})} \tag{6}$$

where $q"$= heat flux, kW/m^2 and h_{LV} =latent heat of vaporization, kJ/kg,

For $B_o \leq 0.3 \times 10^{-4}$,

$$\Psi_{nb} = 1 + 46\,B_o^{0.5} \tag{7}$$

For $B_o < 0.0011$, Froud number, F_r can be calculated as,

$$F_{rl} = 15.4\,B_o^{0.5} \tag{8}$$

Solving for the convection number, C_o,

$$C_o = \left(\frac{1-x}{x}\right)^{0.9}\left(\frac{\rho_G}{\rho_L}\right)^{0.5} \tag{9}$$

where ρ_G=the gas density of R134a,kg/m^3, and ρ_L=the liquid density of R134a,kg/m^3.

For horizontal tubes with $F_r \leq 0.04$,

$$N = 0.38\,Fr_l^{-0.3}C_o \tag{10}$$

So that,

$$\Psi_{cb} = \frac{1.8}{N^{0.8}} \tag{11}$$

and use the result of equation (7) or equation (11) into equation (3) whichever is higher.

3. Description of the Experimental Work

An experimental rig was constructed to house all the refrigeration components and instrumentation sensors. A fabricated mini channel evaporator was used in the experimental test set-up by tapping a 1/8 hp vapour compressor. The dimension of the mini channel is 100 mm *x* 50 mm *x* 20mm copper block. Fins were also fabricated on each face of the copper block and each face contains 24 fins with width of 2 mm, length of 50 mm and thickness of 5 mm.

IOP Conf. Series: Materials Science and Engineering **88** (2015) 012027 doi:10.1088/1757-899X/88/1/012027

The refrigerant passes through the mini channel with hydraulic diameter varied from 1mm, 2mm and 3mm. Leak tests were done to ensure that there will be constant flow of refrigerant in the system.

During the experiments, refrigerant flow rate was controlled using a by-pass valve from the main evaporator. The refrigerant mass flow rate was measured at the inlet of the mini channel evaporator. The two temperature sensors and two pressure transducers were mounted on the refrigerant-side at the inlet and outlet of the mini channel. One temperature sensor was also installed at the middle of the mini channel.

An insulated wind tunnel was fabricated and a forced draft fan was installed to guarantee sufficient supply of air across the mini channel. Two temperature sensors were mounted on the wind tunnel, one at the inlet and one at the outlet of the tunnel. The air velocity over the mini channel was measured using hotwire anemometer.

The temperature sensors, flow meter and pressure transducers were connected to the data logger. The data logger was interfaced with the computer. Lab VIEW software had been used to operate the data logger and to store the data in the computer. The data logger was set to scan the data from the flowmeter, temperature sensors and pressure transducers at an interval of 30 seconds.

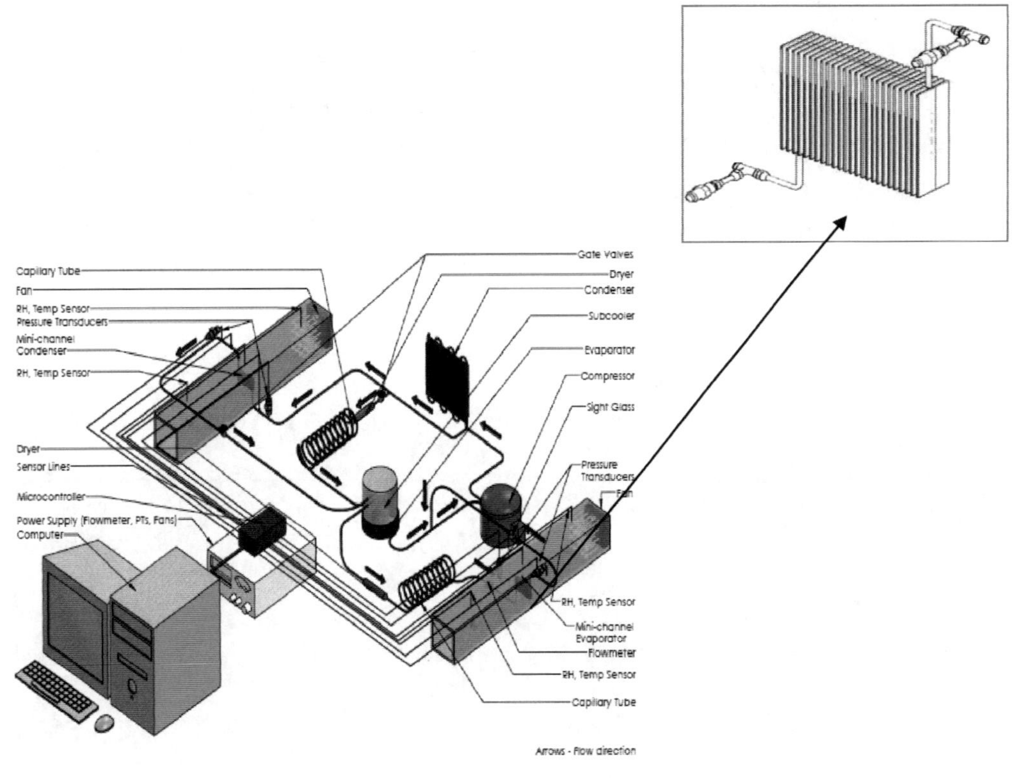

Figure 2. Experimental set-up.

4. Data Analysis and Evaluation of Heat Exchanger Performance

The results related to the heat transfer coefficient and pressure drop were obtained from the readings of three thermocouples, two pressure transducers, a flow meter and an anemometer, installed in the test section. The heat absorbed in the evaporator is calculated by the relation,

$$Q = m_r\,(h_{out} - h_{in}) \tag{12}$$

where m_r is the refrigerant mass flow rate obtained directly from the flow meter, h_{in} is the enthalpy entering to the mini channel evaporator corresponding to a measured pressure and temperature, and h_{out} is the enthalpy leaving the evaporator corresponding to a measured pressure and temperature conditions.

The average values of experimental heat transfer coefficient were calculated based on the average inside surface temperature of the mini channel,

$$Q = h_{TP,Exp}(A_i)(T_{sat} - T_{si}) \tag{13}$$

5. Results and Discussion

The heat exchangers in Figure 3 show the view of the interior part and Figure 4 shows the finned exterior surface of the mini channel evaporator. These mini channels were fabricated and joined together through oxy-acetylene welding. These heat exchangers were capable of operating at maximum pressure of 10 bars.

Figure 3. Interior part of mini-channel.

Figure 4. Exterior Part of mini-channel.

Figure.5, shows the variations of the experimental heat transfer coefficient and the Shah correlation with the length ratio. Shah correlation over predicted the heat transfer coefficient for the D_h = 1.0mm in most part of the total length of the mini channel but approximates each other values near the inlet and outlet of the mini channel.

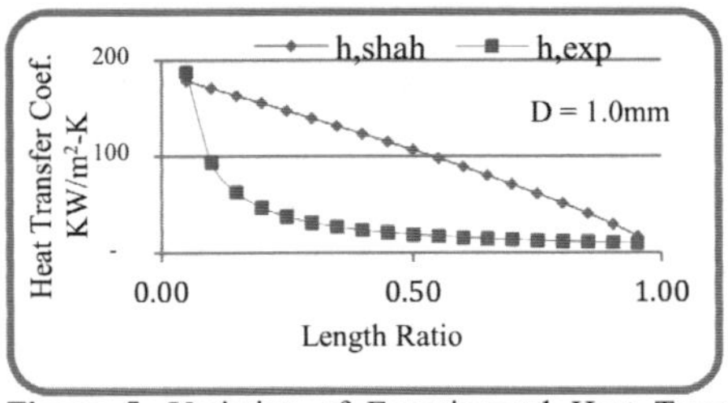

Figure 5. Variation of Experimental Heat Transfer Coefficient & Shah Correlation for 1mm.

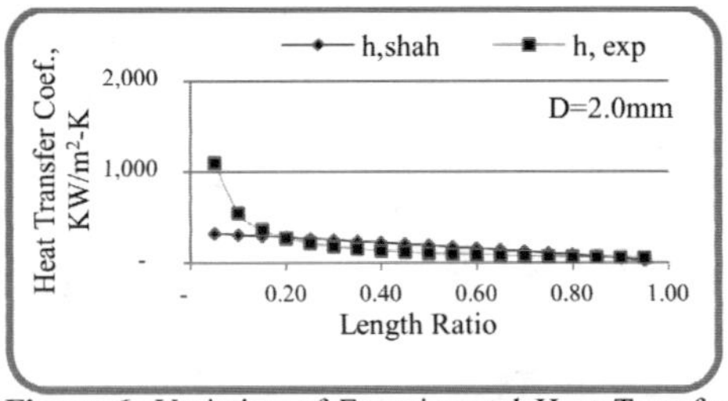

Figure 6. Variation of Experimental Heat Transfer Coefficient & Shah Correlation for 2mm.

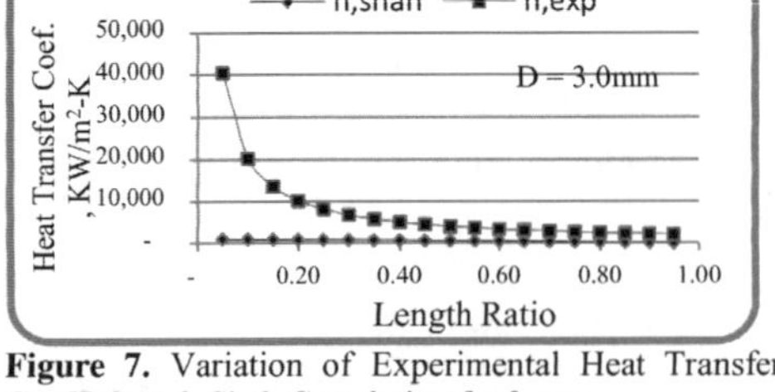

Figure 7. Variation of Experimental Heat Transfer Coefficient & Shah Correlation for 3 mm.

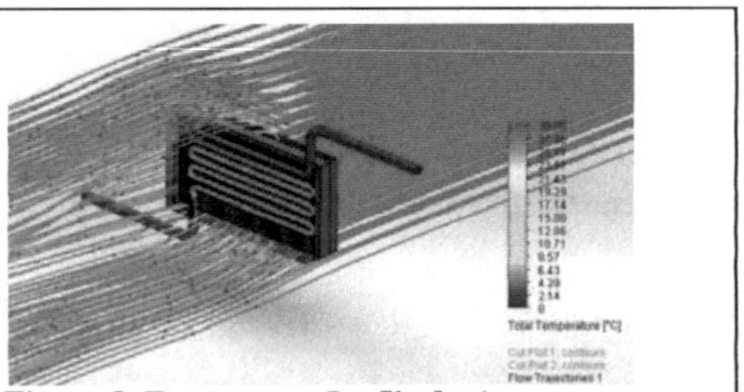

Figure 8. Temperature Profile for 1mm.

Figure 6 shows the variation of the experimental heat transfer coefficient and the Shah correlation with the length ratio. Shah correlation over predicted the heat transfer coefficient for the D_h=2.0mm in some parts of the total length and under predicted in some parts of the mini channel length. Greater variation of the heat transfer coefficients at the inlet is attributed to the sub-cooled liquid condition. Figure 7, shows Shah correlation under predicted the heat transfer coefficient for the D_h=3.0mm in all parts of the length ratio from 0.05 to 0.95.The greater variation occurs below 0.50. Figure 5, Figure 6 and Figure 7, show that the experimental heat transfer coefficient and the Shah correlation increase as the hydraulic diameter of the mini-channel increases. These two methods in calculating the heat transfer coefficient conformed each other as to the magnitude of the heat transfer coefficient.

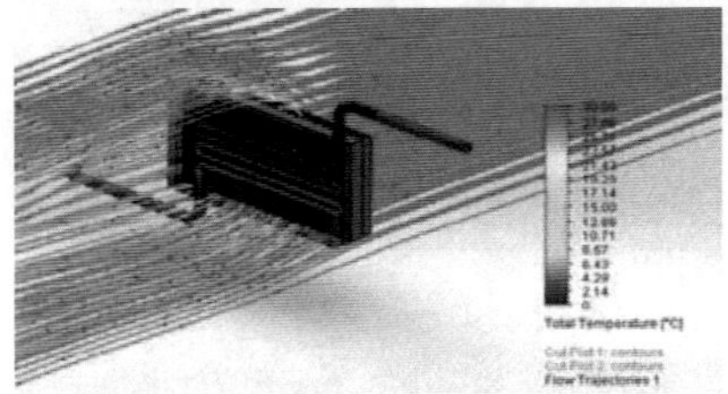

Figure 9. Temperature Profile for 2mm.

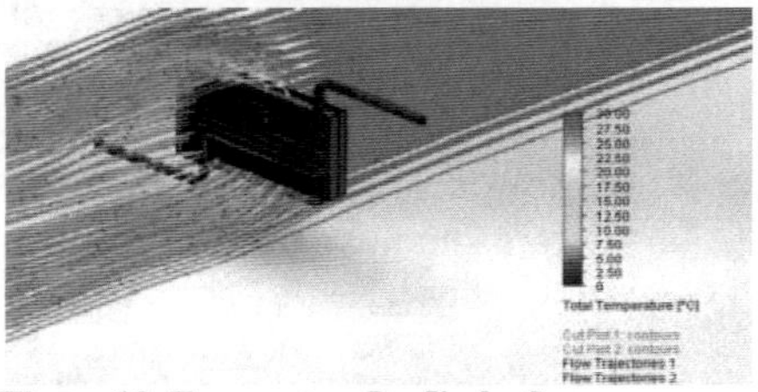

Figure 10. Temperature Profile for 3mm.

Figures 8, 9 and 10, show the temperature profile inside the mini channel. The refrigerant R-134a enters the mini-channel at a sub-cooled liquid condition, represented by the blue color. The temperature at the outlet condition is represented with the red color signifying the superheated condition at the mini-channel outlet.

6. Conclusions and Recommendations

Based on the data obtained in this study, the following conclusions are drawn. The experimental heat transfer coefficient for D_h=1.0mm has the lowest value and D_h=3.0mm has the highest value for the three tested mini channel evaporator. Shah correlation over predicted the values for D_h=1.0mm and under predicted the values for D_h=3.0mm. This variation is caused by the difference of mass flux introduced to the system. Shah correlation approximates the experimental values for 2.0mm near the centre of the length of mini channel.The experimental heat transfer coefficient and the Shah correlation increases as the hydraulic diameter increases. The simulated temperature profile of R-134a approximates the experimental values obtained using temperature sensors. Sub-cooled condition of R-134a results a higher experimental heat transfer coefficient and decreases as the temperature of R-134a increases inside the mini-channel evaporator.

It is recommended that further study will be done by using constant mass flow rate as basis for comparison and to further study the behavior of the mini channel with hydraulic diameter of 0.5mm, 1.5mm and 2.5mm.

7. References

[1] Kandlikar S.G, 2006 *Heat transfer & Fluid Flow in Minichannels & Microchannels*, 1st ed. UK: Elsevier Science; pp. 3

[2] Kandlikar S.G, 2006 *Heat transfer & Fluid Flow in Minichannels & Microchannels*, 1st ed. UK: Elsevier Science; pp. 176,

[3] Shah, M. M., 1976 *A New Correlation for Heat Transfer During Boiling Flow Through Pipes*, ASHRAE Trans., Vol. 82, No. 2, pp. 66-86

[4] Shah, M. M., 1982 *Chart Correlation for Saturated Boiling Heat Transfer: Equations and Further Study*, ASHRAE Trans., Vol. 88, No. 1, pp. 185-196

[5] Dittus, F.W. and Boelter, L.F. K., 1930 University of California, Berkeley, Publication in Engineering, Vol. 2, pp.443.

Experimental Investigation and Flow Process Computer Simulation of the Single Mini Channel Condenser for Vapor Compression Refrigeration System

L L Pabilona[1] and E P Villanueva[2]

[1] Associate Professor II, College of Engineering & Architecture
Mindanao University of Science and Technology
Cagayan de Oro City, Philippines
E-mail: ll_pabilona@yahoo.com

[2] Professor VI,College of Engineering
Iligan Institute of Technology, Mindanao State University
Iligan City, Philippines
E-mail: ep_villanueva@yahoo.com

Abstract. This study is a computer simulation of the temperature profiles and experimental investigation of three 100 mm x 50 mm x 18 mm single mini channel condensers with hydraulic diameters of 3 mm, 2 mm, and 1mm. The mini channels which were made of copper were designed, fabricated and tested. Each unit was connected in a vapor compression cycle with R-134a as the refrigerant. The average refrigerant mass flow rates were varied from 1.296 - 69.471 g/s, and the average inlet and outlet condenser pressure variations were 102.5 - 121.8 kPa and 101.74 -121.23 kPa, respectively. Each condenser was placed inside a mini wind tunnel system where forced draft air was introduced to initiate convective heat transfer. Each condenser was tested and data were gathered every five minute interval for one hour using a Lab View Software. Computer simulations on the flow process were conducted using Solid Works software. The experimental results presented the inlet and outlet condenser pressures, and pressure drops. The experimental heat transfer coefficients were calculated at different mass fluxes during condensation. The values ranged from 3900 to 5200 W/m^2-°K for the 3 mm, 2600 to 9000 W/m^2-°K for the 2 mm, and 13 to 98 W/m^2-°K for the 1 mm. The heat transfer coefficients calculated from experiments were then compared with the computed values using the correlations developed by Dittus-Boelter and Lee-Son. The results showed increasing deviation as the diameter decreased. The discrepancies could be attributed to the appropriateness of the Dittus-Boelter and Lee-Son correlations in small diameter channels, complexities in the flow process which involved two phase flow heat transfer in very small tubes, and the difficulties in attaining very accurate measurements in small channels.

I. Introduction

The condenser in a refrigeration system is the heat sink of the system. The condenser is a major component in the cooling system. Its design, configuration and size greatly affect the performance of the refrigeration system. As the size of the condenser decreases, such as in a mini-channel, the mechanism of fluid flow changes as the capillary effect becomes more prominent. The heat transfer mechanism is also affected since it is very dependent on fluid flow.

Minichannels are increasingly being used with compact heat exchangers. Small hydraulic diameter channels find applications in compact heat exchangers for electronic equipment, in automotive condensers, and in refrigeration applications. The adoption of minichannel also promotes the reduction of the refrigerant charge, which is favorable specially if toxic or flammable refrigerants

[1] To whom any correspondence should be addressed.

are used. Phase change of refrigerants in minichannels may differ from the conventional channels due to differences in the relative influence of gravity, shear stress and surface tension. Design of small heat exchangers is a must as the size of heat generators such as electronic components reduces to miniscale. Fluid flow inside mini channel is one of the efficient processes for small scale cooling. According to Kandlikar [1], mini channels are classified based on their hydraulic diameters (D_h) of 0.5 to 3 mm. There are relatively few studies conducted as regards to mini channels condenser of a refrigeration system.

This study is about the designing, fabrication and testing of a mini channel condenser. Three mini channels of similar configurations but of different channel sizes and used as condensers were experimentally tested. The inlet and outlet pressures, and pressure drops were analysed. Measurements of the heat transfer coefficient were compared to existing correlations developed by Dittus-Boelter and Lee-Son. Finally, Computer simulations on the flow process particularly the temperature distribution inside the single channel condenser were conducted using the Solid Works software.

2. Theoretical Considerations

2.1 Vapor Compression Cycle

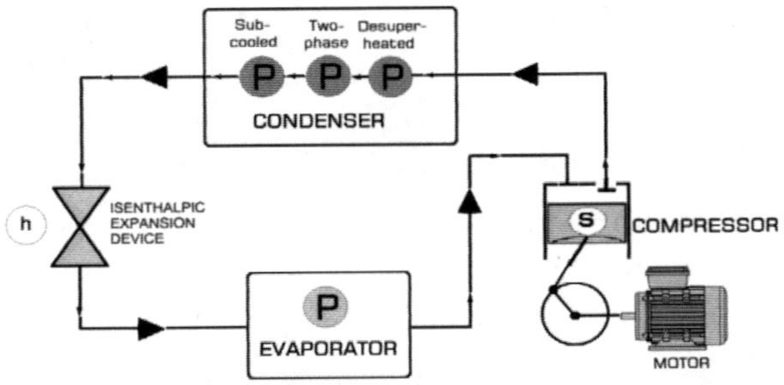

Figure 1. Analysis on the Condenser

The condenser is a heat exchanger that usually rejects all the heat from the refrigeration system. The condenser accept hot , high -pressure refrigerant ,usually a superheated gas, from the compressor and reject heat from the gas to some cooler substance . As energy is removed from the gas it condenses and this condensate is drained so that it may continue its path back through the expansion valve or capillary to the evaporator The condenser is divided into three regions characterized by the thermodynamic state: the de-superheated vapor, the change of phase (two-phase), and the subcooled liquid regions and the refrigerant condenses at a negligible temperature difference from the ambient.

2.2 Heat Transfer Coefficient

The Dittus-Boelter [2] correlation is a common correlation useful for many applications. This correlation is applicable when forced convection is the sole mode of heat transfer. This is an equation quite similar to McAdams [3] equation where the application as presented in the literature is not restricted to small diameters [4]. The equation is expressed below as

$$h_L=0.023(Re)^{0.80}(Pr)^{0.40}\left(\frac{k}{D}\right)$$

Another correlation developed by Son-Lee [5] is the single-phase heat transfer coefficient (h_L), express as

$$h_L=0.034(Re)^{0.80}(Pr)^{0.30}\left(\frac{k}{D}\right)$$

Where k is the thermal conductivity of R-134a, D is the hydraulic diameter, Re is the Reynolds number and Pr is the Prandtl number.

3. Description of the Experimental Work

Three 100 mm x 50 mm x 20 mm copper block mini channel condensers with 1 mm, 2 mm, and 3 mm hydraulic radius for each were fabricated and installed one at a time in the experimental rig as shown in Figure 2. Fins were machined on each face. Each had 24 fins with width of 2 mm, length of 50 mm and height of 5 mm. A fan was installed downstream of the heat exchanger to guarantee sufficient air flow across the condenser inside the insulated long duct used in the rig. Two temperature sensors and two pressure transducers were installed on the refrigerant-side at the inlet and outlet of the mini channel condenser. Another temperature sensor was also installed at the middle of the mini channel condenser. The refrigerant mass flow rate was measured at the outlet of the expansion valve. A valve that passed through the evaporator was used to control the mass flow rate. On the air-side, two thermocouples spaced equally were installed at the inlet and two at the outlet. The air velocity over the mini channel condenser was measured using a hot wire anemometer. Every start of the experiment, the system was vacuumed for about 30-40 min to remove air, followed by leak test before charging the system with the refrigerant R134a. The temperature, flow rate and pressure readings were automatically recorded, stored and processed by a data acquisition system using LabVIEW software. The data logger was set to scan the data at an interval of 5 minutes.

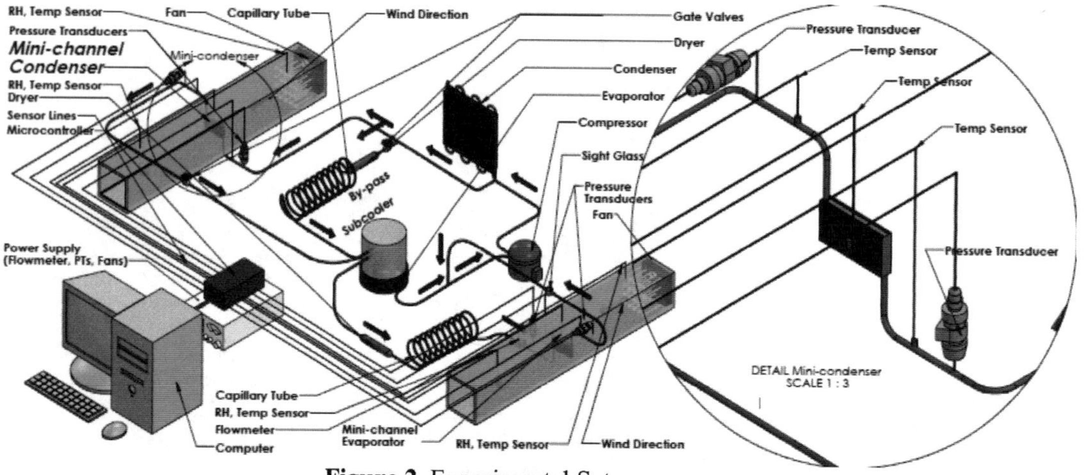

Figure 2. Experimental Set-up

4. Data Analysis and Evaluation of Heat Exchanger Performance

The results related to the heat transfer coefficient and pressure drop were obtained from the readings of thermocouples, pressure transducers, mass flow meter and anemometer, installed in the test section. The heat rejected in the condenser was calculated using the relation

$$Q = m_s (h_{in} - h_{out}) \tag{3}$$

Where m_s is the refrigerant mass flow rate and obtained directly from the flow meter, h_{in} is the enthalpy entering the mini channel condenser corresponding to measured pressure and temperature conditions, and h_{out} is the enthalpy leaving the condenser corresponding to measured pressure and temperature conditions.

The average values of experimental heat transfer coefficient were calculated at the average inside surface temperature of the condenser [6]:

$$Q = h_r A_i (T_{sat} - T_{si}) \tag{4}$$

Where h_r is the refrigerant side heat transfer coefficient, A_i is the inside area of mini channel, T_{sat} is the saturation temperature of refrigerant, and T_{si} is the temperature of inner wall surface.

The refrigerant pressure drops along the mini channel condenser were obtained from the difference in readings of transducer pressures between the inlet and outlet conditions.

$$\Delta P = P_{inlet} - P_{outlet} \tag{5}$$

IOP Conf. Series: Materials Science and Engineering **88** (2015) 012028 doi:10.1088/1757-899X/88/1/012028

5. Results and Discussion

The two parts of the heat exchangers before they were welded together are shown in figures 3. Figure 4 shows the finned exterior surface of the mini channel. The welded assembly was design to withstand pressure of up to 11 bars.

Figure 3. Interior part of mini channel (Unassemble)

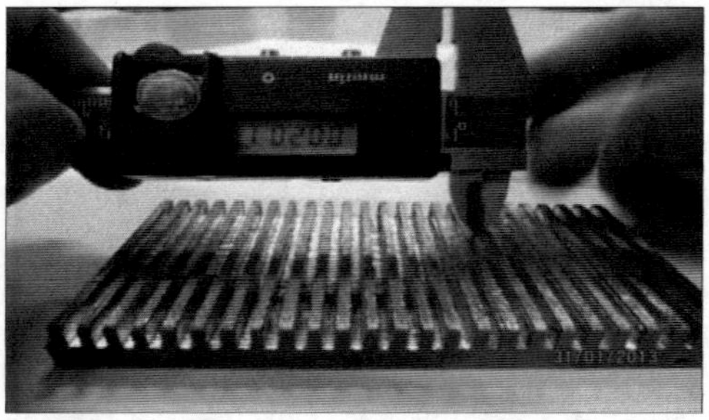

Figure 4. Exterior Part of mini channel Parts

Figures 5, 6 and 7 show the inlet and outlet pressure for 1 hour observation of the different condenser sizes. The pressure readings tend to increase with increase in mass flow rate and decrease in diameter.

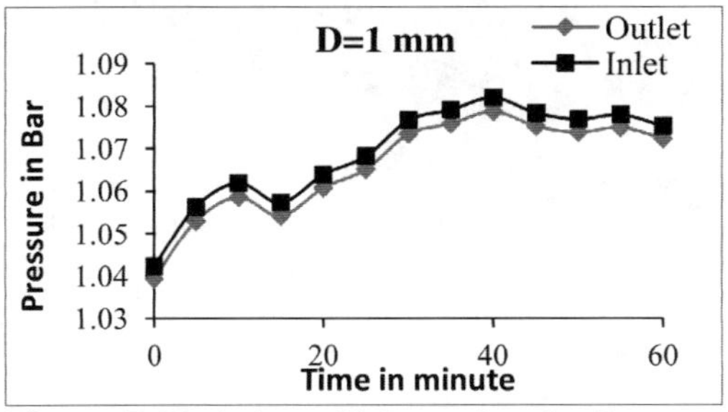

Figure 5. Variation of inlet and outlet Pressure with time for condenser size of 1 mm

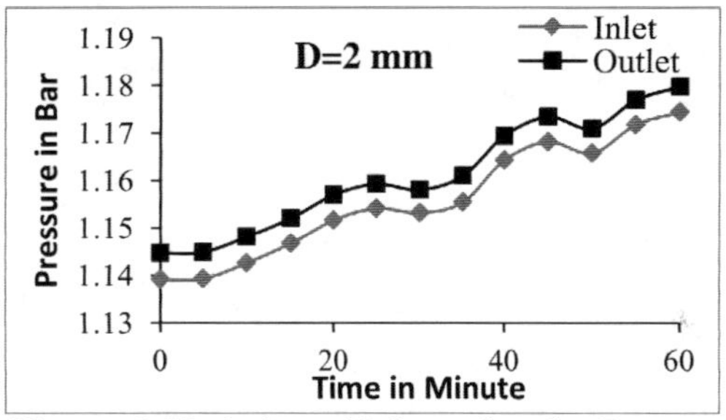

Figure 6. Variation of inlet and outlet Pressure with time for condenser size of 2 mm

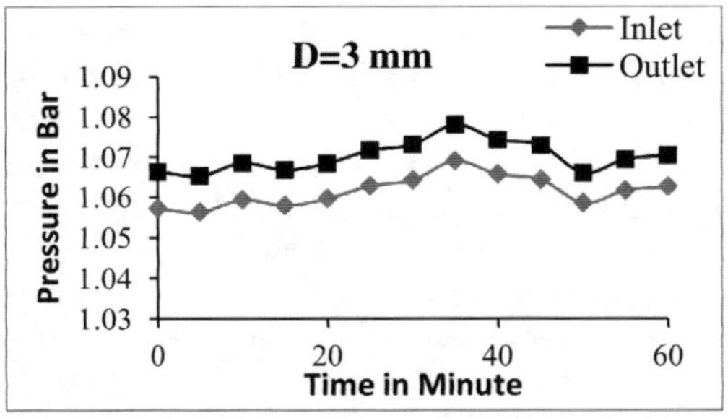

Figure 7. Variation of inlet and outlet Pressure with time for condenser size of 3 mm

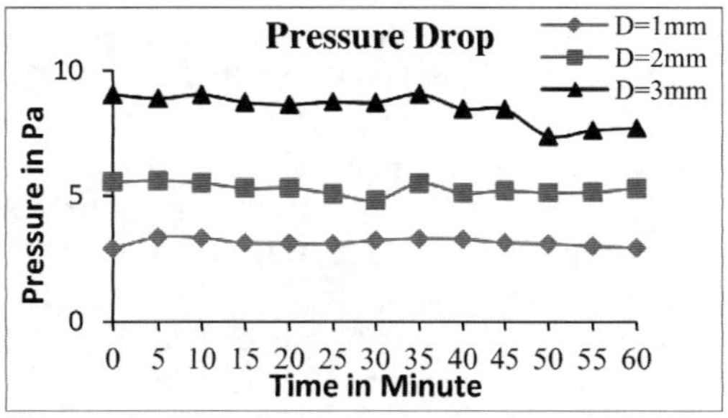

Figure 8. Variation of pressure drop with time for condenser size of 1 mm, 2 mm & 3 mm

Figure 8 shows the pressure drop of the three different sizes of mini channel condensers. The corresponding mass flow rates of size 1 mm, 2 mm and 3 mm are 2 g/s, 52 g/s and 58 g/s, respectively. The pressure drop is quite dependent on the mass flow rates, i.e., the higher the mass flow rate the higher is the pressure drop. However it is also dependent on the diameter. The pressure drop of the 1mm condenser although lower is not very far from the 2 mm and 3 mm condensers as correlated to their mass flow rates differences.

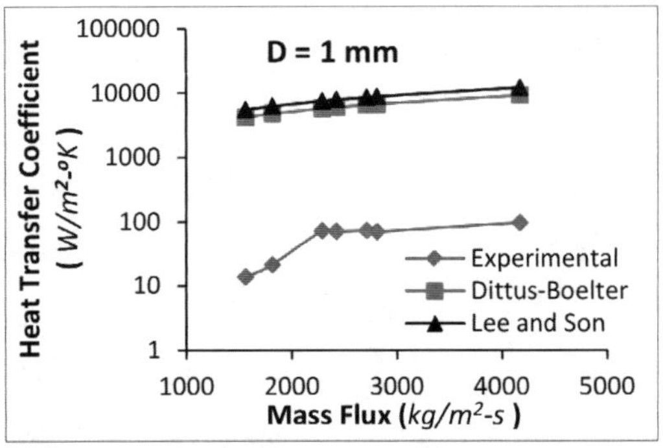

Figure 9. Variation of heat transfer coefficient with mass flux for condenser size of 1 mm

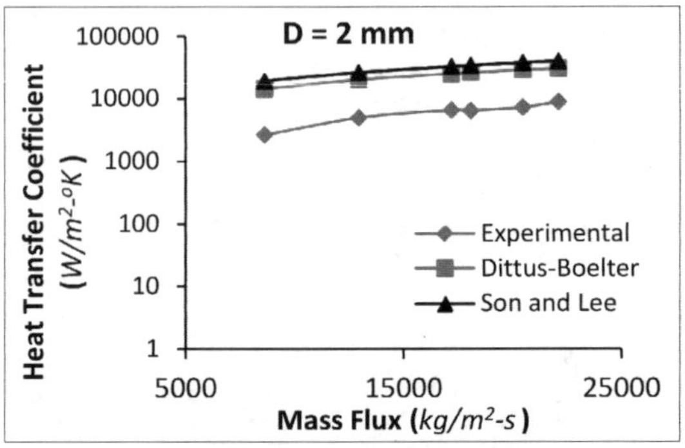

Figure 10. Variation of heat transfer coefficient with mass flux for condenser size of 2 mm

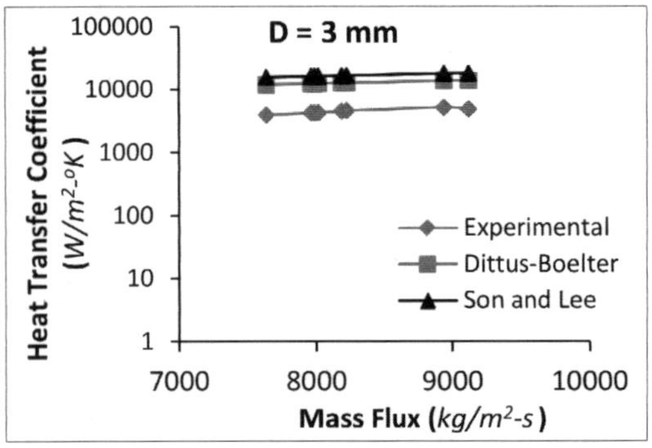

Figure 11. Variation of heat transfer coefficient with mass flux for condenser size of 3 mm

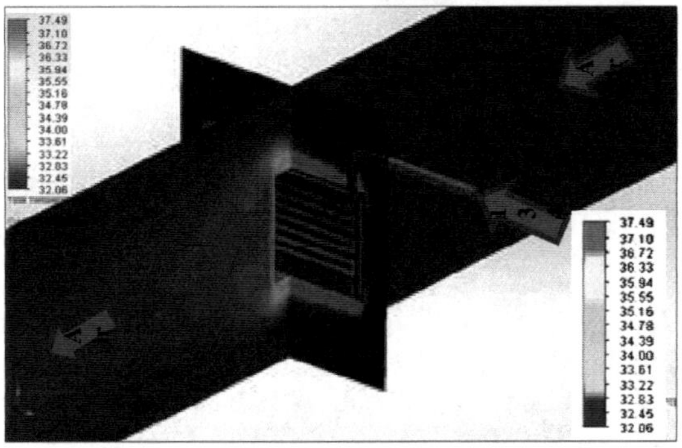

Figure 12. Temperature distribution of the mini channel condenser along the wind

The theoretical values of the average heat transfer in refrigerant side of R134a computed from Dittus-Boelter [2] and Lee-Son [5] correlations at experimental data were compared with experimental values of heat transfer coefficient from equation 4 as shown in figures 9 to 11. As can be seen in figures 9 to 11, the experimental heat transfer coefficients values ranged from 13 to 98 W/m^2-°K for the 1 mm, from 2600 to 9000 W/m^2-°K for the 2 mm, and from 3900 to 5200 W/m^2-°K for the 3 mm. The heat transfer coefficient is affected by the mass flux and the size of channel. The heat transfer coefficient is increased with R134a mass flux rising. Figures 9 to 11consistently show the calculated heat transfer coefficient values of both Dittus-Boelter [2] and Lee-Son [5] correlations which over-predicted the experimental results. Also from the same figures, deviations are shown from the existing correlations of heat transfer coefficient. The deviation of this heat transfer coefficient with this correlation appears to be related to difficulties in accurate measurements [7] particularly the 1 mm in size as shown in figure 9. The primary cause for these deviations are related to difficulties in accurate measurements of small flow rates in small channels, especially the 1 mm channel diameter having a mass flow of 2 g/s. It is worth to note that as the channel dimensions become small, the surface roughness and the measurement error become relatively more significant. Many authors have noted that precise measurements on smaller channels are very difficult and subject to error. Such researchers include Cavallini *et al.* [8], Bergles *et al.* [9], and Koyama *et al.* [10]. The difficulties are caused by very small flow rates and amounts of heat, and small dimensions of the test sections. Other matters to be considered as mentioned by Kandlikar [11] that are frequently neglected is the uniformity of the cross section dimensions along the channel flow length.If unaccounted, these errors significantly alter the results.

The result of the simulation as depicted in figure 12 demonstrates the temperature profile inside the mini channel condenser and the wind tunnel. The inlet and outlet temperatures at the refrigerant side are 37.15°C and 35.22°C, respectively. At the air side, the upstream and downstream temperatures are 32.07 °C and 32.36°C respectively. From the same figure, one can see that the fluid temperature changes along the mini channel condenser and along the wind tunnel during simulation. The simulation results show the temperature profiles along the flow paths of both air and the refrigerant. The colors indicate the increasing temperatures of air as it absorbs heat and the decreasing temperatures of refrigerant as it releases heat.

6. Conclusions and Recommendations

Based on the results of this study, the following conclusions are drawn.

1. The inlet and outlet pressure reading is dependent with the mass flow rates and diameter. The higher the mass flow rate and the smaller the diameter the higher is the pressure reading.
2. The heat transfer coefficient increases with increase in the mass flux in the experimental results and in the calculated values of both Dittus-Boelter [2] and Lee-Son [5] correlations. However the calculated values of both correlations over-predicted the experimental results. This indicates some deficiencies of the correlations in small channels that need further investigations.

The following are recommended for further studies:

1. Conduct more experimental runs using different refrigerants, pressure above 11 bars and hydraulic radii of 1.5 mm, 2.5 mm, and 3.5 mm
2. Conduct similar experiments using condenser made of materials of high thermal conductivity and designed and fabricated to withstand pressures beyond 11 bars.
3. Conduct a boundary layer analysis for small tubes and channels and the result plus the experimental data be used in the formulation of a new empirical equation for the calculation of heat transfer coefficient.

7. References

[1] Kandlikar S G 2006 *Heat transfer & Fluid flow in Minichannels & Microchannels* (1st ed. UK: Elsevier Science)

[2] Dittus F W and Boelter L M K. 1930 *Heat transfer in automobile radiators of tubular type* (University of California, Berkeley, Publications on Engineering Vol. 2, No.13) pp. 443

[3] McAdams W H 1954 *Heat Transmission* (3rd ed. McGraw-Hill Book Company,New York)

[4] Owhaib W 2007 *Experimental Heat Transfer, Pressure Drop, and Flow Visualization of R-134a in Vertical Mini/Micro Tubes* (Doctoral Theses, KTH, School of Industrial Eng'g. and Management , Energy Technology, Applied Thermodynamics and Refrigeration) pp.57-58

[5] Son C and Lee H 2009 *Condensation heat transfer characteristics of R-22, R-134a and R-410A in small diameter tubes* (Heat and Mass Transfer 45:1153–66)

[6] Suhayla Y H 2011 *Experimental Investigation of Condensation of Refrigerants R134a and R12 in Air Cooled Horizontal Condenser* (Journal Of Eng'g. And Dev. Vol. 15,No.4)

[7] Mirza M S 2010 *Heat transfer during condensation inside small channels : Applicability of general correlation for macrochannels* (Proceedings of the 14th International Heat Transfer Conference, August 8-13 2010 Washington, D. C., USA. : 125-134)

[8] Cavallini A *et al* 2006 *Local Heat Transfer Coefficients During Condensation in a 0.8 mm Diameter Pipe* (Proceeding International Conference Nanochannels, Microchannels and Minichannels,Limerick, Ireland) pp. 139-146.

[9] Bergles A E *et al* 2000 *Boiling and Condensation in Small Diameter Channels* (Heat Transfer Engineering) pp. 18-40.

[10] Koyama S *et al* 2003 *Condensation of Refrigerant in a Multi-Port Channel* (First Int. Conf. on Microchannels and Minichannels Rochester, NY,USA) pp. 193-205.

[11] Kandlikar S G 2003 *Microchannels And Minichannels– History, Terminology, Classification and Current Research Needs* (First International Conference on Microchannels and Minichannels)

IOP Conf. Series: Materials Science and Engineering **88** (2015) 012029 doi:10.1088/1757-899X/88/1/012029

Experimental study on the submerged steam jet through side hole nozzle

W Wang[1], F Yuan, Q B Zhao, D T Chong*, W X Chen and J J Yan
State Key Laboratory of Multiphase Flow in Power Engineering, Xi'an Jiaotong University, Xi'an 710049, PR China
Tel:029-82665359, Email:dtchong@mail.xjtu.edu.cn

Abstract: In this paper, the steam shape parameters and the condensation oscillation characteristics of the direct contact condensation of submerged steam jet into subcooled water were investigated through a side hole nozzle. Meanwhile, the dimensionless jet length and the expansion ratio versus pool water temperature and steam mass flux were obtained by measuring the photographs taken by a high speed camera. Both of them increased gradually with rise of pool water temperature and steam mass flux. A comparison of the shape parameters of three different style nozzles was also achieved. And a correlation of the dimensionless jet length was obtained, the predicted data agreed well with the experimental data, while the discrepancy was within ±5%. For the pressure oscillation versus steam mass flux, the peaks and the root mean square value of pressure oscillation were almost consistent in the range of 150-400 $kg \cdot m^{-2} \cdot s^{-1}$, while both of them increase in the range of 400-800 $kg \cdot m^{-2} \cdot s^{-1}$. The positive peak and negative peak were completely symmetrical.

1. Introduction

The direct contact condensation (DCC) has distinct advantages, such as high heat transfer coefficient, without heat transfer surface, and so on. It has been widely used in chemical industry, nuclear power, cryogenic refrigeration, utilization of waste heat, and other fields. The condensation of submerged steam jet is the most important issue of DCC. Many researchers were attracted to investigate this.

Because of the complication of mass and heat transfer and the widely application, many researchers have done large work about this from 1970s. Cho et al.[1] gave a relatively complete condensation regime map in a wide range of steam mass flux and subcool water temperature based on visual observation and acoustics. The condensation pattern was divided into six parts including chugging, transient region, condensation oscillation, bubbling condensation oscillation, stable

[1] To whom any correspondence should be addressed.

condensation and interfacial oscillation condensation. Petrovic et al.[2] developed a three-dimensional condensation regime map of steam jet in water, which includes the effect of injector diameter. Wu et al.[3] suggested a three-dimensional condensation regime map of steam in water for supersonic and subsonic conditions, especially including the pressure ratio. Six shapes of the steam plume were observed, which were expansion-contraction, double expansion contraction, double expansion divergent, contraction, contraction expansion contraction and contraction expansion divergent. Kerney et al.[4] and Chun et al.[5] investigated the size of the steam plume and got the correlation of the dimensionless jet length in the similar pattern respectively. Kim et al.[6] found the relationship between the condensation heat transfer coefficient and the diameter of the nozzle outlet and the steam mass flow flux, and suggested three different models to calculate the condensation heat transfer coefficient. Cho et al.[7] corresponded the pressure oscillation characteristics of different flow patterns to the shape of the steam plume and achieved the effect of the steam mass flow flux on the pressure oscillation pulse. Qiu et al.[8] studied the pressure oscillation of the sonic nozzle including the pressure oscillation amplitude and the main frequency.

The side hole nozzle is a typical simplified model of the multiple holes nozzle. It is significant to study the single hole nozzle to state the mechanism of the multiple holes nozzle. Above all there were few reports to investigate the condensation of submerged steam jet for a side hole nozzle. In this paper, the steam shape and the condensation oscillation characteristics of the DCC of submerged steam jet into subcooled water were investigated for a side hole nozzle.

2. Experimental apparatus and measurement method

The experimental apparatus were schematically shown in figure 1. The experiment apparatus mainly consist of a steam boiler which can supply the saturated steam 0.3t/h continuously, a circular tank with diameter of 2000mm and height of 1500mm. Steam was generated from the boiler and flowed through the adiabatic pipes after being stabilized. Finally it jetted into the water tank through a nozzle and condensed in pool. The steam mass flux was controlled by an electric valve and measured by a vortex type steam flowmeter with accuracy of 0.5%FS. The steam pressure and temperature in the nozzle inlet was measured with a pressure transducer (0.1%FS) and a T-type thermocouple (0.1%FS), respectively. The temperature of pool water was measured by six uniform distributed T-type thermocouples. One fixed pressure sensor was installed on X= 80 mm and R= 50 mm. The pressure sensor accuracy was 0.25% FS with the range of -100kPa–100kPa.

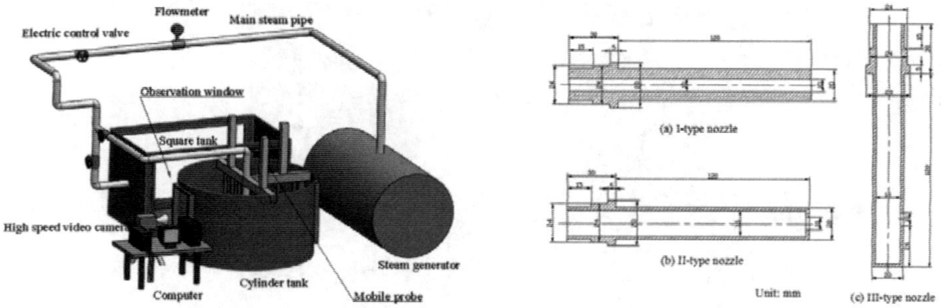

Figure 1. Schematic diagram of experimental system and the nozzle

In order to investigate whether the difference between the side hole nozzle and other kinds of nozzle exists, two different types of nozzle were used in this study as shown in Fig.1, and the side hole nozzle was the Ⅲ-type nozzle. The test of these two kinds of nozzles were taken in the square. They are also widely used in industrial fields.

To obtain steam plume photographs with high quality, a high speed camera coupled with submerged lamps were applied for taking the photograph of plume. And the macroscopic parameters of steam shape (expansion ratio and dimensionless penetration length) versus pool water temperature and steam mass flux were obtained by measuring the photographs. The photograph of steam condensation was obtained with a high speed of 5000 fps. And the test conditions was chosen as table 1. The pressure sensor was arranged at the location in the same plane with the hole.

Table 1 Test conditions

parameters	value
Diameter of hole d_e (mm)	10
Pressure sensor radial distance R(mm)	50
Pressure sensor axial distance X(mm)	80
Pool water temperature t_w (℃)	15-55
Range of steam mass flux G_e (kg·m^{-2}·s^{-1})	150-800

3. Experimental result and discussions

In this article, the jet length was defined as the length of steam plume core which did not include the wake region. The corresponding dimensionless jet length L was the ratio of the jet length l and the hole width D. The jet width was defined as the width of steam plume core. The plume tail which was the steam clumps separated from the steam plume core was not taken into account. The corresponding expansion ratio R_{ex} was the ratio of the jet width d and the hole width D. The two phase mixture region was not measured as the steam jet length either, as shown in figure 2. It was a complete steam plume of the side hole nozzle, and the others had the same shape and measurement standard as this. The length and the width of the jet were obtained by measuring 10 continuous photos and taking an average. During measuring photos, a commercial software was used to obtain the coordinates of the four edge points of the steam plume and the two edge points of the hole. Then a mathematical operation was taken to get the length and the width of the jet.

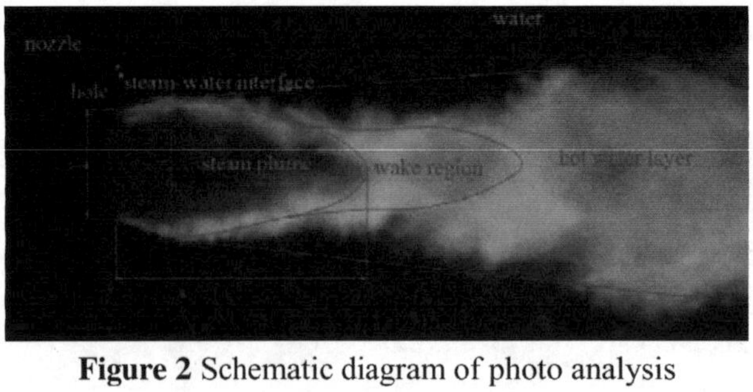

Figure 2 Schematic diagram of photo analysis

3.1 The macroscopic parameters of steam shape

The experiment results showed that the shape of steam jet plume was unstable when the steam flux was under 300 kg·m^{-2}·s^{-1}, and a relative stable steam plume was observed and measured at the steam flux larger than 400 kg·m^{-2}·s^{-1}. According to Cho et al. [1], the change value from condensation oscillation to stable condensation in condensation regime was about 300-400 kg·m^{-2}·s^{-1}. So, in the present paper, the stable shape size of the steam plume was measured when the steam mass flux was larger than 400 kg·m^{-2}·s^{-1}. Firstly, the the dimensionless jet length and the expansion ratio of the side hole nozzle versus pool water temperature and steam mass flux were given. And then a comparison of three different nozzle was suggested.

3.1.1 Dimensionless steam jet length

The corresponding dimensionless jet length was in the range of 1.82 - 4.61 when the pool water temperature ranged from 10 to 55°C and the steam mass flux ranged from 400-800 kg·m^{-2}·s^{-1} as figure 3.

The jet lengths increased gradually with rise of pool water temperature and steam mass flux. There was no noticeable rise under the test condition. This was because the water test condition was in stable condensation region, and no divergent plume occurred.

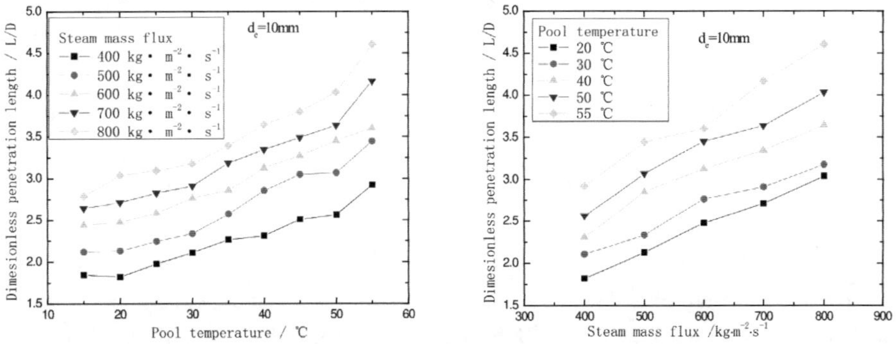

Figure 3 Dimensionless steam jet length versus pool water temperature and steam mass flux

Kerney et al.[4]proposed a theoretical model with the conversation of mass and energy to predict the steam jet length. According to the theoretical model, the jet length was closely related with the steam mass flux (G_0), dimensionless driving potential for the condensation process (B) . Then a correlation of dimensionless jet length was obtained and expressed as $L/D=f(B,G_0/G_m)$. Later, the correlation form, treated as the basic form, was widely used by the later researchers.

Based on the experimental data and the previous work[4], an independent correlation was obtained as follow:

$$\frac{l}{D} = 0.374B^{-0.739}(G_0 / G_m)^{0.628}$$

$$B=c_p\,(T_s - T_\infty)/\,h_{fg},\quad G_m=275\ \text{kg}\cdot\text{m}^{-2}\cdot\text{s}^{-1}$$

Here, G_0 is steam mass flux, kg/m^2·s, B is condensation driving potential, c_p is specific heat of pool, T_s is saturated temperature T_∞ is pool conditions, h_{fg} is latent heat at pool pressure.

And the predicted data and the experimental data agreed well and the discrepancy was within ± 5%

as figure 4.

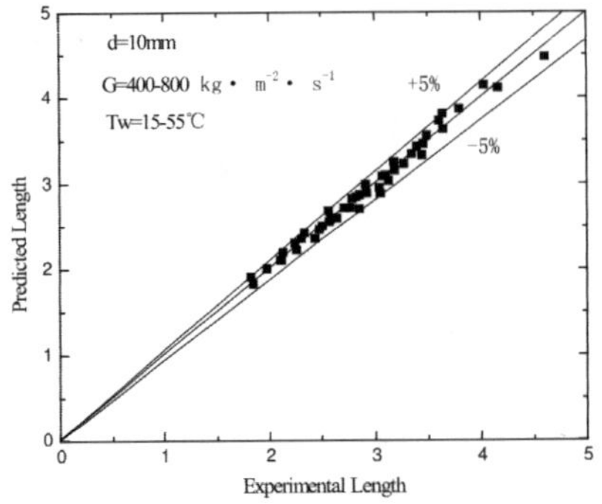

Figure 4 Comparison of predicted and experimental dimensionless jet length

3.1.2 Expansion ratio

The corresponding expansion ratio was in the range of 1.09 - 1.68 when the pool water temperature ranged from 10 to 55℃ and the steam mass flux ranged from 500-800 kg·m^{-2}·s^{-1}. The expansion ratio increased gradually with rise of pool water temperature and steam mass flux as figure 5.

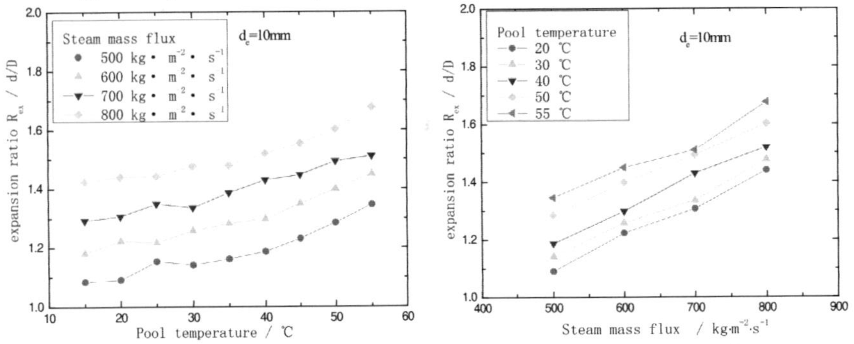

Figure 5 Expansion ratio versus pool water temperature and steam mass flux

It is easy to find that the expansion ratio and the jet lengths increased gradually with rise of pool water temperature and steam mass flux. It is because the capacity of heat transfer is closely related to the pool temperature and the flow flux. When the pool water temperature and steam mass flux increase, the capacity of heat transfer of the interface between steam and cool water decreases. So the larger heat transfer area is needed. Obviously it is efficient to prolong the length of steam plume and widen the width of steam plume.

3.1.3 The comparison of the expansion ratio and the jet lengths of the different style nozzle

The comparison of the jet lengths and expansion ratio of the three different style nozzles was shown as figure 6. In the figure, Ⅲ-type nozzle was also the side hole nozzle . The steam mass flux was 600 kg·m^{-2}·s^{-1}.

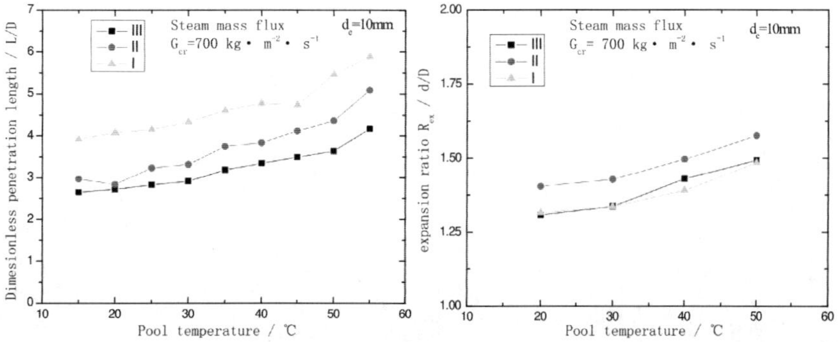

Figure 6 Comparison of the expansion ratio and the jet lengths of the three different style nozzle

From the Fig.6, the expansion ratio and the jet lengths of the side hole nozzle was quite different with the other two nozzles. The jet lengths of the side hole nozzle was smallest, and the expansion ratio was similar with that of Ⅰ-type nozzle but smaller than that of Ⅱ-type nozzle. The smaller jet length and expansion ratio are important parameter for the design of the size of the heat exchanger and have effect on the degree of the pressure oscillation of the whole equipment. So it was meaningful to investigate the pressure oscillation of the side hole nozzle.

3.2 Pressure oscillation

The results show that the condensation of side hole nozzle is really different with the other type nozzle. Pressure oscillation is another important side of the application of the direct contact condensation of submerged steam jet. It will affect the safety and the stability of the system. So it is necessary to study the pressure oscillation of the side hole nozzle.

3.2.1 Root mean square value of pressure oscillation

The pressure oscillation was got through arranging pressure transducer in the water tank. The experimental results show that the root mean square value of pressure oscillation increased gradually with rise of pool water temperature. And with the steam mass flux increasing, the root mean square value of pressure oscillation is almost consistent in the range of 150-400 kg·m^{-2}·s^{-1}, and increases while in the range of 400-800 kg·m^{-2}·s^{-1}as figure 7. This is because of the existence of the condensation transition region.

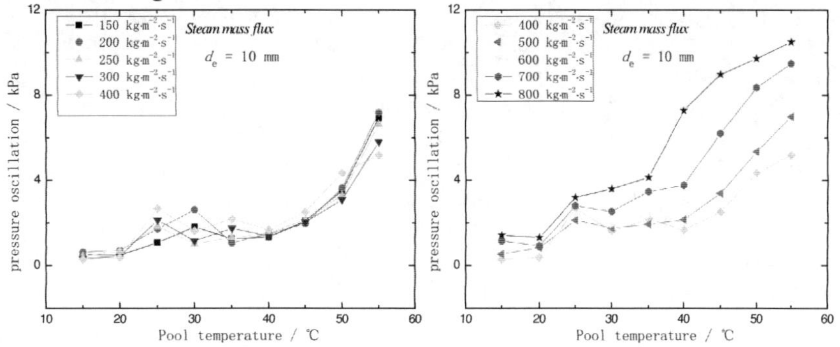

Figure 7 Root mean square value versus pool water temperature and steam mass flux

3.2.2 Positive peak and negative peak of pressure oscillation

The pressure oscillation is also got through arranging pressure transducer in the water tank. The experimental results show that the peaks of pressure oscillation increase gradually with rise of pool water temperature. And with the steam mass flux increasing, the peaks of pressure oscillation are almost consistent in the range of 150-400 kg·m^{-2}·s^{-1}, and increase while in the range of 400-800 kg·m^{-2}·s^{-1} . This is because of the existence of the condensation transition region. And the Positive peak and negative peak are completely symmetrical as figure 8.

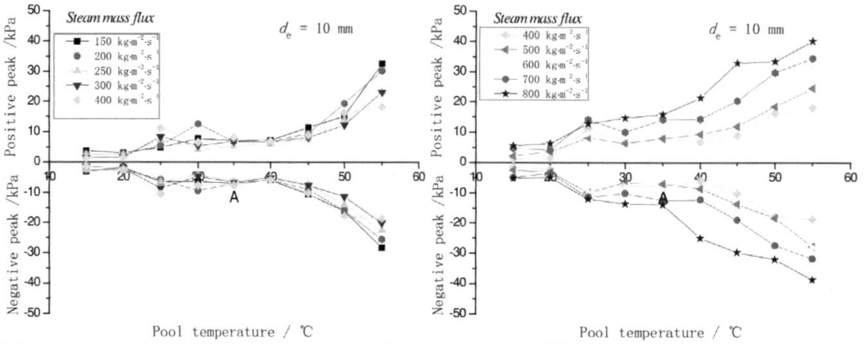

Figure 8 Positive peak and negative peak of the pressure oscillation versus pool water temperature

It is easy to find that with the steam mass flux increasing, the peaks and the root mean square value of pressure oscillation are almost consistent in the range of 150-400 kg·m^{-2}·s^{-1}, while both of them increase in the range of 400-800 kg·m^{-2}·s^{-1} . From the condensation regime map, there is a transient region between condensation oscillation region and stable condensation region. And in the transient region, the range of the steam mass flux was 200-400 kg·m^{-2}·s^{-1}. With the steam mass flux increasing, the pressure oscillation should increase because that the more steam needs to be condensated, but it will also be close to the stable condensation region. So in these two sides, the pressure oscillation are almost consistent in this range of the steam mass .

The law of pressure oscillation suggested in this paper was quite useful to choose the relatively safe pool water temperature and steam mass flux to ensure the safety of the facility. And the conclusion of the study is very meaningful to the investigate of the multiple holes nozzle which is usually used in many industry fields.

4. Conclusions

In this paper, the steam condensation in quiescent subcooled water of the side hole nozzle was investigated experimentally under different steam mass fluxes and water temperatures. The steam shape parameters and the condensation oscillation characteristics were surveyed and studied. The results can be mainly summarized as follows:

(1)The dimensionless jet length and the expansion ratio increased gradually with rise of pool water temperature and steam mass flux and were in different size with other two kinds of nozzles .

(2)A correlation of the dimensionless jet length was obtained, and the predicted data and the experimental data agreed well and the discrepancy was within ± 5%.

(3)In the range of 150-400 kg·m^{-2}·s^{-1} the peaks and the root mean square value of pressure oscillation are almost consistent, while in the range of 400-800 kg·m^{-2}·s^{-1} both of them increase, and the positive peak and negative peak were completely symmetrical.

5. Reference

[1]Cho S, Song C H, Park C K, et al. Experimental study on dynamic pressure pulse in direct contact condensation of steam jets discharging into subccoled water C. In: The 1st Korea－Japan Symposium

on Nuclear Thermal Hydraulics and Safety (NTHAS98), Pusan, Korea, 1998, 291-298.
[2]Petrovic A, Calay R K, With G. Three-dimensional condensation regime diagram for direct contact condensation of steam injected into water J. International Journal of Heat and Mass Transfer, 2007 50(9-10):1762-1770.
[3]Wu XZ,Yan JJ, Pan DD, Liu, GY and Li, WJ. Condensation regime diagram for supersonic/sonic steam jet in subcooled water J. Nuclear Engineering and Design, 2009,239:3142-3150.
[4]Kerney P J, Faeth G M, Olscn D R. Penetration characteristics of a submerged steam jet J . AICHE Journal, 1982, 18(3):548-553.
[5]Chun M H, Kim Y S, Park J W. An investigation of direct condensation of steam jet in subcooled water J . International Communications in Heat and Mass Transfer, 1996, 23(7):947-958.
[6]Kim, H.Y., Bae, Y.Y., Song, C.H. Jong K. Park and Sang M. Cho. Experimental study on stable steam condensation in a quenching tank J . International Journal of Energy Research, 2001, 25(3):239-252.
[7]Cho S, Song CH, Chung HJ, Chun, SY and Chung, MK. Multiple-hole effect on the performance on a sparger during direct contact condensation of steam J . KSME International Journal, 2001,15(4):482-491.
[8]Qiu BB,Tang S,YanJJ, Liu, JP, Chong, DT and Wu, XZ. Experimental investigation on pressure oscillations caused by direct contact condensation of sonic steam jet J . Experimental Thermal and Fluid Science, 2014, 52:p. 270-277.

Heat transfer, erosion and acid condensation characteristics for novel H-type finned oval tube

Y Wang[1], X Zhao, G Tang*
Key Laboratory of Thermal Fluid Science and Engineering of MOE,
School of Energy and Power Engineering, Xi'an Jiaotong University, Xi'an 710049, China

E-mail address: ghtang@mail.xjtu.edu.cn

Abstract. Low efficiency of heat transfer, acid corrosion and erosion of economizers affect the economy and security in coal-fired power plants significantly. The H-type finned oval tube is proposed to alleviate these problems. Based on the H-type finned oval tube, we investigated three novel types of fins, including bleeding dimples, longitudinal vortex generators (LVGs), and compound dimple-LVG. We considered the three aspects together, and obtained the heat transfer, acid condensation rate and erosion loss. The results show that the tube bank with the new structured fins can improve the performance on the three aspects, and the compound dimple-LVG performs the highest comprehensive effect.

1. Introduction

The economizers in coal-fired boilers usually face three significant problems: low heat transfer efficiency of the flue gas, serious tube acid corrosion and erosion. Recently, many researchers have investigated one of the three aspects. Li et al. [1] studied the heat transfer and laminar flow characteristics of a slit finned tube heat exchanger with LVGs numerically. They also optimized the slit fin structure by using the field synergy principle. Bi et al. [2] numerically studied the convective heat transfer inside mini-channels with dimples, cylindrical grooves and low fins. They found that the dimple surface shows the highest performance of heat transfer enhancement. Wilson [3] and Zhang [4] simulated the acid condensation rate and water condensation rate of the laminar flow in a circular tube. The FLUENT solver with laminar model was employed to preliminarily simulate acid condensation rate on the water-cooled wall by Volz [5] and in the concrete chimney in power station by Jin et al. [6], respectively. They found that the acid condensation

[1] To whom any correspondence should be addressed.

rate changes with the physical parameters. Ahmad et al. [7] experimentally studied the gas-liquid two-phase flow in heat exchanger. And they proposed a theoretical model to predict the convection-condensation and two-phase flow distribution. Nagarajan et al. [8] determined the effects of ash particle physical properties and transport dynamics on the erosive wear of three different grades of low alloy steel experimentally. Fan et al. [9-12] made a series of numerical calculation to predict the interaction between particles and tubes and the erosion. In these researches, some anti-wear structures were presented. Few researches, however, take all the three factors into account, and the comprehensive effect of the three aspects is more practical. In particular, for heat exchangers in the coal-fired power plants, it is necessary to consider not only the heat transfer and flow resistance performance, but also the acid anti-condensation and anti-wear performance. In this paper, we study the heat transfer, acid condensation and erosion characteristics of the single H-type finned oval tube, the H-type finned oval tube with bleeding dimples, with -30° rectangular longitudinal vortex generators (LVGs), and with compound dimples and LVGs. The purpose of this paper is to study the mechanism of heat transfer enhancement, acid condensation and erosion reduction of H-type finned oval tube with different vortex generators so that we can achieve a high efficient heat exchanger on heat transfer, acid anti-condensation and anti-wear.

2. Model description and numerical method

2.1. Physical model

A schematic diagram of a novel H-type finned oval tube with compound dimples and LVGs is displayed in Fig. 1. A pair of rectangle winglets is punched out from the fin symmetrically behind each oval tube and the bleeding dimples are penetrated out in the windward side. The geometry parameters of the H-type finned oval tube with dimples and -30° rectangular LVGs are listed in Table 1, and the oval tube minor radius and major radius are 40mm and 60mm, respectively.

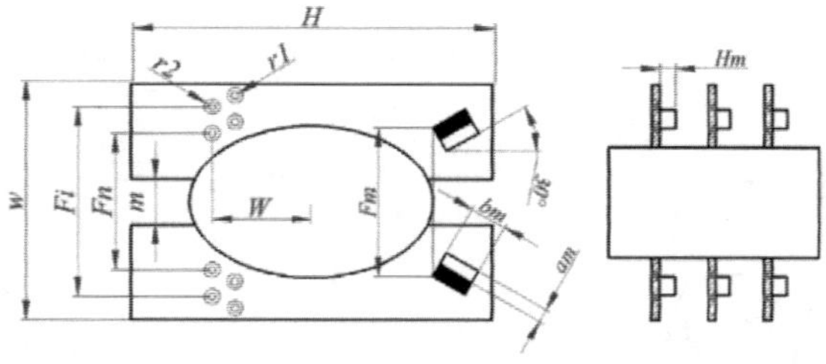

Fig. 1 Schematic diagram of a novel H-type finned oval tube with compound dimples and LVGs.

2.2. Governing equations and boundary conditions

It is assumed that the gas-solid flow in the economizer is a three-dimensional, viscous and steady incompressible turbulent flow. The tube surface is assumed to be at a constant temperature T_w =350 K and the wall thickness is ignored due to the relatively large heat transfer coefficient between the cooling water and inner wall of the tubes and the high thermal conductivity of the tube wall. Moreover, the temperature distribution in the fin surface and in the gas fluid is determined by coupled computation. The solid fin and tube are assumed to have constant thermal conductivity. The condensation of water vapor and sulfuric acid vapor in the flue gas is negligible, but it is computed on the wall. The temperature of liquid film surface is

IOP Conf. Series: Materials Science and Engineering **88** (2015) 012030 doi:10.1088/1757-899X/88/1/012030

equal to the temperature of the wall. The convection caused by the diffusion is negligible.

Table 1 Geometry parameters of the basic H-type finned oval tube with dimples and -30° LVGs.

S_1/mm	S_2/mm	H_m/mm	W/mm	w/mm	Fi/mm
140	100	5	26	72	50
F_t/mm	F_m/mm	F_p/mm	a_m/mm	b_m/mm	Fn/mm
3	40	15	5	10	36
r_1/mm	r_2/mm	r_3/mm	H/mm	m/mm	
1	0.4	1	95	12	

2.2.1 Gas-phase governing equations

The gas-phase governing equations including mass, momentum, energy and composition conservation equations are as follows.

Continuity equation:

$$\frac{\partial}{\partial x_i}(\rho u_i) = 0 \tag{1}$$

Momentum equation:

$$\frac{\partial}{\partial x_i}(\rho u_i u_j) = \frac{\partial}{\partial x_i}(\mu \frac{\partial u_j}{\partial x_i}) - \frac{\partial p}{\partial x_j} \tag{2}$$

Energy equation:

$$\frac{\partial}{\partial x_i}(\rho u_i T) = \frac{\partial}{\partial x_i}(\frac{\lambda}{C_p}\frac{\partial T}{\partial x_i}) + S_e \tag{3}$$

Where ρ is the density, u is the velocity, T is the temperature, μ is the dynamic viscosity, λ is the thermal conductivity, C_p is the specific heat at constant pressure, and S_e is the energy source term.

Species equation:

$$\nabla \cdot (\rho Y_i \mathbf{U}) = \nabla \cdot \left[\left(\rho D_{i,m} + \frac{\eta_t}{Sc_t}\right)\nabla Y_i\right] + S_m \tag{4}$$

Where S_m is the species source term, Y_i is the mass fraction of i component, Sc_t is the turbulent Schmidt number with $Sc_t = 0.7$, and η_t is the turbulent viscosity. The Reynolds number of the flue gas flow in the present study ranges from 15677 to 24705, and the turbulence model is presented below.

The RNG k-ε turbulence model:

$$\rho\frac{Dk}{Dt} = \frac{\partial}{\partial x_i}[\alpha_k \mu_{eff} \frac{\partial k}{\partial x_i}] + \mu_t S^2 - \rho\varepsilon \tag{5}$$

$$\rho\frac{D\varepsilon}{Dt} = \frac{\partial}{\partial x_i}[\alpha_\varepsilon \mu_{eff} \frac{\partial \varepsilon}{\partial x_i}] + C_{1\varepsilon}\frac{\varepsilon}{k}\mu_t S^2 - C_{2\varepsilon}\rho\frac{\varepsilon^2}{k} - R \tag{6}$$

where R is the rate of strain term given by

$$R = \frac{C_{\mu}\rho\phi^3(1-\phi/\phi)}{1+\beta\phi^3}\frac{\varepsilon^2}{k} \quad (7)$$

where $\phi = sk/\varepsilon$, ϕ_0=4.38, β=0.012, $C_{1\varepsilon}$=1.42 and $C_{2\varepsilon}$=1.68. $S^2=2S_{ij}S_{ij}$ is the modulus of the rate of strain tensor expressed as $S_{ij}=\frac{1}{2}(\partial u_i/\partial x_j + \partial u_j/\partial x_i)$.

2.2.2 Vapor condensation model

The mass condensation rate of the water and acid vapor at the wall surface is

$$m_i = -\left(\rho D_{i,m} + \frac{\eta_t}{Sc_t}\right)\nabla Y_i \quad (8)$$

The effective diffusivity $D_{i,m}$ of i component should be calculated from the binary diffusivity coefficients of i component through other components in the flue gas. The binary diffusivities D_{ij} can be calculated using the Fuller correlation [3].

The acid solution concentration at the wall surface is

$$y_a = \frac{R_a}{2R_a + R_w} \quad (9)$$

Where R_a and R_w are the mole condensation rate of the sulfuric acid vapor and the water vapor at the wall surface with $R_a = m_a/M$, and *M* is the mole mass of sulfuric acid.

When the condensation occurs in the grid cells in the fluid region near the wall, the sulfuric acid vapor and water vapor concentration in the gas phase will decrease and the latent heat is released. So the species source and energy source in Eqs. (3) and (4) are redefined as

$$S_e = -m_i r_i A/V, \quad S_m = -m_i A/V \quad (10)$$

Where *A* is the area of the grid on the condensing interface, *V* is the volume of the grid, and r_i is the latent heat of *i* component.

The partial pressure of the vapor at the gas-liquid interface is the key factor affecting the condensation rate. It can be calculated according to the gas-liquid equilibrium data [13].

$$\ln p_i = a_i A + b_i B + c_i C + \Delta H_i^v D + \Delta S_i^v E + \bar{C}_{pi}^l F + \bar{L}_i^l G + \alpha_i H + \ln \bar{a}_i(298) \quad (11)$$

Where p_i is the monomer partial pressure at the gas-liquid interface, *A-H* are the equation coefficients that are affected by the temperature of acid condensate, a_i, b_i, c_i, ΔH_i^v and ΔS_i^v are the thermodynamic parameters for pure components when the temperature is 298 K, and $\bar{C}_{pi}^l$, $\bar{L}_i^l$ and α_i are the parameters dependent on the composition change when the temperature is 298 K [14].

The hydration reaction of the sulfuric acid vapor and the water vapor lead to the production of $H_2SO_4 \cdot H_2O$, so p_i is not the accurate partial pressure. Fugacity coefficient was introduced to adjust the vapor concentration. In the chemical theory of solutions, the apparent partial pressure is equal to the monomer partial pressure divided by the apparent fugacity coefficient [15], given by

$$p_{io} = p_i / \phi_{io} \tag{12}$$

2.2.3 Lagrangian formulation for particle motion

The particles are assumed to be spherical points and we track each individual particle in the Lagrangian framework. In our assumption, the forces acting on the particles such as the gravity, thermophoretic force and Brownian force are ignored, but the Saffman lift [16] caused by the transverse velocity gradient is considered.

The force balance of the particles:

$$\frac{du_p}{dt} = F_D(u - u_p) + \frac{g_x(\rho_p - \rho)}{\rho_p} + F_x \tag{13}$$

where $F_D(u-u_p)$ is the particle drag force per unit mass

$$F_D = \frac{18\mu}{\rho_p d_p^2} \frac{C_D \mathrm{Re}_m}{24} \tag{14}$$

$$Re_m = \frac{\rho d_p \left| u_p - u \right|}{\mu} \tag{15}$$

where u is fluid phase velocity, u_p is particle velocity, μ is dynamic viscosity of the fluid, ρ and ρ_p are density of the fluid and particle, respectively, d_p is the particle diameter, C_D is the drag coefficient and Re_m is the relative particle Reynolds number.

2.2.4 Particle wall collision model and wall wear model

When fly ash particles collide with the surface of the tube and fin, the ash particles will change the trajectory as schematic in Fig. 2. To follow the particle trail, the collision model proposed by Tabakoff et al. [17] for coal ash particle impacting on stainless steel surface is employed. Normal and tangential coefficients of restitution can be assumed as constants during the collision of the particles with the tube and fin surface and presented as

$$\frac{V_{n2}}{V_{n1}} = 1.0 - 0.4159\beta - 0.4994\beta^2 + 0.292\beta^3 \tag{16}$$

$$\frac{V_{t2}}{V_{t1}} = 1.0 - 2.12\beta + 3.0775\beta^2 - 1.1\beta^3 \tag{17}$$

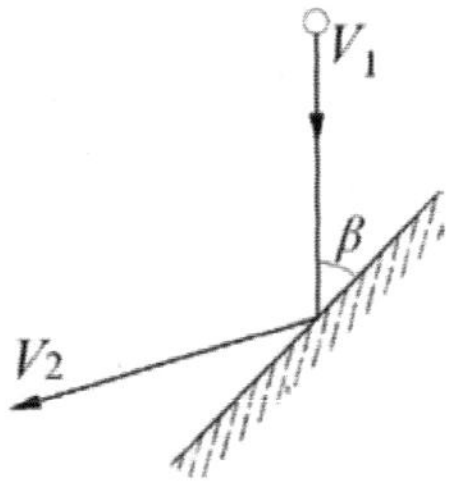

Fig. 2 Schematic of the particle collision with solid surface.

where V_n and V_t represent the particle velocity components normal and tangential to the tube surface, respectively. Subscripts 1 and 2 refer to the condition before and after collision, respectively. β is the angle between the incident velocity and the tangent to the surface as schematic in Fig. 3. In the erosion model by Tabakoff et al. [17], the mass wear rate E is defined as the ratio of the amount of mass loss of the target metal material (milligram) to the collision particle mass (gram).

$$E = K_1\{1 + C_k[K_2 \sin(\frac{90}{\beta_0}\beta)]\}^2 V_i^2 \cos^2\beta(1 - R_i^2) + K_3(V_i^2 \sin\beta)^4 \quad (18)$$

where V_i and β indicate the particle impact velocity and collision angle, respectively. $R_1 = 0.0016V\sin\beta_1$, and $\beta_0 = 25°$ is the angle when the maximum erosion occurs according to the experimental results of Tabakoff et al. [17]. $C_k = 1$ for $\beta_1 \le 3\beta_0$, while $C_k = 0$ for $\beta_1 > 3\beta_0$. K_1, K_2 and K_3 are the empirical constants with $K_1 = 1.505101\times10^{-6}$, $K_2 = 0.296007$ and $K_3 = 5.0\times10^{-12}$. Here, the employed tube material and the particles are the AISI 304 stainless steel and the anthracite, respectively.

2.2.5 Boundary conditions

The computational domain is composed of six boundaries: inlet, outlet, two symmetrical boundary (right and left), and two periodic boundary surfaces (top and bottom). To maintain the uniform inlet velocity and avoid the exit velocity with a recirculation-free flow, the computational domain is extended along the upstream by 3.5 times of the oval tube major radius length at the inlet zone and along the downstream by ten times of the oval tube major radius length at the exit zone, respectively. At the inlet boundary, the flue gas enters the computational domain along the x direction at uniform velocity u_{in}, temperature T_{in} (450K), and the velocity components along y and z directions are set to be zero.

For the particle phase, the diameter distribution of particle group follows the Rosin-Rammler diameter distribution, which can be manifested that the maximum diameter, minimum diameter and mean diameter of particles are 0.03 mm, 0.0001 mm and 0.01 mm, respectively. One-way coupling is applied in the dilute dispersed two-phase flow, and the effect of the discrete phase on the continuous phase is ignored. In this paper, a three-dimensional and steady-state numerical simulation is conducted by using the commercial software FLUENT with a second-order discretization scheme for both convective and diffusive terms. When the residual of each variable for gas phase is below 10^{-4}, the numerical computation is regarded as convergence. The erosion and sulfuric acid vapor condensation of tubes are measured by writing a UDF program. We deal with the mass transfer by solving the species equation and the source term in the equation.

2.3. *Parameter definition*

Some characteristic and non-dimensional parameters are defined as follows.

$$Re = \frac{\rho u_m D}{\mu}, \quad h = \frac{Q}{\Delta T A_t}, \quad Nu = \frac{hD}{\lambda}, \quad Eu = \frac{2\Delta p}{\rho u_m^2 Z}, \quad \Delta T = \frac{(T_{out} - T_w) - (T_{in} - T_w)}{\ln((T_{out} - T_w)/(T_{in} - T_w))}, \quad f = \frac{2\Delta P}{\rho u_m^2}\frac{A_m}{A_t},$$

$$\Delta P = P_{in} - P_{out}, \quad PEC = \frac{Nu/Nu_0}{(f/f_0)^{1/3}} \quad (19)$$

In the above equations, u_m is average velocity in the minimum flow cross-section, Q is the heat transfer capacity, ΔP is the total pressure drop over the whole computational domain, Z is the number of tube rows, A_m and A_t are the minimum flow cross-section area, and the total surface area, respectively, D is the characteristic length, which is the length of the oval tube minor radius [18], and Nu_0 and f_0 are Nusselt number and friction factor of the compared baseline, respectively.

2.4. Grid generation and independence validation

The mesh is generated by using software GAMBIT 2.4.6 and the grid independence test is necessary to make sure the accuracy and validity of the numerical results. Here, we demonstrate the grid independence test of the tube rows with compound dimples and LVGs in detail. Three sets of grid numbers about 190 413, 367 937, and 887 451 cells are examined and the obtained corresponding *Nu* at *Re* = 15677 are 111.52, 113.04 and 115.02, respectively. The relative error of *Nu* between number 2 and number 3 is 1.7%, and the grid number of 113.04 is selected finally considering both the accuracy of the numerical results and the computer resource. Besides, similar tests of the grid independence are also performed for the other cases mentioned above. The detail results are not shown here to save the article's space.

2.5. Validation of the computational model

Apart from the grid independence test, we should verify the computation accuracy of the model and calculation method. Here, the H-type finned tube bank is carried out for the validation. The simulated geometry parameters of the H-type finned oval tube are the same with the ones presented in Ref. [19]. The experimental correlations from Ref. [19] for the Nusselt number and Euler number are presented as $Nu = 0.09152Re^{0.7013}Pr^{0.33}$ and $Eu = 0.2963Re^{-0.449}$, respectively. The comparison of numerical results with the corresponding experimental correlations is displayed in Fig. 3. The average discrepancy and maximum discrepancy between the predicted *Nu* and the experimental correlation are 5.0% and 5.2%, respectively, and the average discrepancy and maximum discrepancy between the predicted *Eu* and the experimental correlation are 8.9% and 10.8%, respectively. However, there are still some obvious deviations between the numerical results and the solutions from experimental correlations, which may be ascribed to the system error in the experiment and the assumptions in the numerical simulation. These deviations could be allowable, and therefore, we can believe that the present numerical method meets the computational requirements for engineering applications.

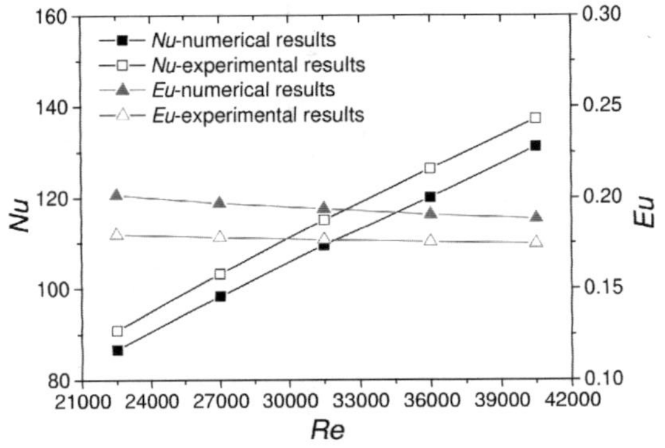

Fig. 3 Model validation with experimental data.

3. Numerical results and discussion

In the following sections, the longitudinal vortices generated by the rectangular winglet and dimples are presented. Then the heat transfer, acid anti-condensation and anti-wear characteristics based on the single H-type finned oval tube with different enhanced structures are discussed.

3.1. *The influence of rectangular winglet and dimples on flow field*

Fig. 4 presents the streamlines starting near dimples for u = 7m·s^{-1} at x= -26mm. When the gas flows over the dimples, the gas develops cyclones inside the dimples as shown in Fig. 4 (b). Zhang et al. [20] pointed that the pits in the surface could decrease both the turbulence intensity near the wall surface and the friction resistance through lowering the velocity gradient. That is because the low speed cyclones inside the pits lead to interaction between the inner flow and the outer flow of the dimples called air cushion, which is the fluid control behavior on the wall boundary layer developed by dimples surface. The friction drag generated from the bottom of dimples acts as an accessional impetus. When flowing over the LVG, the air in the back surface flows to the front fin surface because the pressure in the front surface is lower than that in the back surface. The streamlines traversing across the fin through the bleeding hole and the LVG enhance strongly the fluid transport between the mainstream region and the wake region, which can be observed from Figs. 4 (a)-(c). Fig. 4 (d) shows the velocity vectors of the cross-section at x=50mm behind the fin surface. It can be observed that the vortex is formed behind the fin, which is usually defined as the secondary flow. The structures of secondary flow generated by the LVG have been sufficiently studied and more details can be found in Ref. [21].

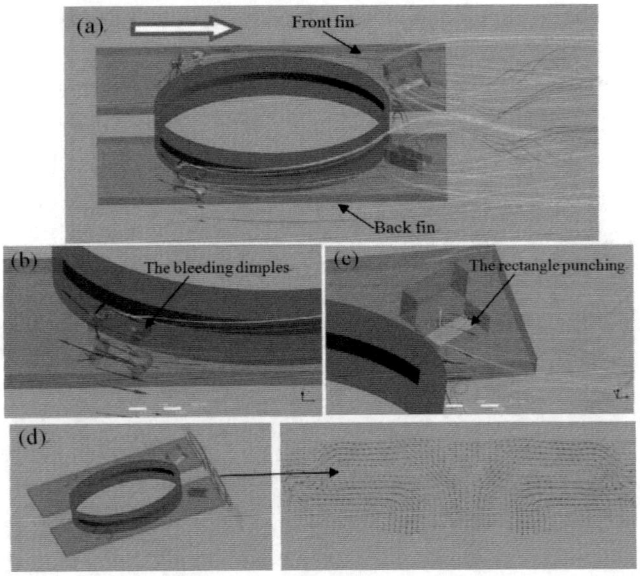

Figs. 4 Streamlines starting from the single tube near dimples in (a), (b) and (c), and velocity vectors of the cross-section at x=50mm behind the fin in (d).

3.2. *Flow and heat transfer characteristics of single tube*

Fig. 5 illustrates the local velocity distribution of the XY-plane in the middle-plane (Z=0) at u =7 m·s^{-1}. It can be seen that behind the tube, because of the obstruction of the tube, the gas separates into two regions, the mainstream region and the wake region. As can be observed from Fig. 5 (a), the flow velocity is very low in the rear of the tube (the wake region). Behind the tube, we can clearly see from the streamlines distribution that the vortices are generated, which are called transverse vortices. The existence of the transverse vortices shows the recirculating of the fluid in the wake region, where the fluid is almost separated from the main flow. The effect of dimples on the velocity field can be observed from Fig. 5 (c). Compared with Fig. 5 (a), the wake region in Fig. 5 (c) is compressed and the streamlines in the centers of the transverse vortices become dense. Besides, comparing with Figs. 5 (b) and (d) with Fig. 5 (a), we can see that the wake region is reduced when the gas flows over the LVG. The LVG is able to increase the flow transport between the mainstream region and wake region so that enhance the flow of wake region and delay the flow separation behind the tube.

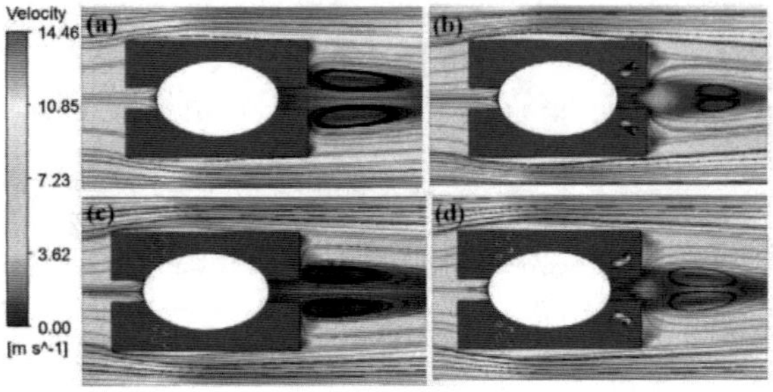

Fig. 5 Local velocity distribution on the middle cross-section in x direction. (a) H-type. (b) LVG. (c) Dimple. (d) Compound dimple-LVG.

Fig. 6 presents the Nusselt number Nu versus Reynolds number in which both the Nusselt number and pressure drop for all the cases increases with the increasing Reynolds number. The Nusselt numbers of the finned oval tubes with different heat transfer enhancement structures are significantly higher than the original H-type finned oval tube. Among the three types of enhanced structures, the dimple, the LVG and the compound dimple-LVG, the enhanced heat transfer of the compound case achieves the highest, followed by the LVG case and the dimple case. Compared to the baseline, the increase of Nu for the three types of enhancement structures from low to high is 25.7-34.8%, 36.4-46.3%, and 44.7-49.9%, respectively.

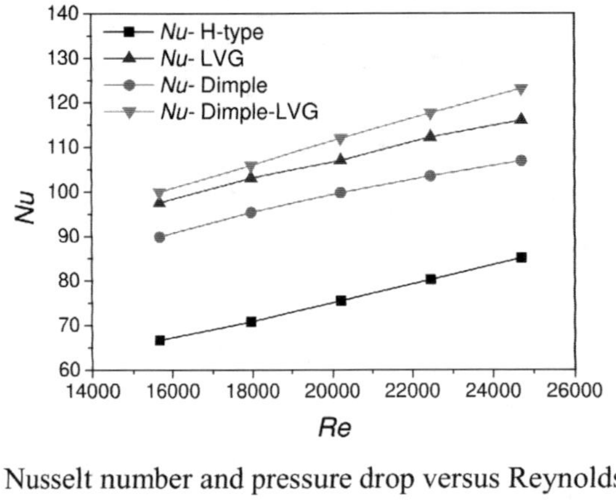

Fig. 6 Nusselt number and pressure drop versus Reynolds number

Fig. 7 shows the relationship between the friction factor and Reynolds number. It can be seen that the friction factor of the H-type finned oval tube with different augment structures are obviously higher than the original H-type finned oval tube. In addition, the friction factor from low to high is the dimple, the LVG, and the compound dimple-LVG cases, and the corresponding friction factor increase compared with the H-type finned oval tube is 34.7-37.5%, 48.5-51.8%, and 48.2-60.2%, respectively. However, at low Reynolds number, the friction factors for the tube with LVG and the tube with dimple-LVG are very close. The results suggest that the dimple could slow the speed of the friction increase at low Reynolds number, and it is in consistent with the conclusion in Ref. [20].

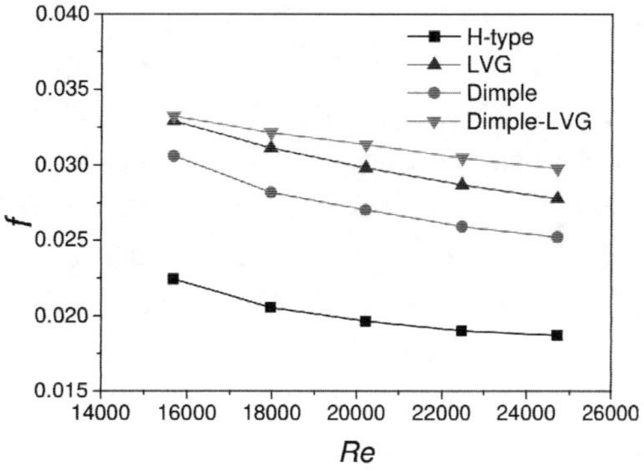

Fig. 7 Friction factor versus Reynolds number

Fig. 8 presents the parameter of PEC for the finned oval tube with different enhanced structures against Reynolds number. The PEC is a universal evaluation parameter which represents comprehensive performance of a heat transfer unit. The H-type finned oval tube is regarded as the compared baseline. From Fig. 10 we can see that the value of PEC from high to low is the compound dimple-LVG, the LVG, and the dimple. The compound dimple-LVG provides higher overall heat transfer performance than the LVG or the dimple, which is resulted from the hybrid effect. These results show that the tube with compound dimple-LVG case achieves considerable augmentation of heat transfer capacity with moderate flow resistance increase.

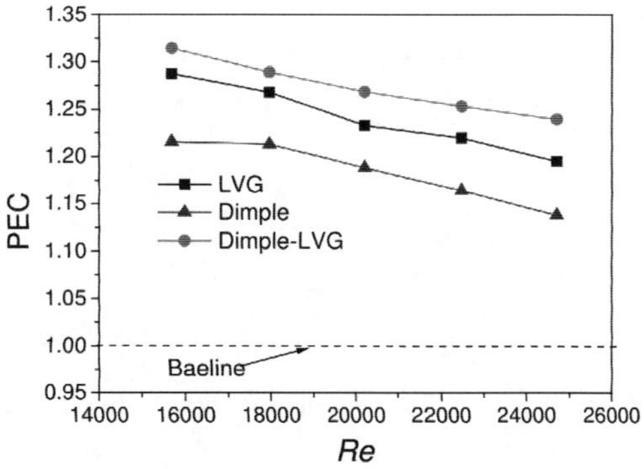

Fig. 8 PEC versus Reynolds number

3.3. Sulfuric acid vapor condensation characteristics of single tube

From Fig. 9, we can find that the distribution of the acid solution concentration is similar to that of the temperature. The high concentration of the acid solution distributes most on the fin surface instead of the tube wall. Due to the cooling of the tube, the acid solution concentration decreases first and then increases along the streamwise direction of the flue gas. Therefore, we can make sure that the fin surface can protect the tube from corrosion significantly. In addition, the acid condensation rate is higher in the central area of the fin and becomes lower at the rear of the fin close to the tube wall. This is because the larger velocity of flue gas in this zone can enhance the mass transfer between the acid vapor in the gas phase and the acid vapor at the gas-liquid interface. We can also find that both the dimples and LVGs have important effect on the acid condensation rate.

In Figs. 10, the inlet velocity is u_{in}=8 m·s^{-1} and the temperature of the tube wall is T_w=350 K. We can see that the acid condensation rate achieves the lowest for the finned surface with only dimples while the highest for the finned surface with only LVGs. Because the low velocity of the fluid near the dimple reduces the mass transfer between the acid vapor in gas phase and the acid vapor at the gas-liquid interface, which will lead to decrease of the acid condensation rate. The LVG structure can destroy the fluid boundary layer, which leads to flow disturbance and reduces the mass transfer resistance, so the LVG structure increases the acid condensation rate in the fin wake zone. It is interesting that in Figs. 10 (a), (b) and (c), the acid condensation rate is all about 1.4% higher for the LVG while about 6.3% and 4.7% lower for the Dimple and the compound Dimple-LVG, compared to the baseline of the H-type finned oval tube. Because the different finned tubes change the acid condensation rate mainly by breaking the flow state of the flue gas, however, the difference in the flow state of the flue gas is negligible among various fin structures at the present range of physical parameters (T_{in}, x_a and x_w).

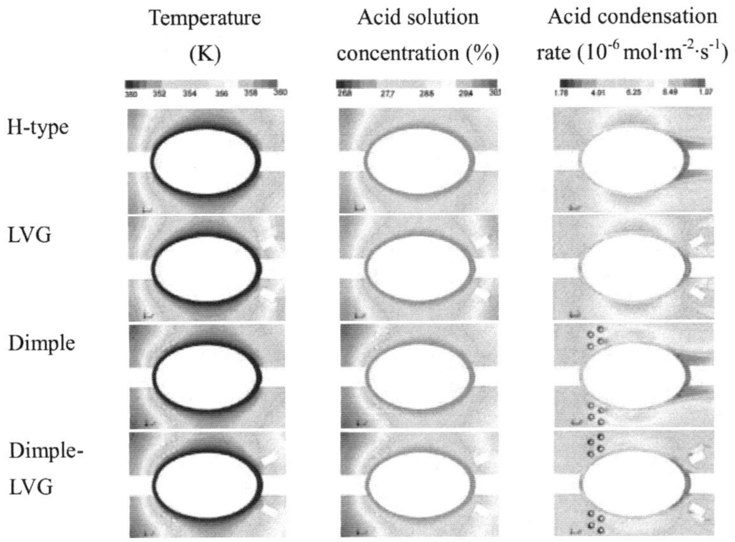

Fig. 9 Contours of temperature, acid solution concentration and acid condensation rate on different fin surfaces.

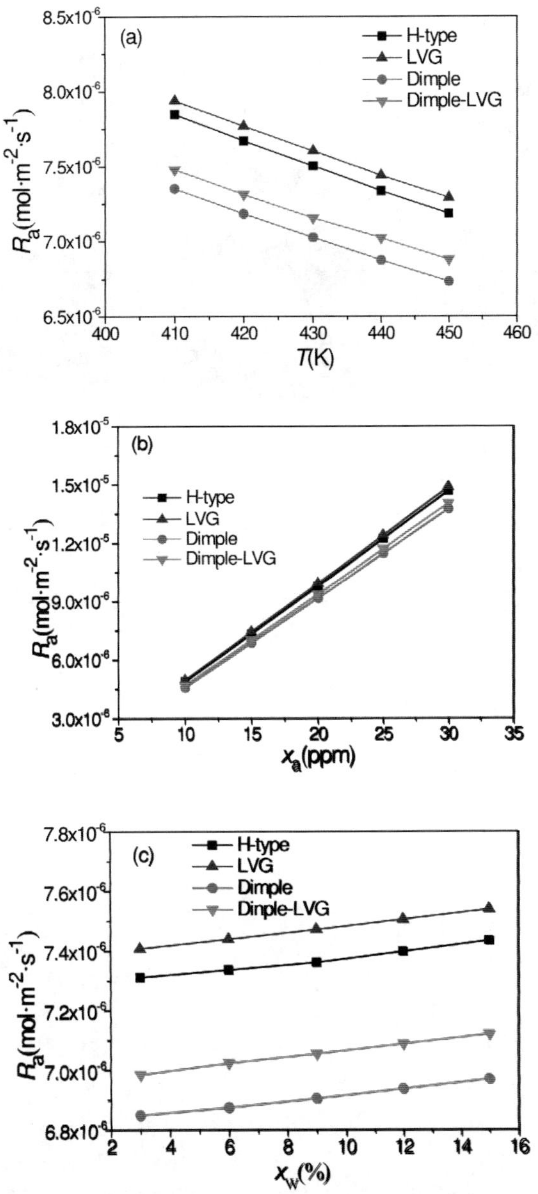

Fig. 10 Effects of physical parameters on acid condensation rate for different finned tubes. (a) Temperature. (b) Sulfuric acid vapor concentration. (c) Water vapor concentration.

3.4. *Erosion characteristics of single tube*

Figure 11 shows the erosion contours of the solid at u = 7 m·s^{-1}. We can see that the fin erosion is mainly distributed on both verges of the fin, which can be explained that those zones have large flow speed and the erosion is proportional to the cube of particle speed. Because the augment structures, namely the dimples or LVGs, change the fluid flied, the erosion loss of the solid tube and fin is decreased compared with the original H-type finned oval tube. Moreover, it can be seen that the maximum erosion of oval tube is concentrated in the range of 20-45° in windward of the oval tube.

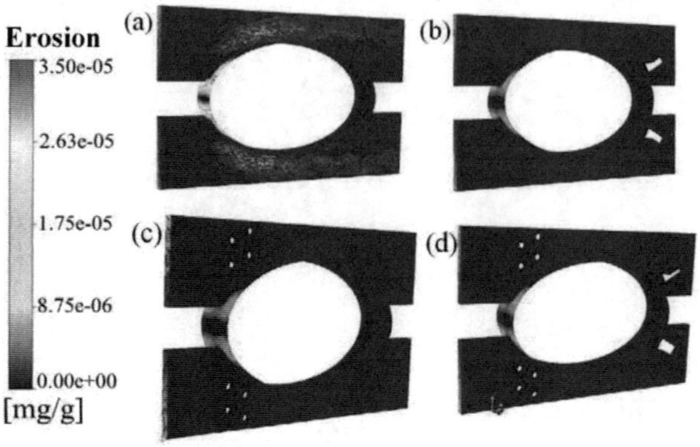

Fig. 11 Erosion contours of solid wall and fin. (a) H-type. (b) LVG. (c) Dimple. (d) Compound dimple-LVG.

From Fig. 12, we know that the tube wall erosion loss for H-type finned oval tube without any enhancement technique is the largest under the same Reynolds number. Specifically, compared with the H-type finned oval tube without any enhancement technique, the tube wall erosion loss of the compound dimple-LVG case and the dimple case is almost the same and reduced by 46.1-49.1%. The tube wall erosion loss for the H-type finned tube with rectangular LVG is reduced by 76.4-76.9%. Furthermore, the anti-wear performance of the enhanced tubes improves as the inlet velocity increases.

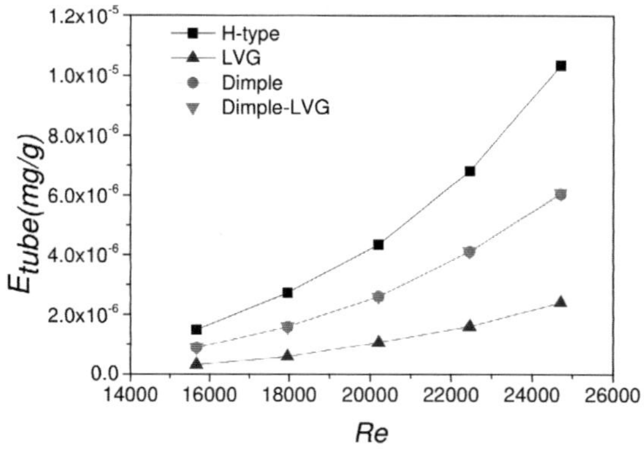

Fig. 12 Tube erosion value versus Reynolds number.

4. Conclusions

In this study, the heat transfer, sulfuric acid vapor condensation and erosion characteristics of the H-type finned oval tube, tube with LVGs, tube with bleeding dimples, and tube with compound dimples and LVGs have been numerically studied. The following conclusions are obtained.

1. The comprehensive performance of the enhanced heat transfer surfaces is evaluated by the parameter PEC. The H-type finned oval tube with compound dimples and LVGs provides the highest overall heat transfer performance. The performance of the tube with LVG is lower than that of the tube with compound dimple-LVG, and the tube with bleeding dimple presents the lowest performance.

2. Both the finned tube with bleeding dimples and the finned tube with compound dimples and LVGs can inhibit acid vapor condensation effectively because the acid condensation rates of them are lower than

that of the H-type baseline under the same conditions. Furthermore the finned tube with bleeding dimples performs slightly higher.

3. The anti-wear performance of the tube with LVGs is the highest, and it is obviously higher than that of the tube with compound dimple-LVG and with the sole dimples. For all the enhanced structures, the erosion loss of the tubes are dramatically less than the original H-type oval tube, and the reduction of erosion loss gets more for a larger Reynolds number.

The H-type finned oval tube with compound dimple-LVG heat exchanger may have significant applications not only to improve the heat transfer efficiency and economy, but also reduce acid condensation and the wear loss of heat exchanger and promote the safe operation level in coal-fired boilers.

5. Acknowledgments

This work was supported by the National Basic Research Program of China (973 Program) (No. 2011CB710702) and the National Natural Science Foundation of China (No. 51222604).

6. References

[1] Li J, Wang SF, Chen JF, Lei YG. Numerical study on a slit fin-and-tube heat exchanger with longitudinal vortex generators. Int J Heat Mass Transfer 2011; 54: 1743-1751.

[2] Bi C, Tang GH, Tao WQ. Heat transfer enhancement in mini-channel heat sinks with dimples and cylindrical grooves. Appl Therm Eng 2013; 55: 121-132.

[3] Wilson RW. Condensation of sulfuric acid and water mixtures in laminar flow. Dissertation for Doctoral Degree. Lehigh University: 1989.

[4] Zhang ZT. Mass transfer processes in flow passages of rotating regenerative air preheaters. Dissertation for Doctoral Degree. Lehigh University: 1991.

[5] Volz W. Condensation of water vapour and acid mixtures from exhaust gases. Dissertation for Doctoral Degree. Berlin University: 2004.

[6] Jin BS, Tang ZY, Sun KQ, et al. Numerical prediction of deposit rate of condensed sulfuric acid from the flue gas of power plant chimney (in Chinese). Proceedings of the CSEE, 2006; 26: 40-44.

[7] Ahmad M, Berthoud G, Mercier P. General characteristics of two-phase flow distribution in a compact heat exchanger. Int Heat Mass Transfer 2009; 52: 442-450.

[8] Nagarajan R, Ambedkar B, Gowrisankar S, Somasundaram S. Development of predictive model for fly-ash erosion phenomena in coal-burning boilers. Wear 2009; 267: 122-128.

[9] Fan JR, Zhou DD, Jin J, Cen KF. Numerical simulation of tube erosion by particle impaction. Wear 1991; 142: 171-184.

[10] Fan JR, Zhou DD, Zeng KL, Cen KF. Numerical and experimental study of finned tube erosion protection methods. Wear 1992; 152: 1-19.

[11] Fan JR, Sun P, Zheng YQ, Zhang XY, Cen KF. A numerical study of a protection technique against tube erosion. Wear 1999; 225: 458-464.

[12] Fan JR, Sun P, Chen LH, Cen KF. Numerical investigation of a new protection method of the tube erosion by particles impingement. Wear 1998; 223: 50-57.

IOP Conf. Series: Materials Science and Engineering **88** (2015) 012030 doi:10.1088/1757-899X/88/1/012030

[13] Gmitro JI, Vermeulen T. Vapor-liquid equilibria for aqueous sulfuric acid. AIChE J, 1964; 10: 740-746.

[14] Giauque WF, Hornung EW, Kunzler JE, Rubin TR. The thermodynamic properties of aqueous sulfuric acid solutions and hydrates from 15 to 300 K. J Am Chem Soc, 1960; 82: 62-67.

[15] Copeman TW, Stein FP. On the derivation of solution thermodynamics for reactive components. Fluid Phase Equilib, 1982; 9: 149-165.

[16] Saffman PG. The lift on a small sphere in a slow shear flow. J Fluid Mech 1965; 22: 385-400.

[17] Tabakoff W, Kotwal R, Hamed A. Erosion study of different materials affected by coal ash particles. Wear 1979; 52: 161-173.

[18] Holman JP. Heat Transfer. 10th ed. China Machine Press: 2011.

[19] Liu YZ, Yuan YC, Xu SY, Wu WB. Experimental study on the characteristics of heat transfer and flow resistance for H-type finned tube banks (in Chinese). J Univ Shanghai Sci Technol 2004; 26: 457-460.

[20] Zhang CC, Ren LQ, Wang J, Zhang YZ. Simulation on flow control for drag reduction of revolution body using bionic dimpled surface (in Chinese). Acta Armamentarii 2009; 30: 1066-1072.

[21] Chang LM, Wang LB, Song KW, Sun DL, Fan JF. Numerical study of the relationship between heat transfer enhancement and absolute vorticity flux along main flow direction in a channel formed by a flat tube bank fin with vortex generators. Int J Heat Mass Transfer 2009; 52: 1794-1801.

Optimising the operational parameters of a spherical steriliser for the treatment of oil palm fresh fruit bunch

D Kumaradevan[1], K H Chuah[1,3], L K Moey[1], V Mohan[2], W T Wan[2]
[1]FOSTEM, INTI International University, Nilai, 71800 Negeri Sembilan, Malaysia
[2]BELL Management Sdn. Bhd., Subang Jaya, 47500 Selangor, Malaysia

E-mail: kenghoo.chuah@newinti.edu.my

Abstract. The extraction of crude palm oil (CPO) begins with the sterilization of oil palm fresh fruit bunch (FFB) in a pressurized, saturated-steam chamber. Sterilization loosens the palm fruits from the stalks and deactivates the free fatty acid (FFA)-producing enzymes. Operational parameters affecting the quality and yield of CPO from an industrial spherical sterilizer are studied at a palm oil mill. The factors are the ripeness of FFB, the number of days before treatment of FFB, and the number of pressure peaks applied in the sterilization process. The results indicate that the degree of ripeness of FFB is the most important parameter affecting the quality and yield of CPO. Ripeness is graded based on the fruits' colour and the presence of loose fruits. Over ripe FFB that goes for the sterilization process has higher FFA content in CPO and more oil loss to the condensate chamber. The spontaneous reaction on FFB due to accumulation at the loading ramp also gives rise to higher FFA content. Oil loss to condensate chamber is reduced using a two-peak sterilization technique for over ripe FFB; the peak refers to the pressure level of stream after a flushing and refilling cycle. Overall, the generated solution improves the quality and yield of the palm oil mill.

1. Introduction

Crude palm oil (CPO) is a globally traded commodity extracted from the flesh or mesocarp of oil palm fruit. The extraction begins at a mill with the sterilization of fresh fruit bunch (FFB), where each bunch contains more than a thousand individual fruits and all bunches are cooked in a pressurized, saturated-steam chamber. The purposes of sterilization are to detach the fruits from the stem, to stop the enzyme reaction that caused free fatty acid (FFA) in the oil, to assist in the breaking of oil cells in the mesocarp, and to condition the nuts for subsequent cracking [1]. As sterilization is the first step in the production of CPO, any problems at this stage will propagate downstream, affecting the successive processes as well as the overall yield and quality of the oil. For instance, an insufficiently cooked FFB may have some of its fruits intact, preventing them from being stripped, or an excessively cooked FFB may have rendered some of its fruits too tender for machineries, creating chances of ripping and causing an oil loss. Sterilization is therefore vital for CPO extraction [2].

[3] To whom any correspondence should be addressed.

This paper is a report to analyse the operational parameters affecting the quality and yield of CPO from an industrial spherical sterilizer. The objectives of the study are:

- to determine the FFB ripeness and to control the FFA level,
- to identify the factors affecting the sterilization process,
- to identify the quality of fruit collected at loading ramp without processing, and
- to provide a solution to overcome the factors affecting the sterilization process and the fruits collected at the loading ramp

Section 2 introduces the operational parameters of spherical steriliser. Section 3discusses the experiments, results and discussions. Section 4 is the conclusions.

2. Operational parameters of spherical steriliser

With the objectives of improving yield and quality of CPO, the independent parameters are the number of pressure peaks applied in the sterilization process, the ripeness of FFB, and the number of days before treatment of FFB. The dependent parameters are the percentage of FFA in oil and the percentage of oil in condensate. The controlled parameters are the steam conditions and the current sources of FFB.

2.1 Sterilization of fresh fruit bunch

Sterilization is performed in a chamber with saturated steam at an elevated pressure. Initially, air is removed from the chamber with steam together with any condensed water. Steam at a lower density can displace air from the top of the chamber, forcing the air to leave at the bottom. Once the air is removed, the pressure in the chamber is raised to a set pressure to start the sterilization process. The process produces low quality steam-water mixture as a result of heat transfers, and the mixture is therefore flushed and replaced with new steam to renew the heating. The process utilizes new steam at 30 – 40 psig and the duration of sterilization is between 75 and 90 minutes; current industrial practice has dropped the time to about 60 and 70 minutes. Overall, sterilization generates about 36% of the total waste water in the mill [1].

Conventional sterilizer is a long cylindrical chamber with a side loading door and rails running all the way into the chamber. A number of carts filled with FFB can be pushed into the chamber for sterilization, and as easily be pulled out at a later time. Steam is piped into the chamber once the chamber is filled, thereby heating both the carts and the FFB.

The mill in this study uses the approach of a spherical sterilizer, a 25 cubic meter enclosure with a top opening for filling the FFB and a bottom condensate drain. The drain is covered with a strainer to block FFB from slipping through the waste water drain. The spherical chamber, pivoting on two horizontally aligned shafts to the chamber's centre, can be rotated 180 degrees to pour out its contents. Steam is supplied in a concentric pipe through each anchoring shaft. In general, spherical sterilizer is more efficient than conventional sterilizer, as the process can be scaled up and transporting carts are not necessary.

In either case, flushing and refilling expend steam energy, which lead to more waste water. Each steam change cycle is called a peak, defined as an increase in chamber pressure until a maximum followed by a decrease in chamber pressure. In practice, two-peak and three-peak sterilizations are common, and the tests in this paper are conducted with these two configurations.

Steam sterilisation is not the only way of treating FFB. The other methods are air sterilisation [3] and microwave sterilisation [4]. Hadi et al explores direct sterilisation with conventional oven, and the result is comparable to steam heating, if not better [3]. Dry heating avoids the generation of waste water, which water ends up as palm oil mill effluent if mixed with oil and dirt in the process.

Similar result is obtained for microwave sterilisation. In addition, microwave irradiation can sterilise FFB in less than 17 minutes with relatively low power while maintaining the CPO quality [4].

IOP Conf. Series: Materials Science and Engineering **88** (2015) 012031 doi:10.1088/1757-899X/88/1/012031

2.2 Classifications of fresh fruit bunch

FFB arrives at the mill in several different conditions: under ripe, unripe, ripe, over ripe, empty and rotten. The quality of FFB affects the yield and quality of the oil produced and is a major factor in the course of optimising the sterilization process [4]. Local plantations in Malaysia practice manual grading of FFB. With human vision, the colour of fruits is the simplest and most direct way of assessing the quality of FFB, although automated grading with machine vision has been developed recently. Jaffar et al. reports an automation method to systematically grade FFB into ripe and unripe grades. The method is a prove-of-concept that correlates colour to ripeness with the use of colour coding and computerized image processing system [5].

Table 1. Classification of FFB into ripeness categories based on mesocarp colour and pattern.

Ripeness	Colour	Loose fruits	Sample
Over ripe	Orange	>10	
Ripe	Orange-red	~10	
Under ripe	Red-purple	<10	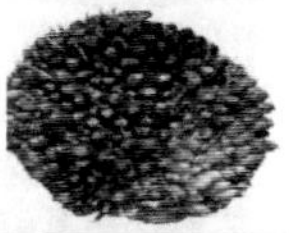
Unripe	Purple-black	<1	

A more comprehensive system for automatic grading, developed by Makky and Soni, shows that machine vision can perform up to 12 tons of grading per hour [6]. The system classifies FFB into 6 categories starting from unripe to over ripe, as well as rejecting empty and rotten bunches with success rate of over 93%. Field tests conducted indicate that just minor adjustment is required.

In this paper, manual grading is performed at the mill for experimental purposes. Random samples of 80 FFB are graded based on mesocarp colour and presence of loose fruits. The classification is divided into unripe, under ripe, ripe and over ripe as shown in Table 1 below. Based on the ripeness distribution sampled, an assessment is made to ascertain the effect of ripeness on the percentage of FFA in oil and the percentage of oil in condensate. For example, the assessment may show that a distribution with more over ripe FFB lose more oil than a distribution with more under ripe FFB.

2.3 Number of days before treatment of FFB

Delay in processing of FFB is sometimes unavoidable primarily due to fluctuations in supply. Storage of FFB before processing is thus a standard practice to buffer away the uncertainty, but the practice is detrimental to the quality of CPO. Zu and Bani notes that stored FFB have higher FFA content than freshly harvested FFB, and the content depends on the duration of storage [7]. The reason is the enzymes in the FFB after harvesting start attacking the mesocarp, making the flesh soft over time with the decomposition of oil and the production of FFA. The FFA content reaches unacceptable level by 72 hours after harvesting.

2.4 Measurement of FFA in oil
FFA is an undesirable content of CPO, and the target is to reduce FFA to below 5%. A simple chemical analysis can be performed to check the content in CPO. The analysis, following MPOB test method [8], mixes 6ml of CPO with 50ml of isopropanol and 3 drops of phenolphthalein in a solution. After boiling the solution, the percentage of FFA is estimated from the amount of sodium hydroxide (ml) required to turn the solution red times the ratio of its concentration and the sample volume.

2.5 Measurement of oil in condensate
Oil in condensate represents the oil loss with waste water. The percentage of oil in condensate is directly measured from the collection tank. The target is to reduce the percentage toward 0%.

3. Experiments, results and discussions

The spherical steriliser is a piece of industrial equipment in full operation at the mill. As such, we adhere to all practical restrictions in performing the process optimisation. Testing and improvement activities are carried out only when possible, and the data collected may not always be complete. In total, six cases of tests are conducted at the mill, and each case represents a single sterilization. Of the six cases, manual grading is performed for three of the cases to assess the effect of ripeness on the FFA and oil loss contents as well as the effect of number of sterilisation peaks on the same two contents. The other three cases assess the effect of storage on FFA content.

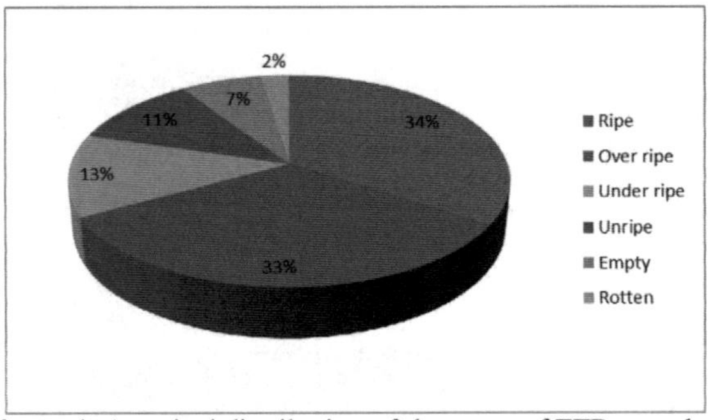

Figure 1. A typical distribution of the types of FFB sampled.

Figure 1 shows a typical distribution of the types of FFB sent to the mill. It is clear from the pie chart that two-third of FFB sampled are either ripe or over ripe, amounting to, respectively, 34% and 33% of all FFB. It is known that these FFB are where most of the oil is derived. In the remaining one-third of FFB, 13% is under ripe and 11% is unripe, respectively, whilst empty FFB (7%) and rotten FFB (2%) account for the rest of the samples. The mill acquires FFB from planters in the vicinity, and thus efforts that dissuade planters from harvesting unripe FFB can reduce these percentages. An effort for example is a system that rewards planters with better prices based on a track record of quality produce.

Table 2. Data collected on site for six test cases for the sterilising process.

Case No.	1	2	3	4	5	6
% over ripe fruits	25	17	58	-	-	-
% ripe fruits	34	48	20	-	-	-
Peak pressure (psig)	35	35	40	35	35	35
Process length (min)	67	67	55	67	67	67
No. of peaks	3	3	2	3	3	3

No. of days stored	-	-	-	2	3	2
% FFA	4.8	3.7	4.3	4.7	5.8	3.7
% Oil	4.6	2.9	3.9	-	-	-

Table 2 summarizes the values and parameters applied to each test case at the mill as well as the respective results. The results indicate that the degree of ripeness of FFB is the most important parameter affecting the quality and yield of CPO. Case 1 and case 2 show that the percentages of FFA and oil decrease with the percentage of over ripe fruits. Percentages of FFA are both below 5%, but the percentage of oil of the first case is the highest observed at 4.6%. It is clear that over ripe FFB that goes for the sterilization process has higher FFA content in CPO and more oil loss to the condensate chamber with the standard three-peak sterilisation process. To test the matter further, a random sample with over 58% of over ripe fruits is treated with a shorter, non-standard, two-peak sterilisation process. The result as case 3 indicates, the oil loss to condensate chamber is reduced to 3.9%. Figure 2 shows the physical appearance of over ripe sterilized fruit. Its physical appearance which is light brownish indicates that it is well sterilized. From these assessments, three-peak sterilization is optimum for ripe bunches and two-peak sterilization for over ripe bunches.

Figure 2. The physical appearance of over ripe sterilized fruits in Case 3.

To assess the effect of storage on the percentage of FFA content, case 4 and case 5 demonstrate that as the time of FFB storage grows from two days to three days, the percentage amount of FFA surges as well, from 4.7% to 5.8%. Clearly, FFA percentage of over 5% is unacceptable. Nevertheless, an interesting factor discovered is that if the stored FFB are spaced out in the open and left for two days, the FFA percentage actually drops to 3.7%. We theorize that the spontaneous enzymatic reaction on FFB to FFA is due to the packed accumulation at the loading ramp, as shown in Figure 3. By simply spacing out the FFB, enzymatic effect is reduced.

Figure 3. The FFB loading ramp at a mill.

4. Conclusions

In conclusion, the solutions to overcome the factors affecting the sterilization process and the fruits collected at the loading ramp are to source the fruits with consistent quality and to properly distribute the fruits at the ramp, conceivably with automation. As the degree of ripeness of FFB is the most important parameter that determines the quality and quantity of the extracted oil, proper grading can ensure that consistent process are always applied for different quality of fruits. To further increase the quality and quantity of oil production, automated handling of harvested fruit should be considered for preventing excessive delays at the ramp. A future study should be conducted on practical ways to incentivize the planters whereby only ripe crops are harvested and sent to mill for processing.

5. Acknowledgements

The authors thank the support of BELL Sri Lingga S/B for the project and the important roles they play in bring real world problems to the academics.

6. References

[1] Thani, M.I., Hussin, R., Ibrahim, W.W.R., and Sulaiman, M.S. (1999). Industrial processes & the environment: crude palm oil industry, Handbook No. 3.

[2] Jusoh, J. M., Rashid, N. A., & Omar, Z. (2013). Effect of Sterilization Process on Deterioration of Bleachability Index (DOBI) of Crude Palm Oil (CPO) Extracted from Different Degree of Oil Palm Ripeness. Int. J. Biosci. Biochem. & Bioinfo., 3(4), 322-327.

[3] Hadi, N.A., Ng. M.H., Choo Y.M., and Ma A.N. (2012). Dry heating of palm fruits: effect on selected parameters. American J. of Engr. & Applied Sci., 5(2), 128-131.

[4] Sarah, M., & Taib, M. R. (2013). Microwave Sterilization of Oil Palm Fruits: Effect of Power, Temperature and D-value on Oil Quality. Journal of Medical and Bioengineering, 2(3).

[5] Jaffar, A., Jaafar, R., Jamil, N., Low, C. Y., and Abdullah, B. (2009). Photogrammetric grading of oil palm fresh fruit bunches. Int. J. Mech. Mechatron. Eng, 9, 18-24.

[6] Makky, M. and Soni, P. (2013). Development of an automatic grading machine for oil palm fresh fruits bunches (FFBs) based on machine vision. Comp. & Electronics in Agriculture, 93, 129-139.

[7] Zu, K. S. A., Adjei-Nsiah, S., and Bani, R. J. (2012). Effect of processing equipment and duration of storage of palm fruits on palm oil yield and quality in the Kwaebibrem District, Ghana. Agric. Res. Rev, 1(1), 18-25.

[8] MPOB, Ed. (2005). MPOB Test Method (A Compendium of Test on Palm Oil Products, Palm Kernel Products, Fatty Acids, Food Related Products and Others, Malaysia Palm Oil Board.

IOP Conf. Series: Materials Science and Engineering **88** (2015) 012032 doi:10.1088/1757-899X/88/1/012032

Characteristics of electricity production by metallic and non-metallic anodes immersed in mud sediment using sediment microbial fuel cell

N Haque[1], D Cho[2], and S Kwon[1*]

[1]Department of Ocean System Engineering, Gyeongsang National University, Cheondaegukchi- Gil 38, Tongyeong, Gyeongnam 650-160, South Korea

[2]Department of Energy and Environmental Engineering, Soonchunhyang University, 22, Soonchunhyang-ro, Asan-si, Chungnam, 336-745,South Korea.

Abstract: Sediment microbial fuel cell (SMFC), equipped with Zn, Al, Cu, Fe orgraphite felt (GF) anodeand marine sediment, was performed. Graphite felt was used as a common cathode. SMFC was single chambered and did not use any redox mediator. The aimof this workwas to find efficient anodic material. Oxidation reduction potential (ORP), cell voltage, current density, power density, pH and chemical oxygen demand (COD) were measured for SMFC's performance.The order of maximum power density was 913 $mWm^{-2}$for Zn, 646 $mWm^{-2}$for Fe, 387.8 $mWm^{-2}$for Cu, 266 $mWm^{-2}$for Al, and 127 $mWm^{-2}$for GF. The current density over voltage was found to be strongly correlated one another in most metal electrodesbutthe graphite felt electrode, in which relatively weaker electricitywas observed because of its bio-oriented mechanism. Metal corrosion reactionsand/or a complicated microbial electron transfer mechanismacting around the anodiccompartment may facilitate to generate electricity. We presume that more sophisticated selection of anodic material can lead to better performance in SMFC.

1. Introduction

Microbial fuel cell (MFC) systems have been of great interest as a potential applicant for future alternative energy production where employ microbes to generate electricity from biochemical reaction of organic and inorganic substances [1].These substances are converted into electricity in the anode of MFC by microbial metabolism (via the action of bacteria as catalysts) [2, 3]. The anodic part of SMFC is important in attempt to (A) determine the most efficient microorganisms, those that can offer the highest rate of oxidation or able to extract the highest number of electrons per mole of the substrate [4-6]; (B) study the effectiveness of redox mediators [7, 8]; (C) select more effective electrode materials [9-11]; (D) determine the most efficient anodic reactions, those producing the highest number of electrons per unit weight of the reactant [12-13]. Specific bacteria, including

[1] To whom any correspondence should be addressed.

Geobacteraceae, were enriched on the anode where organic matter was oxidized and electrons were directly transferred to the anode under anaerobic conditions [14]. Bacteria capable of electricity generation from enriched domestic wastewater [15], ocean sediments [14], animal wastes [16], and anaerobic sewage sludge [17, 18]. Several factors affect MFC performance including the microbial inoculum, chemical substrate (fuel), electrode materials, cell internal and external resistance, solution ionic strength, and electrode spacing, type of proton exchange material (and the absence of this material) [19-21].

However, to increase the anode performance, different chemical and physical strategies have been followed. Park et al. [22] incorporated Mn(IV) and Fe(III) and used covalently linked neutral red to mediate the electron transfer to the anode. Electro-catalytic materials such as polyanilins/Pt composites have also been shown to improve the current generation through assisting the direct oxidation of microbial metabolites [23-25]. Each of these compounds (mediators) has a different potential that possesses in the oxidation-reduction reaction, and their differences result in generate electricity. The difference in the chance of contact between electron shuttles' and anode electrode can also cause the variation in electricity generation. In other views, the higher the concentrations of electron shuttles in the anode phase, the more the electricity generation [26].SMFC design must be made if SMFC systems are to be used for wastewater treatment. In a two chamber SMFC system, anode and cathode separated by proton exchange membrane. Proton exchange membranes (e.g. Nafion) are not suitable if SMFC systems are applied for wastewater treatment processing because of cost. Finally, a general use of the SMFC might be the removal of organic matter from sediments or other polluted sites. As both the bio-electrochemical activity and microbial metal reducing activity appear to share the same or similar electron transport chains, electrochemical activity of microorganisms may play an important role in the biogeochemical cycling of carbon, nitrogen, metal other organic contaminants. Indeed, the concept of mediator-less MFC studies provide new insights into the function of electrochemical activity bacteria directly associated with practical needs of environmental protection. In past years, environmental pollution control has mainly relied on how fast and feasible process could operate to treat environmental pollutants. The challenges facing the community are many. Everyaspect of the SMFC operation can be regarded as sub-optimal at this point, including the anode, the cathode, the membrane, and SMFC design. While these challenges loom as great, the opportunities are equally great. The specific objective in this study was to evaluate the possibility to choose a suitable anode material for field applications by using laboratory experiments as well as to increase power generation. Cost effective SMFC systems can offer a potentiallypromising remediation technology that also earns surplus energy.

IOP Conf. Series: Materials Science and Engineering **88** (2015) 012032 doi:10.1088/1757-899X/88/1/012032

2. Materials and Methods

2.1 Sediment Collection

Estuarine sediment was used for microbial energy generation. The sediment was collected from the bottom (approximately 10 cm from the sediment–water interface) of Lake Sap-Kyo(N 36°52'9.3", E 126°50'29.12") in a southern province in South Korea.The sampling was performed using a Ponar type grab sampler (2.4-L). All samples including surface water were placed into clean polycarbonate jars (Nalgene, Fisher Scientific) with no headspace gas (i.e. no air) and transported to the laboratory in a cooler box with ice packs. All sediments were passed through a 2-mm sieve to remove plant debris, macro fauna, and other large terrestrial leaves and then homogenized by mixing with a stainless steel spatula prior to use. All samples were kept at 4°C before use.

2.2Establishing sediment microbial fuel cell(SMFC)

A one-chambered sediment cell was established to analyze the performance of different electrodes. The body of theSMFC wasa 500-mL Pyrex beaker. Aliquots of 50-g wet sediment and 120-mL of sea-water were loaded into the lower and upper part of sediment MFC, respectively. The SMFCwas monitored for 72 hours.During theruns, water loss due to evaporation was compensated daily by adding distilled water.

2.3Anode and cathode compartments

SMFC, equipping different metallic anodes, was used in respect of graphite felt as common cathode. The electrodes used in this experiment were Zinc (Zn), Aluminium (Al), Copper (Cu), Iron (Fe), and Graphite felt (GF). The dimension of the anode was 150×45×3 mm (length×width×thickness). The cathode (40×35×3 mm) was bare graphite felt and placed parallel to the anode and 4 cm above the sediment–water interface. Both anode and cathode were connected by a platinum wire (internal resistance 20Ω) and an insulated copper wire to an external load. All electrodes had a projected surface area of 0.002 m^2.

2.4Electricity Measurement

Redox potentials were measured using an Ag/AgCl reference electrode (9678BNWP, Thermo Orion, Beverly, MA, USA).The pH was measured using anautomatic calibrated pH meter (Model pH 20, Hanna Instruments, USA). Cell voltage was recorded using a multi-meter and a data acquisition system (Model 2700, Keithley Instruments,Cleveland, OH, USA). Current density, i, was calculated as i=I/A=V/RA, where V(mV) is the voltage, I (mA) the current in electrochemical tests, R(Ω) the external resistance, and A(m^2) the projected surface area of the studied electrode. Power density was calculated according to P (mWm^{-2})=10×iV (10 needed for the given units). Columbic efficiency was calculated as CE $=C_p/C_{th}$•100%, where C_p is the total coulombs calculated by integrating the current over time, and C_{th} is the theoretical amount of coulombs available based on the Chemical Oxidant

Demand (COD) removed in the SMFC. Current density was also calculated by dividing the current by the apparent surface area of anode.

3. Results and Discussion

3.1Redox potentials of SMFC

For each metallic anode has recorded oxidation reduction potential (ORP) of SMFC for both chambers (anode and cathode) with time. Four types(Zn, Al, Cu, Fe) of metals and graphite felt as non-metallic electrode were examined here. Anodic part was not modified any mediatorbut coverednaturally microbial environment condition.As recent studiesshow,a number of bacteria have been found to possess the ability to transfer electrons from oxidized fuel (substrate) to a working electrode without a mediator [27, 28], making it possible to establish mediator-less MFCs. To date, many metal reducing bacteria, which are capable of the reduction of solid metal oxides ability including *Clostridium Butyricum*[29], *Geobactersulfurreducens*[30], *Rhodofoferaxferrireducens*[1] and *Shewanellaputerfaciens*[31]. Furthermore, in the SMFC system, growth and metabolism of *S. puterfaciens* were dependent on the presence of active anode-in essence, it served as the electron acceptor for growth and metabolism. Anodic potentials in our cellranged 184 to -155 mVfor Zn, 155 to -150 mV for Al;148 to -126 mV for Cu;161 to -42 mV for Fe and 178 to -5 mV for Graphite feltfollowed for 72 h, whereas the cathodic potentials showed a slow drift. Metallic anode has shown larger ORP change than that in non-metallic graphite electrode. The results in figure 1 show that as time progresses the potential of the anodetends to decrease.It might be due to corrosion decreases over time for changes in the properties of the electrodes' surfaces resulting from biofilm buildup and polarization effect. Generally biofilm deposit promotes corrosion in aerated waternot in anaerobic water.Oxidation reduction potential changes were higher for metallicanode than for non-metals because of variation in corrosion rate.Sufficient microbial growth guarantees formation of a certain thickness of biofilm. According to electromotive series Al is more reactive than Zn, and Cu metal is much lower than Fein potential value. **The potential of** Al is higher than that of Zn, which is attributed to itsmicrostructure as well as slightchanges with type,concentration, and temperaturearound the metal. In case of Fe and Cu, it might be due to overvoltage (producing hydrogen from H_2S) phenomena in Fe which is not absence inCu. The microbial fuel cell potential reported in the literature varies widely depending on the type of anodic and cathodic efficiencies of the cell used. The anode reaction potentials reported in the literature vary between -300 and -500 mV_{SCE}[19, 30, and 32].The open circuit potential (OCP) of the plain graphite anode, iron containing anodes, and manganese containing electrodes are different from each other, indicating that different electron transfer reactions occur at each of these anode type[24].

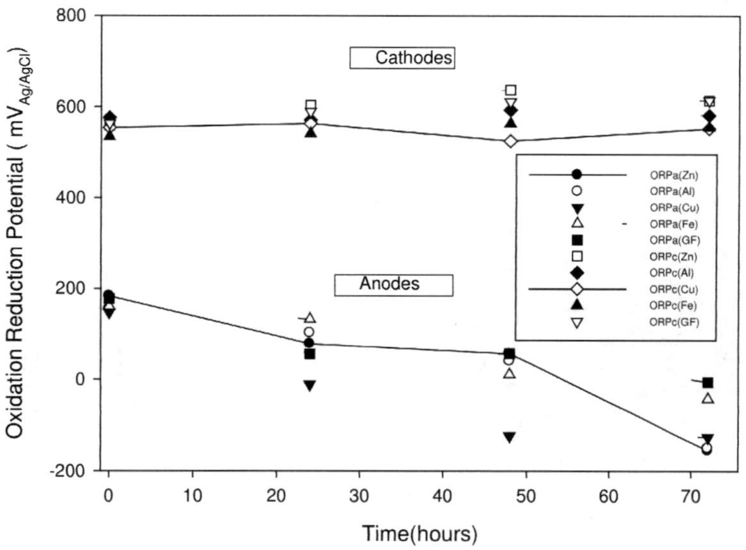

Figure 1. Oxidation reduction potential (ORP) variations in the cathode and anode (against Ag/AgCl)in our sediment microbial fuel cell.

3.2 Power Density of SMFC

From the measurements repeated several times, power density of SMFC using different metallic and non-metallic anode was depicted in figure 2.However, SMFCs were started with variety of anode distinctly and was run till 72 hours. We used Zn, Al, Cu, Fe, and graphite felt (GF) as anodic material. Once operated with metallic and non-metallic anode, the deployed fuel cell has achieved its maximum power density of Zn (913 mWm^{-2}), Al (266 mWm^{-2}), Cu (387.8 mWm^{-2}), Fe (646 mWm^{-2}), GF (127 mWm^{-2}) varying with time. At beginning time of SMFC, the power density decreasesover time significantly (10 h) for all anodes but interestingly Zn anode was found to increase after that time. Increasing trends of power density of Zn anode have sustained till 45 hours after that drop. Power densities were shown to be different for different anodesprobably because of some various catalytic interactions on the electrode surface.Indeed,power densitiesof metallic anodeswere higher than non-metal graphite felt, because corrosion and microbial activity provide electrons to metal whereas only microbial activity provideselectrons to non-metal surface. Bacterial adhesion capacity to electrode might causevariation in power density. And the total area of the electrodes including the internal surface area [22], apparent surface area [32, 33], or projected area [19, 34] has been adopted computationof power density in MFCs. Schroder et al. [10] obtained a power density of 6.0 $Wm^{-2}$for polyaniline–modified platinum as an anode. From literature, it is reported that mediated anode produces higher power than non-mediated one. A similar fuel cell using a plain graphite anode could sustain maximum power of ~20 mW/m^2(~66 mA/m^2) at 0.30 V while AQDS modified graphite electrodes initially produced five-time greater power, and thenthe performance gradually decayed over time[14].

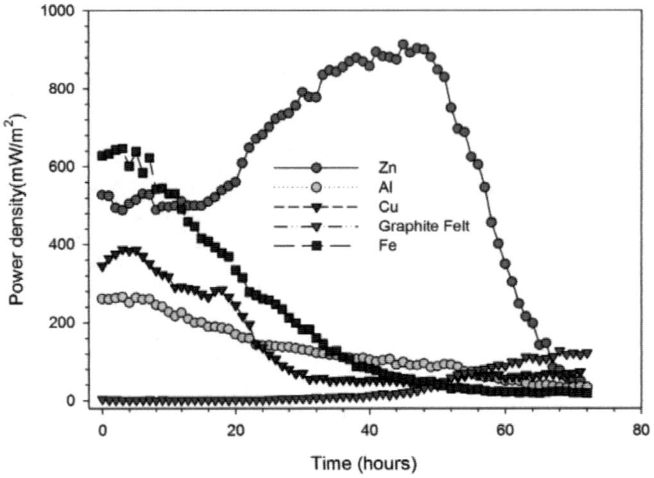

Figure 2. Power density of SMFC according to different anode with respect to time.

3.3 *Cell Voltage and Current density of SMFC*

Figure 3 presents the cell voltage and figure 4 represents the current density of SMFC with adding neither external substrates nor electron transport mediators over 72 hours of operation. Immediately after circuit connection (external circuit resistance fixed at 20 Ω) was established, the current density sharply increased. Maximum current density of Zn (1175 mA/m^2), Al (360 mA/m^2), Cu(610 mA/m^2), Fe (890 mA/m^2), and Graphite felt (255 mA/m^2) were observed at 45 h, 3 h, 5 h, 3h and 68 h, respectively. Current density and cell voltage are varying within a certain range over the next 72 hours.Figure 4 indicate that the majority of electrical charge was generated in the initial stage of the sediment MFC runs: approximately more than 50% of the total amount current density was delivered within the first 45 hours.Corrosion and microbial activity vary over time and condition of solution. The overall pattern of electricity generation was analogous to that observed by Holms et al. [35, 36], who demonstrated microbially-mediated current production using sediment MFC systems: a rapid increase in current production within the first few days of reactor operation with no lag period, followed by a gradual decrease. Wei and Zhang [37] and Mathis et al. [38] isolated the electricigenic (also referred to as anodophilic and electrochemically-active) bacteria from natural marine sediments and proved that the instant electricity generation in their MFC systems augmented with isolated cultures. Current densityamong the metallic anodes of Al, Cu and Fe showed a continuous decrease whereas Zn metal showed somewhat different phase. Zn anode showed rising to falling incidents and non-metallic anode graphite felt showed an elevated trend till end time. These results suggest that catalytic or microbial activity on the surface of Zn or graphite felt was higher than Al, Cu and Fe anode. It is observed that voltage losses as like current density over time (Fig.3). According to anode, the observed maximum cell voltages were Zn(777 mV), Al(751 mV), Cu (645 mV), Fe (734 mV) andGF (499 mV) over different time.A number of different bacteria are able to reduce corrosion rates of different materials in several corrosive media. One type of bacteria can shift corrosion potential (E_{corr}) of one metal in the positive direction while another type can shift corrosion potential (E_{corr}) of some other metal in the negative direction [39,40].

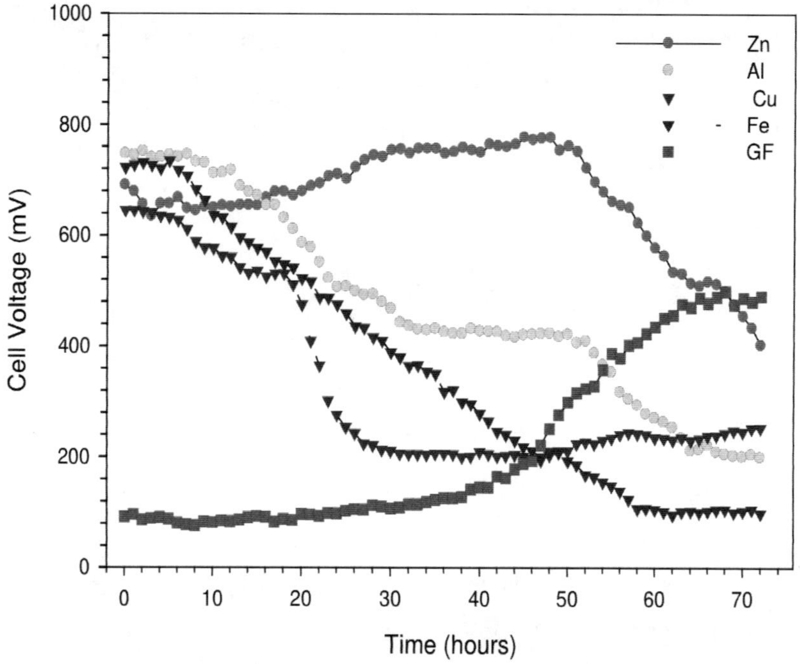

Figure 3. Cell voltages of SMFC with respect to time.

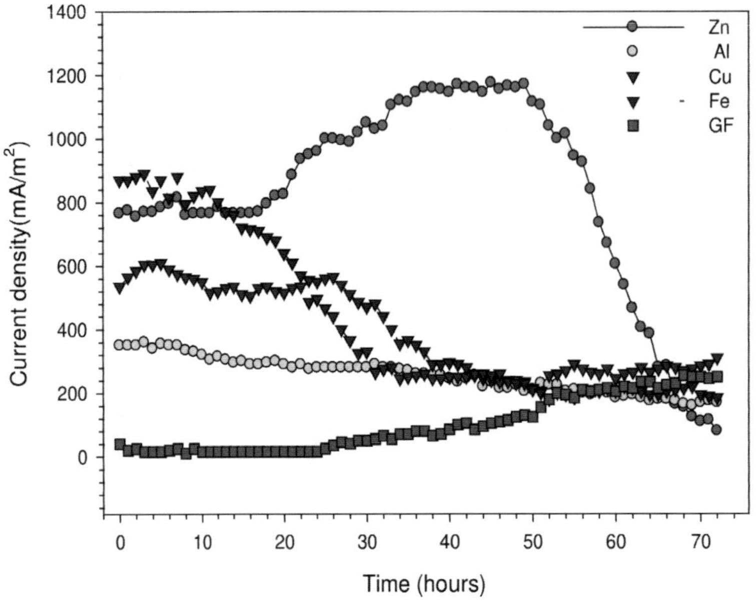

Figure 4. Current density vs time of SMFCoperation (surface area 0.002 m^2 and internal resistance 20Ω).

3.4pH and chemical oxygen demand (COD) of SMFC

pH of SMFCin this workhas changed over time at both in anode part and cathode part in reverse way (Figure 5).When pH of anodic part were decreased, at the same time cathode pH increased. The pH variation ranges were 7.5 to 6.4in anode Zn, 7.3 to 6.8 in Al, 6.6 to 6.4 in Cu, 6.7 to 6.2 in Fe, 7.5 to 6.7 in GFover starting time and ending time, respectively. But, in cathode part, pH changed very little (7.6~7.7). pH reduction indicated that anodic environment is acidic and acidic media reduces the corrosion rate of metal. Metalsulfide or metal oxides are likely to reduce pH value. Acidic environment in anode part is an important issue because MFCs that are able to operate at low pH are technologically advantageous. The proton transport rate from anode to cathode is increases and the kinetic barrier for O_2 reduction to H_2O at the cathode decreases, which leads to the higher current and power densities[41, 42]. In acidic medium acidophilic microorganisms that have colonized the anode and cathode surface can oxidize glucose and otherorganic compounds in the absence of redox mediators [43]. It is well known that*Acidiphilium spp.* is able to use organic compounds such as glucose and glycerol as electron donors [44, 45, and 43]. Chemical oxygendemand of SMFC wasdecreased over time in case of all anodes.But losses of organic matter,shown in COD decreases might be facilitative for long timeelectricity generation. Biomass production could account for additional COD removal, but it was not possible to establish a complete mass balance of COD in SMFC system. Not all the organic matter in a wastewater can be biologically degraded, so that COD removal is typically lower than that for biochemical oxygen demand (BOD) [46].

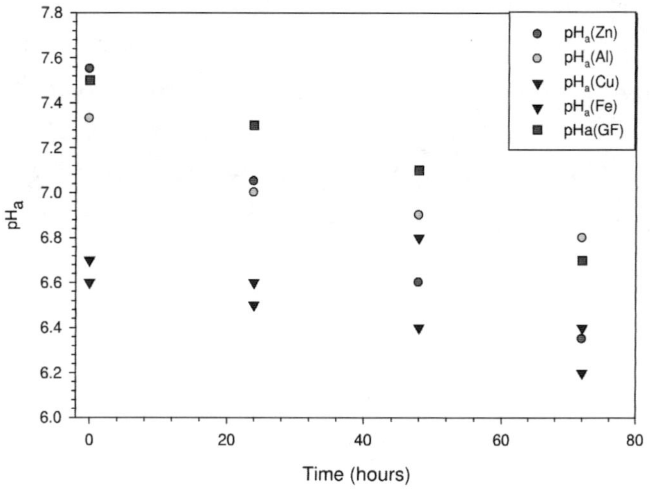

Figure 5.pHchanges in anode chamber over time of SMFC (pHa, anodic pH).

4. Conclusions

This research was aimed to evaluate the feasibility of electricity generation of SMFC using different metal and non-metal anodes. From our study we can assume thata mediator less single chamber sediment microbial fuel cell might produce higher power by changing anode material. As for power density Zn was excellent over other metals. Some interfacial reactions or phenomena on metal or non-metal surface need to be investigated further as mentioned in Section 3. Also time duration of the cell and limitation of current generation could be next issues.

Acknowledgment

This research was supported by BK21 plusprogram and authors acknowledge Soonchunhyang University for its partial finance(No. 20130584).

References

[1] SK Chaudhuri, DRLovley.Electricity generation by direct oxidation of glucose in mediatorless microbial fuel cells[J]. Nat. Biotechnol.,2003, 21:1229-1232.

[2] K Rabaey,W Verstraete.Microbial fuel cells: novel biotechnology for energy generation[J].Trends in Biotechnology, Jun.2005,23:291-298, doi:10.1016/j.tibtech.2005.04.008

[3]K Rabaey, KVDSompel, L Maignien, N Boon, PAelterman, P Clauwaert, LD Schamphelaire, HT Pham, J Vermeulen, MVerhaege, P Lens, WVerstraete. Microbial fuel cells for sulfide removal[J]. Environmental Science & Technology, 40:5218-5224, doi:10.1021/es060382u.

[4] S Tsujimura,AWadano,K Kano,TIkeda. Photosynthetic bioelectrochemical cell utilizing cyanobacteria and watergeneratingoxidase[J]. Enzyme Microb. Technol.,2001, 29:225-231.

[5] HJ Kim, HS Park, MS Hyun,ISChang, M Kim, BHKim. A mediator-less microbial fuel cell using a metal reducing bacterium,*Shewanellaputrefaciense*[J].Enzyme Microb. Technol., 2002, 30:145-152.

[6] Y Choi, J Song,S Jung,S Kim. Optimization of the performance of microbial fuel cells containing alkalophilic*Bacillus sp*[J].J. Microbiol. Biotechnol.,2001, 11:863-869.

[7] DH Park, JGZeikus. Electricity generation in microbial fuel cells using neutral red as an electronophore[J]. Appl. Environ. Microbiol.2000, 66:1292-1297.

[8] JB McKinlay, JG Zeikus.Extracellular iron reduction is mediated in part by neutral red and hydrogenase in *Escherichia coli*[J]. Appl. Environ. Microbiol.2004, 70:3467-3474.

[9] E Katz,I Willner,ABKotlyar. A noncompartmentalized glucose vertical bar O-2 biofuel cell by bioengineered electrode surfaces[J]. J. Electroanal. Chem.,1999, 479:64-68.

[10] U Schroder,J Niessen, F Scholz. A generation of microbial fuel cells with current outputs boosted by more than one order of magnitude[J]. Angew. Chem., Int. Ed.2003, 42:2880-2883.

[11] KB Gregory, DR Bond, DRLovley, Graphite electrodes as electron donors for anaerobic respiration[J]. Environ. Microbiol.,2004, 6:596-604.

[12] SA Lee, Y Choi, SH Jung,S Kim. Effect of initial carbon sources on the electrochemical detection of glucose by *Gluconobacteroxidans*[J].Bioelectrochemistry,2002, 57:173-178.

[13] BE Logan. Extracting hydrogen electricity from renewable resources[J]. Environ. Sci. Technol, 2004, 38:160A-167A.

[14] LMTender, CEReimers, HA Stecher, DE Holmes, DR Bond, DA Lowy, K Pilobello, SJ Fertig, DR Lovley. Harnessing microbially generated power on the seafloor[J]. Nat. Biotechnol., 2002, 20:821–825.

[15] H Liu, R Ramnarayanan, BE Logan. Production of electricity during wastewater treatment using a single chamber microbial fuel cell[J]. Environ. Sci. Technol.,2004, 38:2281-2285.

[16]BMin, BELogan.Continuous electricity generation from domestic wastewater and organic substrates in a flat plate microbial fuel cell[J]. Environ. Sci. Technol.,2004, 38:5809-5812.

[17] JR Kim, B Min, BE Logan.Evaluation of procedures to acclimate a microbial fuel cell for electricity production[J]. Appl. MicrobialBiotechnol.,2005,68:23-30.

[18] K Rabaey, N Boon, SD Siciliano, M Verhaege, W Verstraete. Biofuel cells select for microbial consortia that self-mediate electron transfer[J]. Appl. Environ. Microbiol.2004, 70:5373-5382.

[19] H Liu,BE Logan.Electricity generation using an air-cathode single chamber microbial fuel cell in the presence and absence of a protone exchange membrane[J]. Environ.Sci.Technol.2004, 38:4040-4046.

[20] S Cheng, H Liu,BE Logan. Power densities using different cathode catalysts (Pt and CoTMPP) and polymer binders (nafion and PTFE) in single chamber microbial fuel cells[J].Environ. Sci. Technol.,2006, 40:364-369.

[21] H Liu, SA Cheng,BE Logan. Power generation in fed-batch microbial fuel cells as a function of ionic strength, temperature, and reactor configuration[J]. Environ. Sci. Technol. 2005, 39:5488-5793.

[22] DH Park, JGZeikus. Improved fuel cell and electrode designs for producing electricity from microbial degradation[J]. Biotechnol.Bioeng.,2003,81:348-355.

[23] HAVidela. Manual of Biocorrosion[M], CRC Press, 1996.

[24] PN Froelich, GP Klinkhammer, ML Bender, NALuedtke, GR Heath, D Cullen, P Dauphis, DHammond, B Hartman, V Maynard. Early oxidation of organic-matter in pelagic sediments of the eastern equatorial Atlantic-suboxicdiagenesis[J]. Geochim.Cosmochim.Acta 1979,43 (7), 1075–1090.

[25] GC Gil, IS Chang, BH Kim, M Kim, JK Jang,HS Park,HJ Kim. Operating parameters affecting the performance of a mediator-less microbial fuel cell[J]. Biosens.Bioelectron., 2003, 18:327-334.

[26] BE Logan, B Hamelers, R Rozendal, U Schroder, J Keller,S Freguia, P Aelterman, WVerstraete,KRabaey.Microbial fuel cells: Methodology and technology[J]. Enveronmental Science & Technology, Jul. 2006, 40:5181-5192, doi:10.1021/es0605016.

[27] BH Kim, HJ Kim, MS Hyun, DH Park. Direct electrode reaction of Fe(iii) reducing bacterium, *Shewanellaputerfaciens*[J].J. Microbial. Biotechnol.,1999:127-131.

[28] HJ Kim, MS Hyun, IS Chang, BH Kim. A fuel cell type lactate biosensor using a metal reducing bacterium, *Shewanellaputerfaciens*[J]. J. Microbial. Biotechnol.,1999,9:365-367.

[29] HSPark, BH Kim, HS Kim, HJ Kim, GT Kim, M Kim, IS Chang, YK Park,HI Chang. A novel electrochemically active and Fe(III) reducing bacterium phylogenetically related to *Clostridium butyricum* isolated from a microbial fuel cell[J]. 2001, Anaerobe 7:297-306.

[30] DR Bond,DRLovely.Eelectricity production by *Geobactersulfurreducens* attached to electrodes[J]. Apple. Environ. Microbial.2003, 69:1546-1555.

[31] HJ Kim,HS Park, MS Hyun, IS Chang, M Kim, BH Kim. A mediator-less microbial fuel cell using a metallic bacterium *Shewanellaputerfaciens*[J]. *Enzyme* Microb. Technol., 2002, 30:125-152.

[32] H Liu, SA Cheng, BE Logan, Production of electricity from acetate or butyrate using a single-chambered microbial fuel cell[J]. Environ. Sci. Technol., 2005, 39:658-662.

[33] HMoon, IS Chang, KH Kang, JK Jang,BH Kim. Residence time distribution in microbial fuel cell and its influence on COD removal with electricity production[J]. Biochem. Eng. J.2005, 27:59-65.

[34] BE Logan, C Murano, K Scott, ND Gray,IM Head. Electricity generation from cysteine in a microbial fuel cell[J]. *Water Res.* 2005, 39:942-952.

[35] DE Holmes, DR Bond, DRLovley. Electron transfer by *Desulfobulbuspropionicus*to Fe(III) and graphite electrodes[J]. Applied Environmental Microbiology,2004a,70:1234–1237.

[36] DE Holmes, DR Bond, RA O'Neill, CE Reimers, LR Tender, DRLovley. Microbial communities associated with electrodes harvesting electricity from a variety of aquatic sediments[J]. Microbial Ecology,2004b,48:178–190.

[37] D Wei, X Zhang. Current production by a deep-sea strain *Shewanella sp.* DS1.Current Microbiology2007, 55:497–500

[38] BJMathis, CW Marshall, CE Milliken, RS Makkar, SECreager, HD May. Electricitygeneration by thermophilic microorganisms from marine sediment[J]. Applied Microbiology & Biotechnology,2008, 78:147–155.

[39] E Kus,R Abboud,R Popa,KHNealson,FMansfeld. The concept of the bacterial battery[J]. Corros. Sci., 2005, 47:1063-1069.

[40] F Mansfeld. The interaction of bacteria and metal surfaces; 2007; ElectrochimicaActa 52:7670–7680.

[41] JC Biffinger, J Pietron, O Bretschger, LJ Nadeau, GR Johnson, CC Williams, KH Nealson, BR Ringeisen. The influence of acidity on microbial fuel cells containing *Shewanellaoneidensis*[J]. Biosens.Bioelectron., 2008, 24:900-905.

[42] B Erable, L Etcheverry, ABergel. Increased power from a two chamber microbial fuel cell with a low pH air-cathode compartment[J]. Electrochem.Commun., 2009, 11:619-622

[43] M Malki, AL De Lacey, N Rodríguez, R Amils, VM Fernández. Preferential use of an anode as an electron acceptor by an acidophilic bacterium in the presence of oxygen[J]. Appl. Environ. Microbiol.2008, 74:4472-4476.

[44] DB Johnson, S McGinness.Ferric iron reduction by acidophilic heterotrophic bacteria.Appl.Environ. Microbiol.,1991, 57:207-211.

[45] K Kusel, T Dorsch, G Acker, E Stackebrandt. Microbial reduction of Fe(III) in acidic sediments: Isolation of Acidiphiliuncryptum JF-5 capable of coupling the reduction of Fe(III) to the oxidation of glucose[J]. Appl. Environ. Microbiol.,1999, 65:3633-3640.

[46] Grady CPL Jr,Daigger G T, Lim HC. In Biological wastater treatment[R], 2nd ed., New York,Marcek,1999.

Numerical investigation to study effect of radiation on thermal performance of radiator for onan cooling configuration of transformer

V Chandak[1,*], S B Paramane[1], W V d Veken[2] and J Codde[3]
[1]Global R&D Centre, Crompton Greaves Ltd., Mumbai, 400042
[2]Crompton Greaves Power Systems Belgium NV, Belgium
[3]KU Leuven University Belgium, Belgium

Abstract. In the present work, flow and temperature distribution in the radiator fins of a power transformer is studied numerically with conjugate heat transfer using commercial CFD software to study the effect of radiation on heat dissipation. The approach considered here is a complete 3D geometry of the radiator fins with average height of the flute geometry of the fins for meshing and computational time reduction. Simulations are performed for ONAN (Oil Natural Air Natural) case for one radiator configuration. The simulations also study the effect of radiation and its impact on the overall heat dissipation. These results would give a holistic picture of heat transfer phenomenon to the designers.

1. Introduction

Transformers are one of the primary components for the transmission and distribution of electrical energy. They are extensively used in electric power systems to transfer power by electromagnetic induction between circuits at the same frequency, usually with changed values of voltage and current. The energy losses in the power transformers are proportional to the load. These losses are transformed into heat and consequently temperature of the cooling medium rises and must be cooled.

In most power transformers, the cooling is provided through circulation of oil between ducts in the active parts and radiators outside the transformer tank. The internal flow of oil in the radiator and the external flow of air over the radiator results in a heat transfer by natural or forced convection. Thus, the cooling modes in a transformer are ONAN (oil natural air natural), ONAF (oil natural air forced), OFAF (oil forced air forced) and ODAF (oil directed air forced). Most of the published literature is on internal cooling of the power transformer i.e. cooling of windings. Whereas, for external cooling of the transformer which corresponds to the cooling of radiators, very few work is found.

Radiators-fan constitutes a major portion of thermal system of transformer. It is necessary that this radiator-fan system should be efficient and with lesser weight. Transformer manufacturers are keen to reduce the weight of the system, without compromising on the thermal performance. Due to advanced tools like CFD, now it is possible to visualize virtually the air flow and temperature distribution over the radiators compared to expensive experiments. Furthermore, air flow distribution will give more insight for improving thermal performance of the system.

Limited experimental and simulated theoretical work related to power transformer radiator cooling performance has been reported. In some previous efforts, the focus was on thermal design and methods to predict hot-spot and top-oil temperature [1, 2]. Conventional calculations of internal transformer temperature are not only complicated and difficult, but also lead to conservative estimates obtained on the basis operating condition assumptions [3, 4]. With computational numerical techniques, it is possible to calculate transformer losses and temperature distributions in power transformers and cooling systems [5, 6]. There are still natural convection analysis problems with

[1] To whom any correspondence should be addressed.

IOP Conf. Series: Materials Science and Engineering **88** (2015) 012033 doi:10.1088/1757-899X/88/1/012033

convergence, calculation time, and even accuracy [7]. Nabati et al. [8] carried out numerical modeling of temperature distribution and flow pattern in a block radiator used in power transformer cooling system. They mainly focused on studying the relation between radiator block characteristics and cooling behavior of system. The results indicate that recirculation occurs whenever pressure increase at the end of a radiator block that consequently prevents the enough oil flow through the last radiator of block. Amoiralis et al. [9] studied optimum designs of ONAN transformer cooling system by means of advanced numerical thermal field and computational fluid dynamics techniques. Novel tank designs are examined in conjunction with other crucial parameters, as the number and location of the winding cooling ducts, so as to define the best optimum geometry that ensures maximum efficiency of the transformer cooling system performance. In recent times Kim et al. [10] presents prediction and experimental study on the cooling performance of radiators used in oil filled power transformer applications with non-direct flow (ONAN) and direct-oil-forced flow (ODAN). Radiator temperature distribution and cooling performance was predicted using theoretical calculations, then validated using CFD simulation results. To the best of the author's knowledge there has been no paper found where the effect of radiation on heat transfer from the radiator is reported.

Present study focuses on determination of effect of radiation on various performance parameters of the radiator.

2. Geometry Description

The geometry is composed of one radiator of 520mm width having 2500mm height and 27 fins. The geometry considered for analysis in the present study is shown in figure 1. Oil enters at the top and leaves from the outlet at the bottom of the radiator and air domain surrounding the radiator is shown in figure 1. Close up view of radiator and oil domain is shown in figure 2.

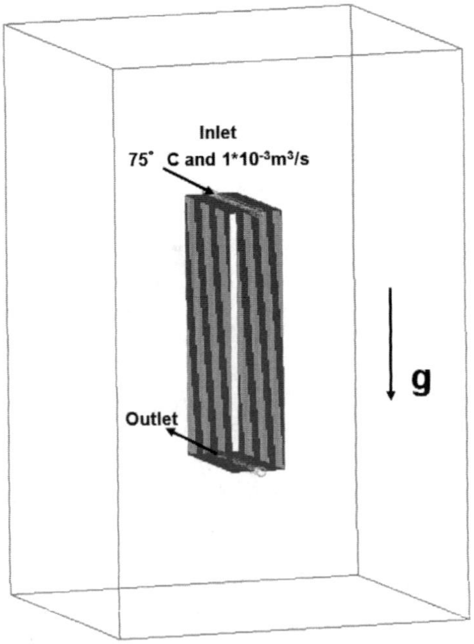

Figure 1. Geometry of the radiator-fan configuration

IOP Conf. Series: Materials Science and Engineering **88** (2015) 012033 doi:10.1088/1757-899X/88/1/012033

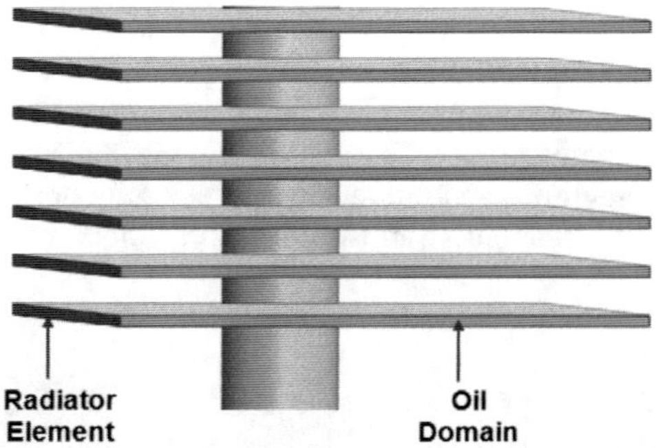

Figure 2. Close up view of radiator and oil domain

3. Methodology Description

The geometry of radiator is made in Solid works modeling software from the geometrical description and imported in Ansys ICEM-CFD mesh generation software. Mesh quality is important for accuracy and stability of the numerical calculation and needs to be of high quality. The domains of the model of the radiator, the inner oil and the surrounding ambient air are discretized into a finite set of control volumes. Hexahedral mesh is used for the radiators for better convergence, accuracy and resolving detailed flow features. By using different options provided in ICEM-CFD, it is possible to have a good control over all parts of the mesh. The image of figure 3 shows a mesh that is used for the CFD analysis. Relatively coarser mesh is used for the regions which are away from the radiators and fine mesh is used near to the radiator and fan. This is because most of the action is happening near to the radiators due to higher gradients of velocity and temperature near these parts. The total size of the mesh used was 2.6 million nodes.

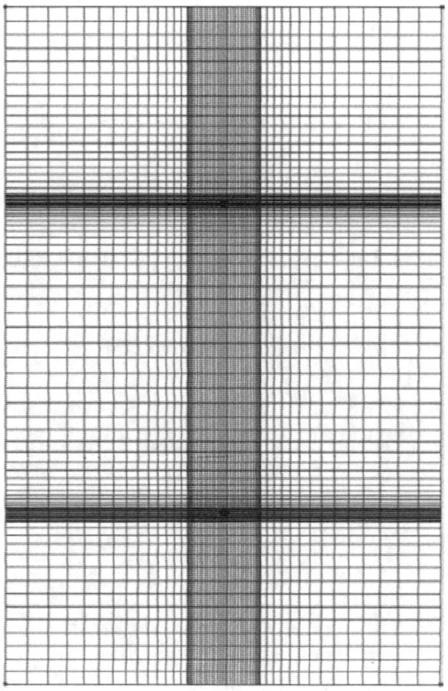

Figure 3. Hexahedral Mesh of radiator

Ansys CFX pre and solver is used for preprocessing and solving, respectively whereas post processing of the results is done in CFX post. In this case, it has been ensured that residuals for all solved quantities are below 10^{-4} and imbalance within domain is less than 1% after which it is considered that convergence is achieved.

Due to natural convection, momentum and energy equations are coupled. Temperature dependent oil properties are used for analysis. Computation time required for completing one simulation on four parallel processors is 120hrs.

4. Configurations studied

Two configurations, both in ONAN, are studied.

1. Radiation is not considered in this case. Oil mass flow rate is specified which is generally observed for heat run test of transformer in ONAN configuration.
2. Radiation is considered in this case. Oil mass flow rate is specified which is generally observed for heat run test of transformer in ONAN configuration.

5. Simulation Parameters

This section summarizes the different parameters and boundary conditions that are used for the numerical simulations

5.1. Material Properties

In the present calculations, the fluid that flows in the transformer winding is a mineral oil having the following properties with the temperature of oil, T_{oil} expressed in Kelvin.

$C_{pf} = 821.19 + 3.563 \times T_{oil}$ (J kg^{-1} K^{-1})

$k_f = 0.15217 - 0.0000716 \times T_{oil}$ (W m^{-1} K^{-1})

$\mu_f = 0.08467 - 0.0004 \times T_{oil} + 0.0000005 \times T_{oil}^2$ (Pa s)

$\rho_f = 1067.75 - 0.6376 \times T_{oil}$ (kg m^{-3})

where C_{pf} is the specific heat, k_f is the thermal conductivity, μ_f is the dynamic viscosity and ρ_f is the density. All the fluid properties, except for the dynamic viscosity, vary linearly with temperature. The material for radiators is taken as steel.

5.2. Boundary conditions

The fluid domain has an inlet boundary condition where a mass flow rate and an oil temperature are specified. At the domain exit, a pressure condition (p_{avg}= 0 Pa) is set. Mass flow rate specified at inlet is $1*10^{-3}m^3/s$ (0.845865kg/s) for both the cases. Flow is considered to be laminar for both cases.

6. Results and Discussions

This section discusses the results obtained from CFD analysis for both the ONAN cases considered in the present study.

Figure 4 below shows the temperature distribution on the radiator for both the cases (without and with radiation). This clearly shows the difference in temperature distribution in the end fins for both the cases. This can be attributed to better heat transfer for the latter case (the one with radiation considered), which thereby results in better cooling leading to lesser oil temperature, hence lesser surface temperature. It was observed that the qualitative temperature plot for rest of the fins in the middle looks similar for both the cases. This is because shape factor is same for the middle fins as they are facing each other.

The better heat dissipation of end fins is further affirmed in figure 5, which compares the heat flux distribution on the radiator surface between the two cases. It is qualitatively evident that the end fins are cooled better for the case with radiation considered. Now, comparison of quantitative results will be made for the two cases along with the results from the theoretical model presented by Kim et al. [10].

IOP Conf. Series: Materials Science and Engineering **88** (2015) 012033 doi:10.1088/1757-899X/88/1/012033

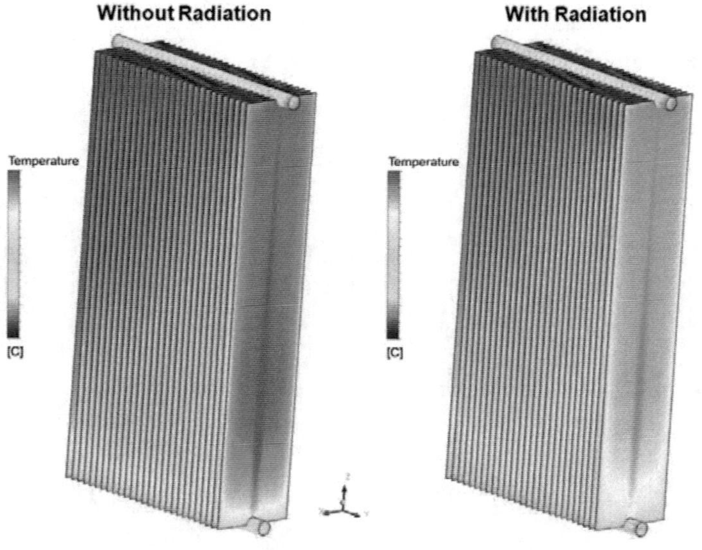

Figure 4. Comparison of Radiator Surface Temperature with and without Radiation

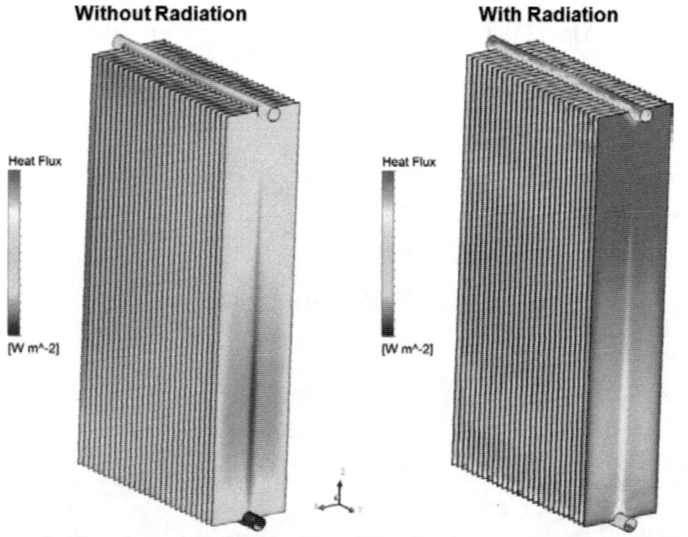

Figure 5. Comparison of Heat Flux Distribution on Radiator Surface with and without Radiation

Figure 6 below shows the percentage oil mass flow distribution among the radiator fins for the two ONAN cases from CFD along with the prediction from the theoretical model. The mass flow distribution predicted from our CFD based model with only convection considered closely matches with the model of Kim et al. [10]. Thereby, the present CFD approach is validated and can be used to compare the convection model (without radiation) with the convection plus radiation model which is shown by the third curve. There is a steep hike in oil flow for the front and the back fins. This can be attributed to the additional radiation heat transfer which provides better cooling at the front and back fins exposed to air. It is to be noted that for the radiation included CFD model, the 1st fin has 29.9%

higher mass flow compared to 2nd fin. The 27th fin has 34.4% higher mass flow compared to 26th fin. The last fin (27th) has 22.6% lower mass flow compared to the 1st fin thereby highlighting the skewed mass flow distribution among the fins. This difference is found even in the other two cases which do not consider radiation.

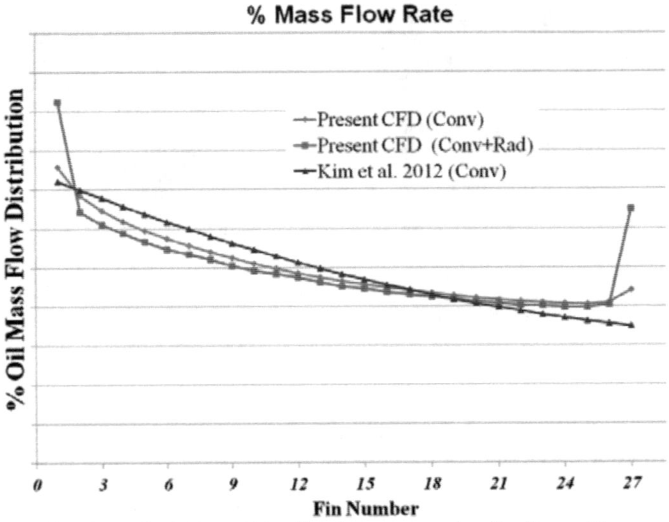

Figure 6. Mass Flow Rate Distribution across the Radiator Fins

Figure 7 shows the distribution of heat dissipation for each fin for the two CFD cases (with, without radiation). It is observed that as compared to without radiation case, the heat dissipation in each of the middle fins with radiation case is at least 100 Watts higher. For end fins, this difference is much higher. This emphasizes the role radiation plays in not just the end fins but also the middle fins for heat dissipation. Also, with radiation case, the 1st fin dissipates 51.3% more heat as compared to the 2nd fin and the 27th fin 55.6% more than the 26th fin.

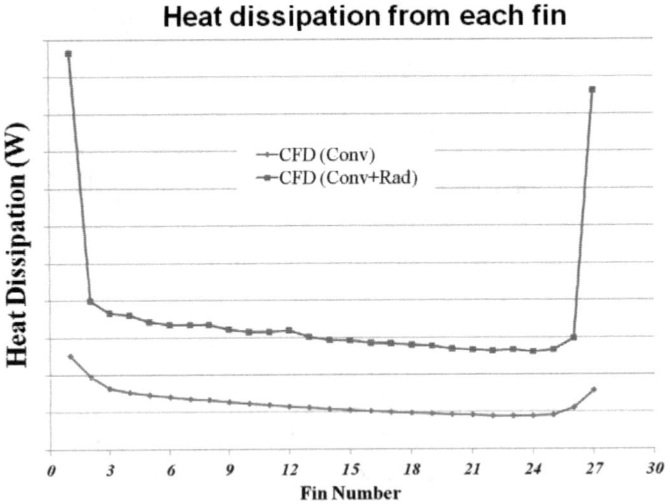

Figure 7. Fin-wise Heat Dissipation Distribution for various cases

Figure 8 shows the comparison of overall heat dissipation from the radiator for the 2 cases considered. Upon including radiation in the CFD model, it is observed that heat dissipation shoots up by 24%. This finally affirms the importance of radiation in the overall heat dissipation from the radiator. Although, it has to be noted that this difference is magnified due to the lesser dominant natural convection being the primary mode of heat transfer. In the cases of forced convection (ONAF), the effect of radiation will not be so high.

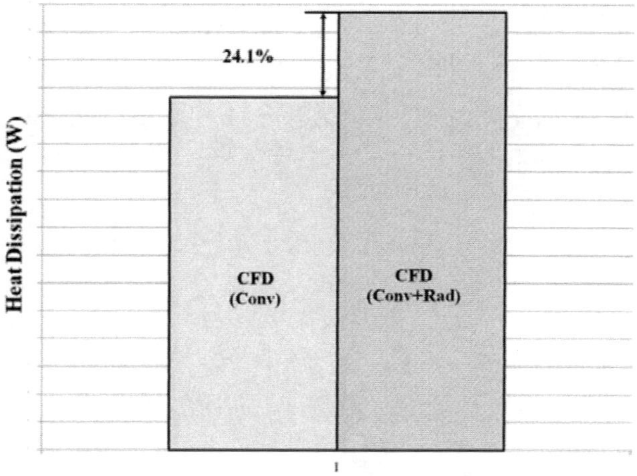

Figure 8. Overall Heat Dissipation for various cases

7. Conclusion

CFD analysis is performed to evaluate the effect of radiation on heat dissipation of transformer radiator. Furthermore, oil flow and heat dissipation distribution among the different fins in the radiators is studied for ONAN cooling configuration without and with radiation. It highlights the significant contribution of radiation in the overall heat dissipation. Detailed analysis shows even the middle fins of the radiator providing higher heat dissipation, the end fins expectedly provide much higher dissipation due to one full face being exposed to the ambient. Thereby, it was observed that radiation needs to be included in the natural convection case (ONAN). These results can be useful for the designers to understand the effectiveness of radiator-fan configuration and heat transfer characteristics for different transformer cooling configurations.

8. Acknowledgements

The authors would like to acknowledge Global R&D Centre, Crompton Greaves Ltd. Mumbai, India for providing the computational resources and permission for publishing this work.

9. References

[1] Zhang J and Li X 2004 Coolant flow distribution and pressure loss in ONAN transformer windings. Part I: Theory and model development *IEEE Transactions on Power Delivery* **19** 186-193

[2] Radakovic Z and Feser K 2003 A new method for the calculation of the hot-spot temperature in power transformers with ONAN cooling *IEEE Trans. Power Del.* **18** 1284–92

[3] Taghikhani M A and Gholami A 2009 Prediction of hottest spot temperature in power transformer windings with non-directed and directed oil-forced cooling *Int. J. Electr. Power. Energy Sys.* **31** 356–364

[4] Tang W H, Wu Q H and Richardson Z J 2004 A simplified transformer thermal model based on thermal–electric analogy *IEEE Trans. Power Del.* **19** 1112–19

[5] Susa D and Lehtonen M 2006 Dynamic thermal modeling of power transformers: Further development - Part II *IEEE Trans. Power Del.* **21** 1971–80

[6] Wakil N E, Chereches N C and Padet J 2006 Numerical study of heat transfer and fluid flow in a power transformer *Int. J. Therm. Sci.* **45** 615–626

[7] Torriano F, Chaaban M and Picher P 2010 Numerical study of parameters affecting the temperature distribution in a disc-type transformer winding *Appl. Therm. Eng.* **30** 2034–44

[8] Nabati H 2008 Optimal pin fin heat exchanger surface *PhD Thesis Dept. Mech. Eng. Mälardalen University Sweden*

[9] Amoiralis E I, Georgilakis P S, Tsili M A and Kladas A G 2009 Global Transformer Optimization Method Using Evolutionary Design and Numerical Field Computation *IEEE Tran. on Mag.* **45** 1720-23

[10] Kim M, Cho S M and Kim J K 2012 Prediction and evaluation of the cooling performance of radiators used in oil-filled power transformer applications with non-direct and direct-oil-forced flow *Experimental Thermal and Fluid Science* **44** 392-397

IOP Conf. Series: Materials Science and Engineering **88** (2015) 012034 doi:10.1088/1757-899X/88/1/012034

Experimental investigation of shell-side steam pressure drop in crossflow in a horizontal falling-film tube bundle

S Q Shen [1], **H Liu, L Y Gong, X S Mu and R Liu**
Key Laboratory for Sea Water Desalination of Liaoning Province, Dalian University of Technology, Dalian 116024, Liaoning, China

E-mail: zzbshen@dlut.edu.cn

Abstract. An experimental investigation of the steam flowing across a horizontal tube bundle with liquid falling film was conducted to simulate flow characteristics in a large desalination plant. Experiments were conducted at pressure ranging from 12 kPa to 31 kPa with the corresponding saturated steam temperature of 50 °C to 70 °C and the falling film spray density of 0.02 $kg \cdot m^{-1} \cdot s^{-1}$ to 0.08 $kg \cdot m^{-1} \cdot s^{-1}$. The influence of the spray density, saturated temperature and tube bundle geometry structure on steam pressure drop was illustrated. The results indicate that the pressure drop increased with increasing spray density, but decreased with increasing saturated temperature. The pressure drop at the maximum spray density of 0.08 $kg \cdot m^{-1} \cdot s^{-1}$ can be three times that in spray density of 0.02 $kg \cdot m^{-1} \cdot s^{-1}$ and the pressure drop at 50 °C is almost twice that at 70 °C. Under the same condition, the pressure drop through a rotated square-arranged tube bundle was smaller than that through a regular triangular-arranged tube bundle.

1. Introduction

The horizontal tube falling-film evaporator is a high-efficiency heat transfer equipment with wide applications in chemical engineering, refrigeration and seawater desalination plants. Most studies on the process of horizontal tube falling film evaporators focus on the heat transfer coefficient [1, 2]. Parameters such as the liquid film Reynolds number, evaporation temperature, liquid properties, tube space and heat flux have been studied as the major parameters that affect heat transfer performance. Ribatski et al. [3] presented a critical review of research on horizontal tube falling film evaporators before 2005. Shen et al. [4] showed the uneven distribution of heat transfer within a large-scale tube falling-film evaporator considering the heat transfer coefficient distribution, which is affected by the spray density and condensation in tubes.

Pressure drop is a critical parameter for the heat transfer performance of the multi-effect distillation (MED) desalination plant for its dominant function in the calculation of temperature loss [5, 6]. The thermal process in horizontal tube falling film evaporators in MED is characterised by the small temperature difference, minimal pressure drop, high sensitivity and saturation state.

The functional relationship between the pressure drop and relevant variables are determined experimentally. Based on an experimental databank, Fujii et al. [7] studied the effect of in-line and staggered arrangements of tube banks on the pressure drop coefficient. The lack of reliable method for predicting the void fraction motivated Grant and Chisholm [8] to establish a new equation for shell-side two-phase pressure drop using their data. Ribatski et al. [9] analysed and summarized previous

[1] zzbshen@dlut.edu.cn.

researches on the two-phase flow. They critically described the frictional pressure drop, the prediction methods of which are generally based on experimental results of the adiabatic air-water flow. Ishihara et al. [10] suggested a Martinelli model-based correlation approach to represent the two-phase pressure drop friction multiplier. Since then, researches on two-phase pressure drop in tube banks are mainly been based on this correlation. Some relative factors, such as the tube pitch [11]; flow pattern [12] and void fraction [13, 14] have been gradually integrated to improve the correlation.

The flow in horizontal tube falling film evaporators is complex and differs from general two-phase flow. Such flow contains a downward falling film flow driven by the gravity and a steam flow across the tube bundle driven by pressure difference. The pressure drop of the steam flowing across tube bundle is caused not only by the tube bundle, but also by the falling liquid. However, the effect of liquid film on pressure drop is not considered in the aforementioned studies. This paper aims to experimentally investigate the characteristics of steam flowing across the falling-film tube bundle and to determine the effect of liquid film on steam pressure drop.

2. Experiment and error analysis

2.1Experimental apparatus and procedure

Figure1 shows the experimental system for the steam pressure drop in the horizontal-tube falling-film evaporator. The system consists of a boiler to provide steam, a temperature & pressure reducer to control the saturated steam pressure and temperature, a test section with a tube bundle and a water spray box, a condenser, water reservoirs, pumps and measuring instruments.

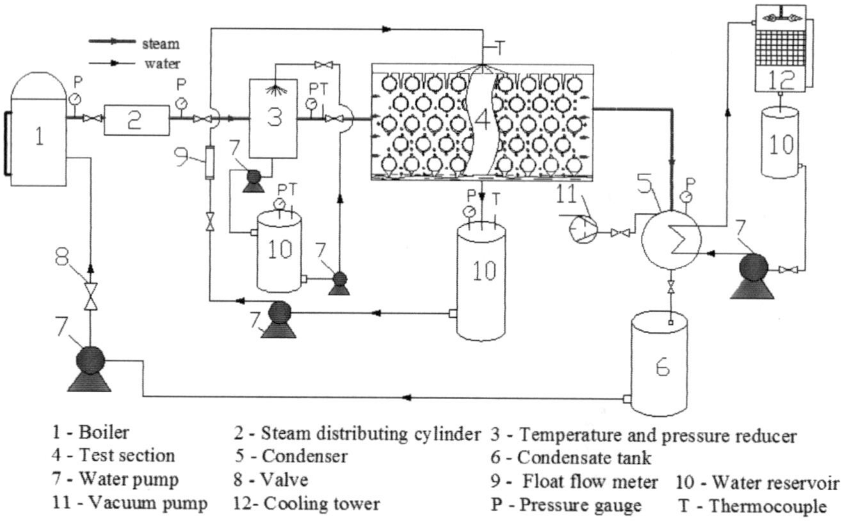

Figure 1. Experimental system diagram.

In each experiment, a designed amount of steam was generated from the boiler and flows through the tube bundle with falling film flow under adiabatic condition. After flowing across the tube bundle, the steam flowed into the condenser, where it was condensed into water and then pumped back to the boiler for recycling. The required steady flow rate of saturated water was controlled by the associated pump at the water reservoir to distribute over the top row of tubes to form falling films. After flowing downward through the tube bundle continuously, the saturated water flowed to the water reservoir for recycling. An electrical bar was employed as the heating source in water reservoir to maintain the temperature of saturated water. The system pressure was maintained by the vacuum pump and condenser.

IOP Conf. Series: Materials Science and Engineering **88** (2015) 012034 doi:10.1088/1757-899X/88/1/012034

The test section is illustrated in figure 2. The inlets and outlets of the test section are equipped with thermocouples to measure the fluid temperatures. Pressure and pressure difference sensors are installed in the inlet and outlet of the tube bundle. The quantity of saturated water supplied by the pump is measured by a rotameter whereas the fluid temperature is measured by a T-type armoured thermocouple.

Figure 2. Image of test section and installation of test instrument.

Tube bundles consist of Al-brass tubes with 0.0254m outer diameter and 0.5 m long; with the tube pitch 1.3 times the outer diameter. As shown in figure 3, two types of arrangement, the regular triangle and rotated square arrangements are selected, for which the number of tubes are 501 and 448, respectively. The minimum cross-section area at the vertical surface across the centre of a tube row is taken as the calculation flow area.

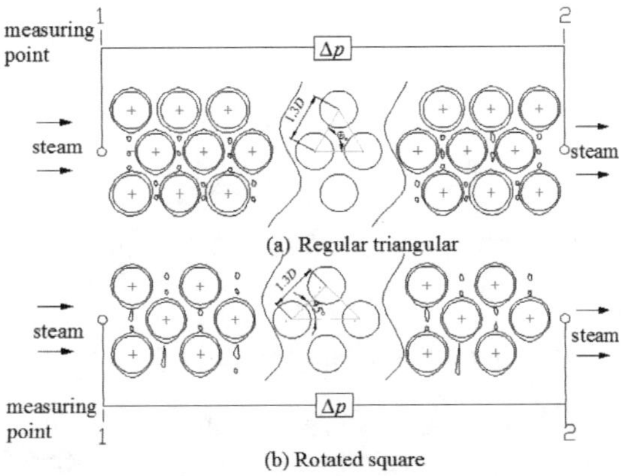

Figure 3. Diagram of tube bundle arrangement.

2.2 Experimental condition and error analysis

In the early stage of each experimental condition, the steam flow rate is changing and not stable, so the measuring result is not steady. The metering begins until the measured parameters are steady to ensure the accuracy. The quantity of falling-film flow is measured by the unilateral spray density Γ. The steam mass flow rate (per area) G is calculated by the minimum cross-section area and employed by the weight of condensed water. The ranges of experimental parameters are given in Table 1. The float of the experimental data is within the estimate range and thus acceptable.

IOP Conf. Series: Materials Science and Engineering **88** (2015) 012034 doi:10.1088/1757-899X/88/1/012034

Table1. Experimental conditions

Name	Regular Triangle	Rotated square
T /°C	50–70	50–70
P /kPa	12–31	12–31
G /kg·m^{-2}·s^{-1}	0.19–1.1	0.39–1.24
Γ/kg·m^{-1}·s^{-1}	0.02–0.08	0.02–0.08

The precision of the pressure difference sensor (GE Druck LPX9381) is ±1 Pa with the measuring range from 0 Pa to 1000 Pa. The precision of the pressure sensor (UNIK PMP5073) is 0.03 kPa with the measuring range from -0.1 MPa to 0 MPa (gauge pressure). The water flow rate is measured by using rotameters (LZB-80) with the measuring range from 2 m3/h to 20 m3/h and 2% precision. The operating temperature from 50 °C to 70 °C is measured by T-type armoured thermocouples with the precision of ±0.05 °C. The sensors are calibrated before the experiment. The measuring signal is collected by a National Instruments (NI) controller PXI-1042. Software, Lab VIEW (NI, Austin, Texas, USA), makes the data visible when they are saved. Measurements are performed under adiabatic conditions.

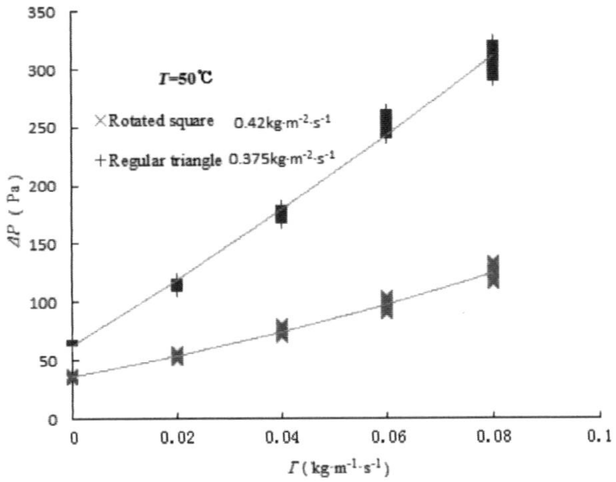

Figure 4. Distribution of data.

Each signal of T, p and Δp is processed by the acquisition system and stored in computer. Under stable experimental conditions, no less than 50 numbers were recorded for each set of parameters. Figure 4 takes some certain conditions to illustrate the experimental data. Owing to the high instrumental sensitivity and the small change in each experimental condition, the metering data varies and hold wave in the vicinity of a value. In each certain condition, the metering data contains more than 50 values. The fluctuate of each set data is within 5%, which shows good steadiness. The average value in each condition will be employed in the following discussion.

Tube bundles consist of Al-brass tubes with 0.0254 m outer diameter and 0.5 m long; with the tube pitch 1.3 times the outer diameter. As shown in figure 3, two types of arrangement, the regular triangle and rotated square arrangements are selected, for which the number of tubes are 501 and 448, respectively. The minimum cross-section area at the vertical surface across the centre of a tube row is taken as the calculation flow area.

3. Results and Analysis

The pressure drop of each column tube $\overline{\Delta p}$ is chosen to characterize the flow resistance in the comparison of different tube arrangements. Figure 5 shows the variation of pressure drop with the spray density and saturated temperature in the regular triangle arrangement.

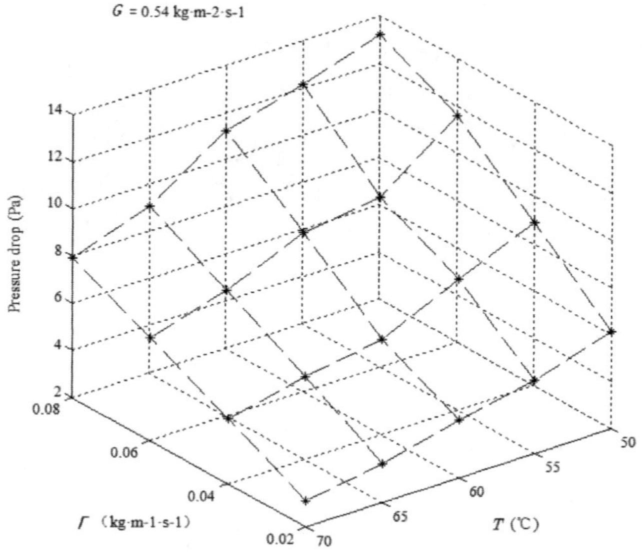

Figure 5. Pressure drop in regular triangle arrangement.

Figure 5 shows that in the range of spray density Γ. On average, $\overline{\Delta p}$ will increase by approximate 2 Pa for every 0.02 kg·m^{-1}·s^{-1} increment of the spray density. Considering different T values, in Γ=0.08 kg·m^{-1}·s^{-1} is two to three times than that of Γ=0.02 kg·m^{-1}·s^{-1}.

The experimental results show that the spray density affects the flow resistance significantly when the steam flows across tube bundle. When liquid sprayed on the tube forms a falling-film outside the tube surfaces, the tube outer diameter increases whereas the tube pitch decreases. This causes the actual flow velocity to become higher than the calculated superficial value. Consequently, the pressure drop, as well as the speed and frequency of droplets between the tubes, will increase with the increasing spray density. When spray density increases to a certain degree, the original falling-film mode of the droplet gradually changes into the jet mode, as illustrated by Armbruster [1] and Hu [16]. This mode further reduces the flow area and increases the actual flow velocity between tubes. Furthermore, the existence of the droplet and liquid film will increase the additional contact area between steam and liquid and thereby increase the frictional resistance. In a word, the pressure drop increases with the increasing of spray density.

Figure 5 also shows that pressure drop decreases with increasing temperature. $\overline{\Delta p}$ at 50 °C is nearly twice that at 70 °C. In a constant G, saturated steam density increases with temperature. Thus, the smaller steam velocity occurs at higher temperature. The impact factor of steam velocity on pressure drop is higher than that of steam density. Therefore, a higher temperature results in a smaller pressure drop. Furthermore, water dynamic viscosity is low at high temperature, which will cause the friction resistance become small at high temperature when steam flows across the falling-film tube. Therefore, pressure drop decreases with increasing temperature.

Figure 6 shows the pressure drop in the rotated square tube bundle arrangement. The tendency of pressure drop is similar to that in regular triangular arrangement as shown in figure 5. However, the value in rotated square tube bundle is obviously small.

The falling film evaporation occurs in an actual evaporator, such that the spray density changes along the vertical direction. Spray density actually decreases gradually along the downward direction. Based on the data of pressure drop illustrated in figures 5 and 6 and the above description of pressure drop change law, we can image the distribution of pressure in evaporator and the coefficient of pressure drop decreases gradually along the tube row direction.

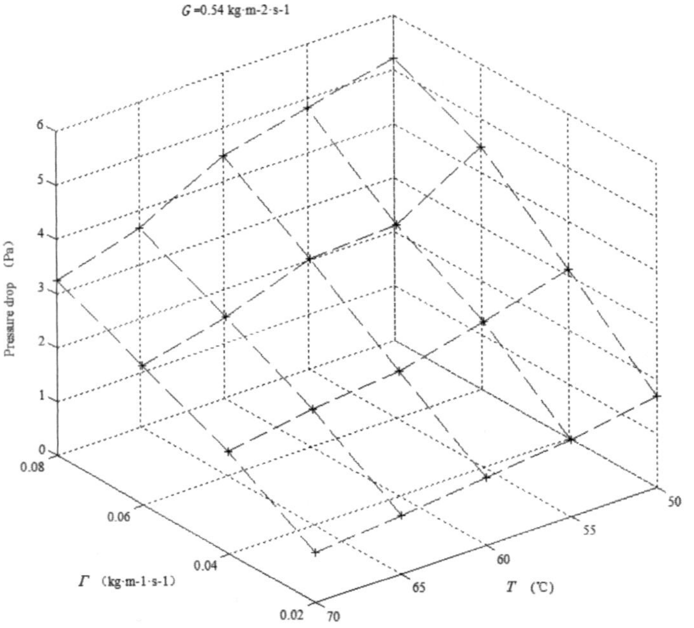

Figure 6. Pressure drop in rotated square arrangement.

A MED system has many effect-evaporators and the steam mass flow rate generated outside of tube in each effect-evaporator is approximately the same. The working pressure and corresponding temperature decrease with the evaporator effect number. Experimental data show that pressure drop decreases with increasing temperature. In other words, the change of temperature in experiment can be regarded as the effect number of the multi-effect evaporation device. The data of pressure drop and its change with the temperature can indicate that the pressure drop in a low-temperature-effect evaporator is higher than that in a high-temperature-effect evaporator.

R is defined to evaluate the effect of tube bundle configuration on pressure drop as follows:

$$R=\frac{\overline{\Delta p}_{\mathrm{RT}}}{\overline{\Delta p}_{\mathrm{RS}}} \tag{1}$$

where $\overline{\Delta p}_{\mathrm{RT}}$ represents the pressure drop in regular triangle arrangement; $\overline{\Delta p}_{\mathrm{RS}}$ represents the pressure drop in rotated square arrangement.

The ratio (R) between $\overline{\Delta p}$ in regular triangle and rotated square arrangements is illustrated in figure 7. The value of R more than 2.5 shows that the pressure drop in regular triangle is obviously high than that in rotated square arrangements. Although same the both relative tube pitch in two arrangements, the adjacent vertical and horizontal tube pitch are different because of the tube bundle geometry structure. The effect of tube bundle geometry structure on pressure drop is analysed in two aspects. 1) Compared with the rotated square arrangement, the vertical adjacent tube pitch of the regular triangle arrangement is longer. Under the same spray density, the velocity of liquid droplet or falling film among the vertical adjacent tubes, as well as the velocity of liquid film outside the tube is higher in regular triangle arrangement. When steam flows through the tube bundle, the friction resistance and shear force of steam occurs with the liquid droplet or falling film among the vertical adjacent tubes as well as with the liquid film outside the tube. Thereby, the higher velocity causes the bigger pressure drop. 2) The horizontal adjacent tube pitch of the regular triangle arrangement is shorter than that of rotated square arrangement, which will cause the steam flow worse in horizontal direction. In comparison, the steam flow is more frequent change in the regular triangle arrangement; consequently the local resistance is much bigger. In all, the pressure drop in regular triangle is higher than that in rotated square arrangements.

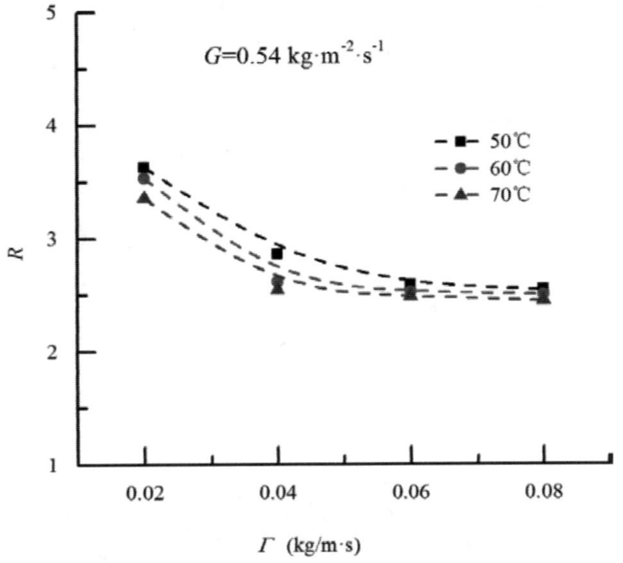

Figure 7. Ratios of pressure drop in two arrangements.

When temperature changes from 50 °C to 70 °C, the change of R is little. With increasing spray density R decreases, nevertheless, the change trend becomes small and it maintains stable. When spray density is 0.02 $kg{\cdot}m^{-1}{\cdot}s^{-1}$, R is more than 3; when spray density increases to 0.04 $kg{\cdot}m^{-1}{\cdot}s^{-1}$, it reduces to 2.8; from 0.06 $kg{\cdot}m^{-1}{\cdot}s^{-1}$ to 0.08 $kg{\cdot}m^{-1}{\cdot}s^{-1}$, R maintains about 2.5. At a small spray density, the flow pattern among the vertical adjacent tubes is drop-wise flow; the development and breakup of liquid pendant and droplet are more intense in long vertical adjacent tube pitch. Thus, the value of R is higher at a small spray density. Given that droplets grow with increasing spray density, the liquid flow pattern turns into the jet mode [1, 15]. The effect of spray density on steam horizontal flow changes little. Thus, the R maintains stability from 0.06 $kg{\cdot}m^{-1}{\cdot}s^{-1}$ to 0.08 $kg{\cdot}m^{-1}{\cdot}s^{-1}$.

4. Conclusions

Experiments were conducted to measure the pressure drop under the condition of steam flow across the horizontal tube bundle with falling-film flow. Experimental results determine the law for pressure drop with tube bundle structure, column number, temperature and spray density. The conclusions are as follows:

1) For a horizontal tube falling film evaporator, spray density has a significant influence on pressure drop. The pressure drop at the maximum spray density of 0.08 $kg \cdot m^{-1} \cdot s^{-1}$ can be three times that in spray density of 0.02 $kg \cdot m^{-1} \cdot s^{-1}$.
2) The saturated temperature also affects the pressure drop. Under the same steam flow rate, the pressure drop at 50 °C is almost twice that at 70 °C.
3) The tube bundle structure has an obvious effect on pressure drop. The rotated square arrangement is more conductive to the steam flowing across a horizontal falling-film tube bundle and the pressure drop in that is smaller.

5. References

[1] Armbruster R and Mitrovic J 1998 *Exp. Therm. Fluid Sci.* 18 183-194
[2] Liu Z H and Yi J 2002 *Appl.Ther. Eng.* **22** 83-95
[3] Ribatski G and Jacobi A M 2005 *Int. J. Refrig.* **28** 635-653
[4] Shen S Q, Liang G T and Gong L Y 2011 *CIESC J.* **62** 3381-85
[5] Hisham T E, Imad A, Bingulac S and Ettouney H 1998 *Chem. Eng. Technol.* **28** 437-451
[6] Hisham T E, Ettouney T, Hisham M E and Faisal M 2000 *Appl. Therm. Eng.* **20** 1679-1706
[7] Fujii T, Uehara H, hirata K and Oda K 1972 *Int. J. Heat Mass Trans.* **15** 247-260
[8] Grant I D R and Chisholm D 1979 *J. Heat Trans.-T. ASME* **101** 38-42
[9] Ribatski G and Thome J R 2007 *Heat Transfer Eng.* **28** 508-524
[10] Ishiraha K, Palen J W and Taborek J 1980 *Heat Transfer Eng.* **1** 23-32
[11] Dowlati R, Kawaji M and Chan A M C 1990 *AIChE J.* **36** 765-772
[12] Xu G P, Tso C P and Tou K W 1998 *J. Fluid Eng.-T. ASME* **120** 140-145
[13] Dowlati R, Kawaji M and Chan A M C 1996 *J. Heat Trans.-T. ASME* **118** 124-131
[14] Feenstra P A, Weaver D S and Judd R L 2000 *Int. J. Multiphase Flow* **26** 1851-73
[15] Hu X and Jacobi A M 1998 *Exp. Therm. Fluid Sci.* **16** 322-331
[16] Hassan A M 2010 *Energ. Convers. Manage.* **51** 703-709
[17] Liu H, Shen S Q, Gong L Y and Chen S 2014 *Appl. Therm. Eng.* **69** 214-220

6. Acknowledgments

This work is supported by the project of Chinese National Natural Science Foundation (No. 51336017). The authors are grateful for the financial support

IOP Conf. Series: Materials Science and Engineering **88** (2015) 012035 doi:10.1088/1757-899X/88/1/012035

Application of calcium chloride as an additive for secondary refrigerant in the air conditioning system type chiller to minimized energy consumption

A Suwono[1], Y S Indartono[1] and M Irsyad[1,2], I C Al-Afkar[1]

[1]Faculty of Mechanical and Aeronautical Engineering, Institut Teknologi Bandung, Indonesia
[2]Mechanical Engineering Department, Lampung University, Bandar Lampung, Indonesia

E-mail: aryadi@termo.pauir.itb.ac.id

Abstract. One way to resolve the energy problem is to increase the efficiency of energy use. Air conditioning system is one of the equipment that needs to be considered, because it is the biggest energy user in commercial building sector. Research currently developing is the use of phase change materials (PCM) as thermal energy storage (TES) in the air conditioning system to reduce energy consumption. Salt hydrates have been great potential to be developed because they have been high latent heat and thermal conductivity. This study has used a salt hydrate from calcium chloride to be tested in air conditioning systems type chiller. Thermal characteristics were examined using temperature history (T-history) test and differential scanning calorimetry (DSC). The test results showed that the thermal characteristics of the salt hydrate has been a high latent heat and in accordance with the evaporator temperature. The use of salt hydrates in air conditioning system type chiller can reduce energy consumption by 51.5%.

1. Introduction

Issues on energy which are encountered in many countries have become a global concern. The problem is that electricity consumption keeps increasing while its energy supply is declining. To narrow this gap, efficiency of energy consumption has to be escalated. Air Conditioning in commercial buildings is recognized as the contributor of the largest quota of its total energy consumption in different countries such as the United States and China 38% [1], Malaysia 57% [2], and Indonesia 65% [3]. Efforts for efficiency of energy on Air Conditioning are applying new air conditioner with higher efficiency, designing integrated and innovative system, or managing control and operation [4].

Innovative design which is being developed is the use of *phase change material* (PCM) as thermal energy storage in air conditioning system. One of its applications is in the secondary refrigerant. To apply this air conditioning system, such criteria should be employed as phase change temperature between 5°C to 12°C, large fusion heat, good thermal stability, good thermal conductivity, stabile chemical features, low corrosions, low environment impact, low viscosity, simple manufacture, and relatively cheap [5]. Some PCMs whose phase change temperature on the range between 5°C up to

[1] To whom any correspondence should be addressed.

12°C are fatty acids and paraffin from organic material and hydrate gas and salt from inorganic ones[6]. Inorganic materials have advantages over organics, i.e. high latent heat and conductivity, low volume change, and available in economical prices. Yet their sub cooling and corrosion characteristics should be considered when the materials are applied [6}.

Calcium chloride is one of materials to build hydrate salt compound. This material is generally utilized to save thermal energy in calcium chloride hexahydrate ($CaCl_2.6H_2O$) compound form with melting point on 29.9°C. The benefits of using this material is it is accessible widely with low-cost, it has high latent heat and low corrosion compared to other similar compound and it is non-toxic[8]. Considering the thermal characteristic displayed on phase diagram in Figure 1, solid to liquid phase change and the reverse can be generated as it is expected by adjusting the $CaCl_2$ concentration. Adjusting concentrate to 39-40% of $CaCl_2$ mass basis into water, the melting and freezing temperature are matched to chiller evaporator working temperature. To identify whether the compound can be applied in secondary refrigerant of chiller type air conditioning system, further study needs to be experimented.

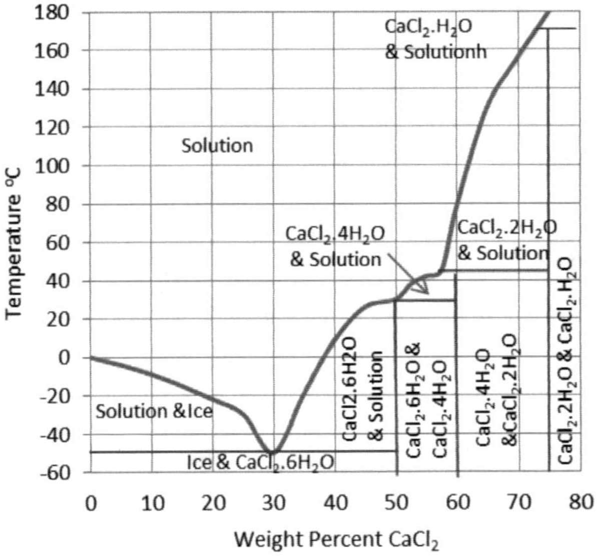

Figure 1. Phase diagram of calcium chloride and water compound [8]

2. Research Methodology

Hydrate salt in the study is the mixture of calcium chloride and water. The calcium chloride found in stores is calcium chloride bi-hydrate ($CaCl_2.2H_2O$) which contains 25% water. The thermal characteristics of the hydrate salt are examined by performing two types of tests. Firstly, temperature history (T-history) method examination is performed to obtain adjusted composition in order to match the melting temperature with evaporator's working temperature. Secondly, *Perkin Elmer's Differential Scanning Calorimetry* (DSC) type 8500 test is used to signify the freezing and melting temperature together with the latent heat value. This test observes thermal characteristics using sample weight 5-7 mg and the applied cooling and heating flow of 2 C/min starting from 25 goes to -30°C for cooling and -30 up to 35°C for heating with flow rate of cooling nitrogen as much as 20 ml/min.

Another test for density and viscosity is executed to support the application of this hydrate salt in the air conditioning system. The study use Hitachi RCU15Y brand chiller specified with compressor power 10.8 kW and secondary refrigerant pump power 1.5 kW. The test is performed for eight hours starting from 08.00 a.m. to 16.00.p.m. Testing equipment scheme is displayed in Figure.2. The test will indicate the energy spent by the pump, compressor, and COP chiller

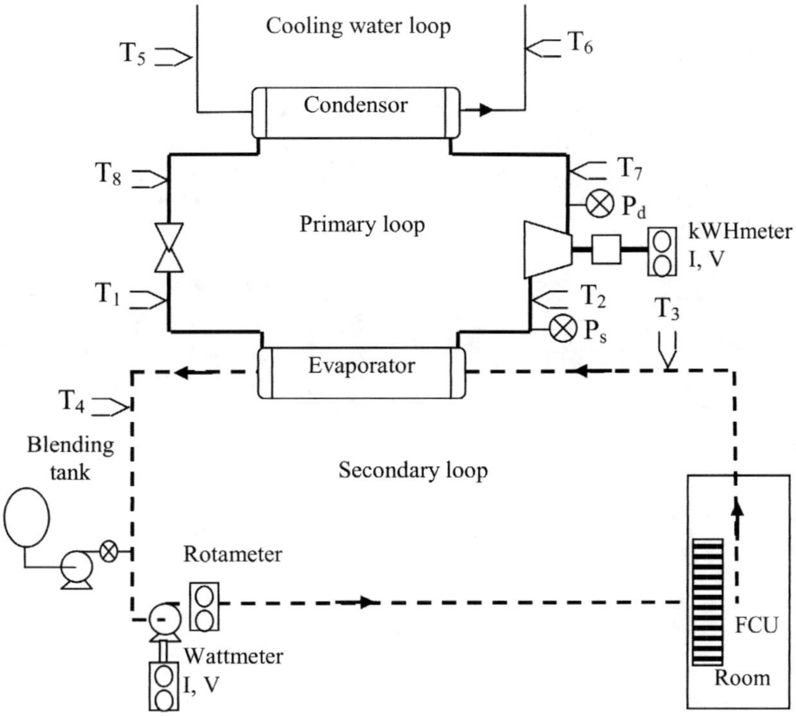

Figure 2 Schematic picture of experimental layout

3. Result and Discussion

3.1. Thermal Characteristics

The result of T-history test indicated that appropriate composition of $CaCl_2$ and water on melting temperature between 5 to 12°C was 55% : 45%. This result is dissimilar from the value obtained from phase diagram curve in Figure 1. This alteration is caused by calcium chloride applied in the test in the form of calcium chloride di-hydrate ($CaCl_2.2H_2O$) compound with water percentage around 25%. The comprehensive result on the thermal characteristics for this hydrate salt is presented in Table 1.

Table 1. Properties of salt hydrate from calcium chloride

Properties	Value
Density (kg/m^3)	1.352
Viscosity (cSt)	3.893
Melting Temperature (°C)	8.81 – 25.68
Freezing Temperature (°C)	0.08 – (-4.10)
Latent Heat (kJ/kg)	315.37

Freezing process started to form on temperature 5.36°C and ended on 5.26°C, just like the T-history curve showed on Figure 3. DSC Test obtained slightly different number on freezing temperature as shown on table 1. Strong sub-cooling character of hydrate salt caused the difference on freezing temperature as well as its melting one. The latent heat value is high as much as 315.37J/g. This is caused by the length of melting temperature range as shown in Figure 5 and the formed solid crystal contained water. The water content in the solid crystal was the result of spectroscopy as explained in Baumgartner, et.al [9].

Freezing and melting process indicated the shaped and the melted solid mass concentration. Solid mass concentration for each temperature on freezing and melting process is shown in Figure 4. In the cooling process, the solid mass is formed in the temperature between 9°C down to 6°C was 10.212%. At the same time, the heating process in the temperature between 6°C up to 10°C, the melting part of the solid mass was 9.942%. Thus, with the energy saved in the form of latent heat, the change flow on temperature in secondary refrigerant can be reduced. Concluding the result of T-history test, DSC and solid mass concentrate test as support, it is signified that the concentration of $CaCl_2$ is 55% and water 45% was fitted to the working temperature of evaporator.

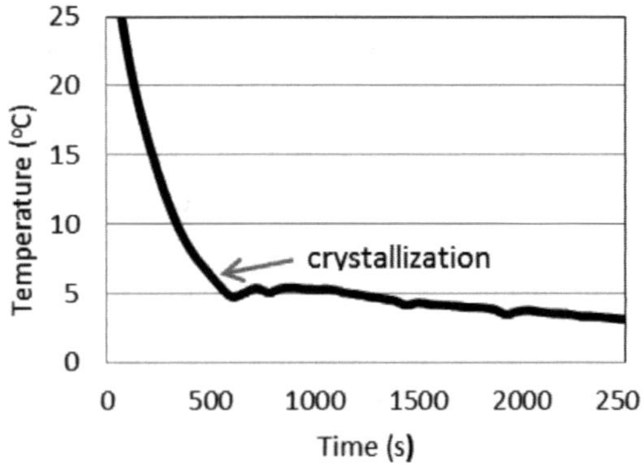

Figure 3. Cooling process of salt hydrate from $CaCl_2$ 55% and water 45% with T-history test

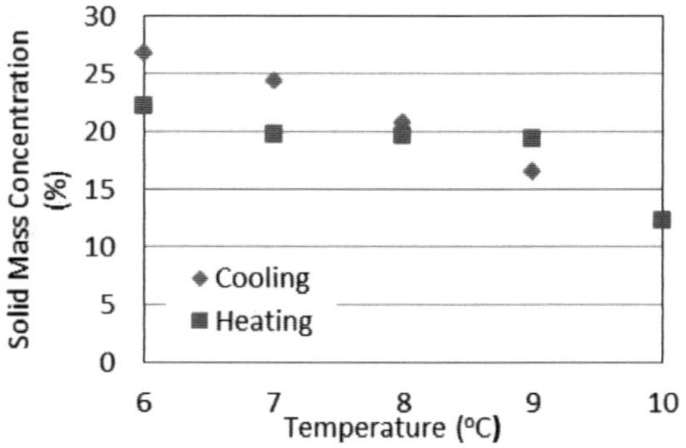

Figure 4. Solid mass concentration of salt hydrate from $CaCl_2$ 55% and water 45% in cooling and heating process

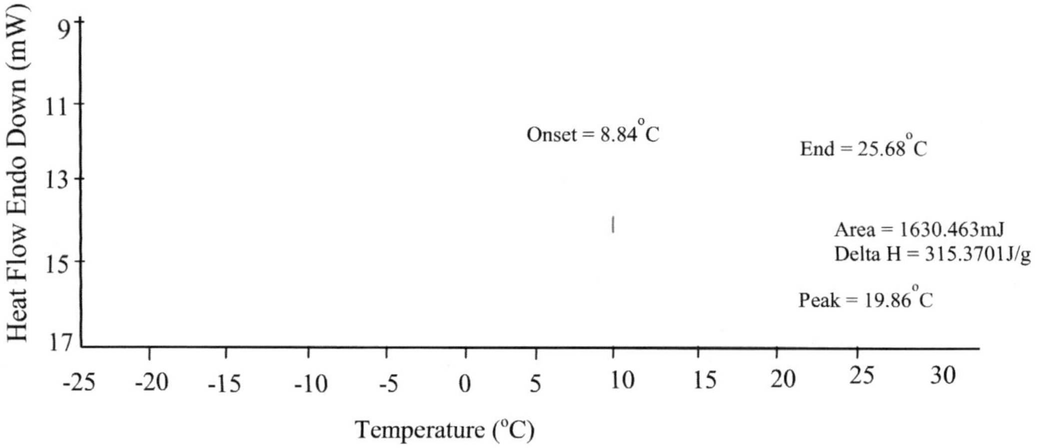

Figure 5. Melting temperature of salt hydrate from $CaCl_2$ 55% and water 45% with DSC test

3.2. Air conditioning performance

To find out the effect of the use of this hydrate salt as refrigerant for air conditioning performance system, energy consumption on chiller's compressor and secondary refrigerant pump had to be identified. The ability to save thermal energy and the effect on COP energy consumption will be discussed on chiller while the effect of hydrate salt use to pump energy consumption

3.2.1. Pump energy consumption

Salt hydrate from calcium chloride has different character from water so that it influenced the performance and energy needed by the pump. The test indicated that during eight hours experiment, the pump energy consumption increased 3.6% as displayed in Figure 6. This increase was not significant because the difference of density and viscosity is not too far. Viscosity value had not exceeded tolerated limit is 5sCt while debit, head and pump power calculation did not need corrective factor [10].

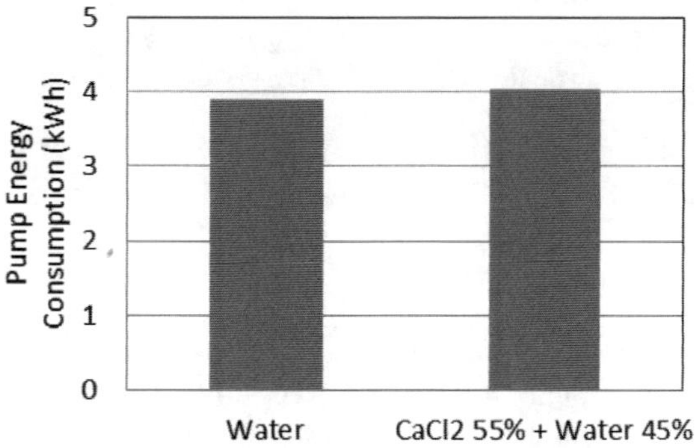

Figure 6. Energy consumption of pump

3.2.2. Chiller energy consumption

The use of salt hydrates can reduce the energy consumption of the chiller compressor, as shown in Figure 7. The total of decline was 51.5%. Thermal energy storage in the form of latent heat in PCM slurry can reduce the frequency of on/off compressor chiller. When the compressor on, some energy is stored as latent heat so that the time required to reach the temperature off of compressor is longer than water as secondary refrigerant. Likewise, when the compressor off, the release of energy in the melting process of solid particles in the slurry can extends the time over to reach the temperature on of the chiller compressor.

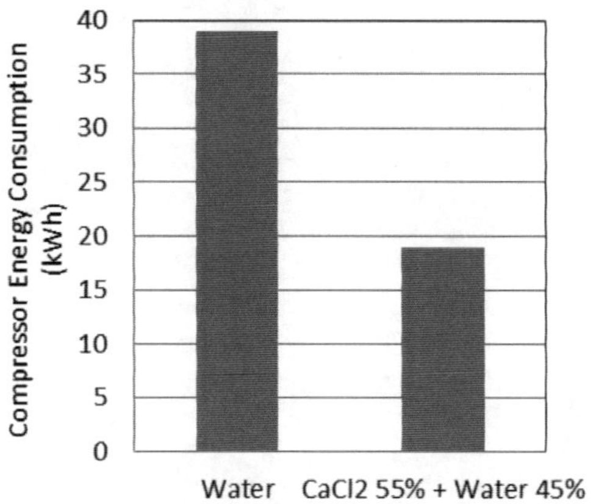

Figure 7. Energy consumption of compressor

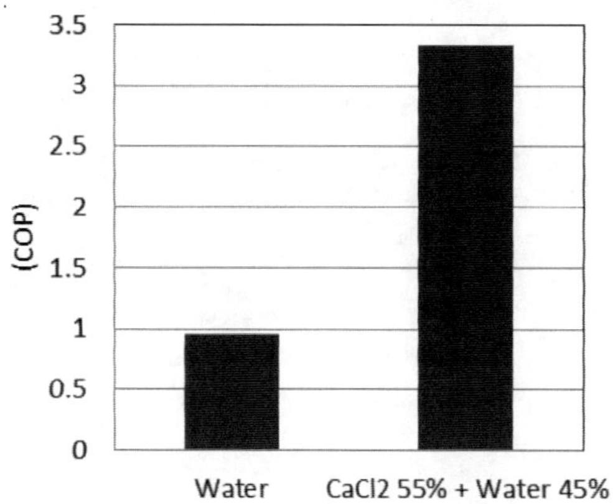

Figure 8. COP of chiller

Daily COP of chiller has been increase for the use of salt hydrates from calcium chloride as a secondary refrigerant as shown in Figure 8. This improved is due to the reduction of energy consumption compressor chiller, and increasing of cooling capacity. With the partly of energy is stored in the form of latent heat, the cooling capacity can be increased compared to water.

4. Conclusion

The use of calcium chloride in the form $CaCl_2.2H_2O$ that is dissolved in water with a composition of 55%: 45% can be used as a thermal energy storage air conditioning system as a secondary refrigerant replaced of water. This material has a high latent heat, and it can reduce the energy consumption of the compressor. In addition, the use of salt hydrates did not significantly affect the increase in electricity consumption of pumps. With the availability of many materials at low prices, and can decrease energy consumption of the air conditioning system makes this material very prospect to be developed.

5. References

[1] Xia J, Hong T, Shen Q, Feng W, Yang L, Im P, Lu A and Bhandari M 2014 *Energy and Buildings* **78** 165–175
[2] Saidur R 2009 *Energy Policy* **37** 4104–4113
[3] *JICA Study on Energy Efficiency and Concervation Improvement in Indonesia 2007 -2008* http://eneken.ieej.or.jp/en/data/pdf/491.pdf, diakses 8 – 10 – 2012
[4] Chua KJ, Chou SK, Yang WM, and Yan J 2013, *Applied Energy* **104** 87-104
[5] Li G, Hwang Y, and Radermacher R 2012 *International Journal of Refrigeration*, **xxx** 1-25
[6] Zhou D, Zhao CY,and Tian Y 2012 *Applied Energy* **92** 593–605
[7] Tsoukpoe KEN, Rammelberg HU, Lele AF, Kormhammer K, Watts BA, Schmidt T, and Ruck WII 2014 *Applied Thermal Engineering* **xxx** 1-19
[8] *Calcium Chloride, A Guide to Physical Properties.* Occidental Chemical Corporation.
[9] Baumgartner M, and Bakker RJ 2009 *Chemical Geology* **265** 335-344
[10] *Centrifugal pumps Selection and Application into the System*, http://jensapardi. files.wordpress.com/2010/02/centrifugal-pump-manual-12.pdf

IOP Conf. Series: Materials Science and Engineering **88** (2015) 012036 doi:10.1088/1757-899X/88/1/012036

Electrical and Joule heating relationship investigation using Finite Element Method

S K Thangaraju, K M Munisamy[1]

Universiti TenagaNasional, Jalan IKRAM-UNITEN, 43009, Kajang, Selangor.

Email: savithry@uniten.edu.my, kannan@uniten.edu.my

Abstract: The finite element method is vastly used in material strength analysis. The nature of the finite element solver, which solves the Fourier equation of stress and strain analysis, made it possible to apply for conduction heat transfer Fourier Equation. Similarly the Current and voltage equation is also liner Fourier equation. The nature of the governing equation makes it possible to numerical investigate the electrical joule heating phenomena in electronic component. This paper highlights the Finite Element Method (FEM) application onto semiconductor interconnects to determine the specific contact resistance (SCR). Metal and semiconductor interconnects is used as model. The result confirms the possibility and validity of FEM utilization to investigate the Joule heating due electrical resistance.

1. Introduction

The current challenge in on-chip interconnects is high temperature due to self-heating or resistive heating or it is also called Joule heating. The quote from Texas Instrument personnel, Mr. Siva (Gurrum, et. al. 2008) supports the motion. Thus, in this research the thermal distribution in semiconductor interconnects due to electrical current will be investigated in Finite Element Method (FEM) and experimental analysis.

There are several literatures reported on Joule heating recently. Read, 2009 has done a research work on Joule heating which was applied onto copper interconnects and the von-Misses strain level was investigated (Read et.al. 2009). When the liner material gets thinner the higher temperature is reached atthe interconnect. And the liner material property influence the temperature profile in the interconnects. But he has not looked at the specific contact resistance effects on the thermal distribution. Pedro in 2011 has reported a study on Joule heating in nanotube (Pedro et. al., 2011). He has made actual-imaging on joule heating but only restricted on the edge and outer surfaces. High current density in interconnects rises the temperatures. Tiedmann uses Scanning Joule Expansion Microscopy (SJEM) and FEM to investigate thermal distribution on aluminum interconnects (Tiedmann et. al., 2009). The expansion in the interconnect is used to validate the FEM model and ANSYS is used for thermal analysis. However, the specific contact resistance and temperature distribution at depletion region is not covered. It is also important to look at Broe's work, who has studied specifically on the ohmic contact resistance but between metal-metal contact temperatures (Broe et. al., 2010). He has the simple geometry of Ruthenium and Gold, and he has experimentally measured temperature of the material and contact resistance. He has highlighted that as the current density increases the temperature and changes the specific contact resistance value. The softening

[1] To whom any correspondence should be addressed.

temperature is made as failure criteria for the test at each current density. This model is a good candidate for the current FEM model validation.

From the literature reviewed, the paper investigate the thermal effect due to Joule heating in 3-Dimensional interconnect via using Finite Element Method (FEM) to improve the thermal distribution through the specific Contact Resistance (SCR) between different material.

A simple geometry is chosen to investigate the effect of SCR in simple semiconductor and metal interface. The geometry of the simple semiconductor to metal structure built in Finite element Abaqus. Different electrical resistivity value applied on the semiconductor and metal. The conductivity value is calculated for the semiconductors and metal for both electrical and thermal. The FEM is used to carry out the electrical analysis using the FourierEquation then the heat analysis to investigate the thermal distribution at the SCR region or the hot spots.

2. Background Theory

The electrical Fourier equation is given below in Eq. 1. The equation is solved using Finite element method. The property of current conductivity, σ is defined and Voltage at one surface is given as well. The current flux distributions from the element are used to calculate the heat flux or joule heat value at each elements using Eq. 2. The calculated heat flux value is then used as input file for heat analysis to obtain temperature distribution by solving Eq. 3. Table 1 shows the similarity of the Fourier Equation for electrical and thermal analysis in FEA.

Equation 1	Equation 2
Electrical analysis $J_x = \sigma_x \frac{\Delta V}{\Delta x}$Where, J_x , Current flux, Amp/μm^2 σ_x, = 1/ρ, Conductivity P, resistivity, Ωμm Δx, spatial coordinate in the direction of current flow, (μm) V, volt	Joule heating calculation $G = J^2\rho$ where G = Joule heat power, (W/μm^3) J = Current density (mA/μm^2) ρ = electrical resistivity, (Ω.μm)
Equation 3	
Heat Analysis $H = k\frac{dT}{dn}$ where H = Heat Flux, (W/μm^2) T = Temperature (K) n = spatial coordinate in the direction of current flow, (μm) k = material thermal conductivity (W/(μm.K))	

3. Model Development

A simple geometry is chosen to investigate the effect of SCR in simple semiconductor and metal interface. The Figure 1, Figure 2, and Figure 3 shows the geometry of the simple semiconductor to metal structure built in FEM platform. The Table 1 illustrates the electrical resistivity used for semiconductor and metal. Table 2 shows semiconductor and metal calculated conductivity value for both electrical and thermal. Next Table 3 shows the SCR value used and Table 4 summarize the electrical and conductivity value applied at the SCR region in FEM (Abaqus).

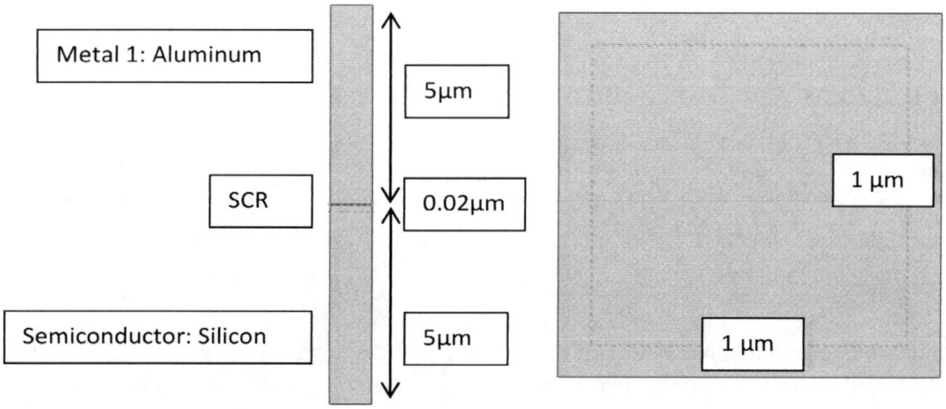

Figure 1. Simple semiconductor to metal contact geometry – view in Z-X and X-Y plane

Table 1: The electrical resistivity used for semiconductor and metal

	Electrical resistivity	
Material	ρ_b (Ω.cm)	ρ_b (Ω.μm)
Silicon	4.5 X 10^{-6}	4.5 X 10^{-2}
Aluminum	4 X 10^{-6}	4 X 10^{-2}

Table 2: The electrical and thermal conductivity used for semiconductor and metal

	Electrical conductivity, σ		Thermal conductivity, k	
Material	σ_b $(\Omega.\text{cm})^{-1}$	σ_b $(\Omega.\mu\text{m})^{-1}$	K (W/(cm.K)	K (W/(μm.K)
Silicon	222.22 X 10^3	22.22	148	1.48 X10^{-4}
Aluminum	250 X 10^3	25	250	2.50 X10^{-4}

Table 3: The Specific Contact Resistance (SCR) value used for Semiconductor to metal interface

Interface	ρ_c (Ω.cm^2)	ρ_c (Ω.μm^2)
Silicon-Aluminum	1 X 10^{-6}	1 X 10^{+2}

Table 4: The electrical and thermal conductivity value applied for SCR - baseline

ρ_c (Ω.μm^2)	1 X 10^{+2}
$R_c = \rho_c$ X area (Ω)	1 X 10^{+2}
$R_c X$ thickness (Ω.μm)	2
SCR electrical Conductivity $(\Omega.\mu\text{m})^{-1}$	0.5
SCR Thermal Conductivity (W/(μm.K)	2.00 X10^{-4}

Table 5: Varying SCR values configuration used

	SCR – 1	SCR-2	SCR-3	SCR-4
SCR Rho-c (Ω.cm^2)	1.00E-06	1.00E-07	1.00E-08	5.00E-08
SCR Rho-c (Ω.μm^2)	100	10	1	5
Area of interface(μm^2)	1	1	1	1
Resistor	100	10	1	5
Thickness (μm)	0.02	0.02	0.02	0.02
Resistor X thickness (Ω.μm)	2	0.2	0.02	0.1
SCR electrical conductivity (Ω.μm)^-1	0.5	5	50	10

Figure 3 shows the computational model of the simple semiconductor to metal geometry with fully meshed and boundary conditions and loads imposed. In Abaqus for electrical analysis the current density at the surface is given as 1 Amp/ $(\mu m)^2$. That makes the total current at the surface is 1Amp for 1μm X μm area. And for the heat analysis the Joule heating will be the heat flux at all the nodes and the boundary temperature is set to be 308 K at both end side of the geometry.

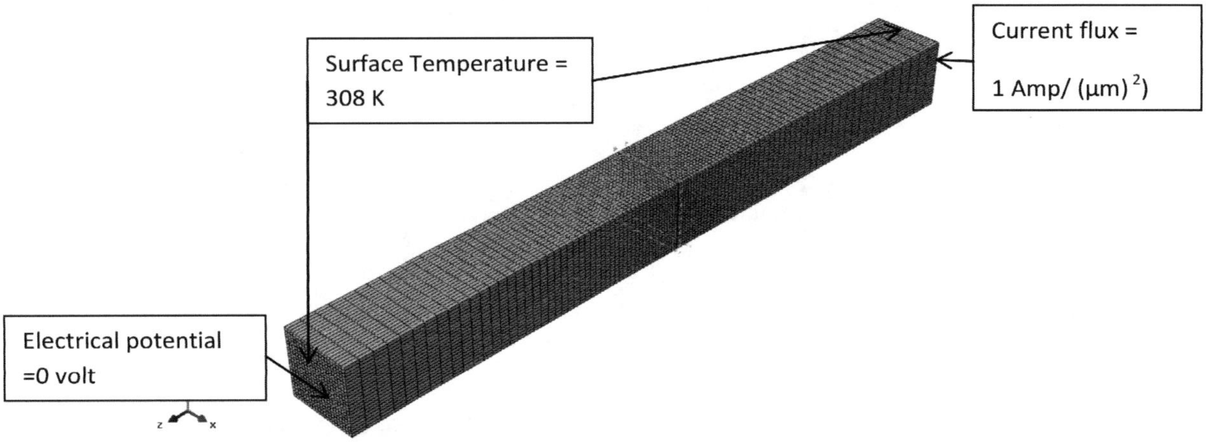

Figure 2. Computational model developed with mesh boundary conditions and Load setting in Abaqus.

Result and Discussion

SCR = 1 X 10^{-6} ($\Omega.cm^2$)

The Figure 5 shows the electrical potential distribution output from electrical analysis. The blue color indicates the zero electrical potential which was set in the boundary condition. The opposite side has the highest voltage of 0.465 Volt. **Figure 6** below is the final heat flux distribution due to heat sink application. The input for the calculation would be the Joule heating as in the Eq. 2. The highest heat flux value is at the bottom part of silicon (semiconductor) with values of 0.254 W/(μm.K). The interface has the lowest heat flux value of 0.00119 W/(μm.K).

IOP Conf. Series: Materials Science and Engineering **88** (2015) 012036 doi:10.1088/1757-899X/88/1/012036

Voltage Distribution

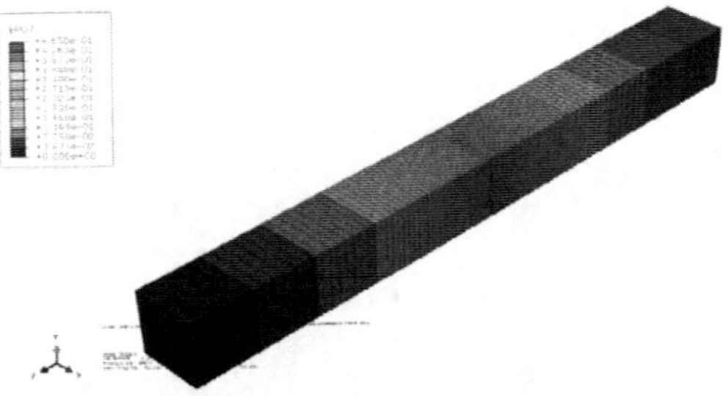

Figure 3 : Output of electrical analysis – Electrical potential in Volts

Heat flux distribution

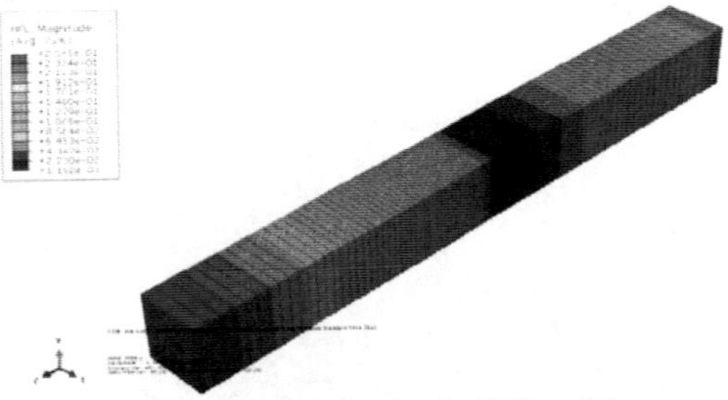

Figure 4 : Heat flux distribution in W/(μm.K)

Temperature distribution from Heat Analysis

The temperature distribution in Figure 6 shows the temperature output from heat analysis, from Eq. 3. The boundary temperature is 308K as set initially. The interface region has recorded highest temperature of 628K, temperature difference of 320K.

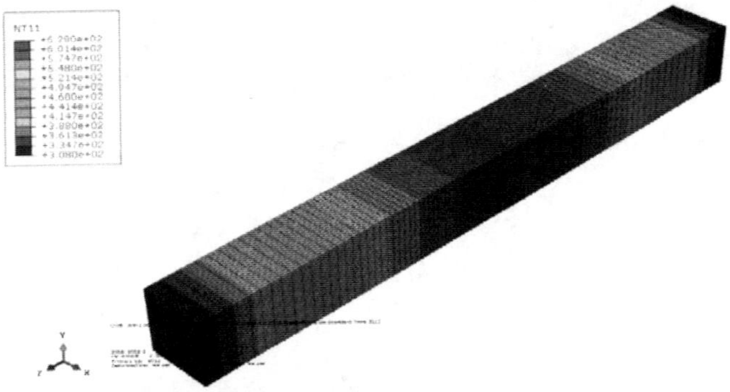

Figure 5. Temperature distribution – from heat analysis in (K).

Table 6: Summary highest Temperature recorded in the model with varying SCR.

SCR value	$1 \times 10^{-6}(\Omega.cm^2)$	$1 \times 10^{-7}(\Omega.cm^2)$	$1 \times 10^{-8}(\Omega.cm^2)$	$5 \times 10^{-8}(\Omega.cm^2)$
Temperature Difference (K)	320	279.9	276	277.7

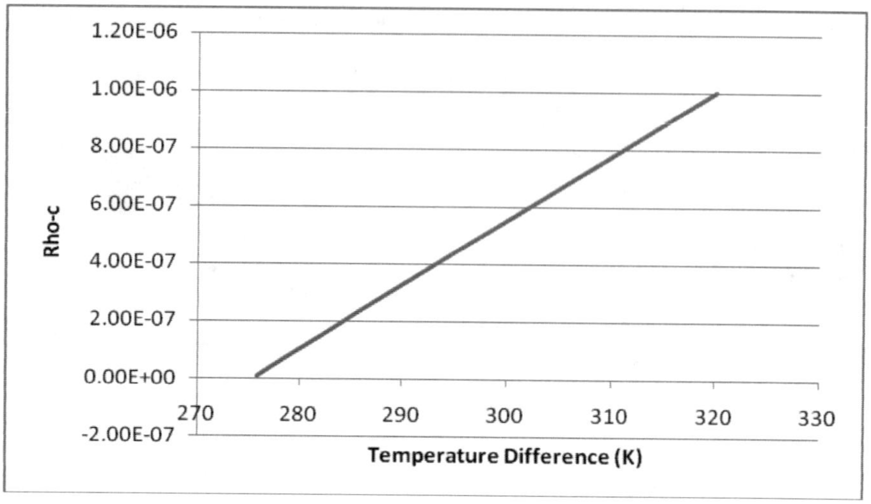

Figure 6. Rho-c of SCR against temperature difference

Conclusion

It can be concluded that the electrical analysis and heat analysis can be done using FEM software. However, a subroutine is needed to input the output value of heat flux from electrical analysis into heat analysis. The Fourier equation for electrical and heat conduction is solved separately to obtain the temperature output

The varying SCR value has confirmed that the temperature distribution is changing with SCR value. And the Figure 6 and Table 6show that the relationship between the Rho-c value of SCR and temperature difference is linear.

References

1. (Breu, 2010) Brue, et. al. Characterization of gold/gold, gold/ruthenium and ruthenium/ruthenium ohmic contacts in MEMS switches imporved by a novel methodology J Micro/Nanolith, MEMS, MOEMS 9(4), 041102 (Dec 2010
2. (Chen, 2011), Chen M et. al., Bonding of carnonnantubes onto microelectrodes by localized, Sensors and Actuators A 170, 2011, 202-204
3. (Gurrum, 2008), Gurrum, S.P, et. al., A compact approach to on-chip interconnect heat conduction modeling using the finite element method, ASME J. Elelctronic Packaging, 2008, Vol 130 (3)
4. (Indukuri, 2012), Indukuri, T.K, et. al., Electrical and reliability characterization of CuMnself forming barrier interconnects on low-k CDO dielectrics, Microelelctronic Engineering 92, 2012, 49-52
5. (ITRS, 2009), International Road-map in Semiconductor 2009.
6. (Khor, 2012), Khor, C.Y. et al. Investigation of the fluid/structure interaction phenomenon in IC packaging. Journal of Microelectronics Reliability 2012; 52, pg 241-252
7. (Lau, 2010), Lau JH., TSV manufacturing yield and hidden costs for 3D IC integration. In: Electronic components and technology conference; 2010. p. 1031-42
8. (Pedro, 2011), Pedro M.F.J.C, Guatam U., Bando Y. Golberg D, Direct imaging of Joule heating dynamics and temperature profiling inside a carbon nanotube interconnect, Nature Communications 1429, 2011.

IOP Conf. Series: Materials Science and Engineering **88** (2015) 012036 doi:10.1088/1757-899X/88/1/012036

9. (Read, 2008), Read, D T, Geiss R.H., Barbosa III N, Constraint effect in deformation of copper interconnect lines subjected to cyclic Joule heating, J. of Strain Analysis Vol 44, 2008, 543-553
10. (Dimitriv, 2011) SimaDimitrijev, Principles of semiconductor Devices, 2nd Ed. 2011 Oxford University Press
11. (Tiedmann, 2009), Tiedmannet. al., Finite element analyses assisted Scanning Joule Expansion Microscopy on interconnects for failure analysis and reliability investigations, Microelectronics Reliability 49, 2009, pp 1165 – 1168.
12. (Tokei, 2010), Tokei, Z, Croes, K, Beyer, G, Reliability of copper low-k interconnects, Microelectronics Eng 87, 2010, 348-354
13. (Tu, 2010), Tu KN, Reliability challenges in 3D IC packaging technology, Microelectronic Reliability 2010; 51(3): 517-523

IOP Conf. Series: Materials Science and Engineering **88** (2015) 012037 doi:10.1088/1757-899X/88/1/012037

Design of Radial Turbo-Expanders for Small Organic Rankine Cycle System

M Arifin[12], A D Pasek[2]
[1]Research Centre for Electrical Power and Mechatronics, Indonesian Institute of Sciences - LIPI
Komp. LIPI Gd.20, Jl. Cisitu No.21/154D, Bandung 4015 Jawa Barat
Telp. 022 2503055, Fax 022 2504773

E-mail: mfilzah@gmail.com / maul004@lipi.go.id

[2]Faculty of Mechanical and Aerospace Engineering – Institute Technology Bandung
Jl. Ganesha No.10 Taman Sari, Bandung 41032 Jawa Barat
Telp. 022 2504243, Fax 022 2534099

E-mail: aripasek@gmail.com

Abstract. This paper discusses the design of radial turbo-expanders for ORC systems. Firstly, the rotor blades were design and the geometry and the perfromance were calculated using several working fluid such as R134a, R143a, R245fa, n-Pentane, and R123. Then, a numerical study was carried out in the fluid flow area with R134a and R123 as the working fluid. Analyses were performed using Computational Fluid Dynamics (CFD) ANSYS CFX on two real gas models, with the k-epsilon and SST (*shear stress transport*) turbulence models. The results analysis shows the distribution of Mach number, pressure, velocity and temperature along the rotor blade of the radial turbo-expanders and estimation of performance at various operating conditions. CFD analysis show that if the flow area divided into 250,000 grid mesh, and using real gas model SST at steady state condition, 0.4 kg/s of mass flow rate, 15,000 rpm rotor speed, 5 bar inlet pressure, and 373K inlet temperature, the turbo expander produces 6.7 kW, and 5.5 kW of power when using R134a and R123 respectively.

1. Introduction

Climate change concerns coupled with high oil prices are driving research and development on renewable energies such as solar and geothermal energy. The Organic rankine Cycle (ORC) uses organic fluid as the working fluid to provide higher thermal cycle efficiency compared to the conventional steam Rankine cycle at resource temperatures below 300^{0}C. ORC has been studied as the utilization of waste heat recovery [1][2], solar energy [3], the combination of heat and power (CHP) [4], geothermal [5], and heat recovery from the exhaust gases from the engine [6]. The results of experimental studies show that the small-scale units ORC has a promising performance for power generation especially in remote areas. The ORC could provide a wide output power range, but consist

[1] To whom any correspondence should be addressed.

of less components, so that the ORC could be more compact and smaller in size compared to conventional power plants.

Most studies were focused on the thermodynamic analysis of the ORC, and the selection of the working fluid, with particular attention to obtain the best efficiency of power generation. On the other hand, only few published papers discuss on the design and geometry optimization of turbo expander. Basically, for output power ranges from 5 to 5000 kW) two types of turbines are proposed, they are axial or radial turbines. The latter turbine type is considered more attractive, because it has better performance at small output capacity. Therefore, it is important to develope knowledge on geometry design and its relation to the performance of radial turbo-expanders in ORC systems. The objectives of this study is to calculate the geometry of a radial inflow Turbo Expander for smalll ORC system and predict its performance for various working fluids and operating conditions.

2. Rotor blade geometry design

Baines' method [7] is normally used for preliminary geometry design of the radial turbine. Load coefficient can be calculated based on the U_4 and can be calculated by using Euler's turbomachinery equation:

$$\psi = \frac{\Delta h_0}{U_4^2} = \frac{C_{\theta 4}}{U_4} - \varepsilon \frac{C_{\theta 6}}{U_4} \quad (1)$$

Where $\varepsilon = r_6/r_4$ is ratio radius of rotor.

Figure 1 shows important parameters those are used for geometry design of the radial turbo-expander rotor. Meanwhile Figure 2 shows the meridional-plane geometry on the radial turbo-expanders. In this model the rotor shroud contour is assumed to be a circle, while the rotor hub to the contours of which are described as elliptical geometry assumptions made by Glassman [8].[8]

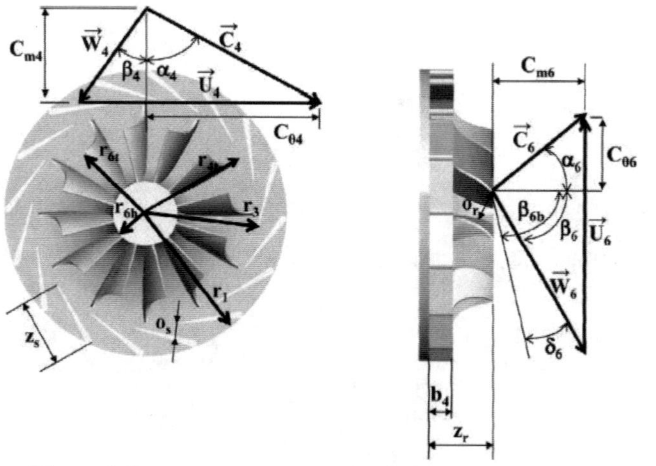

Figure 1 Geometry parameters of radial turbo-ekspander [9].

Magnitude of exit swirl very small, load coefficient can be predicted as:

$$\psi = C_{\theta 4}/U_4 \quad (2)$$

Hence the rotor inlet velocity triangle (station 4) is defined as:

$$C_{m4} = \xi \phi U_4 \quad (3)$$

$$C_4 = \left(C_{m4}^2 + C_{\theta 4}^2\right)^{1/2} \quad (4)$$

$$\alpha_4 = tan^{-1}(C_{\theta 4}/C_{m4}) \quad (5)$$

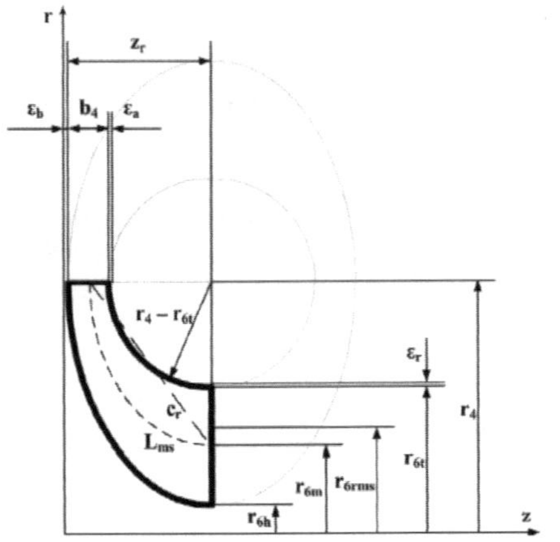

$$\beta_4 = tan^{-1}[(C_{\theta 4} - U_4)/C_{m4}] \quad (6)$$

Figure 2 The meridional-plane geometry on the radial turbo-expanders [8].

The static temperature and pressure at the inlet to the rotor are :

$$T_4 = T_{04} - C_4^2/2C_p \quad (7)$$

$$p_4 = p_{04}(T_4/T_{04})^{k/(k-1)} \quad (8)$$

Where $T_{04}=T_{01}$ dan $p_{04}=p_{01}-\Delta p_0$. Here Δp_0 is the total pressure loss in the stator.

$$A_4 = mRT_4/p_4C_{m4} \quad (9)$$

At exit, the total and static temperatures are :

$$T_{06} = T_{04} - \Delta h_0/C_p \quad (10)$$

$$T_6 = T_{06} - C_6^2/2C_p \quad (11)$$

3. Flow simulation on the rotor of radial turbo-expander

Table 1 Operational condition of the radial turbo-expander

Parameter	Value
Inlet temperature (K)	353-423
Mass flow (kg/s)	0.1-1.0
Outlet pressure (atm)	1.0
Outlet temperature (K)	313-373K

3.1 Number of grid

CFD simulation were done with various number of grids, to know the influence of number of grid on the simulation speed and accuracy. Figure 2 show grid model and Table 2 show the grid number used in the simulation.

Table 2 Grid number for analysis rotor turbo-expander

Model	Grid number
Model 1 (Coarse Grid)	20,000
Model 2 (Medium Grid)	100,000
Model 3 (Fine Grid)	250,000

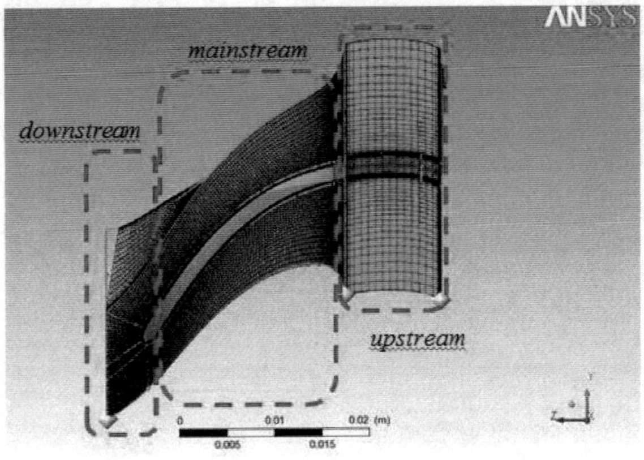

Figure 3 Grid model for CFD analysis for downstream, mainstream, and upstream area

3.2 Turbulence of real gas model

ANSYS CFX turbomachinery software provide for 2 types of turbulence models for real gas, the k-epsilon, and SST (Shear Stress Transport). The influence of these model on the simulation results were investigated. In this paper, simulations were performed at steady state condition at various rotor speeds 15,000 rpm, 20,000 rpm and 30,000 rpm.

4. Result and discussion

4.1 Blade geometry design

Results of geometry design are shown in Table 3. Figure 4 shows model for radial rotor turbo-expander with geometry obtained from the design process.

4.2 CFD analysis on the rotor of radial turbo-expander

Numerical simulations using the finite volume method were done to find the fluid flow charateristics, that able to predict the resulted torque, power and efficiency of the designed turbo-expander. The simulation were done at various of number of grids, rotational speeds and working fluids. As an example. Figure 5 shows three model with different number of grids using R123 as working fluid at 20,000 rpm. Model 3 wih high number of grid (250,000 grid), gives better results. With high number of grid, a strong vortex can be seen. This vortex form a higher pressure area in the region after the leading edge, separating flow from the hub surface and moving it up the blade toward the tip. Other calculation results obtained from the simulation for R-123 and R134 working fluids can be in Table 4.

Table 3 Result for geometry parameter rotor radial turbo-expander

Parameter	Unit	R134a	R123	R245fa	R143a	nPentane
Absolute meridional velocity (C_{m4})(inlet)	m/s	50	44	43	48	49
Blade speed (U_4)	m/s	167	146	143	160	163
Absolute tangential velocity ($C_{\theta 4}$)	m/s	150	132	129	144	147
Absolute flow angle (inlet) (α_4)	Degree	71.57^0	71.57^0	71.57^0	71.57^0	71.57^0
Relative flow angle (inlet) (β_4)	Degree	-18.43^0	-18.43^0	-18.43^0	-18.43^0	-18.43^0
Absolute velocity (C_4)	m/s	158	139	136	151	155
Relative absolute inlet (W_4)	m/s	52.7	46.38	45.33	50.60	51.60
Inlet Temperature (T_4)	K	360.47	361.15	364.37	362.60	367.37

Table 3 (continue)

Parameter	Unit	R134a	R123	R245fa	R143a	nPentane
Inlet Pressure (P_4)	bar	3.60	3.75	3.93	3.85	4.15
Inlet Area (A_4)	m^2	$6.5x10^{-4}$	$4.7x10^{-4}$	$5.3x10^{-4}$	$7.7x10^{-4}$	$8.3x10^{-4}$
Radius rotor (r_4)	m	0.100	0.093	0.091	0.101	0.104
Inlet Blade height (b_4)	m	0.010	0.008	0.009	0.012	0.012
Inlet Density (ρ_4)	Kg/m^3	12.238	19.121	17.384	10.737	9.813
Inlet Mach number (M_4)	-	0.87	0.93	0.85	0.75	0.72

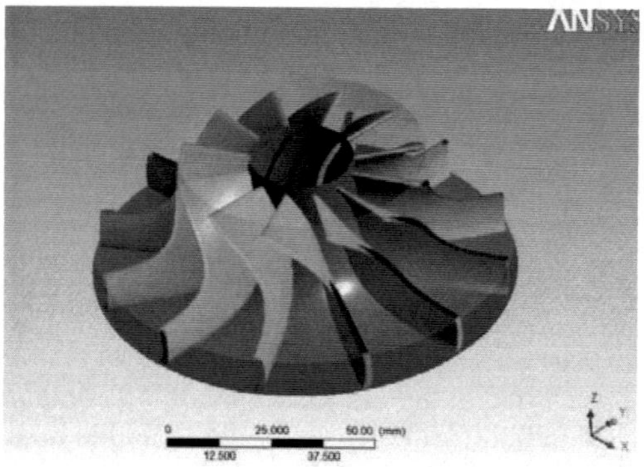

Figure 4 Result geometry radial rotor turbo-ekspander for small organic rankine cycle system

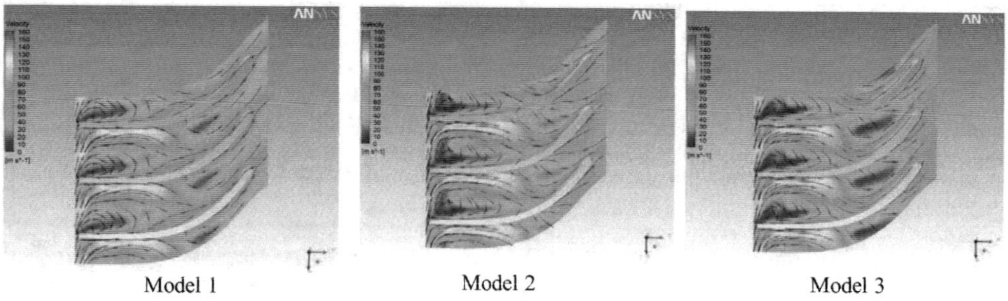

Model 1 Model 2 Model 3

Figure 5 Three model grid flow field predictions for radial turbine using R123 as working fluid with rotational speed 20,000 rpm

Table 4 Power and efficiency for rotor radial turbo-expander with different grid number

Number Grid	Model 1 20,000	Model 2 100,000	Model 3 250,000
Fluid: R123			
20,000 rpm			
Torque(Nm)	0.32	0.35	0.36
Power (kW)	6.56	7.08	7.21
Efficiency	0.63	0.64	0.65
Time consuming	6 minutes 30 seconds	11 minutes 56 seconds	28 minutes 5 seconds
Fluid R134a			
20,000 rpm			
Torque(Nm)	0.44	0.47	0.48
Power (kW)	7.94	8.60	8.53
Efficiency	0.63	0.64	0.65
Time consuming	6 minutes 21 seconds	11 minutes 54 seconds	27 minutes 30 seconds

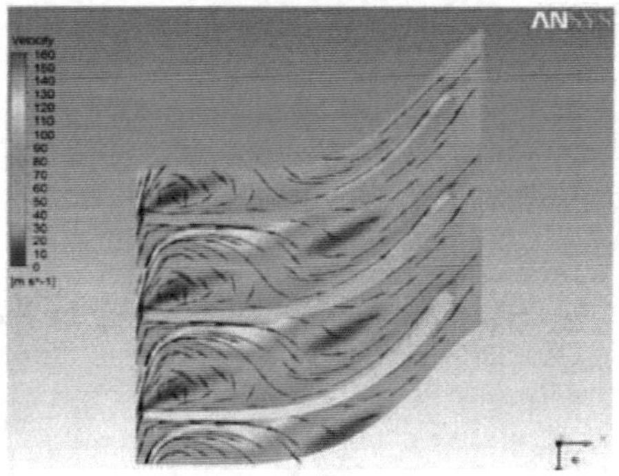

(a)

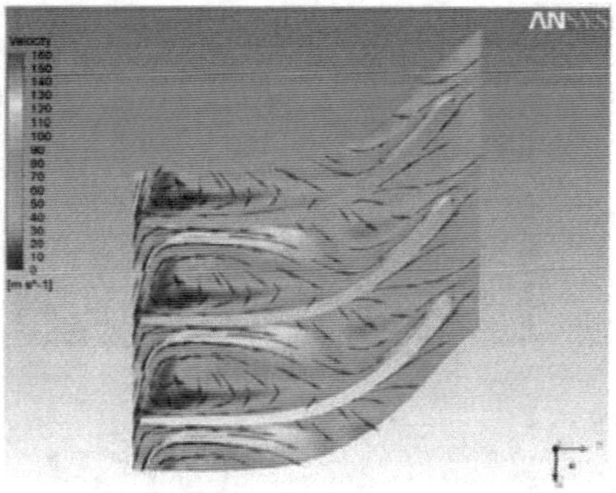

(b)

Figure 6 Velocity contour for turbulence models (a) k-epsilon dan (b) SST

Table 5 Time computation, iteration number on turbulence models for radial turbine small ORC with R123, 20,000 rpm

Model	Time consuming	Iteration	Power (kW)
k-epsilon (k-ε)	28 minutes	200	6.1
SST	28 minutes	200	7.2

Table 5 shows comparison of the computation time and the number of iterations for the k-epsilon and SST turbulence models. From Figure 5 and Figure 6, it can be seen that SST model has good swirl flow prediction on blade passage and resluts higher power output. For k-epsilon turbulence model, the power ouput is 6.1 kW while for the SST models the power output is 7.2 kW, which is 18% higher. A better model between those two cannot be verify in this study, experimental results on the expander model may clarify this discrepancy.

The result of an aerodynamic evaluation of rotor radial turbo-expander using SST turbulence model are shown in Figure 7 and Figure 8. This rotor is designed for a low specific speed of under 1.0. the lower specific speed means that the exducer area is considerably smaller, so that the exit tip to inlet radius ratio is much smaller and the blade height at the exit is smaller fraction of the inlet radius [7].

The results are different with conventional characteristic of rotor radial turbo-expander of this type. At rotor with low specific speed using air as working fluid, the fluid flow is more uniform. Meanwhile, at rotor radial turbo-expander using R123 and R134a as working fluid, a strong vortex can be seen forming on the pressure surface soon after the leading edge, separating flow from the hub surface and moving it up the blade toward the tip. It is interesting to note that the pressure gradient introduced in inlet region of the blade passage affect even the flow upstream of the leading edge quite strongly. The large variations in flow field that are created in the passage cause very significant changes in the exit flow (Figure 9). Contours of both relative and absolute flow angle just downstream of the trailing edge.

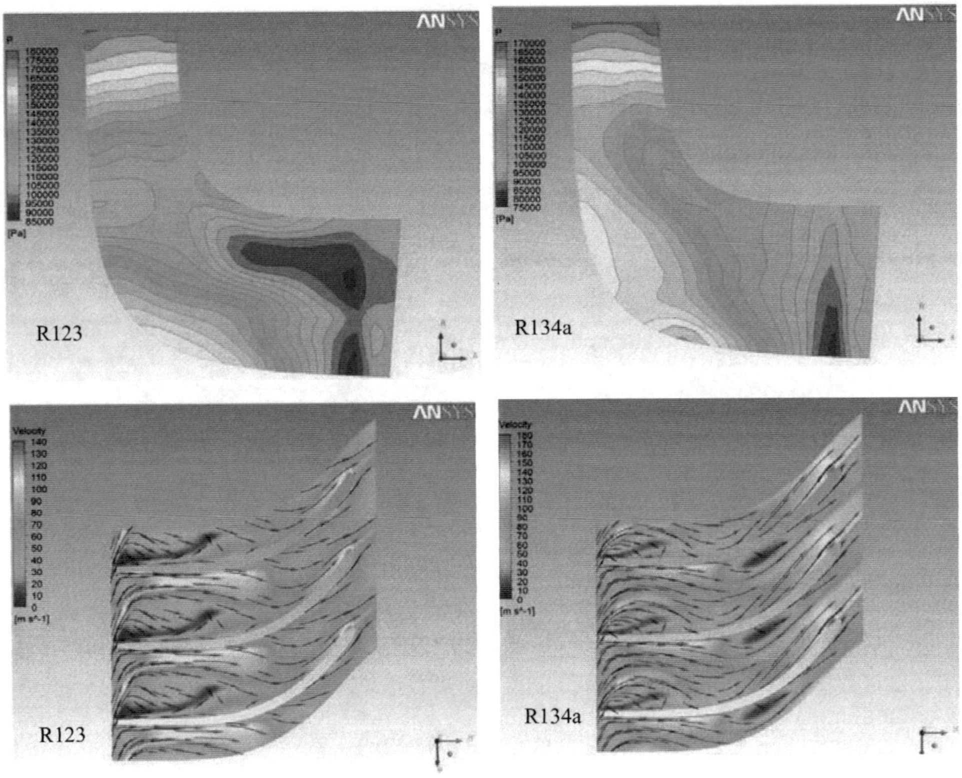

Figure 7 Flow prediction at meridional plane and blade-toblade view for rotor radial turbo-expander using R123 and R134a at 15000 rpm

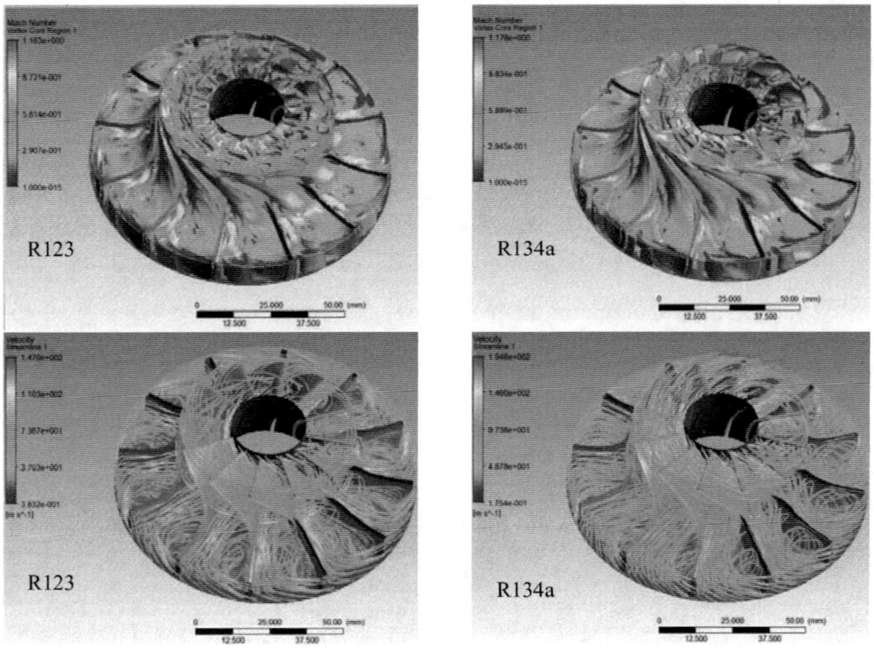

Figure 8 Flow prediction for rotor radial turbo-expander using R123 and R134a at 15000 rpm

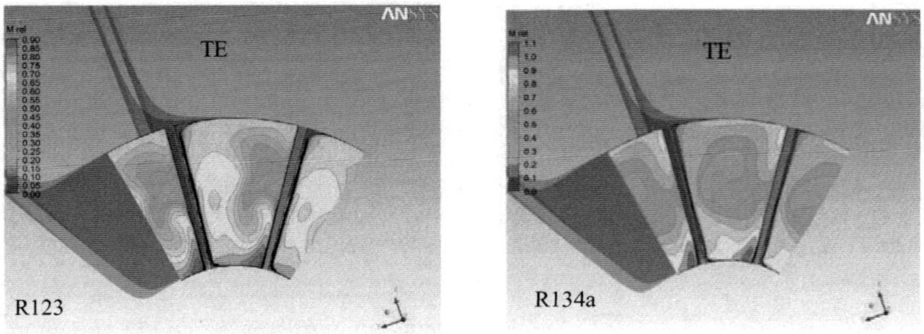

Figure 9 Flow prediction at trailing edge (TE) for rotor radial turbo-expander using R123 and R134a at 15000 rpm

IOP Conf. Series: Materials Science and Engineering **88** (2015) 012037 doi:10.1088/1757-899X/88/1/012037

5. Conclusion

This paper present the geometry calculation of a small radial turbo-expander for small organic rankine cycle system. The performances of the expander have been evaluated using 3D numerical analysis method. From the simulation it can be concluded that the higher tne number of grid will gives better and accurate results. The study also suggested to use SST models for the numerical analysis because it gives higher power output. However this results has to be verify in the experimental study. From the two working fluids use in the analysisi, R134a gives better performance. At mass flow rate 0.4 kg/s, 15,000 rpm, inlet pressure 5 bar, and inlet temperature 373K, the expander with R134a produces 6,7 kW power output , with total efficiency-to-static (ηts) 0,71. On the other hands R123 only produces 5.5 kW, with total efficiency-to-static (ηts) 0,66.

6. Acknowledgement

Authors would like to express gratitude to all the members of the Research Centre for Electrical Power and Mechatronics Indonesia Institute of Science and all members of Thermodynamics Laboratory of Faculty of Mechanical and Aerospace Engineering, Institute Technology of Bandung, for any assistance that have been given.

7. References

[1] T.-C. Hung, "Waste heat recovery of organic Rankine cycle using dry fluids," *Energy Convers. Manag.*, vol. 42, no. 5, pp. 539–553, Mar. 2001.

[2] Z. Gnutek and A. Bryszewska-Mazurek, "The thermodynamic analysis of multicycle ORC engine," *Energy*, vol. 26, no. 12, pp. 1075–1082, Dec. 2001.

[3] D. Manolakos, G. Papadakis, S. Kyritsis, and K. Bouzianas, "Experimental evaluation of an autonomous low-temperature solar Rankine cycle system for reverse osmosis desalination," *Desalination*, vol. 203, no. 1–3, pp. 366–374, Feb. 2007.

[4] A. Schuster, S. Karellas, E. Kakaras, and H. Spliethoff, "Energetic and economic investigation of Organic Rankine Cycle applications," *Appl. Therm. Eng.*, vol. 29, no. 8–9, pp. 1809–1817, Jun. 2009.

[5] M. Kanoglu, "Exergy analysis of a dual-level binary geothermal power plant," *Geothermics*, vol. 31, no. 6, pp. 709–724, Dec. 2002.

[6] M. Talbi and B. Agnew, "Energy recovery from diesel engine exhaust gases for performance enhancement and air conditioning," *Appl. Therm. Eng.*, vol. 22, no. 6, pp. 693–702, Apr. 2002.

[7] N. C. Baines, "Radial Turbine Design," in *Axial and Radial Turbines*, 2003, pp. 199–327.

[8] A. Glassman, *Turbine Design and Application.* 1994.

[9] C. A. M. Ventura, P. A. Jacobs, A. S. Rowlands, P. Petrie-Repar, and E. Sauret, "Preliminary Design and Performance Estimation of Radial Inflow Turbines: An Automated Approach," *Journal of Fluids Engineering*, vol. 134. p. 031102, 2012.

Experimental analysis and FEM simulation of loop heat pipe charged with diamond nanofluid for desktop PC cooling

P Gunnasegaran[1], **M Z Abdullah**[2], **M Z Yusoff**[1]
[1]Centre for Advanced Computational Engineering (CACE), College of Engineering, Universiti Tenaga Nasional, Putrajaya Campus, Jalan IKRAM-UNITEN, 43000 Kajang, Malaysia
[2]School of Mechanical Engineering, Universiti Sains Malaysia, Engineering Campus, 14300 Nibong Tebal, Penang, Malaysia

E-mail: prem@uniten.edu.my

Abstract. This paper discusses the impact of diamond nanofluid on heat transfer characteristics in a Loop Heat Pipe (LHP). In this study, diamond nanoparticles in water with particle mass concentration ranged from 0% to 3% is considered as the operational fluid within the LHP. The experiments are carried out by manufacturing the LHP, in which the setup consists of a water tank with pump, a flat evaporator, condenser installed with two pieces of fans, two transportation lines (vapor and liquid lines), copper pipe sections for attachment of the thermocouples and power supply. The uniqueness of the current experimental setup is the vapor line of LHP which is made of transparent plastic tube to visualize the fluid flow patterns. The experimental results are verified by Finite Element (FE) simulation using a three-dimensional (3D) model based on the heat transfer by conduction where the LHP as a whole is modeled by assuming it as a conducting medium without taking into account the events occurring inside the LHP. The LHP performance is evaluated in terms of transient temperature distribution and total thermal resistance (R_t). The experimental and simulation results are found in good agreement.

1. Introduction

Nowadays, the world is in a constant energy crisis and energy seldom lacks. It would be very useful to find a way to use energy more efficiently. Energy recovery is a very important thing to consider. Loop heat pipe (LHP) is an equipment with high efficiency used to transfer the heat from one place to the other. Normally conventional fluids are used in LHPs to remove the heat based on a temperature range for its particular operating conditions [1]. The lower thermal conductivity of these working fluids limits the thermal performance enhancement of the LHPs. Fluid, with nanoparticles is referred to as nanofluid, a term proposed by Choi [2]. The term 'nanofluid' refers to a two-phase mixture with its continuous phase being generally a liquid and the dispersed phase constituted of 'nanoparticles' i.e., extremely fine metallic particles of size below 100 nm. In other words, the large surface-area-to volume ratio also increases the stability of the suspensions. Thus, the nanofluid is a promising heat transfer fluid in variety of applications.

Numbers of experimental investigations have demonstrated the use as nanofluid with higher effective thermal conductivity and critical heat flux [3 and 4]. These advantage is applications such as heat transfer

[1] prem@uniten.edu.my.

devices. Recently, many researchers have presented the heat transfer characteristics of heat pipe using nanofluids. In contrast, only a few investigations on thermal performance of heat pipe charged with diamond nanofluid are available and LHPs have never used the diamond nanofluid as working fluid and this has motivated the present study. To the best of authors' knowledge; most of the research works are carried out experimentally and hence validating the experimental findings is difficult [5, 6, and 7]. Thus, the present study deals with experimental investigation to evaluate the thermal performance of LHP using diamond (diamond–H_2O) nanofluid for various mass concentrations, which ranged from 0% to 3% under forced convection mode and the results are verified by simulation using FEM (ANSYS 14 software). The results of interests such as total thermal resistance and transient temperature distribution in the LHP are reported to illustrate the effect of nanofluid concentration on LHP performance.

2. Materials and methods

2.1. The experimental setup

Fig. 1 illustrates the schematic diagram of experimental setup for LHP. The main function of this experiment rig is to determine the thermal performance of LHP charged with diamond–H_2O nanofluid with mass concentration ranged from 0%-3% as a working medium. In this experiment, the K-type thermocouples are installed on the pipe/wall in different locations of the loop, including the copper base plate (T_B), the evaporator (T_E), the vapor line (T_V), the condenser section (T_C) and the liquid line (T_L). The temperatures measured by the thermocouples are collected through a data acquisition (Agilent 34970A) with sample rate of 1 Hz and connected to a PC to collect the data. The experiments are conducted under a heating power of 40 W. The airflow velocity is fixed as 4 m/s and the coolant flow rate is 750 liters per hour, controlled by adjusting the flow control valve. In order to study the transient temperature distribution, temperatures at each 750 s from the start of the experiment, are also noted, until the steady state.

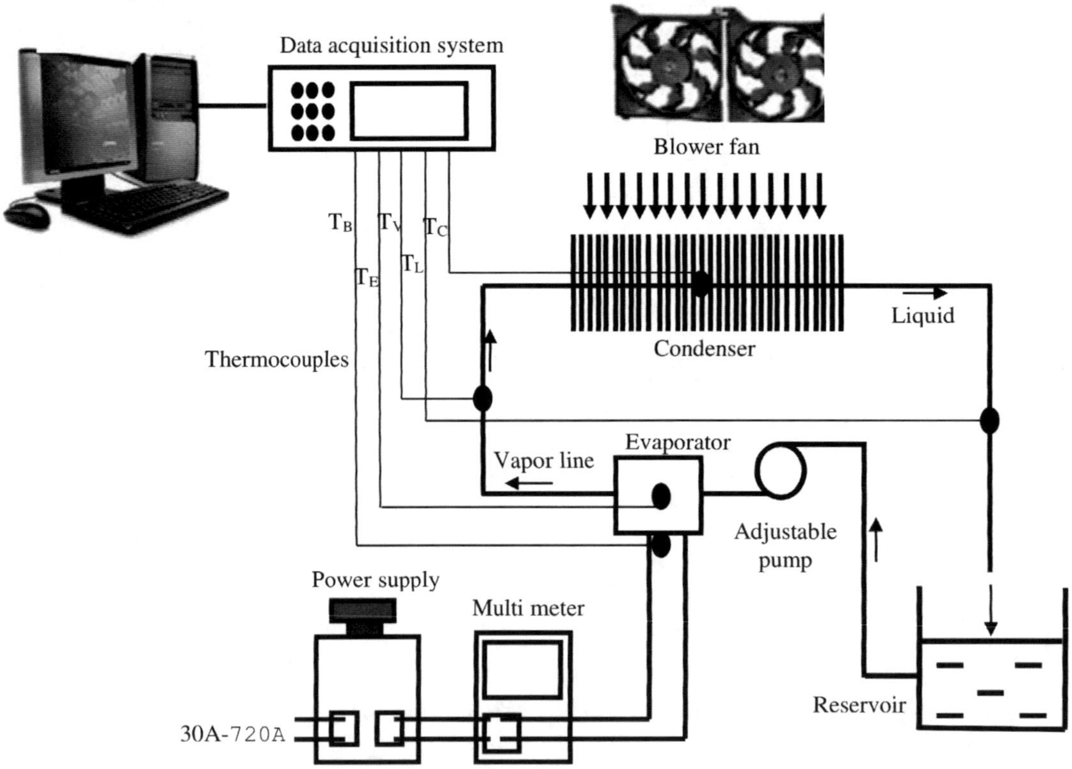

Figure 1. Schematic diagram of experimental apparatus.

IOP Conf. Series: Materials Science and Engineering **88** (2015) 012038 doi:10.1088/1757-899X/88/1/012038

2.2. Nanofluid preparation

In the present study, three mass concentrations Of 0.5%, 1.0% and 3.0% are tested for diamond-H_2O nanofluids. Deionized water is used as the base liquid. The diamond nanoparticles that used for investigation have an average size of 13 nm and density of 3.95 g/cm^3. The diamond nanoparticles with 5 g, 10 g and 30 g are used to prepare 1000 ml of diamond-H_2O nanofluid, which corresponds to 0.5%, 1.0% and 3.0% particle mass concentration, respectively. Diamond nanoparticles are weighted very accurately using a sensitive balance with a 0.1 mg resolution. The particle mass concentration of diamond-H_2O nanofluid in the present study is calculated using Eq. (1) as follows [8].

$$\%\text{ mass concentration} = \frac{W_{diamond}}{W_{bf}} \text{ x } 100\ \% \tag{1}$$

Where,

$W_{diamond}$ = Amount of diamond nanoparticles in gram

W_{bf} = Amount of base fluid in gram

2.3. Thermal analysis

The objective of the current study is to study the total thermal resistance (R_t) and the temperature distributions of the LHP using diamond-H_2O nanofluid as working fluids for heat input of 40 W under steady state and transient conditions. The results obtained from experimental investigation used to verify by FEM simulation. The nanoparticle mass concentration that yields the minimum R_t is then found out, and the effective thermal conductivity (k_{eff}) of evaporator section, vapor line and liquid line is calculated and the various steps to estimate R_t are as follows.

The heat flux ($\dot{q}$) that applied on the bottom of base plate can be expressed as:

$$\dot{q} = \frac{Q}{A_b} \tag{2}$$

where Q denotes the heat input and A_b is the area of base plate. The thermal resistances of the LHP are defined as [9]: The thermal resistance between the copper base plate and the evaporator section (R_b) is:

$$R_b = \frac{T_b - T_e}{Q} \tag{3}$$

where T_b denotes the temperature at the copper base plate and T_e is the temperature at the evaporator.

The thermal resistance of the evaporator section (R_e) is:

$$R_e = \frac{T_e - T_v}{Q} \tag{4}$$

where T_v is the temperature at the vapor line.

The thermal resistance of the vapor line (R_v) is:

$$R_v = \frac{T_v - T_c}{Q} \tag{5}$$

where T_c is the temperature at the condenser section.

The convective thermal resistance of the condenser (R_c) is:

$$R_c = \frac{T_c - T_l}{Q} \tag{6}$$

where T_l is the temperature at the liquid line.

The thermal resistance of the liquid line (R_l) is:

$$R_l = \frac{T_l - T_a}{Q} \tag{7}$$

where T_a is the ambient temperature.

According to the thermal resistance network as shown in Fig. 3, the R_t of the system is given by:

$$R_t = R_b + R_e + R_v + R_c + R_l \tag{8}$$

The heat transfer coefficient (h) for LHP is obtained from Eq. (9) as follows:

$$h = \frac{Q}{50A_f(T_c - T_a)} \tag{9}$$

where 50 is the number of fins and A_f is the surface area of aluminum rectangular fin.

3. 3D numerical heat conduction model

The purpose of simulation in the present study is to verify the experimental observations. Accordingly, a three-dimensional (3D) model is designed by assuming the whole LHP as a conducting medium without taking into account the events occurring inside the LHP. The simulation is performed in commercial FEM software package, ANSYS, for LHP charged with diamond nanofluid of 1% mass concentration. The 3D model of the complex LHP assembly is built in Pro Engineer Wildfire 4.0 and is exported to ANSYS 14. The LHP consists of copper base plate (heat source), evaporator, vapor line, condenser with 50 aluminum rectangular fins and liquid line. The high effective thermal conductivity of LHP is meshed with 1 mm element edge length, whereas the copper base plate and aluminum fins have coarse meshes of 4 mm edge length, forming a total of 205523 triangular elements. The heat flux of 16000 W/m^2 applied on the bottom of the base is calculated by Q/A_b, where Q is 40 W and A_b is the base surface area (0.0025 m^2). The initial temperature is taken as 22 °C and the total time steps (thus the number of iterations) are taken as 1200. The heat transfer coefficient of LHP, calculated using Eq. (9) is taken as 138 W/m^2.K, which corresponds to LHP charged with 1% diamond-H_2O nanofluid. It is to be noted that the value of k for evaporator section, vapor and liquid lines of the LHP charged with 1% diamond-H_2O nanofluid is the effective thermal conductivity (k_{eff}).

4. Experimental results

4.1. Flow pattern in the vapor line

Instead of saturated vapor, two different flow patterns are observed under a heat input power of 40W in the vapor line using pure water (0%) and diamond-H_2O nanofluid with particle mass concentrations of 0.5%, 1% and 3%, which is slug flow and bubbly flow, respectively. It is also observed that the nucleation size of bubble formation is getting reduced and the quantity of bubbles formation are getting increased as the mass concentration increases from 0% to 3%. From the results obtained, the lowest thermal resistance is expected in LHP using 3% diamond-H_2O nanofluid. The reason for reducing the thermal resistance of heat pipe can be explained as follows. A major thermal resistance of heat pipe is caused by the formation of vapor bubbles at the liquid-solid interface. A larger bubble nucleation size creates a higher thermal resistance that prevents the transfer of heat from the solid surface to the liquid [10]. The suspended nanoparticles inside the water tend to bombard the vapor bubble during the bubble formation. Therefore, it is expected that the nucleation size of vapor bubble is much smaller for fluid with suspended nanoparticles than that without them as reported by Tsai et al [11]. According to Huminic et al [12], as the mass concentration of suspended nanoparticles increases within the base fluid, the resultant greater bombardment of vapor bubble has amplified the amount of much smaller nucleation size of vapor bubbles formation. This fact is in good agreement with the achieved results in present study.

4.2. Thermal resistance analysis

Fig. 1 shows the relationship of total thermal resistance (R_t) of LHP charged with diamond-H_2O nanofluid at different mass concentrations (0%-3%). From the graph, it is observed that the R_t of LHP charged with diamond-H_2O nanofluid decreases as the nanoparticle mass concentration increases and thus be responsible for better heat transfer performance instead of pure water (0%). This is due to the fact that the suspended nanoparticles in a fluid flow can increase both the thermal conductivity of fluid and convective heat transfer from fluid flow to the wall as reported by Wang et al. [13] and Xuan and

Li [3], which results in the reduction of R_t of the LHP charged with diamond-H_2O nanofluid as discussed in the present study. A minimum value of total thermal resistance was found at mass concentration of 3.0%, which is 3.15 °C/W.

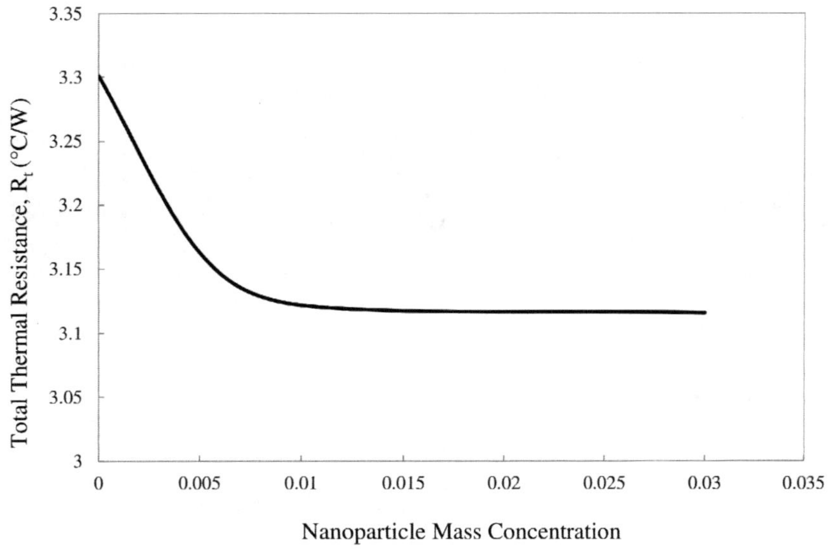

Figure 2. Total thermal resistance of LHP charged with diamond-H_2O nanofluids at different mass concentrations.

5. Comparison of experimental and simulation results

Table 1 illustrates the comparison of experimental and simulation results at steady state for LHP using diamond-H_2O nanofluid with particle mass concentration of 1%. The good agreement indicates the validity of the present methodology for the thermal analysis of the LHP. However, a detailed simulation by incorporating the multi-phase flow within the LHP may yield predictions that are more realistic. It is seen that, the predicted transient temperature distributions for LHP using diamond-H_2O nanofluid in present simulation are also well matching with the experimentally observed trends.

Table 1
Comparison of experimental and simulation temperatures.

Temperature (°C)	Diamond-H_2O (1%)	
	Experiment	Simulation
T_b	148.20	149.51
T_e	75.60	77.47
T_v	61.50	61.98
T_c	24.89	25.09
T_l	23.38	23.29

6. Conclusion

In the present study, extensive experimental and FEM simulation investigations on the LHP are performed. As nanoparticle mass concentration increasing, two different flow patterns are observed under the same heat input power in the vapor line. They are slug and bubbly. The thermal analysis of the LHP charged with diamond-H_2O nanofluid with particle mass concentration ranged from 0% to 3% is studied. The results of the simulation showed the positive influence of nanofluid utilizing as a heat pipe working fluid on the system thermal performance. It is found that the R_t of LHP decreases when particle mass concentration of diamond-H_2O nanofluid increases. However, there existed an optimal concentration of 1%, at which the reduction in R_t is not apparent or remains constant thereafter. Apart from the study on thermal resistance, the transient temperature distributions obtained from experiment and FEM simulation are found in good agreement.

Acknowledgement

The authors would like to sincerely thank the Ministry of Higher Education (MOHE) of Malaysia for the provision of a grant with code no 20120228FRGS to support this work.

References

[1] Rafferty R., *TCE Presentation on 9 December 2009*. Internal document of Thermacore Europe Ltd.

[2] Choi, S. U. S., 1995. "Enhancing thermal conductivity of fluids with nanoparticles," in Developments Applications of Non-Newtonian Flows, FED-vol. 231/MD-vol. 66, ASME: 99-105, edited by D. A. Siginer and H. P. Wang. New York.

[3] Xuan, Y., Li, Q., 2003. Investigation on Convective Heat Transfer and Flow Features of Nano-fluids. Journal of Heat Transfer 125, 151-155.

[4] Das, S. K., Putra, N., Thiesen, P., Roetzel, W., 2003. Temperature Dependence of Thermal Conductivity Enhancement for Nano-fluids. Journal of Heat Transfer 125, 567-574.

[5] Shafahi M., Bianco V., Vafai K., Manca O., 2010. An Investigation of the Thermal Performance of Cylindrical Heat Pipes using Nanofluids. International Journal of Heat and Mass Transfer 53, 376-383.

[6] Alizad, K., Vafai, K., Shafahi, M., 2012. Thermal Performance and Operational Attributes of the Startup Characteristics of Flat-shaped Heat Pipes using Nanofluids. International Journal of Heat and Mass Transfer 55, 140-155.

[7] Do, K. H., Jang, S. P., 2010. Effect of Nanofluids on the Thermal Performance of a Flat Micro Heat Pipe with a Rectangular Grooved Wick. International Journal of Heat and Mass Transfer 53, 2183-2192.

[8] Naphon, P., Thongkum, D., Assadamongkol, P., 2009. Heat Pipe Efficiency Enhancement with Refrigerant-nanoparticles Mixtures. Energy Conversion and Management 50, 772-776.

[9] Franchi, G., Huang, X., 2008. Development of Composite Wicks for Heat Pipe Performance Enhancement. Heat Transfer Engineering 29, 873-884.

[10] Collier, J. G., Thome, J. R., 1996. Convective Boiling and Condensation. Clarendon Press, Oxford.

[11] Tsai, C. Y., Chien, H. T., Chan, B., Chen, P. H., Ding, P. P., Luh, T. Y., 2004. Effect of Structural Character of Gold Nanoparticles in Nanofluid on Heat Pipe Thermal Performance. Materials Letters 58, 1461-1465.

[12] Huminic, G., Huminic, A., Morjan, I., Dumitrache, F., 2011. Experimental Study of the Thermal Performance of Thermosyphon Heat Pipe using Iron Oxide Nanoparticles. International Journal of Heat and Mass Transfer 54, 656-661.

[13] Wang, B. X., Li, H., Peng, X. F., 2002. Research on the Heat-conduction Enhancement for Liquid with Nanoparticle Suspensions. Journal of Therm. Sci. 11, 214-219.

A study on the structural characteristics and shape of outfitting equipment support in 300K DWT crude oil tanker

K Jeong[1], H Jeong[2], M Ji[3], J Kim[4], J Park[5], H Chung[†]

[†]Corresponding Author: Department of Energy and Mechanical Engineering, Gyeongsang
National University, Cheondaegukchi-Gil 38, Tongyeong, Gyeongnam 650-160, Korea,
E-mail: hschung@gnu.ac.kr, Tel: 055-772-9115
[1] Department of Energy and Mechanical Engineering, Gyeongsang National University, Daewoo Shipbuilding & Marine Engineering Corporation Ltd.
E-mail: kwjeong@dsme.co.kr, Tel: 055-735-4851
[2] Department of Energy and Mechanical Engineering, Gyeongsang National University,
E-mail: hmjeong@gnu.ac.kr, Tel: 055-772-9114
[3] Solar E&S Corporation Ltd.
E-mail: mkji@solarens.kr,Tel: 055-263-6868
[4] Daewoo Shipbuilding & Marine Engineering Corporation Ltd.
E-mail: jtkim9@dsme.co.kr, Tel: 055-735-5192
[5] Changshin University,
E-mail: jhoo@cs.ac,kr, Tel: 055-250-3149

Abstract. With the increase in the size and speed of recently built vessels, the output and speed (rpm) of propulsion or generation engines have continuously increased, and the high-output, high-speed engine has become a major cause of excessive vessel noise and vibration. Accordingly, resonance occurs in the equipment and other outfitting equipment installed in a vessel, and thus, periodic requests for correction are received from ship owners or officers. In this study, to resolve this problem, supports that stably fix the outfitting equipment installed in the engine room of a very large crude oil tanker and provide protection from physical or external shock were classified into seven types for three kinds of widely used standard shapes, and an optimized shape was developed and suggested by analyzing the structural characteristics of the shapes of the supports (the maximum bending moment, maximum bending stress, and maximum deformation) using DNV NATICUS HULL 3D BEAM, a structural analysis program, so that it could be used for the outfitting design of a vessel.
[KW] Vessel, Maximum bending stress, Maximum bending moment, Maximum deformation, Outfitting equipment, Support, Structural analysis

[1] To whom any correspondence should be addressed.

IOP Conf. Series: Materials Science and Engineering **88** (2015) 012039 doi:10.1088/1757-899X/88/1/012039

1. Introduction

A vessel has an average life of about 25.3 years, and the performance of its installed devices or equipment should also be continuously maintained and managed until the end of the life of the vessel. With the increase in the size and speed of recently built vessels, the output and speed (rpm) of propulsion or generation engines have continuously increased and have become major causes of excessive vessel noise and vibration [1-2]. Accordingly, resonance occurs in the equipment and other outfitting equipment installed in vessels, and such equipment frequently malfunctions and fails. Therefore, in this study, to resolve this problem, the types, shapes, and structural characteristics of the supports that safely fix the outfitting equipment installed in vessels were investigated, and a standard was made for application to vessels [3-4].

2. Conception of support

A support is a device for attaching outfitting equipment, such as pipes, to a hull structure using various clamps and additional devices through extension, using angles, etc., to fix and support the outfitting equipment. Depending on the position and equipment for installation, the types of support can be classified into the deck mounting type, the wall mounting type, the ceiling mounting type, the vent duct type, and the cable tray type, as shown in Figure 1 [5-7].

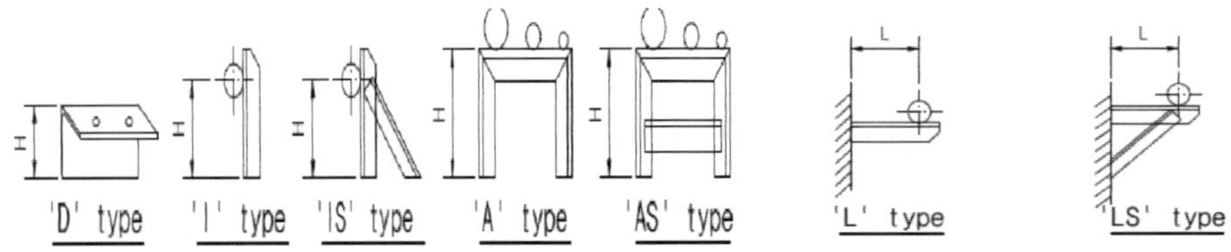

Figure 1. Various types of support

3. Conditions for the structural analysis

The analysis was performed based on the shapes of the supports that are widely used in vessels (Figure 2), the types summarized in Table 2, and the conditions summarized in Table 1, using the NATICUS HULL (DNV) structural analysis program.

Table 1. Model and Load condition

Model	Fig. 2
h	3.0 m
b	2.0 m
Members	200x8+200x12(H)
Load (F)	500 KN
For convenient calculation, only a lateral load was applied	

Table 2. Type of supports

Ser. No.	Description
1	Basic support shape
2	Basic + horizontal lower support structure shape
3	Basic + horizontal middle support structure shape

4	Basic + diagonal support structure shape
5	Vertical + inclined support height structure shape
6	Vertical + inclined support base structure shape
7	Vertical + inclined support structure shape

Figure 2. Model of the support

4. Analysis results

As shown in Figure 3, for Case 1 (the triangular shape), the maximum bending moment was 20 KNm, the maximum bending stress was 37 N/mm^2, and the maximum deformation was 5.74 mm; for Case 2 (the trapezoidal shape), 179 KNm, 339 N/mm^2, and 18.46 mm; and for Case 3 (the quadrilateral shape), 594 KNm, 425 N/mm^2, and 68.99 mm. When the maximum deformation of Case 1 (5.74 mm) was assumed to have been 100%, those of Case 2 and Case 3 were 321% and 1,202%, respectively. Figure 3 shows the changes in the conditions of Cases 1, 2, and 3.

	Case 1	Case 2	Case 3
Max. Bending Moment	20 KNm	179 KNm	594 KNm
Max. Bending Stress	37 N/mm^2	339 N/mm^2	425 N/mm^2
Max. Deformation	5.74 mm (100 %)	18.46 mm (321 %)	68.99 mm (1202 %)

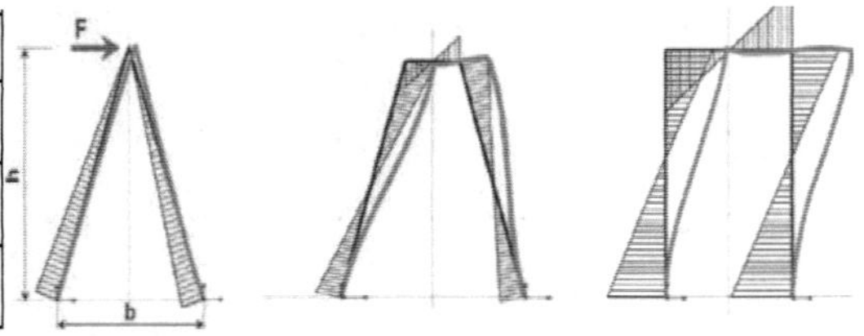

Figure 3. Basic support shapes

As shown in Figure 4, for Case 1, the maximum bending moment was 19 KNm, the maximum bending stress was 37 N/mm^2, and the maximum deformation was 5.74 mm; for Case 2, 178 KNm, 338 N/mm^2, and 18.33 mm; and for Case 3, 402 KNm, 760 N/mm^2, and 67.98 mm. When the maximum deformation of Case 1 in in Figure 3 (5.74 mm) was assumed to have been 100%, those of Case 1, Case 2, and Case 3 were 100%, 319%, and 1,184%, respectively. Figure 4 shows the changes in the conditions of Cases 1, 2, and 3.

	Case 1	Case 2	Case 3
Max. Bending Moment	19 KNm	178 KNm	402 KNm
Max. Bending Stress	37 N/mm^2	338 N/mm^2	760 N/mm^2
Max. Deformation	5.74 mm (100 %)	18.33 mm (319 %)	67.98 mm (1184 %)

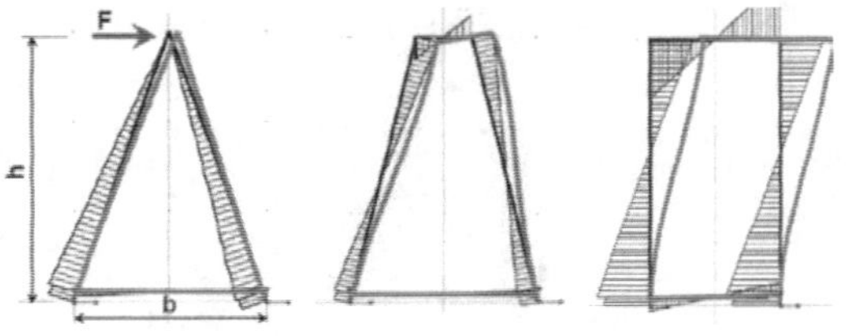

Figure 4. Basic + horizontal lower support structure shape

IOP Conf. Series: Materials Science and Engineering **88** (2015) 012039 doi:10.1088/1757-899X/88/1/012039

As shown in Figure 5, for Case 1, the maximum bending moment was 25 KNm, the maximum bending stress was 47 N/mm^2, and the maximum deformation was 5.65 mm; for Case 2, 137 KNm, 259 N/mm^2, and 12.69 mm; and for Case 3, 275 KNm, 519 N/mm^2, and 38.08 mm. When the maximum deformation of Case 1 in Figure 3 (5.74 mm) was assumed to have been 100%, those of Case 1, Case 2, and Case 3 were 98%, 221%, and 675%, respectively. Figure 5 shows the changes in the conditions of Cases 1, 2, and 3.

	Case 1	Case 2	Case 3
Max. Bending Moment	25 KNm	137 KNm	275 KNm
Max. Bending Stress	47 N/mm2	259 N/mm2	519 N/mm2
Max. Deformation	5,66 mm (98 %)	12,69 mm (221 %)	38,08 mm (675 %)

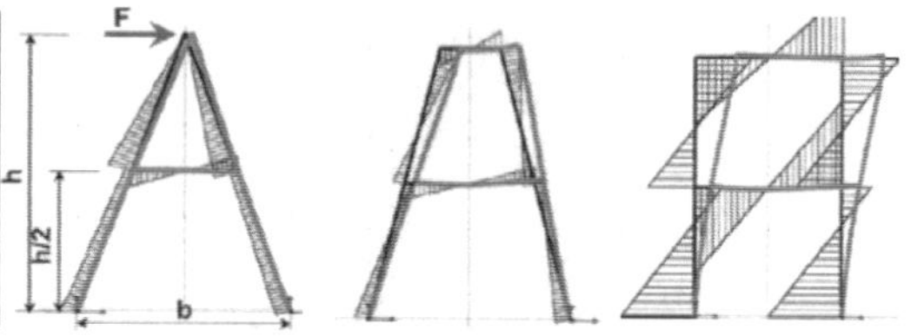

Figure 5. Basic + horizontal middle support structure shape

As shown in Figure 6, for Case 1, the maximum bending moment was 37 KNm, the maximum bending stress was 69 N/mm^2, and the maximum deformation was 5.59 mm; for Case 2, 187 KNm, 345 N/mm^2, and 12.26 mm; and for Case 3, 308 KNm, 581 N/mm^2, and 31.35 mm. When the maximum deformation of Case 1 in Figure 3 (5.74 mm) was assumed to have been 100%, those of Case 1, Case 2, and Case 3 were 97%, 213%, and 541%, respectively. Figure 6 shows the changes in the conditions of Cases 1, 2, and 3.

	Case 1	Case 2	Case 3
Max. Bending Moment	37 KNm	183 KNm	308 KNm
Max. Bending Stress	69 N/mm2	345 N/mm2	581 N/mm2
Max. Deformation	5,59 mm (97 %)	12,26 mm (213 %)	31,35 mm (541 %)

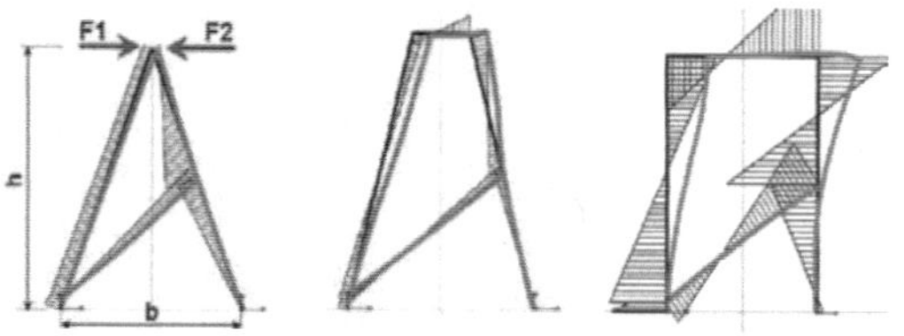

Figure 6. Basic + diagonal support structure shape

As shown in Figure 7, for Case 1, the maximum bending moment was 1,000 KNm, the maximum bending stress was 1,888 N/mm^2, and the maximum deformation was 162.5 mm; for Case 2, 750 KNm, 1,416 N/mm^2, and 90.23 mm; for Case 3, 500 KNm, 944 N/mm^2, and 47.61 mm; and for Case 4, 73 KNm, 138 N/mm^2, and 20.45 mm. When the maximum deformation of Case 4, 20.45 mm was assumed to have been 100%, those of Case 1, Case 2, and Case 3 were 795%, 441%, and 233%, respectively. Figure 7 shows the changes in the conditions of Cases 1, 2, 3, and 4.

	Case 1	Case 2	Case 3	Case 4
Max. Bending Moment	1,000 KNm	750 KNm	500 KNm	73 KNm
Max. Bending Stress	1,888 N/mm2	1,416 N/mm2	944 N/mm2	138 N/mm2
Max. Deformation	162,5 mm (795 %)	90,23 mm (441 %)	47,61 mm (233 %)	20,45 mm (100 %)

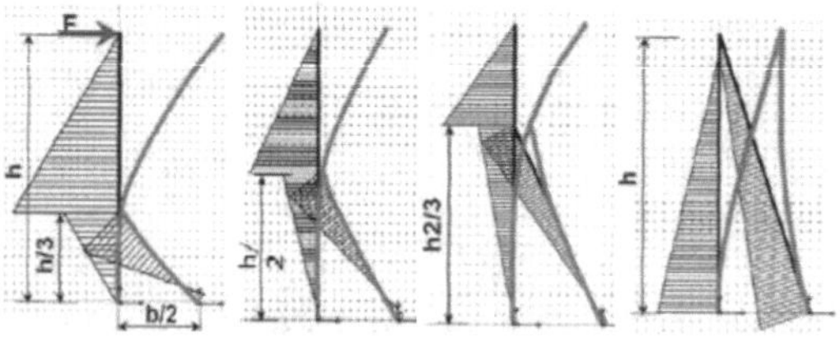

Figure 7. Vertical + inclined support height structure shape

As shown in Figure 8, for Case 1, the maximum bending moment was 1,500 KNm, the maximum bending stress was 2,832 N/mm^2, and the maximum deformation was 372 mm; for Case 2, 500 KNm, 944 N/mm^2, and 79.71 mm; for Case 3, 500 KNm, 944 N/mm^2, and 47.61 mm; and for Case 4, 891 KNm, 1,684 N/mm^2, and 262.7 mm. When the maximum deformation of Case 2 (79.71 mm) was assumed to have been 100%, those of Case 1, Case 3, and Case 4 were 467%, 60%, and 329%, respectively. Figure 8 shows the changes in the conditions of Cases 1, 2, 3, and 4.

	Case 1	Case 2	Case 3	Case 4
Max, Bending Moment	1,500 KNm	500 KNm	500 KNm	891 KNm
Max, Bending Stress	2,832 N/mm2	944 N/mm2	944 N/mm2	1,684 N/mm2
Max, Deformation	372,7 mm (467 %)	79,71 mm (100 %)	47,61 mm (60 %)	262,7 mm (329 %)

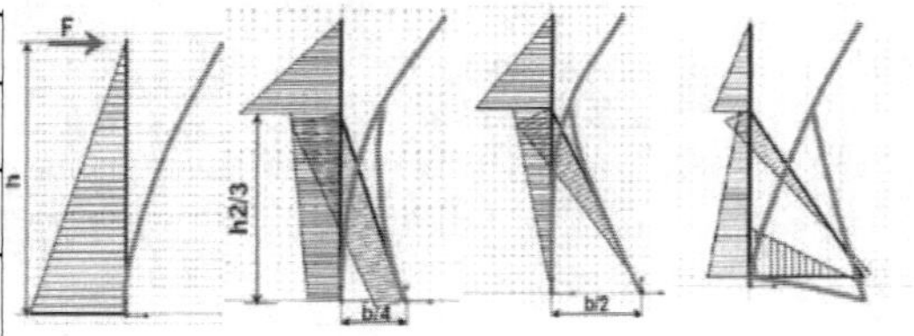

Figure 8. Vertical + inclined support base structure shape

As shown in Figure 9, for Case 1, the maximum bending moment was 500 KNm, the maximum bending stress was 944 N/mm^2, and the maximum deformation was 47.61 mm; for Case 2, 500 KNm, 944 N/mm^2, and 41.183 mm; and for Case 3, 500 KNm, 944 N/mm^2, and 45.91 mm. When the maximum deformation of Case 1 (47.61 mm) was assumed to have been 100%, those of Case 2 and Case 3 were 86% and 96%, respectively. Figure 9 shows the changes in the conditions of Cases 1, 2, and 3.

	Case 1	Case 2	Case 3
Max, Bending Moment	500 KNm	500 KNm	500 KNm
Max, Bending Stress	944 N/mm2	944 N/mm2	944 N/mm2
Max, Deformation	47,61 mm (100 %)	41,18 mm (86 %)	45,91 mm (96 %)

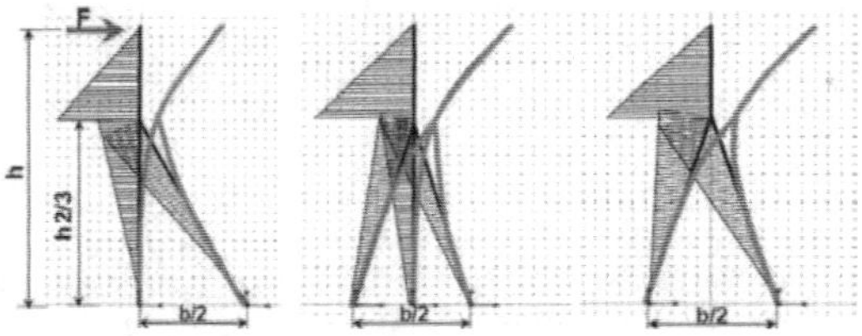

Figure 9. Vertical + inclined support structure shape

5. Conclusion

In this study, when a constant load was applied to outfitting equipment supports that are widely used in very large crude oil tankers, the structural characteristics of the shapes of the supports (i.e., the maximum bending moment, maximum bending stress, and maximum deformation) were examined and analyzed. As a result, the structural characteristics of the following seven typical types of support structure shapes were determined: the basic (triangular, trapezoidal, and quadrilateral) support shape in Figure 3; the basic (triangular, trapezoidal, and quadrilateral) + horizontal lower support structure shape in Figure 4; the basic (triangular, trapezoidal, and quadrilateral) + horizontal middle support structure shape in Figure 5; the basic (triangular, trapezoidal, and quadrilateral) + diagonal support structure shape in Figure 6; the vertical + inclined support height structure shape in Figure 7; the vertical + inclined support base structure shape in Figure 8; and the vertical + inclined support structure shape in Figure 9. The results of the experiment are as follows.

1) Among the seven types of shapes, the basic + horizontal middle support structure shape in Figure 5, which added a horizontal middle support to the basic (triangular, trapezoidal, and quadrilateral)

IOP Conf. Series: Materials Science and Engineering **88** (2015) 012039 doi:10.1088/1757-899X/88/1/012039

support shape in Figure 3, and the vertical + inclined support base structure shape in Figure 8 showed the best performance.

2) In the case of a good support, the shape of the support determines the amount of the required material in terms of the production cost, and the work load and productivity in the manufacturing process should also be considered.

3) In the case of a good support, the shape of the support should be determined considering the structural characteristics (e.g., the strength and rigidity) in terms of mechanics.

4) In the case of a good support, the support should be designed to satisfy the required structural characteristics using the minimum amount of material and so that it can be easily manufactured in terms of the prime cost.

6. Postscript

This research was supported by Basic Science Research Program through the National Research Foundation of Korea(NRF) funded by Ministry of Education, Science and Technology(NO. NRF-2011-0021376) and the Future Leading Project through the Small and Medium Business Administration (No.S2044441).

7. References

[1] Su-Hyun Choi, Kuk-Su Kim, Yeon Cho and Byoung-Gon Kim "A Study on the Evaluation of Vibration Characteristics for Onboard Machinery with Resilient Mountings" Journal of the Society of Navel Architects of Korea Vol. 39, No. 1, pp.73-81, February 2002

[2] Kun-Hee Lee, Jong-Gug Bae and Soo-Mok Lee "Vibration Characteristics of Diesel Generator Set with Resilient Mount and Provention of Vibration on the Design Stage" The Korean Society for Noise and Vibration Engineering (KSNVE) 2005 Spring Conference, pp.921-924

[3] Byung Hyun Kim " Vibration Testing Requirements for Shipboard Equipments (ISO 10055:1996(E) & MIL-STD-167-1)" The Korean Society for Noise and Vibration Engineering (KSNVE) 2013 Spring Conference, pp.459-462

[4] Yun-Hwan Kim, Kook-hyun Kim, Tae-Muk Choi, and Dae-Seung Cho "Numerical Dynamic Stiffness Assessment of Ship Equipment Foundation with FEM" The Korean Society for Noise and Vibration Engineering (KSNVE) Anniversary of 2nd Period, 2010 Spring Conference, pp.212-213

[5] Shipbuilding Technology (Making All Ships) Journal of the Society of Navel Architects of Korea, pp339-343,617-618

[6] DSME Technology Standard DSE-D2H4087 Engine Room Pipe Support Design Standard, pp.4-10

[7] DSME Technology Standard DSE-201006-038 Topside Pipe Support Design Standard (Flxed Platform, FPSO) pp.6-17

IOP Conf. Series: Materials Science and Engineering **88** (2015) 012040 doi:10.1088/1757-899X/88/1/012040

A Study on the compensation margin on butt welding joint of Large Steel plates during Shipbuilding construction.

J Kim[1], H Jeong[2], M Ji[3], K Jeong[4], C Yun[5], J Lee[6], H Chung[†]

[†]Corresponding Author: Department of Energy and Mechanical Engineering, Gyeongsang National University, Cheondaegukchi-Gil 38, Tongyeong, Gyeongnam 650-160, Korea,
E-mail: hschung@gnu.ac.kr, Tel: 055-772-9115
[1] Department of Energy and Mechanical Engineering, Gyeongsang National University, Daewoo Shipbuilding & Marine Engineering Corporation Ltd.
E-mail: jtkim9@dsme.co.kr, Tel: 055-735-5192
[2] Department of Energy and Mechanical Engineering, Gyeongsang National University,
E-mail: hmjeong@gnu.ac.kr, Tel: 055-772-9114
[3] Solar E&S Corporation Ltd.
E-mail: mkji@solarens.kr,Tel: 055-263-6868
[4] Department of Energy and Mechanical Engineering, Gyeongsang National University, Daewoo Shipbuilding & Marine Engineering Corporation Ltd.
E-mail: kwjeong@dsme.co.kr, Tel: 055-735-4851
[5] Changshin University,
E-mail: yck6552@hanmail.net, Tel: 055-250-3145
[6] Daewoo Shipbuilding & Marine Engineering Corporation Ltd.
E-mail: singam@dsme.co.kr, Tel: 055-735-51482

Abstract. This paper examines the characteristics of butt welding joint shrinkage for shipbuilding and marine structures main plate. The shrinkage strain of butt welding joint which is caused by the process of heat input and cooling, results in the difference between dimensions of the actual parent metal and the dimensions of design. This, in turn, leads to poor quality in the production of ship blocks and reworking through period of correction brings about impediment on improvement of productivity. Through experiments on butt welding joint's shrinkage strain on large structures main plate, the deformation of welding residual stress in the form of I, Y, V was obtained. In addition, the results of experiments indicate that there is limited range of shrinkage in the range of 1 ~ 2 mm in 11t ~ 21.5t thickness and the effect of heat transfer of weld appears to be limited within 1000mm based on one side of seam line so there was limited impact of weight of parent metal on the shrinkage. Finally, it has been learned that Shrinkage margin needs to be applied differently based on groove phenomenon in the design phase in order to minimize shrinkage.

1. Introduction

Main plates of hull and off-shore structure are welded by butt welding method and it is quite normal procedure that the 2nd welding shall be performed on the opposite side after cooling and turning over the 1st welding side. During this heating and cooling cycles while welding repeatedly, residual stress occur

[1] To whom any correspondence should be addressed.

IOP Conf. Series: Materials Science and Engineering **88** (2015) 012040 doi:10.1088/1757-899X/88/1/012040

in a welded butt joint and this stress react to produce internal forces, causing shrinkage or length deficiency of main plates in comparison with designed dimension as shown on preceding research data [1]-[3]. The shrinkage around welded joint is one of major concern in the shipbuilding industry and an obstacle that must be overcome to meet the relevant quality standard and production schedule.

In order to avoid butt welding shrinkage effect, Shipyards have been applying 10mm~20 mm of four (4) direction margins or two (2) direction margins at the end of plates and cutting in accordance with actual designed dimension after finishing butt welding. This is one of quality control methods in shipbuilding industry. This paper shows the values of welding shrinkage which were measured around butt welding joint during actual shipbuilding construction and shows how we can control its residual stress and effect on the dimension of main plates assembled by butt weld, applying reasonable margin at design stage, not depending on their experience.

2. Theoretical Analysis of Welding Shrinkage

Atoms being vibrated by getting thermal heat expansion energy act as if they have a larger radius. Therefore, the average distance between atoms and the overall dimensions of material are increased. Such dimensional changes of material, namely the length change (ΔL) per unit length can be represented by a linear thermal expansion coefficient (α) as below equation (1).

$$\alpha = \frac{L_f - L_0}{L_0(T_f - T_0)} = \frac{\Delta L}{L_0 \Delta T} \tag{1}$$

Also, when an isotropic material is heated slowly and evenly, material is equally expanded without any occurrence of residual stress. However, if the material is suppressed to prevent its swelling, its dimensional change does not occur, on the other hand its thermal stress is generated. This thermal stress having thermal expansion coefficient (α), the modulus of elasticity (E) and thermal change (ΔT) has the same as below equation (2) [4].

$$\sigma_{thermal} = \alpha E \Delta T \tag{2}$$

Welding heat input given from the outside. That is to say, electrical heat energy (J) which Arc generates per unit length (cm) of welding can be represented by Arc voltage E(V), Welding current I(A) and Welding speed U(cm/min) as below formula (3) [5].

$$J = \frac{60EI}{v} (joule/cm) \tag{3}$$

Welding deformation and welding residual stress represent an effect opposite to each other that when less restraint is given to welding metal, welding residual stress is rather small, but the welding deformation is large. While if welding restraint is greatly increased to the extent of preventing the free contraction of welding metal, the welding deformation is decreased, but the welding residual stress is increased [4]. The main formulas are the equation (4) and (7) as below.

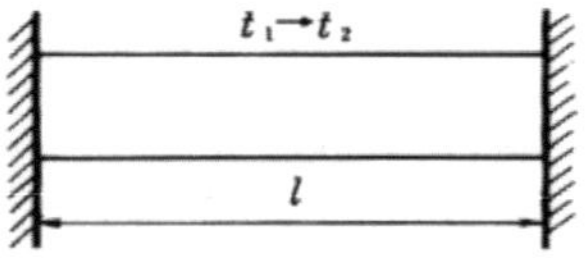

Temperature rise: Compression Temperature drop: Tensile

Figure 1. Compressive strength and tensile strength

Strain $$\delta = l' - l = l \cdot \alpha \cdot (t_2 - t_1) \tag{4}$$

Strain rate $$\epsilon = \alpha \cdot (t_2 - t_1) \tag{5}$$

Thermal stress $$\sigma = E \cdot \alpha \cdot (t_2 - t_1) \tag{6}$$

Tension $$P = AE \cdot \alpha \cdot (t_2 - t_1) \tag{7}$$

$\boldsymbol{\alpha}$ is thermal expansion coefficient and as for mild steel, $\boldsymbol{\alpha}$ is $11.45 \cdot 10^{-6}$ [4].
The way of control of welding deformation is to rate the status of shrinkage and to compensate it by design margin. But, in the workplace, the experienced data is much more utilized rather than theoretical analysis and experimental results of data.

3. Experiment

The shrinkage measurements were carried out on the plates welded in a butt joint using submerged arc welding, during actual shipbuilding construction, at D shipyard in S. Korea. Welding parameters are in the WPS (Welding Procedure Specification) as shown on Table 1. and the dimension of the plates and welding condition including ambient temperatures are shown in Table 2. Groove shapes of the plates are shown on Figure 2.

Table 1. Welding procedure specification

Process	Da.(mm)	Type & Pd.	Amperage(A)	Voltage(V)	Speed (cm/min)	Thickness
SAW	4.0~ 4.8	DCRP	800~ 1050	28~ 38	27~ 45	11≤T 〈15
			900~ 1150	28~ 38	25~ 35	15≤T 〈18
			1000~ 1200	29~ 40	22~ 30	13≤T 〈22
			1100~ 1250	30~ 40	20~ 27	22≤T 〈27

Table 2. Plates dimension and welding condition

Kind	Welding conditions	Remark
Size(L * B)	21.6 M * 2.5～3 M)	
Thickness(t)	11～21.5	Figure 2
Welding Materials	WIRE: H-14(Ø4.8) FLUX:S707T	
Workplace temperature	18 ～ 34℃	
Welding positions	Below	

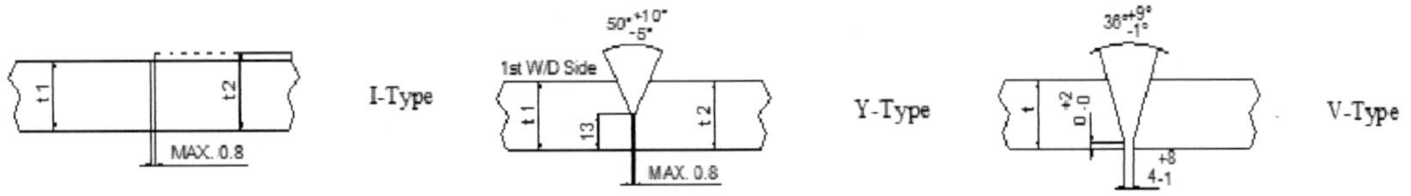

Figure 2. Groove type of large steel plates

3.1 Preparation

Shrinkage measurement was conducted under the following conditions. Run on and run off tab pieces were provided with same edge details as those of the plates before the plates were assembled and tack welded as a normal welding procedure in the shipbuilding industry.

1) Plate thickness : 11mm~21.5mm
2) Plate Length: 15m~21.6m
3) Plate Width: 2.5m~3.5m
4) Steel grade: A, AH, AH36 (Discrepancy between steel grades is not considered)
5) Number of block: 12 blocks
6) Number of butt welding joint: 41 EA

3.2 Measurement

Specially ordered marking device was used for the marking of 150 mm off line from the welding line as shown on Figure3. These 150 mm off line parallel to the welding line was drawn on the plates using the marking device before welding. The length of joint gap was included into 150 mm off marking line and to get the reliability and easy reading its display, Digital Vernier Caliper was used to measure the distance between these lines before and after welding as shown on Figure 3.

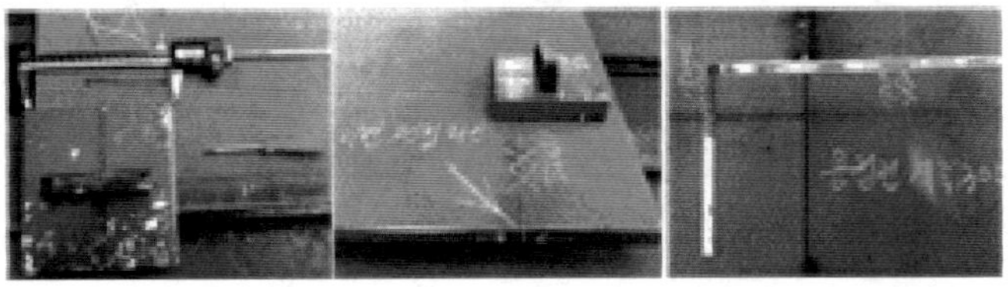

Figure 3. 150mm off marking and measuring tools

After tack welding, 300 mm between two parallel lines were confirmed by Digital Vernier Caliper (150mm + 150mm=300mm). No any arrest devices were used to make the free end of plates during welding. After finishing the first welding side, it takes 2.5~3.0 hours for cooling and the plates were turned over for the second welding and the transverse shrinkage was measured in 3 hours after the second welding considering cooling time as shown on Figure 4.

Figure 4. Shrinkage measurements

4. Experiment results and Discussion

1) The measurement result was obtained from the difference between these 300 mm (150mm+150mm) off lines on the plates and half of this value can be taken as the shrinkage of each individual plate. Measurement results are as shown on Table 3.

Table 3. Summary of shrinkage average collected from the measurement

Block no.	Thickness	Grooves	Measurement(mm) Before welding /	After welding	Average Shrinkage
22XX-538	11.0~ 20.0	I	300	298.8	1.2 mm
51XX-243	20.5~ 21.5	Y	300	298.4	1.6 mm
11XX-122	14.0~ 16.0	V	300	298.7	1.3 mm

2) As a first step, tack welding was carried out and it took about 2.5~3.0 hours for cooling after finishing the first welding and the plates were turned over for the second welding on opposite side. During these twice welding processes, heating and cooling cycles were occurred on the plates. It is assumed that the degree of shrinkage measured by this experiment may differ from the predicted result by the theoretical analysis because of plastic deformation which caused by high heating on the plates during welding.

3) Temperature of welding point just after welding was measured by infrared thermometer during natural air cooling and the result. As it is shown on Figure 5, temperature of welding point is cooled down within 5 minute rapidly and then cooled down gently. Because of the Max capacity of infrared thermometer, it was measured from 650 °C so the temperature of welding point right after welding is predicted about 1300~1500 °C.

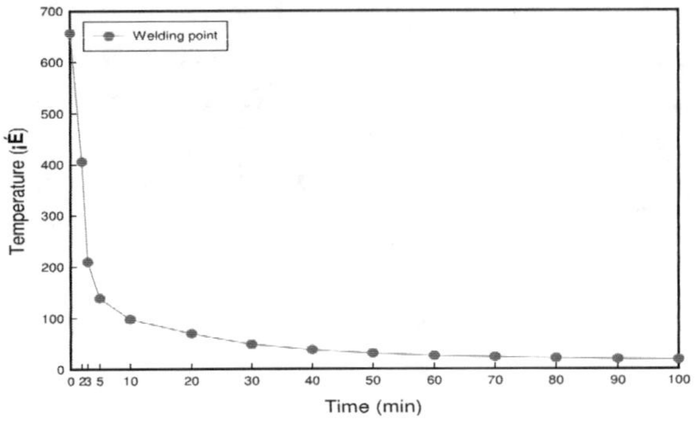

Figure 5. Cooling curve of welding point in 30 Second after welding

As it is shown on Figure 5, temperature of welding point is cooled down within 5 minute rapidly and then cooled down gently. Because of the Max capacity of infrared thermometer, it was measured from 650 °C so the temperature of welding point right after welding is predicted about 1300~1500 °C.

5. Conclusion

The main conclusions, derived from the measurement of transverse shrinkage and investigation in the butt welded joint between two plates during actual shipbuilding construction, are as follows.

The Shrinkage value of Y-Groove is slightly bigger than I-Groove because of the higher welding heat input so the value of compensation to control the dimension of plates shall be applied differently in accordance with groove shape.

In case of I-Groove welding, for the same thickness, the amounts of transverse shrinkages were measured from 1.0 mm ~1.5 mm variously and these results are the same as for the plate thickness of 11.0 mm~20.0 mm. This could be due to no good straightness of the plate edge making no zero joint gaps between two plates. At the welding point of no zero joint gaps, the higher welding heat is inputted during welding and the more shrinkage is occurred vs.

The degree of transverse shrinkage of I, Y and V-Groove welding on the plates thickness of 11.0 mm~21.5 mm, measured almost equally being limited within 1.0 mm~2.0 mm and the effect of heat transfer by welding is limited within 1,000 mm off from the welding line and welding residual stress is occurred due to high cooling velocity. This result suggests that an effect on shrinkage by self-weight of plates is very small.

6. Postscript

This research was supported by Basic Science Research Program through the National Research Foundation of Korea(NRF) funded by Ministry of Education, Science and Technology(NO. NRF-2011-0021376) and the Future Leading Project through the Small and Medium Business Administration (No.S2044441).

7. References

[1] I. S. Kim, H. H Kim, H. K, "A Study about Analysis of Weld Distortion using Genetic Algorithm" Journal of the Korean Welding and Joining Society, Vol. 27, No. 4, pp. 398-403, 2009

[2] J. O. Yun and C. I. Oh, "Trends of Welding Deformation Analysis for Ship Structure" Journal of the Korean Welding and Joining Society, Vol. 30, No. 2, pp. 147-150, 2012.

[3] S. K. Kang and J. S. Yang, "Overview of Research Works Regarding Welding Distortion" Journal of the Korean Welding and Joining Society, Vol. 30, No. 2, pp. 141-146, 2012.

[4] D-J. Yun, H-C. Lee, and M-H. Lee, "Material Properties" Seoul, Korea: Publishing Gold Book, pp.

179-184, 2004 (in korea)
[5] N-J. Kim, "Matal engineering dictionary" Seoul, Korea: Sungandang Book, p. 557, 1998.

Analysis of the Earthquake Impact towards water-based fire extinguishing system

J Lee, M Hur*, K Lee[1]

Dept. of Information and Telecommunications Engineering, Incheon National University, Incheon, KOREA
* Faculty of Fire and Safety, Woosong College, KOREA

E-mail: tnfeco@hanmail.net, mshur@wsi.ac.kr, kylee@incheon.ac.kr

Abstract. Recently, extinguishing system installed in the building when the earthquake occurred at a separate performance requirements. Before the building collapsed during the earthquake, as a function to maintain a fire extinguishing. In particular, the automatic sprinkler fire extinguishing equipment, such as after a massive earthquake without damage to piping also must maintain confidentiality. In this study, an experiment installed in the building during the earthquake, the water-based fire extinguishing saw grasp the impact of the pipe. Experimental structures for water-based fire extinguishing seismic construction step by step, and then applied to the seismic experiment, the building appears in the extinguishing of the earthquake response of the pipe was measured. Construction of acceleration caused by vibration being added to the size and the size of the displacement is measured and compared with the data response of the pipe from the table, thereby extinguishing water piping need to enhance the seismic analysis. Define the seismic design category (SDC) for the four groups in the building structure with seismic criteria (KBC2009) designed according to the importance of the group and earthquake seismic intensity. The event of a real earthquake seismic analysis of Category A and Category B for the seismic design of buildings, the current fire-fighting facilities could have also determined that the seismic performance. In the case of seismic design categories C and D are installed in buildings to preserve the function of extinguishing the required level of seismic retrofit design is determined.

1. Introduction

The definitions for the seismic design category (Seismic Design Categories, SDC) for the seismic design, in KBC2009. And thus to determine the importance of the group and earthquake seismic intensity Seismic Design category. The introduction to the standards based on the recently developed concept of seismic performance seismic design category. The higher the rating the seismic design category is enhanced using the building structure, limitations of the system, analysis methods, and design and the detailed regulations.

The seismic design criteria are established to ensure quality of facilities, including a public service essential to consider arbitrarily large personal accident prevention at the national level is necessary

Corresponding Author [1] Dept. of Information and Telecommunications Engineering, Incheon National University, Incheon, KOREA . E-mail: kylee@incheon.ac.kr leading role. It can be said that the efforts to secure the necessary basis for such publicity for the analysis of earthquake effects on buildings fire safety facilities to be constructed in the study.

The firefighting facilities are mostly installed on the inside of the building. Conduct a seismic design and seismic performance of firefighting facilities classified according to the size of the vibration on the building during an earthquake. Most of the small mass inertia during earthquake size small or not fixed directly to the building fire protection facilities. Cannot be delivered at the time of the earthquake as the Firefighting Facility Construction acceleration and displacement. Therefore, it is reasonable to perform seismic risk for the seismic design category in the seismic design without the need for any seismic design category will be delivered some heavy earthquake loads.

Identify the effect of the fire-fighting water pipe saw the earthquake affected buildings during firefighting facilities in such a perspective. Seismic experiment was conducted by the installation of fire protection facilities of reinforced concrete specimens were determined categories designed for firefighting facilities are required to measure the size of the seismic acceleration and displacement passed the fire test facilities in each step.

When applied to a step-by-step construction of the aqueous digestion experiments seismic equipment and structures in seismic experiment, we measured the size and displacement of the acceleration due to vibration that was added to the building in fire-fighting water pipe water extinguishing system in the building. By analysing the data on the response of fire-fighting water pipe was investigated with respect to the required level of fire-fighting water pipe seismic reinforcement.

2. Seismic performance of water-based fire-fighting equipment (grade) Regulations

2.1. Seismic grade for firefighting facilities and seismic performance objectives

After the earthquake, can continue to operate by checking the equipment and piping, and to maintain the function of the equipment rather than with respect to vibrations transmitted to the earthquake damage occurs as a whole building construction equipment, vibration performance. In the event of falling of a large earthquake is fixed configuration of the installation device, but, it is possible to ensure the maintenance of such specific features. And ensure the safety of human life, is to prevent secondary disasters that accompany this goal. Seismic safety evacuation, etc., secured lifeline, ensuring safety is important after an earthquake, it is necessary to enhance the seismic grade is given.

2.1.1. Wet-piping fire extinguishing facilities

Water based firefighting system is consist of hydrant, sprinkler, pump, pipe, water for fire fighting
The digestion tank plumbing and fire-fighting equipment and fire-fighting water facility such as the main target of seismic design is to be considered a priority. Because they reduce the damage of life and property caused by fire in the building during an earthquake. In addition, it is effective to require seismic design is to be the most robust in the building of other facilities, including evacuation facilities, fire protection facilities, the seismic design of buildings, or include it in the review if necessary. It is reasonable to classify the facility affected by the behaviour of building structures in order to evaluate the seismic performance of water-based fire extinguishing facilities.

2.1.2. Seismic Design Criteria

Seismic design criteria for water-based fire extinguishing equipment facilities (draft), I was prepared to conform to the facility of a water-based fire extinguishing equipment refer to the (KBC2009) Architectural design criteria (II) and the study of seismic design criteria of Construction and Transportation has enacted .

- **Seismic performance objectives**

Set the seismic performance targets by taking into account the importance of the building of the seismic design time, it is determined to understand the intent of the main building and use of the building in the option.

That is, it can be classified to the loss of three forms of the effects of seismic buildings.

① the safety of human life : Injury or death of activists and residents of the neighbourhood and the users of the building
② property damage : Cost for replacing or maintenance of the building or its contents
③ function infeasible : Financial loss due to the fact that it is not possible where the facility to perform the functions of the normal after the earthquake

With regard to the biggest earthquake, for the earthquake to ensure the safety of human life, usually occurring frequently, performance goals of the building, for the purpose of execution of the functions of the building roughly.

2.2. Seismic performance secure firefighting facilities
Enhancing the seismic performance of the facility another is fundamental measures which can minimize the damage of earthquake.

In facilities that are related to lifeline fire extinguishing equipment, power equipment, gas equipment, water and sewage facilities, such as communication facilities, you will need strong earthquake, to be able to allow for quick recovery also earthquake damage, especially after. If the pipe attached to the building, by changing the material with excellent earthquake resistance, it is necessary to consider the structural. You also need to be checked on a regular basis, based on the technical standards for the facility and to ensure a water-based fire extinguishing equipment which operates stably even earthquake disaster.

3. Analytical methods of the impact of the piping of the earthquake
In this section we discuss how to format the title, authors and affiliations. Please follow these instructions as carefully as possible so all articles within a conference have the same style to the title page. This paragraph follows a section title so it should not be indented.

In this study, it is in the grasp of the vibration level that can be to satisfy the seismic performance above a certain level reliability and maintenance of water-based fire extinguishing equipment through the shaking table test.

3.1. Earthquake vibration generator (shaking table test device)
Pusan National University (distributed shared research infrastructure construction projects, Korea Construction Engineering Development Program) KOCED public use Table B to earthquake vibration test centres in three groups was shown in Table1 the following specifications. Figure 1 is Shaking table in Pusan National University Earthquake test Center

Table 1. Shaking Table Specification

Table	**Specifications**	**dimension**
Table B	Degrees of freedom	3DOF (5m×5m, Movable)
	Maximum load weight	60 tonf
	The maximum acceleration	3g
	The maximum speed	100cm/sec
	The maximum displacement	±30.0cm (±20.0cm)*

Figure 1. Earthquake test Center (Pusan National University, korea)

3.2. Seismic experiment body production

Was produced experimental body of reinforced concrete structure of the two-story in order to perform a seismic experiment (Figure2). Installation to aqueous digestion facilities shown in Figure3, in order to determine the effect on the piping of a water-based fire extinguishing system in the earthquake, as shown in Figure4, was placed the measuring sensors.

Figure 2. Make a seismic experiment Structure (Shaking Table)

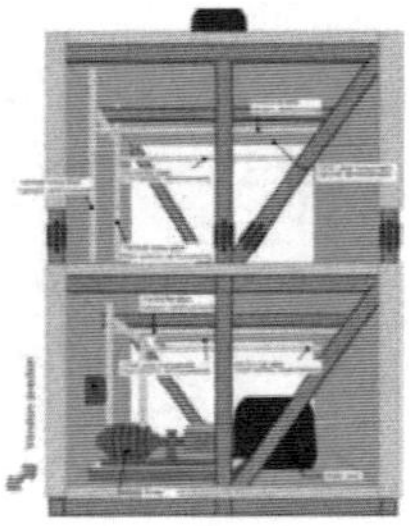

Figure 3. Water-based extinguishing system installation on the shaking. table

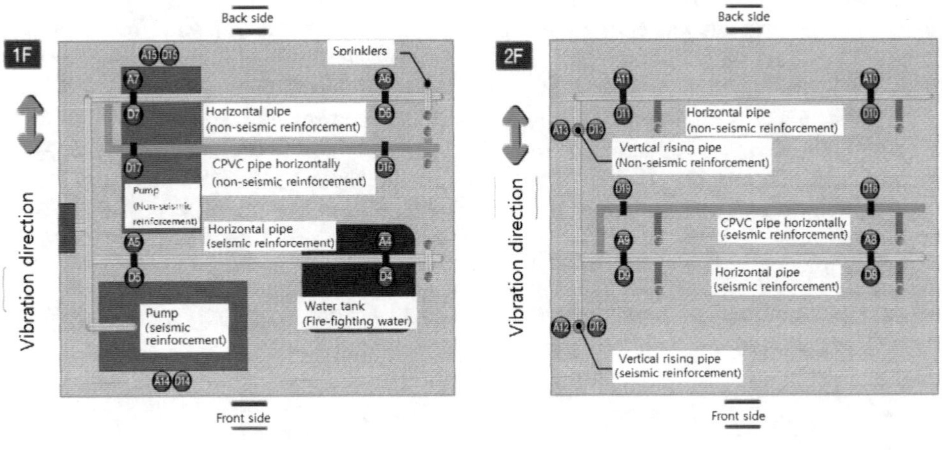

Figure 4. Installing the sensor for measuring the amount of change of the fire extinguishing system

IOP Conf. Series: Materials Science and Engineering **88** (2015) 012041 doi:10.1088/1757-899X/88/1/012041

3.3. production Determination of experimental seismic wave

Seismic wave that has been used to study this time, using artificial seismic waves and El Centro wave. The seismic wave was made on the basis of "Test Method for Seismic telecommunications equipment" announcement No. 2009-3 issue of radio laboratory.(Figure 5,6).

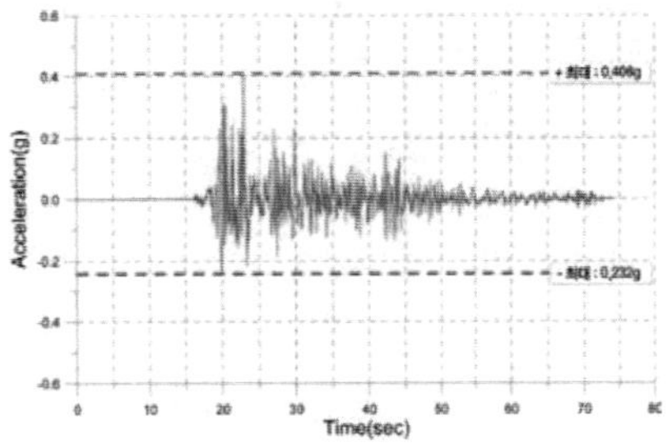

Figure 5. El Centro seismic acceleration input wave for test

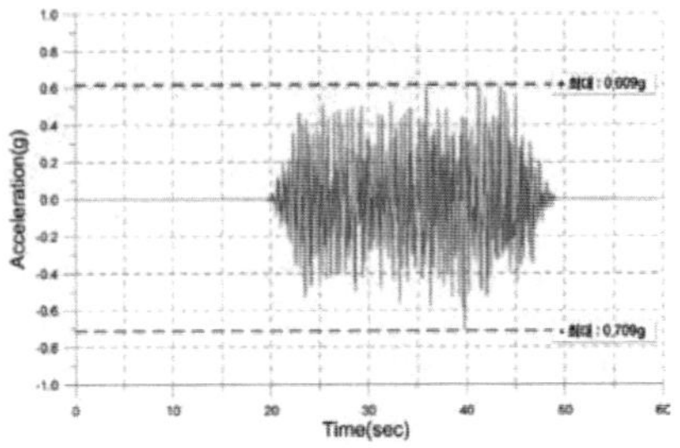

Figure 6. Artificial seismic acceleration input wave for test

3.4. Seismic experimental methods

We assume the acceleration of the acceleration of the bottom surface of the specimens.

Therefore, the user can view the value of the acceleration in accordance with the SDS bottom KBC2009. We have a maximum acceleration level of the bottom to reproduce the acceleration corresponding to the seismic design category in the present study.

- 0.18g corresponding to Seismic Design Category A
- 0.21g corresponding to Seismic Design Category B
- 0.36g corresponding to Seismic Design Category C
- 0.50g corresponding to the seismic design category D.

4. Analysis of the experimental results

4.1. Analysis of experiments results for seismic design categories A and B

In this study, an experiment was carried out (digester, such as pressurized water pumping equipment) for connected devices and piping corresponding to the digestion of water-based facility. The Figure8 and Figure7,8, is the experimental results of the piping and B Seismic Design Category A

Looking at the experimental results, the maximum displacement of the vibration table was attached with about 30mm occurs, but the maximum displacement of the firefighting equipment occurred in less than 1mm. It is determined that the acceleration in the case of acceleration does not occur almost response acceleration and 0.18g, affected by the vibration of the earthquake is very small.

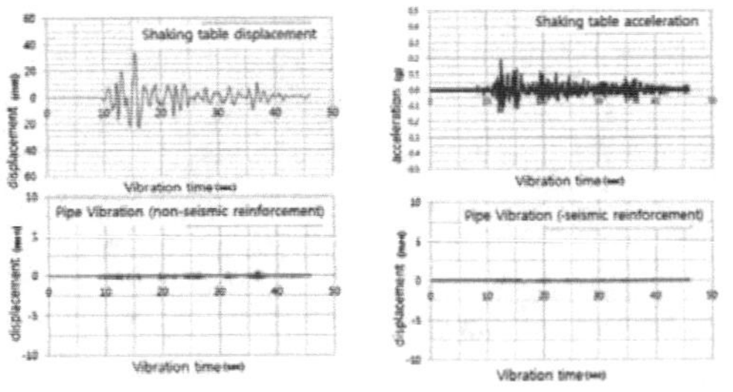

Figure 7. experiments results of fire extinguishing system (Earthquake Wave intensity 0.18g)

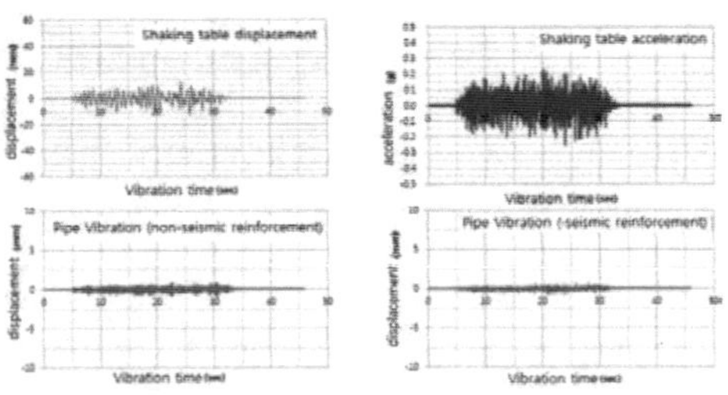

Figure 8. experiments results of fire extinguishing system (Earthquake Wave intensity 0.21g)

4.2. analysis of experiments results for seismic design categories C and D

In Figure9, it is the result of an experiment conducted using 0.50g the base acceleration of the vibration table. Level of 0.50g is the acceleration level is corresponding to the maximum acceleration of the category C of seismic design, corresponding to the seismic design category D. The maximum displacement of the shaking table for about 30mm occur, wall fixing device showed no specificity.

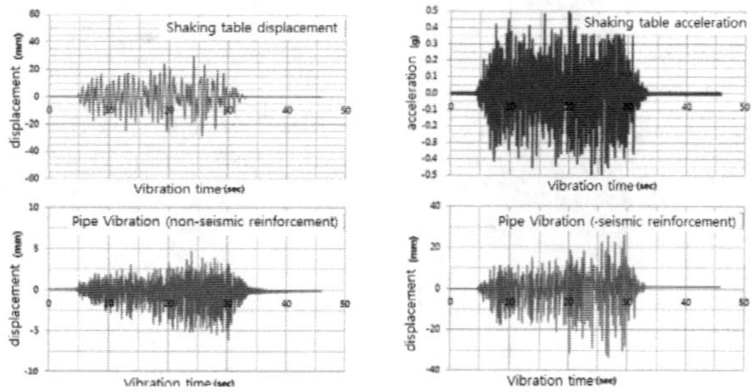

Figure 9. experiments results of fire extinguishing system (Earthquake Wave intensity 0.50g)

5. Conclusion

If the value of the SDS of Category A of seismic design is less than 0.17g. Corresponding to a level of about 0.07g in case 0.19g / 2.5 = 0.068g in terms of the value of this ground acceleration. This is the level at the level corresponding to the collapse prevention level of about 2 grade of seismic design criteria for other facilities, for the ground motion of 0.07g normal level below, has not been carried out seismic design. It is possible to know that for the 0.07g level, damage does not occur at all vibration table test results of this study. Therefore, if applicable to the seismic design category A, it is determined that sufficient even without the seismic design of firefighting facilities.

If the value of the SDS of seismic design category B corresponds to 0.17g~0.33g. As shown in shaking table test results of this study, in the case of a normal type firefighting equipment that are out of seismic design, major damage has not occurred. However, sometimes it causes damage spring anti-vibration of the pump, and the digester, in the inertial force due to the weight of the earthquake, such as water tank. Therefore, when the water tank pressurized pumping water systems, the digester, however, when the firefighting equipment other, it is determined that there is no need to perform seismic designed to be performed the seismic design case of seismic design category B.

If seismic design category C and category D seismic survey was that most of the damage in the fire at the facility during the earthquake spectral acceleration of 0.33g or higher earthquake ground acceleration acting on the occurrence and size of buildings. Therefore, seismic design categories C and Seismic Design Category D is needed for the seismic design of fire protection facilities.

6. References

[1] KBC 2009, Ministry of Land, Infrastructure and Transport.,korea, 2009
[2] Elliott GOODWIN, Emmanuel MARAGAKIS, Ahmad ITANI "Seismic evaluation of hospital piping systems," Earthquake Engineering, 13th World Conference on Earthquake Engineering, Paper No. 1081, 2004.
[3] ICC, International building code (IBC), International Code Council, INC, 2012.
[4] Gimjongbu, Seismic Design Methodology for sprinkler installations, Master Thesis, Seoul City University, 2011.
[5] Namminjun, bakseunghui, yunjonggu, standing chungyo, water-type "by shaking table test Seismic Performance Evaluation of fire-fighting equipment, "Proceedings of the Conference in 2012, South Korea Earthquake Engineering, 121-124, 2012.

IOP Conf. Series: Materials Science and Engineering **88** (2015) 012042 doi:10.1088/1757-899X/88/1/012042

Study of Indonesia low rank coal utilization on modified fixed bed gasification for combined cycle power plant

T Hardianto[1], A R Amalia[1], A Suwono[1], P Riauwindu[1]
[1]*Faculty of Mechanical and Aerospace Engineering, Department of Mechanical Engineering*
Institut Teknologi Bandung
Jalan Ganesa 10, 40132, Bandung, Indonesia

Email: toto@termo.pauir.itb.ac.id

Abstract:
Gasification is a conversion process converting carbon-based solid fuel into gaseous products that have considerable amount of calorific value. One of the carbon-based solid fuel that serves as feed for gasification is coal. Gasification gaseous product is termed as *syngas* (synthetic gas) that is composed of several different gases. *Syngas* produced from gasification vary from one process to another, this is due to several factors which are: feed characteristics, operation condition, gasified fluid condition, and gasification method or technology. One of the utilization of *syngas* is for combined cycle power plant fuel. In order to meet the need to convert carbon-based solid fuel into gaseous fuel for combined cycle power plant, engineering adjustment for gasification was done using related software to create the *syngas* with characteristics of natural gas that serve as fuel for combined cycle power plant in Indonesia.
Feed used for the gasification process in this paper was Indonesian Low Rank Coal and the method used to obtain *syngas* was Modified Fixed Bed Gasifier. From the engineering adjustment process, the yielded *syngas* possessed lower heating value as much as 31828.32 kJ/kg in gasification condition of 600°C, 3.5 bar, and steam to feed ratio was 1 kg/kg.
Syngas characteristics obtained from the process was used as a reference for the adjustment of the fuel system modification in combined cycle power plant that will have the same capacity with the conversion of the system`s fuel from natural gas to *syngas*.

1. Introduction

1.1. Indonesia Coal Reserves Potential and Properties

Indonesia is one of the largest country in the world in term of coal reserves with approximated total reserves of 5.5 billion metric ton and also world`s top coal producer with 237.4 million metric ton mined in 2012 [1]. Medium and low rank coal constitutes the biggest portion of Indonesia coal reserves. It is accounted for 85% of the total Indonesia coal reserves [1]. Medium and low rank coal possesses high level of moisture content and volatile matter with relatively low calorific heating value, therefore the utilization of this type of coal usually involve direct combustion either on coal-fired power plant or domestic application in Indonesia such as cooking and heating.

1.2. Energy Need in Non-coal-fired Power Plant in Indonesia

Having possessed abundant coal reserve, Indonesia doesn`t solely rely on coal to generate the electricity. Several types of power plant were built to meet the electrification target of the nation, for example: diesel, combined cycle, hydropower, and other renewable sources. There`s a possibility to utilize other type of alternative fuel instead of conventional fuel into these existing power plants, especially for combined cycle power plant (PLTGU). The fuel possibility can be derived indirectly from low rank coal due to the abundance presence of this type of fuel in Indonesia.

1.3. Study of utilization of coal through gasification for non-coal-fired power plant

Utilization of coal to supply fuel for combined cycle power plant can be achieved by converting solid coal into gaseous product through process called gasification. This paper focuses on the study of utilization of Indonesia low rank coal for combined cycle power plant (PLTGU) with capacity of 400 MWe with 267 MWe of gas-fired power plant and 133 coal-fired power plant by utilizing modified fixed bed gasification. Reasoning of modified fixed bed utilization will be further discussed in point 2.3. Gasification produces several gaseous products in form of CO, H_2, and CH_4 with different composition based on the coal feed characteristics, gasification method, gasifier fluid condition, and operational condition of the gasification process itself. These gaseous products will be utilized into combined cycle power plant through several adjustments to meet the condition of the existing gas used in Indonesia combined cycle power plant.

2. GASIFICATION TECHNICAL REVIEW

2.1. Combined Cycle Power Plant Fuel Property

Prior to coal conversion process into gaseous fuel for combined cycle power plant, combined cycle power plant fuel characteristics has to be determined. Fuel characteristics in combined cycle power plant is marked as steady aerodynamics and steady flame kinetics [2]. Indonesia combined cycle power plant (PLTGU) utilizes Compressed Natural Gas (CNG) to fuel its gas turbine with the stated characteristics above. CNG used in Indonesia PLTGU with capacity of 400 MWe is mainly dominated by CH_4 gas (70.68%) with molecular weight of 23.21 g/mole and lower heating value of 40630 kJ/kg. Fuel consumption rate in which the fuel is consumed by gas turbine is defined by the following equation [3];

[illegible] (1)

Where: $\dot{m}$ = Fuel consumption rate [kg/s]; [illegible] = Generated power from the gas turbine inside the PLTGU system [MW]; and LHV_{fuel}= lower heating value from gas turbine fuel [MJ/kg]

2.2. Existing Coal Gasification Technology

Gasification process consists of several phases and is described in the Figure 1 below. Existing gasification technologies are grouped and developed based on these three categories, which are: fixed bed gasifier, fluidized bed gasifier, and entrained flow gasifier [4]. The Table 1 presents the comparison between various gasification technologies in the world.

Table 1. Gasification Technologies Comparison

Parameter	Fixed Bed	Fluidized Bed	Entrained Flow
Ash Condition	Dry ash	Dry ash	Slagging
Feed Size	5 - 80 mm	0,5 - 10 mm	<0,1 mm (homogen)
Coal Feed Quality	All type of coal	Low – Medium rank	All type of coal
Oxygen Requirement	Low	Medium	High
Water Steam Requirement	High	Medium	Low
Working Pressure	1 bar-100 bar	10 bar-25 bar	1 bar-40 bar
Reaction Temperature	800 – 1000°C (Below AFT)	950 - 1100°C (Below AFT)	>1400°C
Flue Gas Temperature	Low (400 - 900°C)	Medium (900 - 1100°C)	high (1200 - 1600°C)
Coal Residence Time	Updraft: 1 - 2 hour, downdraft: 20 - 30 minutes	5 - 50 seconds	1 - 10 seconds
Main Gas Product	H_2 = 41-43% mole CO = 14-16% mole CH_4 = 8-9% mole CO_2 = 30-32% mole	H_2 = 6-37% mole CO = 30-58% mole CH_4 = 1-4% mole CO_2 = 2-13% mole	H_2 = 28-34% mole CO = 26-57% mole CH_4 = 0,04-0,5% mole CO_2 = 2-16% mole

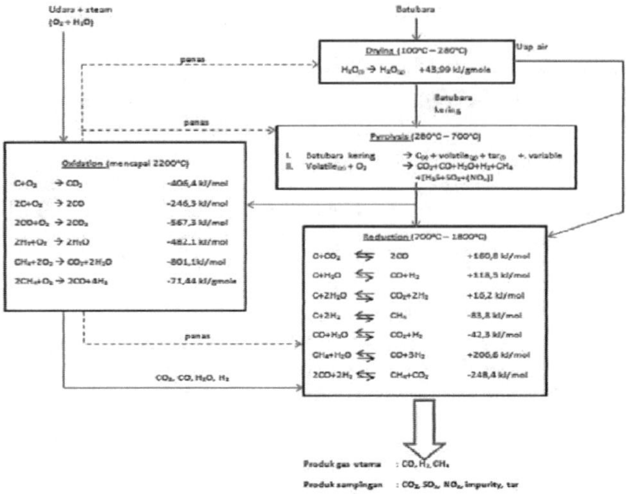

Figure 1. Gasification process phases of coal

2.3. *Modified Fixed Bed Technology Implementatin for Indonesia Combined Cycle Power Plant*

Considering the Indonesia low rank coal potential and the existing gasification technology characteristics, modified fixed bed gasification technology was chosen as the research concern of this paper to study the possibility of utilizing modified fixed bed gasification technology for low rank coal in the Indonesia PLTGU. Modified fixed bed is a fixed bed with modified batch coal feed into continuous feed therefore in the later simulation, the feed was modeled as continuous feed for fixed bed instead of batch feed. Figure 2 shows the schematic of Lurgi Dry Ash Gasifier as an example for modified fixed bed.

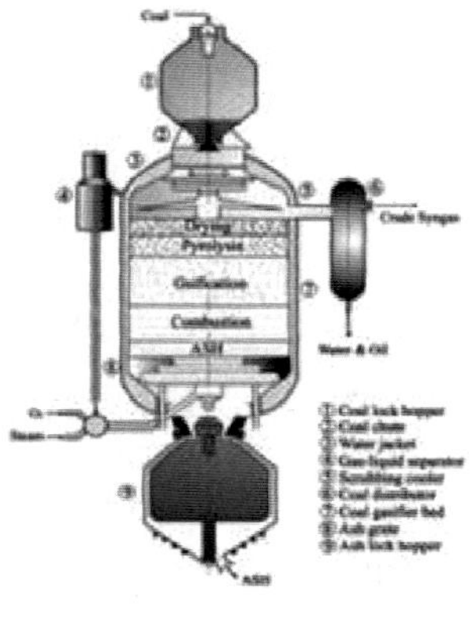

Figure 2. Lurgi dry ash gasifier schematic [7].

Modified fixed bed gasifier is applicable for low rank coal with relatively big size, high moisture content, and volatile matter. Modified fixed bed gasifier is also able to employ either air or oxygen as the gasifier fluid with relatively low consumption of oxygen as its gasifier fluid. With the high water steam consumption, modified fixed bed gasifier is also able to maintain low operating temperature.

For PLTGU application, gasifier with high working pressure is required. Modified fixed bed gasifier is able to operate within wide operational pressure, ranging from 1 – 100 bar. Low operational temperature in this type of gasifier compared to other type of gasifier is capable to produce gaseous product with low temperature thus resulting in high efficiency of the gasification process. Efficiency is defined as;

$$\eta_{cg} = \frac{HV_{syngas} \cdot \dot{m}_{syngas}}{HV_{feed} \cdot \dot{m}_{feed}} \times 100\% \quad (2)$$

Where: η_{cg} = gasification cold gas efficiency [%]; HV_{feed} = heating value (LHV or HHV) from gasification process feed [kJ/kg]; HV_{syngas} = heating value (LHV or HHV) from gasification product [kJ/kmol]; $\dot{m}_{feed}$ = *mass flow rate* from feed [kg/s]

3. FIXED BED GASIFICATION MODELLING FOR INDONESIA PLTGU

3.1. Gasification Process Model and Operational Parameter

In order to meet the defined objectives above, process diagram of gasification was then modeled using thermodynamic numerical software in Figure 3. The diagram is divided into three parts which are left side part, middle part, and right side part. Left side part describe the drying, pyrolysis, reduction, and oxidation process in the gasification. Middle part describe the heat transfer process as a representative from the presence of water jacketing in gasifier. For simplification, in this paper water jacketing was removed because the function of water jacketing is to maintain the operational temperature to prevent over temperature. Due to the fact that the simulation was done in numerical software, the final temperature was still under reasonable value, therefore in this paper the process in the water jacketing part was modeled as adiabatic. On the other hand, the right side part of the model was the result of the process simulation.

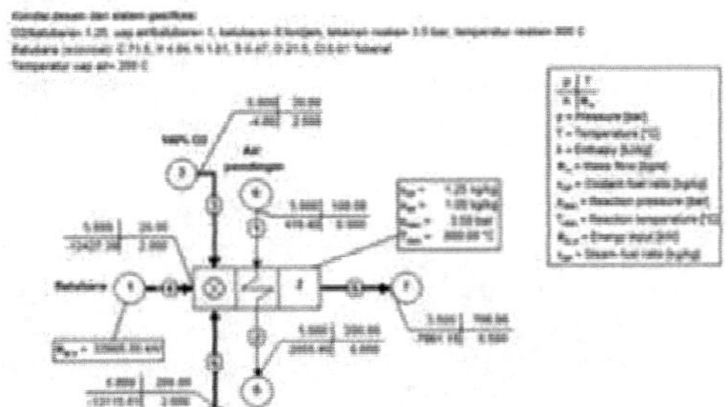

Figure 3. Process diagram of coal gasification

In the initial condition, specific operational condition was determined based on the modified fixed bed gasifier characteristics which were: coal feed rate 8 ton/hour, oxygen to coal ratio 1.25, water steam to coal ratio 1, working pressure 3.5 bar, and operational temperature 800°C. Several variables were also determined. Fixed variables in the simulation were gasifier type (modified fixed bed), coal type (low rank coal), coal feed rate 2 kg/s, gasifier fluid (oxygen), and water steam. The dependent variables were defined as H_2, CH_4, CO_2, and H_2O composition in the gasification gaseous product, heating value of the product, gasification simulation result, gasification efficiency, and final temperature of gasification product.

3.2. Operational Temperature Effect on Low Rank Coal Gasification

There were 6 different simulated operating temperatures in this simulation which were 600°C, 700°C, 800°C. 900°C, 1000°C, and 1100°C. Result shown in Figure 4 shows that CO gas was increasing with the increasing operating temperature. On the other hand, the CH_4 and CO_2 gas composition was decreasing with increasing operating temperature. H_2 was also increased with the increasing temperature but only until maximum peak at certain temperature and was decreasing with the increase in temperature beyond peak temperature. The effect of operating temperature to the main gas LHV, cold gas efficiency, and final temperature of the product is shown in the Figure 5. Increasing temperature increased the gasification process efficiency but was decreasing the LHV of the main gas as well as the product temperature of gasification.

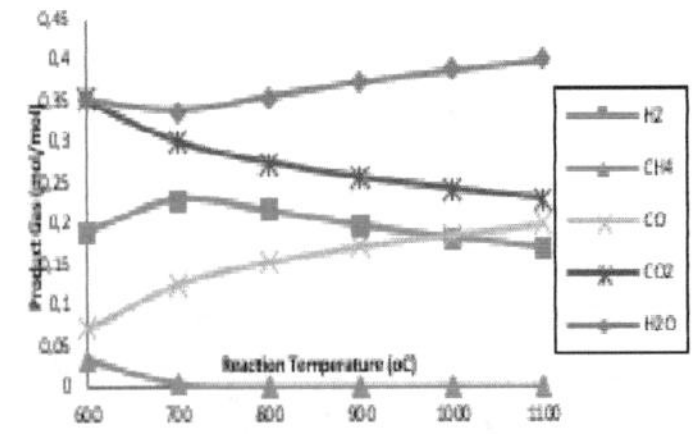

Figure 4 Operational Temperature Effect in the condition of 3.5 bar and H_2O/feed ratio is 1.0 to the gasification gas product

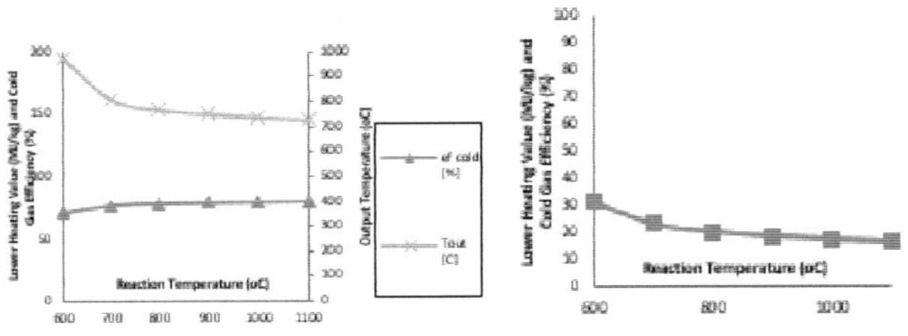

Figure 5 Operating Temperature Effect at 3,5 bar and H_2O/feed ratio 1,0 to LHV, cold gas efficiency, and gasification product temperature.

From 6 operating temperatures simulated, the temperature that yielded the highest yield of CH_4 was at 600°C which is 3.29% of the total composition with LHV at 92.92 kJ/mole and main gas LHV at 92.19 or equal to 31828.32 kJ/kg with gasification cold gas efficiency at 71.20%. At the prescribed condition the highest heating value was also obtained.

3.3. Water Steam to Feed Ratio Effect on Low Rank Coal Gasification

Figure 6 shows that with the increasing water steam to feed ratio, the H_2 and CO_2 composition tend to increase until certain optimum point in which beyond that optimum point the product gas composition was decreasing. On the other hand CO and CH_4 gas decreased with the increase in water steam to feed ratio. The product temperature of the gasifier also decreased with the increase in the water steam to feed ratio. This phenomenon happens due to the ability of water steam to control the reaction process through decreasing the gasification process temperature. The addition of water steam in the water steam to feed ratio also created water shift reaction in the reduction process of gasification that resulted in the increasing composition of H_2 and CO_2. Simulation result shown in Figure 7 predicts that the highest value of gasification process efficiency attained at water steam to feed ratio equal to 0, but the increase of the ratio did not result in a significant decrease in the gasification efficiency. The increase of the ratio instead resulted in the increasing value of gas LHV per kilogram increase of the water steam.

From several simulated ratios, ratio that resulted in the highest yield of CH_4 was on 0.01 that was 0.16% with the gas LHV at 165.42 kJ/mole and the main gas LHV at 164.21 kJ/mole or equal to 13383.15 kJ/kg at calculated gasification efficiency (cold gas) of 81.65%. On the other hand the ratio that resulted in the highest value of LHV, 26315.14 kJ/kg, was at 2 kg/kg of water steam to feed as a result of the increasing production of H_2 gas due to the increasing value of water steam.

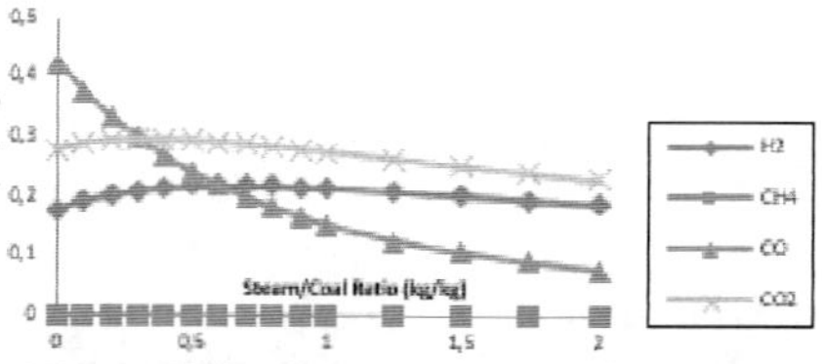

Figure 6 H_2O/feed ratio effeect at 3,5 bar and temperature 800°C to the gas composition of gasification product

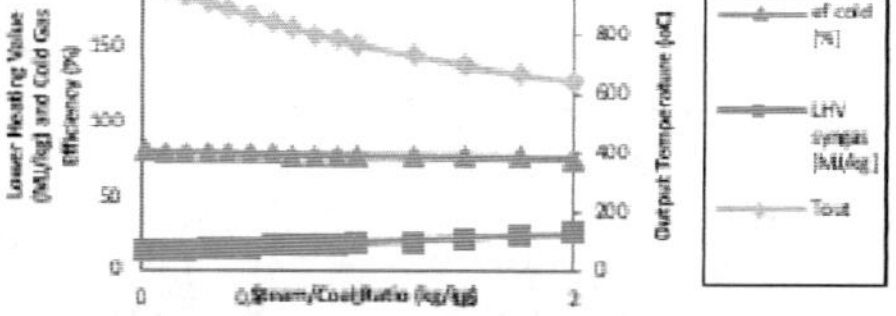

Figure 7 H_2O/feed ratio effect at 3,5 bar and operating temperature 800°C to LHV, cold gas efficiency, and product temperature of gasification.

3.4. Working Pressure Effect on Low Rank Coal Gasification

Working pressure effect on gasification of low rank coal is shown in the Figure 8 and Figure 9. Figure 8 shows that the higher the working pressure the higher the CH_4 and CO_2 composition but resulted in the decreasing composition of CO and H_2. CO and H_2 experienced more drastic decrease with the increasing working pressure and decreasing operating temperature while CO and CH_4 increased.

On the low operating temperature, the change in working pressure significantly affected the gaseous product of gasification. The increase in CO_2 and CH_4 gas composition with the increase in working pressure can be explained based on equilibrium shift of the chemical reaction. The higher the system pressure, the chemical reaction will shift to the reaction with the lowest coefficient number. Another phenomenon that could be observed was that the lower the operating temperature in the process, the increase and decrease in product gas composition resulted in a drastic decrease with the increasing working pressure, while high operating temperature did not significantly affect the composition of the gaseous product of gasification. This is due to the chemical reaction of the process has started to attain the equilibrium state with the increase in operating temperature. It can be concluded that the change in gasification working pressure only had significant impact to the product with the low operating temperature.

The working pressure effect to the gas product LHV, gas product temperature, and gasification efficiency on 3 different operating temperatures is shown on Figure 9. The gasification efficiency decreased with the increase in the operating temperature and product gas temperature and LHV increased with the increase in the working pressure. The relationship between working pressure and gas product temperature can be explained with assuming that the gas product as ideal gas in which the temperature is the function of the pressure, pressure increase will result in the increase of the temperature.

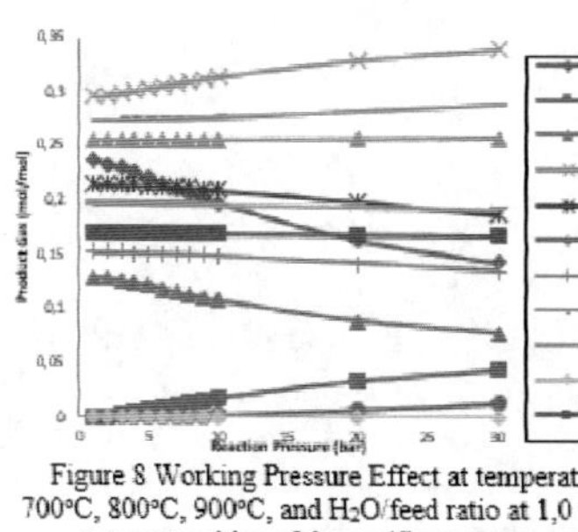

Figure 8 Working Pressure Effect at temperature 700°C, 800°C, 900°C, and H_2O/feed ratio at 1,0 to the gas composition of the gasification process.

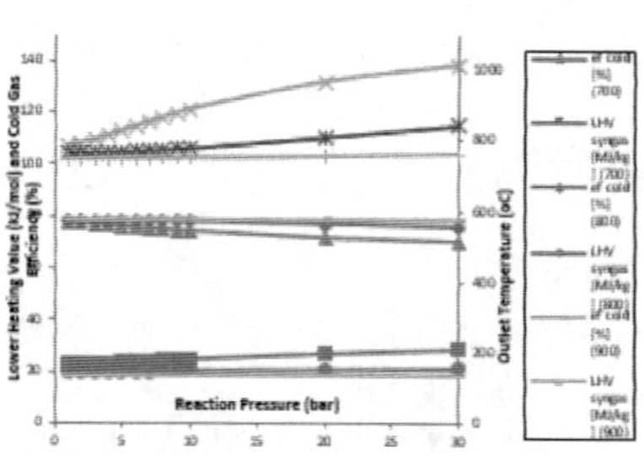

Figure 9 Working Pressure Effect at Temperature 700°C and H_2O/feed Ratio 1,0 to LHV, Cold Gas Efficiency, and Product Temperature of Gasification

From several simulated relation between operating temperature and working pressure, the condition that resulted in the highest yield of CH_4 is at 700°C and 30 bar with gas LHV at 93.41 kJ/mole and main gas LHV at 92.65 kJ/mole with cold gas efficiency at 70,03%. Due to the different composition of CH_4 between gasification products and natural gas that is used in PLTGU, therefore the gasification condition that resulted in highest value of main gas LHV was chosen. Main gas LHV was obtained at the same operating condition that results in the highest composition of CH_4.

4. SYNGAS ADAPTATION FOR UTILIZATION IN EXISTING PLANT

4.1. Lower Heating Value Comparison

From the chosen gasification main gas product LHV, the comparison of the gasification main gas product LHV with the existing LHV of natural gas in PLTGU was conducted. The LHV of natural gas is at 40630 kJ/kg, therefore to generate electricity from PLTGU with capacity of 400 MW will require 267 MW electric power from gas-fired power plant (PLTG) and 133 MW electric power from coal-fired power plant (PLTU). Thermal efficiency of gas turbine in the PLTGU system was assumed to be at 32% then to generate net electric power of 267 MW requires heat as much as;

[illegible] (3)

[illegible] (4)

If natural gas was used, the fuel consumption rate equal to:

[illegible] = 20.53 [illegible]

4.2. Syngas Fuel Consumption Rate

Gasification syngas consumption rate can be obtained by using LHV_{syngas} at 31828.32 kJ/kg, therefore to generate the same power from gas turbine process in PLTGU requires syngas consumption rate at:

[illegible] = 26.20 [illegible]

In order to generate the same electric power with syngas instead of natural gas, adjustment on fuel consumption rate was conducted. Adjustment was to increase the consumption rate by 1 to 2 times of the natural gas consumption. Based on mass conservation law, it was obtained that:

[illegible] = 6.25 [illegible]

Syngas mass flow rate produced from low rank coal gasification was;

[illegible] = 0.8 [illegible]

From the gasification simulation of low rank coal, 8 ton/hour or 2 kg/s coal was needed to produce 0.8 kg/s syngas. Therefore to produce syngas as much as 26.2 kg/s to generate 267 MW with 32% efficiency gas turbine, gasification coal feed required was:

[illegible] = 66.8 [illegible]

In conclusion to generate electric power as much as 267 MW in gas turbine in PLTGU, 26.2 kg/s syngas was needed that was equal to 66.8 kg/s or 240 ton/day gasification coal feed. There was an increase in coal consumption as much as 33 times bigger compared to the simulation conducted, therefore in order to produce required amount of syngas, additional modified fix bed gasifier was needed or by modifying the current system to be able to accommodate required gasification coal feed to produce required amount of syngas for PLTGU.

5. CONCLUSION

Indonesia low rank coal has the potential and characteristics to be used as alternative fuel in form of syngas for combined cycle power plant through gasification process Syngas characteristics produced from gasification that resembles the natural gas characteristics used in PLTGU occurs at gasification

IOP Conf. Series: Materials Science and Engineering **88** (2015) 012042 doi:10.1088/1757-899X/88/1/012042

operating condition at 600°C, 3.5 bar, and water steam to feed ratio 1 kg/kg. LHV characteristics of the syngas at specified condition is at 31828.32 kJ/kg.

CNG required to generate 400 MWe of Indonesia combined cycle power plant (PLTGU) was at 20.53 kg/s. Syngas needed to replace CNG in PLTGU system was 26.2 kg/s or equal to 66.8 kg/s gasification coal feed.

REFERENCES

[1] Handbook of Energy and Economics Statistic of Indonesia, Ministry of Energy and Mineral Resources, Indonesia, 2012
[2] M. Moliere, Expanding Fuel Flexibility of Gas Turbines, Journal of Power and Energy, 219(A2):109-119, 2005 (http://pia.sagepub. com/content/219/2/1 09.full.pdf
[3] Tulus Ruseno, Gas Turbine Performance and Efficiency Analysis for Power Generation, Indonesia Power, Jakarta, 2013.
[4] David A. Bell, Brian F. Towler, and Maohong Fan, Coal Gasification and Its Application, Elsevier, United Kingdom, 2011.
[5] A. Williams, M. Pourkashanian, J. M. Jones, dan N. Skorupska, Combustion and Gasification of Coal, Taylor & Francis, New York, 2000.
[6] Zachary Hoffman, Simulation and Economic Evaluation of Coal Gasification with SETS Reforming Process for Power Production, Thesis, Department of Chemical Engineering Louisiana State University, Baton Rouge, 2005
[7] Chang He, Process Modeling and Thermodynamic Analysis of Lurgi Fixed-Bed Coal *Gasifier* in a SNG Plant, *Applied Energy*

Simulation and Analysis of Temperature Distribution and Material Properties Change of a Thermal Heat sink Undergoing Thermal Loading in a Mobile Computer.

A Xavier[1] and C S Lim[1,2]
[1]University of Nottingham Malaysia Campus, Jalan Broga, Semenyih, 43500, Selangor.
[2]INTI International University, 71800 Nilai, Negeri Sembilan.

Email: chinseong.lim@nottingham.edu.my

Abstract. This paper is aimed at studying the thermal distribution and its associated effects on a thermal heat sink of a mobile computer (laptop). Possible thermal effects are investigated using Finite-Element Method with the help of a FEM software (Ansys Workbench 14). Physical changes of the structure such as temperature change and deformation are measured and are used as the basis for comparison between models of heat sinks. This paper also attempts to study the effect of thermal loading on the materials found in a heat sink hardware in terms of stresses that may arise due to physical restraints in the hardware as well as provide an optimized solution to reduce its form factor in order to be comparable to an Ultrabook class heat-sink. An optimized solution is made based on a cylindrical fin concept.

1. Introduction

Advancement in microprocessor technology as well as manufacturing has enabled the Ultrabook class laptops to be developed. In essence, ultrabooks are mobile computers that are under 21mm thick hence offers a reduced form factor without compromising performance [1]. Comparing typical laptops and ultrabook laptops, an immediate distinction in terms of thermal management could be observed. Ultrabooks, by branding, runs on Intel's Central Processing Unit (CPU) chips and these chips produce a Thermal Design Power, TDP, of less than 15 W whereas typical laptops would use CPU chips that could have a TDP ranging anywhere between 11.5W up to 45W [2]. The TDP of a microprocessor is the thermal power that is generated by it and this is the amount of heat that needs to be rejected so that the microprocessor could function optimally.

The typical permissible operating temperature of the Central Processing Unit (CPU) is 70°C and the reliability decreases by 10% for every increase of 2^{O}C [3]. Ergo, heat generated by CPU and Graphics Processing Unit (GPU) needs to be wicked away and this is achieved by using a heat sink hardware. The heat sink hardware consists of base plates, heat pipes, fins and a fan. At maximum operating conditions, the heat sink needs to be able to dissipate the thermal power generated. Smaller-sized mobile computers require a smaller heat sink assembly. Being smaller; the equivalent stresses in the unit may be higher. In order to study whether or not that is the case, this study also involves modelling a few conceps of heat sinks as well as simulating it at the operating conditions of the initial design.

Current heat sink technology are focussing on reduced dependencies on forced convection, advanced heat pipes as well as better performing and lighter weight materials [4] [5] [6] [7] [8] [9]. The inner workings of the heat pipe will be omitted in this study and will be assumed as a perfect heat conductor [10], conducting heat from the thermal plates to the heat sink fins.

As for the design of heat sink fins, there are two main school of thoughts, straight fins and cylindrical fins which are also known as pin fins [11] [12] [13] .Both have their unique strengths and weaknesses. This study will involve both designs

At elevated temperatures, thermally induced internal stress builds up within a member when it is unable to change length [14]. Material properties also differ at higher temperature as compared to

[1] To whom any correspondence should be addressed.

room temperature and for the most part, mechanical properties decrease with the increase of temperature [15] [14] [16] [17] [18] [19] [17] [18] [20].Seeing that a heat sink assembly is usually made up of more than one type of material, the Coefficient of Thermal Expansion (CTE) also plays a role in inducing stresses [21] .Creep is not considered for this study due to the operating temperature range being at a relatively low temperature (<200°C). Creep only becomes significant at temperatures above 40% of the melting point temperature of the materials [22] [23].
This study will involve the analysis of 7 different models comprising of the initial model and 6 concept models. The table below explains the different characteristics of the concept models.

Table 1. Different physical characteristics of the concept models.

Model	Physical properties
Concept 1	Reduced values of w, H and L (refer toTable 2)
Concept 2	Similar size envelope to the initial model but with angled fins
Concept 3	Similar size envelope to the initial model but with cylindrical fins
Concept 4	Similar size to Concept 3 but with spacing between heat-pipe and fins.
Concept 5	Similar size and spacing as Concept 4 but with increased cylinder diameter,*d* (refer to**Error! Reference source not found.**)
Concept 6	Reduced values of w,H, L and hf and with same cylinder diameter and spacing as Concept 5 (Refer to **Error! Reference source not found.**)

For a similar size envelope, it is found that cylindrical fins could perform better than straight fins. A smaller size cylindrical fin was able to give the same performance as the original design. A cylindrical fin design would be smaller, lighter and material-wise, cheaper than the original design. This warrants further study into the total economic impact

2. Methodology

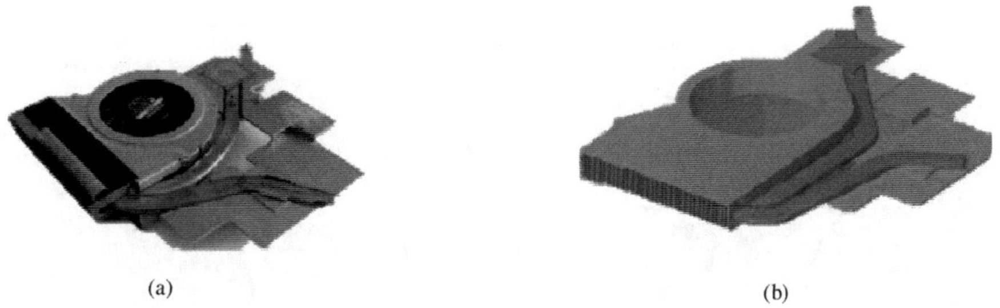

(a) (b)

Figure 1. Figure 1(a) is of the actual heat sink assembly and Figure 1(b) is the rendered image of the Creo Parametric model.

Figure 2. Figures 2(a), (b) and (c) shows where the letters are designated for the types of heat-sink design tested; t is the fin thickness, hf is the fin height and b is the spacing between the fins

Table 2. Shows the dimensions for each letter designation in the straight fin, angled fin and cylindrical fin design 2(a) and Figure 2(b) and 2(c)).

Model	**Straight fins**		**Angled fins**	**Cylindrical fins**			
	Initial	Concept 1	Concept 2	Concept 3	Concept 4	Concept 5	Concept 6
w (mm)	67	55.5	67	67	67	67	59.5
H (mm)	12	11	12	12	12	12	11
L (mm)	26	21	26	26	26	26	22.2
t (mm	0.5	0.5	0.5	-	-	-	-
b (mm)	0.65	1	1.5	2.5	2.5	3.5	3.5
b1 (mm)	-	-	3.7	-	-	-	-
Theta (degrees)	-	-	85	-	-	-	-
hf (mm)	4	3	3.5	3.5	3.4	3.4	2.85
number of fins	55	35	30	225	225	108	80
d (mm)	-	-	-	1.5	1.5	2.5	2.5

1. The parts required for this study is modelled using Creo Parametric and the dimensions used for each part is listed in Table 2. Shows the dimensions for each letter designation in the straight fin, angled fin and cylindrical fin design 2(a) and Figure 2(b) and 2(c)).
2. Simulation is carried out using Ansys Workbench 14. For each simulation, heat is supplied to the fins by the heatpipes. Fixed constraints are at all of the outer wall of the fin body which is meant to emulate the fixed enclosure within a laptop.

3. Results

3.1 Volume reduction and material cost reduction

With reduced volume of materials put into the design, savings in cost could be made.

Table 3. Material Volume reduction in cubic milimeters.

	Material Volume (mm3)		Reduction in Volume
Model	Al	Cu	Al
Initial	8651.25	8558.48	0
Concept 1	4677.75	8558.48	3973.5
Concept 2	6612.45	8558.48	2038.8
Concept 3	7094	8558.48	1557.25
Concept 4	6961.26	8558.48	1689.99
Concept 5	8291	8558.48	360.25
Concept 6	5871	8558.48	2780.25

The reduced volume in turn reduces the mass of materials used. By finding out the mass of the material per unit of heat sink and finding the monetary value of the metals at current market price, a cost comparison between models can be made. All the concept models brought forward into this study are made of less material hence cost cheaper in terms of materials. A smaller heat-sink also translates to a lighter system.

3.2 Comparing simulation results

The maximum permissible stress of Aluminium is 610MPa [24]. As observed in the simulations, the initial design is well below the threshold. However, concepts 3 and 4 have a higher maximum stress which is experienced at the cylindrical fins. Even though the stress is higher, the maximum deformation in concept 3 is still lower than that of the initial design.

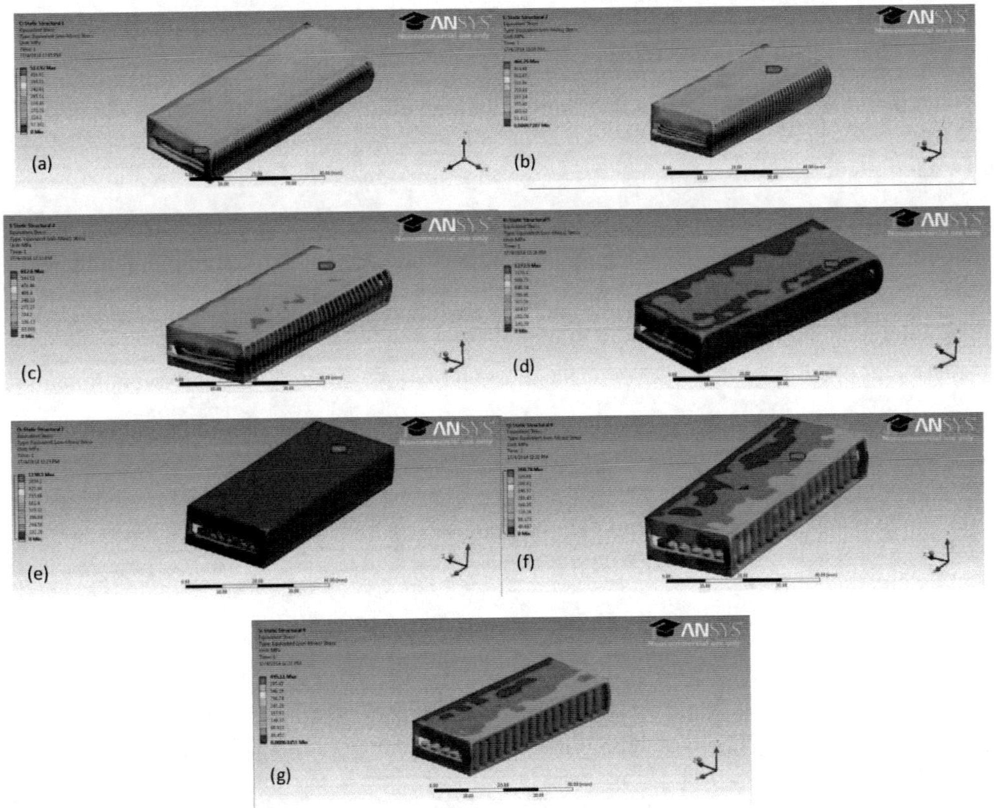

Figure 3 Distribution of stress levels experienced in the models from the first concept model (Figure 3(b)) to the sixth concept model (Figure 3(g).

Concept 5 and 6 is an attempt at reducing the stress levels in the cylindrical fins. This is achieved by using bigger-diameter cylinders. The stress is greatly reduced however the cooling capability is negatively impacted.

Table 4. Maximum stresses and deformation experienced in the models

Model	Maximum Stress (MPa)	Maximum Deformation (mm)
Initial	513.92	0.183
Concept 1	466.29	0.186
Concept 2	612.6	0.188
Concept 3	1272.5	0.181
Concept 4	1247.1	0.188
Concept 5	452.38	0.181
Concept 6	416.66	0.178

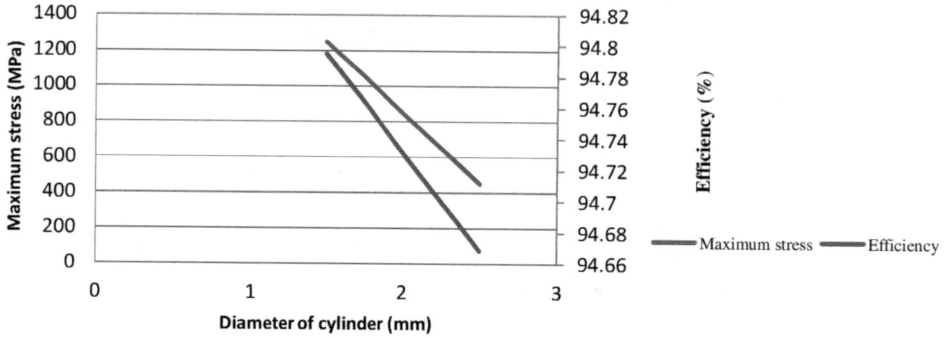

Figure 4. Relationship between cylinder diameter and the stress experienced and thermal efficiency

Trend lines are plotted using Concept model 4 and 5 in an attempt to find a compromise for a suitable cylinder diameter as shown in Figure 4. This results in efficiency higher than the initial model and at the lowest possible stress. Method of obtaining efficiency values will be explained in Section 3.3.

3.3 Formulating measure of comparison between heat-sinks

A measure of comparison based on the ratio of the temperature of the heat sink and the air temperature at the outlet is formulated:

$$\frac{[illegible]}{[illegible]} \times 111 \quad (1)$$

Table 5. Resultant efficiency of heat sinks based on alternative formulation

Model	Maximum temperature at fins (K)	Air temperature at outlet (K)	Comparison value
Initial	370.9	354.3	95.5244
Concept 1	369.7	352.8	95.42873
Concept 2	372.3	356.1	95.64867
Concept 3	373.7	366.2	97.99304
Concept 4	372.7	353.3	94.79474
Concept 5	369.5	349.8	94.66847
Concept 6	368.4	352.7	95.73833

In order to be deemed better than the initial product, the value at the Comparison column has to be of a higher value.

3.4 Cylindrical fins versus Straight fins

The elements that determine the cooling capability of any heat sinks include the surface area, metallurgy and the impact of the fin structure on the incoming air streams. Greater surface area and greater thermally conductive materials do make good heat sinks. However, the impact of the airflow is seldom looked at. Air acts as a good heat insulator if it is wrapped around a surface stagnantly. The function of heat sinks could then be thought of as a means to break the boundary layer by accelerating the air flow. Hence, a faster air stream would result in a higher probability that the boundary layer is broken which in turn results in a more effective heat sink [13].

That being said, the air velocity study at the outlet of all the models could act as gauge for comparison. All models are supplied with the same volumetric flow rate as dictated by the type of fan used. In each case, the same fan is assumed to be used and it supplies 3.28×10^{1} ^{1}ı ^{1}/ı of air.
The Initial Model, Concept Model 2 and Concept Model 3 all have approximately the same air velocity at inlet. However, there are differences in the efficiency of those units. The table below serves as comparison of the velocities.

Table 6. Air velocities and resultant comparison value

Model	Inlet Air Velocity (m/s)	Maximum Outlet Air Velocity (m/s)	Comparison value (%)
Initial	5.07	15.5	95.52
Concept 2	5.05	17.3	95.64
Concept 3	5.05	27.5	97.99

The difference in cylindrical fin design greatly affects the air flow characteristics and in turn affects how effective the heat sink is.The observable stresses are higher in the cylindrical fin design than in the straight fins. This could be due to the greater temperature seen in the cylindrical fins main body. This higher temperature results in greater thermal stresses (Table 6, compare initial model with Concept 3 and 4). For the same amount of force exerted in the straight fins and the cylindrical fins, the stress levels differ according to the surface area. The smaller surface area of each individual cylindrical fins creates for a much higher stress. Because of this, the stress levels can be reduced by designing a heat-sink with bigger cylinder fins.
As for angled fins, the efficiency is higher compared to the initial design as the surface area that is in thermal contact is bigger. However, at angles approaching 0° and 180°, the fins would greatly affect the air flow. Once the fins are perpendicular lengthwise to the heat-pipes, the total surface area of the fins in contact with the airstream is minimal thus decreasing the efficiency of the heat sink.

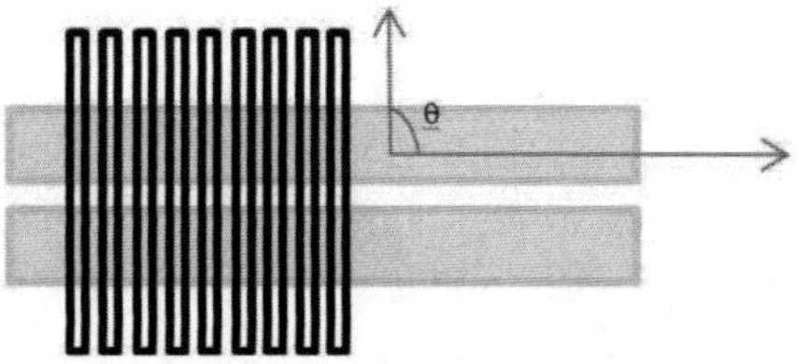

Figure 5. Reference angle for the case of angled fins

4. Conclusion

For a given size envelope, cylindrical fins have a higher efficiency when the cylinder fins are small (Figure 4). However, stress levels in cylindrical fin design are very high. A compromise needs to be made between cylinder diameter and the performance of the heat sink in terms of stress levels and efficiency.

This study has shown that a cylindrical fin design is cheaper, can be made smaller and has a better efficiency for a given size envelope compared to a straight fin design. After identifying a suitable diameter for the cylinder fins, possible limitations may only come from manufacturing methods that are employed in the industry to fabricate the fins. Assuming the cost of machining is higher than the typical straight fins, by economy of scale, the price could be justified reasonably as seeing it from the materials perspective it is actually cheaper. Concept 6 proves that a smaller form factor cylindrical heat-sink with a suitable fin diameter could work. Due to the reduced material needed, it is cheaper, lighter and smaller.

Possible improvement to heat sink design is to incorporate thermocouple materials to transform heat energy in the heat sink into usable electrical energy which could help power the fan in the system in order to improve battery life.

5. Reference

[1] B. Emmet, Intel, 28 July 2011. [Online] http://blogs.intel.com/technology/2011/07/the_world_of_computing_is/.
[2] "Ark.intel," [Online]. http://ark.intel.com/products/family/75024/4th-Generation-Intel-Core-i5-Processors/desktop.
[3] M. J. J. Y. M. S. Jeehon Choi, vol. 44, 2012.
[4] M. H. W. W. K. B. J. B. Kwang Soo Kim, Applied Thermal Engineering, vol. 23, pp. 1137-1144, 2003.
[5] Y. M. Vladimir Pastukhov, Applied Thermal Engineering, vol. 27, pp. 894-901, 2007.
[6] Y. D. X.Yang, Applied Thermal Engineering, Vols. 33-34, pp. 1-14, 2012.
[7] Y. Yamada, Microelectronics Reliability, vol. 52, no. 9-10, pp. 2443-2446, 2012.
[8] P. Wallin, "Heat Pipe, selection of working fluid," Lund University, Lund, Sweden, 2012.
[9] F. Korn, "Heat Pipes and its applications," Lund University, Lund, Sweden, 2008.
[10] S. E.-H. A.Abo El-Nasr, "Effective Thermal Conductivity of heat pipes," vol. 32, no. 1-2, 1996.
[11] Cool Innovations, "Technology: Pin Fin Design," 2014. [Online]. http://www.coolinnovations.com/technology/pin-fin-design.
[12] B. Dagan, ECN Mag, no. January, 2010.
[13] B. Dagan, New Electronics, no. June, 2009.
[14] B. Onouye, Statics and Strength of Materials: Foundation for Structural Design, Prentice Hall, 2004.
[15] M. Vable, Mechanics of Materials, 2nd ed., Oxford University Press, 2007.
[16] M. E. Kassner, Fundamentals of creep in metals and alloys, Second ed., Jordan Hill, Oxford: Elsevier, 2009.
[17] T. R. P. P. R.Murugavel, Materials and Design, vol. 30, no. 2, pp. 221-227, 2009.
[18] H. W. J. Z. S. F. W. Yajun Gao, Computational Materials Science, vol. 50, no. 10, pp. 3032-3037, 2011.
[19] M. M. J.H. You, Journal of Nuclear Materials, vol. 373, no. 1-3, pp. 269-274, 2008.
[20] A. B. Ron Cobden, 1994. http://www.alueurope.eu/talat/lectures/1501.pdf.
[21] H. P.Monard, Dynamic Mechanical Analysis: A Practical Introduction, Boca Raton, Florida: CRC Press LLC, 1999.
[22] L. Michelet, 2013. http://lmm.epfl.ch/files/content/sites/lmm/files/shared/TPDefRupt2013/TPDefFract13_Creep.pdf.
[23] "DoITPoMS," University of Cambridge, http://www.doitpoms.ac.uk/tlplib/creep/stress.php.
[24] R. E.Simmons, "Estimating Parallel Plate-Fin Heat Sink Thermal Resistance," no. February 2003, 2003.
[25] CES Level 2 Material Universe, CES Level 2 Material Universe.

IOP Conf. Series: Materials Science and Engineering **88** (2015) 012044 doi:10.1088/1757-899X/88/1/012044

An Expert Machine Tools Selection System for Turning Operation

C F Tan[1], S N Khalil[2], J Karjanto[3], L S Wahidin[4], W Chen[5] and G W M Rauterberg[6]
[1,2,3,4]Centre for Advanced Research on Energy (CARe)
Faculty of Mechanical Engineering, Universiti Teknikal Malaysia Melaka
Hang Tuah Jaya, 76100 Durian Tunggal, Melaka, Malaysia

[5,6]Designed Intelligence Group
Department of Industrial Design, Eindhoven University of Technology
Eindhoven, the Netherlands

E-mail: cheefai@utem.edu.my

Abstract. The turning machining process is an important process in the manufacturing industry. It is important to select the right tool for the turning process so that the manufacturing cost will be decreased. The main objective of this research is to select the most suitable machine tools with respect to user input requirement. The selection criteria are based on rule based expert system and multi-criteria weighted average method. The developed system consists of Knowledge Acquisition Module, Machine Tool Selection Module, User Interface Module and Help Module. The system capable of selecting the most suitable machine along with its full specification and ranks the machines based on criteria weighted. The main benefits from using the system is to reduce the complexity in the decision making for selecting the most appropriate machine tools to suit one requirement in the turning process for manufacturing industry.

1. Introduction

In metal based industries, the efficiency of the manufacturing system is depending on the selection of machine tools. The machine tools will influence the manufacturing systems such as time to market, precision, productivity, etc. Traditionally, the selection of machine tools in industry is done manually by expert machinist or experience engineers. Thus, the process to select the most suitable machine for an application is a long and complicated process. An expert system is developed to perform the similar task like expert when the he/she is absent. Application of rule-based expert system in manufacturing systems selection has been reported by various investigators. Among others are cutting tools selection, ceramic material selection, rapid prototyping selection, robot selection and material handling selection. Maropoulos and Gill [1] developed a tool selection system for machining cylindrical components. The

[1] To whom any correspondence should be addressed.

tool selection system was developed based on knowledge and logic approach to select the suitable tool to machine the cylindrical components.

Arezoo et al. [2] demonstrated the used of expert system to select the cutting tools and conditions of turning operation. In addition, an expert system for turning and rotating tool selection also developed by Moorkherjee and Bhattacharyya [3]. The aim of the work was to select the most suitable material for the automotive engine components. Another work by Massod and Soo [4] used rule based expert system to select the rapid prototyping. The input for the developed expert system is the commercial available rapid prototyping systems in the market. Chu and Lin [5] developed a fuzzy method for robot selection. It uses various rating alternatives verses subjective criteria and the weights of all criteria are assessed in linguistic terms represented by fuzzy numbers. Work by Chan et al. [6] used expert system to select the material handling equipment selection. Er & Dias [7] developed a rule-based expert system to select the cast components for manufacturing process. Various development of expert system for industry application has been reported [8-10].

This paper describes the development of rule based expert system for selection of CNC machine tools especially turning operation. The system involves the user providing answers to specific questions or entering the required data. The system uses rule based and multi-criteria weighted average method (MCWA) [11] as the decision-making approach. Recommendations and suggestions of the appropriate machine tools based on input requirement are displayed on the screen at the end of the program.

2. Overall System Development

The flow chart of the CNC turning machine tools selection system that used KAPPA-PC expert system shell [12] is shown in Figure 1. The system has for basic modules namely the Knowledge Acquisition Module, CNC Turning Machine Tool Selection Module, User Interface Module and Help Module.

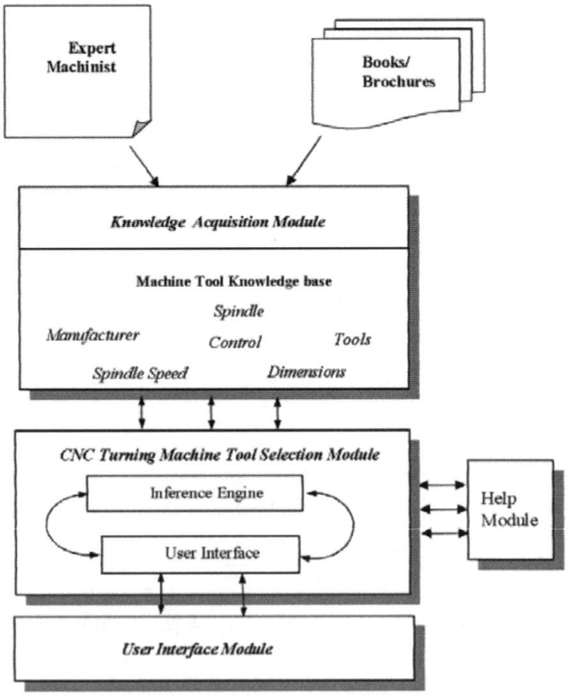

Figure 1. Flow chart of the developed system.

2.1 Knowledge Acquisition Module

Knowledge and information used to develop the system are obtained from brochures, vendors, books, journals and internet. The information related to the CNC turning machines is shown in Table 1. Classification of machine tools are based general view about the machine, spindle, tooling, work support, axis and physical specification. Currently, the database contains 3 machines manufacturers and 17 types of CNC machine tools.

Table 1. Classifications of Machine Tools.

1	General	manufacturer, model, machine type,
2	Spindle	spindle type, no. of spindle, spindle direction type, taper number, max speed, motor power
3	Tooling	standard tool size, number of tools, tool diameter
4	Work Support	table size, rotary table
5	Axis	rapid traverse speed, number of axis
6	Physical	machine dimensions (height, length, width)

2.2 CNC Turning Machine Tools Selection Module

Grouping the detail of machines into the database is the next step in the methodology of development of CNC turning machine tools selection system. Since different machine manufacturers use different terms in describing the machine parameters, the development of the database needs more effort to recognize the same parameters. The database of the machine tools is created into a hierarchy tree in the Kappa-PC as shown Figure 2.

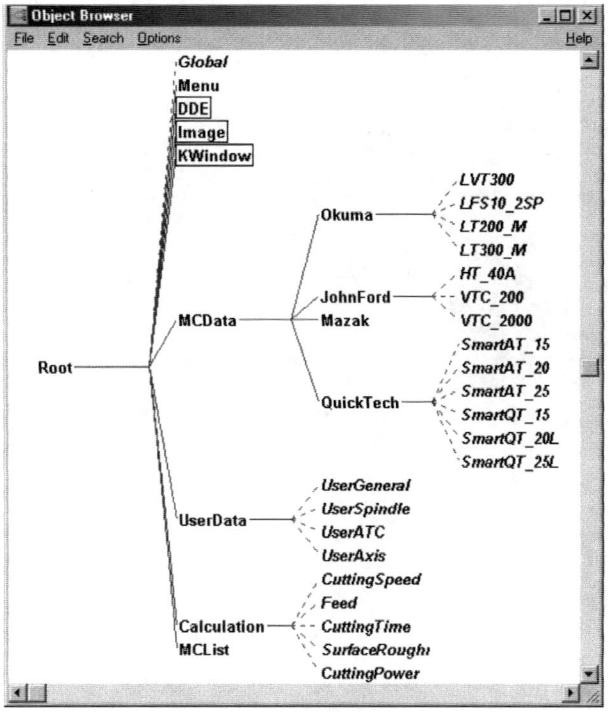

Figure 2. Hierarchy database of machine tools.

IOP Conf. Series: Materials Science and Engineering **88** (2015) 012044 doi:10.1088/1757-899X/88/1/012044

The domain of rule-based expert system is consists of frames or objects. The objects can be classes or instances within classes [13]. The relationship between the objects is linked together by a tree hierarchy. Information related to machine tool manufacturer are represented as objects, and machine tool types are classes. Slot represents important properties of an object or class. The characteristics or specification of machine tools such as number of axis, programming language, number of tool holder are defined as slot in KAPPA-PC.

This system uses rule-based with "IF-THEN" rules and multi-criteria weighted average method (MCWA) is used as the decision-making approach in forward chaining principle. Each turning machine tool in the database is assigned a score according to its properties in relation to the data input. After this, the system will rank the turning machine based on the input scores by the user. The machines are ranked using Multi-Criteria Decision Making (MCDM) Method where the ranking is from best to worst using machine specifications and criteria weights. The alternatives are weighted average and hierarchy tree is used for the selection process.

2.3 User Interface Module

Another important step in the development of expert system is the designing of the user interface. The user interface windows are linked to one another by using functions. A user-friendly interface is created using button, text images and bitmaps. A session window, which is the basic components of KAPPA-PC interface is used to create application interface. Figure 3 shows the main user interface developed in this system. Some of the examples data requires by the system is the general data input, types of spindle, automatic tool changes, axis as shown in Figure 4. Figure 5 shows the sub-menu of "general" data input.

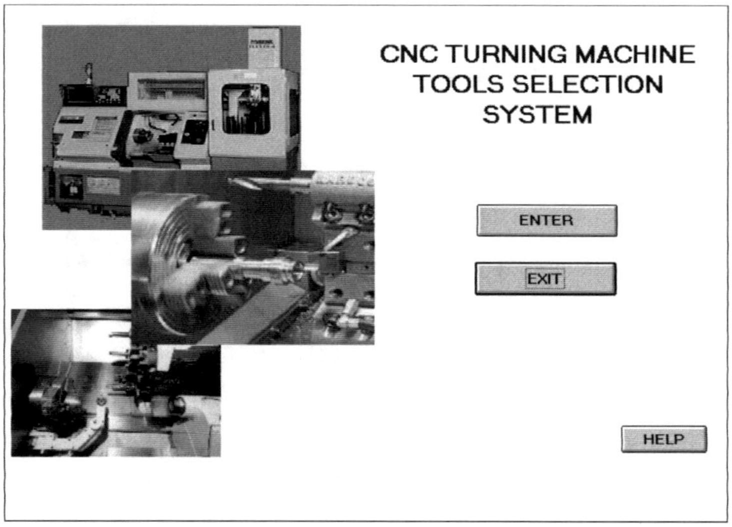

Figure 3. Main menu.

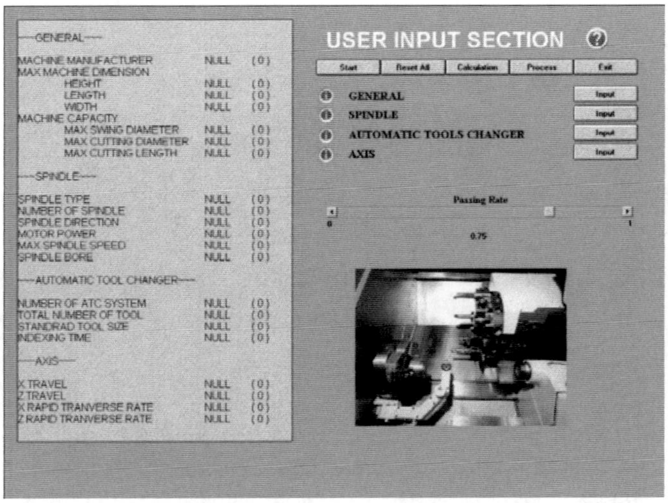

Figure 4. User Data Input interface.

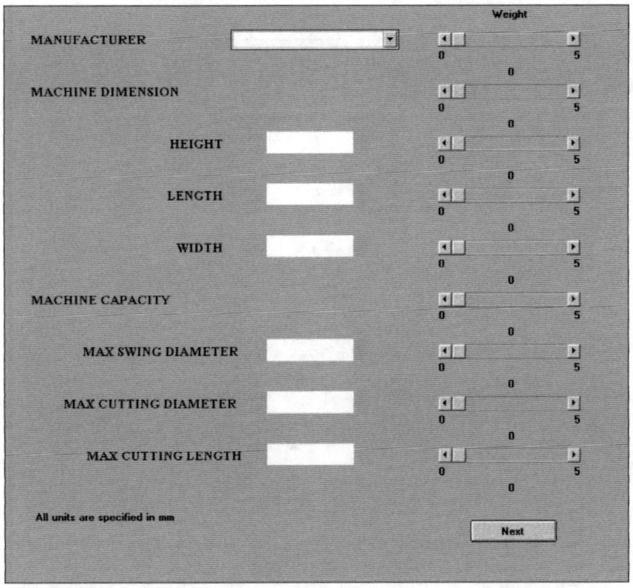

Figure 5. Sub-menu of "General" User Data Input.

2.3 Help Module

User's guide and some extra information are also included in the system. Figure 6 shows some of the examples of help menu in the decision support system. This system provides machinability calculation to assist the user in determining some machining parameters for particular machine tools.

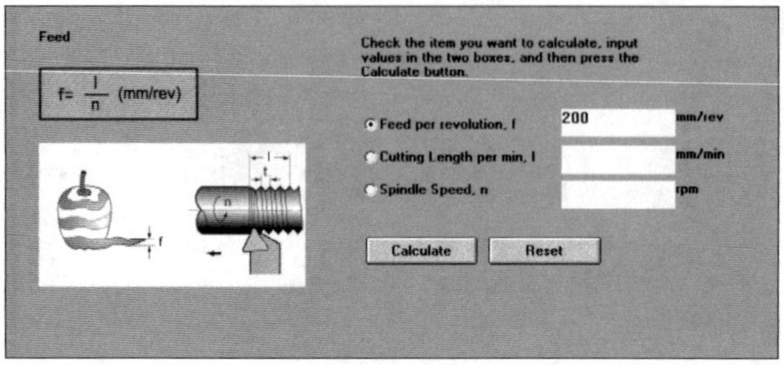

Figure 6. Help module.

3. Result and Discussion

The developed expert turning machine tools system is able to select the suitable turning machine tools based on the ranking by the user. The system is able to show the capability to manipulate the variations of the inputs. Currently, the system has three machine tool manufacturers and seventeen machines inside the database. Figure 7 shows the output menu where the highest score machine will be arranged first according to preferred passing rate. As an example the initial passing rate is 0.75. The details of machine parameters can also be displayed for easy reference.

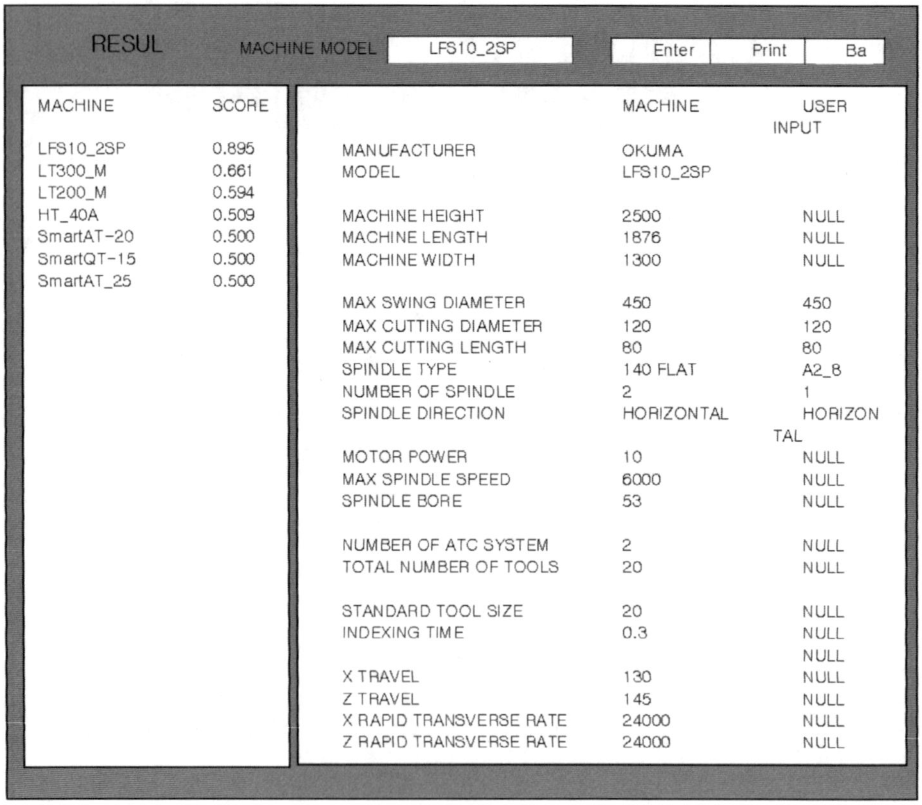

Figure 7. The output window

5. Conclusion

A rule-based expert system for CNC machine tools selection has been developed. The system uses rule based and multi-criteria weighted average method (MCWA) as the decision-making approach. Recommendations and suggestions are displayed on the screen at the end of the program. Furthermore, on line help facilities are also include in the system. Currently the database in the system is not exhaustive for all applications. Nevertheless, it does illustrate the basic concept in developing an expert system for machine tools selection. One of the capability and potential of this system is to become a useful tool for machinist in metal based industries.

Acknowledgment

Special thanks to the Centre of Research and Innovation Management and Faculty of Mechanical Engineering for the support.

References

[1] Maropoulos & Gill 1995 Intelligent tool Selection for Machining Cylindrical Components. Part 1: Logic of the knowledge-based Module *Proceedings of Instn Mechanical Engineers. Part B: Journal of Engineering Manufacture* **209** 173-182

[2] Arezoo B Ridgway K & Al-Ahmalim A 2000 Selection of cutting tools and conditions of machining operations using expert system *Computers in Industry* **42** 43-58

[3] Moorkherjee R & Bhattacharyya B 2001 Development of an Expert System for Turning and Rotating Tool Selection in a Dynamic Environment *Journals of Materials Processing Technology* **113** 306-311

[4] Massod S H & Soo A 2002 A rule based expert system for rapid prototyping system selection *Robotics and Computer Integrated Manufacturing* **18** 267-274

[5] Chu T C & Lin Y C 2003 A Fuzzy TOPSIS Method for Robot Selection *The International Journal of Advanced Manufacturing Technology* **21** 284-290

[6] Chan Ip & Lau 2001 Integration of expert system with analytic hierarchy process for the design of material handling equipment selection system *Journal of Materials Processing Technology* **116** 137-145

[7] Er A & Dias R 2000 A rule-based expert system approach to process selection for cast components *Knowledge-based Systems* **13(4)** 225-234

[8] Urrea C Henriquez G & Jamett M 2015 Development of an expert system to select materials for the main structure crane designed for disabled people *Expert systems with Applications* **42(1)** 691-697

[9] Ipek M Selvi I H Findik F Torkul O & Cedimoglu I H 2013 An expert system based material selection approach to manufacturing *Materials & Design* **47** 331-340

[10] Dobrzanski L A & Madejski J 2006 Prototype of an expert system for selection of coatings for metals *Journal of Materials Processing Technology* **75(1-3)** 163-172

IOP Conf. Series: Materials Science and Engineering **88** (2015) 012045 doi:10.1088/1757-899X/88/1/012045

A Rule-Based Industrial Boiler Selection System

C F Tan[1], S N Khalil[2], J Karjanto[3], B T Tee[4], L S Wahidin[5], W Chen[6], G W M Rauterberg[7], S Sivarao[8] and T L Lim[9]
[1,2,3,4,5,8]Centre for Advanced Research on Energy (CARe)
Faculty of Mechanical Engineering, Universiti Teknikal Malaysia Melaka
Hang Tuah Jaya, 76100 Durian Tunggal, Melaka, Malaysia

[6,7]Designed Intelligence Group
Department of Industrial Design, Eindhoven University of Technology
Eindhoven, the Netherlands

[9]Intellogic Technology Private Limited
Puchong, Malaysia

E-mail: cheefai@utem.edu.my

Abstract. Boiler is a device used for generating the steam for power generation, process use or heating, and hot water for heating purposes. Steam boiler consists of the containing vessel and convection heating surfaces only, whereas a steam generator covers the whole unit, encompassing water wall tubes, super heaters, air heaters and economizers. The selection of the boiler is very important to the industry for conducting the operation system successfully. The selection criteria are based on rule based expert system and multi-criteria weighted average method. The developed system consists of Knowledge Acquisition Module, Boiler Selection Module, User Interface Module and Help Module. The system capable of selecting the suitable boiler based on criteria weighted. The main benefits from using the system is to reduce the complexity in the decision making for selecting the most appropriate boiler to palm oil process plant.

1. Introduction

There are various types of boiler with different specification, fuel base, construction and capacity. Different industries need a specific boiler in order to manufacture its products. In other words, boiler selection is based on the product that is being manufactured. There may be many criteria that can be considered when selecting a boiler to meet the application needs. Some of the criteria are i) steam capacity of boilers, ii) steam pressure of boilers, iii) boiler efficiency, iv) type of fuel, v) weight of boiler, and vi) volume of water.

Currently boiler selection is a manual process in order to fulfil the requirements heat production for a specific product. It is a time consuming and costly affair to determine the specification and requirements of an industry boiler. Boiler specification and selection is very important to produce a good and high quality product. In some cases, some boilers is not able to achieve the optimum

[1] To whom any correspondence should be addressed.

efficiency because the failure to diagnose the criteria needed for boiler due to the lack of knowledge. Boiler that does not fully utilize in capacity or problematic is detrimental to the environment. The Expert System is a technique that can help to improve industry boiler selection by providing guidelines for the optimum boiler selection. By using expert system, the experienced person who is not familiar in determining the type of boiler that is needed for a specific industry is able to gain information and knowledge correctly to select the appropriate industry boiler.

Expert system has been used in different applications, such as selection, diagnosis, gaming and feature recognition. Arezoo et al. [1] demonstrated the used of expert system to select the cutting tools and conditions of turning operation. In addition, an expert system for turning and rotating tool selection also developed by Moorkherjee and Bhattacharyya [2]. The aim of the work was to select the most suitable material for the automotive engine components. Another work by Massod and Soo [3] used rule based expert system to select the rapid prototyping. The input for the developed expert system is the commercial available rapid prototyping systems in the market. Chu and Lin [4] developed a fuzzy method for robot selection. It uses various rating alternatives verses subjective criteria and the weights of all criteria are assessed in linguistic terms represented by fuzzy numbers. Work by Chan et al. [5] used expert system to select the material handling equipment selection.

This paper describes the development of rule based expert system to select the industrial boiler for palm oil process industry. The system involves the user providing answers to specific questions or entering the required data. The system uses rule based and multi-criteria weighted average method (MCWA) [6] as the decision-making approach. Recommendations and suggestions of the suitable boiler based on input requirement are displayed on the screen at the end of the program.

2. Methodology

The expert systems are well-known and widespread applications of artificial intelligent research. There are many methods of expert system developing, but these different methods can be applied efficiently in large, commercial development projects. During our researches a new development process was created, which can be used in experimental projects for rapid development of small, prototype expert systems, considering special characteristics of applied expert system shells namely Kappa PC. The Kappa-PC expert system is a hybrid PC tool that combines critical technologies essential to the rapid development of low-cost and high-impact business applications [1].

2.1 Knowledge Acquisition Module

The knowledge for the developed system was gathered from the guidelines, specification, operators, technician and engineers. The specification of the water tube boiler (Figure 1) consists of the weight of boiler, water volume, fuel type, steam capacity and steam pressure. The fire tube boiler as shown in Figure 2 consists of the different specification from the water tube boiler. The general specification of the industry boiler is shown in the Figure 3.

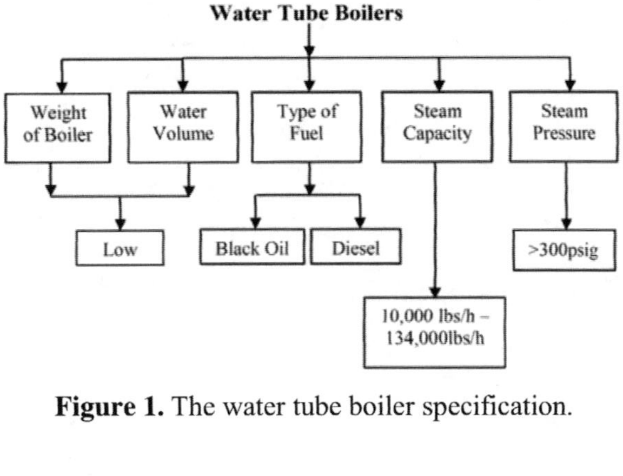

Figure 1. The water tube boiler specification.

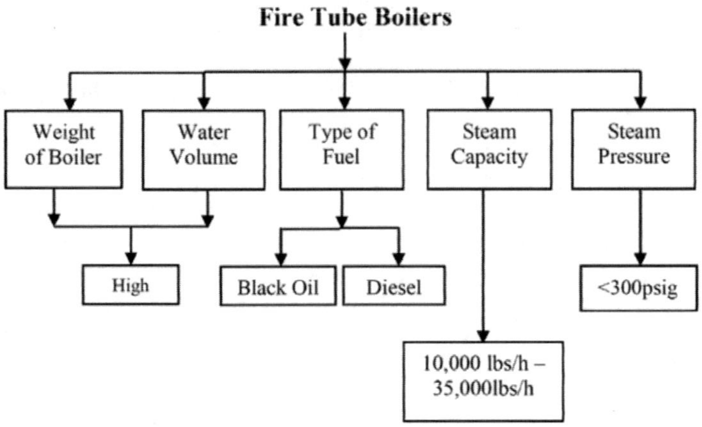

Figure 2. The fire tube boiler specification.

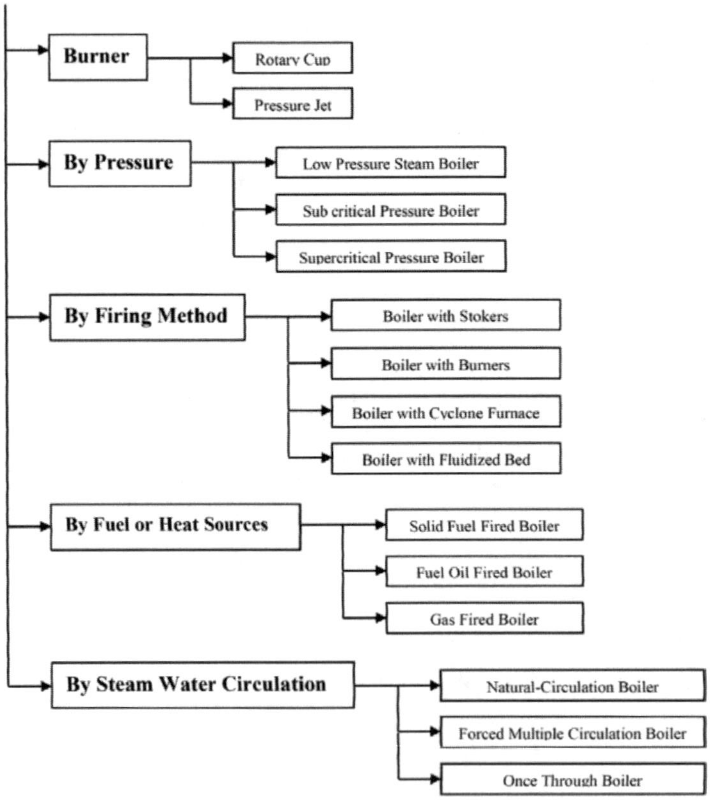

Figure 3. The overall boiler specification.

IOP Conf. Series: Materials Science and Engineering **88** (2015) 012045 doi:10.1088/1757-899X/88/1/012045

2.2 Boiler Selection Module

Grouping the detail of machines into the database is the next step in the methodology of development of boiler selection system (Figure 4). Since different machine manufacturers use different terms in describing the machine parameters, the development of the database needs more effort to recognize the same parameters.

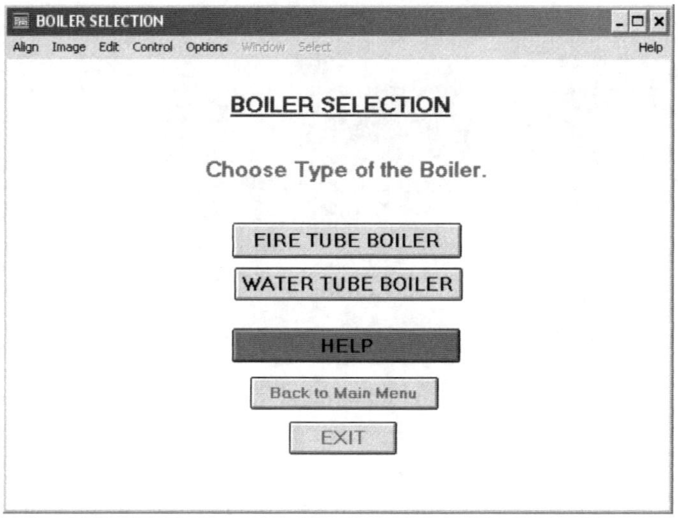

Figure 4. The boiler selection window.

The domain of rule-based expert system is consists of frames or objects. The objects can be classes or instances within classes [13]. The relationship between the objects is linked together by a tree hierarchy. Information related to machine tool manufacturer are represented as objects, and machine tool types are classes. Slot represents important properties of an object or class. The characteristics or specification of machine tools such as number of axis, programming language, number of tool holder are defined as slot in KAPPA-PC.

This system uses rule-based with "IF-THEN" rules and multi-criteria weighted average method (MCWA) is used as the decision-making approach in forward chaining principle. Each turning machine tool in the database is assigned a score according to its properties in relation to the data input. After this, the system will rank the turning machine based on the input scores by the user. The machines are ranked using Multi-Criteria Decision Making (MCDM) Method where the ranking is from best to worst using machine specifications and criteria weights. The alternatives are weighted average and hierarchy tree is used for the selection process.

2.3 User Interface Module

Another important step in the development of expert system is the designing of the user interface. The user interface windows are linked to one another by using functions. A user-friendly interface is created using button, text images and bitmaps. A session window, which is the basic components of KAPPA-PC interface is used to create application interface. Figure 5 shows the main user interface developed in this system.

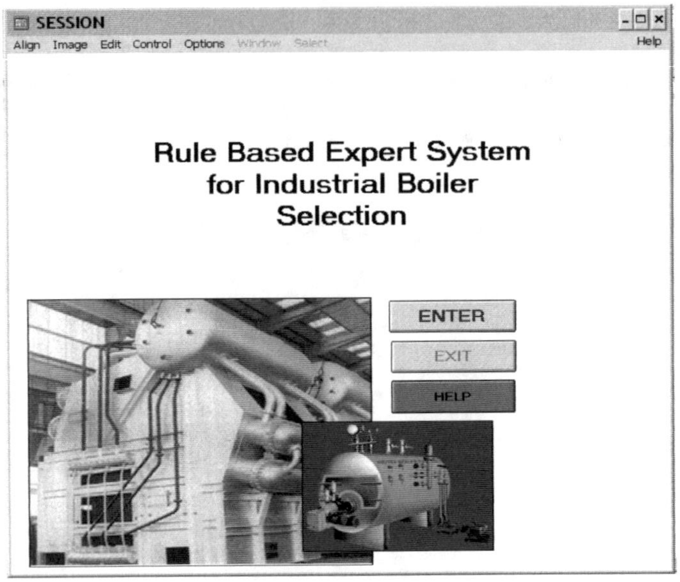

Figure 5. Main menu.

2.3 Help Module
User's guide and some extra information are also included in the system. This system provides machinability calculation to assist the user in determining some machining parameters for particular machine tools.

3. Conclusion

A rule-based expert system for boiler selection has been developed. The system uses rule based and multi-criteria weighted average method (MCWA) as the decision-making approach. Recommendations and suggestions are displayed on the screen at the end of the program. Furthermore, on line help facilities are also include in the system. Currently the database in the system is not exhaustive for all applications. Nevertheless, it does illustrate the basic concept in developing an expert system for boiler selection. One of the capability and potential of this system is to become a useful tool for mechanical engineer in chemical process industries.

4. Acknowledgment

Special thanks to the Centre of Research and Innovation Management and Faculty of Mechanical Engineering for the support.

5. References

[1] Arezoo B Ridgway K & Al-Ahmalim A 2000 Selection of cutting tools and conditions of machining operations using expert system *Computers in Industry* **42** 43-58

[2] Moorkherjee R & Bhattacharyya B 2001 Development of an Expert System for Turning and Rotating Tool Selection in a Dynamic Environment *Journals of Materials Processing Technology* **113** 306-311

[4] Massod S H & Soo A 2002 A rule based expert system for rapid prototyping system selection *Robotics and Computer Integrated Manufacturing* **18** 267-274

[5] Chu C T & Lin Y C 2003 A Fuzzy TOPSIS Method for Robot Selection *The International Journal of Advanced Manufacturing Technology* **21** 284-290

[6] Chan I P & Lau 2001 Integration of expert system with analytic hierarchy process for the design of material handling equipment selection system *Journal of Materials Processing Technology* **116** 137-145

IOP Conf. Series: Materials Science and Engineering **88** (2015) 012046 doi:10.1088/1757-899X/88/1/012046

An Expert Fault Diagnosis System for Vehicle Air Conditioning Product Development

C F Tan[1], B T Tee[2], S N Khalil[3], W Chen[4] and G W M Rauterberg[5]
[1,2,3]Centre for Advanced Research on Energy (CARe)
Faculty of Mechanical Engineering, Universiti Teknikal Malaysia Melaka,
Hang Tuah Jaya, 76100 Durian Tunggal, Melaka, Malaysia.

[4,5]Designed Intelligence Group,
Department of Industrial Design, Eindhoven University of Technology,
Eindhoven, the Netherlands

E-mail: cheefai@utem.edu.my

Abstract. The paper describes the development of the vehicle air-conditioning fault diagnosis system in automotive industries with expert system shell. The main aim of the research is to diagnose the problem of new vehicle air-conditioning system development process and select the most suitable solution to the problems. In the vehicle air-conditioning manufacturing industry, process can be very costly where an expert and experience personnel needed in certain circumstances. The expert of in the industry will retire or resign from time to time. When the expert is absent, their experience and knowledge is difficult to retrieve or lost forever. Expert system is a convenient method to replace expert. By replacing the expert with expert system, the accuracy of the processes will be increased compared to the conventional way. Therefore, the quality of product services that are produced will be finer and better. The inputs for the fault diagnosis are based on design data and experience of the engineer.

1. Introduction

Nowadays, the advancement of computer system has enabled the application of computer to solve the engineering problem such as product development problem, line balancing problem, etc. Computer system is able to support the manufacturing industry especially for the production support function. The production support system is includes the internal customers that represent the end user in terms of cost, quality and delivery. The cost is represented by the robust designs and processes, the quality represented by the ease of manufacture and the delivery is represented by the efficient of communication [1].

Expert system (ES) is a computer program that can used to represent the defined knowledge and experience of expert. ES has been used by various researchers [2-7] to solve the problem of expert absent when needed. ES has been used to diagnose the problem of the gear box of the wind turbine [2]. The ES for the gear box is able to provide a precise and shorter maintenance time. In addition, ES also used to diagnose the faulty of the EAST cryoplant [3]. The developed ES is able to ensure the safe operation of the cyroplant. For the power system, ES has been used to diagnose and assess its

[1] To whom any correspondence should be addressed.

IOP Conf. Series: Materials Science and Engineering **88** (2015) 012046 doi:10.1088/1757-899X/88/1/012046

operation for error reduction [4]. Yang et al. [5] developed an expert system, namely VIBEX, to diagnose the condition of rotating machinery based on vibration approach. The developed ES is able to assist plant operator in rotating machinery diagnosis. In the similar development by Qian et al. [6], an expert system for lubricating oil condition diagnostic, namely LUBRES, has been developed. LUBRES is able to diagnose the refining process of lubricating oil. ES also can be used to select the casting processes [7].

The main objective of the development of an expert fault diagnosis system is to achieve an integrated diagnostic procedure for air-conditioning development. The research was done in a subsidiary of the largest automotive parts manufacturing group in Malaysia. The developed system is to diagnose problems in the product development process in the vehicle air-conditioning manufacturing company. The developed ES will trace the error and defect in the various phases in the development processes. The developed ES is able to shorten the product development time and increase the efficiency in the decision making process.

2. Vehicle Air Conditioners

Vehicles air-conditioning systems is the standard component for vehicles. There are 3 main components in the vehicle air-conditioning system, which is compressor, condenser and evaporator.

2.1 Compressor

The compressor is the heart of the vehicle air-conditioning system. The compressor is attached with the vehicle engine via belt to compress and transfer the refrigerant gas.

2.2 Condenser

The condenser in the vehicle air-conditioning system is used to dissipate the heat from the refrigerant gas. Normally, it is located in front of the radiator. The fan will cool the refrigerant gas by dissipate the heat. Then, the refrigerant gas will flow back to the system.

2.3 Evaporator

The evaporator is located near the dashboard of the vehicle. The evaporator is used to absorb and remove the heat in the vehicle. In addition, the evaporator also used to dehumidify the air in the vehicle.

3. Product Development

In the design methodology, the total design process as shown in Fig. Product development is a step on the quality planning road map to determine the product features required to meet the customer needs. In competitive situations, product development i should comply with the following criteria:

- Meet the need of the customers.
- Meet the needs of the supplier.
- Be competitive,
- Optimize the combined cost of the customers and supplier.

Quality planning is part of the product development activity. Product designers, quality specialists operating personnel, or some combination of personnel in all these categories, may involve in quality planning.

3.1 Air-Conditioner Development Procedures

First of all, Marketing Department will discuss with companies that are interested in developing an air conditioning system for new model vehicle about the requirement of the air conditioners. A meeting will be held as soon as possible between Marketing Department and Product Development Department for the latter to know the customer requirement. The prototype complete-built-up vehicle will be sent for a certain period (normally one month) to allow Product Development Department to develop the air conditioning system for the vehicle. The air conditioners will be based on previous design or new design depending on the requirement. The vehicle may come in two types: i. complete original air-conditioning system, and ii. without air conditioning system. The activity of Product Development Department is shown in Figure 1.

After the air conditioning system has been developed, a complete set of drawings for each newly developed component will be prepared. Then, the vehicle will send back to the customer for testing over the newly developed air conditioning system and evaluated the proposed air conditioning system, or change some components. The engineer will issue the specification for costing. After evaluation, customer will negotiate with the Marketing Department with the price of air-conditioner and decide whether to accept the proposed air conditioning system. After that, the engineer will prepare the installation manual for the proposed air conditioning system. Lastly, the final specification will issue and production will start by Production Department.

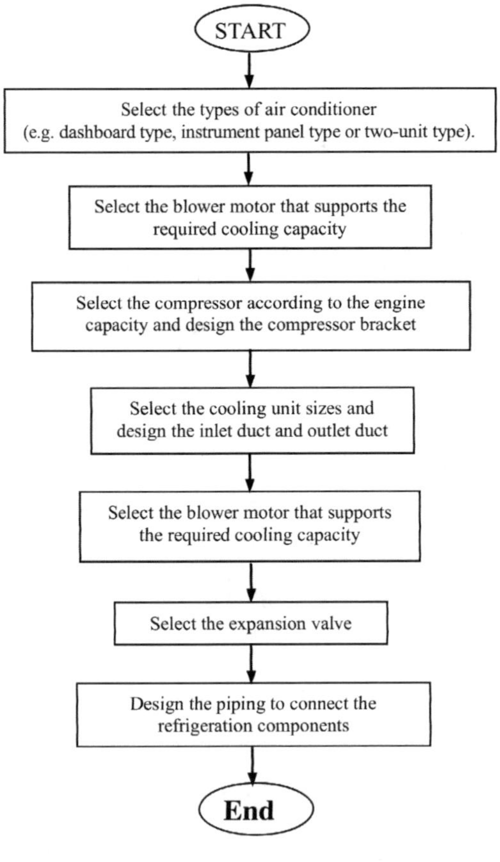

Figure 1. The activities of Product Development Department.

First of all, the Marketing Department will receive the inquiry from the customer for new product. Then the customer will send the original model, sample or drawing to Product Development Department. The executive will check whether the product drawings, specification or sample information sufficient or not. If not, PD department will collect more information from the customer.

Then, the model will be redesigned according to customer specification. The technical drawing for the model will be generated. Executive checks the drawing before it is approved by the manager.

Then, the process of generates the product development schedule & preliminary specification will be done. If the customer requires non-standard component, the component will be fabricated. If not, the sample core and piping will be made by Engineering Department. The sample will be checked by Quality Assurance Department according to the dimension in the drawing. If the sample is not in correct dimension, the checker will refer to the person who draws the drawing or executive for confirmation. If the sample is really not right condition, it has to be remake. Sample in good and right condition will be send to customer for approval. If the sample is not approved, the model will be redesign until the customer is satisfied with the sample. If the model still cannot be redesign, the order for this model will be rejected.

After the customer is satisfied and approved the sample, the final specification will be prepared and released. The Engineering Department will fabricate the jigs and fixtures according to the volume of the product. At last, the model is to be mass-produced.

4. Overall Description of the Expert Fault Diagnosis System

There are six development steps in the Rule Based Expert Fault Diagnosis System for Vehicle Air-conditioner Production System such as knowledge acquisition module, determination of user interface, define the knowledge hierarchy, programming, validating and testing of the program. The expert knowledge is heuristic and difficult to gather. All information regarding the product development processes for the vehicle air-conditioners is compiled from the manager, senior engineers, technicians and operators. Figure 2 shows the framework of the proposed system.

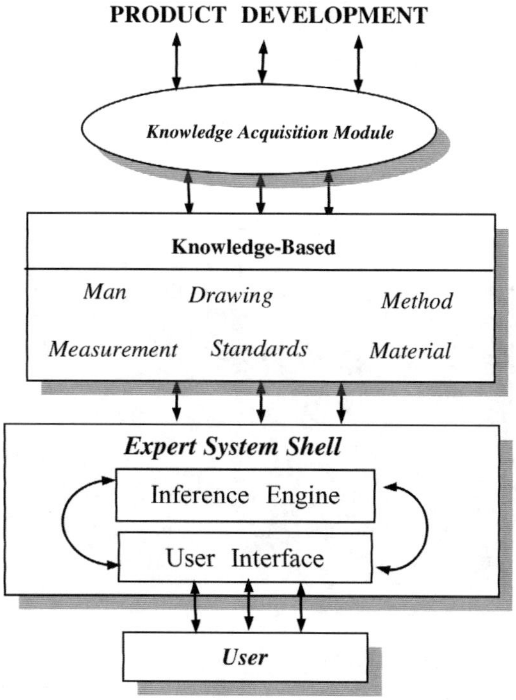

Figure 2. The framework of the developed system.

The knowledge acquisition module is used to store the information about the product development processes. An inference engine will examine the knowledge module. Next, "if-then-rules" will be applied to select the suitable solutions of the problem. The inference engine interacts with knowledge-based in backward chain method to search for right solution. The user interacts with the system

IOP Conf. Series: Materials Science and Engineering **88** (2015) 012046 doi:10.1088/1757-899X/88/1/012046

through user interface as shown in Figure 3 and Figure 4. Figure 5 shows the object browser for original market in the system.

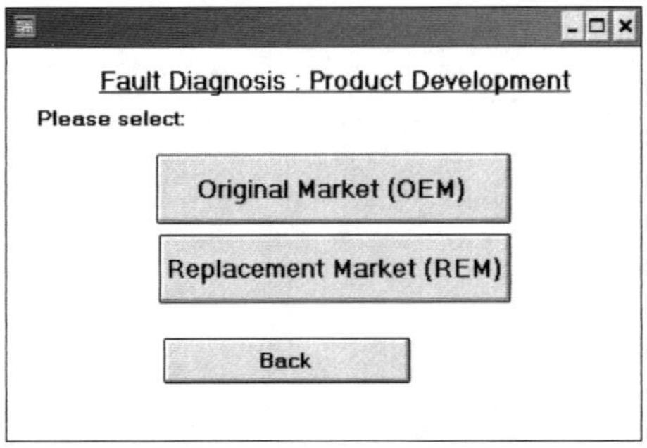

Figure 3. The window of product development.

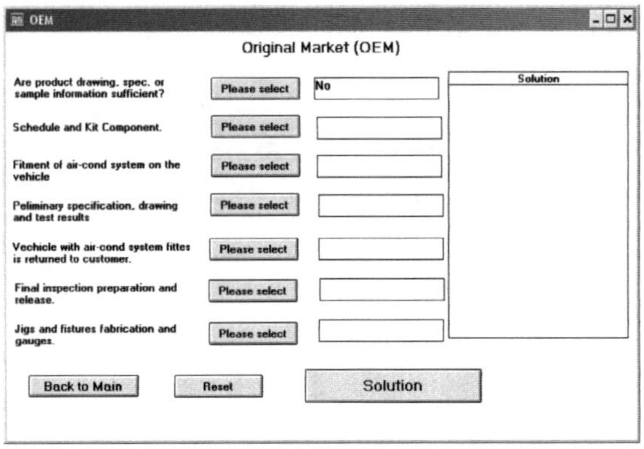

Figure 4. The fault diagnosis window for OEM of product development.

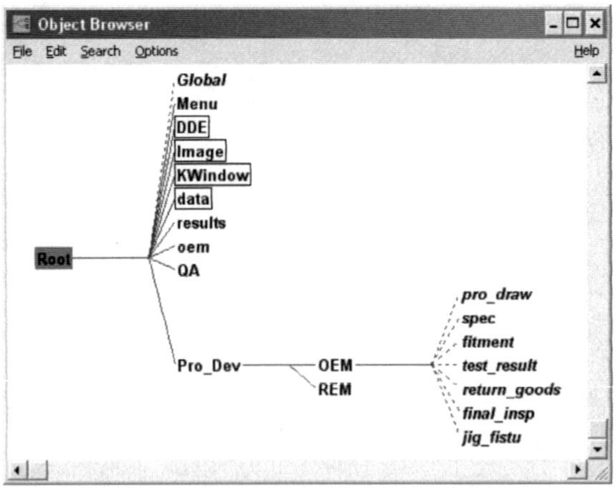

Figure 5. The object browser of the system.

4. Result and Discussion

A prototype expert fault diagnosis system for vehicle air conditioning product development process was developed and tested. The proposed system enables user to diagnose the problem in the development of vehicle air conditioning system. The developed system is able to guide the experienced and non-experienced worker in trouble-shooting the product development process. The database of the proposed system can be changed and upgraded by user easily. Currently, the proposed system interprets 21 rules before giving a solution. Lastly, a few scenarios were used to validate the program for any error. The corrective action will be taken to improve the developed program.

5. Conclusions

The developed rule-based fault diagnosis system for vehicle air-conditioning comprises a product development module. The module includes OEM and REM of the vehicle air-conditioning. In addition, the developed system includes the knowledge-based module and a user interface. The developed system is able to solve the problem of the development process of the vehicle air-conditioning. The system was developed to be flexible and modular-based where the database such as manufacturer, standards, method and materials can be upgraded to make the system more comprehensive.

6. Acknowledgment

Special thanks to the Faculty of Mechanical Engineering, Universiti Teknikal Malaysia Melaka for the support.

7. References

[1] Hetem V 2000 Communication: computer aided engineering in the next millennium *Computer-Aided Design* **32** 389-394

[2] Yang Z Wang B Dong X & Liu H 2012 Expert system of fault diagnosis for gear box in wind turbine *Systems Engineering Procedia* **4** 189-195

[3] Zhou Z Zhuang M Lu X Hu L & Xia G 2012 Design of a real-time fault diagnosis expert system for the EAST cryoplant *Fusion Engineering and Design* **87(12)** 2002-2006

[4] Ma D Liang Y Zhao X Guan R & Shi X 2013 Multi-BP expert system for fault diagnosis of powersystem *Engineering Applications of Artificial Intelligence* **26(3)** 937-944

[5] Yang B Lim D & Tan A C C 2005 VIBEX: an expert system for vibration diagnosis of rotating machinery using decision tress and decision table *Expert System with Applications* **28(4)** 735-742

[6] Qian Y Xu L Li X & Kraslawski A 2008 LUBRES: an expert system development and implementation for real-time fault diagnosis of a lubricating oil refining process *Expert Systems with Applications* **35(3)** 1252-1266

[7] Moorkherjee R & and Bhattacharyya B 2001 Development of an expert system for turning and rotating tool selection in a dynamic environment *Journals of Materials Processing Technology*, **113** 306-311

III Lead ECG Pulse Measurement Sensor

S K Thangaraju , K Munisamy[1]

College of Engineering, Universiti Tenaga Nasional, Jalan IKRAM-UNITEN, 43009, Kajang, Malaysia

Email: savithry@uniten.edu.my

Abstract - Heart rate sensing is very important. Method of measuring heart pulse by using an electrocardiogram (ECG) technique is described. Electrocardiogram is a measurement of the potential difference (the electrical pulse) generated by a cardiac tissue, mainly the heart. This paper also reports the development of a three lead ECG hardware system that would be the basis of developing a more cost efficient, portable and easy to use ECG machine. Einthoven's Three Lead method [1] is used for ECG signal extraction. Using amplifiers such as the instrumentation amplifier AD620BN and the conventional operational amplifier Ua741 that would be used to amplify the ECG signal extracted develop this system. The signal would then be filtered from noise using Butterworth filter techniques to obtain optimum output. Also a right leg guard was implemented as a safety feature to this system. Simulation was carried out for development of the system using P-spice Program.

1. INTRODUCTION

Monitoring of cardiovascular pulse is widely used in health care, sport training; sleep studies and psycho-physiological (polygraph) examinations. Various contact measurement methods have been developed to estimate a subject's cardiovascular pulse. The golden standard for pulse measurement is Electro-Cardio- Graphy (ECG). ECG signal is the measurement of the potential difference of pulse waves that are generated from heart cells called the pacemaker cells. ECG recording requires the use of a signal amplifier and at least three electrodes. ECG readings can be used to monitor the conditions of heart (detect diseases such as Parkinson disease, Myocarditis, etc). There are many type of ECG machine in the market today such as 3 Lead, 12 Lead and 15 Lead ECG machine. However, these machine are expensive, not portable and not user friendly. These ECG machine cost from a minimum price of USD 620 (RM 2200) and are big and bulky therefore not portable. Moreover the 12 lead ECG system usually requires the patient to remove his or her clothes to enable 6 electrode placement on the chest and 3 electrode placement on the limbs, making it difficult an uncomfortable for patients. Hence, this thesis elaborates on the development of three lead ECG pulse measurement system for Electrocardiogram (ECG) signal extraction. The system develop also produces an ECG signal of the same quality however smaller therefore the hardware is more portable than a standard 3 lead ECG machine available in the market today. The Einthoven's 3 Lead Methods is used for ECG signal acquisition and the hardware designed comprises of amplifiers to amplify the signal and filters to remove noise as shown in Figure 1. P-spice simulations are run on all stages of development before any hardware implementation.

[1] To whom any correspondence should be addressed.

IOP Conf. Series: Materials Science and Engineering **88** (2015) 012047 doi:10.1088/1757-899X/88/1/012047

2. METHOD

2.1 Amplification Stage 1

Figure 2 is a circuit diagram used for the 1st stage using an AD620BN instrumentation amplifier A 100mV sinusoidal sign was chosen to be the input voltage, Vin and fed to amplifier. The variable resistor was then adjusted slowly to provide a minimum output voltage, Vol of 10V. The value of the variable resistor for a gain 100 is calculated as below [1]:

$$RV = (R1+R2)/(A1-1)$$

Where R1 and R2 are the internal resistance of the in amp
Assuming R1 = R2 = 24.7kΩ
And choosing A1 = 100
Rv = 500 Ω

The output results are as follows
The gain A1 = Vol / Vin
= 10.78 V / 100.0mV
= 107.8
which is approximate to the minimum desired gain of $A1_{desired} = 100$

2.2 Filter Stage 1

This filter was built using the Sallen Key architecture method[2] in order to have Butterworth 4th order response. In order to achieve the 4th order response, two 2nd order low pass filters were cascaded together. The cut off frequency was chosen to be 0.05 Hz as it the minimum frequency of the ECG signal. To design this filter, considerations on cut off frequency, gain and quality were taken into account. The transfer function of a 2nd order high pass filter is described as below,

$$Vo/Vin = 1/[1-fc^2/f^2 - \sqrt{2}jfc/f]$$

The gain, A_{f1} for the filter is described as,

$$A_{f1} = |(Vo/Vin)|$$
$$= 1/\sqrt{[(1 - fc^2/f^2)^2 + 2(fc2/f2)]}$$

Equation above can further be simplified in term of resistance of R3 and R4,

$$A_{f1} = 1 + (R4/R3)$$

The cut off frequency F_{c1} can be related as per equation below,

$$F_{c1} = 1 / [\, 2\pi\sqrt{(R1*R2*C1*C2)}]$$

The quality, Q of the filter is,

$$Q = \sqrt{(R1*R2*C1*C2)} / [R1*C1 + R2*C1 + R1*C2*(1 - A_{f1})]$$

Thus, the calculated value for 4th Order High Pass Filter parameters are shown in Table 1.

IOP Conf. Series: Materials Science and Engineering **88** (2015) 012047 doi:10.1088/1757-899X/88/1/012047

2.3 Amplification Stage 2

For this stage, an active amplification filter was designed to have a gain of 10 and a cut off frequency of 0.05Hz. the filter was designed using a UA741 operational amplifier. The 2nd Stage Amplifier gain is estimated based on the correlations describe in equations below[2].

$$f_{3dB} = 1/\ 2\ \pi\ R_1C$$

With gain calculation,

$$A2 = R_2/R_1$$

Thus, at $f_{3dB} = 0.05$ Hz, the calculated gain is,

$$A2 = 20$$

2.4 Filtering Stage 2

This filter was built using the Sallen Key architecture method[2] in order to have Butterworth 4th order response. In order to achieve the 4th order response, two 2nd order low pass filters were cascaded together. To design this filter, considerations on cut off frequency, gain and quality were taken into account. The transfer function of a 2nd order low pass filter is described as below,

$$Vo/Vin = 1/[1\text{-}fc^2/f^2 - \sqrt{2}jfc/f]$$

The gain, A_{f1} for the filter is described as,

$$A_{f1} = |(Vo/Vin)|$$
$$= 1/\ \sqrt{[(1 - fc^2/f^2)^2 + 2(fc2/f2)]}$$

Equation above can further be simplified in term of resistance of R3 and R4,

$$A_{f1} = 1 + (R4/R3)$$

The cut off frequency F_{c1} can be related as per equation below,

$$F_{c1} = 1\ /\ [\ 2\pi\sqrt{(R1*R2*C1*C2)}]$$

The quality, Q of the filter is,

$$Q = \sqrt{(R1*R2*C1*C2)}\ /\ [R1*C1 + R2*C1 + R1*C2*(1 - A_{f1})]$$

Thus, the calculated value for 4th Order Low Pass Filter parameters are shown in Table 2.

2.5 Right Leg Guard

Placement of electrodes on the human body is an excellent path for current flow and since the ECG runs on a 15V power supply, the user must be protected by placing a right leg guard in the ECG system. The right leg guard is placed at the input terminal of the instrumentation amplifier and also connected to the right leg of the user. The buffering circuit is chosen is the right leg guard [3]. This is because it is easy to design and low cost. In case of leakage current, the right leg guard will direct the leakage current to the ground and the user would be protected from getting electrocuted. The circuit diagram that could be use s a right leg guard is shown in Figure 3. Furthermore, the right leg guard also improves the impedance of an instrumentation amplifier. This is because the right leg guard reduces capacitive losses that reduce the input impedance of the in-amp. The Right Leg Guard was also being tested using ammeter at output node which gives the reading of

$$Iout = 0\ Amp$$

IOP Conf. Series: Materials Science and Engineering **88** (2015) 012047 doi:10.1088/1757-899X/88/1/012047

This result was matched with the desired current output of 0 Amp.

3. RESULT AND DISCUSSION

The 4th Order ECG System was designed using the 4th order high pass and low pass filters which was integrated with the 1st and 2nd stage amplifiers together with the right leg guard. The designed system was simulated by putting a sinusoidal input of 20 mV 15 Hz into the system. The system then tested with the acquired ECG signal. The output signal then observed to make sure the signal displayed from the ECG system was creating noise. Figure 4 shows the integrated circuit diagram for the 4th Order ECG System.

Figure 5 shows the output result from the oscilloscope output of the integrated 4th Order ECG System. From the output result, we can calculate the heart rate variability in beats per minutes by using Barros and Ohnishi method[4]. They define a compose time series formed by the time difference (in seconds) between two consecutive R-waves as x = [tl, t2,..,tn] it can be written as hrv=[60/t1,60/t2,…,60/tn]. The overall output gain is, A = 929.5.The III lead ECG system design could be used as basis to develop a heart beat detector for monitoring purpose. The 4th order ECG system may not be accurate but may still be used to develop a heart beat detector where the QRS complex is the main concern. Since the QRS complex wave can be seen clearly in the ECG signal produced by the 4th order ECG system, it can be used to calculate the heart beat rate per minute hence be used as a beat detector. The objective to develop a low cost, portable and easy to use III Lead ECG System was successfully accomplished. The 4th Order ECG System output is suitable for application concerning heartbeat detection.

REFERENCES

[1] J. Bhattacarya and P.P Kanjilal, "Assesing determination in ECG signal, "IEEE Trans. Syst. Man. Cybern. Part A, vol. 29, pp. 406-410, July 1999

[2] George Clayton and Steve Winder, Operational Amplifiers, 4th Edition, Philadelphia, PA : W.B Saunders Company, 2000, Chapter 11

[3] Arpad Barna and Dan I. Porat, Operational Amplifiers, 2nd Edition, Boston, MA: Little, Brown, 1989, Chapter 4

[4] Darold Wobschall Circuit Design For Electronic Instrumentation, 2nd Edition, Armonk, NY: Futura, 1995

[5] Charles Kitchin and Lew Counts, A Designer's Guide to Instrumentation Amplifiers, Report

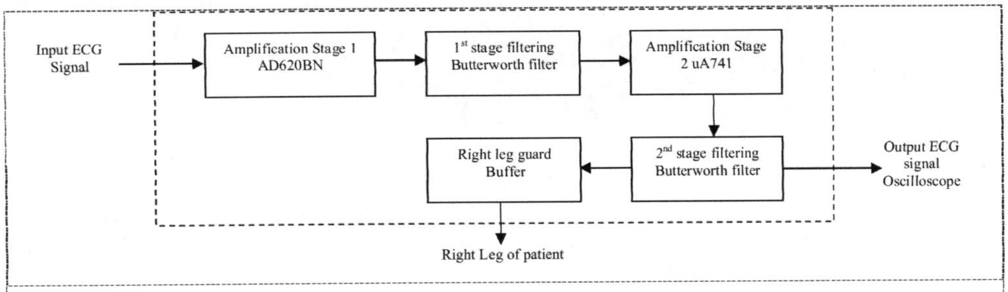

Figure 1. Block Diagram of III Lead ECG Pulse Measurement Sensor.

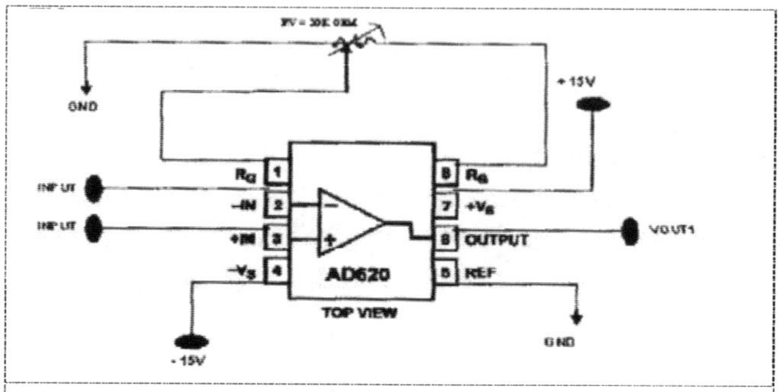

Figure 2. AD620BN Instrumentation Amplifier Circuit Diagram

Table 1: Calculated value of 4th Order High Pass Filter Parameters

	Af1	Q	Fc (Hz)	R1 (kΩ)	R2 (kΩ)	R3 (kΩ)	R1 (kΩ)	C1 (μF)	C2 (μF)
Stage 1	3	0.5412	0.05	5000	10000	1	2	1	1
Stage 2	1	1.3065	0.05	5000	10000	∞	0	1	1

Table 2: Calculated value of 4th Order Low Pass Filter Parameters

	Af2	Q	Fc (Hz)	R1 (kΩ)	R2 (kΩ)	R3 (kΩ)	R1 (kΩ)	C1 (μF)	C2 (μF)
Stage 1	1	0.5412	50	1125	1125	∞	0	1	2
Stage 2	1	1.3065	50	1125	1125	∞	0	1	2

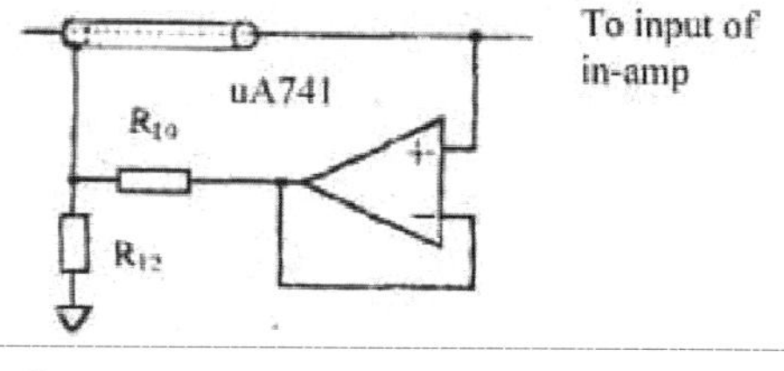

Figure 3. Right Leg Guard Circuit Diagram

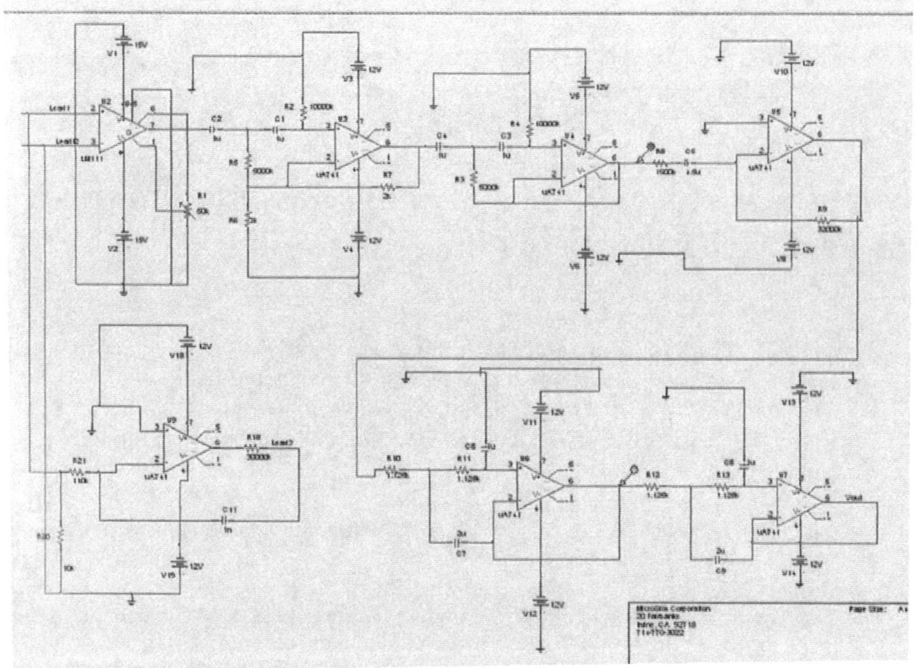

Figure 4. Integrated Circuit Diagram for the 4th Order ECG System

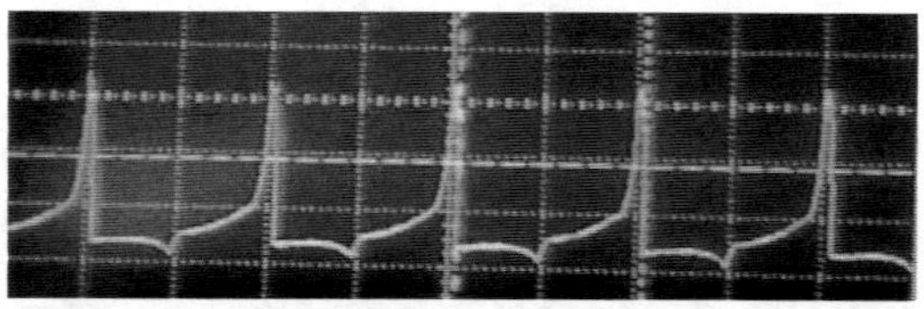

Figure 5. Output Result from Oscilloscope for the 4th Order ECG System

IOP Conf. Series: Materials Science and Engineering **88** (2015) 012048 doi:10.1088/1757-899X/88/1/012048

A study on the surface shape and roughness of aluminum alloy for heat exchanger using ball end milling

E Lee[1],Y Kim[2], H jeong[3] and H Chung[3]
[1]Professor, Department of Korea Polytechnics, Changwon, Korea
[2] President, DAE SUNG AIR TECH. Co., Gimhae, Korea
[3]Professor, Department of Energy and Mechanical Engineering, Gyeongsang National University, Korea

E-mail: wiselej@nate.com, kys@dsat.co.kr, hmjeong@gnu.ac.kr, hschung@gnu.ac.kr

Abstract. Aluminum alloy is a material with a high strength-weight ratio and excellent thermal conductivity. It neither readily corrodes nor quickly weakens at low temperatures, but can be easily recycled. Because of these features, aluminum heat exchangers are widely used in aluminum alloy. In addition, the aluminum alloy used in other areas is expected to gradually increase. As a result, researchers have been continuously studying the cutting patterns of aluminium alloy. However, such studies are fewer than those on the cutting patterns of ordinary steel. Moreover, the research on ball end milling with aluminium alloys has not received much attention. Therefore, in this study, an attempt was made to find the optimal cutting pattern among the seven cutting patterns for the machining of the commonly used aluminum alloy using ball end milling for a heat exchanger. The optimal pattern was found by comparing the different shapes and surface roughness values produced by the seven patterns.

1. Introduction

Aluminum is one of the most abundant metals on Earth. In most cases, it is used in the form of an alloy, which is stronger, except for foil, which is used for wrapping.

Aluminum alloy is used to make heat exchangers such as air conditioners and heaters. Its use is expected to increase in various fields. Thus, studies on the cutting theory relevant to aluminum alloy have been performed.

End milling is the most widely used method of cutting machining. Ball end milling is the most widely used method of finishing a curved surface shape.

In the machining field, demand for free-form surface machining has gradually increased. This demand has led to the abilities to design and machine complicated three-dimensional (3D) shapes, and has driven the development of the CAD/CAM software and machine tool manufacturing techniques. Due to the continuous development of the CAD/CAM software, various cutting pattern methods have been developed, and diverse tool paths can be generated. This study aims to find the optimal cutting pattern for the machining of aluminum alloy using a ball end mill by machining the alloy through the generation of different tool paths based on the seven cutting patterns shown in Figure 2 and by comparing the shapes and surface roughness values of the machined surfaces.

[1] To whom any correspondence should be addressed.

IOP Conf. Series: Materials Science and Engineering **88** (2015) 012048 doi:10.1088/1757-899X/88/1/012048

2. Experimental equipment and method

2.1. Experimental Equipment

2.1.1. Machining Center. The machining center used in this experiment was TNV-40A (Tongil Heavy Industries Co., Ltd.).

2.1.2. Measuring Device . The surface roughness measuring device used in this experiment was model Surfcoder-F3500D (Kosaka Laboratory Corporation).

The shape-measuring device used in this experiment was model Hommel_C8000 (Hommelwerke).

2.2. Experimental tool and material

2.2.1. Experimental tool. The cutting tool used in this experiment was an uncoated general-purpose carbide end mill, which is used to machine free-form surfaces of metal molds.

2.2.2. Experimental material and shape. The specimen used in this experiment had a 70x70x19mm free-form surface shape. In the rough machining, a flat end mill was used, and 0.5 mm was left for finishing. In the finishing, it was machined as shown in Figure 1, using a ball end mill. The chemical composition and mechanical properties of the specimen are summarized in Tables 1 and 2.

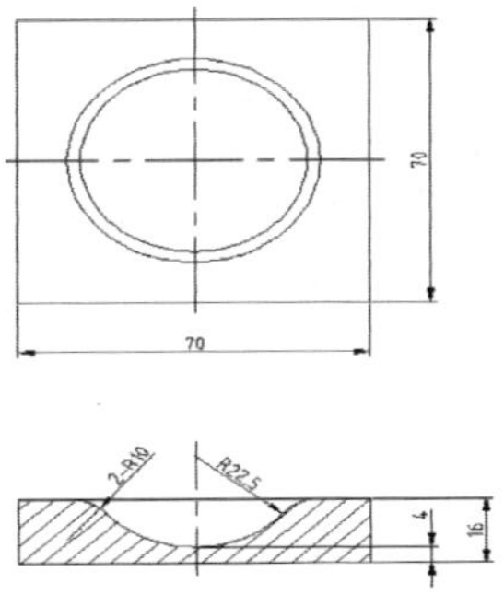

Figure 1. Dimensions of the workpiece

Table 1. Chemical composition

Workpiece material	Chemical composition							
	Cu	Si	Mg	Zn	Mn	Cr	Fe	Al
Al6061	0.15-0.4	0.4-0.8	0.8-1.2	0.25 and below	0.15 and below	0.04-0.35	0.5 and below	The rest

Table 2. Mechanical properties of specimen

Workpiece materia	Tensile strength (N/mm^2)	Yielding strength (N/mm^2)	Elongation (%)	Hardness (HBS 10/500)
Al6061	125	55	25	30

2.3. Experiment method

2.3.1. Specimen Modeling and Tool Path Generation. Before the specimens were machined, 3D modeling was performed using the NX program, which is the CAD/CAM software of Siemens. In the program, the machining conditions were assigned, and tool paths were generated for each cutting pattern. Based on the simulation function of the program, the shapes after the machining were verified in advance, and it was checked if there was a problem. Finally, the NC codes of the rough machining and the finishing were generated.

In the rough machining, all the specimens were identically machined in a '⊏' form along the contour of the machining shape from the outside to the inside. In the finishing, the machining was conducted by generating tool paths using the seven cutting patterns based on down-cutting, which has been verified in a number of surface roughness studies. Figure 2 shows the tool paths and the shapes after the machining.

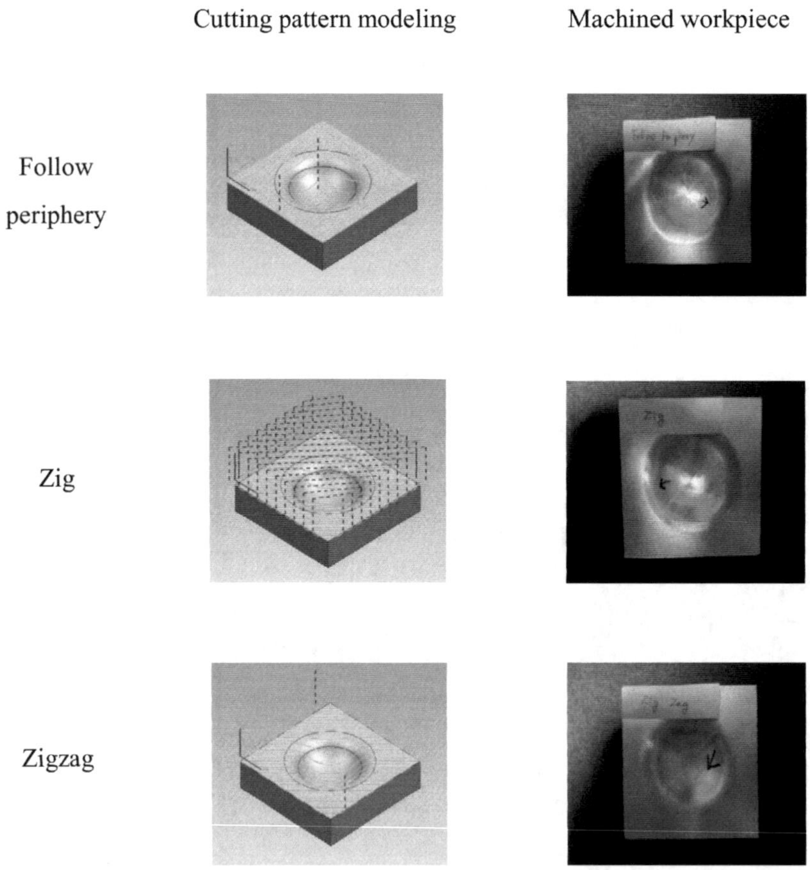

IOP Conf. Series: Materials Science and Engineering **88** (2015) 012048 doi:10.1088/1757-899X/88/1/012048

Concentric zig

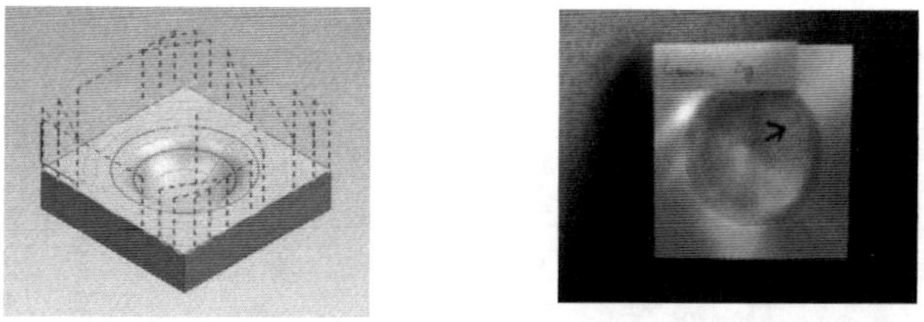

Concentric zigzag

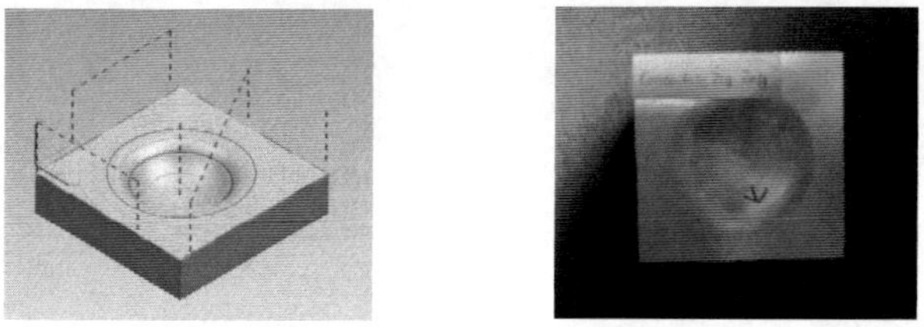

Radial zig

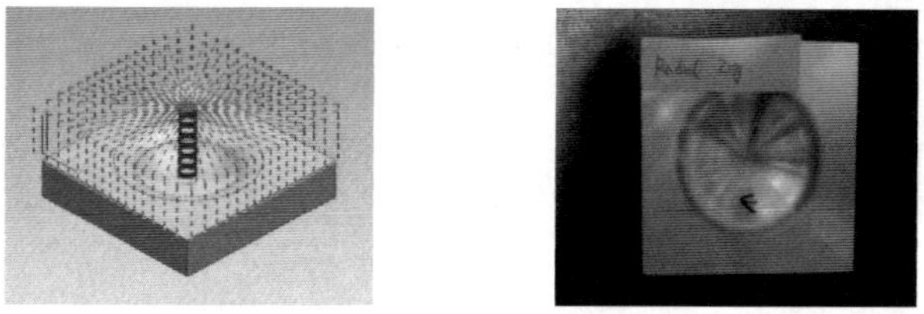

Radial zigzag

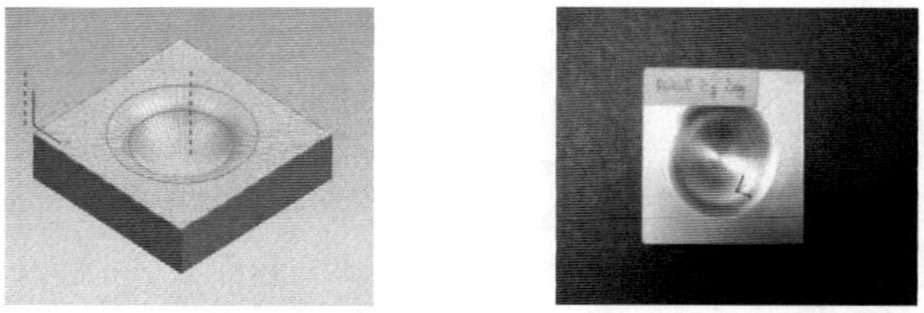

Figure 2. Cutting pattern modeling and machined workpiece

3. Experiment Results and Discussion

3.1. *Surface Roughness and Shape Measurements* The surface roughness was measured at three spots on the flat part of the specimen, among the two parts of the specimen (the flat part and the curved part), and an average value was obtained. Table 3 summarizes the measured surface roughness using the center line average roughness (Ra), the maximum height roughness (Rmax), and the 10-point average roughness (Rz).

To measure the shape, the machined curved part was measured. Table 4 summarizes the values of the measured shape.

Figure 3 shows the graph of the surface roughness data that measured the flat part machined in a radial zig pattern. Figure 4 shows the graph of the shape measurement data that measured the curved part of the material machined in a radial zigzag pattern

Table 3. Surface roughness results of different flat cutting patterns

Cutting pattern	Surface roughness(μm)		
	Ra	Rmax	Rz
Radial zig	1.32	9.30	8.01
Radial zig zag	1.52	11.00	9.06
Concentric zig zag	1.67	14.03	10.44
Concentric zig	1.91	14.42	11.26
Zig	2.19	15.77	12.71
Zig zag	2.20	14.20	11.70
Follow periphery	2.26	16.33	12.90

Table 4. Contour results for different concave cutting patterns

Cutting pattern	Contour(mm)	Error(mm)
	Radius value	
Radial zig zag	22.61	+0.11
Radial zig	22.12	-0.38
Zig	22.11	-0.39
Follow periphery	22.08	-0.42
Concentric zig	22.03	-0.47
Zig zag	21.86	-0.64
Concentric zig zag	21.47	-1.03

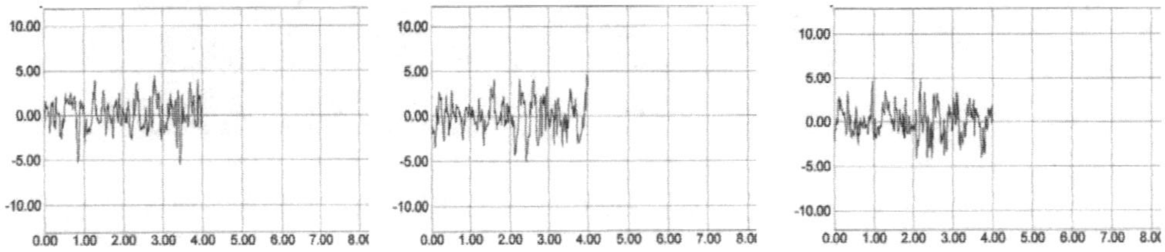

Rmax 9.18(μm) Rmax 8.35(μm) Rmax 10.37(μm)

Rz 7.73(μm) Rz 7.50(μm) Rz 8.81(μm)

Ra 1.34(μm) Ra 1.35(μm) Ra 1.27(μm)

Figure 3. Measured surface roughness characteristics of cutting pattern. (Radial zig).

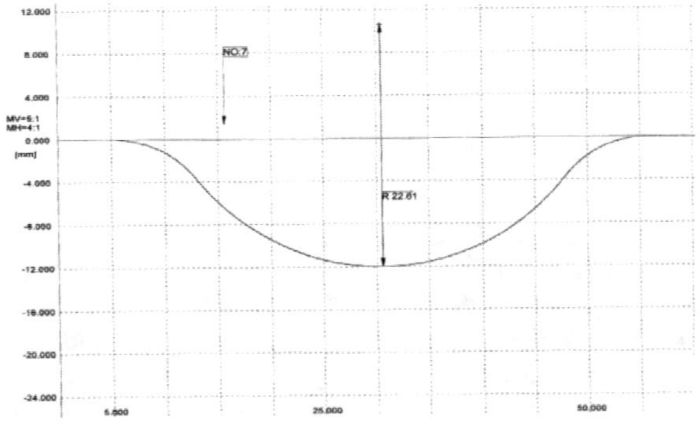

Radius value	22.61(mm)
Error	+0.11(mm)

Figure 4. Measured contour characteristics of cutting pattern. (Radial zig zag)

3.2. *Discussion*

In the case of the machining using a ball end mill, the tool path is determined using two criteria. One is the increase in the machining efficiency, and the other is the increase in the machining precision. This study considered the machining precision. The cutting patterns that had outstanding surface roughness were the radial zig and the radial zigzag. For the radial zig, Ra = 1.32 μm on the flat part; and for the radial zigzag, Ra = 1.52 μm. Thus, their surface roughness values were superior to those of the other cutting patterns.

The cutting patterns that had an outstanding shape were the radial zigzag and the radial zig. For the radial zigzag, the error was +0.11 mm in the curved part; and for the radial zig, -0.38 mm. Thus, their shape values were superior to those of the other cutting patterns.

The question of which cutting pattern is efficient needs to be answered based on the condition of the production requirements, considering the entire process. Studies on the development of a cutting pattern that can simultaneously satisfy the conflicting requirements of machining efficiency and machining precision are expected to be continuously performed

4. Conclusion

In this study, an experiment was conducted to determine the optimal cutting pattern for the machining of aluminum alloy using a ball end mill, by machining the alloy through the generation of tool paths based on different cutting patterns and by comparing the shapes and surface roughness values of the machined surfaces. The following conclusions are drawn.

1) The measurement of the flat-part surface roughness for each cutting pattern indicated that the radial zig cutting pattern, which applied the zig pattern as a linear cutting pattern that extended from an optimal center point, showed the best machined surface, with a center line average roughness (Ra) of 1.32 μm.

2) The measurement of the curved part shape for each cutting pattern indicated that the radial zigzag cutting pattern, which applied the zigzag pattern as a linear cutting pattern that extended from an optimal center point, showed the best shape, with a concave part shape error of +0.11 mm

5. References

[1] Hook S 1989 *High speed cutting in die and mold manufacture* proceedings of the Machine Tool Technologies Assciation Conference (Warwick/Loughborough/Manchester, UK) ISBN_0_M907348_04_1

[2] Aoyama H, Kishinami T and Saito K 1986 *Development of the elliptic ball-endmill* bulletin of JSPE Vol 20 No 4 pp291

[3] Jo, C.Y. *Hemispheric Cutting-edge Optimization of a Ball End Mill for High-speed Machining* Graduate School, Chonbuk National University, Ph.D. Thesis, 2009, pp. 11-13.

[4] Lee, W.H. *A Study on the Ball End Mill Cutting Force for Inclined Surface Angles Using CAM/FEM* Graduate School, Dong-A University, M.S. Thesis, 2012, pp. 8-10.

[5] Park, Y.J. *Evaluation of the Machinability of High-speed Ball End Milling on a Free-form Surface* Journal of the Korean Society of Mechanical Engineers, pp. 66-70 (1999).

[6] Kim, J.S. *Optimization of the Cutting Condition in the High-speed Machining of a Sculptured Surface Using a Ball End Mill* Graduate School, Pusan National University, M.S. Thesis, 2000, pp. 21.

[7] Rundaoz *Study on Non-linear Regression Analysis of the Working Conditions of Aluminium Alloy Using End Milling* Graduate School, Dong-A University, M.S. Thesis, 2011, pp. 20-39.

IOP Conf. Series: Materials Science and Engineering **88** (2015) 012049 doi:10.1088/1757-899X/88/1/012049

Simulation and Experimental Study on Effect of Phase Change Material Thickness to Reduce Temperature of Photovoltaic Panel

Y S Indartono[1,a]; S D Prakoso[1,b]; A Suwono [1,c]; I N Zaini [1,d]; B Fernaldi [1,e]
[1] Faculty of Mechanical and Aerospace Engineering, Institut Teknologi Bandung, Bandung, Indonesia
[1] Research Center on New & Renewable Energy, Institut Teknologi Bandung, Bandung, Indonesia

E-mail: [a]ysindartono@ftmd.itb.ac.id, [b]satrio.dwiprakoso@yahoo.com, [c]aryadi@termo.pauir.itb.ac.id, [d]ilmanuran@students.itb.ac.id, [e]billy.fernaldi@gmail.com

Abstract. Solar energy is promising renewable energy which can be applied in Indonesia. Average solar radiation in the country is 4.8 kWh/day/m^2. Weakness of silicon-based photovoltaic (PV) is efficiency reduction caused by temperature increase. Many attempts have been done to reduce PV temperature. In previous study, palm oil, which is widely available in Indonesia, is suitable to be used as phase change material (PCM) to reduce PV temperature. In this study, thickness of aluminium rectangular-tube containing phase change material oil is varied. The tube is placed at back part of PV. Numerical and experimental study was done to evaluate the effect of tube thickness to the temperature reduction of the PV. Variation of tube thickness used in the experiment is 50.8mm, 76.2 mm, 101.6 mm. Both studies show that increase of PCM thickness reduces PV temperature. Higher PCM thickness cause large reduction on PV temperature. Simulation result shows there is an optimum thickness of the PCM which is applied to the PV.

1. Introduction

Energy plays important role to human's life. Nowadays, fossil fuel is still dominant energy source in the world. However, fossil fuels such as oil, natural gas, and coal are non-renewable energy sources that its amount will always be depleted by use. Hence, the use of renewable energy sources as a substitute for non-renewable energy sources needs to be escalated. Numerous efforts have been carried out to raise the utilization of renewable energy in Indonesia. This effort is also supported by the government to set the application of new & renewable energy in 2025 as big as 17%.[1]

One of the most promising renewable energy sources is solar energy, based on its daily availability. One way to convert solar energy into electricity is by using photovoltaic (PV) cells. The use of PV modules in the world currently has reached 70 GW capacity with a significant grow in the 2010-2011 period amounted to 30 GW. Meanwhile, Germany is the largest user of solar PV technology.

The efficiency of solar modules that are commonly used today is about 16% [2], which is still relatively low. It means that only 16% of the total solar energy that can be converted into electrical energy, though the rest converted to heat and miscellaneous losses.

One of the aspects that closely related with efficiency is the PV module operating temperature. Generally, the rise in temperature will result in a decrease of solar module power output. Therefore, to improve the efficiency of solar PV system, a cooling system is necessary to be installed on the PV module in order to inhibit the rise in temperature at the PV panel.

Phase Change Material (PCM) has the ability as heat storage due to its latent heat capacity. PCM can absorb large amounts of heat in the form of latent heat at its melting temperature. By putting the

[1] To whom any correspondence should be addressed.

PCM in the containment and mounted at the bottom surface of the PV module, this PCM will maintain the temperature at the wall as low as possible. Some examples of this PCM are paraffin, coconut oil, palm oil, and vaseline album [3].

In this study a simulation by using Computational Fluid Dynamic (CFD) (using ANSYS FLUENT package program) is developed in order to examine the temperature profile at the PV wall and to select the suitable PCM substance for PV application, especially for Indonesia condition. Moreover, an experiment is conducted to find the optimum thickness of PCM to inhibit the temperature rise on the PV surface and also to analyse the correlation between temperature, output power, and efficiency by mounting the aluminium tubes filled with PCM at the back surface of the module.

2. Methodology

A simulation and experimental set up was designed to investigate the best PCM thickness in order to reduce temperature at the surface of the solar panel.

2.1 Phase Change Material (PCM)

Phase change material (PCM) is a substance that can undergo solid-solid phase change, liquid-gas and solid-liquid. Solid-liquid PCM type is the most widely used both in research and everyday use applications [3]. PCM absorbs large amounts of energy as latent heat at the phase transition temperature, thus the PCM can be used as passive heat storage. Solid-liquid PCM type has been used as a temperature control on a wide variety of applications. From the few studies that have been done, PCM proved having a significant influence in an attempt to decrease the temperature of the various electronic components and buildings. [3] PCM used in this study is Crude Palm Oil (CPO) and coconut oil.

2.2 Simulation of PV/PCM with Computational Fluid Dynamics (CFD)

Simulation by using ANSYS FLUENT was done to evaluate the influence of PCM thickness to the PV surface temperature. Geometry modelling for the simulation is shown in Figure 1.

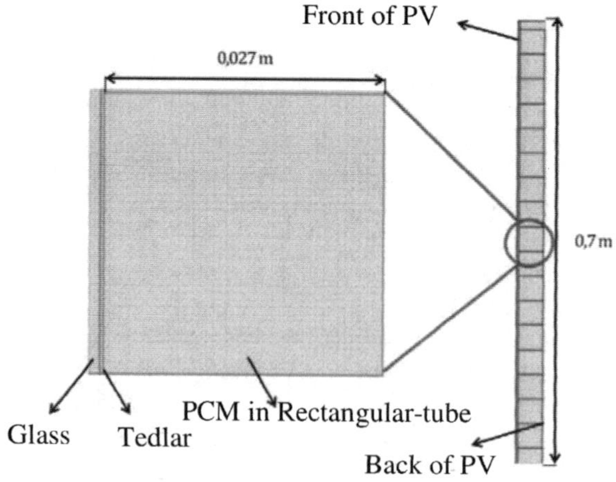

Figure 1. Geometry modelling of aluminium rectangular-tube on back side of PV

While setting of boundary condition for the simulation can be seen in Table 1.

Table 1. Boundary condition used in the simulation

No.	Area	Boundary condition
1	Front surface of PV	*wall* – radiation & convection
2	Side wall of PV	*wall* - adiabatic
3	Back surface of PV, PV material interface, interface between rectangular-tubes, contact between rectangular-tube and back surface of PV	*wall - coupled*
4	Back surface of rectangular-tube	*wall* – constant temperature*

*result of simulation would be compared with experiment of 'PV on roof'. Therefore back surface of rectangular tube is the same with roof temperature (constant).

Heat transfer on PV module

The temperature on the surface of the solar module can be determined by considering the case of heat transfer that occurs between the solar cell modules with the surrounding environment. The three forms of heat transfer that appears in the solar cell module are conduction, convection, and radiation.

- *Convection*

Total convection heat transfer in solar cell modules is formulated with:

$$q_{conv} = h_c \cdot A \cdot (T_{module} - T_{surrounding}) \quad (1)$$

In the equation, T_{module} express the temperature at module surface and $T_{surrounding}$ is the ambient temperature. Convective heat transfer coefficient (CHTC), h depends on the physical conditions around the module. Shao et al. (2009) suggested CHTC for PV panel as follows: [4]

$$h = 6{,}91v + 3{,}9 \quad (2)$$

v in Equation (2) is air velocity.

- Radiation

General form of the equation to express the radiation heat transfer is shown below

$$q_{rad} = \varepsilon\sigma A(T_s{}^4 - T_{module}{}^4) \quad (3)$$

Where q_{rad} is the solar irradiation on the surface of the module (I_{rad}), ε is the emissivity of the module material, σ is the Boltzmann constant (5.67 x 10^{-8} W/m^2.K^4), A is the module area, and T_s express the temperature of radiation source [5].

- Conduction

At steady state condition, heat flux of conduction can be expressed as:

$$q_x'' = -k\frac{\Delta T}{L} \quad (4)$$

In above equation, k is the thermal conductivity of the material, L is the length between sides that has different temperature, and ΔT states the temperature difference between two sides that is being considered [6].

2.3 Experimental set-up with different PCM thicknesses

Based on the simulation result of the correlation between PCM thicknesses with the decrease of temperature, it is appealing to compare this simulation with the experimental result. Therefore, an experimental is set up to analyse PCM thicknesses and PV temperature reduction [7]. Moreover, the influence of CPO as the PCM to the electrical efficiency of the solar module can also be determined.

The experimental set-up is given in Figure 2. The system is installed on the roof top of PAU building at Institut Teknologi Bandung (ITB). Two 10 Watt peak mono crystalline solar modules with dimensions of 39.6 cm x 28.9 cm x 2.3 cm are used in this experiment. The back side of PV panels were directly in contact with the roof top. Therefore, there is no convection at bottom surface of the panels. This situation mimics Building Integrated PV structure on roof. The electricity produced by the solar modules is absorbed by the 10W 27Ω resistors, while the voltage and current are recorded by using multi meter. Five thermocouples connected to the OM-DAQ-USB-2400 data logger are located at top and bottom of both PV panels, and another one recorded ambient temperature every minute. To measure the insolation (incident of solar irradiation), a Lutron solar meter is installed near the PV module location. PV with CPO called as PV/PCM module, while the PV without CPO / normal PV act as a reference.

Table 1 shows the effect of different PCM thicknesses to decrease the operating temperature in the solar module. PCM thickness relation with the operating temperature of the solar modules can be seen in the Figure 4.

Figure 2. Experimental set-up with two 10 Wp solar panels.

3. Result and Analysis

3.1 Simulation results

- The difference between coconut oil and CPO as PCM

Figure 3 shows the curve of the front wall temperature rise of normal modules, modules with coconut oil, and modules with palm oil (CPO). At the beginning, the normal module temperature has increased sharply, while the module with palm oil increases faster than the module with coconut oil. But at 27°C, coconut oil is already melted, and it causes no inhibition of the temperature rise [8]. With the rate of temperature increase is relatively constant, the temperature of coconut oil would exceed the module with palm oil. Therefore, it can be concluded that palm oil is more appropriately used as a phase change material for solar modules operated in Bandung, Indonesia with ambient temperature around 27-30°C.

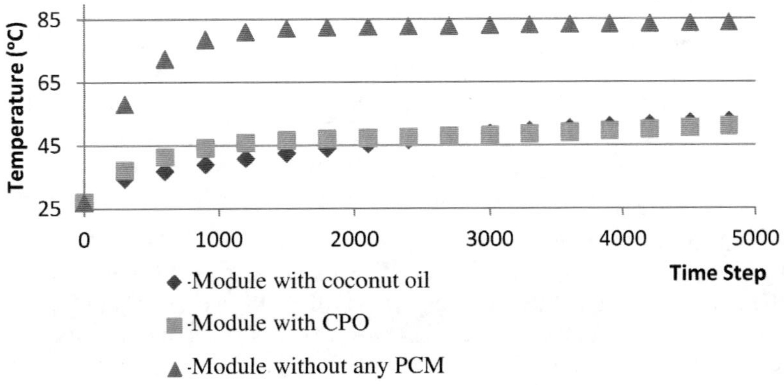

Figure 3. Front wall temperature curves with two type of PCM.

- The impact of PCM thickness

A simulation was then performed to determine the effect of the PCM thickness and to find the optimum PCM volume. Simulation was done by making several models of solar module systems with various PCM thicknesses, i.e. 40, 60, 80, and 90 mm. Result of this simulation is shown in Figure 4.

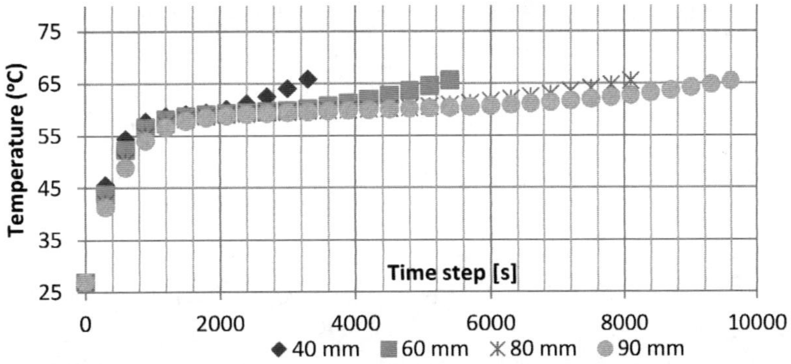

Figure 4. Front wall temperature of PV with various PCM thicknesses

In the temperature curves, it appears that all four modules with different PCM thicknesses have the similar temperature-suppression characteristic. The slightly different temperature-profile was caused by thicker/greater volume of PCM used; while greater volume of PCM resulted longer time it takes to melt the entire PCM. From the Figure 4, temperature-profile of PCM with 80 mm and 90 mm thickness is not significantly different. Therefore, it can be concluded that the optimum thickness of PCM is 80 mm.

3.2 Experiment Result

Table 2 shows temperature difference between reference PV (PV without PCM) and PV/PCM for various PCM thicknesses.

IOP Conf. Series: Materials Science and Engineering **88** (2015) 012049 doi:10.1088/1757-899X/88/1/012049

Table 2. PV/PCM and PV/reference temperature difference comparison.

PCM thickness [mm]	Front surface temperature difference		Back surface temperature difference	
	Maximum (°C)	Average (°C)	Maximum (°C)	Average (°C)
50.8	4,7	2,4	6,0	3,6
76.2	6,5	3,8	8,9	5,3
101.6	12,0	8,3	14,8	10,9

The thicker the PCM is mounted on the solar module, the greater the drop in temperature experienced by the solar modules, especially at the back side of the solar module. This is due to the thickness of the PCM proportional to the total volume of PCM is used. If the PCM volume is larger, then it needs more time to melt all the PCM. Moreover, it will cause the temperature of the PV/PCM panel may rise slowly.

Comparison of the electrical power generated by the PV/PCM with 3 types of PCM thickness variation can be seen in Table 3 below. It can be seen that the maximum power generated by the PV/PCM increases with higher PCM thickness attached to the solar module.

Table 3. PV/PCM and PV/reference power comparison.

PCM thickness [mm]	Power difference between PV/PCM and PV/reference [W]	The increase of average power (%)
50.8	0,9	18,8
76.2	1,1	20,8
101.6	1,3	23,8

Comparison of the PV/PCM efficiency at various PCM thicknesses is described in Table 4 below. It can be concluded that the maximum electrical efficiency of PV/PCM increases with the addition of PCM thickness in solar module.

Table 4. PV/PCM and PV/reference comparison in electrical efficiency.

PCM thickness [mm]	Efficiency difference between PV/PCM and PV/reference [%]	The increase of average efficiency [%]
50.8	1,7	24,1
76.2	1,9	26,2
101.6	2,1	29,3

4 Conclusion

Both from the simulation and experiment, it can be concluded that PCM can reduce the average temperature of PV module. Thickness of the PCM greatly affects the inhibition of temperature rise of the PV surface. From simulation result, the optimum thickness of the PCM is at 80 mm, and Crude Palm Oil (CPO) has better performance than coconut oil in ambient temperature around 27-30°C. Meanwhile from experiment, the biggest temperature difference between reference PV and PV/PCM occurred at 102 mm PCM thickness, i.e. 9.6°C. With 102 mm PCM thickness, output power and efficiency of the PV are higher than the reference, i.e. 23.8% and 2.1%, respectively.

5. Acknowledgement

The authors appreciate full support from Research Center on New & Renewable Energy, Institut Teknologi Bandung (ITB) Indonesia, to conduct this study.

6. References

[1] REN21, *Renewables 2012 Global Status Report*, REN21 Secretariat, Paris, 2012.

[2] J G Ingersoll, *Simplified calculation of solar cell temperatures in terrestrial photovoltaic arrays, ASME J. Solar Energy Eng.* **108** (1986) 95–101.

[3] D Wheelan, *The Use of Phase Change Materials in Building Temperature Regulation Applied Post-Construction*, Thesis, Trinity College, Dublin, Ireland, 2011.

[4] S Armstrong, and W G Hurley, A Thermal Model for Photovoltaic Modules Under Varying Atmospheric Conditions, *Applied Thermal Engineering*, **30** : 1488-1495, 2010.

[5] W T Matthew, W R Santee, L G Berglund, *Solar Load Inputs for USARIEM Thermal Strain Models and The Solar Radiation-Sensitive Components of The WBGT Index*, USAREIM Technical Report, U.S. Army Research Institute of Environmental Medicine, 2001.

[6] F P Incropera, D P Dewitt, T L Bergman, A S Lavine, *Introduction to Heat Transfer*, 5th edition, John Wiley & Sons, New York, 2005.

[7] B Fernaldi., Y S Indartono, A Suwono, *Variation of Phase Change Material Thickness to Prevent Temperature Increase of Photo Voltaic Panel* (*Variasi Ketebalan Bahan Berubah Fasa (Minyak Kelapa Sawit) Untuk Mengurangi Kenaikan Temperatur Dinding Panel Surya)*, Final Work for Undergraduate Degree, Mechanical Engineering Department, Faculty of Mechanical and Aerospace Engineering, Institut Teknologi Bandung, Bandung 2013

[8] I N.Zaini., Y S Indartono, A Suwono, *Simulation on Application of Phase Change Material to Prevent Temperature Increase of Photo Voltaic Panel* (*Simulasi Penggunaan Bahan Berubah Fasa Untuk Menghambat Kenaikan Temperatur Dinding Panel Surya)*, Final Work for Undergraduate Degree, Mechanical Engineering Department, Faculty of Mechanical and Aerospace Engineering, Institut Teknologi Bandung, Bandung 2013.

[9] Y S.Indartono, A Suwono, F Y Pratama, *Improving Photovoltaics Performance by Using Yellow Petroleum Jelly as Phase Change Material*, Proceedings of International Conference on Fluid and Thermal Energy Conversion 2013, 8-11 November, Semarang, Indonesia, 2013.

An experimental study of thermal performance of GN and MWCNTs-based aqueous nanofluids with surfactants SDS and SDBS

A K M Mahmudul Haque[1,5], Tae-Jin Lee[2], Hyo-Min Jeong[3] and Han-Shik Chung[4]

[1]Department of Ocean System Engineering, Gyeongsang National University, Cheondaegukchi- Gil 38, Tongyeong, Gyeongnam 650-160, South Korea
[2]Department of Energy and Mechanical Engineering, Gyeongsang National University, Cheondaegukchi-Gil 38, Tongyeong-si 650-160, Gyeongsangnam-do, Korea
[34] Department of Energy and Mechanical Engineering, Institute of Marine Industry, Gyeongsang National University, 445 Inpyeong Dong, Tongyeong-si 650-160, Gyeongsangnam-do, Korea

E-mail: [1]sana06eee@gmail.com,[2]hansumguy1@nate.com,[3]hmjeong@gnu.ac.kr,[4]hschung@gnu.ac.kr

Abstract: High thermal conductivity enhancement of nanofluid is a promising topic for the recent research fields. And in this regard, GN and MWCNTs based nanofluids with their outstanding properties are examined vastly. Beside this, SDBS and SDS have been concerned for composing better nanofluids. This paper tries to suggest not a solution but a solution approach and deduce a new conclusion by testing thermal conductivity and heat transfer coefficient enhancement ratio of nanofluids with surfactants SDS and SDBS.

1. Introduction

The researches of nano-science filed demonstrated the advantages of nanofluids compared with those fluids containing millimeter or micrometer size particles [1]. Nanofluid with a high thermal conductivity enhancement may be potentially applicable in heat sink applications as coolant [2]. And the concept of naofluid was first introduced by Choi [3] who quantitatively analyzed some potential benefits of nanofluids [4]. Within last few years, many research projects were conducted in order to produce more stable suspensions [1], [5] with well dispersion as well as well homogenization. Some surfactants, such as SDS, SDBS, and so on, are under experiment for using as dispersant for these nanofluids [6] as well as PCA in grinding process [7]. Ultrasonication is used for dispersing nanoparticles in base fluids and grinding method is applied for increasing specific surface area [8] which helps to decrease the sphericity of particles and as a result of this, thermal conductivity is increased [4].

Again, transparent conductive films (TCFs), widely used in transistors [9] as well as in solar cells, are fabricated by indium tin oxide [10]. But the rising cost of indium, high temperature processing in production and brittleness of ITO [11], [12] have introduced some emerging alternatives, such as CNTs [13], GN [14], metal or metal nanowires [15] and hybrids of these [16].

Graphene (GN) with two dimensional extended honeycomb network of sp^2 hybridized carbon atoms [17], high electron mobility [18], excellent mechanical, chemical and thermal properties [19] posses twice specific surface area compared with SWCNTs [11]. However, CNTs(carbon nano tubes) carry some properties such as one of the lightest [20], strongest [21], stiffest [21], electrically [22] and thermally [23] conductive nanoparticles. But, graphene with insolubility and the intrinsic tendency to

[5] **Corresponding author:** email: sana06eee@gmail.com, Tel:(+82)10-4222-1989

agglomerate [24] as well as MWCNTs with insolubility in water due to hydrophobic surface [8] are under research with great challenge to overcome these major obstacles.

Although, ultrasonication is mainly used to disperse nanoparticles in base fluid, there is still agglomeration problem after ultrasoniction (Section 3.1). And with the increasing rpm, there is adhere of powder, especially GN, to grinding media (vial and balls) which increases in grinding process (Section 3.2). This paper chooses wet grinding and ultrasonication in order to minimize these problems during making nanofluid. The reason for wet grinding is to decrease the sticking problem and the possibility of decomposition of surfactants which produces contamination. Again, surfactant controls the welding and fracturing [7] during grinding process and during ultrasonication, it is used as dispersant [1]. So, in order to get the both advantages, surfactants are added with nanoparticles from starting of the process.

In this paper, we do not suggest a solution but a solution approach and try to deduce a new conclusion by giving evidence. This paper tries to show what happens during preparing nanofluid during ultrsonication as well as the effect of SDS and SDBS in making nanofluids (Section 3.1) and in grinding process (Section 3.2). In this paper, the curves of thermal conductivity are analyzed by percent enhancement (Section3.4) and heat transfer coefficient enhancement ratio (Section 3.5) is done for better decision making. Here, it should be noted that the results of this paper are only for comparison and do not suggest the optimum condition for this process.

2. Experimental Setup

2.1 Materials

Raw MWCNTs with a ~20nm diameter, ~5µm length (Carbon Nanomaterial Technology Co., Ltd, South Korea) and graphene nanopowder with 8nm (average flake thickness) flakes, average particle size ~550 nm, specific surface area 100 m^2/g, and 99.9% purity (graphene supermarket) are used in this experimental study. Sodium dodecyl benzene sulfonate (SDBS, $C_{18} H_{29} NaO_3 S$) with hard type, 348.48 molecular weight (Tokyo Chemical Industry Co., Ltd) and Sodium dodecyl sulfate (SDS, $CH_3 (CH_2)_{11} OSO_3 Na$) with 288.38 molecular weight (Junsei Chemical Co., Ltd) surfactants are used as dispersant as well as controller of wielding and fracturing during grinding process. Distilled water (DW) is used as base fluid for making nanofluids.

2.2 Grinding process

A planetary ball mill machine (HPM-700) (Haji Engineering, Korea) is used for grinding process where MWCNT and GN individually as well as with different surfactants and with various weight ratios are ground for one hour with 500rpm rotation speed.

2.3 Ultrasonication

Ultrasonic mixing system (Model 1510E-DTH, 47 kHz is used for making nanofluid of MWCNT and GN in a base fluid distilled water. At first, 5 minute DEGAS then 20 minutes or 40 minutes SONIC time is used for making nanofluid.

In 50ml distilled water surfactant is added and Ultrasonication is performed for 5 minutes. Then, nanoparticles are dissolved in this base fluid and 20 minutes (for sample 7 also 40 minutes) ultrasonication is done for making nanofluid. The surfactants are SDS and SDBS which are added separately. And the nanoparticles are GN and MWCNTs.

2.4 Nanofluids preparation by wet grinding as well as Ultrasonication

By using the planetary ball mill machine and Ultrasonic mixing system as mentioned in section 2.2 and 2.3 respectively, the nanofluids of certain weight ratios (mentioned in section 3) are made. At first, nanoparticles and surfactant are hand mixed and then ground by planetary ball mill machine with 500 rpm for 1 hour. After that, distilled water is added and Ultrasonication is performed for 5 minutes and

IOP Conf. Series: Materials Science and Engineering **88** (2015) 012050 doi:10.1088/1757-899X/88/1/012050

20 or 40 minutes for degassing and ultrasonication respectively. Surfactants are added individually and same is for nanoparticles.

2.5 *Testing apparatus*

In order to get continuous determination of the thermal conductivity as well as heat transfer coefficient enhancement ratio, the LAMDA system measuring instrument is used. Precisely calculation of the thermal conductivity by a theoretical approach is almost impossible [4], so experimental approach is mandatory. And for this, the in-stationary transient hot wire method i.e. LAMDA system is used. At first, hot wire apparatus is calibrated by measuring distilled water (DW) which has known thermal conductivity. The difference between the standard value and the experimental value has been shown in fig.1. And the error is within 1.5% with respect to the standard value.

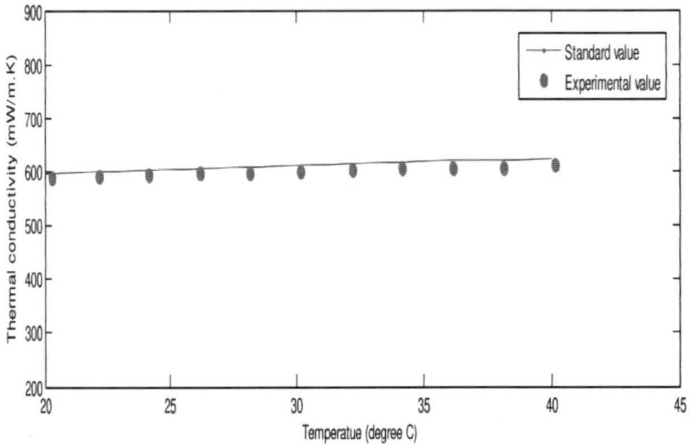

Figure1. Graph of the standard value and the experimental value of distilled water.

3. Results and discussion

3.1 *Result of Ultrasonication*

From fig. 2 and fig. 3, it can be easily seen that after four days of settlement, low added SDBS to GN solutions have shown better dispersed than other weight ratios. Between 2/1 and 3/1 ratios of GN/SDBS solutions, later one is better.

Figure 2. Ultrasonication of GN and SDBS with different weight ratios a) 1/3 b) 3/1 c) 1/1 d) 1/2 e) 2/1

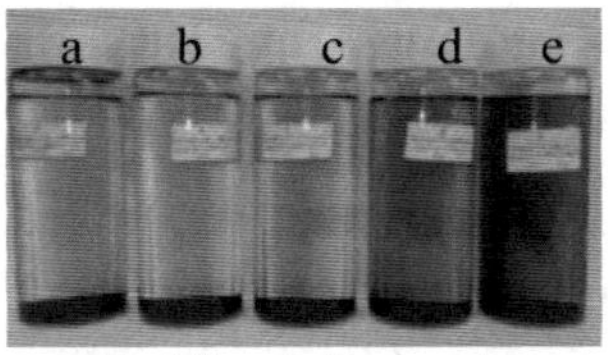

Figure 3. After 4 day settlement of Ultrasonication of GN and SDBS with different weight ratios a) 1/3 b) 1/2 c) 1/1 d) 2/1 e) 3/1

By comparing between fig. 4 and fig. 5, it has been seen that, after 3 day settlement, the three samples (1/1, 2/1 and 3/1ratios) of GN/SDS solution are better than the other two solutions of 1/2 and 1/3 ratios. Among 1/1, 2/1 and 3/1 ratios, 2/1 as well as 3/1 have been more dispersed.

Figure4. Ultrasonication of GN and SDS with different weight ratios a) 1/3 b) 3/1 c) 1/1 d) 1/2 e) 2/1

Figure5. After 3 day settlement of Ultrasonication of GN and SDS with different weight ratios a) 1/3 b) 1/2 c) 1/1 d) 2/1 e) 3/1

3.2 Result of grinding

Fig. 6 has illustrated the adherence problem after grinding of MWCNT as well as GN nanoparticles. This grinding process has been done at 500 rpm. After this experiment, it has seen that the adherence problem of ground GN is more than that of MWCNT. But, by wet grinding this problem has been somewhat reduced.

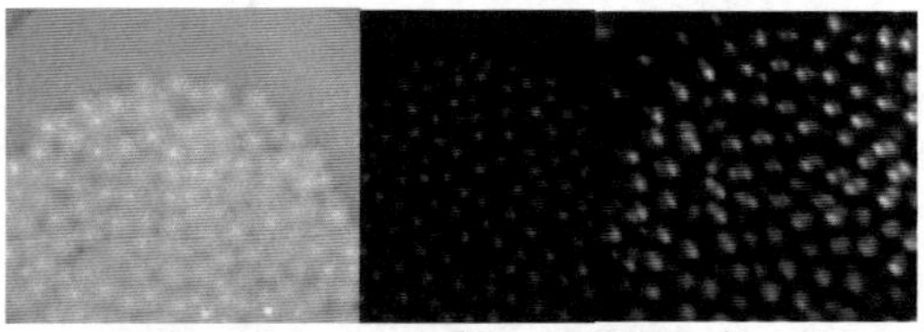

a) b) c)

Figure6. Grinding container and ball a) before grinding b) dry grinding with MWCNT c) dry grinding with GN

3.3 Result of wet grinding and ultrasonication with surfactants

It has been seen that grinding of MWCNTs-surfactant and GN-surfactant have no adhere problem and their nanofluids have no agglomeration (the figures have not been shown).

3.4 Thermal conductivity

The effect of nano particles on the thermal conductivity can be explained from the two aspects: the particles increase the thermal conductivity of nanofluid and the chaotic movement of the particles strengthens energy transport process [4]. The thermal conductivity which is the property of a material to conduct heat has been measured by LAMDA system and the related graph has been shown in fig. 7. It has been seen in fig. 7 that both MWCNTs-SDS and MWCNTs-SDBS have increasing trends and similar to [22], later one has shown better thermal conductivity though in our experiment and both trends are bellow the thermal conductivity of distilled water.

However, for GN nanofluid, it has been seen that GN-SDBS (sample 2) nanofluid trend has gone downward and also below the trend of distilled water. Whereas, GN-SDS nanofluids (sample 4 and 6) have shown significantly good thermal conductivity and 2/1 ratio solution has given better results than 1/1 ratio (sample 4).

In fig. 8, both sample 6 and sample 7 with different ultrasonication times (20 minutes and 40 minutes respectively) have showed thermal conductivity above that of distilled water. And similar to Section: 3.1, it can be said that in our experiment increasing the Ultrasonication time has not showed effective value. Though sample 2 has positive percent enhancement at 25^0 C, it is more negative at 35^0 C. Sample 3 has a negative percent enhancement for both 25^0 C and 35^0 C. At 35^0 C, it is almost 4 times. At 25^0 C, sample 4, sample 6 and sample 7 have positive enhancements (4.202 %, 5.546 %, and 4.706 % respectively) and among them sample 6 has showed better value. Similarly, at 35^0 C sample 6 also has showed better value than other two.

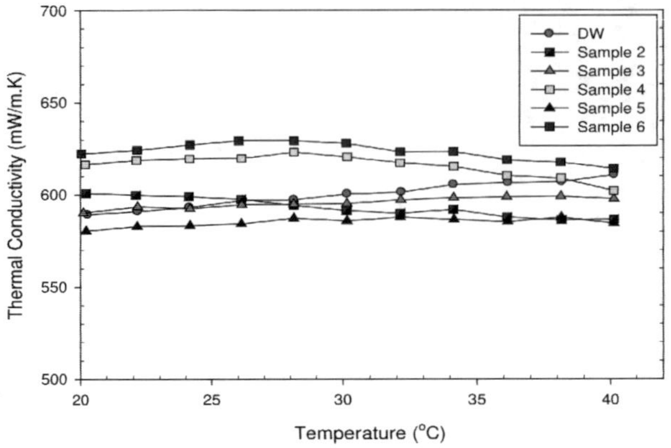

Figure7. Thermal conductivity vs. temperature curve of
Sample 1: Distilled water (DW), used as reference
Sample2: GN (0.5 wt %), SDBS (0.5 wt %), wet grinding, 20 min. Ultrasonication
Sample3: MWCNTs (0.5 wt %), SDBS (0.5 wt %), wet grinding, 20 min. Ultrasonication
Sample4: GN (0.5 wt %), SDS (0.5 wt %), wet grinding, 20 min. Ultrasonication
Sample5: MWCNTs (0.5 wt %), SDS (0.5 wt %), wet grinding, 20 min. Ultrasonication
Sample6: GN (0.5 wt %), SDS (0.25 wt %), wet grinding, 20 min. Ultrasonication

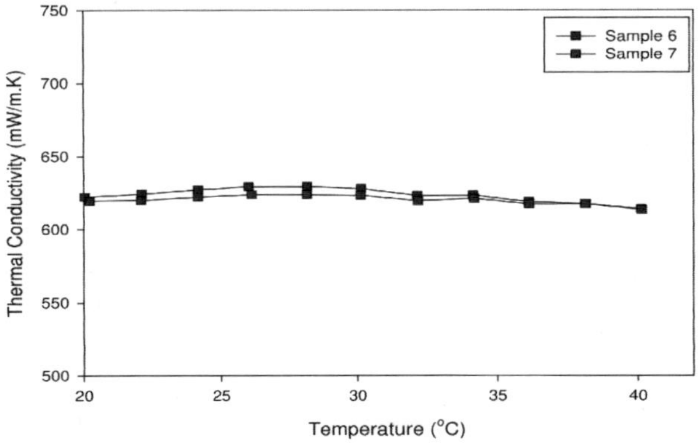

Figure8. Thermal conductivity Vs. temperature curve of
Sample 6: GN (0.5 wt %), SDS (0.25 wt %), wet grinding, 20 min. Ultrasonication
Sample 7: GN (0.5 wt %), SDS (0.25 wt %), wet grinding, 40 min. Ultrasonication

3.5 Heat transfer coefficient enhancement ratio

Heat transfer coefficient is the calculation of heat transfer, typically by convection or phase transition between fluid and a solid and its SI units is watts per square meter Kelvin ($W/m^2.K$). Sadik et al. [27] for constant wall temperature boundary condition and by considering convective heat transfer of laminar nanofluid flow inside a straight circular tube showed that average heat transfer coefficient enhancement ratio is increased by enhancing in volumetric heat capacity and thermal conductivity of nanofluid and there is a pronounced effect of thermal conductivity on heat transfer enhancement.

Heat transfer coefficient enhancement ratio is defined as the ratio of the heat transfer coefficient of the nanofluid to that of the base fluid [27]. From fig. 9, it has been easily seen that sample 6 has higher heat transfer coefficient enhancement ratio from 20^{0}C to 35^{0}C (about) than sample 7 and with the increasing temperature it has decreased.

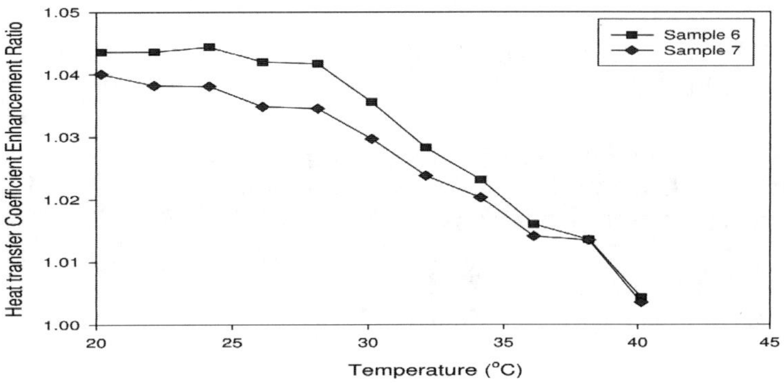

Figure9. Heat transfer coefficient enhancement ratio

4. Conclusion

This experimental approach was done on the basis of thermal conductivity as well as heat transfer coefficient enhancement ratio. After analysis and comparison of these points of view, the key conclusions can be digested as follows:

- Surfactants (SDS, SDBS) are useful for both grinding and dispersing processes.
- In this experiment, the one of the causes of applying wet grinding was to minimize the possibility of contamination by decomposition of surfactants and this was considered theoretically. The presence of surfactant induced contamination in grinding process can be disclosed by the analysis of X-ray diffraction (XRD) spectra where either shifting in peak positions or the appearance of new diffraction peaks will indicate the contamination [20]. The decreasing trends of thermal conductivity were due to the sedimentation of nanoparticles in a base fluid (DW) with time.
- Beside the thermal conductivity, heat transfer coefficient enhancement ratio was calculated and both revealed that GN-SDS with 2/1 ratio, 1 hour with 500 rpm of wet grinding, 20 minute ultrasonication was the better nanofluid in this experiment.

Acknowledgement

This research was supported by Basic Science Research Program through the National Research Foundation of Korea (NRF) funded by the Ministry of Education, Science and Technology (NRF-2014R1A1A4A03005148).

References

[1] K. Wusiman, H. Jeong, K. Tulugan, H. Afrianto, and H. Chung, Feb. 2013, "Thermal performance of multi-walled carbon nanotubes (MWCNTs) in aqueous suspensions with surfactants SDBS and SDS," *Int. Commun. Heat Mass Transf.*, **vol. 41**, pp. 28–33.

[2] M. Kole and T. K. Dey, May 2012, "Effect of prolonged ultrasonication on the thermal conductivity of ZnO–ethylene glycol nanofluids," *Thermochim. Acta*, **vol. 535**, pp. 58–65.

[3] S. U. S. Choi and J. A. Eastman, 1995 ,"Enhancing thermal conductivity of fluids with nanoparticles ".

[4] Y. Xuan and Q. Li, , 2000 "Heat transfer enhancement of nano-fl uids," **vol. 21**, pp. 58–64.

[5] S. M. S. Murshed, K. C. Leong, and C. Yang, , Apr. 2005 "Enhanced thermal conductivity of TiO2—water based nanofluids," *Int. J. Therm. Sci.*, **vol. 44,** pp. 367–373.

[6] M. J. O. Connell, P. Boul, L. M. Ericson, C. Hu, Y. Wang, E. Haroz, C. Kuper, J. Tour, K. D. Ausman, and R. E. Smalley, 2001, "Reversible water-solubilization of single-walled carbon nanotubes by polymer wrapping," **vol. 342**, pp. 265–271.

[7] A. Nouri and C. Wen, Jan. 2014, "Surfactants in Mechanical Alloying/Milling: A Catch-22 Situation," *Crit. Rev. Solid State Mater. Sci.*, **vol. 39,** pp. 81–108.

[8] B. Munkhbayar, S. Hwang, J. Kim, K. Bae, and M. Ji, "Photovoltaic performance of dye-sensitized solar cells with various MWCNT counter electrode structures produced by different coating methods", *Electrochimica Acta.*

[9] Q. Sun, J.-H. Kim, J.-H. Park, and S. Seo, 2010, "Characteristics of a pentacene thin film transistor with periodic groove patterned poly(methylmethacrylate) dielectrics," *Appl. Phys. Lett.*, **vol. 96,** p. 103301.

[10] J. Li, L. Hu, J. Liu, L. Wang, T. J. Marks, and G. Grüner, 2008, "Indium tin oxide modified transparent nanotube thin films as effective anodes for flexible organic light-emitting diodes," *Appl. Phys. Lett.*, **vol. 93,** p. 083306.

[11] Y.-L. Huang, A. Baji, H.-W. Tien, Y.-K. Yang, S.-Y. Yang, S.-Y. Wu, C.-C. M. Ma, H.-Y. Liu, Y.-W. Mai, and N.-H. Wang, Aug. 2012, "Self-assembly of silver–graphene hybrid on electrospun polyurethane nanofibers as flexible transparent conductive thin films," *Carbon N. Y.*, **vol. 50,** pp. 3473–3481.

[12] H. Chang, G. Wang, A. Yang, X. Tao, X. Liu, Y. Shen, and Z. Zheng, Sep. 2010, "A Transparent, Flexible, Low-Temperature, and Solution-Processible Graphene Composite Electrode," *Adv. Funct. Mater.*, **vol. 20,**pp. 2893–2902.

[13] Y.-L. Huang, H.-W. Tien, C.-C. M. Ma, C.-C. Teng, Y.-H. Yu, S.-Y. Yang, M.-H. Wei, and S.-Y. Wu, Oct. 2011, "Effect of cross-linkable polymer on the morphology and properties of transparent multi-walled carbon nanotube conductive films," *Appl. Surf. Sci.*, **vol. 258,**pp. 136–142.

[14] S. J. Wang, Y. Geng, Q. Zheng, and J.-K. Kim, May 2010, "Fabrication of highly conducting and transparent graphene films," *Carbon N. Y.*, **vol. 48,** pp. 1815–1823.

[15] H. Wu, L. Hu, M. W. Rowell, D. Kong, J. J. Cha, J. R. McDonough, J. Zhu, Y. Yang, M. D. McGehee, and Y. Cui, Oct. 2010, "Electrospun metal nanofiber webs as high-performance transparent electrode.," *Nano Lett.*, **vol. 10,**pp. 4242–8.

[16] T.-K. Hong, D. W. Lee, H. J. Choi, H. S. Shin, and B.-S. Kim, Jul. 2010, "Transparent, flexible conducting hybrid multilayer thin films of multiwalled carbon nanotubes with graphene nanosheets.," *ACS Nano*, **vol. 4,**pp. 3861–8.

[17] G. Refael, 2012, "Inducing topological order in a honeycomb lattice," pp. 1–7.

[18] K. I. Bolotin, K. J. Sikes, Z. Jiang, M. Klima, G. Fudenberg, J. Hone, P. Kim, and H. L. Stormer, Jun. 2008, "Ultrahigh electron mobility in suspended graphene," *Solid State Commun.*, **vol. 146,**pp. 351–355.

[19] C. Lee, X. Wei, J. W. Kysar, and J. Hone, Jul. 2008, "Measurement of the elastic properties and intrinsic strength of monolayer graphene.," *Science*, **vol. 321,**pp. 385–8.

[20] X. Zhang, Q. Li, T. G. Holesinger, P. N. Arendt, J. Huang, P. D. Kirven, T. G. Clapp, R. F. DePaula, X. Liao, Y. Zhao, L. Zheng, D. E. Peterson, and Y. Zhu, Dec. 2007, "Ultrastrong, Stiff, and Lightweight Carbon-Nanotube Fibers," *Adv. Mater.*, **vol. 19,**pp. 4198–4201.

[21] F. Inam, A. Heaton, P. Brown, T. Peijs, and M. J. Reece, Jan. 2014, "Effects of dispersion surfactants on the properties of ceramic–carbon nanotube (CNT) nanocomposites," *Ceram. Int.*, **vol. 40,**pp. 511–516.

[22] S. Tawfick, K. O'Brien, and a J. Hart, Nov. 2009, "Flexible high-conductivity carbon-nanotube interconnects made by rolling and printing.," *Small*, **vol. 5,**pp. 2467–73.

[23] S. Berber, Y. Kwon, and D. Tomanek, May 2000, "Unusually high thermal conductivity of carbon nanotubes," *Phys. Rev. Lett.*, **vol. 84,**pp. 4613–6.

[24] S. J. Woltornist, A. J. Oyer, J.-M. Y. Carrillo, A. V Dobrynin, and D. H. Adamson, Aug. 2013, "Conductive thin films of pristine graphene by solvent interface trapping.," *ACS Nano*, **vol. 7,** pp. 7062–6.

[25] D. R. Lide, S. R. Data, E. A. Board, G. Baysinger, S. Chemistry, C. E. Library, L. I. Berger, R. N. Goldberg, B. Division, H. V Kehiaian, K. Kuchitsu, G. Rosenblatt, D. L. Roth, and D. Zwillinger, "CRC Handbook of Chemistry and Physics."

[26] "Thermal effusivity - Wikipedia, the free encyclopedia." [Online]. Available: http://en.wikipedia.org/wiki/Thermal_effusivity. [Accessed: 20-May-2014].

[27] S. Özerinç, A. G. Yazıcıoğlu, and S. Kakaç, "Numerical analysis of laminar forced convection with temperature-dependent thermal conductivity of nanofluids and thermal dispersion, Dec. 2012," *Int. J. Therm. Sci.*, **vol. 62**, pp. 138–148.

IOP Conf. Series: Materials Science and Engineering **88** (2015) 012051 doi:10.1088/1757-899X/88/1/012051

Thermal characteristics of non-edible oils as phase change materials candidate to application of air conditioning chilled water system

M Irsyad[1,2], Y S Indartono[2], A Suwono[2], and AD Pasek[2]

[1]Mechanical Engineering Department, Lampung University, Bandar Lampung, Indonesia
[2]Faculty of Mechanical and Aeronautical Engineering, Institut Teknologi Bandung, Indonesia

Email: irsyad71@yahoo.com

Abstract. The addition of phase change material in the secondary refrigerant has been able to reduce the energy consumption of air conditioning systems in chilled water system. This material has a high thermal density because its energy is stored as latent heat. Based on material melting and freezing point, there are several non-edible oils that can be studied as a phase change material candidate for the application of chilled water systems. Forests and plantations in Indonesia have great potential to produce non-edible oil derived from the seeds of the plant, such as; Calophyllum inophyllum, Jatropha curcas L, and Hevea braziliensis. Based on the melting temperature, these oils can further studied to be used as material mixing in the secondary refrigerant. Thermal characteristics are obtained from the testing of T-history, Differential Scanning Calorimetric (DSC) and thermal conductivity materials. Test results showed an increase in the value of the latent heat when mixed with water with the addition of surfactant. Thermal characteristics of each material of the test results are shown completely in discussion section of this article.

1. Introduction

One of energy crisis' solutions is its efficient consumptions. It is stated that the largest percentage spent for electric consumption on commercial blocks is in their air conditioning systems as demonstrated in United States and China which spend 38 percent [1], Malaysia 57 percent [2], and Indonesia spends more than 65 percent [3]. Methods are being developed to reduce electrical energy consumption in the air conditioning system and one of them is the use of phase change materials (PCM) as thermal energy storage (TES) [4].

Freezing and melting temperature between 5-12°C [5] are important considerations when selecting PCM in order to adjust them to the evaporator of chiller's temperature. Fatty acid's melting temperature ranges from -5.6 up to 79 ° C with latent heat ranging from 126 to 212kJ/kg [6]. Experimented organic PCM from fatty acid like biodiesels from crude palm oil (CPO) [7] and soybean oil [8] are potential candidates material to be mixed into secondary refrigerant. Since these biodiesels are also utilized as edible oils, their functions as PCM should be reconsidered for further development.

[1] To whom any correspondence should be addressed.

On the other hand, many types of non-edible oil PCM from oilseeds of forest trees and plantation plants are promising candidates to be developed as mixing material. Among the seeds, there are *Calophyllum inophyllum, Jatropha curcas L,* and *Hevea braziliensis.* Those tropical plants are widely spread in many countries. The oil content in those three diverse plants is relatively high as it is displayed in Table 1.

Table 1. Non edible oils potential

	Seed production tonnes/(ha.year)	oil content in seeds (%)	kg oil/ (ha.year)	Density (kg/m^3)	Viscosity (cSt)
Calophyllum inophyllum	16[9]	40 - 73 [10]	4680[10]	896[11]	71.98[11]
Jatropha curcas	0.1 - 15[10]	37 - 45[12]	1900 - 2500[10]	920[11]	18.2[11]
Hevea braziliensis	0.15[13]	40 - 50[11]	40 - 50[11]	910[11]	66.2[11]

Vegetable oil is composed of diverse fatty acids, both saturated and unsaturated fats. The most fatty acid composition is palmitic and stearic from saturated group; oleic, linoleic, and linolenic from unsaturated ones. Fatty acids percentage from seeds of *Calophyllum inophyllum*, *Jatropha curcas L*, and *Hevea braziliensis* are shown on Table 2.

Table 2. The content of fatty acids from nonedible oil

Fatty acid	Name Structure	Fatty acid composition (%)				Melting point (^{0}C)[c]	Latent heat (kJ/kg)[d]	Types of fatty acids
		CISO[a]	JCSO[a]	CPO[b]	HBSO[a]			
Palmitic	Hexadecanoic (C16:0)	12.01	16.00	43.50	10,20	60,45	221.42	saturated fatty acid
Stearic	Octadecanoic (C18:0)	12,95	6.50	4.30	8,70	66,87	242.15	saturated fatty acid
Oleic	cis-9-Octadecenoic (C18:1)	34.09	43.50	39.80	24,60	13 s/d 14	138.07	monounsaturated fatty acid
Linoleic	cis-9,cis-12-Octadecadienoic (C18:2)	38.26	34.40	10.20	39,60	-9 s/d -8	n/a	polyunsaturated fatty acid
Linolenic	cis-9,cis-l2,cis-15-Octadecatrienoic (C18:3)	0,30	0.80	0,30	16,30	-17	n/a	polyunsaturated fatty acid

a [13], b [14], c [15], d [16]

Based on the potency and composition of fatty acids within the plant *Calophyllum inophyllum, Jatropha curcas L*, and *Hevea braziliensis*, these oilseeds can be further researched as candidates of phase change materials for air conditioning system. It still is necessary to test the thermal characteristics such as freezing and melting temperature, latent heat and the concentration of solids.

2. Experimental

Thermal characteristics had been obtained using the T-history test and differential scanning calorimetry (DSC).The T-history testing was done on cooling rate ranged from 0.10 to 0.25°C/min with a sample mass of 70 g. DSC test used a Perkin Elmer type 8500. Criteria referenced in this DSC testing were as follows; mass of the sample is 5 -7 mg, cooling and heating rate was 2°C/min, and the sample was cooled from 25 to -30°C and heated from -30 to 35°C. The nitrogen flow rate of cooling was 20 ml/min. Testing of the solid mass concentration was done by taking samples at temperatures of

IOP Conf. Series: Materials Science and Engineering **88** (2015) 012051 doi:10.1088/1757-899X/88/1/012051

6 to 10 °C during cooling and heating. Mass of solids was separated from the liquid and then a solid mass was measured.

3. Result and Discussion

The result of T-history testings for three vegetable oils indicated several findings. First, their freezing temperatures were able to achieve evaporator' performing condition as displayed on Table 3. Second, different treatment for *Jatropha curcas* seed oils (JCSO) was executed. It needed *Crude palm oil* (CPO) added in such composition of 93% : 7%. Third, the highest fatty acid composition percentage on *Calophyllum inophyllum* seed oil (CISO) is oleic; 48.49% and linoleic 20.70% with melting temperatures 13 – 14°C and -9 to -8°C respectively. These two fatty acids contributed to freezing temperature value of CISO. Similar compositions were signified in JCSO and *Hevea braziliensis* seed oil (HBSO), in which oleic and linoleic had the highest percentages. Having different treatment using CPO, JCSO is composed mainly of 46.9% palmitic in melting temperature 6045°C. Consequently, CPO's melting temperature is crucially high; 20.13°C and 46.36°C [17]. The melting temperatures on fatty acids were governed by hydrocarbon chains. These linear hydrocarbon chains were in saturated fatty acids like palmitic and stearic. On the other hand, one or more chains with double carbon ties were in unsaturated fatty acid like oleic and linoleic. Unsaturated fatty acid was formed by monounsaturated fatty acid and polyunsaturated fatty acid. Another controlling factor is that melting temperature is parallel with the chain length in saturated fatty acid, while is opposite to unsaturated level [18]. Melting temperature of saturated fatty acid is higher than unsaturated one because saturated fatty acid has stabile thermodynamic feature [19].

Table 3. Freezing temperature of nonedible oil with T-history test

Phase change material	Composition (%)	Freezing temperature (°C)
CISO	100	12.77 – 5.26
JCSO + CPO	93 + 7	12.02 – 2.44
HBSO	100	11.57 – 1.73

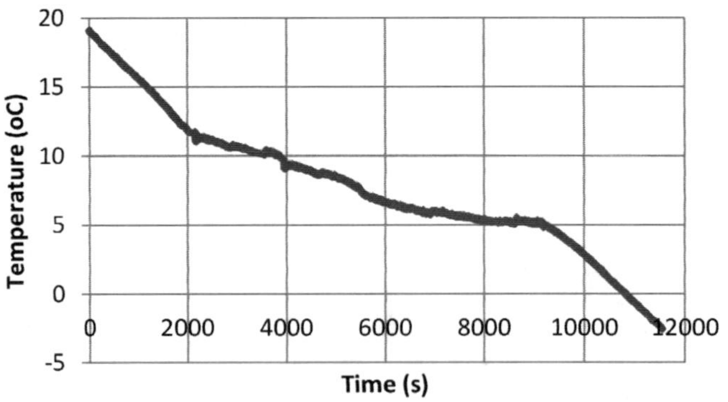

Figure 1. T-history test of CISO for cooling process

DSC test results of nonedible oil are shown in Table 4. Temperature and HBSO CISO crystallization from DSC test is not quite distinctive from the T-test history. This difference was caused by different cooling rates between the two tests. HBSO melting temperature was almost equal to the melting temperature of Malaysian rubber seed oils (MRSO) which were tested by Abdullah et al, but HBSO freezing temperature is much higher than MRSO [20]. This difference could be caused by the content of fatty acid composition.

Table 4. Melting and freezing characteristic of nonedible oil with DSC test

PCM	Freezing				Melting			
	T_{onset} (°C)	T_{peak} (°C)	T_{endset} (°C)	Laten heat (kJ/kg)	T_{onset} (°C)	T_{peak} (°C)	T_{endset} (°C)	Laten heat (kJ/kg)
CISO	11.85	9.41	4.12	32.8039	-6.42	10.35	14.29	184.4739
JCSO + CPO	-14.81	-14.98	-15.59	21.1714	-11.19	-1.79	3.57	8.4387
HBSO	6.26	5.84	5.37	0.3872	-11.41	-4.52	-0.82	9.3000
CISO + water	-11.57	-11.66	-12.26	188.3105	11.08	15.23	16.12	219.7453
JCSO +CPO + water	-11.09	-11.20	-11.69	78.2715	10.48	12.70	13.54	95.0304
HBSO + water	-8.45	-8.53	-9.02	153.2356	9.80	12.69	13.44	171.4922

The freezing and melting process of the nonedible oils had occurred on a large temperature range. Figure 2 shows the melting process for the CISO, where it started at a temperature of -6.42°C and finished at a temperature of 14.29°C. These results were consistent with the research results of Indartono et al to coconut oil and JCSO with the ratio of 1: 1 [8]. DSC test results for a solution of oil in water by adding anionic Texapon surfactant as much as 500ppm showed latent heat higher than the seed oil. Temperature range for freezing and melting processes are very small. Figure 3 shows the melting process for solution of CISO in water, where it started at a temperature of 11.08°C and finished at a temperature of 16.12°C. The addition of seed oils affected the temperature of crystallization and melting of water in the mixture. This finding was consistent with the results of the research of Tippets and Martini, where increasing oil concentration of 20% to 40% could increase the melting temperature [21].

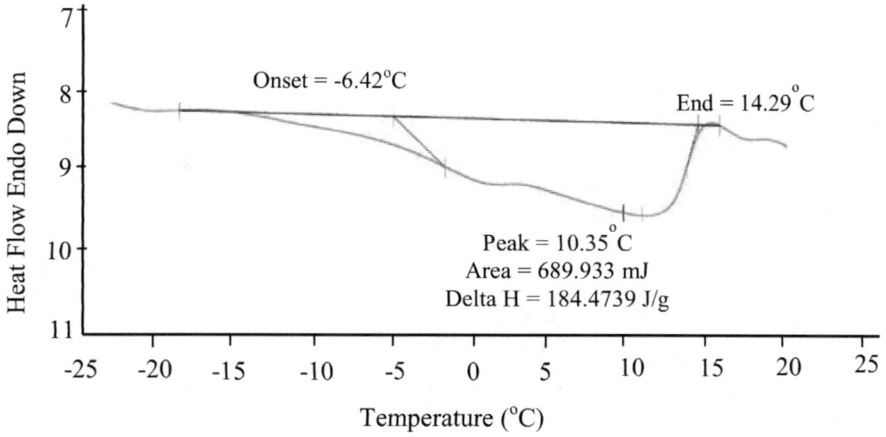

Figure 2. DSC curve of CISO in heating process

IOP Conf. Series: Materials Science and Engineering **88** (2015) 012051 doi:10.1088/1757-899X/88/1/012051

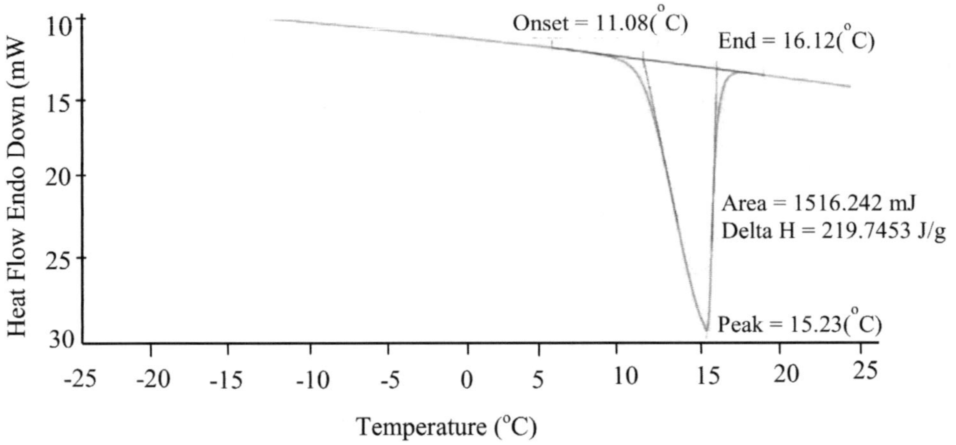

Figure 3. DSC curve of CISO slurry in heating process

Solid mass concentrations for PCM slurry in the heating and cooling process is shown in Figure 4 In the heating process, the concentration of solid mass was reduced from 24 359% at temperatures 7°C, to 22 60% at temperatures 9°C. While in the process of cooling, solid mass concentration increased from 33.13% at the temperature 9°C, to 37.22% at temperatures 7°C. These changes were caused by the melting and freezing processes. Energy from the utilized latent heat was proportional to change in the solid mass concentration of PCM slurry.

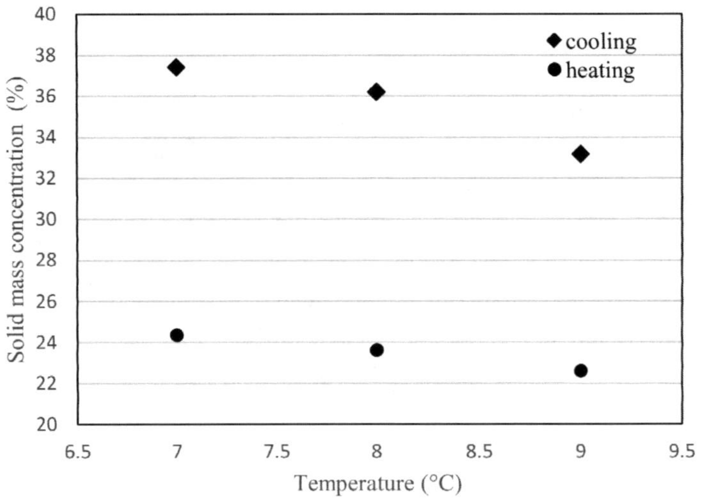

Figure 4. Solid mass concentration of CISO slurry

Referring to the results of observations on the mass concentration of the solid test and t-test of history, in which the process of freezing and melting of non-edible oil for the third is still in accordance with the evaporator temperature, they can be used as candidates of material mixing in secondary refrigerant. CISO and HBSO were more potential to be developed because after they were dissolved in the water, latent heat was high. CISO slurry had a latent heat of 219.745kJ/kg and HBSO slurry had 171.4922kJ/kg.

4. Conclusion

Several conclusions could be drawn from the experiments. Firstly, CISO and HBSO had the prospect to be developed into mixing material in the secondary refrigerant as they were highly potential in availability. Secondly, melting and freezing temperatures were matched with evaporator temperature. Latent heat of two seed oils increased after they were mixed into the water. CISO latent heat of slurry on the melting process was 219.75kJ/kg and the freezing process was 188.31kJ/kg. Besides, the latent heat of HBSO slurry on the melting process was 171.49kJ/kg and the freezing process was 153.24kJ/kg. These mean that both seed oils are good thermal energy storage. Thirdly, utilized latent heat value is proportional to the change in solid mass concentration on the freezing and melting process. Since these materials were insoluble in water, it was necessary to conduct research to get surfactant that less affect the thermal properties of the materials.

5. References

[1] Xia J, Hong T, Shen Q, Feng W, Yang L, Im P, Lu A and Bhandari M 2014 *Energy and Buildings* **78** 165–175
[2] Saidur R 2009 *Energy Policy* **37** 4104–4113
[3] *JICA Study on Energy Efficiency and Concervation Improvement in Indonesia 2007 -2008* http://eneken.ieej.or.jp/en/data/pdf/491.pdf, diakses 8 – 10 – 2012
[4] Al-Abidi A, Mat SB, Sopian K, Sulaiman MY, Lim CH and Th A 2012 *Renewable and Sustainable Energy Reviews* **16** 5802–5819
[5] Li G, Hwang Y and Radermacher R 2012 *J. of Refrigeration* **XXX** 1-25
[6] Sarier N and Onder E 2012 *Thermochimica Acta* **540** 7– 60
[7] Indartono YS, Usui H, Suzuki H, Komoda Y and Mujahidin D 2010, *Hydrodynamics Characteristics of Water Based Mixture of Coconut Oil and FAME-Rich Mixture to be used as Secondary Refrigerant in Air Conditioning System*, Region Conference on Mechanical and Aerospace Technology, Bali, 9-10,2010
[8] Indartono YS, Suwono A, Pasek AD, Mujahidin D and Rizal I 2010 *Jurnal Teknik Mesin*, **12 2** 119-124
[9] Jahirul MI, Brown JR, Senadeera W, Ashwath N, Laing C, Taylor JL and Rasul MG 2013 *Procedia Engineering* **56** 619 – 624
[10] Kumar A and Sharma S 2011 *Renewable and Sustainable Energy Reviews* **15** 1791–1800
Atabani AE Cesar AS 2014 *Renewable and Sustainable Energy Reviews* **37** 644–655
[11] Rodrigues J, Miranda I, Gominho J, Vasconcelos M, Barradas G, Pereira H, Aguuiar FB and Dias FS 2013 *Industrial Crops and Products* **50** 828– 837
[12] Zhu Y, Xu J, Li Q and Mortimer PE 2014 *Energy* **69** 837-842
[13] Tan CH, Ghazali HM, Kuntom A, Tan CP and Ariffin AA 2009 *Food Chemistry* **113** 645–650
[14] Ping BTY and Yusof M 2009, *Characteristics and Properties of FTTY Acid Distillates from PLM Oil*, Oil Palm Bulletin 59 (November 2009) 5-11
[15] Yuan Y, Zhang N, Tao W, Cao X and He Y 2014 *Renewable and Sustainable Energy Reviews* **29** 482–498
[16] Narvaes PC, Rincon SM, Castaneda LZ and Sanchez FJ 2008 *Latin American Applied Research* **38** 1-6
[17] Scrimgeour C 2005 *Chemistry Bailey's Industrial Oil and Fat Products, Sixth Edition* Fereidoon Shahidi, sixth volume John Wiley & Sons
[18] Rustan AC and Drevon CA 2005 *Fatty Acids: Structures and Properties, Encyclopedia of Life Science* John Wiley and Sons
[19] Abdullah BM and Salimon J *2009 European Journal of Scientific Research* **31 3** 437-445
[20] Tippetts M and Martini S 2009 *Food research International* **42 7** 847-855

IOP Conf. Series: Materials Science and Engineering **88** (2015) 012052 doi:10.1088/1757-899X/88/1/012052

Heat-Assisted Machining for Material Removal Improvement

A B Mohd Hadzley[1], **S Muhammad Hafiz**[1], **W Azahar**[1], **R Izamshah**[1], **K Mohd Shahir**[1] **and A Abu**[1]

[1]Faculty of Manufacturing Engineering, UniversitiTeknikal Malaysia Melaka, Hang Tuah Jaya, 76100 Durian Tunggal, Melaka, Malaysia.

E-mail: hadzley@utem.edu.my

Abstract. Heat assisted machining (HAM) is a process where an intense heat source is used to locally soften the workpiece material before machined by high speed cutting tool. In this paper, an HAM machine is developed by modification of small CNC machine with the addition of special jig to hold the heat sources in front of the machine spindle. Preliminary experiment to evaluate the capability of HAM machine to produce groove formation for slotting process was conducted. A block AISI D2 tool steel with100mm (width) x 100mm (length) x 20mm (height) size has been cut by plasma heating with different setting of arc current, feed rate and air pressure. Their effect has been analyzed based on distance of cut (DOC).Experimental results demonstrated the most significant factor that contributed to the DOC is arc current, followed by the feed rate and air pressure. HAM improves the slotting process of AISI D2 by increasing distance of cut due to initial cutting groove that formed during thermal melting and pressurized air from the heat source.

1. Introduction

The demand for higher strength and high wear resistance materials is increasing, particularly in high perfromance application such as in tool and die, oil and gas and aerospace industries. However these materials are often difficult to machine due to their hardness and strength, which make the cutting tool easy to wear or damage within a short time [1]. Because the hardness and strength of materials normally decrease with increasing temperature, heat assisted machining becomes a one of the alternative techniques when performed material removal process of these hard-to-machine materials [1]. Heat assisted machining (HAM), is the combination of heating and machining actions in one material removal process. During HAM process, the workpiece material is heated and softened by a focused heated source and then removed by the cutting tools to perform required shape [1, 2]. Some example of HAM are plasma asissted machining, laser assisted machining and thermaly enhanced machining. This technique have been reported efficient in machining of materials that are difficult to cut, elimination cooling lubricants, geometrically flexible and feasible manufacture of complex component as compared to grinding.

Slotting process is material removal process to form the material in a shape of groove. It is considered one of the milling process that requires high toughness and high wear resistance cutting tools.In mold and die making, slotting is considered one of difficult task ecspecially when machining hardened steel with high ratio between widht and depth of cut. On the other hand,plasma cutting is the process that used a plasma torch to heat and cut steel based materialsin different thickness [3].The plasma exiting from the torch nozzle will melt localized area and remove the molten material by the force of the plasma jet. The intensity and velocity of the plasma are determined by a several factors such asthe pressure &volume the gas, type of gases, the flow pattern, the amount of arc current, the size and design of the nozzle orifice and the distance between nozzle tip to workpiece [3]. In addition, for slotting process, the efficiently to produce groove is also depended on the feed rate of the machining setup.

In this paper, the combination of air plasma cutting machine and CNC mini milling machine were carried out to investigate the effect of plasma cutting on slotting process.The plasma is used as a heat source to soften the layer of material and also to remove part of material by air pressure from plasma nozzle. Several major parameters such as arc current, feed rate and air pressure have been evaluated to

[1] Faculty of Manufacturing Engineering, Universiti Teknikal Malaysia Melaka, Hang Tuah Jaya, 76100 Durian Tunggal, Melaka.

determine their significant factors that contributed to the HAM performance, measured by distance of cut.The results from this paper is an extension of previous findings [4,5].

2. Experimental Setup

The experiment was conducted by using mini CNC machine, modified with the addition of plasma torch shown in Figure 1.A plasma cutting machine with maximum air pressure of 1MPa and arc current of 50Ampere was used. A special jig was used to hold the plasma torch for better flexibility to control of plasma torch movement [5]. For the preliminary experiments, the AISI D2 tool steel was used as the workpiece in this study. Table 1 shows the mechanical properties of AISI D2. The raw material was prepared around the size of 100mm (width) x 100mm (length) x 20mm(height)as shown in Figure 2. Three factors, arc current, air pressure and feed rate were selected as preliminary tests for the experiment in order to study the effects of HAM parameters on distance of cut (DOC). The parameter setup is shown in Table 2. Figure 3 shows the illustration of HAM process. The DOC is measured based on the peak-valley distance of the groove (Figure 4) using optical microscope.

Figure 1. Mini CNC Machine with plasma torch attached in front of the spindle.

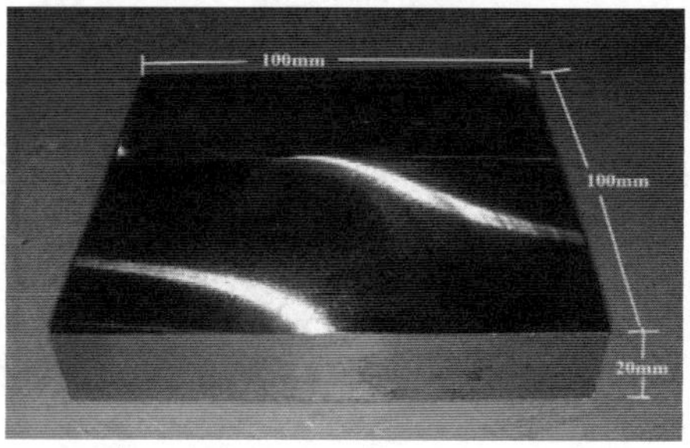

Figure 2. AISI D2 as workiece speciment

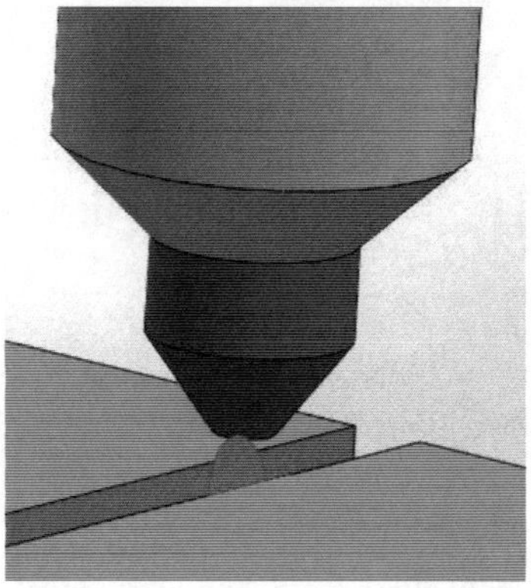

Figure 3. Illustration of HAM process

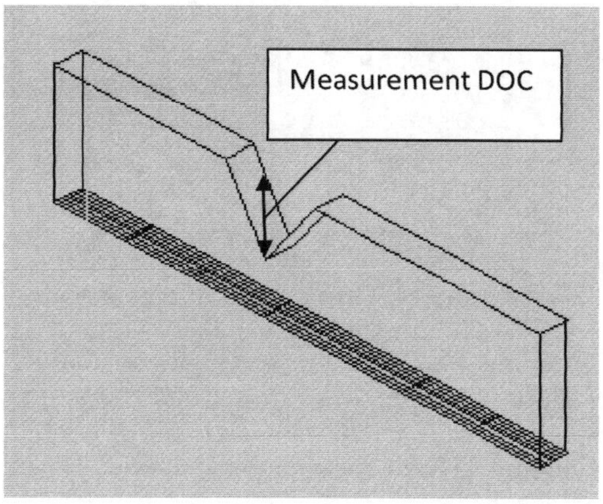

Figure 4. Measurement of DOC based on the peak-valley distance of the groove

IOP Conf. Series: Materials Science and Engineering **88** (2015) 012052 doi:10.1088/1757-899X/88/1/012052

Table 1. Mechanical properties of work material (AISI D2) [6]

Mechanical Properties	AISI D2 Steel
Tensile Strength (Mpa)	710 – 1260
Yield Strength (Mpa)	≥ 320
Hardness (HRC)	52-62
Elongation	≥ 16
Delivery condition	Soft annealed to 3pprox. 210HB

Table 2. Parameter Setup

Control Factor	Minimum Level	Maximum Level
Arc Current (A)	20	30
Air Pressure (MPa)	0.3	0.4
Feed Rate (mm/min)	200	400

3. Results and Discussion.

In this study, the implementation of HAM process has been commenced with the modification of CNC machine by combination between milling and plasma cutting processes in one machine system. Plasma cutting is a process that utilizes an ioned compressed air with high temperature to melt and cut metal in different thickness [6]. In plasma cutting, a compressed gas is blown out from a nozzle at high speed condition. On the same time an electrical arc is formed between the nozzle tip to the surface being cut to excite a powerful amount of heat [7]. The temperatures from the plasma that formed in front of the nozzle can rise up to 16000°C with maximum 6000 m/s air exits speed, which sufficiently high enough to melt and blow the molten metal away from the cutting zone [8]. As the plasma torch attached in front of the machining spindle, the feed rate can be set in automatic mode to provide the consistent plasma movement. The high temperature and high pressure of air plasma capable to produce a groove which can be partly prepared as a pre-slotting process during machining, as shown in Figures 5 and 6. The depth of groove, defined as distance of cut (DOC) depended on HAM parameters (arc current, feed rate and air pressure).

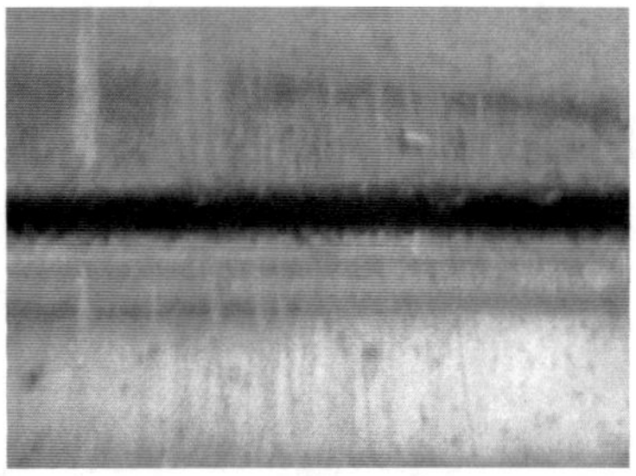

Figure 5. Groove formation after HAM process (top view)

Figure 6. Groove formation after HAM process

Table 3 shows the results of distance of cut, DOC for each parameter investigated.The results of analysis of variance (ANOVA) based on the output DOC is shown in Table 4. From the analysis, it was established that the most significant factors that contributed to the DOC isarc current, followed by the feed rate and air pressure.

For plasma arc current, it shows that the DOC increased and the arc current increased, as shown in Figure 7. The arc current within plasma welding controls the axial electrical field at the nozzle exit [9]. The current level determine the energy density required to get to heat the steel. Higher arc currents provided higher temperature and therefore increasedintensity to melt the steel, hence increased DOC.

For feed rate, it shows that the DOC decreased as the feed rate increased, as shown in Figure 8. In this study, feed rate was controlled by the CNC machine whereits defined by the time of plasma travel with respect to the work piece in certain lenght. As the plasma torch travels across the workpiece, plasma jet have adequate time to provide heat energy to melt the steel. This consequently provide greater the heat affected zone and consequently create larger area of molten metal.

On the other hand, air pressure gave less effect to the DOC as shown in Figure 9. The function of air pressure is to blow the molten metal away from the main workpiece. Even though the pressure was set at high level (0.4 Mpa), the affected area only focussed on the molten region where dominat energy was controlled by the arc current dan and feed rate. This is explained why air pressure not significantly contributed to the efficientcy of grooving process. However in this study, air pressure must be set at minimum 0.2 MPa as the electrical arc will not be formed at a lower air pressure.

Table 3. Experimental results of parameters investigated

Exp No	Arc Current (A)	Feed Rate(mm/min)	Pressure(MPa)	DOC(mm)
1	20	200	0.3	1.865
2	20	200	0.4	1.441
3	20	400	0.3	1.151
4	20	400	0.4	0.438
5	30	200	0.3	4.877
6	30	200	0.4	4.260
7	30	400	0.3	3.602
8	30	400	0.4	4.743

Table 4. Analysis of variance (ANOVA) based on the output DOC

Source of variance	Sum of square	DF	Means square	F-value	P-value Prob>F	Percentage contribution
Model	21.16	4	5.29	22.00	0.0147	20
A-Air Pressure	0.047	1	0.047	0.20	0.6885	0.17
B-Current	19.80	1	19.80	82.34	0.0028	74.87
C-Feed Rate	0.79	1	0.79	3.27	0.1682	2.99
D-Standoff height	0.52	1	0.52	2.18	0.235	1.97
Residual	0.72	3	0.24			
Cor Total	21.88	7				

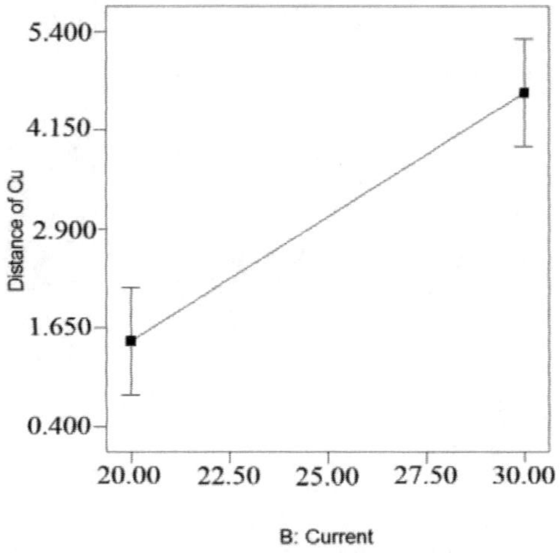

Figure 7. Effect of arc current on DOC

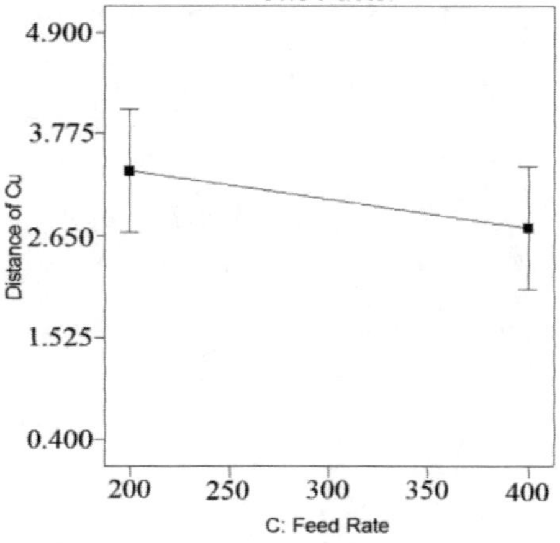

Figure 8. Effect of feed rate on DOC

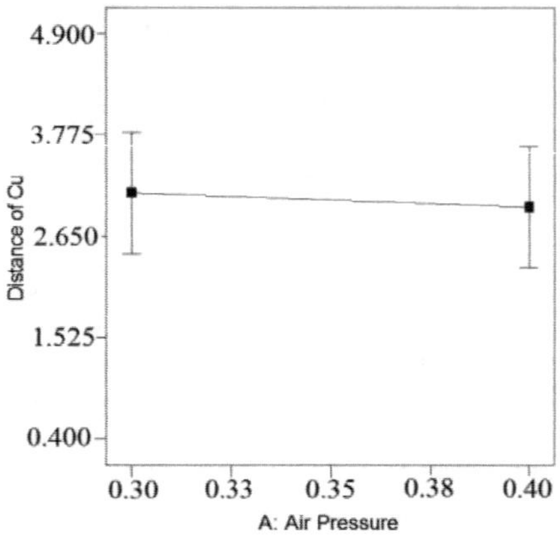

Figure 9. Effect of air pressure on DOC

4. Conclusion

From the experiment, the following findings have been determined:

1. The most significant factors that contributed to the distance of cut is arc current, followed by the feed rate and air pressure.
2. Distance of cut increased as the arc current increased
3. Distance of cut decreased as the feed rate increased
4. Air pressure not significantly contributed to the efficientcy of HAM process.

5. Acknowledgement

The authors would like to thank Ministry of Higher Education Malaysia(MOHE) and Universiti Teknikal Malaysia Melaka (UTeM) for their support that enabled this work to be carried out through the Grant RAGS/201/FKP/TK01/1 B00012.

6. References

[1] Sun S., Brandt M and Dargusch M S 2010 Thermally enhanced machining of hard-to-machine materials—A review, *International Journal of Machine Tools and Manufacture* **50(8)** 663-680

[2] Hongtao D and Yung C S 2010 Laser-assisted machining of hardened steel parts with surface integrity analysis *International Journal of Machine Tools and Manufacture* **50(1)** 106-114

[3] www.millerwelds.com

[4] Mohd Hadzley A B, Muhammad Hafiz S, Mohd Najib Z, Raja Izamshah R A, Mohd Shahir K and Abu A 2014 Development of 3-Axis Mini Milling Machine for Small Scale Production, *International Conference on Design and Concerent Engineering,* 22-23 September 2014, Melaka, Malaysia

[5] Mohd Hadzley A B, Muhammad Hafiz K , Wan Azahar W Y, Raja Izamshah R A and Abu A 2014 Fabrication of Jig as an Assisted Tool in Hybrid Machining Technique, *International Symposium on Research in Innovation and Sustainability 2014 (ISoRIS '14)* 15-16 October 2014, Melaka, Malaysia

[6] http://www.bucorp.com/

[7] Anderson C A 1987 Comparison between Laser, Plasma, Waterjet, Oxygen Cutting and other Mechanical Cutting Processes for Low Thickness (0.5 up to 5mm), *Welding In the World* **25(5/6)** 88-99

[8] Moarrefzadeh A 2011 Numerical Analysis of Thermal Profile in Plasma Arc Cutting *International Journal of Multidisciplinary Sciences and Engineering* **2(6)** 22-25

[9] Tewari S P, Ankur G and Jyoti P 2010 Effect of Welding Parameters on the Weldability of MateriaL, *International Journal of Engineering Science and Technology* **2(4)** 512-516

The empirical correlations for natural convection heat transfer Al_2O_3 and ZrO_2 nanofluid in vertical sub-channel

K Kamajaya, E Umar and Sudjatmi

Center for Applied Nuclear Science and Technology
National Nuclear Energy Agency, Bandung, Indonesia
Jl. Tamansari 71 Bandung 40132, Indonesia
kamajaya@batan.go.id

Abstract. Study on convection heat transfer using water-Al_2O_3 nanofluid as the working fluid in the vertical sub-channel has been conducted. The results of the study have been compared with the water-ZrO_2 nanofluid and pure-water as the working fluid. The equipment used in this experiment is a vertical triangular sub-channel, equipped by primary cooling system, heat exchanger and a secondary cooling system. As a heating source used three vertical cylinders that have a uniform heat flux with a pitch to diameter ratio (P/D) 01:16. Cooling is used is water-Al_2O_3 colloid at 0.05 wt. %. Heat transfer from heating to cooling would occur in natural or forced convection. However, in this study will be discussed only natural convection heat transfer. The results showed that the natural convection heat transfer of water-Al_2O_3 nanofluid in a triangular sub-channels depending on the position. The results of the correlation as follows,

$$Nu = 15.97\left(Ra\frac{D_h}{x}\right)^{0.2778}$$

1. Introduction

Nanofluid is a fluid containing nano-sized particles. This liquid is made of a colloidal suspension of nanoparticles in the base-fluid[1]. The nanoparticles used in nanofluid are usually made of metal, oxide, carbide, or carbon nanotubes. Nanofluid has special properties that make them potentially useful in many applications in heat transfer[2]. Nanofluid has the properties of thermal conductivity and convective heat transfer coefficient is better than base-fluid.[3] In the analysis using computational fluid dynamics (CFD), nanofluid can be assumed to be single phase fluid. According to the classical theory of single-phase fluid, the physical properties of nanofluid are taken as a function of the constituents and concentrations.[4] Many researchers have conducted research on the heat transfer performance and flow characteristics of various nanofluid with different nanoparticles and base-fluid materials. Several articles published relating to the use of nanofluid are described in the following sections. Chein and Chuang reported experimentally on micro channel heat sink performance using CuO–water nanofluid as coolants. The results showed that the presence of nanoparticles creates greater energy absorption than pure-water at a low flow rate and that there is no contribution from heat absorption when the flow rate is high [5]. Duangthongsuk and Wongwises observed the effect of thermophysical properties models on prediction of the heat transfer coefficient and also reported the heat transfer performance and friction characteristics of nanofluid, respectively. The results also indicated that the heat transfer coefficient of nanofluid is slightly greater than that of pure-water by approximately (6–11) % [6]. Hwang et al. through experiments found of flow and convective heat transfer characteristics of Al_2O_3-water nanofluid, with convective heat transfer characteristics of Al_2O_3-water nanofluid with particles varying in the range of 0.01–0.3% in a circular tube with the constant heat flux in fully developed laminar regime reported improvement in convective heat transfer coefficient in the thermally fully developed regime[7]. Li and Xuan studied experimentally the convective heat transfer and flow features for Cu–water nanofluid flowing through a straight tube under laminar and turbulent flow regimes with a constant heat flux. The experimental results showed that addition of nanoparticles into the base liquid remarkably enhanced the heat transfer performance of the base liquid. They also proposed new convective heat transfer correlations for calculating the heat transfer coefficients of the nanofluid for both laminar and turbulent flow conditions [8]. Mirmasoumi and Behzadmehr have studied the effects of nanoparticle mean diameter on the heat transfer and flow behavior into a horizontal tube under laminar mixed convection condition. Their calculated results demonstrate that the convection heat transfer coefficient significantly increases with decreasing the nanoparticles means diameter. However, the hydrodynamics parameters are not significantly changed [9]. Sudjatmi et al. has conducted studies to use of ZrO_2-water nanofluid as coolant in experimental triangular vertical sub-channel with low concentrations of 0.05 to 0.1%. Their results showed that for low concentrations of the heat transfer coefficient of nanofluid is slightly higher than pure water. But for higher concentrations can reduce the heat transfer coefficient. [10]. Zamzamian et al._investigated the effects of forced convective heat transfer coefficient with Al_2O_3 nanofluid in double pipe and plate heat exchangers. Their results indicate that increasing the nanoparticle concentration and temperature could enhance the convective heat transfer coefficient of nanofluid, leading to a 2–50% enhancement in convective heat transfer coefficient of the nanofluid [11].

This paper focuses on experimental research, particularly with respect to the natural convection of water-Al_2O_3 nanofluid in vertical sub-channels. Then the results of this study compared with the results of other experiments using water-ZrO_2 nanofluid and pure-water as the working fluid.

2. Methodology of the experimental

The experimental equipment is built to conduct the experiments on the vertical triangular sub-channel. This equipment is designed for natural convection flow however it can also be used for forced convection. In addition, this equipment is also equipped with plate type heat exchanger, the primary tank, secondary tank as well as primary and secondary pumps.

At each stage of this experiment, each heating providing electric power is 250 W, 350 W, 500 W, 650 W, 750 W, and 850 W. So the experiment was conducted for six stages of power. Coolant circulation by the primary pump is in constant flow rate of 3 liters/minute and a secondary pump constant flow rate is 20 liters/minute. Heat of the primary cooling system is transferred to the secondary cooling system through a plate type heat exchanger. The amount of heater power at a heater power test section changed by adjusting the voltage on the heater. For each data collection, the surface temperature of the heating and cooling fluids are measured and recorded at 5 points at different height positions. Recording temperature for each experiment was performed after heating run for three hours on the expected steady state has been reached. Thermocouples used for temperature measurement directly connected to the data acquisition equipment. For the experiment associated to this paper, the main test section was configured in the natural convection mode with the triangular configuration as shown in figure 1a and 1b. The main experiment test section and the test box are made of glass sheets so that they are transparent as shown in figure 1.c. Cylinder assembly consists of three vertical cylinders, which is equipped with an electric heater and the type-K thermocouple sensors. Each cylinder has an outer diameter of 25.4 mm and length of 340 mm, and arranged with a pitch of 29.5 mm between the two cylinders. In this experiment, the input variable is the power dissipated by each heater cylinder. The output consists of the heater surface temperature and the temperature of the cooling water-Al_2O_3 nanofluid in a sub-channel.

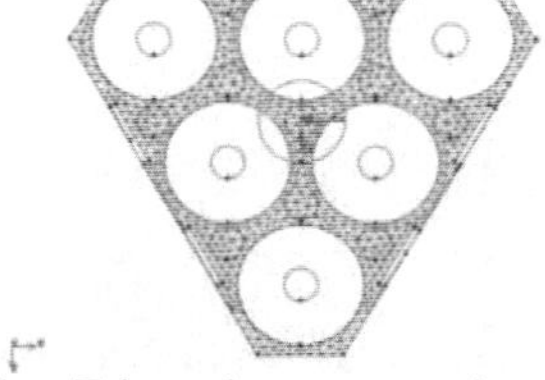

1.a. Triangular cross section

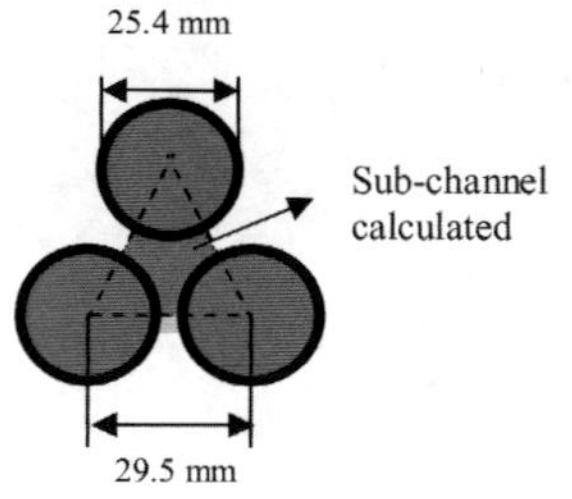

1.b. Calculated flow sub-channel

1.c. Experiment test section

Figure1. Cross section of the triangular configuration as main test section

3. Data analysis

Independent input variables for this experiment is the power of electric heater or heat flux on the surface of the heating cylinder, and position of measurement locations along the surface of the heater. The values of the input variables are shown in table 1. Meanwhile, the output variables are measured or calculated during the experiment is the temperature of the heater surface, the temperature of the water-Al_2O_3 nanofluid in a sub-channel at five points at different height positions the heat transfer coefficient.

Table 1. The values of the input variables

Variable Name	**Variable Type**	**Input Variable Values**
Electric Power, Q	Input	250, 350, 500, 650, 750, and 850 W/cylinder
Surface Heat Flux, q''	Input	9.3, 13,0 18.6, 24.2, 27.9, and 31.5 kW/m^2
Distance, x	Input	1.0, 9.0, 17.0, 25.0, and 33.0 cm
Surface Temperature, T_s	Output	-
Nanofluid bulk Temperature, T_b	Output	-
Heat Transfer Coefficient, h	Output	-

The main objective of this study is to obtain an empirical equation for natural convective heat transfer correlation of water-Al_2O_3 nanofluid, in the form of equation Nusselt Number, Nu, as a function of the Rayleigh number, Ra, and the non-dimensional position, x/D_h, as in the equation below,

$$Nu = f(Ra\frac{D_h}{x}) = a\left(Ra\frac{D_h}{x}\right)^b \quad (1)$$

With x is the measured position of the upstream and D_h is the diameter hydraulic cylinder of the sub-channels, whereas ***a*** and ***b*** are empirical constants to be determined from experiment.

Modified Nusselt number is defined by the following equation,

$$Nu = \frac{q^{"} D_h}{k(T_s - T_b)} \quad (2)$$

with q'', k, T_s , and T_b are surface heat flux, thermal conductivity of fluid, cylinder heater surface temperature, and bulk temperature of water-Al_2O_3 nanofluid, respectively.

The modified Rayleigh number is defined by the following equation,

$$Ra_q = \frac{g\rho\beta\, c_p q^{"} D_h^{\,4}}{k^2 \upsilon} \quad (3)$$

Where g is gravity, ρ is density, β is coefficient of expansion, c_p is heat capacity, and υ is kinematic viscosity of the nanofluid water-Al_2O_3 film near the cylinder surface.

All physical and transport properties are evaluated at the film temperature, T_f, as in below,

$$T_f = \frac{T_s + T_b}{2} \quad (4)$$

By using the heating cylinder geometry data, all input variables in table 1, and all measured temperatures, T_s and T_b, modified Nusselt and Rayleigh numbers for all measurement locations can be calculated. The relationship among the Nusselt number, Rayleigh number, and the location of the measurement can be determined using linear regression analysis.

4. Experiment results

The Al_2O_3 nanoparticles of about 20 nm diameter are used in the present study. Distribution of average wall temperature and water-Al_2O_3 nanofluid temperature distribution along the heater cylinder length at several of heat flux for triangular sub channel at various heat fluxes are shown in figure 2 and 3, respectively. The trend lines shown on these figures are just intended to emphasize the trends of the data. From the figures shown very clearly that the heating cylinder wall temperature tends polynomial shape and temperature of the water-Al_2O_3 nanofluid always increasing along sub channel, so that the line of linear approach. The general trend shows that the temperature curves in the figure are realistic and acceptable.

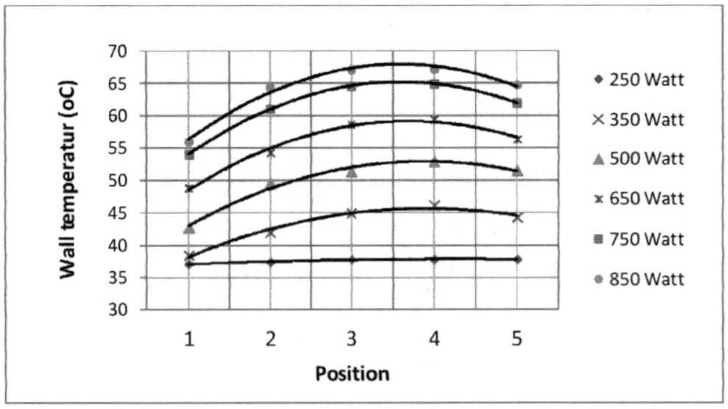

Figure 2. Wall temperature distribution at various heat fluxes

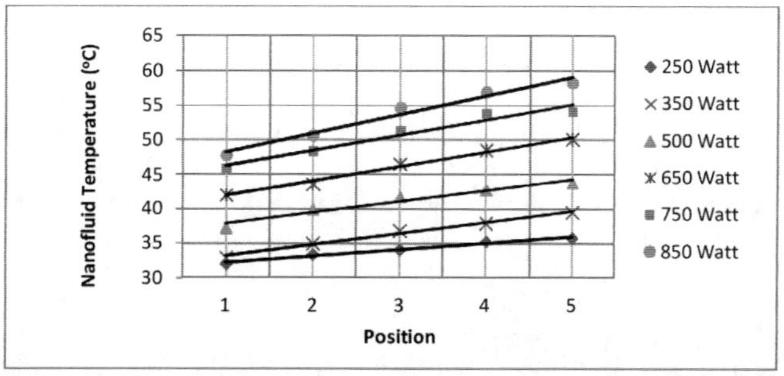

Figure 3. Nanofluid temperature distribution at various heat fluxes

From these figures it is clear that the water-Al_2O_3 nanofluid temperature triangular sub channel increase as the heat flux increases. Since the heater cylinder surface temperature and the water-Al_2O_3 nanofluid temperature were known from the measurements, then the film temperature can be calculated by using Equation 4, so that thermal conductivity and other physical properties of the water-Al_2O_3 nanofluid can be determined. These physical properties were evaluated at the film temperature. Meanwhile, the hydraulic diameter can be calculated from the know geometry of the heater cylinder arrangement. By knowing these variables and using eequation 2 and 3 the modified Nusselt and Rayleigh number can be calculated. A linear regression analysis gave an empirical correlation of natural convective heat transfer for water-Al_2O_3 nanofluid in the triangular sub channel between the Nusselt number, Rayleigh number and non-dimensional position as expressed by the following equations:

$$Nu = 15.97\left(Ra\frac{D_h}{x}\right)^{0.2778}$$ or 111 ⍰1 11 = 0.2778111 1 $\frac{1\,1}{1}$1 + ⍰.2035⍰

Meanwhile, a linear regression analysis gave an empirical correlation of natural convective heat transfer for water-ZrO_2 nanofluid in the sub channel between the Nusselt number, Rayleigh number and non-dimensional position as expressed by the following equations[10]:

$$Nu = 16.22\left(Ra\frac{D_h}{x}\right)^{0.0696}$$ or 111 ⍰1 11 = 0.0696111 1 $\frac{1\,1}{1}$ + ⍰.2102⍰

If the result of the experiment using water-Al_2O_3 nanofluid were compared with the results of the experiments using water-ZrO_2 and pure-water as the working fluid for the triangular sub-channel obtained graphs in figure 3 and 4. Based on the result, it was found that, for water Al_2O_3 and water-ZrO_2 the heat transfer coefficient can be up to 5 – 10 % higher than that of pure-water.

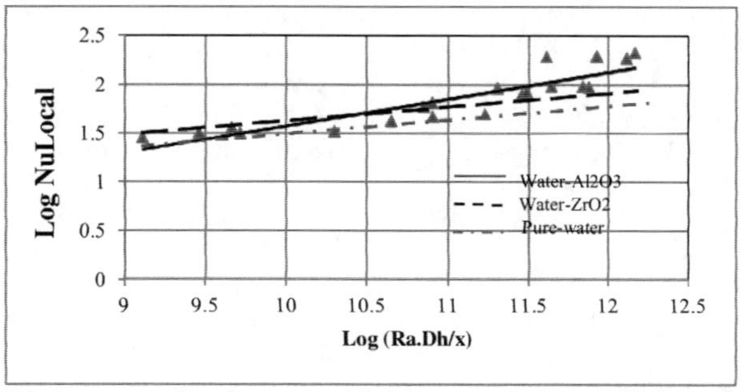

Figure 4. Linear regression for water-Al_2O_3, water-ZrO_2 nanofluid and pure-water

5. Conclusions

Based on the current experimental study, the following important points need to be highlighted as conclusions of this study.

1. It was obtained that on the vertical triangular sub-channel geometry, water-Al_2O_3 nanofluid heat transfer coefficient can be up to 5 - 10 % higher than that of pure-water.
2. A natural convective heat transfer equation for water-Al_2O_3 nanofluid in the vertical triangular sub-channel that depends on the position was obtained from the current study. The equation can be written as: $Nu_q = 15.97\left(Ra_q \frac{D_h}{x}\right)^{0.2778}$
3. The linear regression between water-Al_2O_3 and water-ZrO_2 nanofluid as working fluid is not much different, and the slope of the graph for the linear regression of water-Al_2O_3 slightly steeper.
4. The linear regression graph of water-Al_2O_3 and ZrO_2 nanofluid are above of that of the pure-water as a working fluid.

Acknowledgement

Authors are grateful to Dr. Dani Gustaman Sharif, Mr. Adis Bajarzali, Mr. Tata Kusmayadi and Mr. Budy Darmono for their assistance to prepare nanofluid and construct experiments. The project is funded by the National Nuclear Energy Agency of Indonesia.

6. References

[1] Buongiorno, J. March 2006 Convective Transport in Nanofluid, *Journal of Heat Transfer* (American Society of Mechanical Engineers), **128** (3) 240.

[2] Minkowycz, W., et al. 2013 Nanoparticle Heat Transfer and Fluid Flow, *CRC Press*, Taylor & Francis.

[3] Kakaç, Sadik; Anchasa Pramuanjaroenkij 2009 Review of convective heat transfer enhancement with nanofluid, *International Journal of Heat and Mass Transfer* (Elsevier) **52** 3187–3196.

[4] Maiga, Sidi El Becaye; Palm, S.J.; Nguyen, C.T.; Roy, G; Galanis, N, June 2005 Heat transfer enhancement by using nanofluid in forced convection flows, *International Journal of Heat and Fluid Flow* **26** 530–546.

[5] R. Chein, J. Chuang 2007 Experimental microchannel heat sink performance studies using nanofluid *International Journal of Thermal Sciences* **46** 57.

[6] W. Duangthongsuk, S. Wongwises 2008 Effect of thermophysical properties models on the prediction of the convective heat transfer coefficient for low concentration nanofluid, *International Communications in Heat and Mass Transfer* **35** 1320

[7] K.S. Hwang, S.P. Jang, S.U.S. Choi 2009 Flow and convective heat transfer characteristics of water based Al_2O_3 nanofluid in fully developed laminar flow regime, *International Journal of Heat and Mass Transfer* **52** 193–199

[8] Y. Xuan, Q. Li, 2003, Investigation on convective heat transfer and flow features of nanofluid *ASME Journal of Heat Transfer* **125** 151

[9] S. Mirmasoumi, A. Behzadmehr 2008 Effect of nanoparticles mean diameter on mixed convection heat transfer of a nanofluid in a horizontal tube, *International Journal of Heat and Fluid Flow* **29** 557–566

[10] Sudjatmi K.A, K. Kamajaya, and Efrizon Umar 2013 The effect concentration ZrO_2 heat transfer correlation nanofluid water-ZrO_2 for reactor coolant, *Journal of Nuclear Reactor Technology (Indonesia)* **15** (3) 172-182

[11] A. Zamzamian, S.N. Oskouie, A. Doosthoseini, A. Joneidi, M. Pazouki 2011 Experimental investigation of forced convective heat transfer coefficient in nanofluid of Al_2O_3/EG and CuO/EG in a double pipe and plate heat exchangers under turbulent flow *Experimental Thermal and Fluid Science* **35** 495

IOP Conf. Series: Materials Science and Engineering **88** (2015) 012054 doi:10.1088/1757-899X/88/1/012054

Physical Properties Variation of Graphene and Multi Wall Carbon Nanotubes by Planetary Ball Mill and Heat Treatment

G Jo[1], A K M Mahmudul Haque[2], S Kim[3], H Jeong[4], H Chung[5]
Production Editor, *Journal of Physics: Conference Series,*
[13]Department of Energy and Mechanical Engineering, Gyeongsang National University, Cheondaegukchi-Gil 38, Tongyeong-si 650-160, Gyeongsangnam-do, Republic of Korea
[2]Department of Ocean System Engineering, Gyeonsang National University, Cheondaegukchi-Gil 38, Tongyoung-si 650-160, Gyeonsangnam-do, Republic of Korea
[45]Department of Energy and Mechanical Engineering, Gyeongsang National University, Institute of Marine Industry, 445 Inpyeong-dong, Tongyeong-si 650-160, Gyeongsangnam-do, Republic of Korea

E-mail: handsomeboy1119@gmail.com

Abstract. The article reports the experimental study of graphene, multiwall carbon nanotubes and their composite where samples are made by applying dry grinding and wet grinding as well as heat treatment processes. The investigation information is recorded and compared by SEM analysis, EDS spectrum and density measurement which give the informative evidence. And finally the article tries to find the basic difference in physical properties of these samples.

1. Introduction

The rising cost of indium and high temperature processing [1)] and brittleness of ITO (indium tin oxide) [2)] engendered thinking alternatives which are GN (graphene) [3)] as well as CNTs (carbon nanotubes) [4)]. And, it is the extensive concern to evaluate which one is better one by comparing among GN, CNTs and their composite in various applications, for example, in heat sink applications as coolant [5)]. Within a few decades there are several researches have been done in this field and are still going on. And the outcomes are not only positive but also informative, alluring and practically applicable.

GN, two dimensional extended honeycomb network of sp^2 hybridized carbon atoms [6)], high electron mobility [7)], excellent mechanical, chemical and thermal properties [8)], has been used broadly in super capacitors [9)], optoelectronic devices [10)] as well as in several high performance sensors [11)]. And, CNTs, one of the lightest [12)], strongest [13)], stiffest [13)], electrically [14)] and thermally [15)] conductive nanoparticles, is also applied almost in same fields as that of GN. Now a days, metal or metal nanowires [16)] and

[1] To whom any correspondence should be addressed.

hybrids of these [17] are also used for making composites with GN and CNTs for getting better performance.
This paper studies the comparison among GN, MWCNTs and their composite i.e. GN-MWCNTs and tries to report as follows:

- SEM reports give the surface analysis information (Section 3.1) by which roughness and porosity are revealed.
- EDS spectrum gives the information of the presence of carbon as well as other particles (Section 3.2) which also confirms the presence of any contamination.
- Density measurement informs the amount of particles within a given area (Section 3.3) of each sample.

It should be noted that the procedure and the results of this paper are only for comparison and do not suggest the optimum conditions.

2. Experimental Details

2.1. Materials
Raw MWCNTs with a ~20nm diameter, ~5μm length (Carbon Nanomaterial Technology Co., Ltd, South Korea) and grpahene nanopowder with 8nm (average flake thickness) flakes, average particle size ~550 nm, specific surface area 100 m^2/g, and 99.9% purity (graphene supermarket) are used in this experimental study.

2.2. Grinding process
A planetary ball mill machine (HPM-700) (Haji Engineering, Korea) is used for grinding where MWCNT, GN and their composite are ground for one hour with 400rpm according to the experimental approach.

2.3. Density measurement
A gas Pycnometer (AccuPye1340, micromeritics) is used for measuring density of 0.10g of GN, MWCNTs and GN-MWCNTs composite samples.

2.4. Furance
An electric box furnace (Dongwon scientific Co.) is used for applying thermal treatment of the samples at 90°C for a day.

2.5. Samples

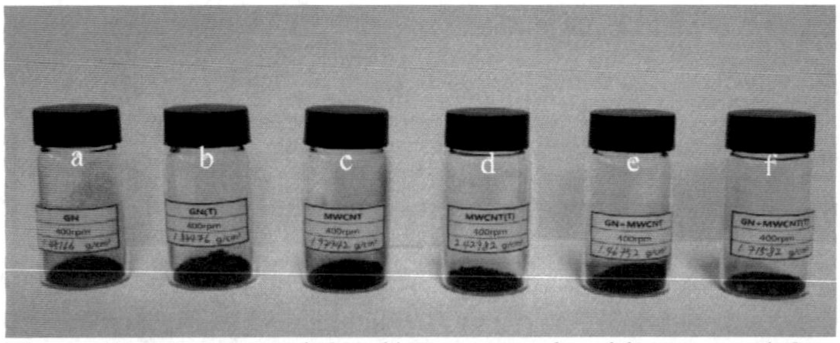

Figure 1. a)Dry ground GN, b)Wet ground and heat treated GN, c) Dry ground MWCNT, d) Wet ground and heat treated MWCNT, e) Dry ground GN-MWCNT, f) Wet ground and heat treated GN-MWCNT(T)

Total six samples are made where GN, MWCNTs and GN-MWCNTs composite are three main samples. And each sample is prepared in two ways: one is dry grinding and another one is wet grinding and furnace under some given parameters as mentioned above.

3. Experimental results and discussion

3.1. SEM results

Figure 2. SEM image of dry grinding of GN.

Figure 3. SEM image of wet grinding and heat treatment of GN.

From Fig. 2. and Fig. 3, it has seen that there is no eloquent information of substantial difference between without and with heat treatment of ground GN. So, in this experiment heat treatment has not been effective for GN samples.

Figure 4. SEM image of dry grinding of MWCNTs.

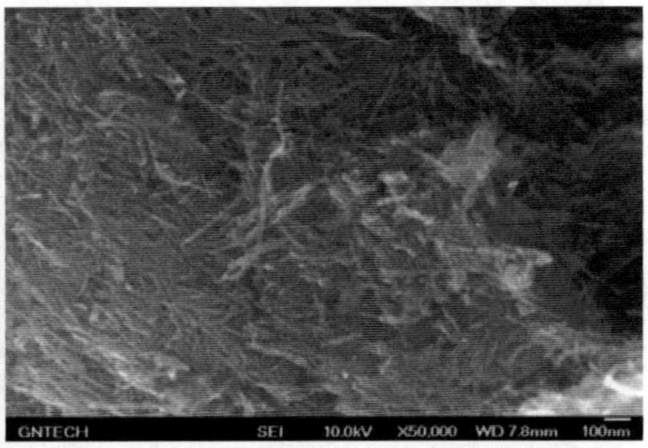

Figure 5. SEM image of wet grinding and heat treatment of MWCNTs.

Without heat treating the surface of ground MWCNTs has been rough, whereas, heat treatment (Fig.5.) has ameliorated the surface.

Figure 6. SEM image of dry grinding of GN-MWCNTs.

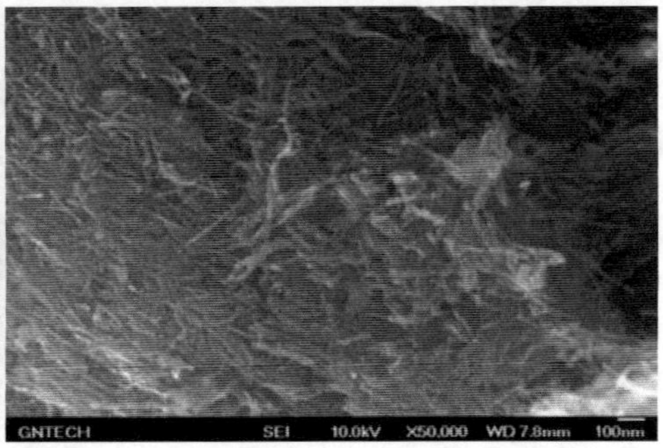

Figure 7. SEM image of wet grinding and heat treatment of GN-MWCNTs.

Fig. 6. and Fig. 7. have revealed that same as ground MWCNTs, heat treatment has upgraded the surface of GN-MWCNTs incomparably.

3.2. EDS spectrum

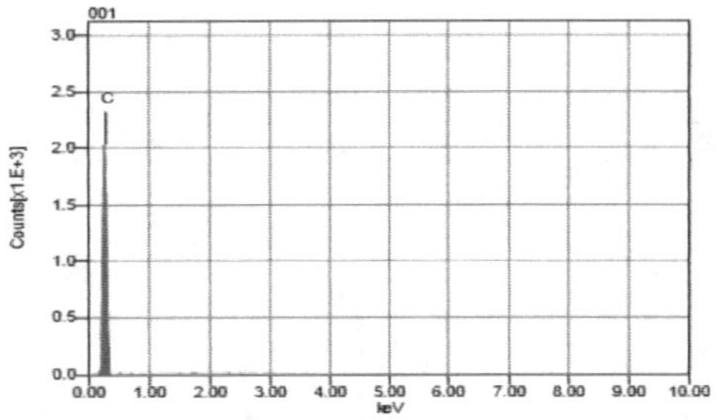

Figure 8. SEM image of dry grinding of GN-MWCNTs.

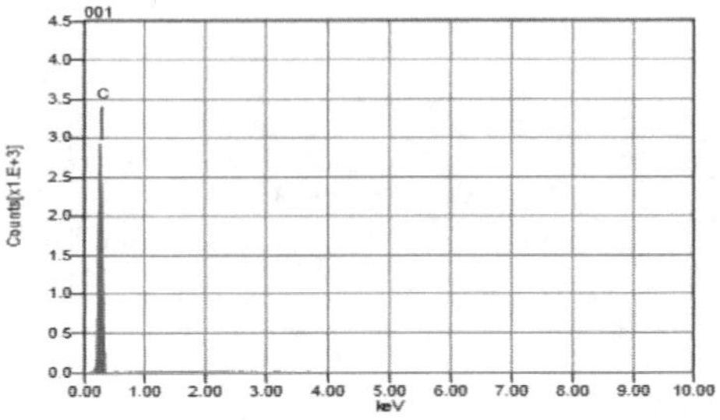

Figure 9. EDS spectrum of wet grinding and heat treatment of GN.

Table 1: Dry grinding of GN.

Chemical	mass%	Atom%
C	100.00	100.00
Total	100.00	100.00

Table 2: Wet grinding and heat treatment of GN.

Chemical	mass%	Atom%
C	100.00	100.00
Total	100.00	100.00

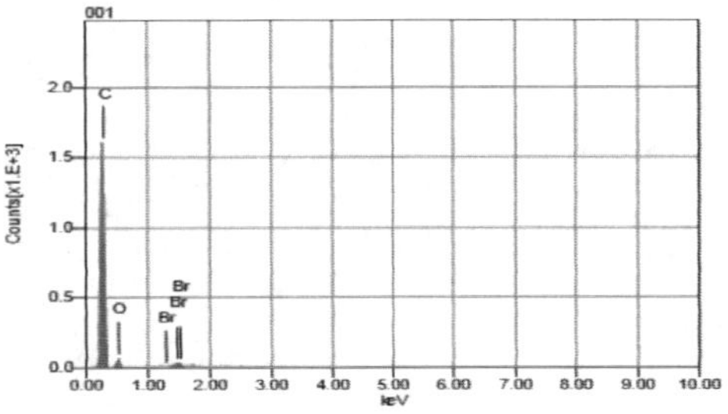

Figure 10. EDS spectrum of dry grinding of MWCNTs.

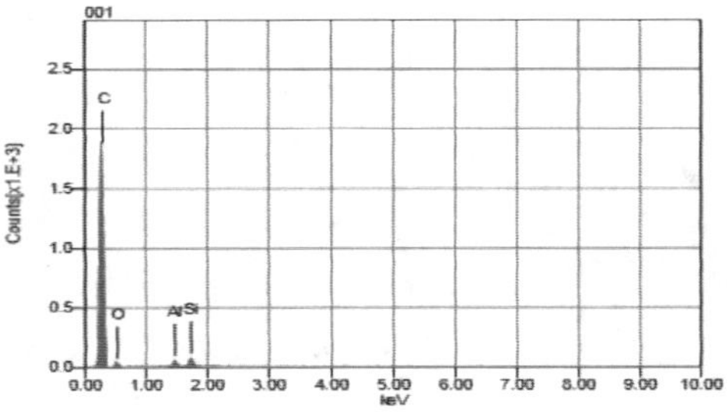

Figure 11. EDS spectrum of wet grinding and heat treatment of MWCNTs.

Table 3: Dry grinding of MWCNTs.

Chemical	mass%	Atom%
C	88.47	93.06
O	8.10	6.40
Br	3.43	0.54
Total	100.00	100.00

Table 4: Wet grinding and heat treatment of MWCNTs.

Chemical	mass%	Atom%
C	91.45	94.23
O	5.96	4.61
Al	1.11	0.51
Si	1.48	0.65
Total	100.00	100.00

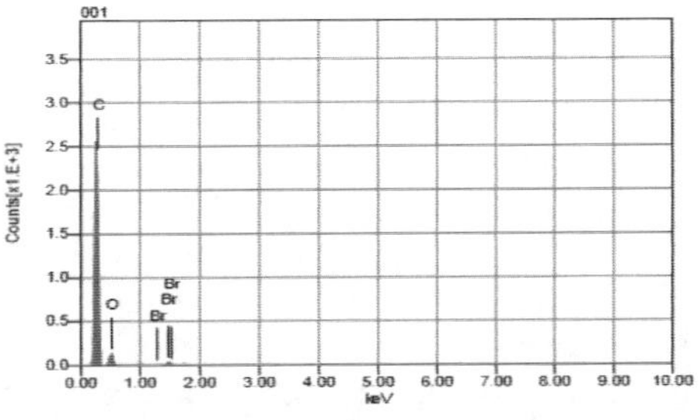

Figure 12. EDS spectrum of dry grinding of GN-MWCNTs.

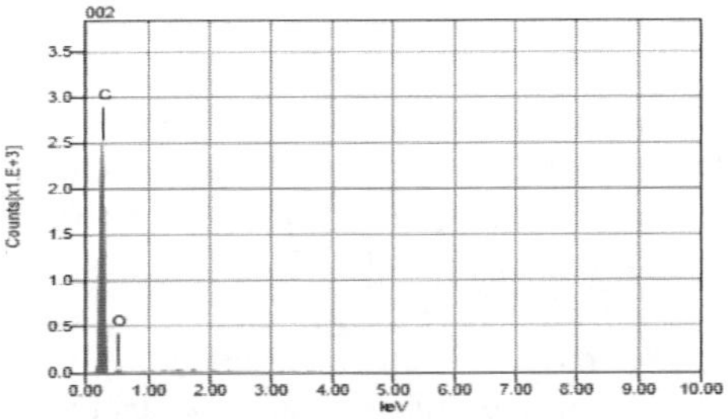

Figure 13. EDS spectrum of wet grinding and heat treatment of GN-MWCNTs.

Table 5: **Dry grinding of GN-MWCNTs.**

Chemical	mass%	Atom%
C	86.15	90.74
O	11.17	8.84
Br	2.68	0.42
Total	100.00	100.00

Table 6: **Wet grinding and heat treatment of GN-MWCNTs.**

Chemical	mass%	Atom%
C	97.13	97.83
O	2.87	2.17
Total	100.00	100.00

3.3. Density

For without heat treatment, the densities of GN, MWCNTs and GN-MWCNTs samples are 1.45166g/cm^3, 1.97942 g/cm^3 and 1.46752 g/cm^3 respectively and for with heat treatment, the values are 1.83476 g/cm^3, 2.42982 g/cm^3 and 1.71582 g/cm^3 respectively (Fig: 14). In both cases, density of MWCNTs is higher than other two. And heat treatment has given better density. The adherence problem of making GN sample has occurred severely than other that of other samples. And because of this problem the sample of GN has not been collected properly after dry grinding. There may be a side effect of this on the density of GN sample which is the lowest value of this density measurement.

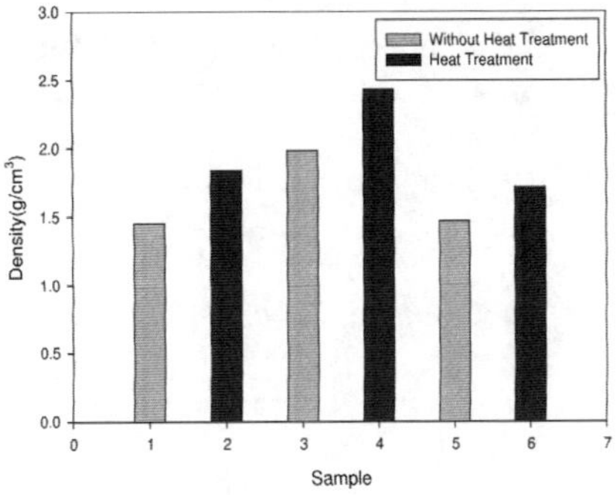

Figure. 14. Density of sample1: dry grinding of GN, sample 2: wet grinding and heat treatment of GN, sample 3: dry grinding of MWCNTs, sample 4: wet grinding and heat treatment of MWCNTs, sample 5: dry grinding of GN-MWCNTs, sample 6: wet grinding and heat treatment of GN-MWCNTs.

4. Conclusions

At first, the surface analysis was done with the help of SEM. After that, EDS spectrums were analyzed in order to analyze the surfaces broadly. And finally the values of the density of the six samples were measured. After analyzing these data, the key conclusions can be digested as follows:

- There is no significant change or improvement by applying heat treatment of ground GN.
- Heat treatment is effective for decreasing the surface roughness of MWCNTs as well as GN-MWCNTs.
- The amount of carbon in these samples increased noticeably by heat treatment.
- There was an adherent of particles to grinding media, i.e., vial and balls during dry grinding which was substantially reduced by wet grinding process.
- Wet grinding and heat treatment approaches represented the considerably higher density than only dry grinding process. And for the dry grinding and the wet grinding as well as heat treatment processes, the density of MWCNTs is noticeably higher than others.

Acknowledgment

This research was supported by the Basic Science Research Program through the National Research Foundation of Korea (NRF) funded by the Ministry of Education, Science and Technology (NRF-2014R1A1A4A03005148).

5. References

[1] Y.-L. Huang, A. Baji, H.-W. Tien, Y.-K. Yang, S.-Y. Yang, S.-Y. Wu, C.-C. M. Ma, H.-Y. Liu, Y.-W. Mai, and N.-H. Wang, "Self-assembly of silver–graphene hybrid on electrospun polyurethane nanofibers as flexible transparent conductive thin films," *Carbon N. Y.*, vol. 50, no. 10, pp. 3473–3481 (2012).

[2] H. Chang, G. Wang, A. Yang, X. Tao, X. Liu, Y. Shen, and Z. Zheng, "A Transparent, Flexible, Low-Temperature, and Solution-Processible Graphene Composite Electrode," *Adv. Funct. Mater.*, vol. 20, no. 17, pp. 2893–2902 (2010).

[3] S. J. Wang, Y. Geng, Q. Zheng, and J.-K. Kim, "Fabrication of highly conducting and transparent graphene films," *Carbon N. Y.*, vol. 48, no. 6, pp. 1815–1823 (2010).

[4] Y.-L. Huang, H.-W. Tien, C.-C. M. Ma, C.-C. Teng, Y.-H. Yu, S.-Y. Yang, M.-H. Wei, and S.-Y. Wu, "Effect of cross-linkable polymer on the morphology and properties of transparent multi-walled carbon nanotube conductive films," *Appl. Surf. Sci.*, vol. 258, no. 1, pp. 136–142 (2011).

[5] M. Kole and T. K. Dey, "Effect of prolonged ultrasonication on the thermal conductivity of ZnO–ethylene glycol nanofluids," *Thermochim. Acta*, vol. 535, pp. 58–65 (2012).

[6] G. Refael, "Inducing topological order in a honeycomb lattice," pp. 1–7 (2012).

[7] K. I. Bolotin, K. J. Sikes, Z. Jiang, M. Klima, G. Fudenberg, J. Hone, P. Kim, and H. L. Stormer, "Ultrahigh electron mobility in suspended graphene," *Solid State Commun.*, vol. 146, no. 9–10, pp. 351–355 (2008).

[8] C. Lee, X. Wei, J. W. Kysar, and J. Hone, "Measurement of the elastic properties and intrinsic strength of monolayer graphene.," *Science*, vol. 321, no. 5887, pp. 385–8 (2008).

[9] S. R. C. Vivekchand, C. S. Rout, K. S. Subrahmanyam, a. Govindaraj, and C. N. R. Rao, "Graphene-based electrochemical supercapacitors," *J. Chem. Sci.*, vol. 120, no. 1, pp. 9–13 (2008).

[10] X. Wang, L. Zhi, and K. Müllen, "Transparent, conductive graphene electrodes for dye-sensitized solar cells.," *Nano Lett.*, vol. 8, no. 1, pp. 323–7 (2008).

[11] P. K. Ang, W. Chen, A. T. S. Wee, and K. P. Loh, "Solution-gated epitaxial graphene as pH sensor.," *J. Am. Chem. Soc.*, vol. 130, no. 44, pp. 14392–3 (2008).

[12] X. Zhang, Q. Li, T. G. Holesinger, P. N. Arendt, J. Huang, P. D. Kirven, T. G. Clapp, R. F. DePaula, X. Liao, Y. Zhao, L. Zheng, D. E. Peterson, and Y. Zhu, "Ultrastrong, Stiff, and Lightweight Carbon-Nanotube Fibers," *Adv. Mater.*, vol. 19, no. 23, pp. 4198–4201 (2007).
[13] F. Inam, A. Heaton, P. Brown, T. Peijs, and M. J. Reece, "Effects of dispersion surfactants on the properties of ceramic–carbon nanotube (CNT) nanocomposites," *Ceram. Int.*, vol. 40, no. 1, pp. 511–516 (2014).
[14] S. Tawfick, K. O'Brien, and a J. Hart, "Flexible high-conductivity carbon-nanotube interconnects made by rolling and printing.," *Small*, vol. 5, no. 21, pp. 2467–73 (2009).
[15] S. Berber, Y. Kwon, and D. Tomanek, "Unusually high thermal conductivity of carbon nanotubes," *Phys. Rev. Lett.*, vol. 84, no. 20, pp. 4613–6 (2000).
[16] H. Wu, L. Hu, M. W. Rowell, D. Kong, J. J. Cha, J. R. McDonough, J. Zhu, Y. Yang, M. D. McGehee, and Y. Cui, "Electrospun metal nanofiber webs as high-performance transparent electrode.," *Nano Lett.*, vol. 10, no. 10, pp. 4242–8 (2010).
[17] T.-K. Hong, D. W. Lee, H. J. Choi, H. S. Shin, and B.-S. Kim, "Transparent, flexible conducting hybrid multilayer thin films of multiwalled carbon nanotubes with graphene nanosheets.," *ACS Nano*, vol. 4, no. 7, pp. 3861–8 (2010).

Effect of pervaporation plate thickness on the rate of methanol evaporation in a passive vapor-feed direct methanol fuel cell

N F I Fauzi, U A Hasran[1], S K Kamarudin
Fuel Cell Institute
Universiti Kebangsaan Malaysia, 43600 UKM Bangi, Selangor, Malaysia

E-mail: umi.h@ukm.edu.my

Abstract. In a passive vapor-feed direct methanol fuel cell (DMFC), methanol vapor is typically obtained using a pervaporation plate in a process by which liquid methanol contained in the fuel reservoir undergoes a phase change to vapor in the anodic vapor chamber. This work investigates the effect of pervaporation plate thickness on the rate of methanol evaporation using a three-dimensional simulation model developed by varying the plate thickness. A. The rate of methanol evaporation was measured using Darcy's law. The rate of methanol evaporation was found to be inversely proportional to the plate thickness, where the decrease in thickness inevitably lowers the resistance along the plate and consequently increases the methanol transport through the plate. This shows that the plate thickness has a significant influence on the rate of methanol evaporation and thereby plays an important role in improving the performance of the passive vapor-feed direct methanol fuel cell.

1. Introduction

Methanol in a direct methanol fuel cell (DMFC) can be used as fuel in its liquid or vapor form. However, the vapor feed DMFC has an advantage over the liquid feed system as it has the potential to have a higher operating temperature, increase the reaction rates and power outputs, enhance the mass transfers, and reduce methanol crossover. It is reliable for high methanol concentration and therefore can increase the fuel cell performance [1, 2].

In terms of fuel and oxidant delivery schemes, DMFCs can be classified either as active-feed or passive-feed. An active DMFC operates with the help of external devices such as a liquid pump for methanol delivery and a gas blower or fan for oxygen delivery, while a passive DMFC can perform stand-alone operations that eliminate the need for any external devices [3, 4].

Since early 1990s, methanol evaporation is achieved in an active vapor feed DMFC by utilizing an external device, such as an electric vaporizer or heater. However, these devices can suffer from excessive complexity that causes significant parasitic power losses, along with increased system volume and cost, especially in the case of small scale power sources. A passive vapor feed DMFC is therefore preferable for portable application devices because of its simplicity, small volume and weight and a more compact design. Vapor phase operation at high temperatures is incompatible with

[1] To whom any correspondence should be addressed.

the need for passive approaches for portable DMFCs [5] and thus pervaporation plate can be used as a medium to vaporize methanol fuel in order to operate passively at ambient temperature.

Abdelkareem et al. investigated the effect on the performance of a passive DMFC under closed circuit conditions due to the usage of a porous carbon plate (PCP) as the vaporizer with different methanol concentration ranging from 1M to neat methanol. As a result, a maximum power density of 24 mW cm^{-2} was achieved at room temperature with a 2M solution without the PCP, and the result can be reproduced at 16M with the PCP. A novel PCP with better durability than other pervaporation medium was also developed [6, 7]. In this study, a model for a pervaporation plate as the vaporizer in a passive vapor feed DMFC that utilizes high concentration methanol was developed. The effect of the plate thickness on the rate of methanol evaporation is investigated by using COMSOL Multiphysics 4.3 software (COMSOL Inc., USA).

2. Methods

A schematic of a half-anode side layer in a passive vapor feed DMFC system operating with neat methanol is depicted in figure 1. Generally, the function of the vaporizer is to adsorb liquid methanol from the methanol reservoir and evaporate it as methanol vapor. The model utilized to describe mass transport in a pervaporation process in this simulation work is the pore flow model. This model assumes that the concentrations of solvent and solute within a plate are uniform and a pressure gradient is used to express the chemical potential gradient across the plate [8]. According to the pore flow model, three steps are included to describe mass transport in the pervaporation process: (i) the permeant is transported through the liquid-filled portion of the pore, (ii) a liquid-to-vapor phase change takes place inside the pore and (iii) the permeant is transported through the vapor-filled portion of the pore [9].

Liquid methanol reservoir
Vaporizer
Vapor transport layer

Figure 1. Schematic of half-anode side layer.

2.1. Model Development

The pervaporation process is a phenomena that affects the phase change of methanol from liquid to vapor. A three dimensional pervaporation plate model was developed using COMSOL Multiphysics. The effects of plate thickness on the performance of the vaporizer were investigated by measuring the rate of methanol evaporation. The range of thickness values used is 0.08-0.30 mm and the plate active area is 10 x 10 mm.

2.1.1. Model Assumptions. Model assumptions are as given below:

- The system is isothermal.
- The system operates under steady state condition.
- There is no chemical reaction through the plate.
- The pervaporation plate is a porous media.
- Vaporization of methanol from the liquid phase to the vapor phase occurs due to the pressure gradient in plate.
- Methanol vapor generated at the vapor phase will not condense.

- Loss of methanol vapor through the vent during the CO_2 releasing process is neglected.
- Only methanol species exists in the vaporizer.

To make the complicated model more tractable, the following simplifications are listed below:

- Methanol fed into the system is neat methanol with vapor pressure 13.02 kPa at 20°C.
- For all the thickness values in the model, the inlet pressure going into the plate is 101.325 kPa, as methanol in the reservoir is at atmospheric conditions. The outlet pressure going away from the plate is 12.0 kPa, which is assumed to be below the vapor pressure of methanol.

2.1.2. Model Description. The model consists of two compartments separated by a plate of thickness, L, as shown in figure 2 below. The following three phases have been considered: (i) the pure methanol in liquid phase (phase 1) in the upstream compartment is at atmospheric pressure (P1 = 1 atm) and well-stirred so that no concentration gradient exists, (ii) the plate (phase 2) is assumed to be filled with homogeneous permeants within the media and (iii) the methanol vapor (phase 3) in the downstream compartment is maintained at a pressure lower than the vapor pressure of methanol.

In liquid phase, methanol permeates from the reservoir tank to the surface of the pervaporation plate that interfaces the methanol reservoir. In the plate phase, methanol diffuses through the pervaporation plate and follows a path defined by pressure gradient, which is the major driving force. Methanol is transported from the high pressure region to the low pressure region, which leads to evaporation. Finally, methanol vapor desorbs from the pervaporation plate surface.

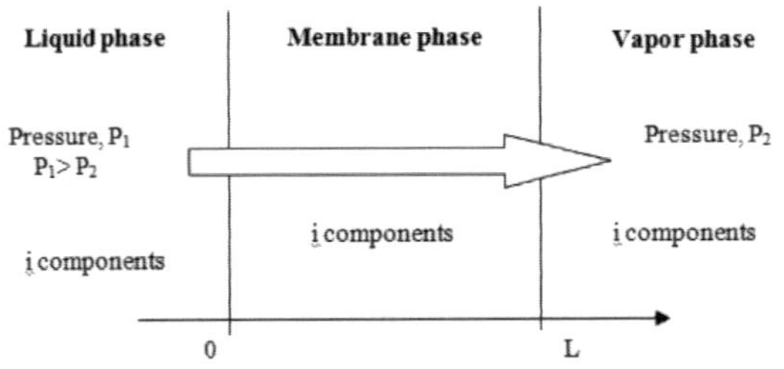

Figure 2. A schematic of the pervaporation plate.

3. Results and discussion

3.1. Rate of methanol diffusion

Figure 3shows the profile of the rate of methanol diffusion into the plate for varying thickness values with neat methanol as fuel. The inlet pressure is higher than the outlet pressure as the pressure in the plate gradually decreases until the end of the plate thickness. As the outlet pressure is lower than the methanol vapor pressure, the methanol can diffuse completely through the plate and converted to vapor. The rate of methanol diffusion is seen to increase from the inlet to the outlet of the plate.

At low pressure, the distance between the plate molecules is larger. Thus, the space that allows for methanol molecules going through the plate becomes larger. Therefore, the rate of methanol diffusion through the plate is also higher. In contrast, the pressure at the beginning of the plate is high. Therefore, the distance between the plate molecules is closer and results in a smaller space for the methanol molecules to go through the plate. Thus, the rate of methanol diffusion through the plate is lower.

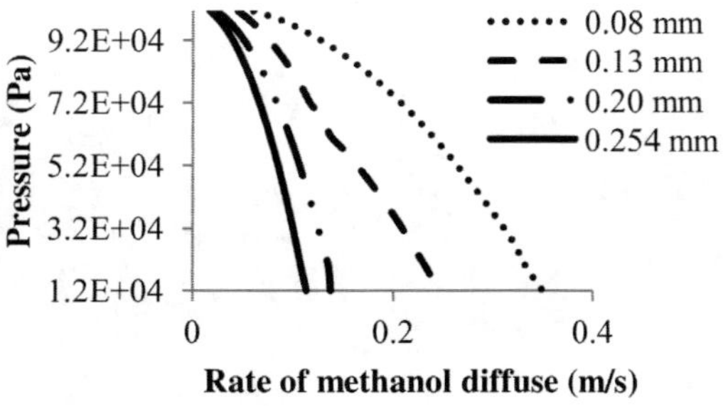

Figure 3. Effect of pressure on the rate of methanol diffusion varies thickness of plate.

From figure 3, methanol diffuses through the plate and starts undergoing a phase transition from the liquid phase to the vapor phase at 13.02 kPa as a result of the methanol vapor pressure. At 12.0 kPa, all the liquid methanol has been transformed to the vapor phase. Hence, the rate of methanol evaporation was started at end of the plate when the pressure was at 12.0 kPa. The rates of methanol evaporation for the plate thickness 0.08, 0.13, 0.20 and 0.254 mm are 0.348, 0.245, 0.137 and 0.113 m/s, respectively.

Figure 4 - 7 show the visual profile of the rate of methanol diffusion through the plate for the thickness values of 0.08, 0.13, 0.20 and 0.254 mm, respectively. The pressure difference between the upstream (inlet) and downstream (outlet) compartments is kept constant. It induces methanol to flow through the plate in a perpendicular direction to its surface, i.e. in a positive x-axis direction. The color indicators show a similar trend for all the profiles, i.e. the methanol diffusion starts at a very slow rate and increases gradually as it goes from inlet to outlet.

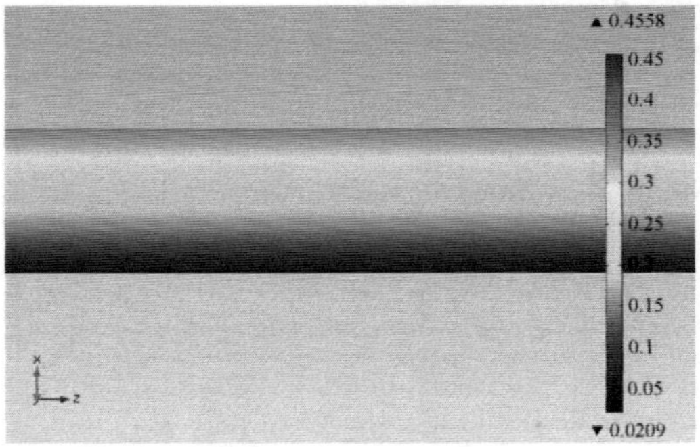

Figure 4. Profile of rate of methanol diffusing through plate for 0.08 mm thickness.

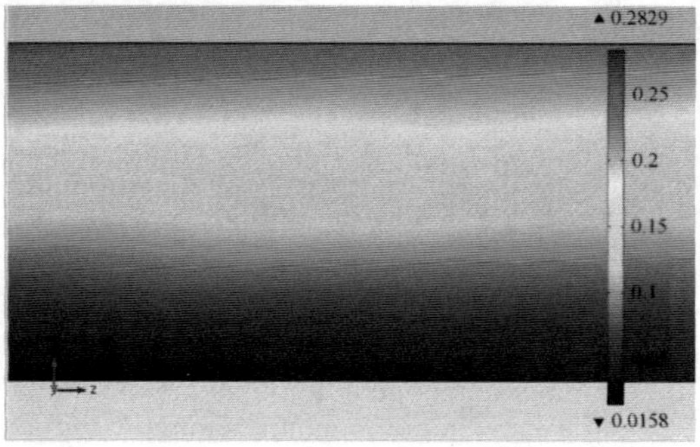

Figure 5. Profile of rate of methanol diffusing through the plate for 0.13 mm thickness.

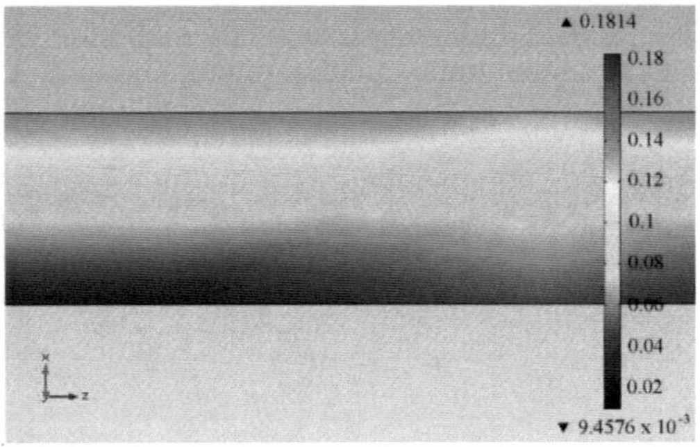

Figure 6. Profile of rate of methanol diffusing through the plate for 0.20 mm thickness.

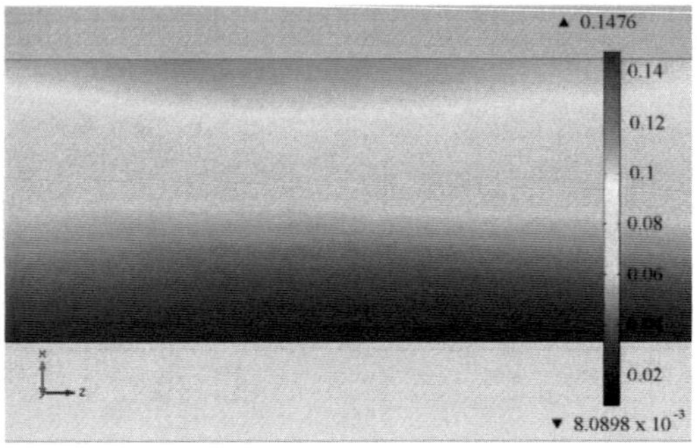

Figure 7. Profile of the methanol diffusion rate through a plate with 0.254.

3.2. Comparison between the rates of methanol diffusion

The rate of methanol diffusion for the plate thickness value of 0.13 mm in figure 8 is shown to have a total difference of 0.21 m/s, while the difference before methanol evaporation is 0.10 m/s. Whereas, the rate of methanol diffusion through a 0.254 mm thick plate in figure 9 is shown to have a total difference of 0.10 m/s and before methanol evaporation of 0.02 m/s.

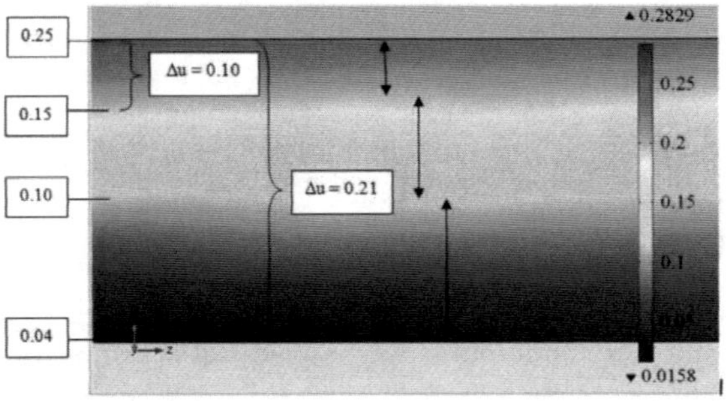

Figure 8. Detail information for the rate of methanol diffuse through the plate for 0.13 mm thickness.

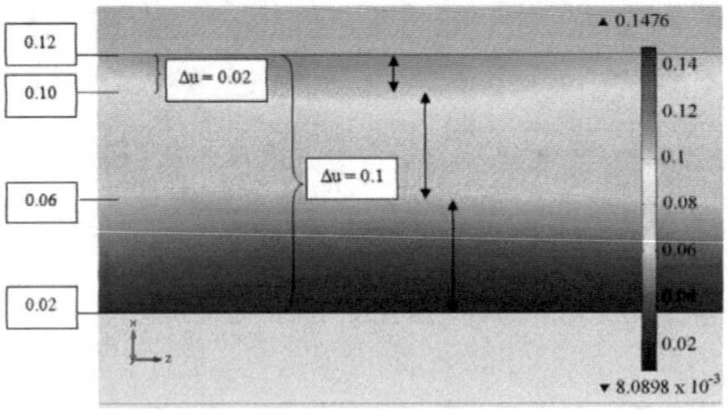

Figure 9. Detail information for the rate of methanol diffuse through the plate for 0.254 mm thickness.

From figure 8 -9, it can be deduced that the total difference in the rate of methanol diffusion and the difference in the rate of methanol evaporation for the thinner plate is larger. This indicates that the rate of methanol evaporation increases as the methanol diffusion along the thinner plate proceeds at a higher rate. Therefore, it can be concluded that plate thickness has a significant influence on the rate of methanol evaporation.

Figure 10 shows the relationship between the rate of methanol evaporation and the plate thickness. The graph shows that the rate of methanol evaporation is indirectly proportional to the plate thickness and the rate of methanol evaporation will increase with decreasing thickness of the plate. This trend is probably because the thinner the plate is, the less resistance is experienced by the methanol molecules that are being transported through the plate.

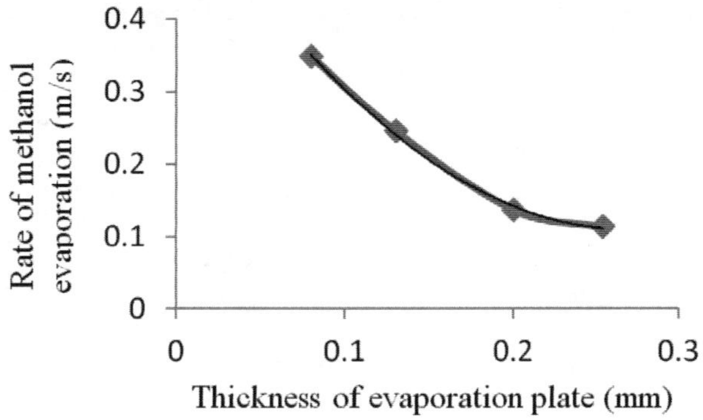

Figure 10. Rate of methanol evaporation versus the thickness of plate.

4. Conclusion

This paper reports the simulation study based on the performance of passive vapor feed DMFC operated with neat methanol as a fuel. The effect of the plate thickness on the rate of methanol evaporation was carried out. The rate of methanol evaporation is found to be inversely proportional to the plate thickness. Therefore, as the thickness of the plate decreases, the rate of methanol evaporation increases, and this will in turn improve the methanol vapor feed into the direct DMFC.

5. Acknowledgements

This work was supported by the Ministry of Science and Technology Innovation under the Science Fund Grant Number of ERGS/1/2012/TK07/UKM/03/2.

6. References

[1] Rice J and Faghri A 2008 *International Journal of Heat and Mass Transfer* **51**(3–4) 948-959
[2] Xu C and Faghri A 2010 *Journal of Power Sources* **195**(20) 7011-7024
[3] Chen R and Zhao T 2005 *Journal of Power Sources* **152** 122-130
[4] Fauzi N F I, Hasran U A and Kamarudin S K 2013 *IOP Conference Series: Materials Science and Engineering* **50**.
[5] Eccarius S, Krause F, Beard K and Agert C 2008 *Journal of Power Sources* **182**(2) 565-579
[6] Abdelkareem M A and Nakagawa N 2006 *Journal of Power Sources* **162**(1) 114-123
[7] Abdelkareem M A and Nakagawa N 2007 *Journal of Power Sources* **165**(2) 685-91
[8] Wijmans J G and Baker R W 1995 *Journal of Membrane Science* **107**(1–2) 1-21
[9] Sukitpaneenit P, Chung T -S and Jiang L Y 2010 *Journal of Membrane Science* **362**(1–2) 393-406

IOP Conf. Series: Materials Science and Engineering **88** (2015) 012056 doi:10.1088/1757-899X/88/1/012056

Effect of surface tension on SiO_2–methanol nanofluids.

M.H.U. Bhuiyan [1, 2], R. Saidur[1], M.A. Amalina[1], R.M. Mostafizur[1]
[1]Department of Mechanical Engineering, Faculty of Engineering, University of Malaya, 50603 Kuala Lumpur, Malaysia.

E-mail: mhub2k2@yahoo.com

Abstract. Surface tension, the cohesive energy of an interface dominated the transportation behaviour of the liquids play an important role in the heat transfer performance. A new class of heat transfer fluid denoting "Nanofluids" with impressive thermo-physical properties, proved its promising potentiality in the heat transfer performance. However, very few numbers of studies observed for the effect of nanoparticles on the surface tension of liquids, also noted controversial results. In the present study, SiO2 nanoparticles dispersed in methanol solution to investigate the effect of surface tension with the change of concentration and their sizes. The most common Du–Noüy ring method was used to measure the surface tension of methanol based nanofluids by an automatic surface tensiometer.The results denote that the surface tension of the nanofluids increases with increase in concentration. On the other hand, the results indicate that the surface tension decreases with the increase in temperatures. Besides, the surface tension of SiO2–methanol nanofluids enhances compared to pure methanol. All in all, the enhancement observed 1.7% to 8.9% of the variation of volume fractions (0.05 Vol % to 0.25 Vol %) and the temperature change of 25 °C to 50 °C.

1. Introduction:

Enhancing the heat transfer is one of the foremost apprehensions in the research community to save the energy of a thermal system due to it is anticipated that all the energy of the universe will spring up over 50 % in the next 25 years. A colloidal mixture denoting "Nanofluids" is the suspension of nanoparticles (normally less than 100 nm) in the conventional fluids introduced by Choi et al. [1]. It has shown dramatic improvements of its thermophysical properties and used in various applications in the field of heat transfer [2]. Besides all other properties, surface tension of the fluids has significantly influenced the heat transfer performance of a thermal system. It is the amount of surface free energy per unit area of the liquid droplets [3]. Surface tension is the force which influences the surface wettability [4], bubble growth, and enhances the boiling heat transfer coefficient [5]. Moreover, the capillary action causes by surface tension of the working fluid transfer the liquid from the condenser to evaporator, which influences to increase the thermal performance of the heat pipes [6]. However, most of the experimental and theoretical studies investigated thermal conductivity, viscosity, density and specific heat of the nanofluids [7, 8]. Besides, some studies were executed to measure the surface tension of the nanofluids[9]. Therefore, it is imperative to examine the effect of nanoparticles on the surface tension of the conventional heat transfer fluids.

There is no significant change on surface tension observed while measuring the Al2O3–H2O nanofluids[10]. Zhu et al. [11] reported surface tension of Al2O3–H2O nanofluids decrease with the increase in concentration and temperatures. Moreover, they observed that smaller nanoparticles showed lower surface tension than a larger one. However, Tanvir et al. [12] found surface tension of the nanofluids enhances with the increase in both the concentration and particle sizes. Similarly, Majid

Moosavi et al. [13] demonstrated the ratio of surface tension of ZnO/EG nanofluids increases with the increasing the volume fraction. Further, Murshed et al.[14] pointed that the surface tension of TiO2- DIW nanofluids reduced linearly with the change of temperature 25°C to 55°C at 0.1 Vol%. Furthermore, surface tension of Bi2Te3 nanoparticles in aqueous solution at first decreases, and, again rises with the increase in concentration observed by Saeid Vafaei et al. [15]. Therefore, the earlier studies demonstrated that the effect of surface tension of the nanofluids governed by the concentration, temperatures and nanoparticle sizes.

Besides water, methanol is the most widely used heat transfer fluids used in the different applications [16, 17]. Morover, Kim et.al[18] observed mass transfer enhancement in Al2O3-methanol nanofluids. Further, methanol based nanofluids showed their performance in the absorption of carbon dioxide (CO2) gas [19, 20], HVAC system [21], heat pipes and air-conditioning system [22]. Therefore, methanol based nanofluids is potential to enhance the heat transfer rate of heat exchangers and save on energy. Additionally, some literatures were found about methanol based nanoluids. Nevertheless, most of them were discussed only the thermal conductivity and rheological properties [23, 24]. In the present experimental study, the authors investigate the effect of SiO2–methanol nanofluids on surface tension for the change of volume fractions (0.05 Vol % to 0.25 Vol%) along with the variation of temperatures (25 °C to 50 °C) .

2. Experimental process

2.1 Materials and sample preparation

SiO2 (5~15nm) nanoparticles purchased from Sigma Aldrich, Malaysia. The base fluid, methanol (CH3OH) with 99.9% purity was purchased from R&M Chemical, Malaysia. In this study, the two-step method applied to prepare the nanofluids with the volume fraction of 0.05 Vol % to 0.25 Vol %. At first, the required mass of the nanoparticle calculated by a digital weight gauge and dispersed in the measured amount of methanol (CH3OH) for the preparation of an individual volume fraction of nanofluids. Then, the solution vibrated in an incubator type mechanical shaker for 1 h at 150 RPM to homogenize the mixture. After that, an ultrasonic homogenize used to distribute the nanoparticles uniformly and evenly in the solution by producing high amplitudes and as well breaking the particle clusters. The sonication process continued for 2 h at the frequency of 10 kHz (50% amplitude) in a series of 2–second long pulses in 2 seconds apart along with a cooling bath to maintain a constant temperature (10°C) within the mixture.

2.2 Surface tension measurement

An automatic surface tensiometer (DCAT 11EC, Data Physics Instruments GmbH, and Germany) was used to measure the surface tension of SiO2–methanol nanofluids. The measuring range of the instrument is 1 to 1000 mN/m; ± 0.001 mN/m. Platinum-iridium RG11 Du–Noüy rings (Part no≠2000321), 70 mm diameter sample vessels, and electric hand dryer were used to conduct the experiments. The experimental results were evaluated by SCAT 31 software. And, the traditional Du–Noüy ring method (DIN 53915 and ASTM-971) applied for this experiment. In this method, the ring emerged and dipped from or into the sample vessels by an inward pushed or outward pulled lamella. The surface tension was measured by the changes of the ring weight during pushing and pulling. Figure 1 shows the surface tension measurement by pulling and pushing of the ring during the interval of time. Three readings took for a particular volume fraction, and the average value was considered for the final calculation. All the measurements were executed at an atmospheric pressure.

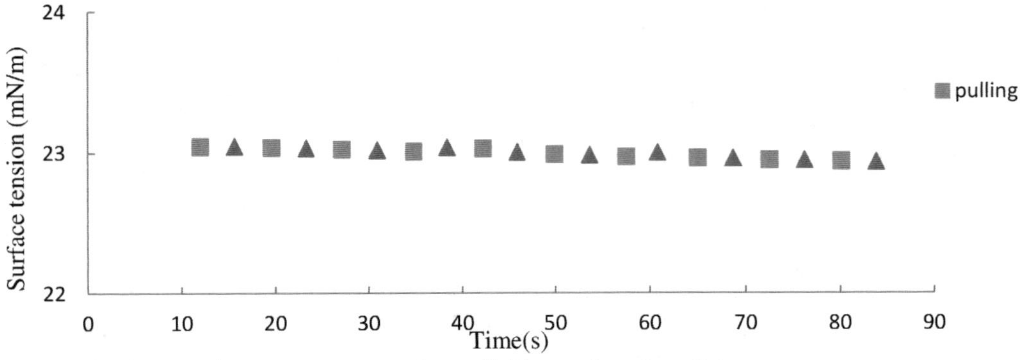

Figure 1. Surface tension measurements of nanofluids as a function of time.

3. Result and discussion

3.1 Effect of volume fractions on the surface tension of the nanofluids

The effect of surface tension of SiO2-methanol nanofluids is shown in Figure 2 with the variation of volume fractions (0.05 Vol % to 0.25 Vol %). The diagram clearly represents that the surface tension of the nanofluids increases with the increase in volume fractions. For instance, the value of the surface tension changes from 22.84 mN/m to 23.17 mN/m for the variation of volume fractions (0.05 Vol % to 0.25 Vol % at 25 °C. Moreover, the results represent the similar trends for other's temperatures. The results of this study are analogus with the experimental study of Tanvir et al. [14]. The results might be happened, as the volume fraction increases, more nanoparticles gathered into the solution.So, the nanoparticles are trying to get close with the surface molecules. As a result, a strong, cohesive force was exerted between the nanoparticles and liquid molecules, and hence, the surface tension of the nanofluids increases with the increase in volume fractions [14]. Moreover, the mean distance between the liquid molecules and the nanoparticles reduces with the increase in volume fractions. Thus, the surface to volume ratio increases,and consequently, increases the surface energy, and accordingly, the surface tension of the nanofluids increases.

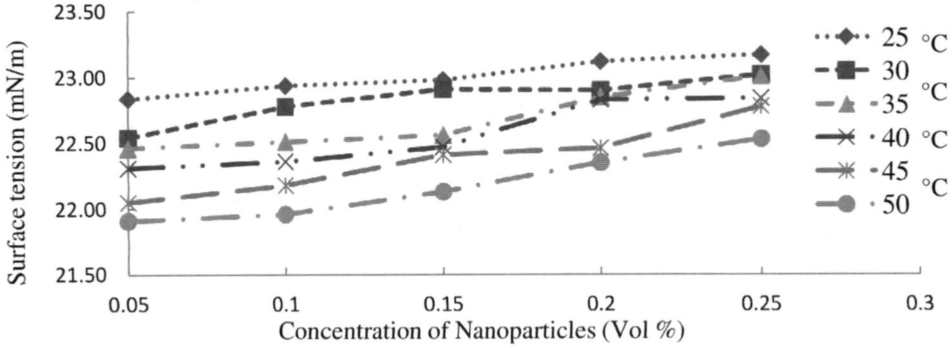

Figure 2. Surface tension of SiO_2–methanol nanofluids as a function of volume fractions.

3.2 Effect of temperature on the surface tension

Apart from the change of volume fractions, the variation of temperatures from 25 °C to 50 °C influences the outcomes of the surface tension of the nanofluids. In the figure 3, the results of the surface tension of SiO2–methanol nanofluids are exhibits with respect to the change of temperatures. The results clearly indicate that the surface tension of the nanofluids decreases linearly with the increase in temperatures. For example, the plot demonstrates that the surface tension of the SiO2methanol nanofluids dropped from 22.84 mN/m to 21.91 mN/m for the temperature change from 25 °C to 50 °C at 0.05 Vol %.The similar trends were observed in the experimental study of Murshed et.al [16]. As the temperature increases, the kinetic energy increases, and hence, rapidly accelerated the molecules of the nanofluids. As a result, the molecules are expanded from each other, and thus, weakening the strength of the intermolecular forces, which reduces the surface tension. Moreover, the density of the liquid molecules decreases with the increase in temperatures. Therefore, the shearing forces of the liquid layers reduce and resulting a lower value of the surface tension. Further, nanoparticles increases the Brownian motion in the liquid molecules, which lessen the cohesive energy and the resultant reduces the surface tension of the nanofluids.

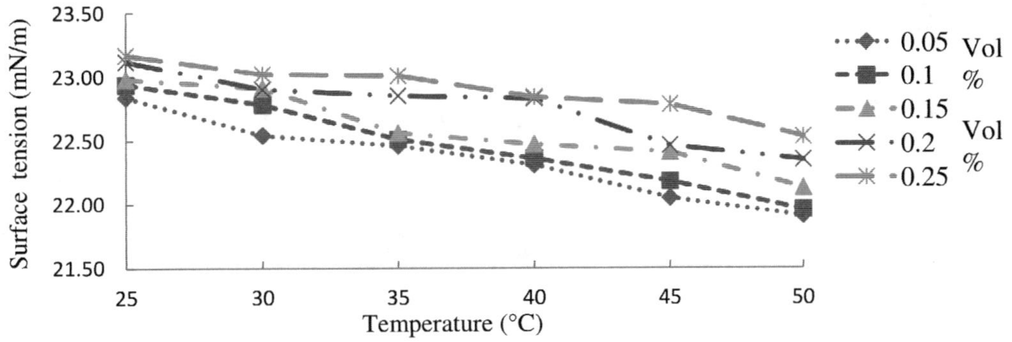

Figure 3. Surface tension of nanofluids as a function of temperatures.

3.3 Enhancement of surface tension of the nanofluids

Though the surface tension of SiO2–methanol nanofluids increases with the increase in concentration and decreases with an increase in temperature, the surface tension of the nanofluids enhances in all cases incomparison with the base fluid (methanol). The figure 4 shows that the values of the surface tension of pure methanol are measued as a function of temperature. Besides, the figure 5 exhibits, the enhancement of the surface tension of the nanofluids as function of temperature and volume fractions.The highest enhancement was observed about 4.27% to 8.92% for 0.25 Vol% at the temperature variation of 25°C to 50 °C. Moreover, the similar results are shown in other's volume fractions. Overall, the lowest improvements exhibit1.72% for 0.05 Vol% at 30 °C. The surface tension of nanofluids will increase or decrease depends on the forces of solid-liquid are attractive or repulsive. As the nanoparticles added to the liquid, the molecules are interact with each other. As a result, the forces of nanoparticles with its adjacent nanoparticles along with the liquid molecules are exerted in the liquid surfaces. Hence, an attractive Van der Waals force is employed, which resulting the enhancement of surface tension. The findings are similar with the outcomes of Moosavi et al. [15]

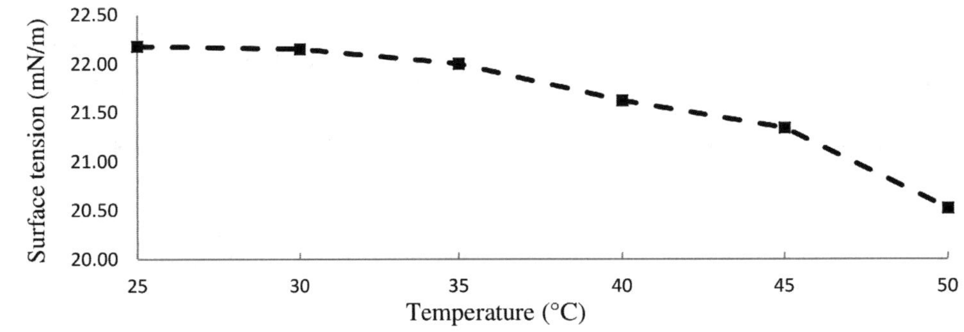

Figure 4. Surface tension of pure methanol as a function of temperatures.

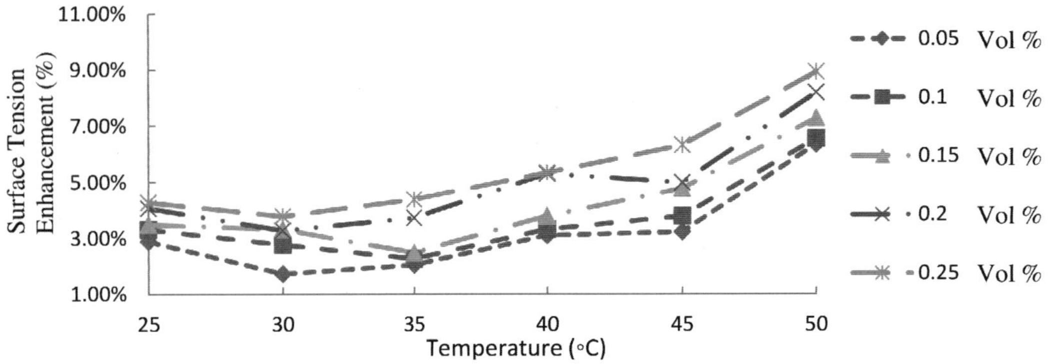

Figure 5. Surface tension of nanofluids as a function of temperatures.

4. Conclusion

The results of the experimental study inferred that surface tension of nanofluids changes with the change of concentration and temperatures. The following observations are the outcomes of this study.

a) The surface tension of the SiO_2-methanol nanofluids increases with the increase in concentrations due to nanoparticles increased the surface energy by increasing the surface to the volume ratio in the nanofluids.
b) Surface tension of nanofluids decreases with the increase in temperatures due to Brownian motions of the nanofluids lessens the cohesive energy.
c) All in all, surface tension of the nanofluids enhances with that of the pure methanols.

More studies are required to develop the models or correlations for prediction of the surface tension effects on nanofluids.

Acknowledgement

"This research is supported by High Impact Research MoE Grant UM.C/625/1/HIR/MoE/ENG/40 from the Ministry of Education Malaysia".

References

[1] Choi S U and Eastman J 1995 Argonne National Lab., IL (United States).
[2] Taylor R, Coulombe S, Otanicar T, Phelan P, Gunawan A, Lv W, Rosengarten G, Prasher and R, Tyagi H 2013 *J. Appl.Phys.* **113** 011301
[3] Gräfe W, 2013 *J. Mater. Sci.*, **48** 2092
[4] Vafaei S, Wen D and Borca-Tasciuc T 2011 *Langmuir* **27** 2211
[5] Raveshi MR, Keshavarz A, Mojarrad MS and Amiri S 2013 *Exp.Therm. Fluid Sci.*, **44** 805
[6] Jiang L, Ling J, Jiang L, Tang Y, Li Y, Zhou W and Gao J 2014 *Ener.g Conver. Mana.***81** 10
[7] Barbés B, Páramo R, Blanco E, Pastoriza-Gallego M, Piñeiro M, Legido J and Casanova C 2013 *J. Therm. Anal .Calorim.* **111** 1615
[8] Yiamsawas T, Mahian O, Dalkilic AS, Kaewnai S and Wongwises S 2013 *Appl.Energy* **111** 40
[9] Khaleduzzaman SS, Mahbubul IM, Shahrul IM and Saidur R 2013 *Int. Commun. Heat Mass Trans.* **49** 110
[10] Das SK, Putra N and Roetzel W 2003 *In.t J. Heat Mass Trans.* **46** 851
[11] Zhu BJ, Zhao WL, Li JK, Guan YX and Li DD 2011 *Mat. Sci. Forum.* Trans Tech Publ. 266
[12] Tanvir S and Qiao 2012 L *Nanoscale Res. Lett.* **7** 1
[13] Moosavi M, Goharshadi EK and Youssefi A 2010 *Int. J. Heat Fluid Flow* **31** 599
[14] Murshed SS, Tan S-H and Nguyen N-T 2008 *J. Phys.D: Appl.Phys.*, **41** 085502
[15] Saeid V, Arup P, Abhishek J, Ganapathiraman R and Theodorian B-T 2009 *Nanotechnology* **20** 185702.
[16] Shwin-Chung W, Siao-Fang H and Kuo-Chun H 2010 *IMPACT 5th International 20-22 Oct.* **2010** 1
[17] Wong S-C, Lin Y-C and Liou J-H 2012 *Int.J. Therm. Sci.* **52** 154
[18] Kim JH, Jung CW and Kang YT 2014 *Int. J. Heat Mass Trans.* **76** 484
[19] Pineda IT, Choi CKand Kang YT2014 *Int. J. Greenhouse Gas Cont.* **23** 105
[20] Jung J-Y, Lee J and Kang Y 2012 *J. Mech. Sci .Technol.* **26** 2285
[21] Firouzfar E, Soltanieh M, Noie SH and Saidi SH 2011 *Appl. Therm. Eng.* **31** 1543
[22] Firouzfar E, Soltanieh M, Noie SH and Saidi MH 2012 *Int. J. Environ. Sci. Technol.* **9** 587
[23] Mostafizur RM, Abdul Aziz AR, Saidur R, Bhuiyan MHU and Mahbubul IM 2014 *Int.J. Heat Mass Trans.* **77** 765
[24] Mostafizur RM, Bhuiyan MHU, Saidur R and Abdul Aziz AR 2014 *Int. J. Heat Mass Trans.*, **76** 350

IOP Conf. Series: Materials Science and Engineering **88** (2015) 012057 doi:10.1088/1757-899X/88/1/012057

Investigation on stability and density of methanol based TiO_2 nanofluids

R M Mostafizur[1,3], R Saidur[1], A R Abdul Aziz[2], M H U Bhuiyan[1]
[1]Department of Mechanical Engineering, Faculty of Engineering, University of Malaya, 50603 Kuala Lumpur, Malaysia.
[2]Department of Chemical Engineering, Faculty of Engineering, University of Malaya, 50603 Kuala Lumpur, Malaysia.

E-mail: mostafiz.rahman@live.com

Abstract. This work has investigated the enhancement of heat transfer considering the density of methanol based nanofluids with TiO_2 nanoparticles at various volume fractions and temperatures. Nanoparticles are suspended into methanol base fluid at four different concentrations which are 0.01, 0.05, 0.10 and 0.15 vol% and the nanofluids preparation were completed through sonication using ultrasonic homogenizer. The density of prepared nanofluids was measured by KEM-DA 130N density meter at four different temperatures (5, 10, 15and 20 °C). The results show that the density increases with increase of volume concentrations while the density decreases with increasing temperature and the increment is observed about 2.2% over the base fluid at 0.15 vol% and 20 °C temperature.

1. Introduction:

Nanofluid is a fluid suspension containing metallic or non-metallic nanoparticles with a typical size of (1-100 nm) dispersed into the base fluid. The concept of nanofluids was first introduced by Choi, in 1995 [1].It is a new class of heat transfer fluids due to its attractive enhancements on thermo–physical properties and heat transfer. For heating and cooling process, cooling is an essential part of multitudinous industrial and civil application such as power generation, chemical process, microelectronics, transportation, air condition and micro–sized application etc. [2, 3]. However, existence of solid particles leads to interesting characteristics in the fundamental thermo-physical properties of nanofluids. Thermal conductivity, viscosity, density and stability have been investigated throughout the recent years by many researchers [4, 5].

Density of fluid is an important thermophysical property. Like viscosity, density of any fluid also has direct impact over pressure drop and pumping power and it also affects on Reynolds number, Nusselt number, thermal diffusivity as well as heat transfer coefficient of a heat transfer fluid. However, only a limited number of studies have been conducted on density of nanofluids. Sommers et al. [6] observed a linear relationship between density and particle concentration for Al_2O_3–propanol nanofluid. Correspondingly, Teng et al. [7] found that the density of Al_2O_3–water nanofluid increased with the increase of volume concentrations (0.5–1.5 wt %) and decreased with increase in temperature (10–40 °C).

In a temperature range of 200–500 K methanol was used as a working fluid of a different heat pipe [8]. Different type of heat pipes like conventional, vapor–dynamic thermosyphons, sorption and micro/miniature heat pipe have recently been used for heat transfer applications [9]. The methanol based

[3] mostafiz.rahman@live.com

nanofluids were also used as a working fluids in HVAC systems [10]. However, there is no such study done about density measurement of methanol based nanofluids. Only few literature focuses on thermal conductivity measurement and rheology behavior of methanol based nanofluids [11-13]. Therefore, the objective of this study is to investigate the stability and density of TiO_2 nanoparticles suspended in methanol at different volume fraction of nanoparticles with different temperatures. This study will help the researchers to get the idea about the effect of nanoparticles on the density of methanol nanofluids which will encourage the researchers to apply nanoparticles in different kind of heat pipes and thermoshypons heat exchangers in HVAC system.

2. Experimental process

2.1 Materials and preparation of nanofluids

The nanoparticles, TiO_2 were purchased from Sigma-Aldrich, Malaysia. Field emission scanning electron microscope (FE-SEM) was used to obtain the morphological characterization of the nanoparticles with SIGMA Zeiss instrument (Carl Zeiss SMT Ltd., UK). SEM pictures of the respective nanoparticles are shown in Figure 1 (a), where the actual dimensions of the nanoparticles can be determined to be ~21 nm. The base fluid, methanol was purchased from R&M Chemicals (Malaysia).

Then the two-step method was applied to prepare methanol based nanofluids at different volume concentrations. The nanoparticles were suspended into the base fluid (methanol) followed by shaking in the incubator for 30 min at 150 rpm. The suspension was mixed using ultra-sonication homogenizer. The details of experiment conditions are summarized in table 1. Figure 1(b) shows the TEM images of nanoparticles mixed in methanol which was obtained using Libra 120 transmission electron microscope. The images obtained show that the particles are indeed very small, whereas some of the particles match the manufacturer's specified size, while some exceed the specified size.

Table 1: Experimental conditions

Base fluid	Methanol	
Nanoparticle	TiO_2	
Nanoparticle type	Spherical	
Nanoparticle size (nm)	21 nm	
Volume concentration (vol%)	0.01, 0.05, 0.10 and 0.15	
Ultra–sonicator	Time [min]	120
	Power [W]	500
	Frequency [kHz]	20
	Pulse [s]	2
	Term [s]	2

IOP Conf. Series: Materials Science and Engineering **88** (2015) 012057 doi:10.1088/1757-899X/88/1/012057

(a) (b)

Figure 1. (a) FESEM image of TiO_2 nanoparticles (b) TEM image of TiO_2–methanol nanofluids

2.2 Stability characterization of nanofluids

In this investigation, the UV–Vis. spectrophotometer, Lambda 35 model, Perkin Elemer make, absorption range of 190 nm to 1100nm was used to study the stability of nanofluid. The inspection range is from 200nm to 800nm. The U-V vis. spectrometer works under the principal of Beer–Lamberts law. Beer–Lamberts law relates that an absorbance of light and proportion of material through is passing. The lesser the suspended particles in the solution makes the light absorption lesser.

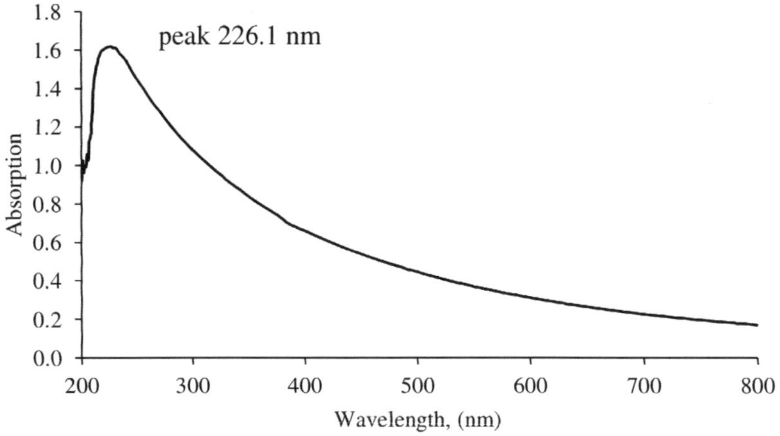

Figure 2. UV–Vis spectrum of TiO_2–methanol nanofluids

Figure 2 shows the stability of TiO_2–methanol nanofluids after ultrasonic agitation. It also shows that the peak absorbance of TiO_2 nanoparticle suspension in methanol occurred at 226 nm. The absorption range of the nanofluid was 1.62. The UV–visibility was measured after seven days and there was no distinguished difference between the density of freshly prepared nanofluids and the old one of seven days.

2.3 Measurement of density

The density of methanol based nanofluids was measured by using a density meter, KEM-DA 130N (Kyoto, Japan). This device measures a density range of 0 to 2000 kg/m^3. The accuracy of the equipment is ± 0.001% kg/m^3. The density was measured at different temperatures and volume concentrations in this study. All data were recorded for three times and the corresponding average values were plotted. since, there is no existing literature related to density of methanol based nanofluids, the experimental values were then compare with that obtained from the most used Pak and Choi [14] model, which is as follows:

$$\rho_{nf} = (1-\phi)\rho_{bf} + \rho_{np}\phi \quad (1)$$

where ρ is the density, ϕ is the volume fraction and subscript nf–nanofluid, bf–base fluid and np–nanoparticle.

3. Result and discussion

3.1 Visualization effect

Figure 3 shows the photos of TiO_2–methanol nanofluids for four volume fractions by shaking with an orbital incubator shaker and sonication. Photos of TiO_2–methanol nanofluids were taken just after preparation and after one week. Figure 3 (b) shows that the sedimentation starting rate was very slow. Generally, the sedimentation of mixtures is measured from the bottom of the specimen. It could be possible when there are slurries obvious at the bottom of the sample. This happened especially for the low concentration in the suspension.

(a) Just after preparation (b) One week later

Figure 3: Photos of visualization (a) just after preparation and (b) one week later for TiO_2–methanol nanofluids

3.2 Density

Figure 4 shows the comparison between the experimental values obtained from the current study with the existing literature data [15]. The results show relatively acceptable consensus with the existing data. The uncertainties in the measurements of density are approximately 1.2% for pure methanol.

The density of TiO_2–methanol nanofluids, as a function of different temperatures and volume concentrations is shown in Figure 5. The results shows that the density of methanol based nanofluids increases with volume concentration but decreases with temperature. Besides, the density of nanofluids was higher than the base fluids. For example, at 5 °C, the density was 811.83 kg/m^3 for 0.01 vol% and 818.74 kg/m^3 for 0.15 vol%. At volume concentration of 0.15 vol%, the density was 809.13 kg/m^3 at 20 °C and 818.74 kg/m^3 at 5 °C. The highest increment was observed for 0.15 vol% and 5 °C. Similarly, the same result was found for all volume concentration in this experiment. This was due to dispersion of high density nanoparticles in the base fluids.

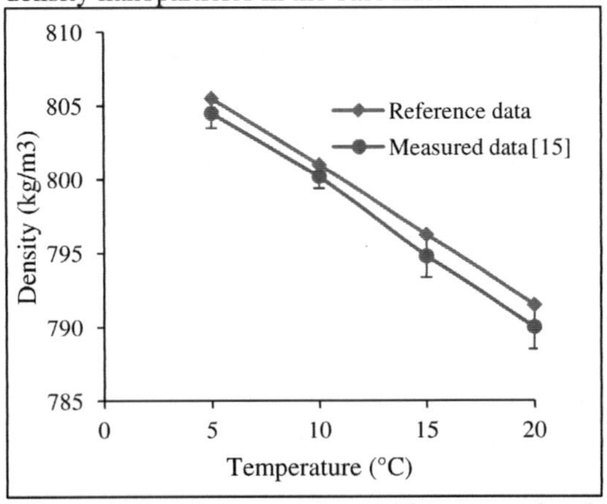

Figure 4. Density comparison with measured and reference data

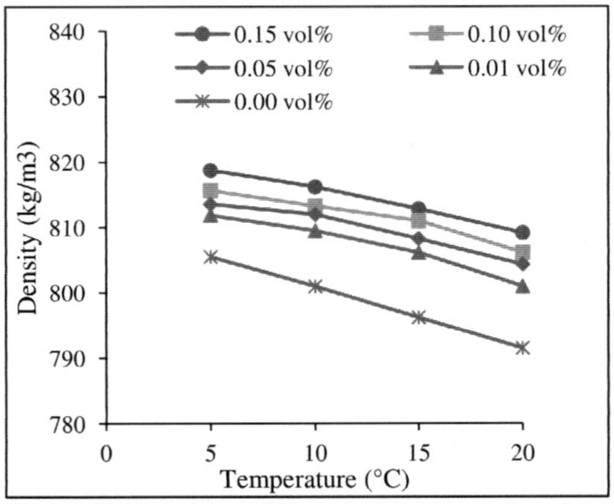

Figure 5. Density of TiO_2–methanol as a function of temperature and volume fraction

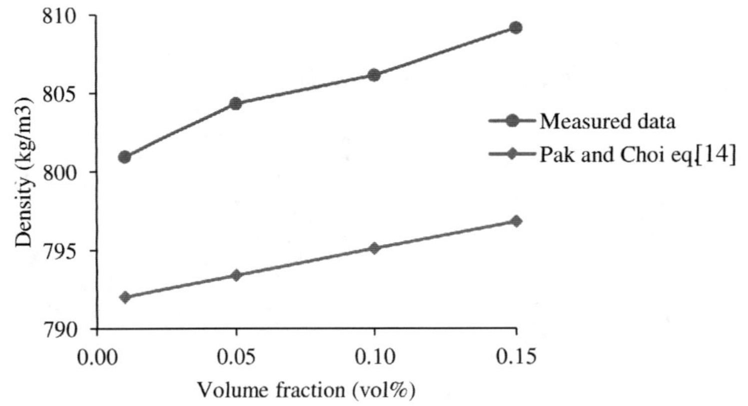

Figure 6. Comparison of measured density with the values obtained from existing correlations

The experimental value of density of methanol based nanofluids at 20 °C was compared with the model given by Pak and Choi [14] shown in Figure 6. The experimental value was slightly higher than the existing correlation. The highest deviation was observed about 1.5% from the existing correlation. This was due to the difference in the density of the base fluids and water based nanofluids whereas Pak and Cho model was proposed latter.

4. Conclusion

In this study, density of TiO_2–methanol nanofluids is investigated at different volume concentrations and temperatures. The experimental results are then compared with existing correlations. The findings of this study are summarized as follows:

(a) Density increases with the increase of volume concentrations and decreases with the increase of temperatures.

(b) Density increases about 2.2% with nanoparticle volume concentration 0.15 vol% and at 20 °C temperature.

(c) The existing correlations for predicting the density of methanol base nanofluids gave lower values than the experimental values.

Therefore, the density and stability of the above mentioned nanofluids will be further investigated with different surfactants such as, PEG, HTAB, SDS and PEG.

Acknowledgement

"This research is supported by High Impact Research MoE Grant UM.C/625/1/HIR/MoE/ENG/40 from the Ministry of Education Malaysia".

References

[1] Choi S U and Eastman J 1995 Argonne National Lab., IL (United States).
[2] Nuntaphan A, Vithayasai S, Vorayos N, Vorayos N and Kiatsiriroat, T 2010 *Int. Comm. Heat Mass Tran.* **37** 287
[3] Saidur R, Leong K Y, Mohammad H A 2011 *Renew. Sustain. Energy Rev.* **15** 1646
[4] Chandrasekar M and Suresh S 2009 *Heat Trans. Eng.* **30** 1136
[5] Lee S, Choi S U S, Li S, Eastman J 1999 *J. Heat Trans.* **121** 280
[6] Sommers A D, and Yerkes K L 2010 *J. Nanopart. Res.* **12** 1003
[7] Teng T-P, and Hung Y-H, 2012 *J. Exp. Nanosci.* 2012 **12** 1
[8] Vasiliev L L 1998 *Appl. Therm. Eng.* **18** 507
[9] Yang X, Yan Y, and Mullen D 2012 *Appl. Therm. Eng.* **33** 1
[10] Firouzfar E, Soltanieh M, Noie S H and Saii S H 2011 *Appl. Therm. Eng.* **31**1543
[11] Mostafizur R M, Bhuiyan M H U, Saidur R and Abdul Aziz A R 2014 *Int. J. Heat Mass Tran.* **76** 350
[12] Mostafizur R M, Abdul Aziz A R, Saidur R, Bhuiyan M H U and Mahbubul I M 2014 *Int. J. Heat Mass Tran.* **77** 765
[13] Pang C, Jung J-Y, Lee J W and Kang Y T 2012 *Int. J. Heat Mass Tran.* **55** 5597
[14] Xuan Y and Li Q 2000 *Int. J. Heat. Fluid. Fl.* **21** 58
[15] *Thermophysical Properties - Methanol* 11.09.2013]; Available from: http://www.engineeringtoolbox.com.

IOP Conf. Series: Materials Science and Engineering **88** (2015) 012058 doi:10.1088/1757-899X/88/1/012058

A PID de-tuned method for multivariable systems, applied for HVAC plant

A B Ghazali [1]

Department of Electronics & Communication, College Of Engineering, Universiti Tenaga Nasional, 43009 Kajang, Selangor, Malaysia.

E-mail: Abakar@uniten.edu.my

Abstract. A simple yet effective de-tuning of PID parameters for multivariable applications has been described. Although the method is felt to have wider application it is simulated in a 3-input/ 2-output building energy management system (BEMS) with known plant dynamics. The controller performances such as the sum output squared error and total energy consumption when the system is at steady state conditions are studied. This tuning methodology can also be extended to reduce the number of PID controllers as well as the control inputs for specified output references that are necessary for effective results, i.e. with good regulation performances being maintained.

1. Introduction

The three-term or PID (Proportional, Integral and Derivative) controller is still by far the most widely used industrial controller due to its simplicity and robustness. The controller has traditionally been developed for SISO applications and various tuning and setting up procedures (see for example in [1]) have been developed to assist the commissioning of PID loops. Advances in auto-tune the PID parameters in an adaptive manner have been developed and commercially used [2]; both analogue and digital versions exist that also include intelligent methods [3][4]. The PID controllers have also widely been used in multivariable applications where their commissioning is less well formalised. The normal approach is to use a robust and adhoc de-tuning philosophy to reduce the gains so that the multivariable application can operate [5][6] but no overall optimality is considered. It would be useful if a more generic methodology is available which allows PID controllers to be setup in such multivariable situations. In this way the classical merits of PID controllers can be extended to general multi loop implementations. A solution for addressing this difficulty is to apply state-space methods where the matrices of PID gains are obtained by observing the outputs of the system when step inputs are applied in turn [7]. In this manner the system inputs and outputs can be "connected" to give rise to the dominant "diagonal" systems as well as the most important cross coupling interactions. However establishing the most diagonal form is difficult and is based on trial and error approach. This methodology has been improved [8] so that the PID gains are obtained in a more unified manner but the trial and error foundations persist. A more recent method to auto-tune the multi-loop PID controllers is to combine sequential loop closing and relay tuning, which is proposed [9] but this method only applies to square systems, i.e. the number of inputs equals the number of outputs and hence it cannot be used for wider applications. There are many other methods and regimes to obtain the PID gains that based on generic and intelligent approaches, for example as discussed in [10].

In this paper, we propose a far simpler PID tuning methodology also based on open-loop step responses but focusing on the performance of individual PID controllers. The method is felt to be application independent but for convenience it is discussed here through it implementation to a

[1] To whom any correspondence should be addressed.

building energy management system (BEMS) with three inputs and two outputs (non-squared system).

2. Multi PID-loop controller

In this multivariable system, we assume that the number of output and input are m and n respectively. The traditional method for designing the control system for this situation is to use m PID controllers for each input. Therefore, we require to utilise $m \times n$ or mn PID controllers for the overall system. The block diagram for this configuration is shown in figure 1 where PID#11, PID#12, ..., PID#mn are the PID controllers and M_{ij}, i = 1, 2, .., m and j = 1, 2, .., n are their respective outputs to drive the control inputs u_i. The PID output signals are given by:

$$\text{PID\#}ij{:}M_{ij} = \text{Kp}_{ij}\, e_j + \text{Ki}_{ij}\, I_j + \text{Kd}_{ij} \frac{de_j}{dt} \tag{1}$$

where output errors, e_j are given by $e_j = y_{Rj} - y_j$. Here, y_j and y_{Rj} are the actual and desired outputs respectively. Kp_{ij}, Ki_{ij} and Kd_{ij} are the proportional, integration and derivative gains for controllers PID#11, ..., PID#mn respectively. For a digital PID implementation, the integration term with respect to time t can be approximated by the Euler method, which employs

$$\int_0^t e_j(t)dt \equiv I_j(t) = I_j(t-T) + e_j(t)\, T \tag{2}$$

where T is the sampling period, and the derivative term is given by

$$\frac{de_j(t)}{dt} = \frac{e_j(t) - e_j(t-T)}{T} \tag{3}$$

Each control input u_i, to drive the plant is obtained by summation of the m PID outputs

$$u_i = M_{i1} + M_{i2} + \ldots + M_{in}, \tag{4}$$

with constraints: $u_i(\min) \leq u_i \leq u_i(\max)$. The objective here is to tune the PID gains for optimum performance so that the outputs remain at required setpoints over the control period. The performance of the tuned controller will be judged by looking at the regulation errors as well as the number of inputs consumed. Clearly a good controller design should have small errors and consume less inputs as well as the number of controllers used.

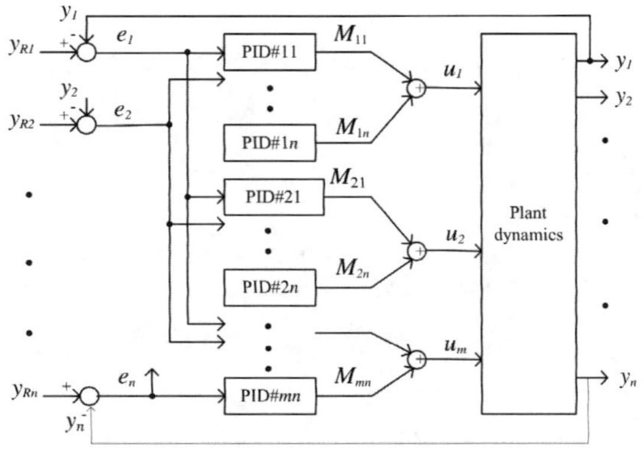

Figure 1. A PID controller structure for multivariable system

A. Ziegler and Nichols (Z-N) tuning rules

There are many methods of PID tuning and a well-known one is the classical Ziegler and Nichols tuning rules [11] from which were followed on the open loop step test and the PID gains are given by

$$\text{Kp}_{ij} = \frac{1.2}{K_{ij}} \frac{\tau}{\alpha}, \quad \text{Ki}_{ij} = \frac{\text{Kp}_{ij}}{2\alpha} \text{ and } \text{Kd}_{ij} = \text{Kp}_{ij} \frac{\alpha}{2} \tag{5}$$

where $K_{ij} = \frac{\Delta y_{sj}}{\Delta u_{si}}$ is the slope of change in process value y_{sj} against the step change in manipulated value, u_{si}. α and τ the output delays as shown in figure 2. The integration term, I_j from equation (2) will make the output M_{ij} increasing or decreasing against time depending on either positive and negative error, e_j respectively, so the limits, $I_j(\min) \leq I_j \leq I_j(\max)$ shall be imposed in the setting up of the controller to avoid these winding up effects.

This PID controller is then tested on a 3 input/ 2 output model of building energy management system (BEMS), also known as HVAC plant at Loughborough University, UK with known plant dynamics as reported in [12]. The three inputs are the heater, cooler and humidifier with maximum power, $u_i(\max)$ of 5.0, 2.7 and 2.6 kW respectively. The outputs are the room temperature and its relative humidity. They have presented the plant dynamics in a linear difference equation with sampling period of 5 minutes. Other parameters such as the surrounding temperature and humidity inside and outside the room/ building, solar irradiation, occupancy and the door opens and closes are treated as the disturbances and noises respectively to this plant dynamics. The results at various output setpoints have shown that the control performances are poor with large output error and majority of the 6 PID act like an on/ off controller where the output fluctuations are large and majority of the control inputs swing between off and on (maximum) limits due to the above imposed constraints. In another word, the Ziegler-Nichols tuning rules are not suitable for multivariable system because a good controller shall vary only between the control inputs ranges in order to regulate the outputs with acceptable error. The Z-N method is only applicable if all the loops are independent of each other and this does not apply to this multivariable case. Example of the result is shown in figure 3 where the required room temperature (y_{R1}) and relative humidity (y_{R2}) were set at 25°C and 50%rh respectively and found that the temperature and humidity sum squared error for the last 100 samples were 68.5 $^{o}C^2$ and 8181.0 %rh^2. In view of this performance it appears beneficial to investigate whether it was possible to develop a methodology so that the PID gains could be de-tuned in multivariable situations so that the resulting controllers yield better performances.

B. A PID de-tuning method

Based on these simulation studies, it is felt that the PID gains obtained by Z-N for each controller needs to be modified. We propose the procedure as follows: assume that the first set of PID controllers nos. 11, 12, …, 1*m* is used in the first control loop (u_1), the set of second with PID nos. 21, 22, …, 2*m* for u_2, etc. The proposed de-tuning algorithm is as follows:

Step 0: Set *m*, *n* and obtain the Z-N gains Kp_{ij}, Ki_{ij} and Kd_{ij} for $i = 1, 2, ..., m, j = 1, 2, ..., n$ from the open loop step responses as given in equation (5). Set $-1 \leq I_j \leq 1$ for convenience.

Step 1: Simulate the results for a period where the responses have reached a steady state conditions (ssc) as shown in figure 3.

Step 2: Records M_{ij} maximum, $M_{ij}(\max)$ and minimum, $M_{ij}(\min)$ at ssc for all combination of *i* and *j*. Compute the maximum swing for each PID, $M_{ij}^s = \text{abs}[M_{ij}(\max) - M_{ij}(\min)]$

Step 3: Compute the de-tuning (reducing) factor, ε_{ij} for each PID; $\varepsilon_{ij} = \frac{u_i(max)-u_i(min)}{M_{ij}^s}$. Thus, the PID output in equation (1) becomes: PID#*ij*:

$$M_{ij}^d = \varepsilon_{ij}[Kp_{ij}\, e_j + Ki_{ij}\, I_j + Kd_{ij}\, \frac{de_j}{dt}] \tag{6}$$

and the control input in equation (4) is replaced by:

$$u_i = M_{i1}^d + M_{i2}^d + \ldots + M_{in}^d \text{ where } i = 1, 2, .., m \tag{7}$$

with similar input constraints.

IOP Conf. Series: Materials Science and Engineering **88** (2015) 012058 doi:10.1088/1757-899X/88/1/012058

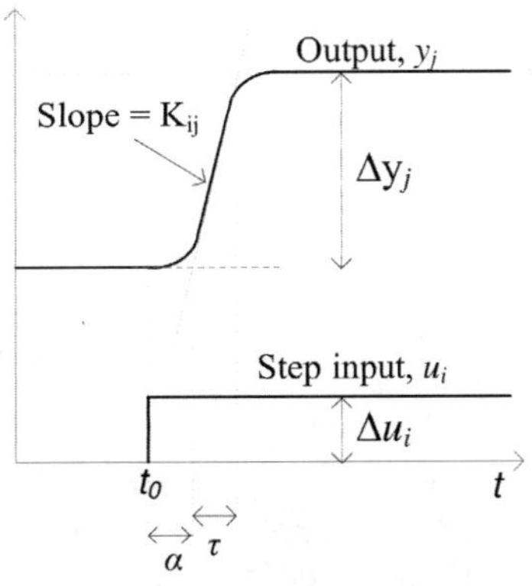

Figure 2. Open-loop system step response

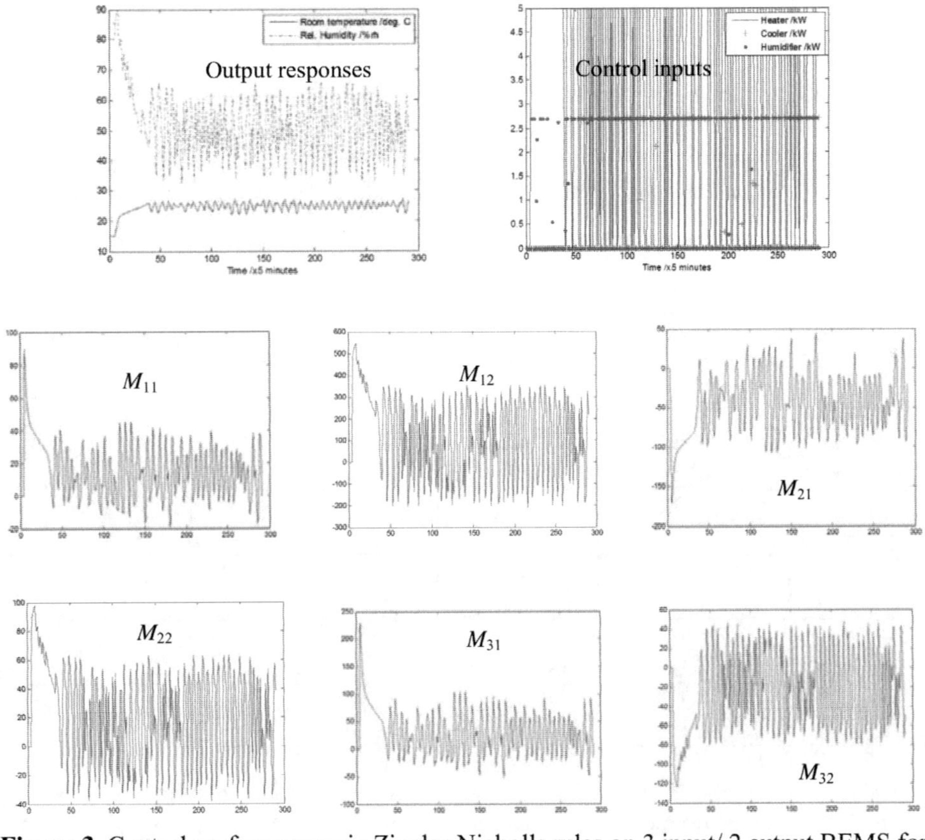

Figure 3. Control performance via Ziegler-Nicholls rules on 3 input/ 2 output BEMS for references y_{R1} = 25°C and y_{R2} = 50%rh and without noise

C. Simulation Results

For setting up the initial Z-N PID controller parameters, the open loop step responses were carried out on the model as described in Section 2 where the control inputs were step increased from 0% to 100% of their maximum powers and then calculate for Kp_{ij}, Ki_{ij} and Kd_{ij}, i = 1, 2 and 3 and j = 1, 2; these parameters are given in table 1. The performance results for required setpoints (y_{R1} = 25°C and y_{R2} = 50%rh) as shown in figure 3 were obtained. But, when the detuned PID gains were used, i.e. Kp_{ij}, Ki_{ij} and Kd_{ij} were reduced by ε_{ij} factor as indicated in equation (6), the performance results as shown in Figure 4 were improved significantly.

We can see that a good regulation has been achieved and the performance over the last 100 sampling intervals is as follows:

- output sum squared errors are 0.0003 °C^2 and 0.001 %rh^2 for the temperature and the relative humidity respectively; and
- required control input energies are 29.5, 5.8 and 10.9 kWh for the heater, cooler and humidifier respectively.

To add a degree of realism into our simulation studies, we can insert some white noise processes V_1 and V_2 to the room temperature and relative humidity respectively. For convenience a pseudo random binary sequences (PRBS) generated using 6 shift registers [13] with magnitudes V_1 = ±0.5°C and V_2 = ±0.5%rh, were used as the noise processes. These represent large disturbances to the plant dynamics and consequently should effect the system quite significantly. These terms were then included into the system of identical settings as used in figure 4. The simulated result is shown in figure 5 where the control performances for the last 100 sampling intervals are as follows:

- output sum squared errors are 1.8 °C^2 and 96.2 %rh^2 for the temperature and the relative humidity respectively; and
- required control input energies are 29.5, 5.8 and 10.9 kWh for the heater, cooler and humidifier respectively.

Moreover, the settling periods are 340 minutes and 135 minutes for the temperature and the relative humidity respectively.

We also found that the controller is capable to regulate the output for required setpoint of y_{R1} = 20°C and y_{R2} = 40%rh and this is shown in Figure 6

Table 1. PID gains and their reduction factors applied for The BEMS plant

Input	M_{ij} (kW)	Kp_{ij}	Ki_{ij}	Kd_{ij}	M^S_{ij} (kW)	u_i(max)- u_i(min) (kW)	ε_{ij}
Heater, u_1	M_{11}	6.41 (kW/°C)	84.34 (kW/°C_h)	0.12 (kW_h/°C	64.9	5.0	0.0770
	M_{12}	-7.88 (kW/%rh)	-231.71 (kW/%rh_h)	-0.077 (kW_h/%rh	566.6	5.0	0.0088
Cooler, u_2	M_{21}	-9.04 (kW/°C)	-265.92 (kW/°C_h)	-0.077 (kW_h/°C	152.2	2.7	0.0177
	M_{22}	-1.41 (kW/%rh)	-41.34 (kW/%rh_h)	-0.0119 (kW_h/%rh	101.2	2.7	0.0494
Humidi fier, u_3	M_{31}	18.31 (kW/°C)	152.58 (kW/°C_h)	0.55 (kW_h/°C	158.0	2.6	0.0316
	M_{32}	1.78 (kW/%rh)	52.35 (kW/%rh_h)	0.015 (kW_h/%rh	127.3	2.6	0.0392

This simulation result can also be used to analyse for possible components reduction in a control system so that the overall cost of installation is reduced. Moreover, this can also reduce the fighting among inputs for similar objective, i.e. to regulate the outputs but contribute to the waste of energy. As shown in figure 4, the cooler requirement for the prescribed setpoint is lowest at steady state conditions and it may not require to regulate the output. If so, the cooler as well as PID#21 and PID#22 controllers can be eliminated. A simulation was done for this case and the result is shown in figure 7 where it is found that the controller is still capable to regulate the output. Similarly, it is found that the humidifier is least required for other setpoints such as for y_{R1} = 20°C and y_{R2} = 40%rh since the heater can also function as to reduce the relative humidity. So, when the humidifier as well as PID#31 and PID#32 are excluded from this controller, we found that the controller can well perform its function and this is shown in figure 8.

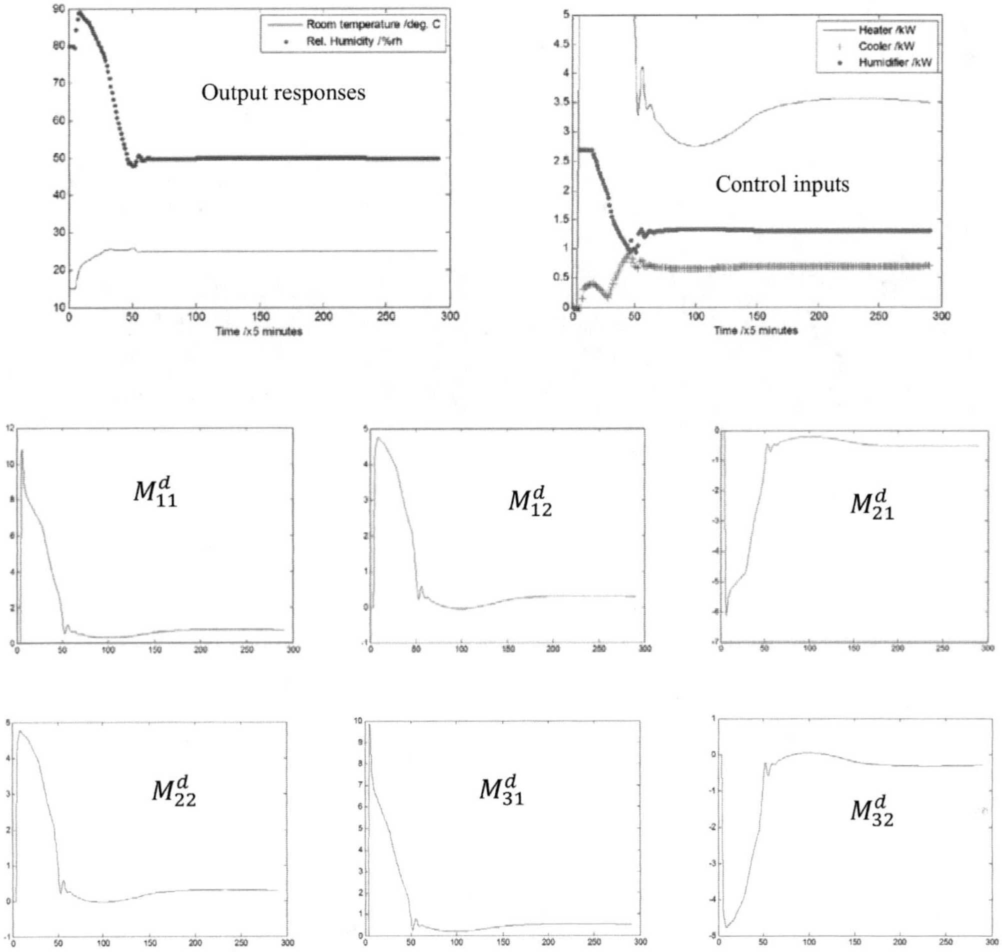

Figure 4. Control performance via modified gains on 3 input/ 2 output BEMS for y_{R1} = 25°C and y_{R2} = 50%rh (no noise)

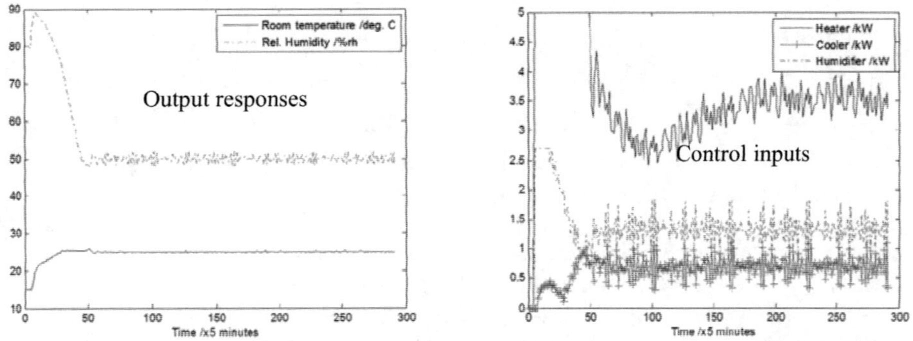

Figure 5. Control performance with PRBS stochastic noises of magnitude ±0.5

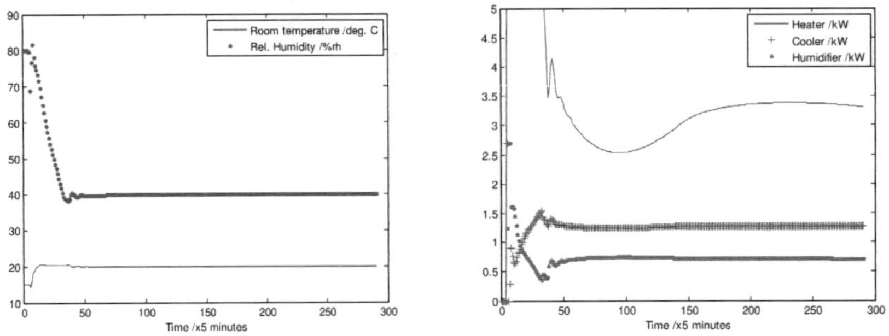

Figure 6. Control performance for y_{R1} = 20°C and y_{R2} = 40%rh

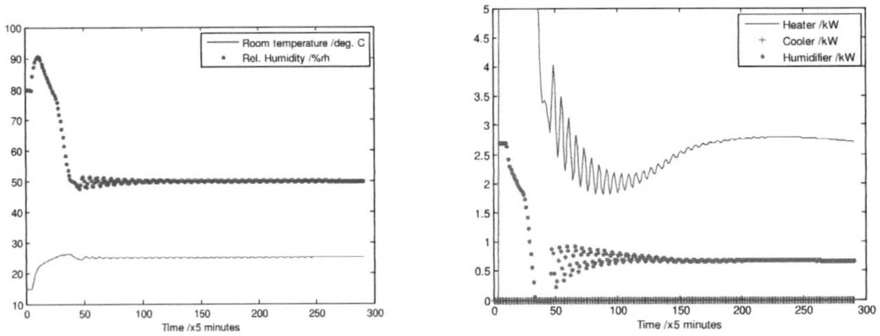

Figure 7. Control performance without cooler for y_{R1} = 25°C and y_{R2} = 50%rh

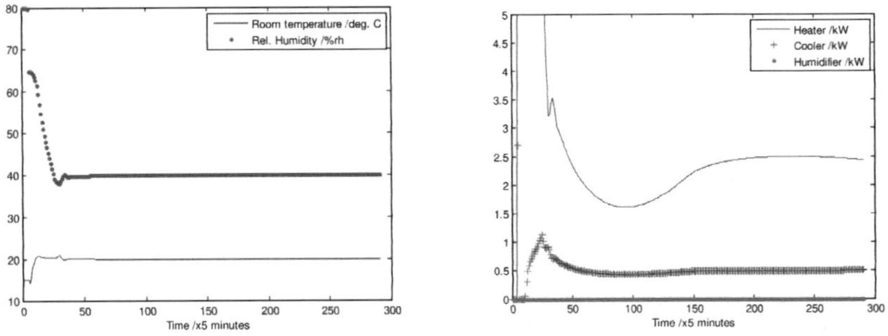

Figure 8. Control performance without humidifier for y_{R1} = 20°C and y_{R2} = 40%rh

3. Conclusions

In this paper a simple yet effective de-tuning methodology for PID controllers in multivariable applications has been described. It has been shown that good control performances can be achieved by using the de-tuning factors to modify the Z-N proportional, integration and derivative gains. The methodology has also been useful for reducing the number of PID controllers as well the control inputs that are necessary for the selected setpoint in order to reduce the overall cost.

The result has also shown that this tuning methodology is suited to the control of HVAC systems and may be applicable to the effect PID based control of general multivariable systems.

4. References

[1] Hang C C, Astrom K J and Ho W K 1991 Refinement of the Ziegler Nichols tuning formula. *IEE Proc. Part D* **138**.

[2] Astrom K J, Hang C C, Persson P and Ho W K 1992 Towards intelligent PID control *Automatica*, **28**, 1 pp 1-9.

[3] Savran, A 2013 A multivariable predictive fuzzy PID control system *Applied Soft Computing*, vol 13 Issue 5 pp 2658-2667.

[4] Chang W D 2007 A multi-crossover genetic approach to multivariable PID controllers tuning. *Expert Systems with Applications* vol 33 Issue 3 pp 620-626.

[5] Luyben W L 1986 Simple method for tuning SISO controllers in multivariable systems. *Ind. Eng. Chem. Process Des. Dev*. **25** pp 654-660.

[6] Dittmar R, Gill S, Singh H and Darby M 2012 Robust optimization-based multi-loop PID controller tuning: A new tool and its industrial application. *Control Engineering Practice*, vol 20 Issue 4 pp 355-370.

[7] Penttinen J and Koivo H N 1980 Multivariable tuning regulators for unknown systems. *Automatica* vol 16 pp 393-398.

[8] Koivo H N and Pohjlainen S 1985 Tuning of multivariable PID-controllers for unknown systems with input delay. *Automatica*, **21**, 1 pp 81-91.

[9] Loh A P, Hang C C, Quek K and Vasnani U 1993 Autotuning of multiloop proportional-integral controllers using relay feedback *Ind. Eng*. Chem. Res. **32** pp 1102-1107.

[10] Helon V H A and Coelho L D S 2012 Tuning of PID controller based on a multiobjective genetic algorithm applied to a robotic manipulator *Expert Systems with Applications* vol 39 Issue 10 pp 8968-8974.

[11] Ziegler J G and Nichols N B 1942 Optimum settings for automatic controllers. *ASME Trans*. **64** pp 759-768.

[12] Loveday D L and Virk G S 1992 Final Report to the Science and Engineering Research Council, Research Contract GR/F/02014: The Effectiveness of Predictive Control Applied to Warm Air Heating Systems For Commercial Building.

[13] Godfrey K R 1980 Correlation Methods *Automatica* Pergamon Press Ltd vol 16 pp 527-534.

IOP Conf. Series: Materials Science and Engineering **88** (2015) 012059 doi:10.1088/1757-899X/88/1/012059

Theoretical Study of New Combined Absorption-Ejector Refrigeration System

A M Abed[1,2], K Sopian[1], M A Alghoul[1], A N Al-.Shamani[1], M H Ruslan[1], S Mat[1]

[1] Solar Energy Research Institute (SERI), Universiti Kebangsaan Malaysia, 43600 Bangi, Selangor Malaysia
[2] Department of Air conditioning and Refrigeration, Al-Mustaqbal University College, Babylon, Iraq

E-mail:azhermuhson@gmail.com

Abstract. An improved system of the new combined single stage absorption cycle operated with NH_3/H_2O as working fluid was performed. In order to enhance performance the cycle a new configuration of absorption system was utilized. The performances of two configurations of the combined absorption cycle were compared; a) with common solution heat exchanger and b) divided the streamline of solution heat exchanger to recover the internal heat. Based on the analysis, it has been shown that the second configuration a significant reduction of the required generator and absorber loads by about 20% and 17% respectively, with increased coefficient of performance (COP) about 12 % compared to the first configuration. This improvement in the overall COP is found due to improve energy utilization efficiency significantly.

1. Introduction

Absorption cycles in cooling systems represent an alternative to predominant compressor refrigerating cycles. Its application to solar cooling makes this technology attractive due to the synchronicity between solar thermal energy and air-conditioning demand. The advantage of the absorption cycles is their much lower consumption of electricity achieved by replacement of a compressor by so called a thermo-chemical compressor. A thermo-chemical compressor ensures absorption of the refrigerant vapor in an absorption liquid, pumping of the created solution to a higher pressure level, and absorption of the refrigerant vapor from the solution. The main benefits of the absorption cycles are the following: very low consumption of electricity (only up to 5 % of the cooling capacity), possible utilizing of thermal energy obtained from renewable sources (solar radiation, biomass combustion) or whatever source of waste heat energy with temperature above 80 °C for the production of cooling. Another positive effect of the absorption cycles is utilizing of ammonia as refrigerant. Disadvantage of the absorption cycles is the overall performance of the absorption cycle in terms of cooling effect per unit of energy input is generally poor. For this reason, the researchers have been devoted to improvement the absorption cooling systems. However, the COP of the NH_3/H_2O systems is limited due to various constraints posed by the component temperatures and pressures. Combining of the ejector with the single stage absorption cycle showed potential improvement in the COP of the cycle

[1] To whom any correspondence should be addressed.

IOP Conf. Series: Materials Science and Engineering **88** (2015) 012059 doi:10.1088/1757-899X/88/1/012059

[1]. It was showed that COP's as high as 0.86 to 1.04. However, this system must be operated with a high temperature heat source (190 to 210°C) and acceptable surrounding temperature. The advantage of using an ejector refrigerator cycle can utilize thermal energy at temperature levels upward from 333 K [2]. Heat energy at these temperatures is available from the absorber of the absorption refrigerator. Previous researches demonstrated that the COP of absorption and ejector refrigerators are in the ranges 0.5–0.8,while the basic absorption cycle 0.2–0.6 [3, 4]. Many researchers have improved the performance of the single effect absorption cooling system in recent years [5, 6]. It was showed further enhancement by adding flash tank to the cycle with the presence of the ejector. Moreover, experimental studies shows that COP can be increased up to 60% when a solution heat exchanger is used [7]. The effect of the generator –absorber temperature on the system performance has been studied [8-12]. They informed that the system has a low generator temperature limit which that the cycle cannot be operated at generator lower than its limit. The optimum generator temperature determines the optimum solution concentration leaving the generator. Therefore, there is thermal energy wasted within the generator and the absorber, when the solution temperature at the generator higher than its optimum value. From this reasons, for thermal energy saving, the modifications are made in the system to save the energy wasted and to improve the COP of the cycle. In the present study, the effect of streamline of strong solution and heat exchanger effectiveness on the thermal loads of components, entrainment ratio (ω) and coefficient of performance are investigated.

2. System Description

Figure1 shows schematically the main part of (combined ejector –absorption refrigeration system) was done by [13] and [5]. The schematic diagram of the proposed new design is shown in Figure 2. In this new design the ammonia mixture , leaves the absorber (state 11) as a saturated solution at the low pressure with a relatively high ammonia concentration. It is pumped to the system high pressure (state 12) and divided into two parts. : One part flows to the solution heat exchanger (state 12A) it used to recovers the heat from the returning weak liquid solution in the recovery heat exchanger before entering the generator, mainwile the remaining part flows to the rectifier. In the rectifier the vapor is cooled by the rich solution pumped from the absorber to the generator (state 12B).As the generator operates from high temperature source to separate the binary solution of water and ammonia (strong solution comes from absorber) , the basic solution is partially heated to produce a two – phase mixture : a liquid (state 14) , which is relatively weak in ammonia , and a vapor (state 1) with a high concentration of ammonia. This two-phase mixture is separated, and the weak liquid transfer heat to the high concentration stream before it is throttled to system low pressure and sprayed into the absorber. The rectifier cools the saturated ammonia vapor (state 2) to condense out any remaining water (state 16) . It is assumed that the ammonia vapor concentration leaved the rectifier passed to the ejector is 99.6%. On the ejector, the secondary flow (ammonia vapor from the mixing of the evaporator and flash tank (state 17) and the primary flow of ammonia vapor from the rectifier are mixed and passed to the condenser (state 3). In the combined flash –ejector absorption cooling system done by Sirwan et al.[5], the flash tank still serves as an accumulator storing excessive refrigerant for test conditions requiring smaller mass in the system's high pressure side. In order to mix the vapor flow with the evaporator exit flow the Ft valve is necessary to create an identical pressure drop across the vapor. The condensed refrigerant from the condenser is separated into two streams: one stream is sucked into the refrigerant heat exchanger (state 5), meanwhile, other flows into the flash tank (state 6). When refrigerant heat exchanger is employed , the refrigerant entering the absorber (state 10A) and the ejector (state10B) has been superheated by heat exchange with mix liquid exiting from condenser and flash tank (state 7) which causes the liquid to enter the expansion device in a subcooled state (state7'). This subcooling reduces the fraction of flash gas produced at the downstream side of expansion valves.

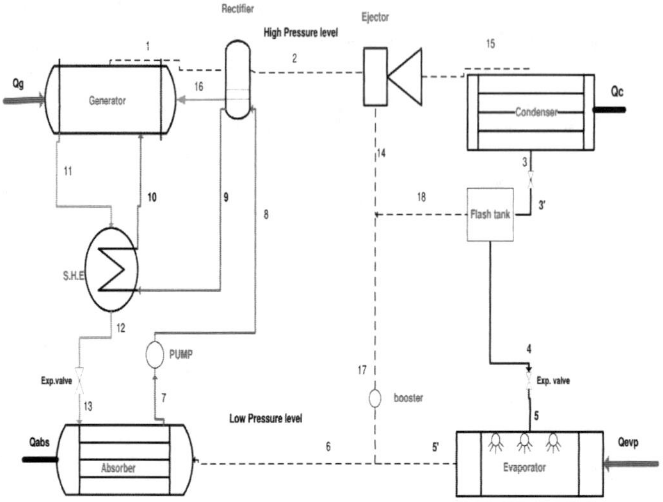

Figure 1. First configuration of combined cycle – with common solution heat exchanger.

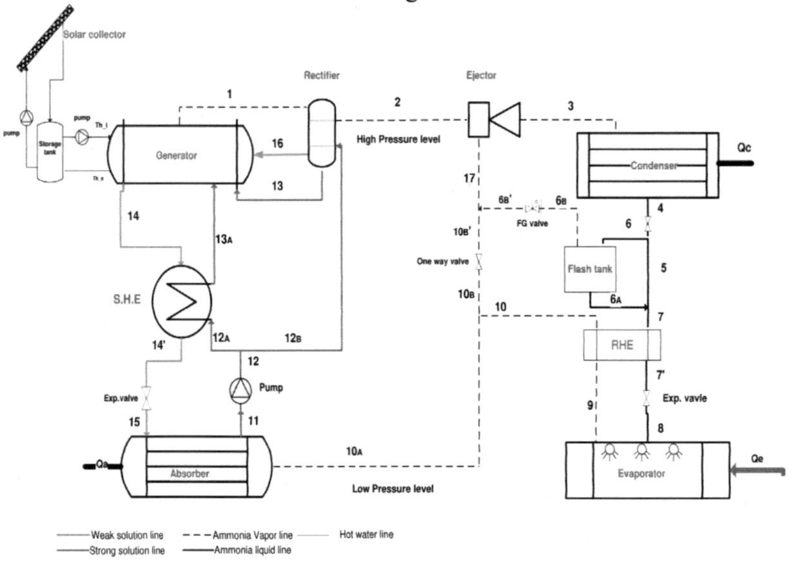

Figure 2. Second configuration of combined cycle - through utilizing the streamline of strong solution.

3. Thermodynamic analysis of absorption cooling system

The mass balances, energy balances, and the equations of state for NH_3/H_2O solution, and refrigerant at each component involved in the cycle are required for system study. In order to calculate the heat and mass balance for the proposed cycle, the thermodynamic properties (pressure, temperature, concentration, enthalpy and density) are necessary for the simulation. The binary mixture of NH3/H2O and pure NH3 are used in the proposed system. The detailed thermodynamic property equations of NH3/H2O are found by [14] and [15]. To evaluate the COP for the cycle, the governing equation required through the each component.

- **Mass balance and Energy balance**

$$\sum_e \dot{m}_e = \sum_i \dot{m}_i \qquad (1)$$

$$Q_K = \sum \dot{m}_e h_e - \sum \dot{m}_i h_i \quad (2)$$

where Q_K is the heat added to component K at temperature T_k .

- **Pump**

The electrical power needed for the pump

$$w_p = \frac{\dot{m}_{11} v_{11} (P_{12} - P_{11})}{\eta_{hed} \eta_{mec}} \quad (3)$$

$$w_p = \dot{m}_{12} h_{12} - \dot{m}_{11} h_{11} \quad (4)$$

- **Heat exchanger**

The solution and refrigerant heat exchanger performance, expressed in terms of an effectiveness ε .

$$T_{7'} = T_9 \varepsilon_{RHE} + (1 - \varepsilon_{RHE}) T_7 \,, \; \dot{m}_{7'} = \dot{m}_7 \,, \; X_7 = X_{7'} \,, \; h_{7'} = h_r(T_{7'}, P_{7'}) \quad (5)$$

$$T_{14'} = T_{12A} \varepsilon_{SHE} + (1 - \varepsilon_{SHE}) T_{14}, X_{14} = X_{14'} \,, \; P_{14} = P_g \,, \; h_{14'} = h_s(T_{14'}, X_{14'}) \quad (6)$$

The component's thermal loads of the combined –ejector cooling system per unit of refrigerant mass are expressed as follows:

$$Q_g = \dot{m}_1 h_1 + \dot{m}_{14} h_{14} - (\dot{m}_{13} h_{13} + \dot{m}_{16} h_{16} + \dot{m}_{13A} h_{13A}) \quad (7)$$

$$Q_e = \dot{m}_9 (h_9 - h_8) \quad (8)$$

$$Q_c = (1 + \omega) \dot{m}_2 (h_3 - h_4) \quad (9)$$

$$Q_a = \dot{m}_{10A} h_{10A} + \dot{m}_{15} h_{15} - \dot{m}_{11} h_{11} \quad (10)$$

The coefficient of performance COP is defined as

$$COP = \frac{Q_E}{Q_G + w_p} \quad (11)$$

- **Equation of state**

Finally, The thermodynamic properties at states (1)-(10),and (17) in Figure2 are determined by NH_3 and other properties at states (11)-(16) can be calculated based on the binary mixture of NH_3/H_2O which are given by the correlations of [14] and [15]:

4. Results and Discussion

EES software is used to evaluate the performance of the system using NH_3/H_2O as working fluid. The operation temperatures range of the proposed cycle are selected as Tgen = 60 – 120 °C, Tcond=Tabs = 20-50 °C, Tevp=-15-15 °C , and mass flow rate of refrigerant = 0.0166 kg/s. The effectiveness of the solution and refrigerant heat exchangers is assumed 0.5 and 0.6 respectively.

4.1 Evaluation of utilizing the streamline of strong solution

In this work, the effects of the streamline of strong solution on the performance of the proposed cycle are analyzed. The effects of the streamline of strong solution on the thermal loads of generator and absorber are shown in Figs.3and 4.It can be seen that the heat load of the generator decreases with an increase temperature of the strong solution entering the generator. Similarly, with a decrease in the weak solution temperature entering the absorber, the heat rejected from the absorber also decreases.

IOP Conf. Series: Materials Science and Engineering **88** (2015) 012059 doi:10.1088/1757-899X/88/1/012059

For this reason, the reduction ratios of both generator and absorber thermal loads increase with increases the generator temperature by about 20 % and 17% respectively as shown in Figure3. Also it is found that the decreasing ratios in both generator and absorber thermal loads increase with the increases of SHE effectiveness and reach up to 38% as shown in Figure4.

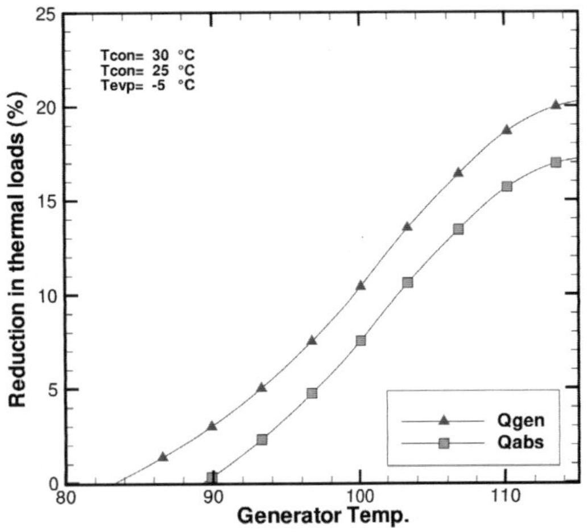

Figure 3. Reduction in thermal loads versus the generator temperature.

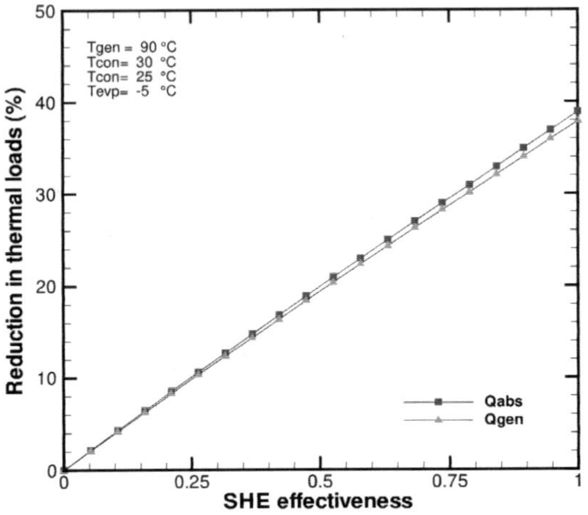

Figure 4. Reduction in thermal loads versus the effectiveness of SHE

As it can be seen from Figure5, when the generator temperature increases, the thermal loads of the generator and absorber decrease. The optimum generator temperatures corresponding to the minimum thermal loads of the generator and absorber after this optimum value the thermal loads remain unchanged. If the generator temperature gets higher, the concentration of the solution leaving the generator increase, and hence, the circulation ratio decreases. The thermal loads of the condenser and the evaporator remains constant, as the generator temperatures increases. Moreover, the entrainment ratio ω remains almost constant under generator temperature range.

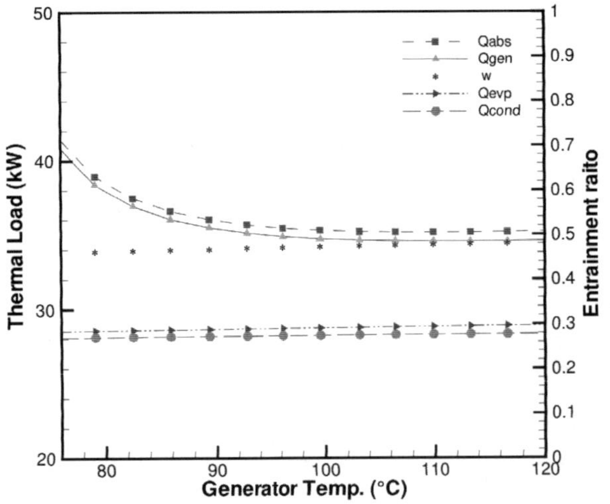

Figure 5. Variation of thermal loads with the generator temperature at (Tc=30 °C , Ta=25 °C ,Te=-5 °C).

4.2 Comparison between the first and second configuration

Figures 6 a & b illustrate the variation of the coefficient of performance with generator temperature for five scenarios of different operating conditions (condenser, absorber and evaporator temperature) for combined cycle before and after the modification. These five scenarios can be used to evaluated the COP of the systems under generator temperature range (70-110 °C), condenser temperature range (25-45 °C) to operate under water –cooled or air cooled condenser . The range of the evaporator temperature is (-10 to 10 °C) that shows the best cooling effect under the employed generator and ambient –condensation temperature ranges. The overall COPs are found higher at the new modified combined cycle (second configuration) about 12 % than that of combined cycle (first configuration). This improvement in the overall COP is found due to minimize the energy consumption at the generator, and absorber. As well as, the RHE is used to assist the flash tank to improve the quality of the refrigerant that enters the evaporator.

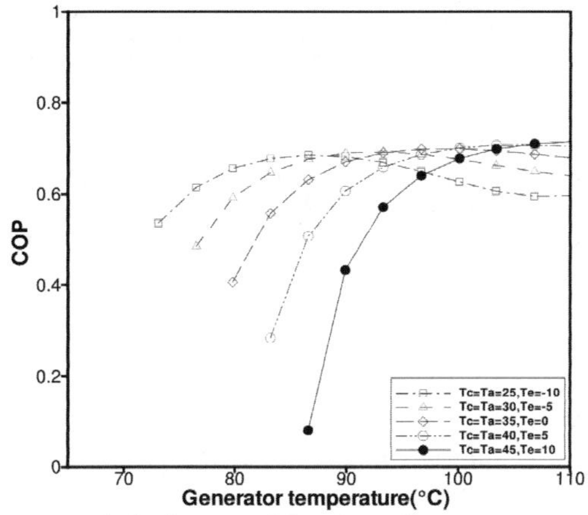

a). Cycle before modification (first configuration).

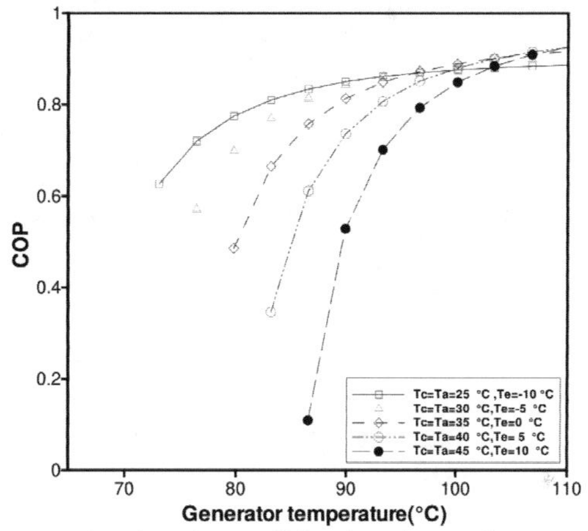

b) Cycle after the modification (second configuration).

Figure 6 (a&b) variation of COP vs generator temperature at different operating conditions for combined cycle: a) first configuration, and b) second configuration.

5. Conclusion

In this study, a single stage combined ejector–absorption cooling system using NH_3/H_2O as working fluid was analyzed and then a new improved configuration based on this system was proposed. The effects of the streamline of strong solution on the performance of the proposed cycle were investigated. It has been shown that the second configuration a significant reduction of the required generator and absorber loads by about 20% and 17% compared to the ordinary combined cycle. It was also found that the SHE decreasing the thermal loads of generator and absorber by about 38%.The thermal loads of the condenser and the evaporator remains constant, as the generator temperatures increases. Moreover, the entrainment ratio ω remains almost constant under generator temperature range. The overall COPs are found higher at the new modified combined cycle about 12 % than that of ordinary combined cycle. This improvement in the overall COP is found due to utilize the internal heat recovery in the new modified cycle.

IOP Conf. Series: Materials Science and Engineering **88** (2015) 012059 doi:10.1088/1757-899X/88/1/012059

6. References

[1] Aphornratana S and Eames I W 1998 Experimental investigation of a combined ejector-absorption refrigerator *International Journal of Energy Research* **22** 195-207

[2] Sun D-W, Eames I W and Aphornratana S 1996 Evaluation of a novel combined ejector-absorption refrigeration cycle — I: computer simulation *International Journal of Refrigeration* **19** 172-80

[3] Srikhirin P, Aphornratana S and Chungpaibulpatana S 2001 A review of absorption refrigeration technologies *Renewable and Sustainable Energy Reviews* **5** 343-72

[4] Ward D S 1979 Solar absorption cooling feasibility *Solar Energy* **22** 259-68

[5] Sirwan R, Alghoul M A, Sopian K, Ali Y and Abdulateef J 2013 Evaluation of adding flash tank to solar combined ejector–absorption refrigeration system *Solar Energy* **91** 283-96

[6] Sirwan R, Alghoul M A, Sopian K and Ali Y 2013 Thermodynamic analysis of an ejector-flash tank-absorption cooling system *Applied Thermal Engineering* **58** 85-97

[7] Aphornratana S 1995 Theoretical and experimental investigation of a combined ejector-absorption refrigerator. University of Sheffield, Department of Mechanical and Process Engineering)

[8] Kang Y, Akisawa A and Kashiwagi T 1999 An advanced GAX cycle for waste heat recovery: WGAX cycle *Applied Thermal Engineering* **19** 933-47

[9] Kang Y T, Chen W and Christensen R N 1996 Development of design model for a rectifier in GAX absorption heat pump systems *Ashrae Transactions* **102** 963-72

[10] Grossman G, DeVault R C and Creswick F A 1995 Simulation and performance analysis of an ammonia-water absorption heat pump based on the generator-absorber heat exchange (GAX) cycle. Oak Ridge National Lab., TN (United States))

[11] Sun D-W 1997 Computer simulation and optimization of ammonia-water absorption refrigeration systems *Energy Sources* **19** 677-90

[12] Sun D-W 1998 Comparison of the performances of NH3-H2O, NH3-LiNO3 and NH3-NaSCN absorption refrigeration systems *Energy Conversion and Management* **39** 357-68

[13] Abdulateef J M, Sopian K, Alghoul M A and Sulaiman M Y 2009 Review on solar-driven ejector refrigeration technologies *Renewable and Sustainable Energy Reviews* **13** 1338-49

[14] Bourseau P and Bugarel R 1986 Absorption-diffusion machines: comparison of the performances of NH3H2O and NH3NaSCN *Réfrigération par cycle à absorption-diffusion: comparison des performances des systèmes NH3H2O et NH3NaSCN* **9** 206-14

[15] Pátek J and Klomfar J 1995 Simple functions for fast calculations of selected thermodynamic properties of the ammonia-water system *International Journal of Refrigeration* **18** 228-34

Applications of Jet Ejectors for efficient refrigeration and modelling of Multi Phase Multi Fluid flow in Ejectors

D Konar[1]
[1] Research Scholar, Dept of Mechanical Engg., IIT Kharagpur, Kharagpur – 721302. W.Bengal. India.

E-mail[1]: dibyendukgp@gmail.com

Abstract. The uses of ejector for efficient refrigeration are manifold – it has been used, among other applications, in the VCRS to reduce the compression ratio, in the combined ejector absorption cycle to enhance the refrigeration capacity and in the ejector absorber cycle to obtain lower evaporator pressures, higher absorber pressures and pre-absorption of the refrigerant in the ejector. Hence, modeling of flow which may be two phase two fluid as in ejector absorber cycle or two phase single fluid as in VCRS in an ejector assumes utmost importance. However, much work has not been done in this field. The primary objective of the present work is to discuss about the role of ejectors in various refrigeration systems and to model the two phase two fluid flow in the nozzle and the diffuser of an ejector under suitable assumptions. The equations of conservation of mass, momentum and energy have been solved to find the different flow properties like pressure, temperature and velocity of the two phases as function of the length in the diffuser. Different cases pertaining to different flows have been taken care of by appreciating what type of phenomena can actually occur at the interface of the two phases. Higher pressure rise was obtained for a given diffuser length with higher diffuser angles, smaller droplet diameter, higher inlet velocity of the gaseous phase and higher drag coefficients. Among other results, it was also seen that the two phases reached thermal equilibrium faster with higher diffuser angle, smaller droplet diameter and higher heat transfer coefficient.

KEYWORDS: Vapour Absorption Refrigeration Systems, Ejector, Multiphase-flow.

NOMENCLATURE

COP	Coefficient of Performance
η	2nd Law efficiency of the system
ρ	Density(Kgm^{-3})
u, v, w	Velocites in the x,y and z direction $\left(ms^{-1}\right)$
A	Cross-sectional area $\left(m^2\right)$
$\dot{m}$	Mass flow rate $\left(Kgs^{-1}\right)$
P	Pressure $\left(Pa\right)$

[1] To whom any correspondence should be addressed.

IOP Conf. Series: Materials Science and Engineering **88** (2015) 012060 doi:10.1088/1757-899X/88/1/012060

F	Force (N)
Q	Heat (KJ)
W	Work (KJ)
H	Enthalpy (KJ)
μ	Coefficient of viscosity $(Kgm^{-1}s^{-1})$
K	Turbulent Kinetic energy (KJ)
ε	Turbulent dissipation
Re	Reynold's number
D	Diameter (m)
C_D	Non-dimensional Drag coefficient
C_{vm}	Virtual mass coefficient
ξ	Concentration(Kgm^{-3})
h	Convective heat transfer coefficient $(Wm^{-2}K^{-1})$
h_m	mass transfer coefficient
u_r	Relative velocity (ms^{-1})
T	Temperature (K)
N_d	Number of droplets per unit volume (m^{-3})
t	time (s)
f, α	volume fraction
ν	Kinematic viscosity (m^2s^{-1})
γ	interfacial tension (Nm^{-1})
h_l	head loss (m)
ε	roughness of pipe inner surface (m)
f	Darcy Weisbach friction factor

Subscripts

e	Evaporator
g	Generator/Gaseous phase
a	Absorber
p	Pump
c	Condenser
μ_t	Coefficient of turbulent or eddy viscosity
l	Liquid phase

IOP Conf. Series: Materials Science and Engineering **88** (2015) 012060 doi:10.1088/1757-899X/88/1/012060

1. INTRODUCTION

The importance of conservation of energy in the context of growing global population, increasing per capita power consumption and depleting fossil fuel resources cannot be overemphasized. Energy can be conserved by using it more efficiently which depends not only on the amount of the energy used for the given application but also on the form of the energy. In other words, as far as possible the energy used for a given application should be of correct amount and type. As an example, it would make more sense thermodynamically to spend thermal energy for heating rather than using high grade energy such as electricity. The same holds good for refrigeration and air-conditioning, which can be obtained by thermal energy, whose quality is far inferior to mechanical or electrical energy.

1.1 VAPOR ABSORPTION SYSTEMS

Refrigeration by mechanical vapor compression system uses shaft work which is a high grade of energy and therefore expensive. Vapor absorption system however makes use of waste heat which is a low grade of energy and readily available. The amount of work required is also less and so is the maintenance cost owing to very few moving parts. Since the compressor gets hot we have to have means of cooling it and this puts a constraint on the refrigerant we can use. Conventional VCRS unlike VARS use CFCs which are facing opposition owing to high GWP & ODP. Hence, studies on VARS assumes utmost importance and methods to increase their performance has been an area of concern.

In vapor absorption refrigeration system, the function of the compressor is accomplished by a 3 step process by the use of absorber, pump and generator. In absorber, absorption of the refrigerant vapor coming from the evaporator by its weak or poor solution in a suitable absorbent occurs forming a rich solution of the refrigerant in the absorbent. The rich solution is pumped to the generator raising its pressure to the condenser pressure. In the generator or desorber, distillation of the vapor from the rich solution occurs leaving the poor solution for recycling through an expansion valve back to the absorber where it is again ready to absorb fresh refrigerant. A simple vapor absorption system thus consists of a condenser, an expansion device and an evaporator as in the vapor compression system and in addition an absorber, a pump, a generator and a pressure reducing valve to replace the compressor.

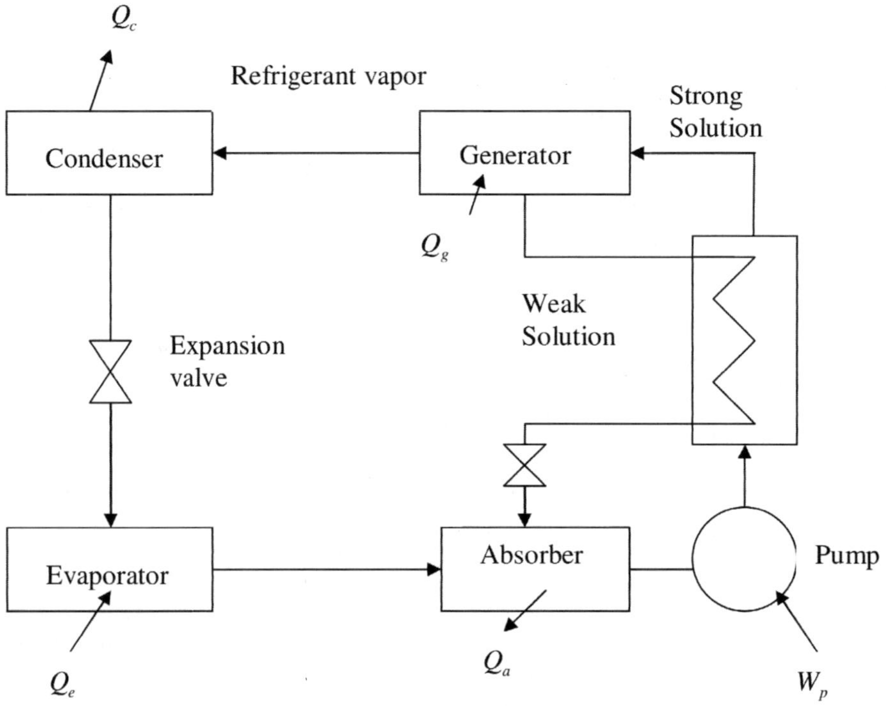

Fig 1 A simple vapor absorption cycle

An ejector is a device where a motive stream expands through a nozzle creating a region of low pressure which entrains the secondary stream which may evaporate in the process thus causing refrigerating effect. The primary objective of the present work is twofold – to discuss the role of ejectors in improving the performance of absorption systems and to model the flow in the nozzle and the diffuser of an ejector and find the different flow properties like pressure, temperature and velocity as function of the length in the diffuser. Different literature surveys have been made in this connection. An attempt to model various one dimensional flows through ejectors like two phase two fluid which is present in different refrigerating systems has been made. For this an one dimensional model has been developed and the different cases pertaining to different flows has been taken care of by appreciating what type of phenomena can actually occur at the interface of the two phases. The equations for conservation of mass, momentum and energy have been solved and flow properties like pressure, temperature and velocity have been plotted as function of length in the diffuser.

2. APPLICATIONS OF EJECTORS IN REFRIGERATION SYSTEMS

The applications of ejector and their role in various systems has been discussed in this section.

2.1.1 VAPOR COMPRESSION REFRIGERATION SYSTEM

An experimentally validated approach to recover some of the expansion losses of the transcritical carbon dioxide cycle is based on replacing the expansion valve by an ejector as shown below.

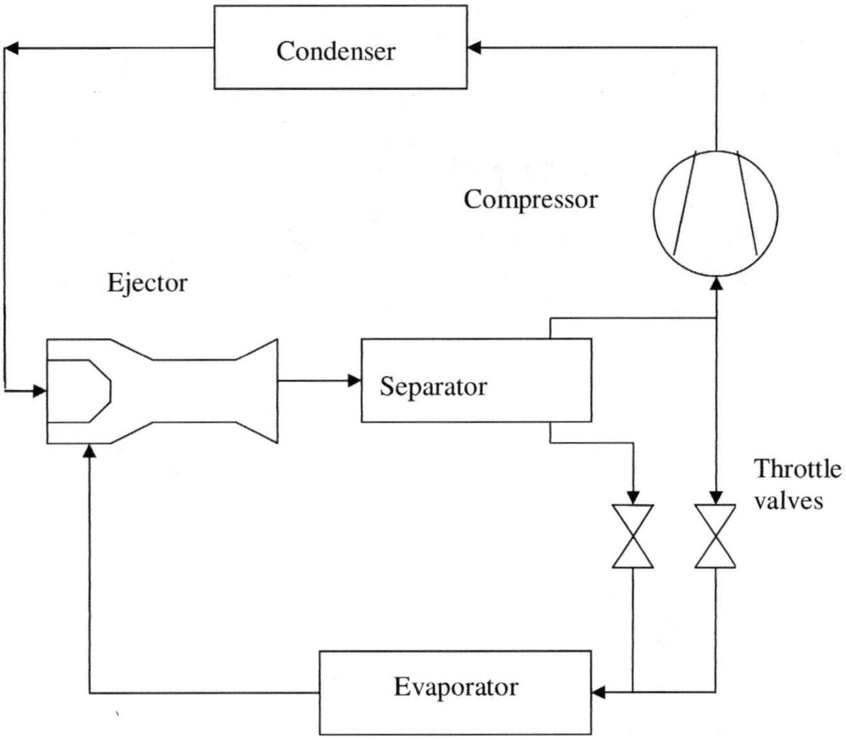

Fig 2 Schematic of trans-critical CO_2 cycle with ejector expansion device

The working processes of the ejector can be explained as follows. The motive stream in the form of high-pressure liquid leaves the condenser and enters the ejector. The motive stream expands in the ejector nozzle from the high pressure to the receiving chamber pressure. The enthalpy of the motive stream reduces and the velocity increases. At the same time the suction stream is drawn from the evaporator at the evaporator pressure and enters the ejector in the receiving chamber where its velocity increases. The suction stream mixes with the motive stream in the mixing section and become a single stream at an intermediate mixing pressure. This stream further increases its pressure in the diffuser by converting its kinetic energy into potential energy. The stream leaves the ejector at a pressure slightly higher than the evaporator pressure and enters a separator to separate saturated liquid from saturated vapor. The saturated liquid expands to the evaporator pressure and enters the evaporator while the saturated vapor enters the compressor. The cycle efficiency increases because the compressor operates at a reduced pressure ratio compared to the cycle without the ejector. In the above system the ejector operates with a 2 phase single fluid.

2.1.2 COMBINED EJECTOR ABSORPTION CYCLE

Eames et al [1] investigated an ejector-absorption cycle both theoretically and experimentally. This cycle shown in the figure below consists of 2 sub-cycles: the steam ejector cycle and LiBr-water single-effect absorption cycle. The ejector cycle is driven by water vapor from the generator of the absorption cycle.

IOP Conf. Series: Materials Science and Engineering **88** (2015) 012060 doi:10.1088/1757-899X/88/1/012060

Because the steam ejector utilized the energy, otherwise lost in a conventional absorption cycle, to enhance the vaporization process in this novel cycle, a higher COP was expected. The computer simulation of this novel cycle was reported by Sun et al[2]. The experimentally-measured COP of this cycle was reported in the range between 0.8 to 1.04 for $5^{\circ}C$ cooling temperature [3]. However, this cycle requires a generator temperature of at least $200^{\circ}C$ and this may result in increased corrosion rates, which may be problematic[4]. In this system the ejector operates with a single phase single fluid.

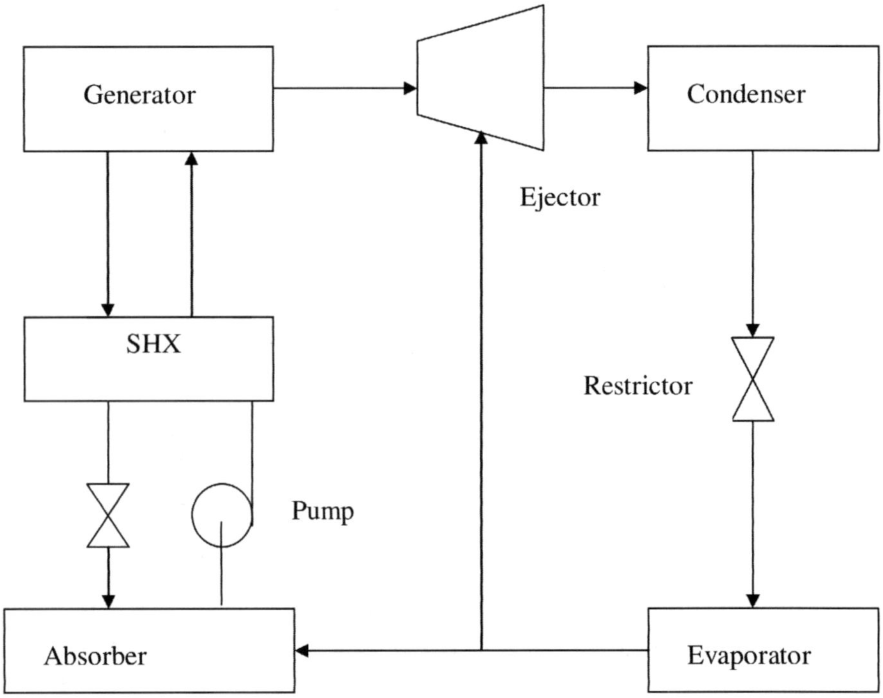

Fig 3 Combined ejector-absorption cycle

2.1.3 EJECTOR ABSORBER CYCLE

Chung et al[5] and Chen [6] used the returning solution from the generator as the primary fluid to entrain the refrigerant vapor from the evaporator. The mixture fluid was discharged into the absorber. Since this arrangement allowed the absorber pressure to be higher than the evaporator pressure, the circulation ratio of the solution could be reduced. Therefore, the COP was improved. However, this system can only be operated using high density refrigerant vapor, because a liquid driven ejector is not suitable for low density vapor such as water vapor (which would involve huge volumetric flow rates). The flow in the ejector is that of 2 phase 2 fluid.

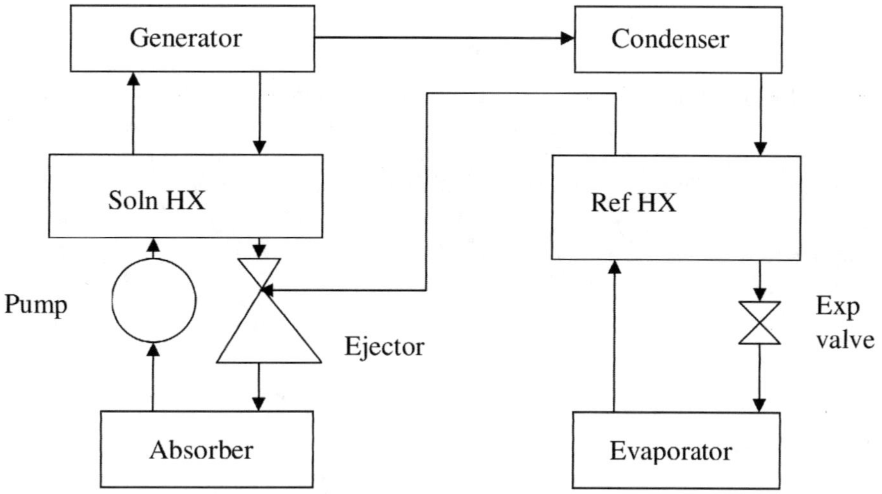

Fig 4 An ejector absorber cycle

The above cycle has the following advantages:

(i) For a given absorber pressure it is possible to achieve a lower value of evaporator pressure, which would enable lower temperatures.

(ii) For a given evaporator pressure we can have a higher value of absorber pressure and hence a higher temperature and hence the absorber can reject heat to a hotter sink.

(iii) There is pre-absorption of the refrigerant in the ejector.

2.2 CONCLUSIONS FROM LITERATURE SURVEY

The application and the use of ejector in different systems has been looked into. In vapor compression refrigerating systems the introduction of the ejector reduces the work of compression by reducing the compression ratio. In combined ejector absorption cycle, the ejector by sucking vapor from the evaporator increases the refrigerating capacity subject to the condition that the generator has to be operated at a higher temperature. In ejector absorber cycle, the introduction of the ejector enables us to operate the absorber at a higher pressure for a given evaporator pressure which implies that the absorber can reject heat to a hotter sink. It also makes the evaporator work at a lower pressure for a given absorber pressure which means we can get lower temperatures. Besides, the ejector also improves pre-absorption which implies that the size of the absorber is reduced.

A generalized model for 3 dimensional flow in ejectors has been looked into. A model in which the liquid phase has turbulent flow has also been studied. The book on " 1 dimensional 2 phase Flow" by Wallis gives different modeling of flows namely the homogeneous model, the separated flow model, the drift-flux model which is applicable to flow of different fluids under different operating conditions. From the studies it is apparent that not much work has been done in the field of two phase flows. Hence, there is large scope for work in this field which is relevant considering that there is so much application of such flows in different systems.

3. MODELING OF 1-D TWO PHASE FLOW IN AN EJECTOR

Levy et al (9) models two phase two fluid flow in ejector which is present in vapor absorption systems as discussed in the literature survey under a set of assumptions given below.

IOP Conf. Series: Materials Science and Engineering 88 (2015) 012060 doi:10.1088/1757-899X/88/1/012060

3.1 DESCRIPTION

The gaseous phase is the pure refrigerant which comes from the evaporator after passing through the refrigerant heat exchanger. The dispersed phase is the solution which comes from the generator after passing through the solution heat exchanger and is injected with the help of a nozzle into the diverging section. The two phases pass through the diverging section of the ejector whose other end is connected to the absorber. The flow in the diverging section assumed 1 dimensional has been modeled in this section.

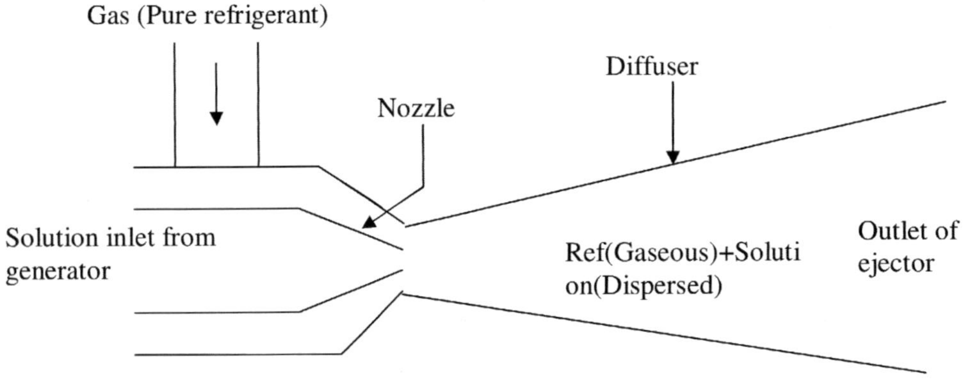

Fig 5 Schematic of the ejector

3.2 ASSUMPTIONS

(i) Steady state.
(ii) One Dimensional flow with u=u(x).
(iii) Pressure P=P(x) which is reasonably good assumption as flow is in X-direction.
(iv) Properties at a particular section are constant.
(v) Mass, momentum and heat transfer only between the 2 phases which implies adiabatic diffuser and frictional work with walls of diffuser negligible.
(vi) Dilute phase flow
(vii) Droplets consist of a binary mixture of absorbent and refrigerant
(viii) Constant droplet diameter
(ix) Electrical forces and surface tension neglected.
(x) All relevant functions are continuous.
(xi)The dispersed phase is fully dispersed i.e. there is no coagulation of any sort.
(xii) The two phases share the same pressure field i.e. at a particular section $P_g = P_d = P$.

3.3 THE MODELLING OF 2 PHASE FLOW IN THE DIVERGING SECTION ASSUMING ABSORPTION

We apply the Reynold's Transport theorem to a control volume of length dx and cross sectional area same as that of the diffuser at that particular section. We can apply Reynold's Transport theorem to both the dispersed and liquid phase individually (separated flow model) or as a whole to the entire fluid flowing through the diffuser (homogeneous model) which obviously depends on the conditions prevalent in the

IOP Conf. Series: Materials Science and Engineering **88** (2015) 012060 doi:10.1088/1757-899X/88/1/012060

diffuser. But it should be seen that the number of differential equations obtained by the conservation laws equal the number of unknowns.

CONSERVATION OF MASS

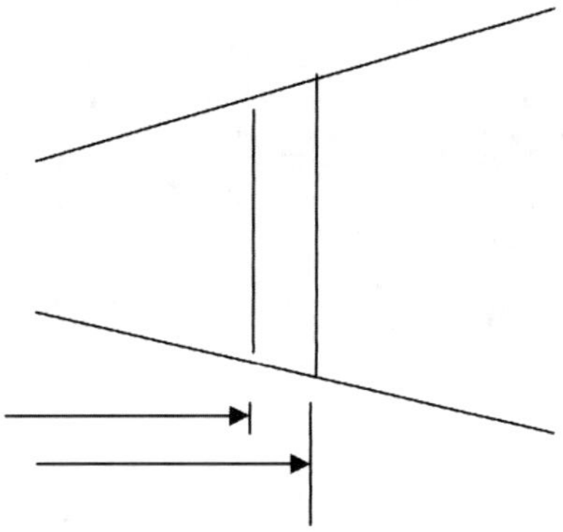

Applied to the gaseous phase, $\left(\rho_g u_g A_g\right)\big|_x - \left(\rho_g u_g A_g\right)\big|_{x+dx} = d\dot{m}_{gd}$ where $\dot{m}_{gd}$ is the rate of absorption of gaseous phase to dispersed phase which is assumed positive i.e. mass is absorbed into the dispersed phase. Here A_d, A_g and A_{tot} are the areas of cross-section occupied by the dispersed phase, gaseous phase and total diffusion cross-sectional area respectively.

$$\frac{d\left(\rho_g u_g A_g\right)}{dx} = -\frac{d\dot{m}_{gd}}{dx} \quad \text{.....................................(3.1)}$$

Applied to the dispersed phase, $\left(\rho_d u_d A_d\right)\big|_x - \left(\rho_d u_d A_d\right)\big|_{x+dx} = -d\dot{m}_{gd}$

$$\frac{d\left(\rho_d u_d A_d\right)}{dx} = \frac{d\dot{m}_{gd}}{dx} \quad \text{.....................................(3.2)}$$

CONSERVATION OF MOMENTUM

Applied to the gaseous phase,

$\rho_g u_g^2 A_g\big|_{x+dx} - \rho_g u_g^2 A_g\big|_x = dF_{gd} - A_g P\big|_{x+dx} + A_g P\big|_x - d\dot{m}_{gd} u_d$ where F_{gd} is the force on the gaseous phase by the dispersed phase which is a surface force.

$$\frac{d\left(\rho_g u_g^2 A_g\right)}{dx} = \frac{dF_{gd}}{dx} - \frac{d\left(A_g P\right)}{dx} - \frac{d\dot{m}_{gd}}{dx} u_d \quad \text{.........................(3.3)}$$

Applied to the dispersed phase,

$$\rho_d u_d^2 A_d\big|_{x+dx} - \rho_d u_d^2 A_d\big|_x = -dF_{gd} - A_d P\big|_{x+dx} + A_d P\big|_x + d\dot{m}_{gd} u_d$$

$$\frac{d\left(\rho_d u_d^2 A_d\right)}{dx} = -\frac{dF_{gd}}{dx} - \frac{d\left(A_d P\right)}{dx} + \frac{d\dot{m}_{gd}}{dx} u_d \quad \text{.....................(3.4)}$$

where the last term of the above 2 equations implies that the momentum that is lost by the gaseous phase is gained by the dispersed phase.

CONSERVATION OF ENERGY

Applying it for gaseous phase,

$$\left\{\rho_g u_g A_g (h_g + \frac{u_g^2}{2})\right\}\Bigg|_x + dQ_{gd} - dW_{gd} + d\dot{m}_{gd}\left\{h_{gd} + \frac{u_d^2}{2}\right\} = \left\{\rho_g u_g A_g (h_g + \frac{u_g^2}{2})\right\}\Bigg|_{x+dx}$$

$$\Rightarrow + \frac{dQ_{gd}}{dx} - \frac{dW_{gd}}{dx} + \frac{d\dot{m}_{gd}}{dx}\left\{h_{gd} + \frac{u_d^2}{2}\right\} = \frac{d}{dx}\left\{\rho_g u_g A_g (h_g + \frac{u_g^2}{2})\right\} \quad \text{..............(3.5)}$$

where h_{gd} is the enthalpy of the refrigerant at the droplet surface and the last term on the L.H.S denotes the energy transfer from the gaseous to dispersed phase due to absorption.

Applying it for dispersed phase,

$$\left\{\rho_d u_d A_d (h_d + \frac{u_d^2}{2})\right\}\Bigg|_x - dQ_{gd} + dW_{gd} - d\dot{m}_{gd}\left\{h_{gd} + \frac{u_d^2}{2}\right\} = \left\{\rho_d u_d A_d (h_d + \frac{u_d^2}{2})\right\}\Bigg|_{x+dx}$$

$$-\frac{dQ_{gd}}{dx} + \frac{dW_{gd}}{dx} - \frac{d\dot{m}_{gd}}{dx}\left\{h_{gd} + \frac{u_d^2}{2}\right\} = \frac{d}{dx}\left\{\rho_d u_d A_d (h_d + \frac{u_d^2}{2})\right\} \quad \text{..................(3.6)}$$

The last term on the L.H.S of the above two equations denotes that the energy lost by the gaseous phase due to absorption is gained by the dispersed phase. Since we assume the ejector to be adiabatic, we have only internal heat exchange. We will use different empirical relations relating heat transfer, work transfer and force to other properties of the gas or solution and other unknowns like velocities so that we have same number unknowns as the number of equations.

THE MODELLING OF 2 PHASE FLOW IN THE DIVERGING SECTION ASSUMING NO ABSORPTION e.g. AIR-WATER

This can be easily obtained from the previous set of equations by putting $\dot{m}_{gd}$=0 i.e. the masses of the dispersed and gaseous phase throughout the entire diffuser remain constant. Several relations to close the system of equations have been developed. These include equations for number of droplets per unit volume, interface forces including forces due to drag and forces due to virtual mass, work transfer at the interface, heat transfer at the interface and mass transfer. The correlation for estimating the overall mass transfer was obtained from an approximation of the theoretical solution of Kronig and Brink for one drop which fits the experimental results of Garner for a cloud of drops (*Clift et al* (1978))

4. RESULTS AND DISCUSSION

A sample set of input includes the following: velocity of air at inlet to diffuser = 45 ms^{-1}; velocity of water at inlet to diffuser = 60 ms^{-1};mass flow rate of water = 24 Kgs^{-1};density of water = 1000 Kgm^{-3};density of air = 1 Kgm^{-3}; non-dimensional drag coefficient = .44; convective heat transfer coefficient= 250 $Wm^{-2}K^{-1}$; diameter of droplets= 0.0002 m; specific heat of water = 4180 $JKg^{-1}K^{-1}$;

specific heat of air = 1005 $JKg^{-1}K^{-1}$; radius of diffuser at exit=0.2 m; radius of diffuser at inlet=0.1295 m; length of the diffuser=.4 m

Equations 3.7 to 3.12 for air as the continuous phase and water as the dispersed phase have been solved with the above input data using MATLAB 6.5 and the results are shown.

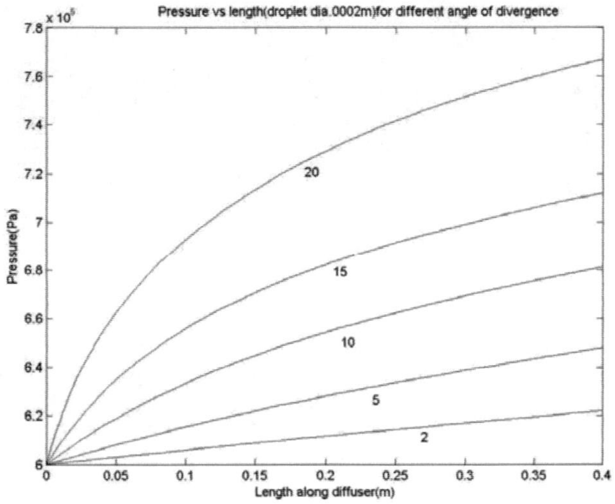

Figure 6 Pressure vs length (droplet diameter=.0002m)for different diffuser angles

From the above plot(Fig 6), it is seen that higher pressure recovery was obtained for higher diffuser angles over a given diffuser length. Also, the rise in pressure progressively becomes more and more as the diffuser angle is increased. Beyond 15°, the diffuser pressure rises suddenly to high values and problems of flow separation may occur with consequent loss of energy. For this purpose, diffuser angles are generally kept much below this value. Apart from same conclusion which can be reached by the following plot (Fig 7), the later shows that for lesser droplet diameters we have higher pressure rise.

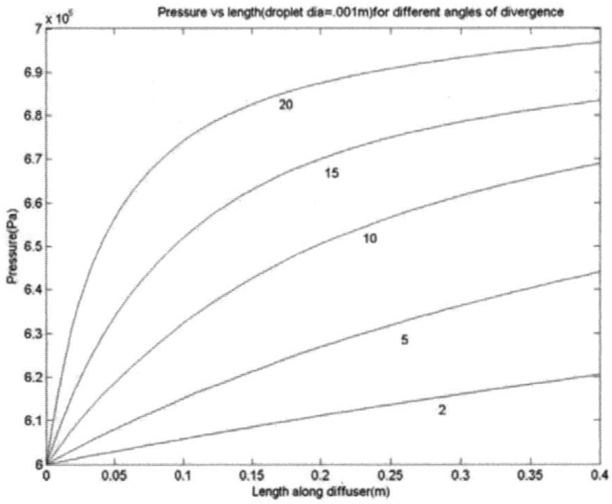

Figure 7 Pressure vs length (droplet diameter=.001m)for different diffuser angles

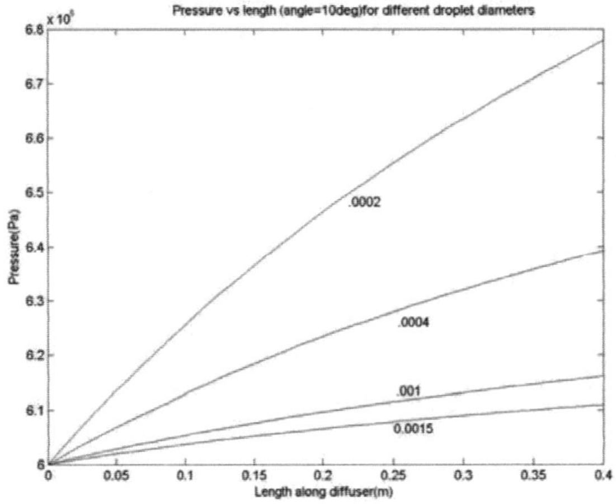

Figure 8 Pressure vs length (angle= 10°)for different droplet diameters

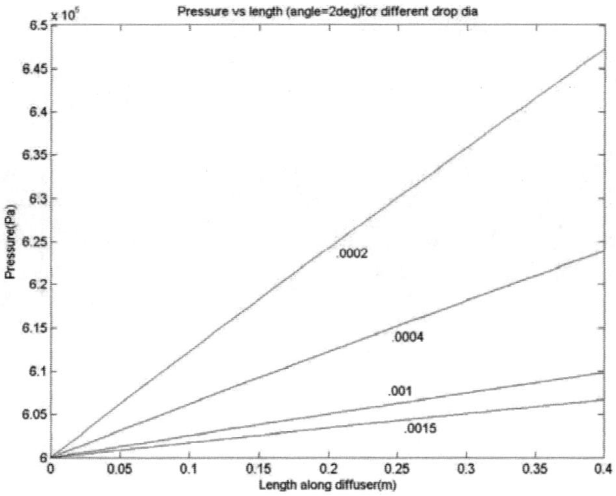

Figure 9 Pressure vs length (angle= 2°)for different droplet diameters

The plot in the previous page(Fig 8) as well as the plot (Fig 9) above shows that a higher pressure rise was obtained when the droplet diameter was reduced. This can be explained from the fact that when the droplet diameter was reduced a higher total surface area was obtained for the same volume and hence there was larger momentum transfer area.

Fig 10 shows that a higher value of pressure is reached if the inlet velocity of the gaseous phase is higher. This is obvious from the fact that with a higher velocity there is a higher velocity head which can be converted to pressure head.

Fig 11 shows that a higher value of pressure is reached if the drag coefficient between the two phases is higher. With more drag there will be more momentum transfer resulting in a higher exit pressure.

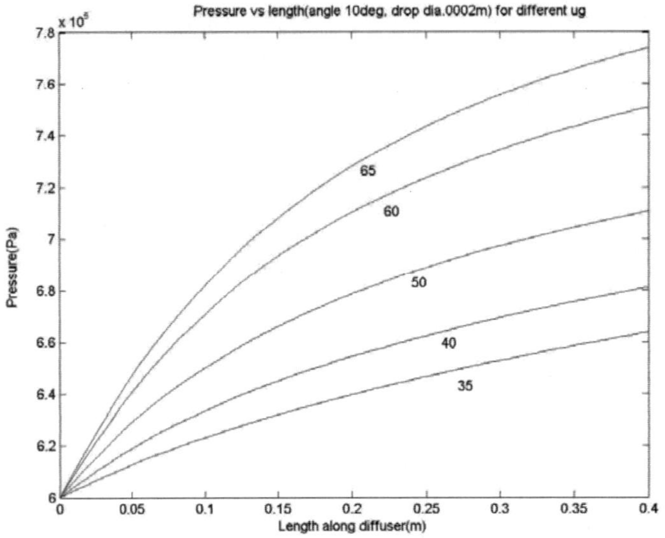

Figure 10 Pressure vs length for different inlet velocities $\left(ms^{-1}\right)$ of the gaseous phase.

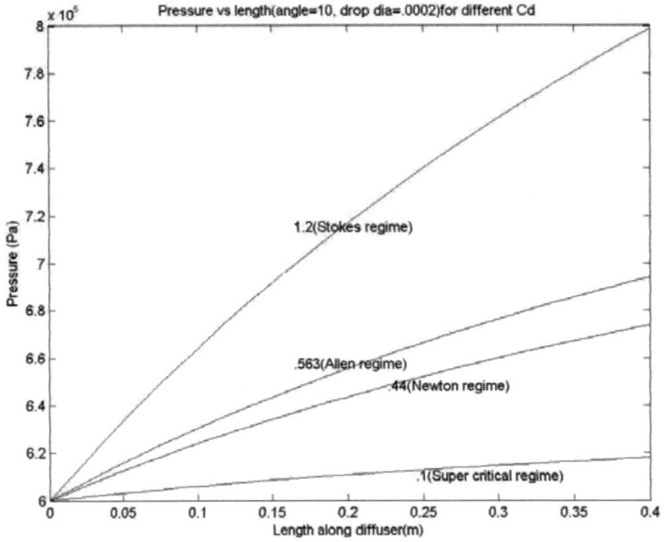

Figure 11 Pressure vs length for different drag coefficients

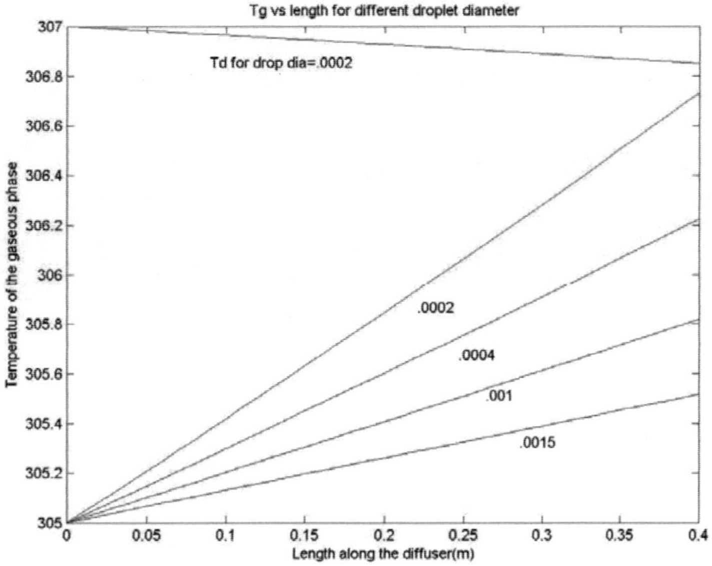

Figure 12 Temperature vs length for different droplet diameters

From Fig 12, it is seen that the temperature rise of the gaseous phase is more than the temperature fall in the dispersed phase. This is because the specific heat of the dispersed phase (water) is about four times more than that of the gaseous phase (air). Also as the droplet diameter is decreased the temperature of the gaseous phase increases implying more heat transfer which is obvious from the increased surface area for heat transfer available due to reduced droplet diameter. Also as the droplet diameter is reduced there is a tendency of the two phases to reach thermal equilibrium over a smaller length.

Figures13 and14 show that the tendency of the two phases to reach thermal equilibrium is higher for larger diffuser angles.

Figure 15 shows that the temperature rise of the gaseous phase and hence the tendency to reach thermal equilibrium is more for larger values of convective heat transfer coefficient.

Figure 16 shows that the velocity of the gaseous phase fell faster as the diffuser angle grew.

IOP Conf. Series: Materials Science and Engineering **88** (2015) 012060 doi:10.1088/1757-899X/88/1/012060

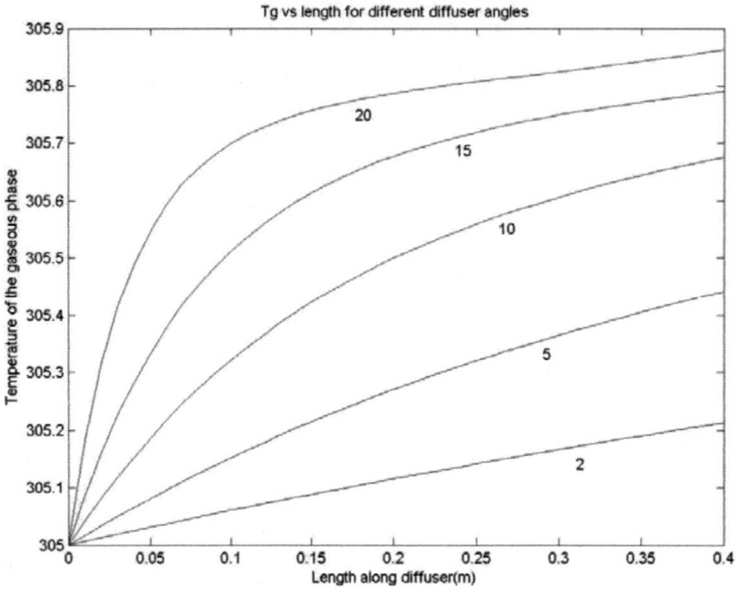

Figure 13 Temperature of the gaseous phase vs length for different diffuser angles

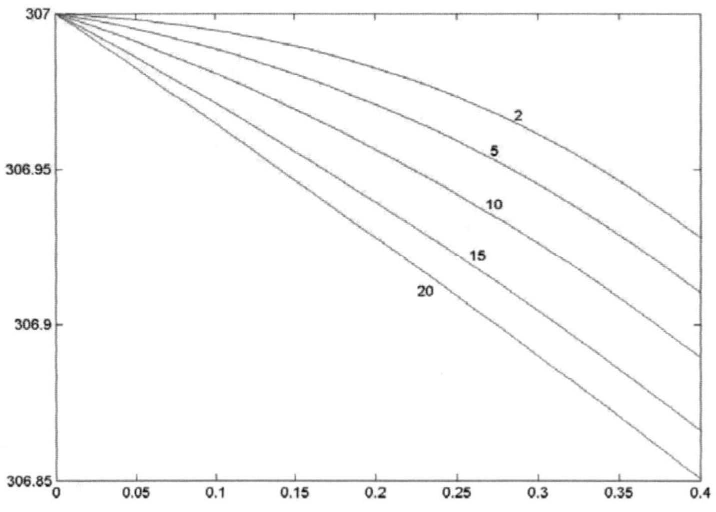

Figure 14 Temperature of the dispersed phase vs length for diff diffuser angles

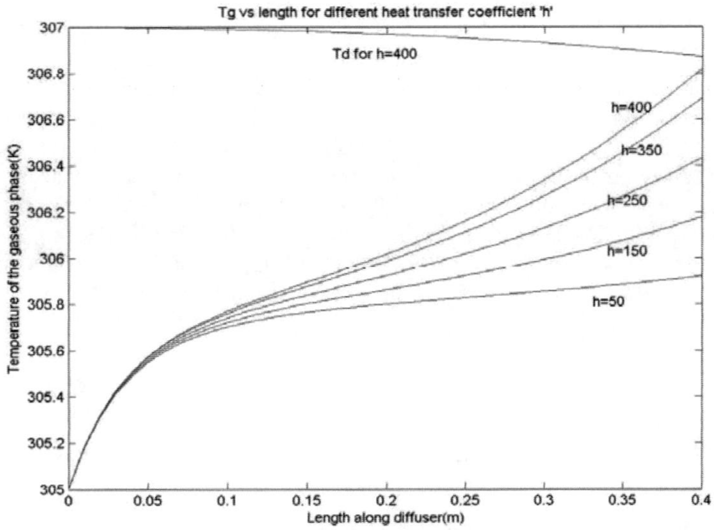

Figure 15 Temperature of the gaseous phase vs length for different heat transfer coefficient

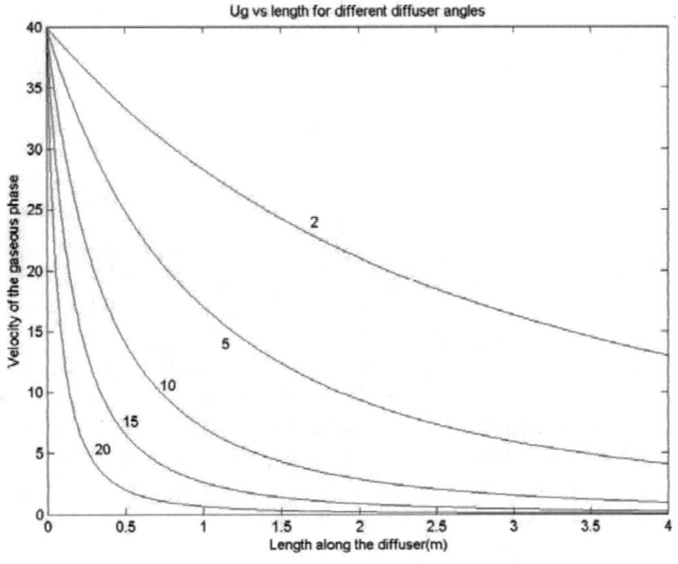

Figure 16 Velocity of the gaseous phase vs length for different diffuser angles

5. CONCLUSION

5.1 SUMMING UP

The results of the one dimensional flow model developed and solved using air water mixture are quite interesting. The trends of the basic model may be used to design diffuser for actual refrigerant absorber pairs. It is seen that the pressure rise is more for larger diffuser angles but beyond a certain value this may not be advisable for design purposes owing to problems of flow separation and consequent loss in energy. A larger pressure recovery is also observed for smaller droplet diameter and hence the nozzle has to be suitably designed to produce droplets of the required diameter depending on the required pressure at the diffuser exit. The pressure rise is also large if the drag coefficient is more and hence suitable substances may be added to the solution to increase the drag coefficient. A higher pressure rise is also noted for larger inlet velocities of the gaseous phase. It is seen that the temperature rise of the gaseous phase was higher than the temperature fall of the dispersed phase and the two phases tend to reach thermal equilibrium faster for smaller droplet diameters owing to larger surface area for heat transfer. The temperature rise is also more for larger diffuser angles and higher values of heat transfer coefficient. The fall in velocity is also faster for wider diffuser angle.

5.2 FUTURE SCOPE OF WORK

This work can be extended to practical refrigerant absorbent pairs and using the model developed, it can be found which pair is suitable for which application. For this the general one dimensional model developed assuming mass absorption can be used and studies have to be made for getting the appropriate constitutive relations at the interface and the properties of the respective mixture and the refrigerant. We can also develop a two dimensional model.

Though we have focused our attention on the flow in the diffuser part of ejector only, we can also analyze how the ejector improves system performance by comparing the COP with and without the ejector. The results of the present study can be used to predict the performance of complex systems having ejectors as one of the components.

REFERENCES

[1] Eames IW, Aphornratana S. *A novel ejector/absorption cycle refrigerator for building air conditioning.* Proc. CIBSE National Conference, Eastbourne, 1995

[2] Sun D, Eames IW, Aphornratana S.*Evaluation of a novel combined ejector-absorption refrigeration cycle - I: computer simulation.* Int J Refrig 1996;**19(3):**172-180

[3] Aphornratana S. *A theoretical and experimental investigation of a combined ejector- absorption refrigerator.* PhD thesis, The University of Sheffield, 1994

[4] Eames IW, Georghiades M, Tucker RJ, Aphornratana S, Wu S. *Combined ejector – absorption cycle technology for gas - fired air conditioning.* Proc. Absorption 96, **Vol I**, 17-20 September, 1996. p 201-8.

[5] Chung H, Huor MH, Prevost M, Bugarel R. *Domestic heating application of an absorption heat pump, Directly fired heat pumps*, Procs. Int. Conf., Uni. Of Bristol, 1984

[6] Chen LT. *A new ejector – absorber cycle to improve the COP of an absorption refrigeration system.* Energy 1988; **30**: 37 – 51

[7] Wu S, Eames IW. *Innovations in vapor – absorption cycles.* Applied Energy **66** (2000):251-266

[8] Jelinek M, Levy A, Borde I. *Performance of a triple-pressure-level absorption cycle with R 125- N, N'-dimethylethylurea.* Applied Energy **71** (2002): 171-189

[9] Jelinek M, Levy A, Borde I. *Study of Design Parameters of a jet ejector for absorption systems*

IOP Conf. Series: Materials Science and Engineering **88** (2015) 012060 doi:10.1088/1757-899X/88/1/012060

[10] Bai H, Thomas BG. *Two phase flow in tundish nozzles during continuous casting of steel.* Materials Processing in the Computer Age III, TMS Annual Meeting, TN, March 12-16 2000,pp 85-99

[11] Apte SV, Mahesh K, Lundgren T. A *Eulerian- Lagrangian model to simulate 2 phase/ particulate flows*. Center for Turbulence Research. Annual Research Briefs 2003

[12] Al-Ansary HAM, Jeter SM. *Numerical and experimental analysis of single-phase and two phase flow in ejectors.* **Vol 10**, Number 4. HVAC & R Research

[13] Chapra S C, Canale RP. *Numerical Methods for Engineers*. 4th Edn, TMH

[14] Groll E A. *Recent Advances in the transcritical* CO_2 *Cycle Technology*, Proc. XVIII National and VII ISHMT- ASME Heat and Mass Transfer Conference, IIT Guwahati, Jan 4-6, 2006

[15] Wallis GB. *One dimensional two phase flow*

[16] Baehr HD, Stephan K. *Heat and Mass Transfer*. Springer

[17] Menegay P,PhD dissertation submitted to the Faculty of the Virginia Tech

IOP Conf. Series: Materials Science and Engineering **88** (2015) 012061 doi:10.1088/1757-899X/88/1/012061

A study on the ventilation method for a factory exposed to high temperature

Y Kim[1], E Lee[2], H Jeong[3] , H Chung[3]

[1] President, DAE SUNG AIR TECH. Co., Gimhae, Korea
[2] Professor, Department of Korea Politechnics, Changwon, Korea
[3] Professor, Department of Energy and Mechanical Engineering, Gyeongsang National University, Korea

E-mail: kys@dsat.co.kr

Abstract. On this work, the importance of industrial ventilation was investigated and examined the theoretical point and problems about general ventilation of factory exposed on high temperature during summer. As a case study, the ventilation planning of the PCB etching process for an electronic company was carried out and each of those characteristics were compared by installing actual ventilation systems and measuring the changing state of the working environment in accordance with ventilation method during summer. The purpose of the study is to present an efficient ventilation method for a factory with a closed structure under high temperature environment. In summary, for a factory with a sealed structure such as the target PCB manufacturing factory in this study, the forced supply and exhaust method was the most appropriate ventilation method for maintaining a low indoor air temperature and for keeping the contaminated air of the factory below the hazardous level.

1. Research Background and Purpose

With the rapid economic growth and industrialization of Korea, industrial complexes have been built all over the country, and such industrialization has rapidly progressed since the 1970s. In factories, all sorts of hazardous factors are produced during the manufacture of products using various materials and methods, unlike in houses or office spaces. Accordingly, the occurrence of industrial accidents in Korea has increased, as has the occurrence of work-related diseases (e.g., occupational diseases).
Of the different types of occupational disease that occurred in 2004,[1] pneumoconiosis accounted for 77%, which suggests that a poor working environment is important. The working environment of a factory is directly related to the health of its workers, and thus, methods of minimizing the production of hazardous factors and the exposure of the workers to them should be established.
In the target factory in this experiment, an etching process for copper clad laminate is performed among the printed circuit board (PCB) manufacturing processes.[2] In such process, circuit boards are manufactured by etching copper clad laminate, on which electronic circuits are printed, in a chemical

[1] To whom any correspondence should be addressed.

IOP Conf. Series: Materials Science and Engineering **88** (2015) 012061 doi:10.1088/1757-899X/88/1/012061

solution. A scrubber is installed in the interior and locally collects dust and performs exhaust treatment of the corrosive harmful gases (e.g., sulfur dioxide) produced during the process. However, ventilation[3] is absolutely required because the building is a sealed area without any air flow and because the leak in the scrubber is problematic.

In this study, the importance of ventilation was examined through a theoretical review of general ventilation;[4] and the characteristics of a high-temperature environment in summer, which has recently been problematic in many factories, were investigated through the changes in the ventilation method of a factory with a sealed structure.

2. Experiment Outline

2.1 Experiment Subject

The target factory in this study is a PCB manufacturing factory of an electronics company located in Chungcheongbuk-do. It is an outermost building of a five-span structure with a medium-height roof, and its interior has partition walls on all sides. It is box-shaped with a 4m-high ceiling, an automatic door on the southeast side, and another door on the northwest side that is open at all times.

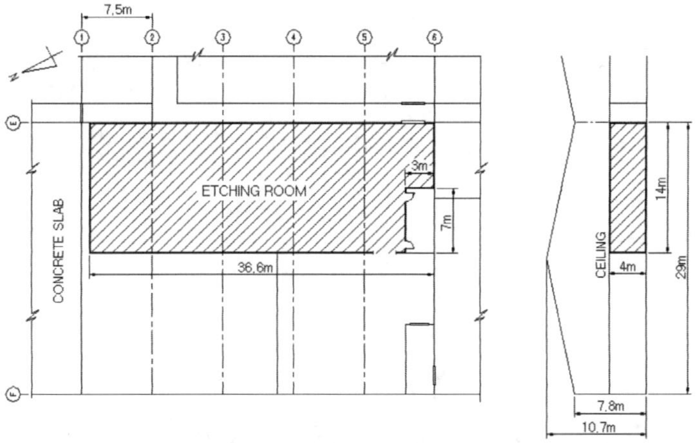

Figure 1. Ventilation target area

Table 1. Experiment cases

Experiment Case	Exhaust	Supply	Etc.	Time
Case 1	Operation	Operation		13:00-15:00
Case 2	Stop	Operation		15:35-17:20
Case 3	Operation	Stop		17:55-19:20
Case 4	Stop	Stop	Nature ventilation	20:00-20:40

2.2 Experiment Method

The experiment was performed based on four cases, as summarized in Table 1. The experiment site was a factory for manufacturing products, and thus, the experiment was conducted within a range that had no effect on the quality of the products.

Figure 2 shows the format of the check sheet used to measure the experiment results.
For the measurement points, tags were attached to the floor at four outdoor points and at a total of 57 indoor points (five points along the width and 12 points along the length, excluding the three points where the doors were positioned), as shown in Figures 3 and 4. For the point where the measurement was difficult due to the apparatus or equipment, the measurement spot was selected within 30 cm of the corresponding point at the breathing line (1.5 m above the floor).
As for the measuring instrument, the Model 6531 Climomaster (hot wire anemometer) of Kanomax, Inc. (Japan) was used to measure the dry-bulb temperature, the relative humidity, and the wind speed. Table 2 summarizes the measurement ranges, and Figure 5 shows the measurement of the temperature, relative humidity, and wind velocity.
It was difficult to read the continuously changing digital data with the naked eyes during the measurement of the temperature, relative humidity, and wind velocity. Thus, the calculation mode of the instrument was used. In other words, three consecutive measurements were taken for 10 seconds each at a point and averaged.
To measure the wind direction, the Model AS-1 of Komyo Rikagaku Kogyo Co., Ltd. was used. Figure 6 shows the measurement. To determine the air flow direction, the average wind direction was measured three times at a point using an air flow indicator.[5]

환기상태 CHECK SHEET (환기방식 : , 측정일시 : 2006. . . ~ :)

No	풍속(m/s)			온도(℃)			습도(%)			비 고
	1	2	3	1	2	3	1	2	3	
A1										
A2										
A3										
A4										
A5										
A6										
A7										
A8										

Figure 2. Check sheet

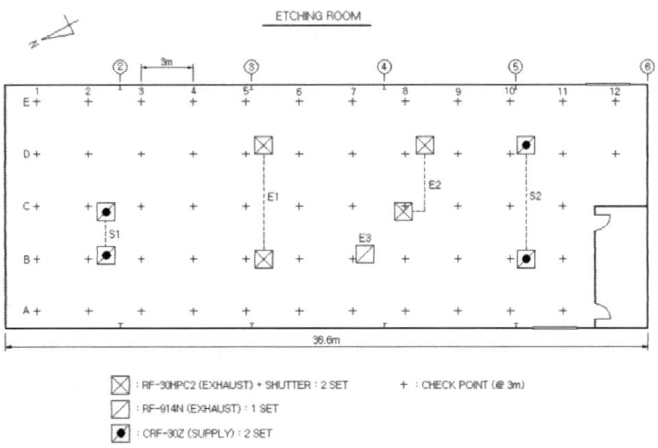

Figure 3. Indoor measurement point

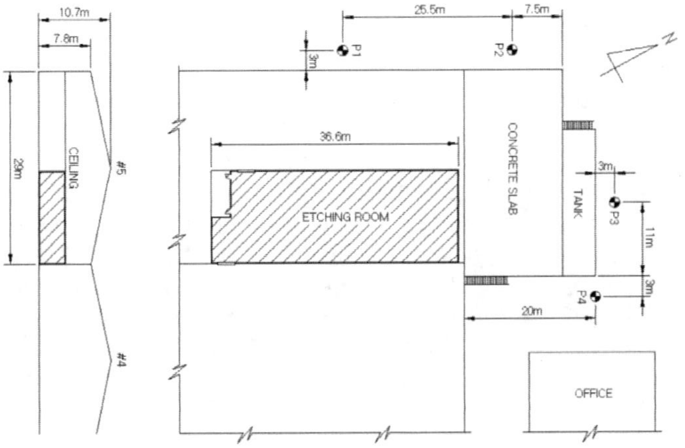

Figure 4. Outdoor measurement point

Table 2. Climomaster

Function	Measurement Extent	Tolerance	Indication Unit	Response Time
Wind velocity	0.00-9.99 m/s	±(3%+0.1) m/s	0.01 m/s	About 1 second
	10-30 m/s		0.1 m/s	
Temperature	0.0-60.0°C	±0.5°C	0.1°C	About 30 seconds
Humidity	2-98% RH	2-80% RH: ±2% RH	0.1% RH	About 15 seconds
		80-98% RH: ±3% RH		

Figure 5. Measurement image

Figure 6. Measurement image (air flow)

IOP Conf. Series: Materials Science and Engineering **88** (2015) 012061 doi:10.1088/1757-899X/88/1/012061

3. Experiment Results

3.1 Indoor Temperature Distribution

Figures 7-10 show the distributions of the indoor temperatures measured by operating the ventilation system using diverse ventilation methods based on the four experiment cases (↔: average outdoor air temperature).

The results of the experiment showed that in all the experiment cases, the average indoor air temperature was higher than the average outdoor air temperature. The ventilation method in Case 4 was natural supply and natural exhaust, and all the indoor air temperatures were higher than the outdoor air temperature. The average outdoor air temperature was 29.1°C and the average indoor air temperature, 36.8°C, which showed a 7.7°C difference. The ventilation method in Case 3 was natural supply and forced exhaust. The average outdoor air temperature was 30.9°C and the average indoor air temperature, 33.8°C, which showed a 2.9°C difference. The ventilation method in Case 2 was forced supply and natural exhaust. The average outdoor air temperature was 31.6°C and the average indoor air temperature, 34.3C, which showed a 2.7°C difference. The ventilation method in Case 1 was forced supply and forced exhaust. The average outdoor air temperature was 31.5°C and the average indoor air temperature, 32.5°C, which showed a 1°C difference. Thus, Case 1 showed the smallest air temperature difference among the four cases.

The cases with the smallest to the greatest difference between their average outdoor air temperature and their average indoor air temperature were as follows: Case 1 < Case 2 < Case 3 < Case 4. The fact that the indoor air temperature distribution in Case 1 was the closest to the average outdoor air temperature indicates that the ventilation method in Case 1 was superior to the other ventilation methods.

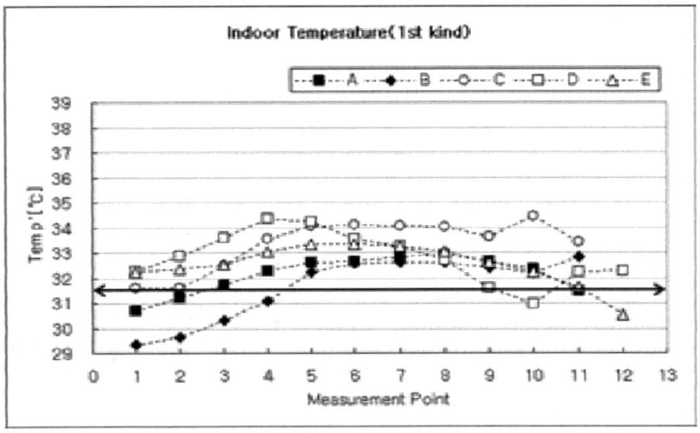

Figure 7. Temperature (Case 1)

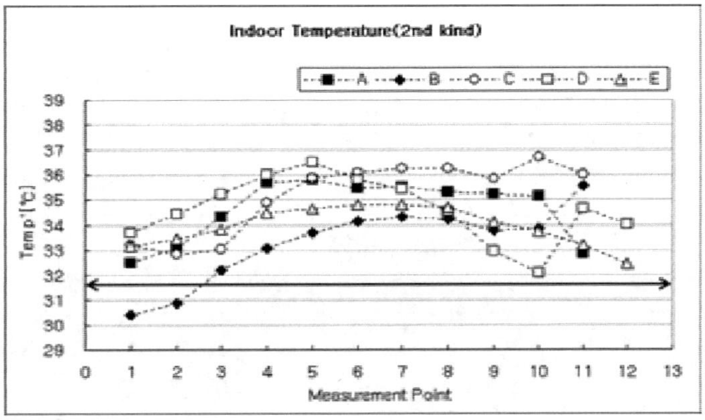

Figure 8. Temperature (Case 2)

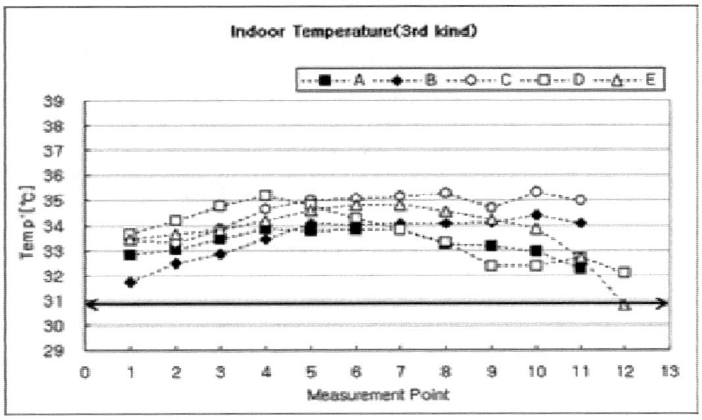

Figure 9. Temperature (Case 3)

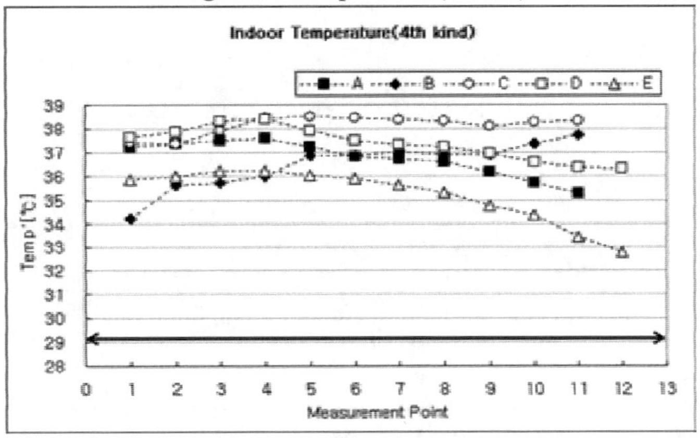

Figure 10. Temperature (Case 4)

Also, the fact that the deviation of the indoor air temperature in Case 2 was larger than that of the other ventilation methods indicates that only the area adjacent to the air inlet was affected by the ventilation system due to the forced supply.

3.2 Indoor Absolute Humidity Distribution

The amount of saturated water vapor changes depending on the temperature, and thus, the relative humidity, has different values depending on the temperature, although the amount of water vapor in the air is constant. Figures 11-14 show the indoor absolute humidity distribution in each experiment case, to examine the indoor humidity distributions depending on the various ventilation methods(↔: average outdoor air absolute humidity).

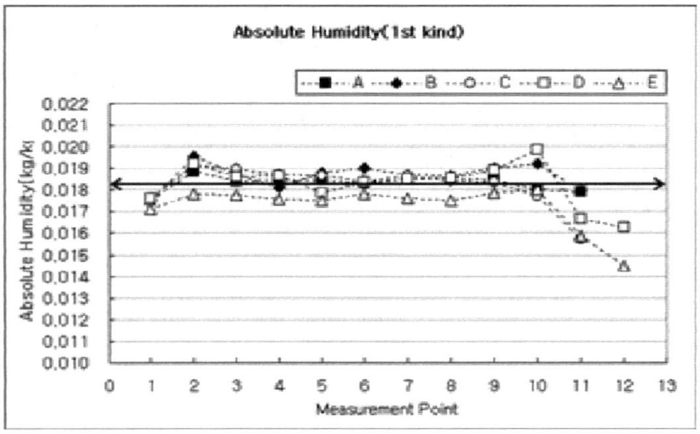

Figure 11. Absolute humidity (Case 1)

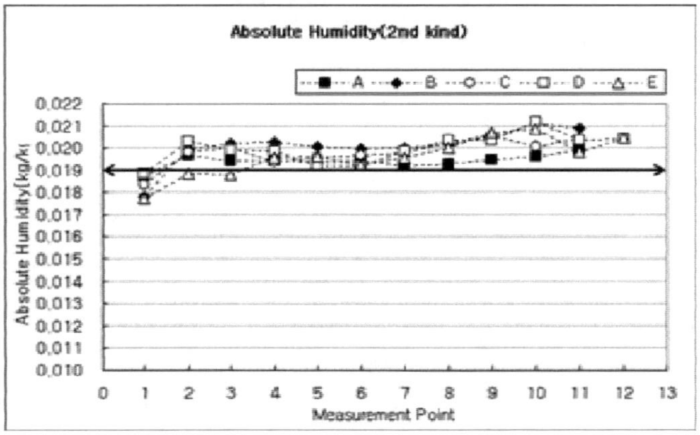

Figure 12. Absolute humidity (Case 2)

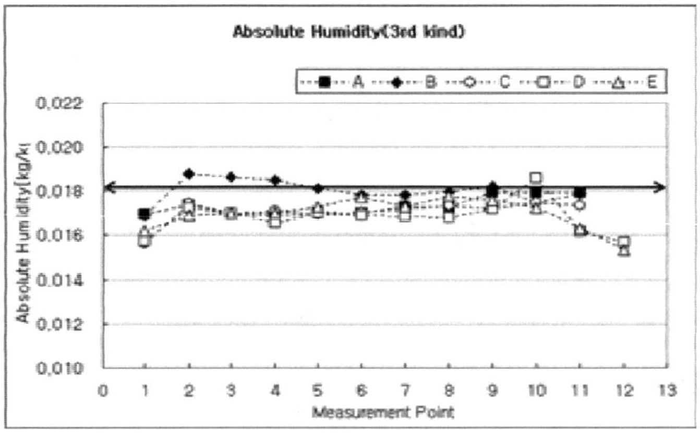

Figure 13. Absolute humidity (Case 3)

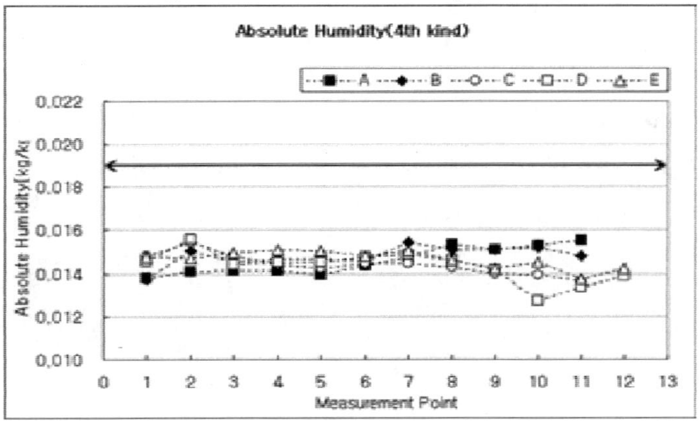

Figure 14. Absolute humidity (Case 4)

As shown in Figures 11-14, the indoor air absolute humidity in Case 2 was slightly higher than the outdoor air absolute humidity, and the indoor air absolute humidities in Case 1, Case 3, and Case 4 were lower than the outdoor air absolute humidity.

In Case 1, the average outdoor air absolute humidity was 0.01821 kg/kg, and the average indoor air absolute humidity was 0.01813 kg/kg; in Case 2, 0.01901 kg/kg and 0.01978 kg/kg; in Case 3, 0.01816 kg/kg and 0.01722 kg/kg; and in Case 4, 0.01897 kg/kg and 0.01455 kg/kg.

The differences between the average outdoor air absolute humidity and the average indoor absolute humidity were 0.00008 kg/kg in Case 1, 0.00077 kg/kg in Case 2, 0.00094 kg/kg in Case 3, and 0.00442 kg/kg in Case 4.

The difference between the average outdoor air absolute humidity and the average indoor absolute humidity in Case 1 was the smallest, followed by those in Case 2, Case 3, and Case 4. Therefore, the measurement of the absolute humidity also showed that the ventilation was most efficient in Case 1.

4. Conclusion

To establish the most appropriate ventilation plan for the target factory, the distributions of the temperature and the relative humidity were measured based on four experiment cases. The following conclusions were drawn.

(1) In all the experiment cases, the average indoor air temperature was higher than the average outdoor air temperature. The cases with the smallest to the greatest difference between the average outdoor air temperature and the average indoor air temperature were as follows: Case 1 < Case 2 < Case 3 < Case 4.

(2) The comparison of the average outdoor air absolute humidity and the average indoor air absolute humidity by converting the relative humidity into the absolute humidity showed that the cases with the smallest to the greatest difference between the average outdoor air absolute humidity and the average indoor air absolute humidity were as follows: Case 1 < Case 2 < Case 3 < Case 4.

In summary, for a factory with a sealed structure such as the target PCB manufacturing factory in this study, the forced supply and exhaust method (i.e., in Case 1) was the most appropriate ventilation method for maintaining a low indoor air temperature and for keeping the contaminated air of the factory below the hazardous level. Also, the ventilation method in Case 2 (forced supply and natural exhaust) was more advantageous than that in Case 3 (natural supply and forced exhaust).

5. References

[1] D.W. Park · T.H. Kim · H.W. Kim, “Control of Work Environment Ventilation,” Korea

National Open University Press, 2004

[2] H.R. Kim, “A Study on the Additive Method for the Formation of the Ultrafine Pattern of Printed Circuit Boards (PCB),” M.S. thesis, Graduate School of Information Industry, Chungwoon University, 2005.

[3] (社) 空氣調和. 衛生工學會,『空氣調和. 衛生用語辭典』, オーム社, 1996.

[4] D.W. Park・N.W. Baek, “Industrial Hygiene," Korea National Open University Press, 2002.

[5] 皆川皆豊,「夏季對策の換氣について」,『空氣調和と冷凍』Vol. 9, No. 6, 1969.

IOP Conf. Series: Materials Science and Engineering **88** (2015) 012062 doi:10.1088/1757-899X/88/1/012062

ACMV Energy Analysis for Academic Building: A Case Study

R Hywel[1], B T Tee[1*], M Y Arifin[1], C F Tan[1], C K Gan[2] and CT Chong[3]
[1]Centre for Advanced Research on Energy (CARe)
Faculty of Mechanical Engineering, Universiti Teknikal Malaysia Melaka,
Hang Tuah Jaya, 76100 Durian Tunggal, Melaka, Malaysia.

[2]Centre for Robotic and Industrial Automation (CeRIA)
Faculty of Electrical Engineering, Universiti Teknikal Malaysia Melaka,
Hang Tuah Jaya, 76100 Durian Tunggal, Melaka, Malaysia.

[3]Faculty of Mechanical Engineering, Universiti Teknologi Malaysia,
81310 Skudai, Johor.

E-mail: tee@utem.edu.my

Abstract. Building energy audit examines the ways actual energy consumption is currently used in the facility, in the case of a completed and occupied building and identifies some alternatives to reduce current energy usage. Implementation of energy audit are practically used to analyze energy consumption pattern, monitoring on how the energy used varies with time in the building, how the system element interrelate, and study the effect of external environment towards building. In this case study, a preliminary energy audit is focusing on Air-Conditioning & Mechanical Ventilation (ACMV) system which reportedly consumed 40% of the total energy consumption in typical building. It is also the main system that provides comfortable and healthy environment for the occupants. The main purpose of this study is to evaluate the current ACMV system performance, energy optimization and identifying the energy waste on UTeM's academic building. To attain this, the preliminary data is collected and then analyzed. Based on the data, economic analysis will be determined before cost-saving methods are being proposed.

1. Introduction

Building energy auditing comprises of evaluating and indicating actual energy consumption, at site, which completed and occupied of particular building and essentially for the purposes to reduce and minimize current energy usage. It is useful for analyzing patterns of energy consumption, monitoring on how the energy used varies with time in the building, how the system elements interrelate and study the effect of external environment towards building[1]. Nowadays, high demands for efficient ways of building operation such as better energy conservation and optimization usage in purpose to cut off energy cost. Many studies has been made about the building energy audit for this particular purpose such Yu & Chow [2] [3], Ramli et. al.[4]. Even for the previous studies, there are

* To whom any correspondence should be addressed.

several researches have been conducted in the other academic buildings regarding the energy and ACMV system analysis such as Rosli [5], Tee et. al. [1], Ismail [6], and Roslan [7].The authors are therefore interested to further conduct a building energy audit and analysis focusing on the ACMV system at FKM's academic building located in UTeM's Technology Campus.

2. Methodology

The building comprises of seven-storey with central air-conditioning system and the case study covered the ground and first floor as shown in Figure 1 and Figure 2 respectively. The operating hours of the ACMV system is from 8.00 am to 6.00 pm on Monday to Friday. This is the preliminary audit and energy analysis which consists of 5 aspects described as following:

Figure 1. Ground Floor corridor.

Figure 2. First Floor corridor.

(a) Walk – Through Audit
An initial facility tour was conducted to observe the major operational and equipment features of the facility. This can assist in identifying areas where have potential energy saving so that a simple and inexpensive action can provide immediate energy use and/or operating cost savings. The main purpose of the tour is to obtain general information and understanding on the ACMV system of the building. In fact, more specific data or information about ACMV system, equipment and occupancy should be obtained from the maintenance and operational officer during the second visit with detailed data collection tour [8]. In addition, it was also conducted to obtain electricity power and cooling load capacity (tones refrigerant-TR) for the water-cooled chiller, water pump, cooling tower and air-conditioning system units.

(b) Utility Cost Analysis
The main purpose is to analyze and evaluate the operating cost of the facility. This is important for several reason: the bills show the proportionate use of each different energy source when compared to the total energy bill; an examination of where energy use can point out previously unknown energy wastes; and, the total amount spent on energy puts an obvious upper limit on the amount can be saved [8]. Besides that, the electricity bills are evaluated to identify the pattern of energy use, peak demand, and potential energy savings.

IOP Conf. Series: Materials Science and Engineering **88** (2015) 012062 doi:10.1088/1757-899X/88/1/012062

(c) Physical Parameters Measurement
To obtain the best information for successful energy cost control program, some measurement must be conducted during the audit. In this case study, the physical parameters involved in the measurement are air temperature, air velocity and relative humidity. The purpose of this measurement is to determine whether ACMV system performance compromise the comfort level as outlined in MS1525 [9]. Air velocity meter was used to measure those parameters. The time and height of measurement are based on the previous works [1][10]. The air velocity meter is set at the height of 1.1 meter for both pilot and actual measurements. At any typical point of measurement, the time to record the readings is every 1 minute for one hour. The case study of measurement areas is located in lecture rooms and main lobby of the ground and first floor.

(d) Cooling Load
Cooling load calculation is calculated for each of case study floor of the building in order to identify the overall load that to removed for the building. The overall results are shown in Table 1.

(e) Economic Analysis
Economic analysis is provided the costing estimation for the energy usage of the building. Suitable measures are proposed based on the analysis and data for energy saving purpose.

3. Result & Discussions

(a) Cooling Load
The total cooling load can be estimated based on the information obtained through the walk-in survey and data measurement. Table 1 shows the total cooling load for ground and first floor based on the specific time taken. The total load minimum and maximum total load for ground floor are 76.9 TR and 109 TR respectively while the first floor are 62.4 TR and 91 TR.

Table 1. Cooling Load Summary for Ground and First floor.

Level	Time (hours)	Total Load (kW)	Total Load/Unit (TR)
Ground Floor	0700	270.4	76.9
	1000	326.5	92.8
	1300	378.7	107.7
	1600	383.4	109
	1800	327.2	93
First Floor	0700	219.3	62.4
	1000	261.1	74.24
	1300	306.4	87.1
	1600	319.3	91
	1800	261.7	74.4

(b) Physical Parameter Measurement
All the data are shown in Table 2. The data then are compared with recommended indoor design conditions of an air condition space for cooling comfort as stated in the Malaysian Standard MS1525:2007 (Air velocity: 0.15-0.50 m/s, Air temperature: 23° C – 26° C and Relative Humidity: 55-70%) to indicate whether the current ACMV system is operating within the recommended comfort level or compromise.

Table 2. Physical Parameters Measurement Data for Ground and First floor.

Location	Measurement Area	Average Flow (cfm)	Average Operating Temperature (°C)	Average Relative Humidity (%)	Average Air Velocity (m/s)
Ground Floor	Lecture Room 3	17.82	23.1	73.8	0.09
	Lecture Room 5	12.74	22.1	74	0.06
	Lobby	16.04	27.8	62.7	0.08
First Floor	Lecture Room 10	32.27	22.1	72.9	0.16
	Lecture Room 11	2.61	24	72.9	0.01
	Corridor	17.74	29.9	60.7	0.09

Based on Table 2, most of the areas in ground and first floor are operating not within the recommended standard. As indicated in the table, air velocities were lower than the requirement by MS1525 except for the lobby area. Relative humidity seems to be slightly higher than the standard except for lobby and corridor area. Therefore, it is shown that the current ACMV systems for the ground and first floor are not operating within the expected comfort requirement.

(c) Energy Data Analysis
Electricity is the main sources of energy that generated in the case study building. All the electricity consumption of the building is charged directly by the local electricity utility company. An electricity meter or energy meter is a device that measures the amount of electric energy consumed and supposed to be installed in every building for billing purposes. Unfortunately, detail billing data in this case study is unavailable for the area. Specific assumption and consideration has to be made for determine cooling load and economic analysis based on the building section. The analysis of the overall monthly energy used pattern is based on the electricity consumption as shown in Figure 3.

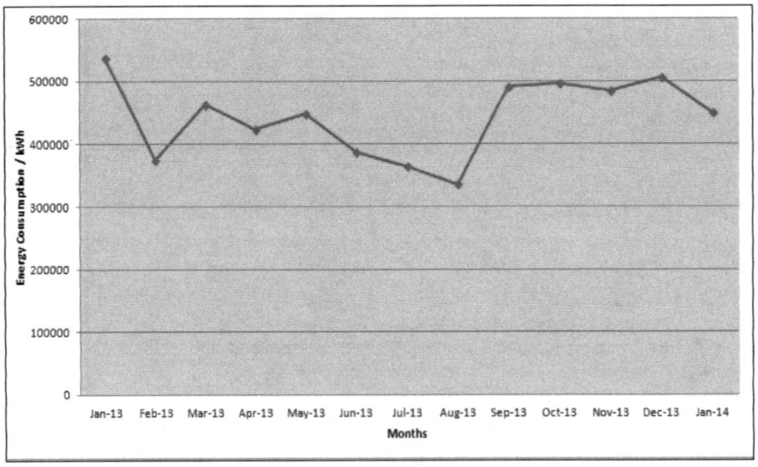

Figure 3. Energy consumption for one year.

(d) Retrofit Analysis

The aim of conducting retrofit analysis is to reduce energy consumption in the study case building. Retrofit installations sometimes could be more expensive but it is better to know the long term benefits of energy efficient technologies from cost-benefit perspective. The result from inspection and after measurement work indicates that the current ACMV system is not working efficiently within the standard requirement probably due to lack of system maintenance and less control equipments. Retrofit installations can be considered include control setback thermostat temperature, off hour control timer, heat recovery system and variable air volume. Table 3, 4 and 5 summarize the total cost and savings estimation based on the retrofit implementation.

Table 3. Total Cost of Electricity for Ground floor and First floor

Location	Total Cost of Electricity			
	Per monthly		Per year	
	Maximum	Minimum	Maximum	Minimum
Ground Floor	RM 18771.26	RM 13238.78	RM 225255.17	RM 158865.41
First Floor	RM 15632.93	RM 10736.93	RM 187595.136	RM 128843.14

Table 4. Total Energy Savings after Retrofit for Ground floor and First floor.

Location	Total Energy Saving after Retrofit			
	Per monthly		Per year	
	Maximum (kW)	Minimum (kW)	Maximum (kW)	Minimum (kW)
Ground Floor	115.02	81.12	1380.24	973.44
First Floor	95.79	65.79	1149.48	789.48
Total	**210.81**	**146.91**	**2529.72**	**1762.92**

Table 5. Total Cost of Electricity after Retrofit for Ground floor and First floor.

Location	Total Cost of Electricity after Retrofit			
	Per monthly		Per year	
	Maximum	Minimum	Maximum	Minimum
Ground Floor	RM 13139.88	RM 9267.15	RM 157678.62	RM 111205.80
First Floor	RM 10943.05	RM 7515.84	RM 131316.59	RM 90190.19

4. Conclusions

Building energy audit is the suitable approach method or tool that used to measure and obtains information regarding actual energy consumption in site or the buildings by purposes to purposes of reducing and minimizing energy usage in order to improve the energy performance. Based on the preliminary building energy audits on ACMV system installed in study case areas, it was found that that the ground and first floor of the FKM's building are contributed about 13 % - 18% of the total energy costs. Based on the retrofit saving analysis, it is estimated about 30% can be saved from the actual amount of the electricity and energy cost.

Acknowledgment
The authors would like to thank Universiti Teknikal Malaysia Melaka for the financial support through Short Term Research Grant award (PJP/2012/FKM(6A) S01077.

References
[1] Tee B T, Gan C K & Rashid M R A 2006 Energy Consumption of HVAC System at KUTKM's Academic Building *Proc ICEE 2006 (Bangi, Malaysia, 28-30 August 2006)* (UNITEN)
[2] Yu P C H & Chow W K 2006 *Energy* **32(2)** 83–94
[3] Yu P C H & Chow W K 2001 *Energy* **69** 243-255
[4] Ramli J, Nor Imrah Y, Jeefferie A R & Mahat M M 2011 *J. Mech. Eng. & Sci.* **1** 25-36
[5] Rosli C M 2012 *The Study of Energy Saving on ACMV (Air–Conditioning and Mechanical Ventilation) System for FKM Building* Universiti Teknikal Malaysia Melaka: Bachelor Degree Thesis
[6] Ismail A 2009 *Study on Predictive Maintenance of HVAC System* Universiti Teknikal Malaysia Melaka: Bachelor Degree Thesis
[7] Roslan N E 2010 *Design and Develop Test Unit Split Air Conditioning System* Universiti Teknikal Malaysia Melaka: Bachelor Degree Thesis
[8] Krarti M 2000 *Energy Audit of Building Systems: An Engineering Approach 2nd Ed.* (Bota Raton, FL: CRC Press)
[9] Malaysian Standard 2007 *Code and Practice on Energy Efficiency and Use of Renewable Energy for Non-Residential Buildings* (Putrajaya:Department of Standards Malaysia)
[10] Tee B T 2001 *Effects of Airflow Towards Thermal Comfort in Lecture Rooms* Universiti Teknologi Malaysia:Bachelor Degree Thesis

Thermal environmental case study of an existing underfloor air distribution (UFAD) system in a high-rise building in the tropics

Y H Ya[1] and K S Poh[2]

[1]Associate Professor, Department of Mechanical Engineering, University of Malaya, 50603 Kuala Lumpur, Malaysia

[2]Postgraduate Student, Department of Mechanical Engineering, University of Malaya, 50603 Kuala Lumpur, Malaysia

Email: yhyau@um.edu.my

Keywords: Underfloor Air Distribution (UFAD), Indoor Environmental Quality (IEQ), Thermal Comfort, Sick Building Syndrome, Energy Efficiency

Abstract. The performance of an existing underfloor air distribution (UFAD) system in a renowned high-rise office tower in Malaysia was studied to identify the root cause issues behind the poor indoor air quality. Occupants are the best thermal sensor. The building was detected with the sick building syndrome (SBS) that causes runny noses, flu-like symptoms, irritated skin, and etc. Long period of exposure to indoor air pollutants may increase the occupant's health risk. The parameters such as the space temperature, relative humidity, air movement, air change, fresh air flow rate, chilled water supply and return are evaluated at three stories that consist of five open offices. A full traverse study was carried out at one of the fresh air duct. A simplified duct flow measurement method using pitot-tubes was developed. The results showed that the diffusers were not effective in creating the swirl effect to the space. Internal heat gain from human and office electrical equipment were not drawn out effectively. Besides, relative humidity has exceeded the recommended level. These issues were caused by the poor maintenance of the building. The energy efficiency strategy of the UFAD system comes from the higher supply air temperature. It may leads to insufficient cooling load for the latent heat gained under improper system performance. Special care and considerations in design, construction and maintenance are needed to ensure the indoor air quality to be maintained. Several improvements were recommended to tackle the existing indoor air quality issues. Solar system was studied as one of the innovative method for retrofitting.

[1] To whom any correspondence should be addressed.

IOP Conf. Series: Materials Science and Engineering **88** (2015) 012064 doi:10.1088/1757-899X/88/1/012064

1. Introduction

Indoor air quality problems can be non-obvious and do not always generate straightforwardly recognized impacts on health, well-being or the physical plant. Symptoms such as headache, fatigue, shortness of breath, sinus congestion, coughing, sneezing, dizziness, nausea, and irritation of the eye, nose, throat and skin, are not really due to the air quality deficiencies, but can also be caused by other factors such as the poor lighting, stress, noise, and others [1].

A complete indoor environmental quality (IEQ) assessment is necessary to find out the alarming factors that influence the poor IEQ. Unfortunately, occupants moved out from the site and the demolishment work started. Hence, a complete IEQ assessment could not be performed under the current site situation. For this reason, alternative assessment procedures focusing on indoor air quality (IAQ) has been taken in place to identify the possible causes to the poor IEQ.

The assessments were conducted with the objectives of identifying the design and the performance of the existing underfloor air distribution (UFAD) system, and to obtain the systems' data and thermal environment data prior to the refurbishment as a baseline to verify the future enhancement to be made. Figure 1 shows the schematic diagram of the underfloor air distribution system (UFAD).

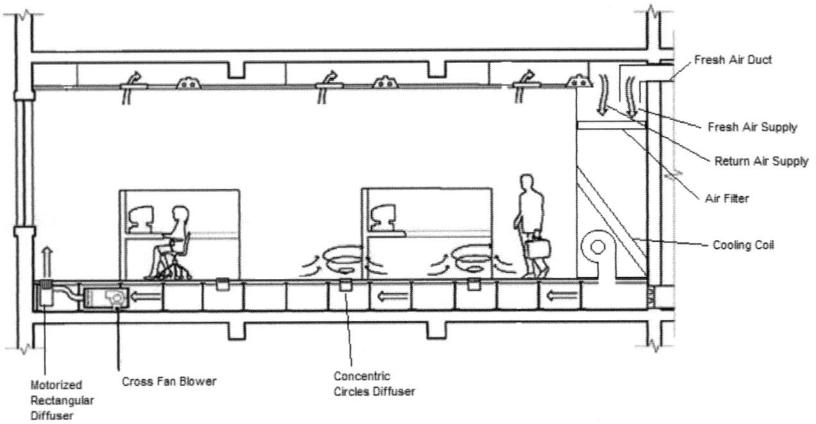

Figure 1. Schematic Diagram of Underfloor Air Distribution System (UFAD) [2].

2. Methodology

The assessment methodology was modified from conventional IAQ assessments methodology [3]. The procedures are divided into 4 stages: preliminary, sampling, evaluation and recommendation.

2.1 *Preliminary Stage*

A walkthrough process is necessary to identify the possible areas where the problem mostly occurs. Occupants are the best thermal sensor. Feedbacks from the occupants are essential during the walkthrough process as their long hours of exposure in the environment. It was evident that the occupants were feeling sneezing, dizziness, irritation of the eye and skin in a long period of time.

2.2 *Sampling Stage*

The sampling data was emphasized on the understanding of the room air physical parameters and the performance of the HVAC system. The assessed parameters were the space temperature, relative humidity, air movement, air change, fresh air flow rate, chilled water supply and return flow rate and temperature. The sampling locations were Levels 33N, 33S, 34N, 34S and 35S. Each location is divided into five thermal zones. Physical parameters were measured at each individual thermal zone to identify the overall indoor air quality.

Fresh air is drawn from outdoor and precooled using a Precool Air Unit (PAU) before it is supplied to each Underfloor Supply Unit (USU) located at each thermal zone. The fresh air flow rate was measured using a pitot tube Traverse Method (Log-Tchebycheff Method). In order to minimize the damage impact on the existing duct, a simplified traverse method was developed and compared with data collected from the full Traverse Method.

Figures 2 and 3 show the Traverse Method in the fresh air flow rate measuring in the ducting.

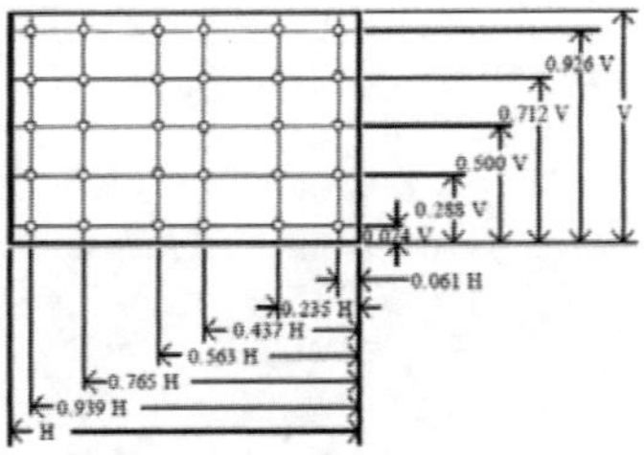

Figure 32: Location of Measuring Points for Traversing a Rectangular Duct using Log-Tchebycheff Method

For this duct, a 30-36" horizontal dimension requires 6 points (or 6 traverse lines). For this duct, a vertical dimension less than 30" requires 5 points (or 5 traverse lines).

# of Points or Traverse Lines per Side	Position Relative to Inner Wall
5	0.074, 0.288, 0.500, 0.712, 0.926
6	0.061, 0.235, 0.437, 0.563, 0.765, 0.939
7	0.053, 0.203, 0.366, 0.500, 0.634, 0.797, 0.947

Figure 2. Traverse Methodology [4].

Figure 3. Measuring fresh air flow rate inside the ducting.

2.3 Evaluation Stage

Based on the parameters measured, data were compared to the local and international standard and regulation. The local and international standard and regulation used in the present study are Industry Code of Practise on IAQ 2010 [5], Malaysian Standard MS 1525 [6] and ASHRAE Standard 62.1 [7].

2.4 Recommendation Stage

Due to the site was under renovation, recommendations have to be realistic, and it was crucial to have the rectifications implemented on time to eliminate the future repeating IAQ problems.

3. Results and Discussion

3.1 Indoor Air Temperature, Relative Humidity (RH) and Air Velocity

The results as shown in Figure 4 indicate that the air temperatures are in the average of 21-23°C. As compared to 23-26°C recommended, the low average temperatures throughout all spaces are due to without occupancy conditions. The heat load from humans and electronics are not available. The temperatures are expected to rise within the recommended range during full occupancy conditions based on the current data obtained.

Relative humidity has exceeded the 70% threshold limit recommended. The relative humidity at the just mentioned level will encourage the growth of mold. During the walkthrough stage, mold was seen growing seriously especially at the return air grill. The data shown have proven the real condition of mold growth due to the high humidity.

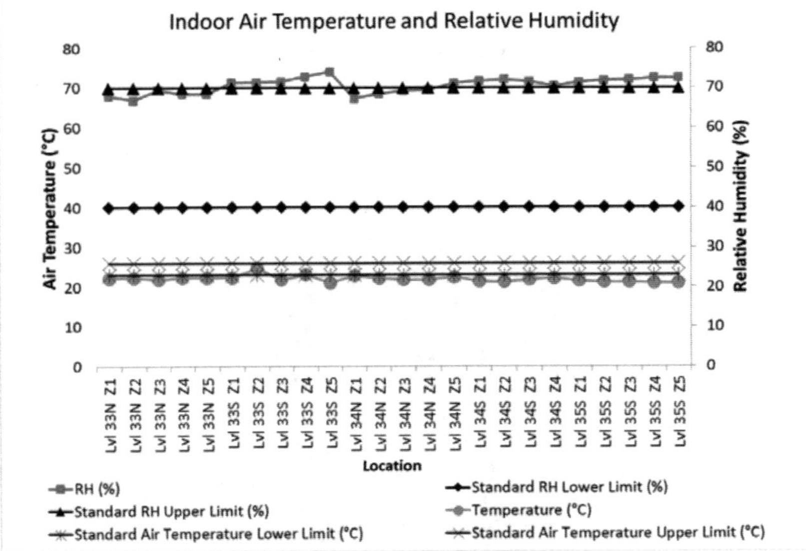

Figure 4. Indoor air temperature and relative humidity compared to recommended threshold limit based on Malaysian Standard MS 1525.

3.2 Fresh Air Intake

The fresh air flow rate was measured using a full Traverse Method at Level 35S. The full data at Level 35S is shown in Table 1.

Table 1: Traverse Measurements in CFM at the Fresh Air Duct at Level 35S.

2839	2840	2758	3413	4308	4632	4480
3037	2954	2924	3550	4183	4488	4265
3024	2721	3033	3352	4034	4129	3907
3048	2834	2968	3338	3541	3661	3585
2440	2521	2861	2909	2939	3156	3220

A simplified method was developed using a correction factor on the flow rate measurements at the center point of the ducting. The following measurements at Levels 33N, 33S, 34N and 34S were measured using the simplified method. The data and calculation are shown in Tables 2 and 3.

Table 2. Fresh Air Data for Level 35S.

Air Temperature (°C)	19.30
Relative Humidity (%)	89.50
Average Volumetric Flow Rate (cfm)	3368.34
Correction Factor with Centre Point	1.0049
Area of L35S (ft^2)	8676.00
Floor height (ft)	10.00

Table 3. Fresh Air Data for Levels 33N, 33S, 34N and 34S.

Level	33N	33S	34N	34S
Air Temperature (°C)	18.80	15.60	16.50	19.20
Relative Humidity (%)	89.60	92.10	88.90	89.30
Volumetric Flow Rate at Centre-Point (CFM)	4059.00	2735.00	1951.00	2167.00
Corrected Volumetric Flow Rate (CFM)	4078.79	2748.33	1960.51	2177.57
Floor Area (ft^2)	9310.50	9310.50	9171.00	9171.00
Floor height (ft)	10.00	10.00	10.00	10.00

A comparison with the ASHRAE Standard 62.1 is shown in Figure 5. The results show that the fresh air intake at Levels 33N, 33S and 35S are higher than the minimum fresh air rate required by the ASHRAE Standard 62.1. The outdoor fresh air is normally high in temperature and humidity. High latent heat load from the outdoor fresh air may cause the indoor relative humidity to exceed the threshold limit because the USUs are not tasked to accomplish such a load. Hence, the optimum fresh air intake closes to the minimum fresh air intake is needed.

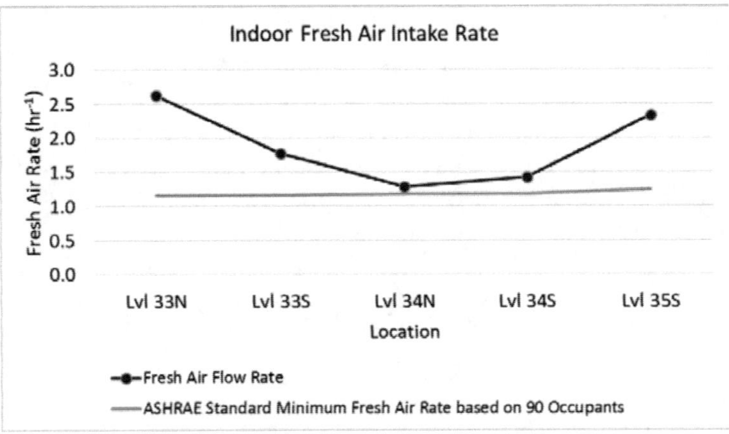

Figure 5. Indoor fresh air intake rate.

3.3 Walkthrough Inspection

There were some findings during the walkthrough stage. First, the valve actuator was not functioning, and hence, the chilled water was running at the full load at all time. Second, the underfloor diffusers were not properly maintained. Most diffusers were broken or parts were missing. These issues must be rectified to ensure that the air-side and water-side HVAC systems function properly.

Figure 6. Walkthrough inspection shows improper underfloor diffusers and malfunctioning valve actuator.

4. Recommendations and Conclusions

The recommendations to rectify the poor indoor air quality are as follows:

1. It is suggested that all broken underfloor diffusers to be replaced to ensure a proper air flow. So that the air stagnancy can be eliminated.
2. It is recommended that the malfunctioning valve actuators to be replaced to ensure a proper control of the chilled water flow rate.
3. The CO2 sensors are recommended to be installed to control the optimum fresh air supply flow rate.

The current study reveals that the maintenance team has a crucial role to ensure the indoor air quality to be maintained up to the standard. A scheduled checking and maintenance work must be taken in place. So that HVAC&R systems can function based on the intended design. Solar energy is studied as one of the possible methods to enhance the humidity control. It is noted that the façade of the high-rise building has a huge surface area and can be installed with solar panels to harvest solar energy for the use of air heating in the USUs. This will improve the relative humidity of the air before it is delivered to the space.

5. Acknowledgements

The authors will like to acknowledge the financial support from the Ministry of Higher Education (MOHE), Malaysia for providing PPP Grant PG111-2012B. Also, special thanks are extended to Ir. Den Low Han Guan, Mr. Jefferson Chuah, Mr. Chan Wai Chuan, Mr. Tam Jun Hui and Mr. Low Heng Kit for their technical helps in the present project.

References

[1] Syazwan AI, Juliana J, Norhafizalina O, Azman ZA and Kamaruzaman J 2009 Indoor air quality and sick building syndrome in Malaysian buildings *G. J. of Health Sci.* **1** 126-135

[2] Publishing, T.C. Pressurized plenum underfloor air distribution (UFAD) system http://what-when-how.com/energy-engineering/underfloor-air-distribution-ufad-energy-engineering/ Retrieved on 15th November, 2013

[3] Cheong KWD and Lau HYT 2003 Development and application of an indoor air quality audit to an air-conditioned tertiary institutional buildings in the tropics *Building and Environment* **38** 605-616

[4] TSI – Alnor 2012 *Electronic Balancing Tool Models EBT730/EBT731 Owner's Manual*

[5] Mokhtar THA et al 2010 *Industry Code of Practice on Indoor Air Quality* (Malaysia: Department of Occupational Safety and Health) 40-42

[6] Department of Standards Malaysia 2007 *Code of Practice on Energy Efficiency and Use of Renewable Energy for Non-Residential Buildings* (Malaysia: Department of Standards Malaysia)

[7] ASHRAE, Inc 2007 *ASHRAE Standard on Ventilation for Acceptable Indoor Air Quality* ANSI/ASHRAE Standard 62.1-2007

Cooling channels design analysis with chaotic laminar trajectory for closed cathode air-cooled PEM fuel cells using non-reacting numerical approach

W Mohamed W A N[1]
[1] Alternative Energy Research Centre, Faculty of Mechanical Engineering
Universiti Teknologi MARA, Malaysia

E-mail: wanajmi@salam.uitm.edu.my

Abstract. The thermal management of Polymer Electrolyte Membrane (PEM) fuel cells contributes directly to the overall power output of the system. For a closed cathode PEM fuel cell design, the use of air as a cooling agent is a non-conventional method due to the large heat load involved, but it offers a great advantage for minimizing the system size. Geometrical aspects of the cooling channels have been identified as the basic parameter for improved cooling performance. Numerical investigation using STAR-CCM computational fluid dynamics platform was applied for non-reacting cooling effectiveness study of various channel geometries for fuel cell application. The aspect ratio of channels and the flow trajectory are the parametric variations. A single cooling plate domain was selected with an applied heat flux of 2400 W/m^2 while the cooling air are simulated at Reynolds number of 400 that corresponds to normal air flow velocities using standard 6W fans. Three channel designs of similar number of channels (20 channels) are presented here to analyze the effects of having chaotic laminar flow trajectory compared to the usual straight path trajectory. The total heat transfer between the cooling channel walls and coolant were translated into temperature distribution, maximum temperature gradient, average plate temperature and overall cooling effectiveness analyses. The numerical analysis shows that the chaotic flow promotes a 5% to 10% improvement in cooling effectiveness, depending on the single-axis or multi-axis flow paths applied. Plate temperature uniformity is also more realizable using the chaotic flow designs.

1. Introduction

The application of hydrogen-fuelled PEM fuel cells is growing positively from test scales to market research scale as its fast initiation and reliable operation proves to be an attractive factor. A significant effort to model the behaviour of fuel cell operation has been made due to the fact that the electrochemical and thermo fluid reactions occur internally and non-visible for direct analysis. Numerical models have proven highly useful to optimize the design of various components such as the reactant flow channels and the cooling channels.

[1] To whom any correspondence should be addressed.

The generation of electrical power within the fuel cell system is always accompanied by a near equivalent amount of heat energy and increases in an exponential behaviour as the electrical power increases. Therefore, continuous active cooling of the cells is required. Thermal management of a PEM fuel cell keeps cell temperatures at levels where excessive membrane drying leading to increased ionic transport resistance is avoided [1]. It also allows near-uniform temperature distribution, stabilizes the power density [2] and prevents propagation of thermal stresses [3].

Air as a coolant requires minimal components and merely increases the overall system size by less than 50% whereas water-cooled systems generally increase the system size by more than 200%. For portable and limited space applications, air-cooled fuel cells are very much desirable. Technically, the design of the coolant flow path is very important in allowing the proper air quantity to flow smoothly and interact with the hot surfaces.

Cooling channels provides an interface for the interaction of the cooling fluid with the hot surface. Conventional air-cooling channels for PEMFC are normally straight path channels integrated between the anode and cathode sides of the bipolar plates. The temperature difference between the anode/cathode and cooling channel surfaces are typically very small due to the high plate conductivity and small distance between them.

Conventional cooling channel designs of a PEM fuel cell have hydraulic diameters larger than 1 mm and are still within the mini-channel definition where the classic theory of fluid behaviour is still valid. For liquid cooling, a variety of cooling channel designs has been proposed such as multi-pass serpentine oblique-fins, coiled, parallel-serpentine, wavy and a novel hybrid parallel-serpentine-oblique-fins design [4]. Lasbet et al. [5] proposed chaotic flow channels to enhance water and surface interactions. From the technical perspective of air-cooling, the chaotic flow channels are more suitable than multi-pass serpentine channels. As the specific heat of air is much lower than water, it has to be allowed to flow in and out of the heated domain swiftly to avoid acting as a heat carrier to the other parts of the cell. At low Reynolds number, the cooling effectiveness could be unsatisfactory as effective cooling would be severely constrained within the inlet region, resulting in large temperature gradients relative to the air outlet region.

Very limited numerical studies on air-cooled closed cathode PEM fuel cell, which utilizes separate fluids as reactants and coolant, are available compared to open cathode thermal investigations. Previously, a non-reacting numerical investigation at stack level was published for a simple iseal case of singular air cooling channel system [6]. Then, Matian et al. [7] investigates three different air cooling channel designs for closed cathode PEM fuel cells in a non-reacting fuel cell numerical domain. The cooling channel designs all are rectangular with a straight flow path but with different aspect ratios. Their main conclusion stated the capability of the cooling plate to balance the temperature variation with a strong inclination towards wider cooling channel designs.

Here, a simplified non-reacting domain is applied to investigate three rectangular cooling channel designs of equal numbers but with variations in aspect ratio and flow path, specifically for closed cathode air-cooled PEM fuel cell application. The computational domain consists of a single cooling plate imbedded with the cooling channels and a constant heat flux was applied on one side of the plate based on the actual heat generation of an experimental fuel cell stack. The results are analyzed from the perspectives of cooling effectiveness, average plate temperature, plate temperature distribution and specific cooling with the purpose for identifying the optimal design for physical development.

The principle question of the analysis was on the cooling effectiveness (Eq. 1) of each cooling plate, calculated from the ratio of the plate cooling rate (total boundary heat transfer) to the thermal power (heat generated) acting on the plate. The cooling effectiveness and average plate temperature are the main optimization parameters for design evaluation.

$$\varepsilon_c = \frac{Q_c}{P_{th}} \tag{1}$$

2. Cooling channel designs

A commercial closed cathode PEM fuel cell stack originally designed for water cooling was chosen as the reference design. The commercial stack is rated at 3 kW power output, consisting of 73 bipolar plates, each 5 mm thick, and the overall dimension of the stack is 365 mm in length x 149.5 mm in width x 23 9 mm in height. Technically, it can be considered as a large-sized fuel cell. Of the total plate height of 239 mm, only 179 mm is available for cooling area designation due to the requirements of the reactant gas ducts and manifolds construction (Figure 1). The length of the channels follows the exact width of the plate at 149.5 mm (refer Table 1).

The number of cooling channels is limited to the width of the channels design. In this work, the number of channels for each design are fixed, n = 20 channels. Ribs between the channels must be at least 1 mm distance as the ribs assist electron flow in actual operation. Three cooling channel designs are covered in this brief report. The first design is a standard straight flow channel (Ms20). As n figure 3, the second design is termed as the square wave mixer channel with 11 bend and single-axis horizontal fluid flow (Ch20-CF). The third design explores a more complex geometry which is the C-shaped channels with 11 bend and two-axis horizontal-vertical flows (Ch20-CV). The Ch20 designs theoretically promotes more fluid-surface interactions as it allows the fluid to be in a more non-linear (chaotic) trajectory even in laminar phase flow.

The practical inlet air velocity is taken at 2 m/s, which translates to a Reynolds number of approximately 400 for the conventional Ms20 straight channel design. This Reynolds number is taken as a reference for the other designs. At a Reynolds number of 400, the coolant air inlet velocity varies for each design due to the differences in hydraulic diameter of the channels. Consequently, the air mass flow rate differs slightly.

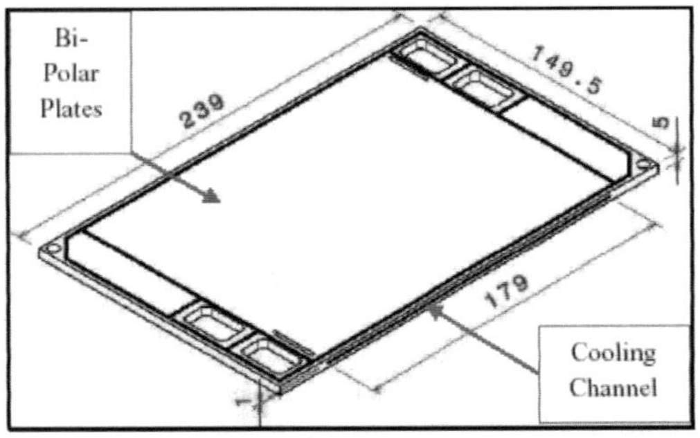

Figure 1. Cooling plate geometry and base dimension.

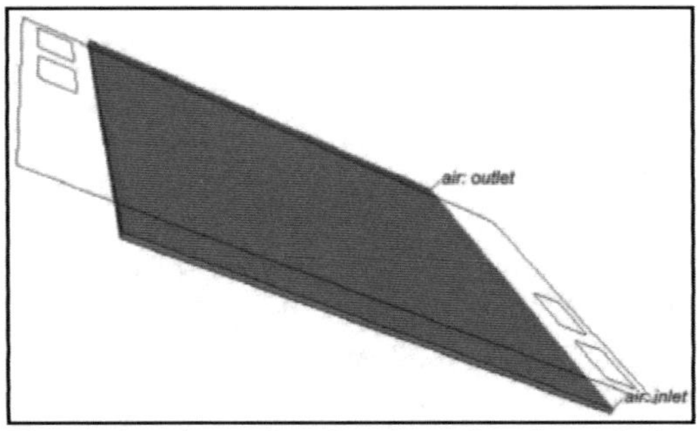

Figure 2. Inlet and outlet air surface boundaries.

Table 1. Cooling plate design specifications.

	Design label	Number of channels	Width (mm)	Height (mm)	Length (mm)	Aspect ratio
1.	Ms20	**20**	**8.0**	**2**	**149.5**	**4**
2.	Ch20-CF	**20**	**4.0**	**2**	**180.0**	**2**
3.	Ch20-CV	**20**	**4.0**	**1**	**200.0**	**4**

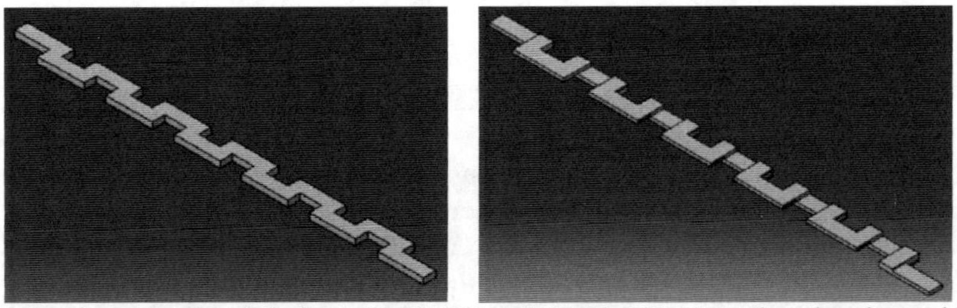

Figure 3. Channel geometry of designs Ch20-CF and Ch20-CV. Note that the channel height of design CV is only 1 mm, or half, of CF design.

3. Computational domain, models and boundaries

The applied commercial software was STAR-CCM. There are only two regions in the work domain which are the plate (graphite) region and the fluid (air) region. The interface layers between the fluid region and plate region was designated as the boundary heat transfer layer where the heat flow rate between the regions, or the cooling rate, would occur and its values captured.

In practice, the nominal (incoming) cooling air velocity range for a closed cathode depends on the fan model and operation setting. Usually, a smooth flow of air is induced into the stack by negative pressure (suction) flow to avoid high fluid entry resistance into the channels. Hence, the typical air velocity range is between 1 m/s to 2 m/s if the fuel cell is assumed to apply a set of standard 6W mini fans. Here, the cooling performance at the highest air velocity is reported where the flow is a fully developed laminar flow.

The projected heat generated is the assumed thermal energy of the fuel cell that the cooling plate needs to remove. Based on the 3 kW stack of reference which is made up of 73 cells, each cell would theoretically generate 41W of thermal energy at 50% conversion efficiency. The cooling plate in this simulation is assumed to handle thermal energy generated from two adjacent cells simultaneously, or applying a thermal safety factor of 2. The heat is applied to one of the plate surfaces with an effective surface area of 0.0342 m^2. This approach is similar to [8] in their work on a water cooled fuel cell using a similar computational platform. The other sides of the plate were set as adiabatic surfaces to totally reflect the cooling effectiveness of the coolant by internal convection. External forced and free convection as well as radiation cooling effects are neglected as a simplification to the simulated model. Tables 2 and 3 lists the other main boundary conditions of the simulation.

Table 2. Boundary conditions for bipolar plate.

	Properties / parameters	Values
1.	Material	Carbon graphite
2.	Specific heat	710 J/kg.K
3.	Density	2240 kg/m^3
4.	Thermal conductivity	20 W/m.K
5.	Projected heat generated	82 W
6.	Applied heat flux	2400 W/m^2

Table 3. Boundary conditions for cooling air.

	Properties / parameters	Values
1.	Flow phase and model	Laminar (Re= 250 -750)
2.	Flow source – Uniform plane	50 mm from channel inlets
3.	Inlet specifications	3 m/s, 30°C
4.	Outlet specifications – pressure outlet	1 atm, 30°C
5.	Thermal conductivity	Constant
6.	Specific heat, Cp	Constant

In the cooling channels, the air flow field and internal forced convection was obtained by solving the steady state Navier-Stokes equations, which includes the continuity equation, momentum equation and the mass and energy conservation equation. The computational physics models selected are ideal gas, laminar flow model, segregated flow and the segregated solid energy model. Iterations in the range of 800-1200 are needed to obtain a satisfactory convergence with residuals less than 0.05% than the previous iteration value.

The air inlet and outlet surface boundaries are set as in figure 2. At the inlet, air velocity inlet is specified. In practice, one or more air fans are used to induce the cooling air. Simplification was made here by neglecting the presence of the momentum source (fan blowers), hence eliminating complex transitional flows between the fan area and the channel inlet area. Inlet air flows are assumed uniform with a constant incoming velocity; therefore, the cooling rates would be theoretically optimal at the stated boundary conditions. At the outlet surface, the exit velocity could not be directly specified; so, the main boundary conditions are pressure and the static temperature. The static fluid temperature of both inlet and outlet were set at the normalized ambient condition of 30°C and 1 atm.

4. Results and discussion

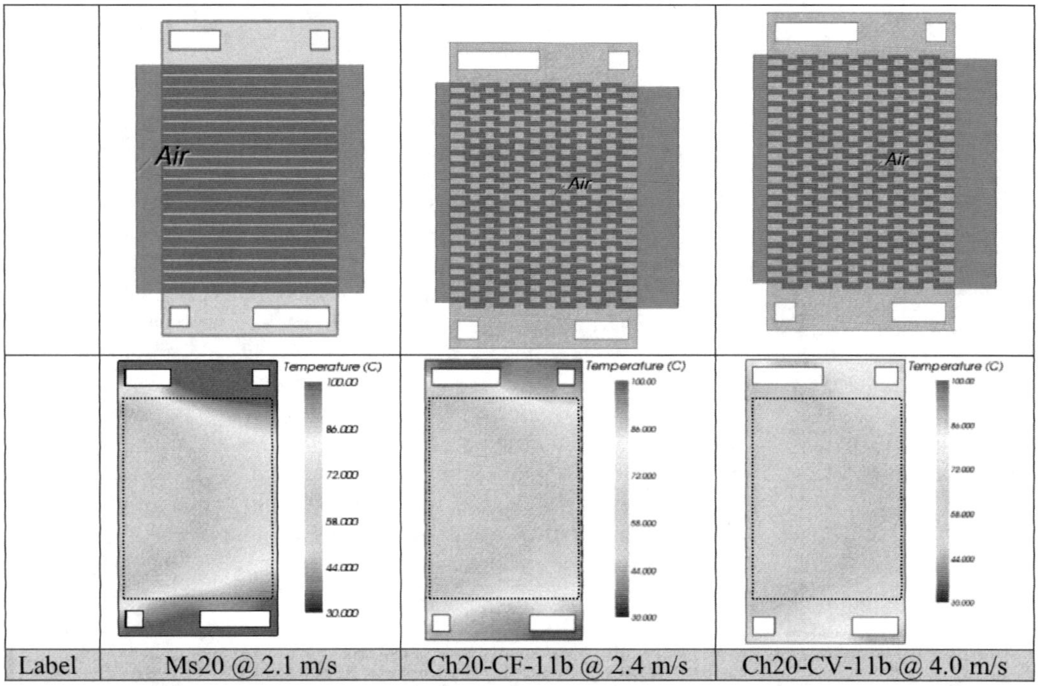

Figure 4. Scalar temperature distribution at Reynolds number of 400.

The main numerical analyses for a fuel cell cooling plate design are temperature distribution, average plate temperature and the cooling effectiveness. Figure 4 shows the plate temperature distribution of each cooling channel design. The dotted box within the plate indicates the boundary of the fuel cell active area. The general profile for all three cases are similar; lowest plate temperature are found near the inlet (approximately 50°C) and increases in a semi-ellipsoidal profile along the coolant path towards the exit. The near-inlet temperatures are more uniform compared to other parts of the active area. This trend is normally encountered for bulk solids with internal convection effects as the coolant gradually loses its cooling capacity along the flow path. In practice, this steady-state thermal profile is achieved when both the conduction heat flow within the plate and the convection heat transfer from the plate to fluids in contact reaches an asymptotic saturation level.

The straight path design (Ms20 design) leads to a highly heterogeneous temperature profile compared to the channels with chaotic flow trajectory. Within the active area, the maximum temperature gradient between the inlet and exit is 40°C. The maximum temperature gradient for designs CF and CV are approximately 30°C and 20°C respectively. As the cooling effect is based on a similar Reynolds number for all cases, the differences in nominal coolant velocity of each design have a big impact on the overall temperature distribution. Here, the temperature profile as shown by design CV is most desirable for fuel cell operation as the plate temperature distribution approaches homogeneity with high temperature areas merely located at the edges of the exit section.

Figure 5 displays the combined parametric comparison analysis on cooling effectiveness and averaged plate temperature. These parameters contribute to an inversely proportional effect to one another. At a nominal coolant flow of Reynolds number equals to 400, the CV design reached 90% cooling effectiveness which is the minimal performance criteria for a sustained bulk cooling effect. This in turn translates to a much lower average plate temperature of 64°C which is still in the normal range of operating temperatures for PEM fuel cells.

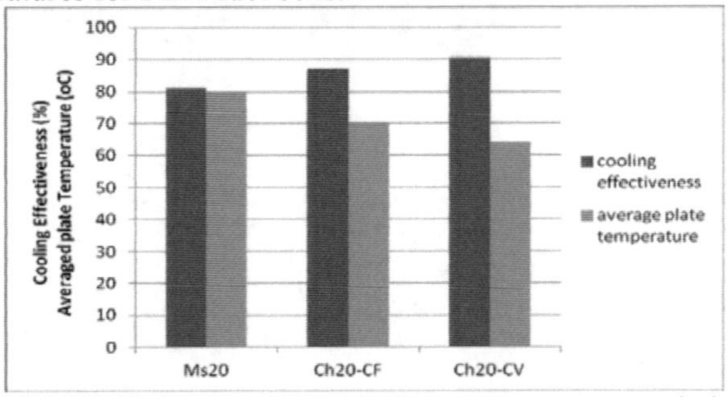

Figure 5. Cooling effectiveness and averaged plate temperature relation at Re 400

The results on cooling effectiveness indicate an improved cooling capability when laminar flows are allowed to have a more interactive trajectory with the surfaces. At similar nominal Reynolds number, the cooling improvement is nearly 10% when the flows are multi-axial while a 5% increase can be expected for single axis chaotic flow. Based on Nusselt number definitions, the heat transfer coefficient of the three cases might not show any significant differences as the Reynolds number is similar. Therefore, the increase in cooling effect is highly due to the contact surface factor in the classic Newton's law of cooling theory.

The translation of this result to actual fuel cell cooling plate design depends on the plate structural integrity due to the more complex geometry of the CV channel designs. Further study in optimizing the cooling plate thickness (from the base case of 5mm thickness here) while simultaneously applying better cooling channel geometries is a challenge in fuel cell thermal engineering. The main results here show that laminar coolant flows with lower fan power consumption performs better when the flow trajectory that depends on the cooling channel design are built to promote a chaotic flow path throughout the channel.

5. Conclusions

The thermal engineering aspect in PEM fuel cell operation is critical for performance reliability and sustainability. The use of air cooling for closed cathode PEM fuel designs is a rare option and was studied here to identify the basic design effects on cooling capability. The three designs proved that introducing a slightly different coolant trajectory within the cooling channels can have positive cooling effects and is useful for size-constrained systems such as for vehicle applications. The results obtained here at a practical Reynolds number of 400 provided a platform for more detail studies in cooling plate design optimization for air-cooled closed cathode PEM fuel cells.

References

[1] Park J and Li X 2006 *J. Power Sources* **162** 444
[2] Didierjean S, Lottin O, Maranzana G and Geneston T 2008 *Electrochimica Acta* **53** 7313
[3] Sadiq al-Baghdadi M A R and Shahad al-Janabi H A K 2007 *Proc. of IMechE 2221 Part A: J. Power and Energy*
[4] Sasmito A P, Kurnia J C and Mujumdar A S 2012 *Energy* **44** 278
[5] Lasbet Y, Auvity B, Castelain C and Peerhossaini H 2006 *J. Power Sources* **156** 114
[6] W Mohamed W A N, Ismail A A and Atan R 2010 *IEEE 2010 Int. Conf. on Sci. and Soc. Research (CSSR), IEEE Xplore Digital Library* 91
[7] Matian M, Marquis A and Brandon N 2011 *Int J Hyg Energy* **36** 6051
[8] Yu S H, Sohn S, Nam J H and Kim C-J 2009 *J. Power Sources* **194** 697

Temperature Dependences on Various Types of Photovoltaic (PV) Panel

I A Audwinto[1], **C S Leong**[1], **K Sopian**[1], and **S H Zaidi**[1]
[1] Solar Energy Research Institute (SERI), Universiti Kebangsaan Malaysia, Bangi, Malaysia.

E-mail : IvanOu.1991@yahoo.com

Abstract. Temperature is one of the key roles in PV technology performance, since with the increases of temperature the open-circuit voltage will drop accordingly so do the electrical efficiency and power output generation. Different types of Photovoltaic (PV) panels- silicon solar panels and thin film solar panels; mono-crystalline, poly-crystalline, CIS, CIGS, CdTe, back-contact, and bi-facial solar panel under 40°C to 70°C approximately with 5°C interval have been comparatively analyzed their actual performances with uniformly distribution of light illumination from tungsten halogen light source, ±500W/m^2. DC-Electronic Load and Data Logger devices with "Lab View" data program interface were used to collect all the necessary parameters in this study. Time needed to achieve a certain degree of temperature was recorded. Generally, each of the panels needed 15 minutes to 20 minutes to reach 70°C. Halogen based light source is not compatible in short wave-length in response to thin-film solar cell. Within this period of times, all the panels are facing a performance loss up to 15%. Other parameters; P_{max}, V_{max}, I_{max}, V_{oc}, I_{sc}, $R_{serries}$, R_{shunt}, Fillfactor were collected as study cases. Our study is important in determining Photovoltaic type selection and system design as for study or industrial needed under different temperature condition.

1. INTRODUCTION

Increasing the performance of the photovoltaic technology has been becoming a main issue of most researchers; increasing the capability of solar cell to absorb more light photons, increasing the life span of the devices, improving its sustainability of heat, and to reduce the cost of the material. There are some ways to do that, such as improving the morphology of the devices, using different method of how to produce the solar cell, inventing a new technology to assist the production of the cell, using low-concentration technology to reduce the whole system cost, and even using lots of methods to reduce the ambient and panel temperature. To put aside of others perspective, our team in this study was working on the temperature dependence from various material or type of PV panels. Including silicon based solar panel and also thin film solar cell; mono-crystalline, poly-crystalline, CIGS, CIS, CdTe, back-contact, and bi-facial solar panel. Our aim is to provide some informations from all the panels regarding how it works under different temperature condition. But of course to do this, an indoor solar simulator was set up to produce 500W/m^2 of light uniformity through the whole area of the panel. The importance of having an indoor experiment is that for us to control the surrounding condition which will affect the performance of the PV system, such as the light illumination, temperature control, humidity control, etc. Different type of material used on the production of PV panel also affecting not a small portion on the impact. In general, typical PV panel converts only around 20% of the light to electricity and the rest of the incident solar radiation converted to heat [1]. Increasing the temperature of the cell modulates the spectral response of the solar cell and the spectral response is found to be dependent with the temperature [2].

Photovoltaic (PV) technology behaves differently under different condition; climate, temperature, humidity, irradiance, etc. A rise of temperature module, leads to the decreasing of electrical generating capability of the solar PV panels due to the big reduction of open-circuit voltage, V_{oc} overwhelms the slightly increases in short-circuit current, I_{sc} [3]. Andreev et al. [7] calculated that the photocurrent increases with the temperature at 0.1%/°C due to the decreasing of the gap of the solar cell and that the open-circuit voltage decreases at -2mV/°C. Many researchers did various kind of experiment to examine the power loss due to the temperature rises. Radziemska and Klugmann [4] reported that crystalline based of solar panel facing a reduction of -0.65%/°C, Taguchi et al. [5] reported that HIT thin-film cell based hetero-junction structure cell has been reduced from -0.33% to -0.25%/°C by a new process of manufacturing, and Nishioka et al.[6] reported InGaP/InGaAs/Ge triple-junction solar cell having the temperature coefficient of conversion efficiency to be -0.248%/°C, etc [8-14]

2. MATERIAL AND METHODS

2.1. Materials and Systems Installation

In this indoor experiment, we were using tungsten halogen bulbs as solar simulator with each bulb being controlled by dimmer. The Dimmers are manually controlled to ensure we get the uniformity out from the bulbs through the whole surface are of panels. Before the experiment began, light radiation was measured by using pyranometer. Our solar simulator was managed to get a uniformity of 500W/m^2 with only 20% miss match at the corner. All parameters were measured by using "DC Electronic Load" and "Data Logger" devices with Lab View program system interface. Thermocouples K type were used to collect the temperature reading.

2.2 Uniformity Light Source Measurement

The intensity was measured by pyranometer. It was tested in 12.5cm gap in each point as the size of the solar cell. Basically, the intensity ready was measured by putting the pyranometer on top of the solar cell in the middle. Red dot is where we put the pyranometer at with 12.5cm gap from each dot to the other as shown in table 1 below. With the 20% of mismatch at the corner area, we still could get a linear data with panels. All the panels were producing half of its capability to produce electrical power out-put within this table range measurement. Which means the 20% of mismatch at the corner area will not affect much of the performances.

Table 1. Uniformity of light radiation emits by the solar simulator. All the values in the table are in W/m^2 units.

430	500	460	435	423	441
458	500	468	440	430	458
455	500	480	440	440	465
458	500	475	445	440	457
468	510	490	450	435	452
470	514	500	455	436	446
475	510	500	460	440	445
440	440	460	497	500	470
440	470	481	440	425	420

2.3 Experiment Methodology

As all the equipment used at this experiment, here we will describe about the whole work on how the study being progressed. As we mentioned that we are using lots quantity of tungsten halogen bulbs which being controlled by dimmer as light source, hence the panels are basically we put on below perpendicular to the incoming light one by one to be tested. The negative and positive wires from the panels were connected to the DC Electronic Load to receive its current-voltage signal, few thermocouples were attached at the middle, rare and behind of the panels and then connected to the Data Logger device. After the program finished its job to calculate all the parameters we could see the results on the computer screen. Temperature differences between the front middle and back side are in between 5% to 10%, so we decided to take the middle front data as to make a comparative study. The whole picture of this experiment is shown in the figure 1 below.

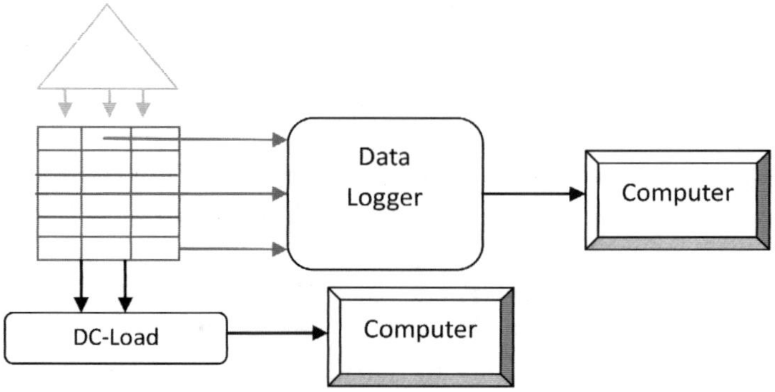

Figure 1. Schematic diagram of how was the experiment being conducted. Yellow represent the solar simulator, red represent thermocouples, blue represent solar panel.

3. RESULTS AND DISCUSSIONS

Seven different types of solar panels in table 2 are based on crystalline silicon and thin film were used to examine the temperature differences effect on each of the panels, such as mono-crytalline (175W), poly-crystalline (110W and 200W), CIGS (85W), CIS(75W), CdTe (72.5W), back-contact (30W), and bi-facial solar cell (45W).

Table 2. Characterization of solar panels.

Type	Area, m2	1000W/m2 Watt Power at 25°C, Wp	500W/m2 Watt Power at 40 °C, Wp	500W/m2 Watt Power at 70°C, Wp
Mono-crystalline	1.31	175	77.97	69.04
Poly-crystalline	0.97	110	48.22	39.81
Poly-crystalline	1.41	200	94.88	84.97
CIGS	0.702	85	42.31	36.27
CIS	0.72	75	44.98	38.69
CdTe	0.613	72.5	20.39	17.89
Back-Contact	0.153	30	15.03	13.35
Bi-Facial	0.12	(Front) 45	20.09	(60°C) 18.59
		(Back) 45	21.21	(60°C) 17.67

A study of how fast the temperature increases with a certain amount of time is important. Some panels were able to reach 75°C after being illuminated by the light radiation for about ±1200 seconds but some others only able to reach 60°C.

Table 3. Percentage of losses in power, V_{oc} and fill factor in between time to reach 65°C from 40°C.

	Pmax(Watt)	Time(Sec)	% Lost (40°C - 65°C)		
			Wp	Voc	FF
Poly-cystalline	200	780	10.08	12	3.14
Poly-Crystalline	110	1050	10.52	9.92	5.91
Mono-Crystalline	175	950	11.45	12.78	3.58
CIS	75	827	13.98	13.32	2.06
CIGS	85	1281	14.27	5.62	9.79
CdTe	72.5	1183	12.26	plus 2.82	14.82
Back-Contact	30	745	11.18	10.39	6.05
Bi-Facial					
Front	45	1048	11.85	8.94	1.24
Back	45	1350	16.96	10.83	12.73

From the results showed in table 2 and 3, we found that the electrical power generated by our solar simulator was linear to any of the panels accept the CdTe panel due to its mismatch to the solar responses of halogen light source. On average, all those panels are losing 10%-15% of its power output after being heated up to 65°C. Every 5°C of temperature raises, each of the panels respond differently; mono(175W) losing ~2%, poly(200W) losing ~2%, poly(110W) losing ~3%, CIGS losing ~2%, CIS losing ~ 2.5%, CdTe losing ~2%, back-contact losing ~2%, and bi-facial losing ~3%.

Under the same circumstances, thin film based of solar panels were facing a greater power lost than silicon based solar panel. Thin films generally are made of tempered glass on both side which will absorb more heat in a longer time. After being heated up in a longer time, solar panel will greatly reduced its performances. We could conclude that mono-silicon based solar panel is heated faster in compare to bi-facial and thin film solar panel. Open-circuit voltage, Voc is the most affected parameter under different temperature. Voc is the voltage across the sytem when the current is zero. Temperature changes will affect the band-gap of the solar cell[7]. In our study, we found out that after temperature increased about 30°C which is from 40°C to 70°C, most of the Voc from the panels will decrease by ~10% to ~15%. With this reduction, it is enough to disturb the whole performance of the system. This is also one reason why we are facing a power losses in high temperature condition.

The fill factor as a parameter which shows how the solar panel performs was facing a great negative impact. There were no big impact toward the current flow in the system since the intensity remains stable at 500W/m^2. While the experiments were conducted indoor under ambient temperature of 21 °C.

4. SUMMARY

Understanding the effect of temperature on to various types of photovoltaic panels will give us a wide picture in industrial, manufacturing, and study field to choose a better PV panel in certain condition. Power lost after being heat up for 30°C is about 15% and 2% reduction in average with 5°C increment of temperature for each panel under a uniform light radiation, stable surrounding condition (indoor experimental study). Halogen light source is not a good choice for Thin-Film based solar cell because

IOP Conf. Series: Materials Science and Engineering **88** (2015) 012066 doi:10.1088/1757-899X/88/1/012066

of the ban gap and light wavelength and spectrum mismatch.The open-circuit voltage is affected mostly from the temperature changes. Improvement of design and material for the PV panels is important to avoid the power lost due to high temperature.

REFERENCES

[1] Dubey Swapnil et al. 2013 Temperature dependent Photovoltaic (PV) Efficiency and Its Effect on PV Production in the World *PV Asia Pasific Conference, Energy Procedia* **33** 311-321

[2] Mosalam Shartout,M.A et al. 2000 The Temperature Dependence of the Spectral and Efficiency Behavior of Si Solar Cell Under Low Concentrated Solar Radiation *Renewable Energy* **21** 445-458

[3] Lorenzo, E., 1994 Solar Electricity. Engineering of Photovoltaic Systems, *Artes Graficas Gala* Sevilla.

[4] Radziemska, E and Klugmann, E 2002 Thermally Affected Parameters of the Current-Voltage Characteristics of Silicon Photocell *Conversion Management* **43** 1889-1900

[5] Taguchi, M, Terakawa A, Maruyama E and Tanaka M 2005 Obtaining a Higher Voc in HIT cells. Progress in Photovoltaics: Research and Applications **13** 481-488

[6] Nishioka K, Takamoto et al. 2006 Annual Output Estimation of Concentrator Photovoltaic Systems Using High-Efficiency InGap/InGaAs/Ge Triple-Junction Solar Cells Based on Experimental Solar Cell's Characteristics and Field-test Meteorological Data *Solar energy Materials and Solar Cells* **90** 57-67

[7] Andreev VM and Grilikhes VA et al. Photovoltaic Conversion of Concentrated Sunlight London: Wiley;1967 [ISBN:0471967653].

[8] Khamid P and Ketjoy N et al. 2012 Investigation on Temperature coefficients of Three Types Photovoltaic Module technologies Under Thailand Operating Condition *Procedia Engineering* **32** 376-383

[9] Emery K, et al. Temperature Dependance d Photovoltaic Cells, Modules, and Systems Proceeding *of the 25th IEEE PV Specialist Conference, Washingtion, USA*; 13-19 May 1996 1275-8

[10] King DL and Eckert PE. Characterizing (rating) The Performance of Large Photovoltaic Arrays for All Operating Conditions *Proceeding of the 25th IEEE PV Specialist Conference, Washington, USA*; 13-19 May 1996 1385-9.

[11] Wilshaw AR, Bates JR and Pearsal NM. Photovoltaic Module Operating Temperature effects *Proceeding of Eurosun'96, Munich, Germany*; 1996 940-4.

[12] Krauter S, Araujo RG, Schroer S, et al. 1999 Combined Photovoltaic and Solar Thermal System for Façade Integration and Building Insulation *Solar Energy* **67(4-6)** 239-48

[13] Moshfegh B, et al. 1998 Flow and Heat Transfer in the Air gap behind Photovoltaic Panels *Renewable Sustainable Energy Rev* **2** 287-301

[14] Del Cueto JA. Comparison of energy Production and performance from flat-Plate Photovoltaic Module technologies Deployed at Fixed tilt *Proceeding of the 29th IEEE PV Specialist Conference, New Orleans, USA*; 20-24 May 2002

IOP Conf. Series: Materials Science and Engineering **88** (2015) 012067 doi:10.1088/1757-899X/88/1/012067

Viscosity Model for Predicting the Power Output from Ocean Salinity and Temperature Energy Conversion System (OSTEC) Part 2: Computer Simulation

SK Lee[a], Ag Sufiyan Abd Hamid[a], Jedol Dayou[a, 1], Rubena Yusoff[a], Fauziah Sulaiman[a], Ejria Saleh[b] and HLH Chong[c]
[a] Energy, Vibration and Sound Research Group (e-VIBS), School of Science and Technology, Universiti Malaysia Sabah, Jalan UMS, 88400 Sabah, Malaysia.
[b] Borneo Marine Research Institute (BMRI), Universiti Malaysia Sabah, Jalan UMS, 88400 Sabah, Malaysia.
[c] Water Research Unit (WRU), School of Science and Technology, Universiti Malaysia Sabah, Jalan UMS, 88400 Sabah, Malaysia.

E-mail: jed@ums.edu.my

Abstract. The paper presents the computer simulation of the improved prediction model in Part 1 and compared with the classical density model. The prediction model in Part 1 is improved by incorporating the effect of fluid dynamic viscosity when there are salinity and temperature differences between two fluids, for better prediction of kinetic power output. The viscosity prediction model takes account of the water head losing causing by frictional effect and viscous dissipation which in turns provides an analytical predicted outcome. Computer simulations are presented in this paper to assess the system as the parameters of system are varied using the Viscosity Model.

1. Introduction

The present study is the continuation of the Part 1 [1] for performing the computer simulations from the improved prediction model. In Part 1, a new refined formulation known as Viscosity Model is developed for predicting the kinetic power output of Ocean Salinity and Temperature Energy Conversion System (OSTEC). Viscosity Model is formulated by considering the effect of fluid dynamic viscosity for specifically determining the flow velocity of incoming water from on-land reservoir. It consists of a series of fluid dynamic equations to estimate the flow velocity. In a different way, Density Model [2] uses one direct equation to determine the respective flow velocity. Despite of the lengthy formulation, Viscosity Model is found relatively accurate than Density Model in Part 1. To gain more insight of the model performance, several important parameters that influencing the overall power output are identified and their corresponding relationships are simulated and compared with classical density model.

The main resources of OSTEC concept for harnessing energy are the salinity and temperature difference between two fluids. The mixing of sea water and incoming water from reservoir at different salinity and temperature in a vertical tube submerged in sea water, produces a rising water mixture due

[1] To whom any correspondence should be addressed

IOP Conf. Series: Materials Science and Engineering **88** (2015) 012067 doi:10.1088/1757-899X/88/1/012067

to buoyant force. Kinetic power can be derived from the upward motion of the mixture by using appropriate turbine rotor. For clarity, the conceptual design of OSTEC experiment as shown in figure 1 is given which is similar to figure 1 in Part 1.

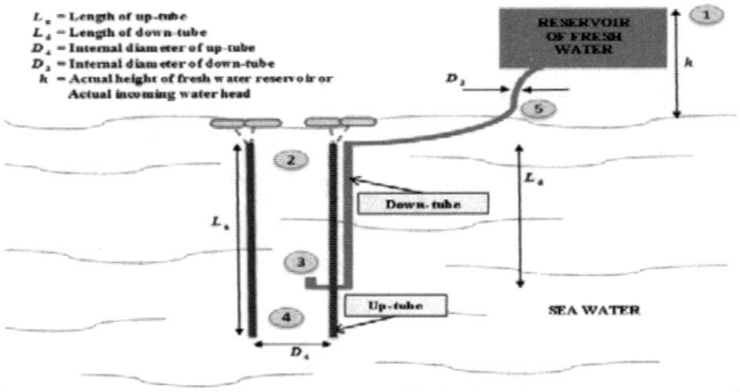

Figure 1. Conceptual design for OSTEC theoretical experiment

To simulate the kinetic power output, several governing equations are summarized. Basically, the maximum kinetic power that can be harvested due to the rising water mixture can be written as

$$P = 8\frac{Q_2^{\ 3}\rho_2}{\pi^2 D_4^{\ 4}} \tag{1}$$

where Q_2, ρ_2, and D_4 are the volume flow rate of water mixture and density of the water mixture at Point 2, and internal diameter of up-tube, respectively. Volume flow rate at Point 2 is very important in determining the kinetic power. Since the only inlets to Point 2 are from Point 3 and Point 4, the volume flow rate at Point 2 is the summation of the flow rate from Point 3 and Point 4 or

$$Q_2 = Q_3 + Q_4 \tag{2}$$

Volume flow rate at Point 3 which is the volume flow rate of incoming water from ground elevated reservoir help exciting the amount of sea water going into the bottom of up-tube due to density difference between the two fluids. It relies on the internal cross sectional area of down-tube, A_3 and the velocity of the incoming water at Point 3, V_3 where it is written as

$$Q_3 = A_3 V_3 \tag{3}$$

To quantify the velocity of incoming water at Point 3, two methods which are Density Model [2] and Viscosity Model [1] are used.

1.1. Density Model

As discussed in Part 1, V_3 in Density Model considers the incoming fluid density directly which can be expressed as

$$V_3 = \sqrt{\frac{2gh_1\rho_{PW}}{\rho_3}} \tag{4}$$

where h_1, ρ_{PW}, ρ_3 and g are the total height of water reservoir, density of pure water at standard temperature and pressure (STP), density of incoming water and gravitational constant, respectively. It can be seen that V_3 is affected by the density of incoming water. As the independent parameters of fluid density, both the fluid salinity and temperature are thus varied and simulated to examine the corresponding effects on kinetic power output (Equation (1)).

IOP Conf. Series: Materials Science and Engineering **88** (2015) 012067 doi:10.1088/1757-899X/88/1/012067

1.2. Viscosity Model

In Viscosity Model, the dynamic viscosity of the incoming fluid is considered indirectly to determine the frictional head loss, h_L of flowing fluid through down-tube, and therefore the effective water head, h_e and velocity of incoming water which can be written as

$$h_e = h_1 - h_L \tag{5}$$

and

$$V_3 = \sqrt{2g(h_e)} \tag{6}$$

respectively. However, from the formulation of Viscosity Model in Part 1, V_3 is determined based on the dynamic viscosity (Equation (8)) of incoming water in a series of equation (Equation (3) to (10)). It can be noticed that the V_3 is influenced by the dynamic viscosity of incoming water. As the independent parameters of fluid dynamic viscosity, both the fluid salinity and temperature are thus varied and simulated to study their corresponding effects to the kinetic power output. Detail examination of Viscosity Model also shows that fluid dynamic viscosity is used in the early stage to model the frictional head loss of the incoming water flowing through down-tube as shown in equation (5) in Part 1. It can be seen that the frictional head loss is affected by the down-tube diameter and the actual height of reservoir (equation (6) in Part 1). The down-tube diameter and the actual height of reservoir are thus varied and simulated to visualize the corresponding effects to the remaining effective head as shown in equation (5) in this paper.

As the incoming water moves upwards from down-tube outlet to mix with sea water at Point 3, by assuming that there is complete transfer of kinetic power from incoming water to the surrounding sea water to enable continuous flowing of fluid in the up-tube and hence the volume flow rate of sea water at Point 4 can be determined as

$$Q_4^{\,3} = Q_3^{\,3}\left(\frac{\rho_3}{\rho_4}\right)\left(\frac{D_4^{\,4}}{D_3^{\,4}}\right) \tag{7}$$

where ρ_4 and D_3 are the density of sea water and internal diameter of down-tube, respectively. It can be seen that Q_4 is influenced by the ratio between fluids density and tube diameter. Due to that the effect of density to the overall kinetic power output is examined, the interest now would be the effects of the diameter ratio and the up-tube diameter to kinetic power output. They are varied correspondingly to examine the impacts to the kinetic power predictions.

2. Computer Simulation Results and Discussions

Computer simulations from Viscosity Model are presented for examining the relationship of important parameters with the potential kinetic power output and further compare with the simulations from Density Model. The main parameters of the Ocean Salinity and Temperature Energy Conversion system (OSTEC) are temperature and salinity difference between sea water and incoming water from reservoir. During the simulation, temperature and salinity of sea water are fixed at 20 °C and 35 psu respectively whereas temperature and salinity of incoming water are varied to increase the kinetic power output possibly. Figure 2(a) and 2(b) show the effects of temperature and salinity to the predicted kinetic power output using Viscosity Model and Density Model respectively, with the setting given in table 1 of Part 1. For clarity, the parametric setting of OSTEC as shown in table 1 is presented here which is similar with table 1 of Part 1.

IOP Conf. Series: Materials Science and Engineering **88** (2015) 012067 doi:10.1088/1757-899X/88/1/012067

Table 1. Parametric Dimensions of OSTEC.

Parameter	Value (m)
Internal diameter of up-tube, D_4	0.150
Internal diameter of down-tube, D_3	0.018
Length of up-tube, L_u	1.500
Length of down-tube, L_d	1.000
Height of reservoir from mean sea level, h_1	0.550

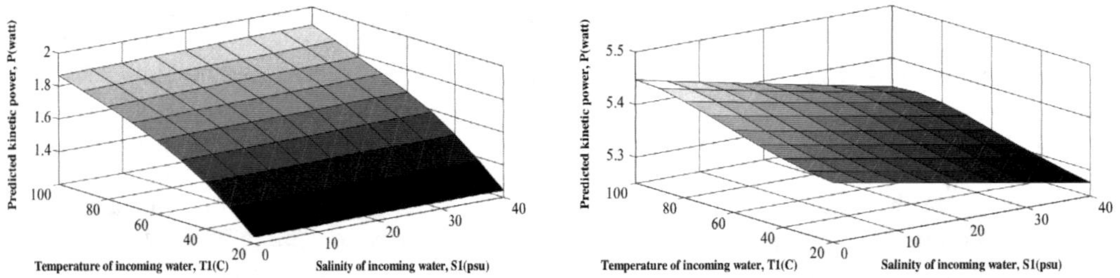

(a) Viscosity Model (b) Density Model

Figure 2. Comparison of the temperature and salinity effect to the kinetic power output using different models.

From figure 2(a) and 2(b), It can be observed that by using Viscosity Model, the overall maximum kinetic power output is 1.89 w whereas it is 5.45 w when Density Model is used. The overall power prediction using Viscosity Model is lower if compared to when Density Model is used. This outcome is due to that in Viscosity Model, dissipation of energy (head loss) from frictional effect is considered based on the coefficient of fluid viscosity at specified fluid temperature and salinity, where these are not considered in Density Model.

It is nevertheless that in Viscosity Model, the predicted power increases more with higher temperature at a given salinity. Similar ascending trend of power output is observed for other tested salinity as in figure 2(a). When the salinity of incoming water is set at 0 psu, power increases for 50.4 % from temperature of 20 °C to 100 °C using Viscosity Model. It is however at the similar conditions, power increases for only 1.5 % using Density Model. The discrepancy happens due to dynamic viscosity reduces with higher temperature. This results to the reducing of energy loss due to frictional effect when incoming fluid temperature increases, and therefore contributes to higher increment of power output.

The effect of salinity to kinetic power output is not so obvious in Viscosity Model but it is quite apparent in Density Model. Theoretically higher fluid salinity contributes to higher fluid density and fluid viscosity and therefore results to lower kinetic power output of the water mixture. This can be observed in figure 2(b) where kinetic power output increases with lower salinity at a given temperature and similar trend is observed for other tested temperature also. Regarding to the effect of fluid salinity, the different trend in figure 2(a) results from the property of fluid dynamic viscosity also where it is more temperature dependent if compared to salinity. This can be seen from the reduction of fluid salinity from 40 psu to 0 psu at given temperature of 20 °C results to reduction of dynamic viscosity of only 7.8 % whereas changes of fluid temperature from 20 °C to 100 °C at given salinity of

0 psu results to reduction of dynamic viscosity of 71.9 % [3]. This explains the dominancy of temperature in affecting power output prediction using Viscosity Model.

It is specifically can be observed that in figure 2(a), the kinetic power output is strangely having a slight increment with higher salinity of incoming water at a given temperature, and the similar trend is observed for other tested temperature also. This suggests that possibly smaller salinity difference between sea water and incoming water may be used for increasing the kinetic power output slightly. The prediction may be explained by the incoming water requires certain amount of mass to have sufficient momentum for flowing through down-tube, up-tube and towards ocean surface. It is in addition reported by [4] that unpublished experiments show that smaller salinity differences raise the efficiency. However, smaller salinity difference between fluids seemed does not produce significant power increment in figure 2(a) and therefore this finding requires further experimental validation and optimization to be applied in the full-scale experiment. As a result, incoming water with zero salinity or fresh water is preferable at the current stage of investigation.

The percentage of tube diameter ratio (D_3/D_4) at constant up-tube diameter has been found useful to determine the flow rate of incoming water along the down-tube from reservoir as in equation (3). By using the similar setting as the experimental setting [4], the percentage of diameter ratio is varied from 2.4 % to 28 % where up-tube diameter is fixed at 0.15 m. Figure 3(a) and 3(b) present the effective water head of incoming water when percentage of diameter ratio and the height of reservoir are varied using Viscosity Model and Density Model respectively.

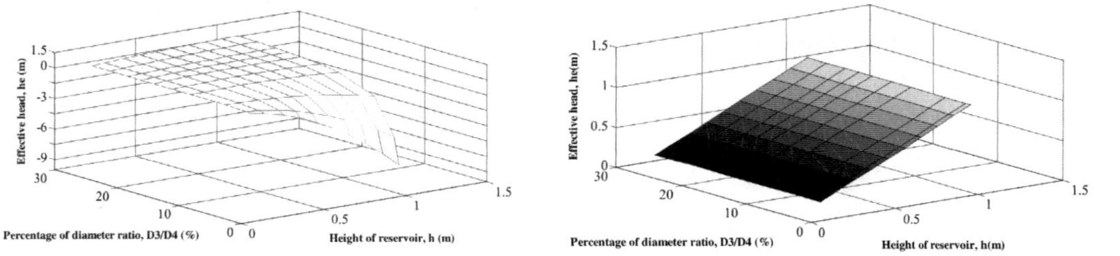

(a) Viscosity Model. (b) Density Model.

Figure 3. Comparison of the effective head from incoming water at different percentage of diameter ratio and height of reservoir using different models.

It is presented that in figure 3(a), the percentages of diameter ratio less than 10 % have negative effective head of incoming water and these heads tend to become more negative with higher height of reservoir. This is causing by the increasing of head loss beyond the cutoff water head when down-tube diameter is reduced to less than 10 % of up-tube diameter. The increasing of reservoir height does not help in adding up the effective head but in reverse results to enlarging of head loss. This may due to that the frictional head loss is velocity dependent and proportionate with higher heights. This finding suggests that tube diameter ratio of less than 10% might not work if for the sake of minimizing the input incoming water, neither it worked by elevating the incoming water reservoir at this absolutely lower diameter ratio.

Also, from figure 3(a), effective head increases with higher diameter ratio and tend to become larger with higher elevation of reservoir. This can be explained through equation (5) in Part 1 where head loss reduces with larger down-tube diameter. Due to that the frictional head loss is velocity dependent at various heights, the head loss is in fact increasing with higher heights but its amount is not so significant at larger down-tube diameter and therefore ends up with increasingly positive effective water head with higher heights. Compare to the effective head predicted using Density Model as in figure 3(b), it is always similar with the theoretical height of reservoir h_1 since it has not considered any energy dissipation or water head losing.

Up-tube diameter is proportionate to certain order of sea water flow rate going into the bottom of up-tube as in equation (7). As a result, various percentages of diameter ratio more than 10 % at different up-tube diameters are used to simulate the corresponding effects to the kinetic power output as presented in figure 4 using Viscosity Model.

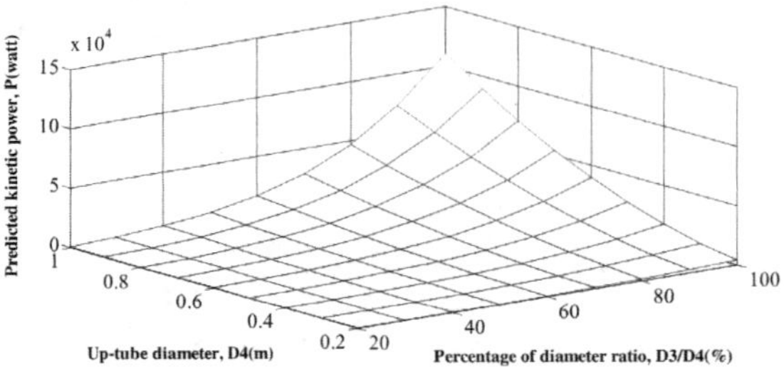

Figure 4. Prediction of kinetic power at different diameter ratio and up-tube diameter using Viscosity Model.

Further examination of figure 4 and figure 4 in [2] for Density Model shows that it is desirable to have diameter ratio as high as possible so that higher power output can be produced. However, there should be a practical ratio constrained by physical argument. From equation (2), the volume flow rate at Point 2 is the summation of the volume flow rate from Point 3 and also Point 4. Thus, adequate space to allow the sea water drift upwards from Point 4 to Point 2 is required. Based on this argument, diameter ratio of 40 % is chosen for further investigation.

Another important point to investigate is on how the height of reservoir (refer figure 1) affects the power output of the system. This is important when the investigation is performed in full-scale experiment at the specific location reported in [2]. In order to have further insight to this, figure 5 is simulated with tube diameter ratio is fixed at 40 % and the temperature of sea water and incoming water (S_1 = 0 psu) from reservoir are set at 25 °C and 32 °C, respectively. This temperature selection is based on the ambient properties where the future full-scale experiment is located. Comparison of figure 5 and figure 6 in [2] for Density Model exhibits that higher kinetic power output can be acquired when the height of reservoir is increased. It is furthermore showing that higher increase rate of power output can be obtained with larger diameter of up-tube.

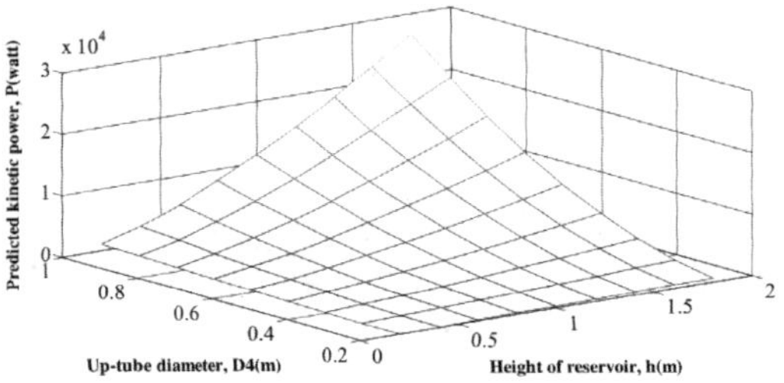

Figure 5. Prediction of kinetic power at different up-tube diameter and height of reservoir using Viscosity Model. The tube diameter ratio (D_3/D_4) is fixed at 40%.

As discussed in Density Model, it is found that up-tube diameter and length of D_4 = 0.6 m and L_u = 7 m are capable to construct a 10kW electrical power generator, at reservoir elevation of h_T = 1.8 m to meet the power demand of a specified amount of households. Meanwhile in Viscosity Model, a slightly larger setting is predicted through figure 5, the required up-tube diameter is 0.7 m with the similar up-tube length and height of reservoir as Density Model for the similar power output. In overall both proposed settings from different model are not differ much except for the up-tube diameter (Viscosity Model: 0.7 m; Density Model: 0.6 m) and down-tube diameter which is 40 % of up-tube diameter (Viscosity Model: 0.28 m; Density Model: 0.24 m). In view that part of the simulation results using Viscosity Model [1] have relatively closer agreement with reported experimental measurements, it is decided to pursue the actual experiment with the setting from Viscosity Model, with the additional intention that the system is still capable to achieve the expected outcome after undergoing minimum unavoidable energy loss.

Based on the project scale assessment, the expected power output of 10 kW might involve big structure, large volume of incoming water (fresh water) and therefore big project costing. For the purpose of fundamental investigation, it would be desirable to reduce the expected outcome to 1 kW. As mentioned in the system mechanism, the rising of water mixture due to buoyant force is excited by the injection of incoming water at different salinity and temperature. It is therefore expected to have higher normalized power output over the volume flow rate of incoming water. Based on this argument, tube diameter ratio of 15 % is selected for additional investigation. In order to visualize this, and take into account of the previous argument, figure 6 is simulated with tube diameter ratio of 15 % at the respective ambient temperature of sea water and incoming water using Viscosity Model.

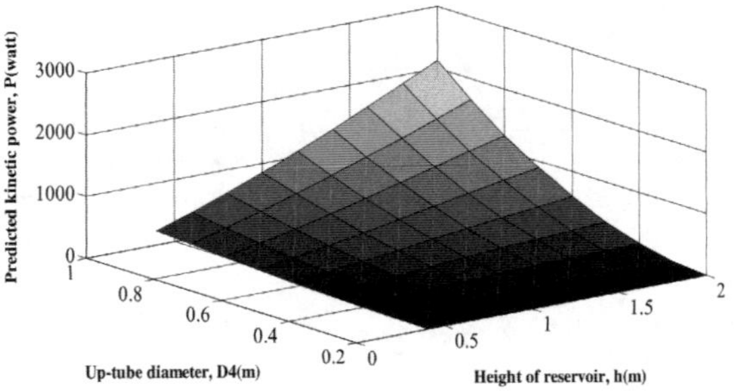

Figure 6. Prediction of kinetic power at different up-tube diameter and height of reservoir using Viscosity Model. The diameter ratio (D_3/D_4) is fixed at 15%.

Detail examination of figure 6 shows that up-tube diameter and down-tube diameter of $D_4 = 0.8$ m and $D_3 = 0.12$ m respectively, with reservoir elevation of 1.8 m from the mean sea level are sufficient to meet the new power demand. Compared to the previous expected power outcome and therefore the predicted down-tube diameter ($D_3 = 0.24$ m) by Viscosity Model, the current predicted down-tube diameter is reduced by half which indicates lesser usage of incoming water in generating useful power. Although for the new power output, the required up-tube diameter is increased by 0.1 m, but in practical the increment of up-tube size does not consume large costing of input resources.

4. Conclusions

The paper of Part 2 presents the computer simulations of Viscosity Model which is derived in Part 1 and later compare with the classical simulations from Density Model. Overall comparison of simulation results reveal that there is power increment with higher temperature as simulated by both models but particularly the increment rate is higher in Viscosity Model. This is due to the reduction of fluid dynamic viscosity where it is relatively temperature dependent. Larger percentage of diameter ratio, bigger up-tube diameter and higher reservoir elevation are found capable to increase the kinetic power output by both models. However in specifically, a limiting boundary is found by Viscosity Model where those tube diameter ratios less than 10 % are unable to produce any useful incoming water flow, neither has it worked by the increasing height of incoming water reservoir at this lower diameter ratio. Parametric setting of the future full-scale experiment predicted by both models to achieve the similar expected outcome are not differ much but Viscosity Model predicts slightly larger setting. It is preferable to pursue the full-scale experiment with the setting from Viscosity Model, with the additional intention that the system is still capable to achieve the expected outcome after undergoing minimum unavoidable energy loss.

Acknowledgments
This study was supported by the Malaysia Ministry of Higher Education under Research Acculturation Grant Scheme (RAGS) no RAG0011-TK-2012, and Ministry of Science, Technology and Innovation under ScienceFund no SCF0089-IND-2013, and are greatly acknowledged.

References
[1] Abd Hamid A S, Lee SK, Dayou J, Yusoff R and Sulaiman F 2013 Viscosity Model for Predicting the Power Output from Ocean Salinity and Temperature Energy Conversion System (OSTEC) Part 1: Theoretical Formulation *Int. J. of Environmental, Ecological, Geological and Mining Engineering* **7(10)** 698-702
[2] Lee SK, Dayou J, Abd Hamid A S, Saleh E and Ismail B 2012 A Theoretical Investigation on the Potential Application of Ocean Salinity and Temperature Energy Conversion (OSTEC) *Int. J. of Renewable Energy Research* **2(2)** 326-31
[3] Franzini J B and Finnemore E J 1997 *Fluid Mechanics with Engineering Applications* (New York: McGraw-Hill)
[4] Pscheidt E and Finley W 2003 *Deriving useful power from the osmotic potential between solutions* Internal Report <http://www.waderllc.com> (retrieved on January 13, 2013)

Study of Solar Driven Silica gel-Water based Adsorption Chiller

K Habib[1], **M K Assadi**[1] **and M H B Zainudin**[1]
[1]Department of Mechanical Engineering, Universiti Teknologi PETRONAS, Bandar Seri Iskandar, 31750 Tronoh, Perak, Malaysia

*E-mail: khairul.habib@petronas.com.my

Abstract. In this study, a dynamic behaviour of a solar powered single stage four bed adsorption chiller has been analysed designed for Malaysian climate. Silica gel and water have been used as adsorbent-refrigerant pair. A simulation program has been developed for modeling and performance evaluation of the chiller using the meteorological data of Kuala Lumpur. The optimum cooling capacity and coefficient of performance (COP) are calculated in terms of adsorption/desorption cycle time and regeneration temperature. Results indicate that the chiller is feasible even when low temperature heat source is available. Results also show that the adsorption cycle can achieve a cooling capacity of 14 kW when the heat source temperature is about 85°C.

1. Introduction

Environmental concerns and the rising energy costs necessitate looking for renewable energy driven environmentally benign cooling systems to meet the ever–increasing demand of refrigeration and air–conditioning. Heat driven sorption based cooling systems are gradually emerging as environmentally friendly alternatives to conventional vapour compression based refrigeration cycles. Since the detection of ozone holes in the stratosphere, caused by the CFCs and HCFCs of vapour compression based coolers, interest on adsorption cooling systems have increased a lot [1]. Adsorption cooling cycles are considered as environmentally benign, having zero ozone depletion potential (ODP) due to the use of natural refrigerants or alternative refrigerants. The other advantages of adsorption refrigeration systems are that they are free of vibration as they do not have any moving parts, simple control and lower operation costs.

Solar adsorption refrigeration system seems to be a promising alternative refrigeration device since cooling load of buildings is roughly in phase with solar energy availability. Habib et al. [2] designed a solar driven dual mode chiller which can provide cooling load throughout the year. Zhai et al. [3] designed a solar-powered air-conditioning system which consisted of a 150 m^2 solar collector and two adsorption chillers with a capacity of 8.5 kW each. Pons and Guilleminot [4] worked on an activated carbon-methanol based solar adsorption refrigeration system to produce ice. Habib et al. [5] presented a solar driven combined adsorption refrigeration cycle which can deliver refrigeration load at -10C. Habib et al [6] used activated carbon fiber-ethanol pair to design a solar driven adsorption chiller. A cooling capacity of 12 kW can be achieved with a heat source temperature of 85C.

IOP Conf. Series: Materials Science and Engineering **88** (2015) 012068 doi:10.1088/1757-899X/88/1/012068

From the above perspective, the present study deals with the transient analysis of a single stage four-bed silica gel-water based adsorption cooling cycle. The novel cycle is driven by solar energy and the analysis of solar thermal system is carried out with TRNSYS software using the solar data of Kuala Lumpur (3.16^{0}N, 101.7^{0}E), Malaysia. A cycle simulation computer program of the novel adsorption cooling system is developed to analyze the cooling capacity and coefficient of performance (COP) variations by varying adsorption/desorption cycle times and regeneration temperatures. The present study can be useful to fabricate environmental friendly cooling cycle which can tackle global warming and ozone depletion problems.

2. Working principle of silica gel-water adsorption cooling cycle

Figure 1 shows the schematic diagram of the four bed adsorption cooling cycle which consists of a condenser, an evaporator, four sorption beds filled with silica gel. The details of the adsorption cooling system have been presented by Habib et el. [2].

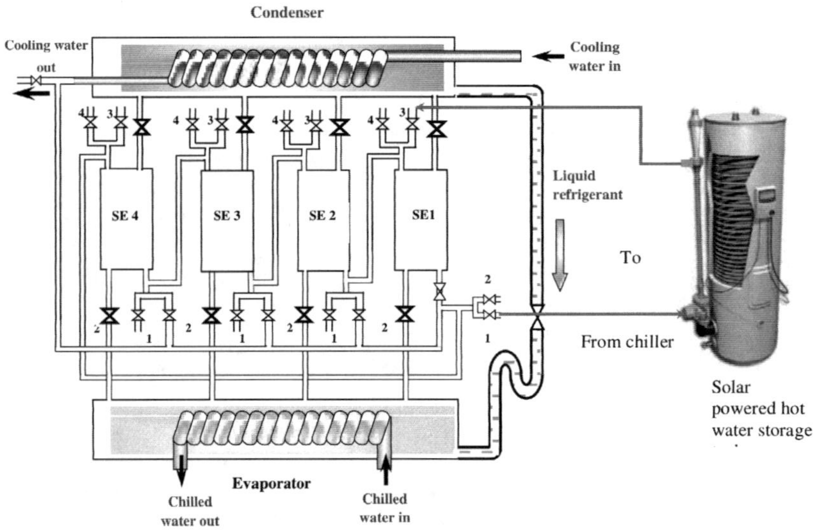

1: Hot water outlet; **2**: Coolant inlet to the adsorber; **3**: Hot water inler; **4**: Coolant outlet from the adsorber

Figure 1. Schematic diagram of four bed adsorption chiller [2].

3. Mathematical modeling

3.1 Adsorption isotherms

The modified Freundlich model, which is expressed by equation (1), is used to estimate the equilibrium uptake of silica gel-water pair [2].

$$x^* = \alpha(T_{sg})\left[\frac{P_s(T^{ref})}{P_s(T_{sg})}\right]^{\beta(T_{sg})} \tag{1}$$

Where,

$$\alpha(T_{sg}) = \alpha_0 + \alpha_1(T_{sg}) + \alpha_2(T_{sg})^2 + \alpha_3(T_{sg})^3$$

$$\beta(T_{sg}) = \beta_0 + \beta_1(T_{sg}) + \beta_2(T_{sg})^2 + \beta_3(T_{sg})^3$$

3.2 Adsorption kinetics

The rate of adsorption or desorption is governed by the linear driving force model [2].

$$\frac{dx}{dt} = 15D_{so}\frac{\exp\left(-\frac{E_a}{RT}\right)}{R_p^2}(x^* - x) \tag{2}$$

3.3 Adsorption and desorption energy balance

Using the lumped approach for the adsorption bed, which comprises silica gel, the heat exchanger fins and tubes, the energy balance equation is given by [2],

$$\left[m_{sg}C_{p,sg} + m_{sg}C_p^{ref}x^{ref} + m_{Al}C_{p,Al} + m_{Cu}C_{p,Cu}\right]\frac{dT^{ref}_{ads/des}}{dt} = \delta m_{sg}\left\{h_g\left(P^{ref}_{eva/cond}, T^{ref}_{ads/des}\right) - h_g\left(T^{ref}_{eva/cond}\right) + \Delta h^{ref}_{st}\right\}\frac{dx^{ref}_{ads/des}}{dt} + \dot{m}_w C_{p,w}\left(T^{ref}_{w,in,ads/des} - T^{ref}_{w,o,ads/des}\right) \tag{3}$$

The outlet temperature of the source is to be modeled by the log mean temperature difference (LMTD) method and it is given by,

$$T^{ref}_{w,o,bed} = T^{ref}_{bed} + \left(T^{ref}_{w,in,bed} - T^{ref}_{bed}\right)\exp\left[-\frac{(UA)^{ref}_{bed}}{(\dot{m}C_p)_w}\right] \tag{4}$$

3.4 Evaporator energy balance

The evaporator energy balance of adsorption cycle can be expressed as [2]:

$$\left[m^{ref}_{eva}C_{p,eva} + m_{hex}C_{p,Cu}\right]\frac{dT^{ref}_{eva}}{dt} = -\delta\left\{h_g\left(P^{ref}_{eva}, T^{ref}_{ads}\right) - h_g\left(T^{ref}_{eva}\right) + h^{ref}_{fg}\right\}m_{sg}\frac{dx^{ref}_{ads}}{dt} + (\dot{m}C_p)_{chill}\left(T_{chill,in} - T_{chill,o}\right) \tag{5}$$

The water outlet temperature can be expressed as,

$$T_{chill,o} = T^{ref}_{eva} + \left(T_{chill,in} - T^{ref}_{eva}\right)\exp\left[-\frac{(UA)^{ref}_{eva}}{(\dot{m}C_p)_w}\right] \tag{6}$$

3.5 Condenser energy balance

The condenser energy balance of adsorption cycle can be expressed as [2]:

$$(m^{ref}_{cond}C_{p,cond}+m^{ref}_{hex}C_{p,Cu})\frac{dT^{ref}_{cond}}{dt}=-\delta\{h_g(P^{ref}_{cond},T^{ref}_{des})-h_g(T^{ref}_{cond})+h^{ref}_{fg}\}m_{sg}\frac{dx^{ref}_{des}}{dt}+\dot{m}_wC_{p,w}(T^{ref}_{w,in,cond}-T^{ref}_{w,o,cond})$$

The condenser outlet temperature can be expressed as:

$$T^{ref}_{w,o,cond}=T^{ref}_{cond}+(T^{ref}_{w,in,cond}-T^{ref}_{cond})\exp\left[-\frac{(UA)^{ref}_{cond}}{(\dot{m}C_p)_w}\right] \tag{7}$$

3.6 Cooling Capacity and coefficient of performance

The cooling capacity is obtained at the evaporator of adsorption cycle.

$$Q^{cycle}_{eva}=\frac{1}{t_{cycle}}\int_0^{t_{cycle}}(\dot{m}C_p)_w(T_{chill,in}-T_{chill,o})dt$$

The coefficient of performance (COP) of adsorption cycle can be expressed as:

$$COP=\frac{\int_0^{t_{cycle}}(\dot{m}c_P)_{chill}(T_{chill,in}-T_{chill,o})dt}{\int_0^{t_{cycle}}(\dot{m}c_P)_{des}(T_{h,in}-T_{h,o})dt} \tag{8}$$

4. Weather data

An accurate climatic database is necessary in solar energy technologies. Habib et al. [5] showed the variation of monthly average solar radiation of Kuala Lumpur, Malaysia.

Table 1. Rated conditions.

Hot water inlet		Cooling water inlet		Chilled water inlet	
Temperature (°C)	Flow rate (kg/s)	Temperature (°C)	Flow rate (kg/s) (ads+cond)	Temperature (°C)	Flow rate (kg/s)
85	1.28	30	(1.25+1.25)	14	0.71
Adsorption/desorption cycle time: 600s			**Switching time: 30s**		

Table 2. Values adopted for simulation [2].

Symbols	Values	Units
m_{sg}	40	Kg
UA_{bed}	3500	W/K
UA_{eva}	4870	W/K
UA_{cond}	15330	W/K
m_{cond}	24.28	Kg
m_{eva}	12.45	Kg
h_{fg}	2.8x106	J/Kg
$C_{p,hex}$	950	J/Kg-K
$C_{p,sg}$	960	J/Kg-K

5. Results and discussion

5.1 Hot water temperature variation

Figure 2 depicts the variation of hot water supply from the tank to the adsorption chiller Kuala Lumpur climate throughout a day (1st May, 2013). It can be seen from Fig. 2 that the maximum temperature of hot water temperature is achieved from 12-2 pm and the temperature is around 85°C.

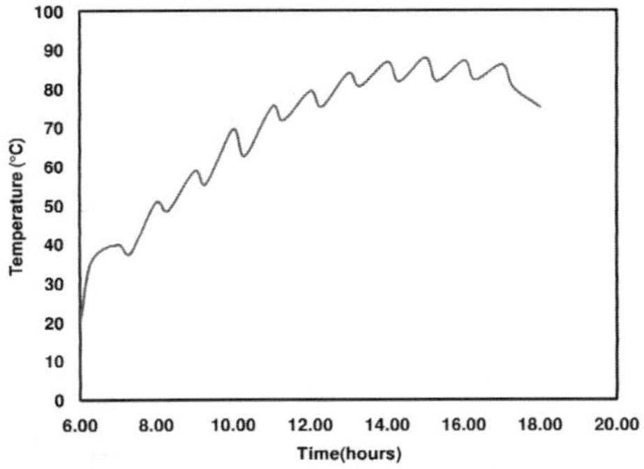

Figure 2. Variation of hot water supply from the tank to the chiller in Kuala Lumpur in May 2013.

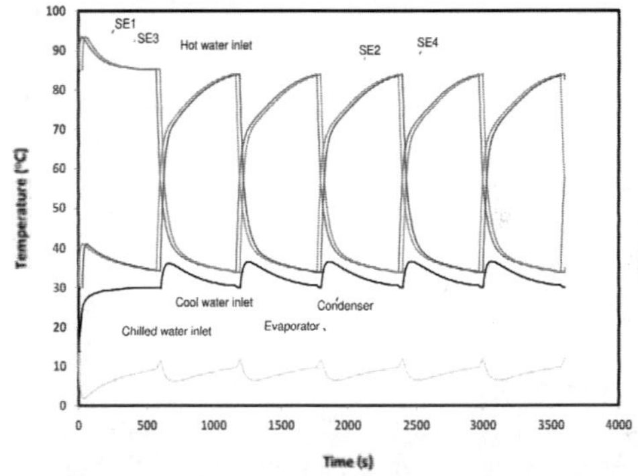

Figure 3. Temperature profiles for various components of silica gel-water cycle at rated conditions

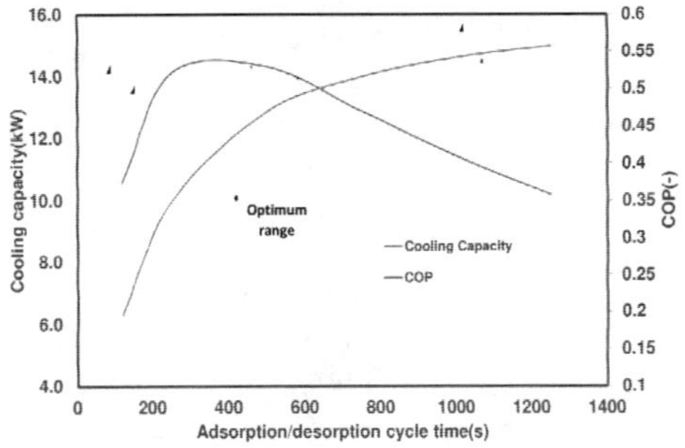

Figure 4. Effects of adsorption / desorption cycle time on cooling capacity and COP

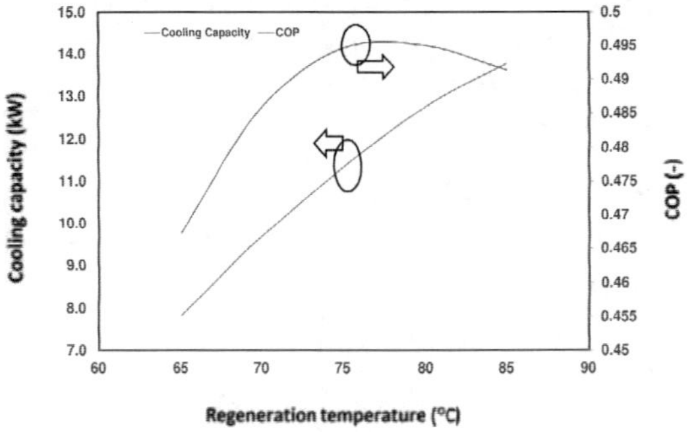

Figure 5. Effects of regeneration temperature on cooling capacity and COP

5.2 Chiller transient response

Figure 3 shows the chiller temporal histories for all the components (adsorber, desorber, evaporator and condenser) by using the mathematical model presented herein. Table 1 depicts the rated conditions for adsorption cycle. Table 2 lists the values adopted for simulation [2]. It can be seen from Fig. 3, the four bed silica gel-water based adsorption chiller is able to reach from transient to nearly steady state within three half cycles or 1900s, where adsorption/desorption cycle time is taken as 600 s and switching time is taken as 30s.

5.3 Adsorption/desorption cycle time

The simulated results of cooling capacity and COP variations with adsorption/desorption cycle time for silica gel-water cycle are shown in Fig. 4. From the Fig. 4, it is observed that for four bed silica gel-water cycle, the highest cooling capacity is around 14 kW for cycle times between 460 and 600s. From the Fig. 4 it is noted that the COP increases uniformly with the increase of adsorption/desorption cycle time and after reaching a certain value (around 600 s) it becomes steady. When the cycle time is less than 450s, there is not enough time for adsorption/desorption to occur satisfactorily. As a result, the cooling capacity decreases abruptly. On the other hand, when the cycle time is longer than 650s, the cooling capacity decreases gradually due to the less intention of adsorption after the first 10 min as the adsorbent reaches towards equilibrium or near equilibrium.

5.4 Regeneration temperature

Figure 5 shows the effects of heat source temperature on cooling capacity and COP for silica gel-water cycle. From Fig. 5, it is noticeable that the cooling capacity increases linearly from 8 to 14 kW with heat source temperature varies from 65 to 85°C. This is due to the fact that the amount of refrigerant circulation increases with the amount of desorbed refrigerant increases with the higher driving heat source temperature. The optimum COP value is around 0.495 for silica gel-water system when the regeneration temperature is between 80 to 85°C.

6. Conclusion

The paper investigates the transient analysis of a solar heat driven four-bed single stage silica gel-water based adsorption cooling cycle. The solar system is modeled with TRNSYS program using the meteorological data of Kuala Lumpur. Simulation results indicate that both the cooling capacity and COP increase with the regeneration temperature. The proposed four bed adsorption system can achieve cooling capacity of 14 kW and COP of 0.49 when the driving heat source temperature is 85°C. The present study can be useful to fabricate environmental friendly cooling cycle which can tackle global warming and ozone depletion problems.

7. Nomenclature

Symbols		**Subscripts**	
A	Area (m^2)	sg	Silica gel
m	Mass (kg)	eva	Evaporator
COP	Coefficient of performance (-)	cond	Condenser
C_p	Specific heat capacity (J/kg-K)	W	water
h	Enthalpy (J/kg)	in	inlet
$\dot{m}$	Mass flow rate (kg/s)	out	outlet
P	Pressure (Pa)		
T	Temperature (K)	**Subscript**	
t	Time (s)	ref	refrigerant
Q	Power (W)		

U	Overall heat transfer coefficient (W/m^2-K)
h_{fg}	Isosteric heat of adsorption (J/kg)

References

[1] Wang D C, Wang Y J, Zhang J P, Tian X L and Wu J Y 2008 *Energy Conv. Manage.* **49** 1063
[2] Habib K, Choudhury B, Chatterjee P K and Saha BB 2013 *Energy* **49** 1063
[3] Zhai X Q, Wang R Z, Wu J Y, Dai Y J and Ma Q 2008 *Appl. Energy* **85** 297
[4] Pons M and Guilleminot J J 1986 *J. Solar Energy Eng. (Trans. ASME)* **103** 332
[5] Habib K, Saha B B, Chakraborty A, Oh S T and Koyama S 2013 *Appl. Therm. Engg.* **50** 1582
[6] Habib K and Saha B B 2013 *Asian J. Scientific Research* **6** 146

7th International Conference on Cooling & Heating Technologies (ICCHT 2014) IOP Publishing
IOP Conf. Series: Materials Science and Engineering **88** (2015) 012069 doi:10.1088/1757-899X/88/1/012069

A Study on the Performance of the Split Reaction Water Turbine with Guide Ribs

Deuel H Allen[1] and Eliseo P Villanueva

[1] Dept. of Mechanical Engineering, College of Engineering,
MSU-Iligan Institute of Technology, Iligan City, Philippines

E-mail: deuel.allen@g.msuiit.edu.ph

Abstract. The development of technologies that make use of renewable energy is of great significance presently. A new kind of turbine called Split Reaction Water Turbine (SRWT) using PVC pipes as material is a major contribution towards harnessing the energy potentials of small stream low head water resources. SRWTs of diameter to height ratio (D/H = 110 cm/160 cm) were tested at the MSU-IIT College of Engineering Fluid Engineering Laboratory. Data on volumetric flow and pressure head at the turbine inlet of the SRWT were recorded using National Instrument Data Processing System using LabView software. In later experiments, guide ribs were installed at the vane of the exit nozzles in order to determine the difference in the performance of the ribbed and the non-ribbed SRWT. Simulations of the running SRWT were conducted using SOLIDWORKS software. Results of the simulations aided in the thorough analyses of the data from the experimental runs. A comparison of data from the ribbed and non-ribbed SRWT shows that guide ribs were effective in directing the momentum of the exiting water to improve the speed of rotation. In this study, the increase in the speed of the Split Reaction Water Turbine was as much as 46%.

1. Introduction.

In developing countries like the Philippines, there are few fossil fuel resources and the effort towards an industrialized economy is hindered by the great need for cheap and reliable power. The correct solution to the growing energy demand is to develop renewable energy technologies. Research for turbine that can be operated to utilize small stream and low head water resources is a contribution to the development of renewable energy technologies. A new kind of reaction turbine, fabricated by splitting a PVC pipe, has a great potential in renewable energy development. The turbine is called Split Reaction Water Turbine (SRWT) [1,2].

2. Statement of the Objectives

The objectives of the study are:

[1] To whom any correspondence should be addressed.

1. To investigate the effect of guide ribs in the vanes at the exit nozzles of the Split Reaction Water Turbine (SRWT);
2. To compare the performance of SRWT with guide ribs on the vanes with the Non-ribbed vanes SRWT;
3. To identify some parameters that influences the performance of SRWT.

3. Scope and Limitations

This study focused on the evaluation of the performance of SRWTs with and without guide ribs at the exit nozzles. Before fabricating the SRWT simulations of the running SRWT were conducted using SOLIDWORKS computer software.

As the objective of the study was to apply SRWTs in small stream low head water resources, the experimental rig used a pump that can deliver 2 liters per second at an average hydraulic head of 1.5 meters at the turbine inlet port. Thus, findings in this study are by no means complete but important parameters were identified, paving the way for better designed and fitted SRWT to operate at any given low flow and low head water stream.

There was only one SRWT diameter to height ratio (*i.e.* D/H = 110 cm/160 cm) that was investigated in this study.

4. Theoretical Considerations

The bases of this study are the basic theory in fluid mechanics and the accepted principles in fluid machinery.

The *continuity equation* in the non-conservative form, as applied to an infinitesimal mass of fluid moving through the flowfield is:

$$\frac{D\rho}{Dt}+(\rho\nabla\bullet V)=0 \tag{1}$$

where $\frac{D\rho}{Dt}=\frac{\partial\rho}{\partial t}+(V\bullet\nabla)\rho$ Substantial Derivative of ρ

and $V=u\vec{i}+v\vec{j}+w\vec{k}$ Velocity vector

For a constant density (ρ) equation (1) is expanded and the *momentum equations* for the *x, y and z* directions are:

In the ***x*-direction** only

$$\rho\frac{\partial u}{\partial t}-\frac{f_x}{\beta}+\left(\rho u\frac{\partial u}{\partial x}+\frac{\partial p_x}{\partial x}\right)+\left(\rho v\frac{\partial u}{\partial y}-\frac{\partial\tau_{yx}}{\partial y}\right)+\left(\rho w\frac{\partial u}{\partial z}-\frac{\partial\tau_{zx}}{\partial z}\right)=0 \tag{2}$$

In the ***y*-direction** only

$$\rho\frac{\partial v}{\partial t}-\frac{f_y}{\beta}+\left(\rho u\frac{\partial v}{\partial x}-\frac{\partial\tau_{xy}}{\partial x}\right)+\left(\rho v\frac{\partial v}{\partial y}+\frac{\partial p_y}{\partial y}\right)+\left(\rho w\frac{\partial v}{\partial z}-\frac{\partial\tau_{zy}}{\partial z}\right)=0 \tag{3}$$

In the ***z*-direction** only

$$\rho\frac{\partial w}{\partial t}-\frac{f_z}{\beta}+\left(\rho u\frac{\partial w}{\partial x}-\frac{\partial\tau_{xz}}{\partial x}\right)+\left(\rho v\frac{\partial w}{\partial y}-\frac{\partial\tau_{yz}}{\partial y}\right)+\left(\rho w\frac{\partial w}{\partial z}+\frac{\partial p_z}{\partial z}\right)=0 \tag{4}$$

where $f=f_x+f_y+f_z$ is the total body force

The momentum in *z*-direction is converted to the momenta in ***x*** and ***y***-directions when the water hits the walls of the turbine. Thus, summing the momenta in the ***x*** and ***y***-directions:

$$\rho\left(\frac{\partial u}{\partial t}+\frac{\partial v}{\partial t}\right)+(\rho g)+\left(\rho u\frac{\partial u}{\partial x}+\frac{\partial p_x}{\partial x}\right)+\left(\rho u\frac{\partial v}{\partial x}-\frac{\partial\tau_{xy}}{\partial x}\right)+$$

$$\left(\rho v\frac{\partial u}{\partial y}-\frac{\partial\tau_{yx}}{\partial y}\right)+\left(\rho v\frac{\partial v}{\partial y}+\frac{\partial p_y}{\partial y}\right)=0$$

IOP Conf. Series: Materials Science and Engineering **88** (2015) 012069 doi:10.1088/1757-899X/88/1/012069

(5)

where $f_x = 0$; (assuming no body force on the fluid along the x-direction)
$f_y = - mg$ (negative since the weight of the fluid is directed downward)

The momentum equation (5) shows that the total momentum is composed of the momenta along the x and y directions. The momentum along the z-direction adds to the momenta in the x and y directions.

Guide ribs placed at the exit nozzles of the vanes of SRWT would allow the optimum transfer of momentum from the water to the turbine. That is, the guide ribs placed along the x-direction will limit the movement of water along the y-direction. Further, the momentum due to the gravitational force along the y-direction can be redirected towards the x-direction. Thus, the scheme is an innovation to improve the non-ribbed vane Split Reaction Water Turbine developed at RMIT University, Melbourne, Australia [1, 2].

5. Design of SRWT

The result of simulating a running SRWT in SOLIDWORKS software indicated that there is counter impulse at the nozzles created by the exiting water. The impulse is due to water that impinges on the small area equivalent to two times the thickness of PVC pipe times the height of the turbine, see figure 1. To reduce the counter impulse, the lips of the nozzles were sharpened, see figure 2.

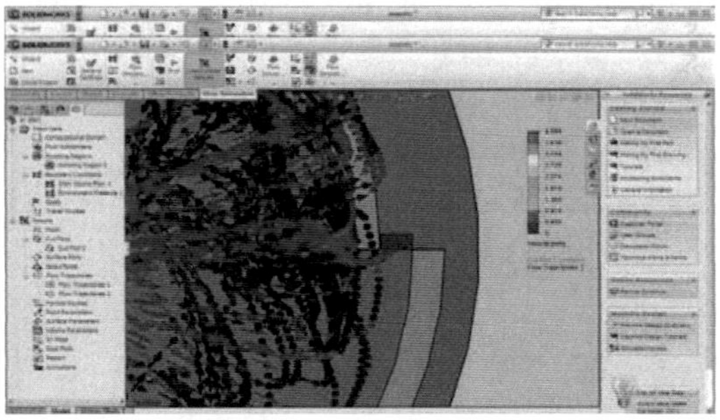

Figure 1. Top View of SRWT from Computer Simulation using SOLIDWORKS Software.

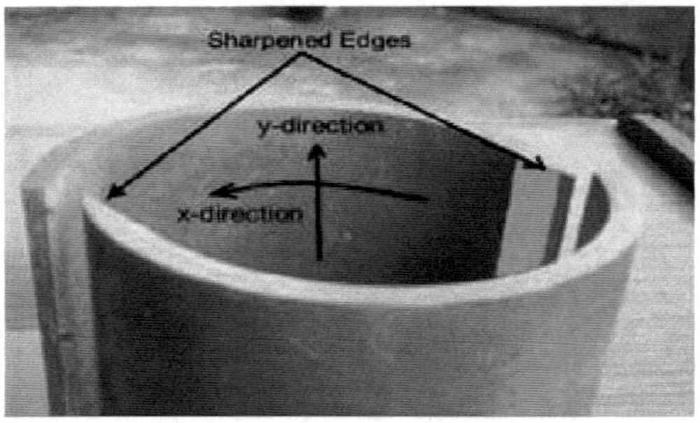

Figure 2. Sharpened Inner Lips of the Nozzles.

6. Experimental Test Rig and Test Procedure

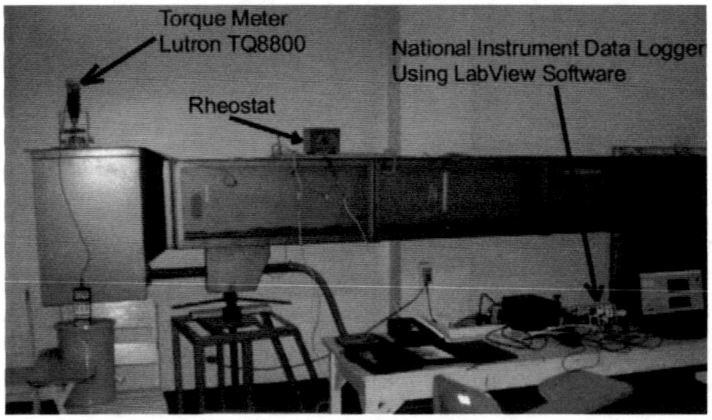

Figure 3. The Experimental Set-up.

Figure 4. SRWT Without Guide Ribs.

The SRWTs were tested in the Mechanical Engineering Fluids Laboratory using the Hydraulic Flume as the platform. The Hydraulic Flume is equipped with an electric pump that can deliver 2.0 liters/sec with a head of 1.5 meters to the SRWT. The SRWT was installed at one end of the Hydraulic Flume inside the water source tank and over it was the Lutron TQ8800 torque meter. Figure 3 is a photo of the experimental set-up. Water was introduced to the inlet port of SRWT located at its bottom as shown in figure 4.

The water jet caused the SRWT to rotate and the exiting water from the turbine falls to the bottom of the water source tank. Then it flows through the hydraulic flume back to the storage tank located below the hydraulic flume. Figure 6 shows the installed rotating Ribbed-SRWT inside the water source tank of the Hydraulic Flume.

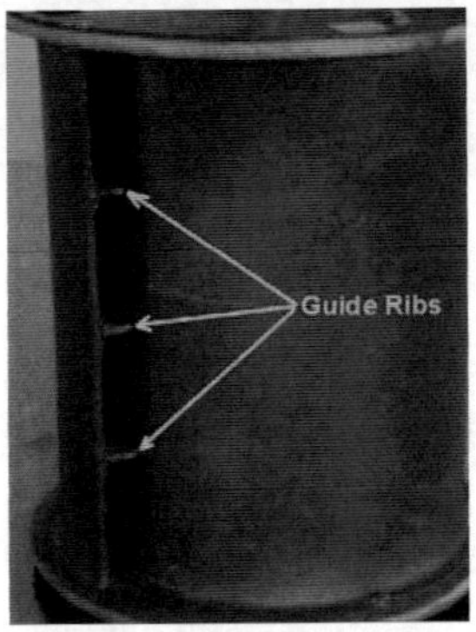

Figure 5. Photo of the SRWT With Guide Ribs Installed.

Figure 6. Photo of a running SRWT With Guide Ribs.

Figure 7. Photo of a static SRWT Without Guide Ribs.

The Hydraulic Flume is also equipped with a Venturi Meter of 44 mm diameter by 16 mm diameter at the throat. To control the flow of water through the Venturi Meter, a gate valve was installed before the Venturi Meter. Pressure difference at the Venturi Meter was measured using a differential pressure transducer and data were recorded through a National Instrument Data Processing System using LabView software. The logged data were recorded in a format that is Microsoft Excel compatible. Likewise, the pressures at the inlet of the SRWTs were recorded in the same manner.

The power outputs of the SRWTs were measured using Lutron TQ8800 torque meter and a non-contact digital tachometer manufactured by TT T-ECHNI-C. A prony brake was fabricated so as to adapt to the torque measuring procedure of the Lutron torque meter. Measuring the torque output of the SRWTs using Lutron TQ8800 and the digital tachometer combined, resulted to more accurate measurements of the actual power output of each SRWT. With this procedure, experimental runs were conducted to measure static torque with the prony brake absorbing the power output of the turbine.

The experiments on the SRWTs were executed by varying the volumetric flow into the turbine. This was carried out by opening and closing the gate valve located before the Venturi meter. In order to test the SRWTs uniformly on different volumetric flows the number of turns in opening and closing the valve were set. There were 7¾-turns from the closed position to the full open position of the valve. However, it was observed that the number of turns from closed to open do not result to the same volumetric flow if the number of turns were executed from open to closed.

Recorded data were analyzed using the basic principles of fluid mechanics and fluid machineries. In most cases where data were needed to complete the analyses, values were deduced from the recorded data using statistical tools. For example, since there is enormous data that were recorded for one gate valve opening, average values were used for analyses. Further, since the recorded data were discrete values and often values between them are needed for analysis, regression analyses were applied to the data.

7. Results and Discussion

Table 1. Representative results from running Non-ribbed-SRWT.

Non-Ribbed SRWT	Data Using Lutron TQ-8800								
Q (L/sec)	Hj (m)	Power (watts)	N (rev/min)	T (N-cm)	Ts (N-cm)	Dynamic Ø	Static Øs	Input Power Pi (watts)	Hydraulic Eff. (%)
1.87	1.42	0.118	105.0	1.08	7.7	0.27	0.28	25.98	0.46
1.85	1.42	0.277	97.9	2.70	7.6	0.27	0.27	25.70	1.08
1.86	1.41	0.385	91.9	4.00	7.6	0.28	0.28	25.84	1.49
1.84	1.41	0.336	87.3	3.68	7.6	0.27	0.27	25.54	1.32
1.72	1.43	0.267	94.4	2.70	7.7	0.25	0.25	24.17	1.10
1.43	1.38	0.189	93.9	1.93	6.7	0.21	0.21	19.29	0.98
0.45	0.83	0.003	32.2	0.10	1.5	0.09	0.09	3.68	0.09
1.08	1.34	0.037	93.1	0.38	5.8	0.16	0.16	14.16	0.26
1.88	1.43	0.222	93.3	2.28	7.8	0.28	0.28	26.35	0.84
1.88	1.42	0.345	100.7	3.28	7.7	0.28	0.28	26.25	1.32
1.88	1.43	0.229	97.3	2.25	7.8	0.28	0.28	26.24	0.87
0.43	1.06	0.065	66.6	0.93	2.6	0.07	0.07	4.51	1.43
1.82	1.41	0.114	104.1	1.05	7.5	0.27	0.27	25.19	0.45
1.72	1.39	0.102	102.6	0.95	7.1	0.26	0.26	23.43	0.44

Referring to tables 1 and 2, a comparison of the results from the experimental tests on the Non-ribbed and Ribbed-SRWTs is made. The results show that the Ribbed-SRWT had faster speeds at comparable flow-rate, Q and hydraulic head, H_j. Although, the differences in speeds of the two turbines are very pronounced, the torque T, of the Ribbed-SRWT was lower and uniform. The torques resulting from running the Non-ribbed-SRWT were averages of the minimum and maximum readings during one particular experimental run.

Static torque T_s, were obtained by tightening the contact of the prony brake and the turbine in such a way that the turbine was prevented from rotating. The torque reading from the torque meter was the maximum torque attainable by the turbine at such a given flow-rate Q and hydraulic head H_j. With dynamic torque T and static torque T_s known, the frictional losses can be calculated. However, since the objective of the study is to investigate the effect of guide ribs, frictional losses were not included in table 1 and table 2.

The differences in the static torques T_s, of the Non-ribbed and Ribbed SRWTs are indicative of the effect of guide ribs at the nozzles of the SRWT. Although the differences in the static discharge coefficients φ_s, of the Non-ribbed and Ribbed SRWTs are small it is noticeable that the Ribbed-SRWT had higher static discharge coefficients. This also indicates the effect of guide ribs at the nozzles of the SRWT. The same is noticeable with data from the two turbines on their dynamic discharge coefficients φ.

Table 2. Representative results from running Ribbed-SRWT.

Ribbed SRWT	Data Using Lutron TQ-8800								
Q (L/sec)	Hj (m)	Power (watts)	N (rev/min)	T (N-cm)	Ts (N-cm)	Dynamic Ø	Static Øs	Input Power Pi (watts)	Hydraulic Efficiency (%)
1.96	1.37	0.14	151.7	0.90	7.9	0.29	0.29	26.36	0.54
1.94	1.37	0.14	150.7	0.87	8.1	0.29	0.29	26.16	0.52
1.94	1.37	0.14	151.4	0.87	8.7	0.29	0.29	26.16	0.53
1.90	1.36	0.14	149.7	0.88	8.3	0.28	0.29	25.31	0.55
1.80	1.35	0.13	148.1	0.87	8.6	0.27	0.27	23.84	0.56
1.47	1.32	0.13	146.8	0.85	7.8	0.22	0.23	19.06	0.69
1.09	1.30	0.13	140.8	0.90	2.2	0.17	0.17	13.99	0.95
1.98	1.37	0.15	152.1	0.92	6.6	0.29	0.30	26.59	0.55
1.97	1.37	0.15	153.9	0.90	7.9	0.29	0.30	26.54	0.55
1.97	1.37	0.14	153.3	0.87	7.4	0.29	0.30	26.43	0.53
0.27	1.07	0.09	101.6	0.87	8.1	0.05	0.05	2.83	3.26
1.92	1.36	0.14	151.8	0.90	3.8	0.29	0.29	25.74	0.56
1.81	1.34	0.14	148.6	0.88	7.9	0.27	0.28	23.79	0.58

IOP Conf. Series: Materials Science and Engineering **88** (2015) 012069 doi:10.1088/1757-899X/88/1/012069

Data from table 1 and table 2 were regressed and the resulting equations plotted to produce figure 8 and figure 9. Figure 8 is a plot of Torque versus Speed (*T vs. N*) for both the Non-ribbed SRWT and the Ribbed-SRWT. The plot for the Non-ribbed-SRWT has a slope that is steeper than the slope of the plot for the Ribbed-SRWT. This indicates that a small increase in speed of the Non-ribbed-SRWT will result to a significant decrease in its torque output. This condition resulted to the erratic speed of the Non-ribbed SRWT. On the other hand, the plot of the Ribbed-SRWT has lesser slope. Thus, torque output of the Ribbed-SRWT was easily controlled by controlling its speed resulting to a smooth operation of the turbine.

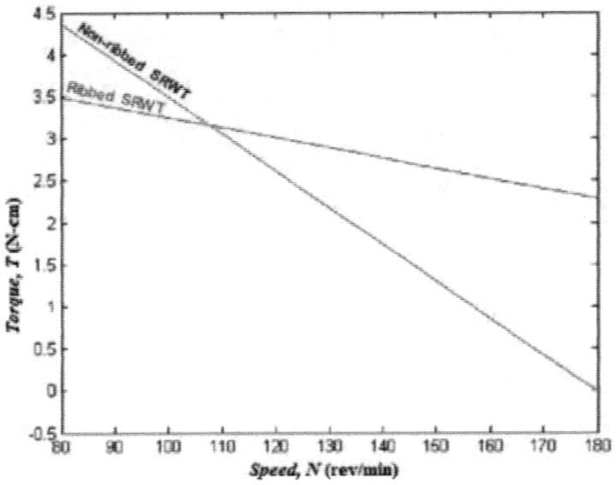

Figure 8. Plot of the Torque vs. Speed @ $Q = 2$ L/sec & $H_j = 1.5$ m.

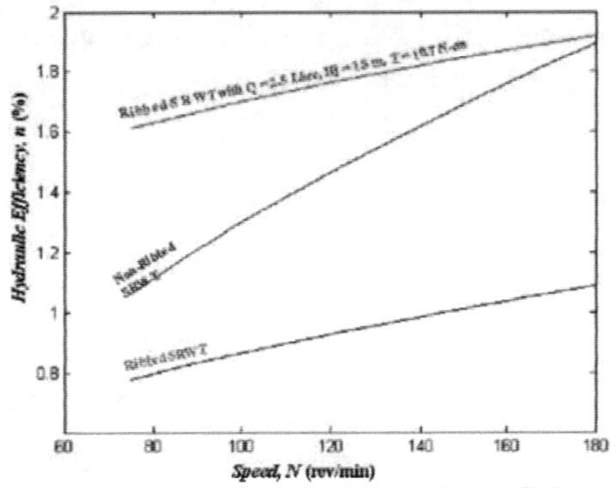

Figure 9. Plot of the Hydraulic Efficiency vs. Speed @ $Q = 2$ L/sec & $H_j = 1.5$ m.

Figure 8, shows clearly the effect of guide ribs in increasing the speed of the Ribbed-SRWT. At a torque of 3.0 *N-cm* the speed of the Ribbed-SRWT was 11% faster than the Non-ribbed-SRWT. On the other hand, at a torque of 2.4 *N-cm* the speed of the Ribbed-SRWT was 41% faster than the Non-ribbed-SRWT. At a much lower torque of 1.0 *N-cm* as shown in table 1 and table 2, the Ribbed-SRWT indicates faster speed than the Non-ribbed SRWT by as much as 46%.

Figure 9, shows that the Ribbed-SRWT attained lower hydraulic efficiencies compared to the Non-ribbed-SRWT. This is due to the requirement for hydraulic reaction turbines to have their runner passages completely filled with water [3]. In this case, the maximum flow-rate of 2 liters/sec of the circulating pump of the hydraulic flume was a limitation in attaining higher hydraulic efficiencies of the Ribbed-SRWT. Besides, at faster speeds of the Ribbed-SRWT the centrifugal effect contributed to demand higher flow-rate in order to completely fill its runner. Upon extrapolation from the data in Table 2 of the Ribbed-SRWT, higher efficiencies can be attained at a flow-rate $Q = 2.5$ liters/sec, hydraulic head $H_j = 1.6$ m resulting to a torque $T = 10.7$ N-cm. The plot is shown in figure 9.

8. Conclusions

The following conclusions were reached based on the data gathered during this study:

1. Guide ribs improved the coefficient of discharge from 0.27 to 0.29 as shown by the data from Non-ribbed SRWT listed in table 1 and from Ribbed-SRWT listed in table 2;
2. Guide ribs improved the speed of the SRWT by as much as 46% as shown by comparing the data from Non-ribbed SRWT listed in table 1 and from Ribbed-SRWT listed in table 2;
3. Installation of guide ribs at the exit nozzles although resulted to increase the speed of the turbine, required high flow-rate Q in order to completely fill the runner of the Ribbed-SRWT;
4. Sharpened lips at entrance to the nozzles of the SRWT reduced the impulse counter to the direction of the rotation of the SRWT. This conclusion is consistent with the results of the computer simulations using Solidworks software.

9. References

[1] Date A and Akbarzadeh A 2010 Design and Analysis of a Split Reaction Water Turbine *Renewable Energy* **35(9)** 1947-1955

[2] Date A and Akbarzadeh A August 2012 (http://www.SciRP.org/journal/sgre) Performance Investigation of a Simple Reaction Water Turbine for Power Generation from Low Head Micro Hydro Resources *Smart Grid and Renewable Energy* 239-245

[3] Daugherty, R L, Franzini J B and Finnemore E John 1989 Fluid Mechanics with Engineering Applications 507 McGraw-Hill Book Company, Singapore

[4] Lipson C and Sheth N 1979 Statistical Design of Engineering Experiments **(3rd Ed.)** McGraw-Hill Book Co., New York

[5] Sayers A T 1992 Hydraulic and Compressible Flow Turbomachines McGraw-Hill Book Company, Singapore

IOP Conf. Series: Materials Science and Engineering **88** (2015) 012070 doi:10.1088/1757-899X/88/1/012070

Experimental Study of Additives on Viscosity biodiesel at Low Temperature

Berkah Fajar[1], Sukarno[2]
[1]Department of Mechanichal Engineering, Diponegoro University, Jl.Prof.Sudarto SH, Tembalang, Central Java, Semarang 50275, Indonesia
[2]Department of Mechanichal Engineering, Akademi Teknik Wacana Manunggal, Plamongsari Raya, Pucang Gading, Semarang 502193, Central Java, Indonesia

E-mail : berkahfajar@undip.ac.id

Abstract. An experimental investigation was performed to find out the viscosity of additive and biodiesel fuel mixture in the temperature range from 283 K to 318 K. Solutions to reduce the viscosity of biodiesel is to add the biodiesel with some additive. The viscosity was measured using a Brookfield Rheometer DV-II. The additives were the generic additive (Diethyl Ether/DDE) and the commercial additive Viscoplex 10-330 CFI. Each biodiesel blends had a concentration of the mixture: 0.0; 0.25; 0.5; 0.75; 1.0; and 1.25% vol. Temperature of biodiesel was controlled from 40°C to 0°C. The viscosity of biodiesel and additive mixture at a constant temperature can be approximated by a polynomial equation and at a constant concentration by exponential equation. The optimum mixture is at 0.75% for diethyl ether and 0.5% for viscoplex.

1. Introduction

Biodiesel is an alternative to diesel fuel in diesel engines made from vegetable oils or animal fats by the transesterification process. The viscosity of biodiesel becomes higher at low temperatures due to crystallization or a gel formation which can cause problems in the fuel line. At low temperature, the gel will clog the fuel filter so the fuel injection pump needed more energy or will be failed. Solutions to reduce the viscosity of biodiesel is to add the biodiesel with some additives.

This study is a continuation of the studies that have been conducted by previous researchers. Udomsap, P. e*t al,* (2008) studied a commercial additive, which was mixed with the biodiesel from CPO. The results showed that the use of commercial additives FA 205 with a concentration of 0.5% vol could reduce the pour point of 4°C (from 18°C to 14°C) [5].

Refaat et al, (2008) also have examined the oil-based biodiesel blending cotton (cottonseed oil methyl ester) with four commercial additives consisting of Technol B100 biodiesel cold flow improver, Gunk premium anti-gel diesel fuel, diesel fuel anti Heet -gel, and Howe's diesel treat conditioner and anti-gel, in which the concentration of each mixture was: 0.2; 0.5; 0.75; and 1% vol. Joshi, H. et al, (2008) stated that the Tecnol B100 was the best additive. The results showed that the use of commercial additives Technol B100 biodiesel cold flow improver at a concentration of 0.2; 0.5; and 0.75 vol% could reduce the pour point by 3°C (from 7 °C to 4°C) [5].

[1] To whom any correspondence should be addressed.

IOP Conf. Series: Materials Science and Engineering **88** (2015) 012070 doi:10.1088/1757-899X/88/1/012070

Rohmax Company (in 2010) is a manufacture of commercial additives Viscoplex 10-330 CFI, which was carried out a research using the commercial additives Viscoplex 10-330 CFI and the biodiesel based from palm oil with various concentrations of 0, 0.2, 0.4, 0.6, 0.8, 1, and 1.2% vol. At a concentration of 10-330 Viscoplex of 1% vol could reduce the pour point the by 5°C (from 11°C to 6°C) [4].

Effect of diethyl ether generic mixing additives on biodiesel JME (Jojoba Methyl Ester) with an average concentration: 5, 10 and 15 vol% has been investigated by Selim, M.Y.E. (2009). Generic additive diethyl ether was selected for very low viscosity (diethyl ether kinematic viscosity of 0.32 cSt at 40 °C). The results showed that the viscosity of the mixture add 5% diethyl ether could reduce viscosity of 16.6 cP to 11.3 cP, adding 10% diethyl ether reduced to 7.7 cP and 15% diethyl ether added declined from 16.6 cP viscosity becomes 5.3 cP.

Sivalakshmi, S. et al, (2010) studied the addition of diethyl ether (DEE) to reduce emissions in particular soot, NOx and particulate emissions. The results showed a significant reduction in NOx emissions primarily to the addition of DEE over 10% vol and a slight decrease compared with JOME pure smoke. It has been found that the addition of 15% volume DEE on JOME improved the performance and the emission [4].

2. Experimental Methods

Biodiesel, diethyl ether and Merck Viscoplex 10-330 CFI (Cold Flow Improver) were used to study the viscosity blends. Table 1 shows the specification of the biodiesel.

Table 1. The specification of Biodiesel.

	Specification	
Density at 15°C	kg/m^3	815-870
Kinematic viscosity at 40°C	mm^2/s	2.0-5.0
Cetane number		≥48-45
Flash point 40°C	°C	≥60
Pour point	°C	≤18
Carbon residue	% mass	≤30
Water content	mg/kg	≤50
T90/95	°C	<370
Oxidation stability	g/m^3	-
Sulphur	%m/m	≤0.35
Total acid number	mg-KOH/g	≤0.6
Ash content	% m/m	≤0.01
Sediment	>% m/m	≤0.01
FAME	% m/m	≤10
Particulate	mg/l	-

The diethyl ether was made by Merck. The diethyl ether specification can be seen on table 2.

Table 2. The specification of diethyl ether.

Specification	
Grade	ACS,ISO,Reag. Ph Eur
Chemical formula	$(C_2H_5)_2O$
HS code	2909 11 00
EC code	200-467-2
Molar mass	74.12 g/mol
Index number EC	603-022-00-4
Number CAS	60-29-7
Ignition temperature	180°C
Solubility	69 g/l (20°C)
Melting point	-116.3°C
Massa molar	74.12 g/mol
Density	0.71 g/cm^3
Boiling point	34.6 °C (1013 kPa)
Vapour pressure	587 kPa (20°C)
Flash point	-40°C
Heating value	8096.32 cal/gr
Kinematic viscosity	0,32 cSt at 40°C

The Cold Flow Improver Viscoplex 10-330 (CFI) is designed to control the formation of wax crystals from various sources of biodiesel feedstock, so it has a very good performance at low temperature. The Viscoplex 10-330 CFI can be used in mixtures with the biodiesel concentration of 0.5% - 1% to increase the cold filter plugging point (CFPP) - (DIN EN 116), pour point (PP) (ASTM D97) and cloud point (CP).

Table 3. The composition of the mixture of biodiesel with diethyl ether and the biodiesel with viscoplex.

Concentration of Generic Additive Diethyl Ether or Viscoplex (%)	Volume of Generic Additive Diethyl Ether or Viscoplex (ml)	Volume of Biodiesel (ml)
0.25	0.04	15.996
0.50	0.08	15.992
0.75	0.12	15.988
1	0.16	15.984
1.25	0.20	15.980

Biodiesel was produced by transesterfication from a mixture of crude palm oil (CPO) and methanol, whereas KOH was as a catalyst. The biodiesel was blended with diethyl ether with the concentration of 0.25%, 0.5%, 0.75%, 1.0% and 1.25%. The Biodiesel was also blended with the

IOP Conf. Series: Materials Science and Engineering **88** (2015) 012070 doi:10.1088/1757-899X/88/1/012070

viscoplex with the concentration the same with the concentration of diethyl ether. The volume of biodiesel, the volume of diethyl ether and the volume of viscoplex can be shown in table 3.

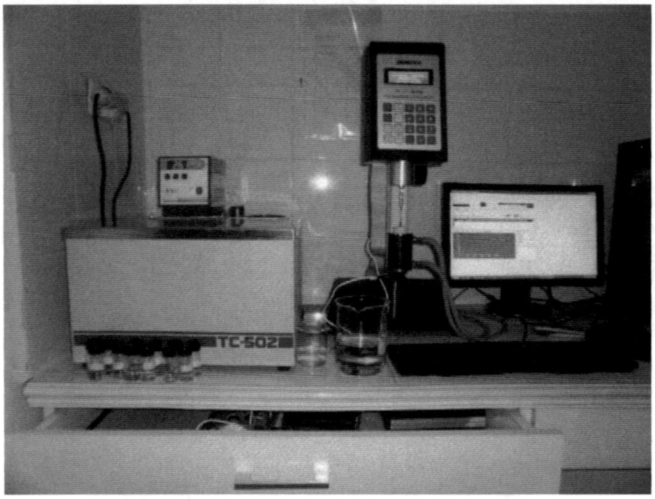

Figure 1. Brookfield DV-III Ultra Rheometer.

The Brookfield DV-III Programmable Rheometer measures fluid parameters of Shear Stress and Viscosity at given Shear Rates. Viscosity is a measure of a fluid's resistance to flow. The principle of operation of the DV-III is to drive a spindle (which is immersed in the test fluid) through a calibrated spring. The viscous drag of the fluid against the spindle is measured by the spring deflection. Spring deflection is measured with a rotary transducer. The measuring range of a DV-III (in centipoise) is determined by the rotational speed of the spindle, the size and shape of the spindle, the container the spindle is rotating in, and the full-scale torque of the calibrated spring.
The specification of the Brookfield III as follows:

- Speed Range: 0-250 RPM, 0.1 RPM increments
- Viscosity Accuracy: ± 1.0% of full-scale range for a specific spindle running at a specific speed.
- Temperature sensing range: - 100°C to 300°C (-148°F to 572°F)
- Temperature accuracy: ± 1.0°C from -100°C to 150°C

  ± 2.0°C from +150°C to 300°C

- Analog Torque Output: 0 - 1 Volt DC (0 - 100% torque)
- Analog Temperature Output: 0 - 4 Volts DC (10mv / °C)

3. Results and discussion

Figure 2 shows the dynamic viscosity as a function of the concentration viscoplex. The unit of the kinematic viscosity is cP (centi Poise) while concentration is a ratio of the volume of viscoplex to the total volume in%. The study was carried out in the range temperature from 282 K up to 318 K and in the range viscoplex concentraion from 0% up to 1.25 %. At first, the viscosity decreases when the concentration decreases and then increases again with the increase in the viscoplex concentration. This case occurs when the temperature is 282 K. Minimum viscoplex viscosity occurs at a concentration of 0.5%. In the range temperature of 283 K and 284 K, the viscosity decreases up to the viscoplex concentration of 1.25. Then these characteristics change when the temperature is greater than 284 K. In this case the viscosity rises with increase in the viscoplex concentration. When the viscoplex concentration reaches 0.5%, the viscosity decreases again. This characteristic can be approximated by a polynomial equation [6]. General polynomial equation can be seen in equation 1.

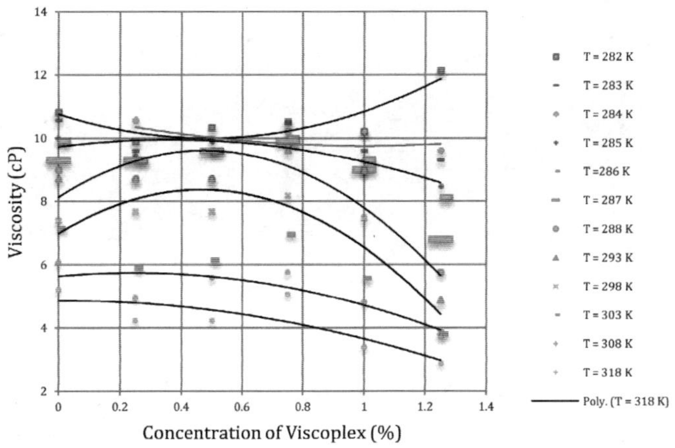

Figure 2. Viscosity blends vs Concentration of Viscoplex.

$$\mu = p_1 C^2 + p_2\, C + p_3 \qquad (1)$$

where μ is the dynamic viscosity in cP, C is the volume percentage (vol.%) of the biodiesel blend; p_1, p_2 and p_3 are the adjustable parameters. The adjustable parameter can be seen on table 4. The transition characteristic of biodiesel blends can be be seen in table 4. The changing of p1 indicates the characteristic transistion. The value of p_1 changes from negative to positive value. If the p1 are negative, the each curve of polynomial equation has a maximum viscosity. Whereas the p1 are positive, the polynomial curves have a minimum viscosity.

Figure 3 shows the dynamic viscosity as a function of the diethyl ether concentration. The each polynomial curve is indicated that has a minimum viscosity. Those minimum viscosities occur at the diethyl ether concentration of 0.75 %. The parameter of p_1, p_2 and p_3 can be seen on table 4. The p1 is always positive therefore the each curve has a minimum viscosity.

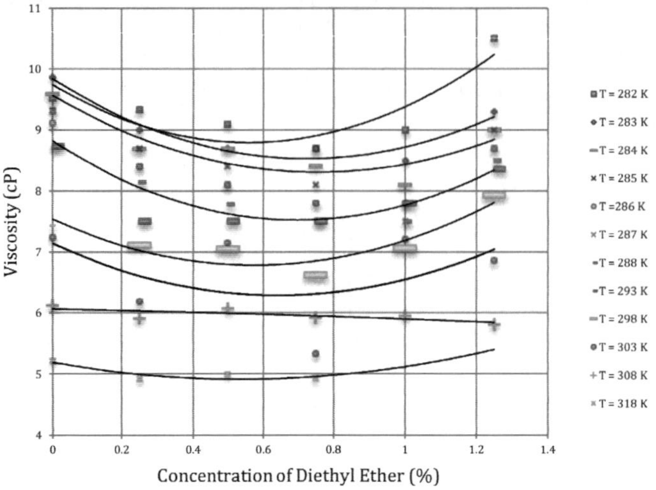

Figure 3. Viscosity blends vs Concentration of Diethyl Ether.

Table 4. The empirical correlation of biodiesel blends at constant temperature.

$\mu = P_1 C^2 + P_2 C + P_3$

No	T (K)	Diethyl Ether P_1	Diethyl Ether P_2	Diethyl Ether P_3	Viscoplex P_1	Viscoplex P_2	Viscoplex P_3
1	282	3.02	-3.378	9.7416	3.229	-3.1406	10.761
2	283	2.4943	3.6127	9.8399	-0.5314	0.0026	10.208
3	284	2.2143	-3.345	9.5704	-0.3429	0.3943	10.243
4	285	2.604	-3.7946	9.4301	-1.7314	1.2489	9.7364
5	286	2.6543	-3.7967	9.1719	-1.9886	1.5017	9.5657
6	287	3.0171	-4.464	9.3634	-4.5371	4.1651	8.9769
7	288	2.6863	-3.7412	8.8242	-5.5857	5.4027	8.3848
8	293	2.9086	-3.7477	8.57533	-6.5743	6.323	8.1315
9	298	2.2674	-2.617	7.538	- 6.3983	5.9567	6.9821
10	303	2.0566	-2.6491	7.1418	-2.6097	1.3346	6.5794
11	308	0.046	-0.1199	6.0641	-1.789	0.8696	5.6369
12	318	0.96	1.0336	5.1906	-1.22	0.061	4.8688

The study was carried out using a mixture of biodiesel concentration from 0.25 % up to 1.25%. From figures 1 and 2 can be seen that there are two difference characteristics. Biodiesel blends and diethyl ether has a viscosity minimum of around 0.6 %, while a mixture of biodiesel and viscoplex have a maximum of 0.5 %. But viscoplex biodiesel blends have the same characteristics as biodiesel and diethyl ether mixture at low temperatures (less than 284 K), therefore a mixture of biodiesel with diethyl ether selected for a concentration of 0.5% and 0.75%, while the mixture of biodiesel with viscoplex selected for the concentration of 0.5% and 1.25%.

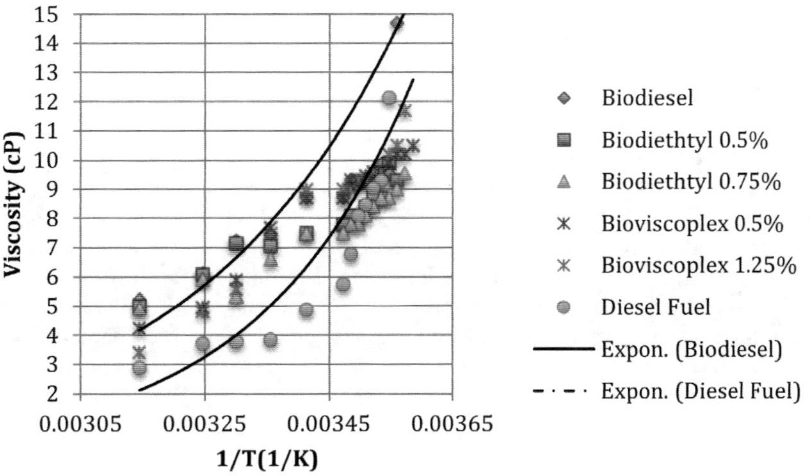

Figure 4. Viscosity of biodiesel blends as function temperature.

The dashed line in Figure 4 shows the trend line for diesel oil and the continous line shows the trendline for biodiesel. When the temperature is above 298 K or less than 0.0345 K^{-1}, all the mixture has a viscosity of biodiesel greater than the diesel fuel viscosity. Even a mixture of biodiesel and diethyl ether has viscosities greater then the pure biodiesel viscosity. But overall viscosity value is still smaller the allowable viscosity of 4.8 cP.

When the temperature is below 298 K or greater than 0.0345 K^{-1}, all the viscosities of the mixture are below the trendline diesel fuel line. The Lowest viscosity of the biodiesel blends and viscoplex occurred at concentrations of 1.25%, while for the biodiesel and diethyl ether mixture occurs at a concentration of 0.75%.

The empirical correlation of biodiesel blends at constant concentration could be written as [6]:

$$\mu = p_1 \exp\left(\frac{p_2}{T}\right) \quad (2)$$

where μ is the dynamic viscosity in cP, T is the absolut temperature in K of the biodiesel blend, p1and p_2 are the adjustable parameters. The empirical correlation of biodiesel blends at constant concentration can be seen on table 5.

Table 5. The empirical correlation of biodiesel blends at constant concentration.

Concentration of Diethyl Ether (%)	$\mu=p_1\exp(p_2/T)$	
	p_1	p_2
0,25	2.10^{-5}	3807.7
0,5	6.10^{-5}	348,7
0,75	0.0019	2428.4
1,0	8.10^{-6}	4093.6
1,25	7.10^{-6}	4154.3
Concentration of Viscoplex (%)	$\mu=C_1\exp(C_2/T)$	
	C_1	C_2
0,25	9.10^{-6}	4091.1
0,5	0.006	2097.2

IOP Conf. Series: Materials Science and Engineering **88** (2015) 012070 doi:10.1088/1757-899X/88/1/012070

0,75	8.10^{-5}	3418.9
1,0	0,0002	3219.1
1,25	0,0003	3003,6

4. Conclusions

The addition of additives into biodiesel can improve the viscosity at low temperatures. In this study was used two kinds additivetive, namely the diethyl ether and the viscoplex (commercial additive). The mixtures have a different character to the change in concentration at a constant temperature. Biodiesel blends and diethyl ether have a minimum viscosity at a concentration of 0.75%, while the mixture of viscosity of biodiesel and viscoplex have maximum at concentration of 0.5%. But mixtures of biodiesel and viscoplex have the same characteristics with the characteristics of biodiesel and diethyl ether mixture at temperatures below 284 K. This mixture has a minimum value at a concentration of 0.5%. This characteristic can be approximated by a polynomial equation: $\mu = p_1 C^2 + p_2 C + p_3$.

The empirical correlation to predict the viscosity of biodiesel blends at various Temperatures, which is characterized by exponential equation. The optimum mixture is at 0.75% for diethyl ether and 0.5% for viscoplex. This characteristic can be approximated by a exponential equation: $\mu = p_1 \exp\left(\frac{p_2}{T}\right)$

Reference

[1] Refaat, Joshi M, and Michael J 2007 Flow properties of biodiesel fuel blends at low temperatures *J.Fuel* **86** 143–151.
[2] Thomas TP, Birney BM and Auld DL 2011 Viscosity reduction of castor oilesters by the addition of diesel safflower oil esters and additives *J Industrial Crop and Product* **36** 267-270
[3] Selim, MYE 2009 Reducing the viscosity of Jojoba Methyl Ester diesel fuel and effects on diesel engine performance and roughness *J. Energy Conversion and Management* **50** 1781–1788
[4] Sivalakshmi S and Balusamy T 2010 Research on Di-ethyl Ether as an Oxygenated Additive with Biodiesel in CI Engine PEA-AIT *International Conference on Energy and Sustainable Development: Issues and Strategies (ESD 2010)*, The Empress Hotel, Chiang Mai, Thailand, 1–7.
[5] Udomsap P, Sahapatsombat U, Puttasawat B, Krasae P, Chollacoop N and Topaiboul S 2008 Preliminary Investigation of Cold Flow Improvers for Palm-Derived Biodiesel Blends Journal of Metals, Materials and Minerals **18(2)** 99–102
[6] Verduzo LPR, Flores BEG, Rodrigues ER, Jacob ARJ 2011 Prediction of the density and viscosity in biodiesel blends at various temperature *J Fuel* **90** 1751-1761

Feasibility study on pliant media drying using fluidized bed dryer

J H Zakaria[1], M H H M Zaid[1], M F M Batcha[1,2] and N Asmuin[1,3]
[1] Centre for Energy and Industrial Environment Studies,
Universiti Tun Hussein Onn Malaysia (UTHM), Parit Raja,
86400 Batu Pahat, Johor, Malaysia.

Email: [2]mfaizal@uthm.edu.my, [3]norzela@uthm.edu.my

Abstract. The usage of pliant media for blasting in surface preparation has gained substantial interest in various industries, particularly oil and gas. Being a clean technology, this relatively new method of surface preparation has become an alternative to conventional abrasive blasting technique which lowers fugitive emissions from blasting process and hence lowering risk to workers in the industry. Despite proven to be effective and cost efficient, the usage of pliant media in tropical climate poses a new challenge due to the torrential rain in the monsoon season. During rainy and wet conditions, the pliant media was literally soaked and the recovery rate of the pliant media for a continuous blasting becomes retarded. A viable technique for drying of this pliant media has then become imperative. The present study proposes to dry water laden pliant media in a Swirling Fluidized Bed Dryer (SFBD). In this preliminary study, three bed loadings of 1.7, 2.0 and 2.3 kg of pliant media was dried in the SFBD at 80°C, 90°C and 100°C. The experimental works revealed that the SFBD has shown excellent potential to dry the pliant media with a relatively short drying time. The behaviour of moisture ratio and drying rate against time are discussed. The findings conclude that the SFBD is a feasible technique for wet pliant media drying and can be extended for continuous processing system.

Keywords: Pliant media drying, Swirling Fluidized Bed Dryer (SFBD), moisture ratio, drying rate

1. Introduction

Surface preparation using pliant media is gaining substantial interest in many industries in Malaysia where process is required on annual basis, such as the oil rig platform, material processing plants and during maintenance of heavy machines. Unlike traditional practice in surface preparation where abrasive materials were used, utilization of pliant media results in reduction of air borne dust to almost 99% [1]. Hence, operator gains better visibility during blasting, especially while working in constrained areas. More importantly, the physically soft blasting medium ensures less rebound energy which reduces surfaces damages especially sensitive ones. As a result, overall productivity of blasting was increased while less impact on the environment was produced.

[1] To whom any correspondence should be addressed.

The concept of pliant media blasting is shown in Figure 1. First, high pressure air from compressor was directed into the blasting nozzle containing pliant media. When this blasting nozzle was directed towards the surface to be prepared, the pliant media was released at high velocity. As a result, the chunks of pliant media hit the surface, and transfer the impact energy. In a matter of few micro-seconds, the pliant media flattens and suppresses the surface to scrap the unwanted surface before falling on the ground with the debris.

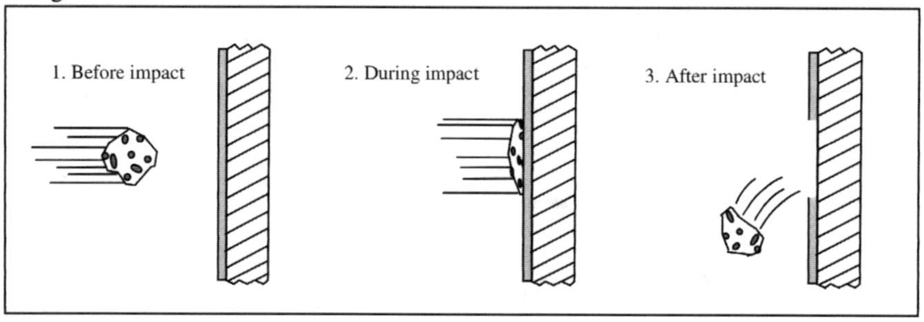

Figure 1. The pliant media blasting concept.

Despite the benefits and efficiency of the blasting process, pliant media is strictly limited for usage on dry conditions only. Being very high in porosity, the pliant media readily absorbs moisture hence prohibiting the usage in wet conditions. These limitation causes surface preparation during monsoon season becomes very difficult and challenging as blasted pliant media could not be recycled readily before drying. Currently, operators use large inventory of dry pliant media for surface preparation in wet conditions since wetted media cannot be used in the existing equipment. This leads to loss of precious operation time as well as increase in pliant media usage. Treating this drawback as a problem to solve, the present study explores the feasibility of using a Swirling Fluidized Bed Dryer (SFBD) to dry the wetted pliant media. Batch experiments were conducted using three bed weights of 1.7 kg, 2.0 kg and 2.3 kg at three drying air temperatures of 80°C, 90°C and 100°C. Prior to this, the hydrodynamics of the SFBD operating with pliant media was investigated. Regimes of operation were classified and minimum fluidization velocity, U_{mf} were recorded.

2. The swirling fluidized bed dryer (SFBD) system

The swirling fluidized bed (SFB) dryer which was proposed in this study for pliant media drying, is a relatively new method in drying. The SFBD system was made by an array of 60 blades which resembles a stator configuration in a gas turbine system. These blades which were inclined at 12° from the horizontal plane form an annular bed as a centre body is placed in the core of the system as shown in Figure 2.

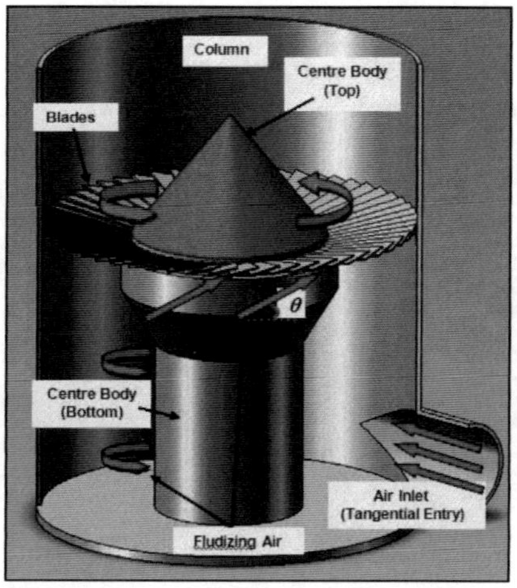

Figure 2. SFB configuration [2].

The centre body was necessary to avoid dead zone in the bed [3]. When hot air flows through the inclined distributor, two jet velocity components created in the bed; the vertical component which fluidizes the bed while the horizontal component which induce swirling. The magnitude of each component depends on the blade inclination. Lower blade angles were preferred as they provide larger horizontal momentum and hence faster swirling with reduced elutriation due to smaller vertical momentum [3,4]. Previous study by [4 – 6] reported that there can be three or four operational regimes in this bed. The most apparent of this bed is its low distributor pressure drop and vigorous mixing during operation. A study on wheat grains due to swirling effect in fluidized bed dryer (FBD) was reported by [7]. Moisture extraction rate and dryer efficiency were increased when swirl motion up to 5-25% and 38% when compared with non-swirl motion. Thomas and Varma [8] studied the drying behaviour of granular materials experimentally for different temperatures, velocity, particle sizes and mass of solids and concluded the critical moisture content was depending on all the experimental parameters especially temperature. Abid et. al [9] studied the effect of humidity and velocity on corn kernels drying rate and reported both effects were less significant in comparison to drying temperature. In this study, batch experiments were conducted using three bed weights of 1.7 kg, 2.0 kg and 2.3 kg at three drying air temperatures of 80°C, 90°C and 100°C. Prior to this, the hydrodynamics of the SFBD operating with pliant media was investigated. Regimes of operation were classified and minimum fluidization velocity, U_{mf} were recorded

3. Methodology

Experiments were started with loading column or drying chamber or bed with wet sponge media with 30% till 40% moisture content. Then, frequency converter (model Holip-HL PA07543B) was switched on to supply air flow and must satisfy the required flowrate for U_{mf} for each loadings. The air flow were accelerated to the SFBD system by blower and heated by electrical heater. This means that, hot jet air flows were used to dry the sponge media which following prepared design of experiment as shown in Table 1. About 3 – 5 grams of the sponge media samples were taken from the bed and were weighed using electronic weight balance model Perkin-Elmer with an accuracy of ± 10 mg before putting it into industrial oven model Memmert to dry the samples at 105°C. The samples were taken at 0th until 20th minute with 2 minutes time interval. After 3 hours, the samples were taken out from the oven and were weighed once again to determine the samples final weight. Figure 3 shown the schematic diagram of the experimental and SFBD setup while Table 1 below shown the design of experiment in order to conduct the experiments.

IOP Conf. Series: Materials Science and Engineering **88** (2015) 012071 doi:10.1088/1757-899X/88/1/012071

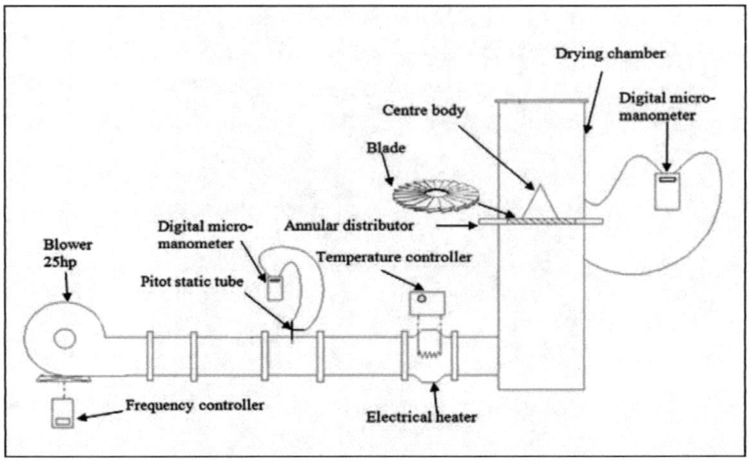

Figure 3. Schematics of the experiment apparatus setup and SFBD setup

Table 1. The pliant media drying experiment parameters.

Temperature, T (°C)	Weight (kg)	Minimum Fluidization Velocity, U_{mf} (m/s)
80, 90, 100	1.7	1.25
80, 90, 100	2.0	1.29
80, 90, 100	2.3	1.37

4. Results and discussion

4.1 Hydrodynamics of pliant media

Hydrodynamic characteristics are important to be determined in any system involving as it gives an insight to the system behaviour as well as its resistance on flow. Both of this finding are useful in estimating the system's performance as well as related cost involved, which will finally analyzed through benefit-cost analysis. In the present study, three most important hydrodynamic characteristics are studied, namely the distributor pressure drop,ΔP_d, bed pressure drop,ΔP_b and minimum swirling velocity, U_{ms}. ΔP_d indicates the amount of kinetic energy loss (converted the potential energy) when the system operates without any pliant media inside (base resistance towards flow) as presented in Figure 4. From the figure, the ΔP_d for the SFBD increases exponentially with superficial velocity which is typical for fluidized bed's distributor. The interesting finding however, is the magnitude of pressure drop which is relatively smaller in comparison with conventional distributors. For instance, Agarwal et. al [11] proposed that ΔP_d should never be less than about 3400 Pa are required to ensure good fluidization characteristics. In the present study, the highest recorded value of ΔP_d yet capable of providing desired fluidization condition.

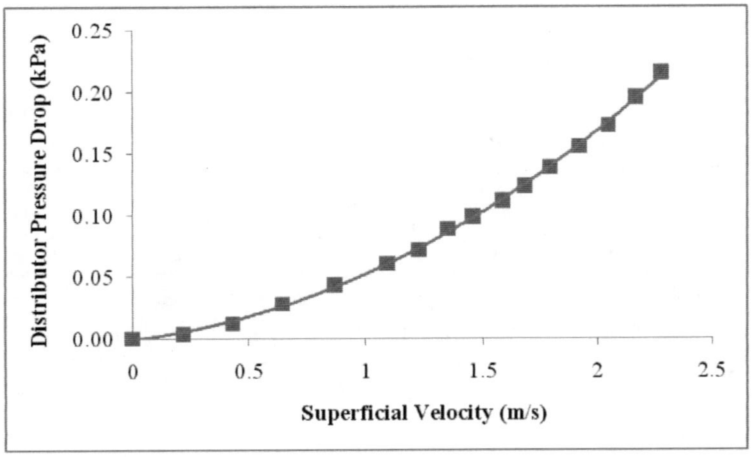

Figure 4. Distributor pressure drop, ΔP_d, against superficial velocity, V_s

Apart from ΔP_d, ΔP_b and U_{mf} in are shown in Figure 5 where ΔP_b is plotted against superficial velocity. In the packed condition (region I), the ΔP_b increased exponentially before achieving U_{mf} condition in region II at about 1.25 m/s. The exponential trend is because the pliant media is made by coarse particles and thus large voidage between particles. As a result, the flow inside the voidage becomes turbulent and resulting the exponential trend, as discussed by [10].

Upon U_{mf}, pressure fluctuations are recorded in region II where random bubbles were observed. Increasing the superficial velocity of air even further, the whole begins to swirl as a single mass at about V_s = 2.15 m/s and finally elutriation occurred beyond V_s = 2.52 m/s.

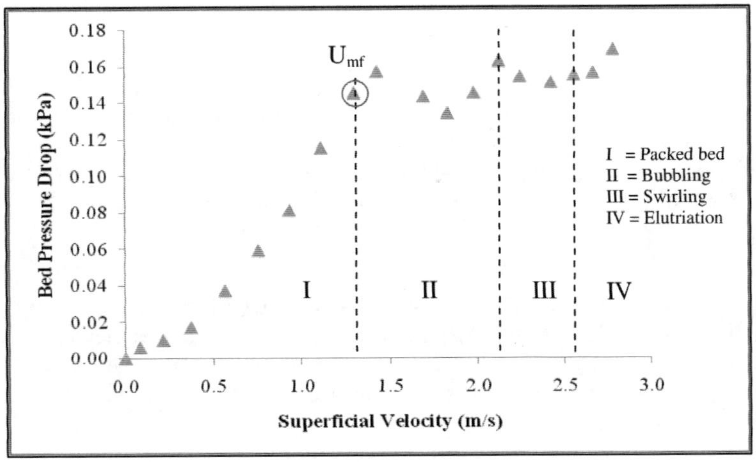

Figure 5. Bed pressure drop, ΔP_b against superficial velocity, V_s for 1.7 kg bed

The regimes of operation are constant and changes as the particle becomes lighter and lighter caused by continuous drying process acting on it. The differences of swirling pattern were caused by the weight changes with the respected to the time. The swirling patterns observed such as below [3,5]:

a) Bubbling: in the beginning of the drying process, after packed condition, the bed was to have small bubbles at certain location and the bubbles grow at this location only with the increase in the drying air flowrate. This phenomena is known as channelling as per described by [10].

b) Two layer fluidization: at this phase, lower layer particle having swirling motion while upper layer still having bubbling condition. At this regime, velocity tangential velocity component was decayed and at certain height, it is insufficient to create any swirling in the bed.
c) Fully swirling: when the particle becomes drier, stable swirling pattern was finally reached. In this phase, swirling velocity eventually becomes higher as a result of lower bed weight as a result of continuous moisture transfer.

4.2 Moisture ratio and drying rate

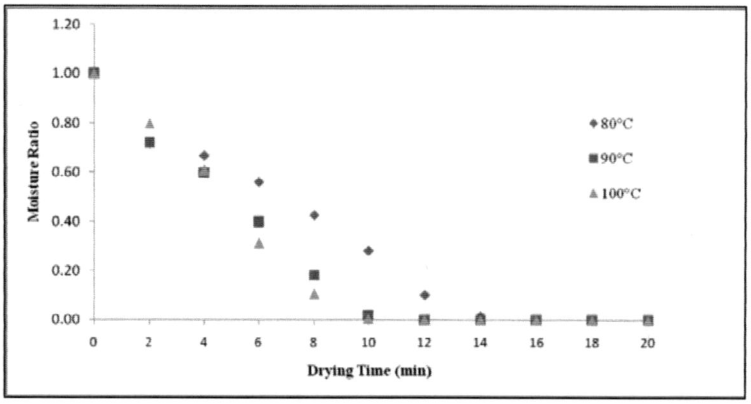

Figure 6. Temperature comparison of moisture ratio versus drying time for 2.0kg bed loads.

Figure 6 above shown temperature comparison between 80°C, 90°C and 100°C of moisture ratio against drying time for 2.0kg bed loads graph. The plotted graph has shown that, higher temperature extracted more moisture from the particle than less temperature. The plotted graph in the figure has shown most contrast comparison started 6th minute till the particle have moisture loss. In short, hotter air flow leads to the higher moisture loss of the sponge media. Apart from that, fast moisture removal was observed for all drying temperatures, with typical drying type of about 10 minutes for all drying temperatures. This is due to the high porosity of the sponge media apart from good contact with air through vigorous mixing in the bed. The drying rate however, remained almost steady throughout the drying process. The total drying time was ideal for actual industrial process which requires rapid drying to ensure high productivity in the surface preparation processes.

4. Conclusion

Experimental results and discussion proves that the capabilities of drying the sponge media through SFBD method. It also shown that drying the particle became faster when higher temperature was used. In the other hand, the sponge media swirling pattern depends on the particle moisture content. The bed undergoes several notable regions, designated at the packed region, minimum fluidization, swirling region and finally elutriation.

5. Acknowledgement

Authors would like to thanks to Deleum Primera Sdn. Bhd. for providing sponge media for the experiments.

6. References

[1] http://www.spongejet.com/index.php (accessed: 23/03/2014)
[2] Mohideen M F, Sreenivasan B, Sulaiman S A and Raghavan V R 2012 Heat Transfer In A Swirling Fluidized Bed With Geldart Type-D Particles *Korean J. Chem. Eng.* **29(7)** 862-867

[3] Mohideen M F, Faiz M, Salleh H., Zakaria H and Raghavan V R 2011 Drying of Oil Palm Frond via Swirling Fluidization Technique *Proc. of World Congress on Engineering* **3** 2375-2380
[4] Mohideen M F, Md Seri S and Raghavan V R 2012 Fluidization of Geldart Type-D Particles in a Swirling Fluidized Bed *Applied Mechanics and Materials* **110-116** 3720-3727
[5] Kaewklum R and Kuprianov V I 2010 Experimental Studies on a Novel Swirling Fluidized Bed Combustor Using an Annular Spiral Air Distributor *Fuel* **89** 43-52
[6] Shu, J, Lakshmanan, V I and Dodson CE 2000 Hydrodynamic Study of a Toroidal Fluidized Bed Reactor *Chemical Engineering and Processing* **39** 499-506
[7] Özbey M and Söylemez M S 2005 Effect of Swirling Flow on Fluidized Bed Drying of Wheat Grains *Energy Conversion and Management* **46** 1495-1512
[8] Thomas P P and Varma YBG 1992 Fluidized Bed Drying of Granular Food Materials *Powder Technology* **69** 213-22
[9] Abid M, Gilbert R and Laguire C 1990 An Experimental and Theoretical Analysis of the Mechanisms of Heat and Mass Transfer During the Drying of Corn Grains in a Fluidized Bed *International Chemical Engineering* **30(4)** 632-42
[10] Kunii D and Levenspiel O 1991 *Fluidization Engineering, 2nd Ed.* Butterworth-Heinemann
[11] Agarwal, J C, Davis, W L and King, D T 1968 Fluidized-bed coal dryer *Chemical Eng. Proc.* **58** 85

Characteristics of electricity production by metallic and non-metallic anodes immersed in mud sediment using sediment microbial fuel cell

N Haque[1], D Cho[2], S Kwon[1]*
[1]Department of Ocean System Engineering, Gyeongsang National University, Cheondaegukchi- Gil 38, Tongyeong, Gyeongnam 650-160, South Korea
[2]Department of Energy and Environmental Engineering, Soonchunhyang University, 22, Soonchunhyang-ro, Asan-si, Chungnam, 336-745, South Korea

E-mail: shkwon@gnu.ac.kr

Abstract. Sediment microbial fuel cell (SMFC), equipped with Zn, Al, Cu, Fe or graphite felt (GF) anode and marine sediment, was performed. Graphite felt was used as a common cathode. SMFC was single chambered and did not use any redox mediator. The aim of this work was to find efficient anodic material. Oxidation reduction potential (ORP), cell voltage, current density, power density, pH and chemical oxygen demand (COD) were measured for SMFC's performance. The order of maximum power density was 913 mWm^{-2} for Zn, 646 mWm^{-2} for Fe, 387.8 mWm^{-2} for Cu, 266 mWm^{-2} for Al, and 127 mWm^{-2} for GF. The current density over voltage was found to be strongly correlated one another in most metal electrodes but the graphite felt electrode, in which relatively weaker electricity was observed because of its bio-oriented mechanism. Metal corrosion reactions and/or a complicated microbial electron transfer mechanism acting around the anodic compartment may facilitate to generate electricity. We presume that more sophisticated selection of anodic material can lead to better performance in SMFC.

1. Introduction

Microbial fuel cell (MFC) systems have been of great interest as a potential applicant for future alternative energy production where employ microbes to generate electricity from biochemical reaction of organic and inorganic substances [1]. These substances are converted into electricity in the anode of MFC by microbial metabolism (via the action of bacteria as catalysts) [2, 3]. The anodic part of SMFC is important in attempt to (A) determine the most efficient microorganisms, those that can offer the highest rate of oxidation or able to extract the highest number of electrons per mole of the substrate [4-6]; (B) study the effectiveness of redox mediators [7, 8]; (C) select more effective electrode materials [9-11]; (D) determine the most efficient anodic reactions, those producing the highest number of electrons per unit weight of the reactant [12-13]. Specific bacteria, including *Geobacteraceae,* were enriched on the anode where organic matter was oxidized and electrons were directly transferred to the anode under anaerobic conditions [14]. Bacteria capable of electricity generation from enriched domestic wastewater [15], ocean sediments [14], animal wastes [16], and anaerobic sewage sludge [17, 18]. Several factors affect MFC performance including the microbial inoculum, chemical substrate (fuel), electrode materials, cell internal and external resistance, solution ionic strength, and electrode spacing, type of proton exchange material (and the absence of this material) [19-21].

[1] To whom any correspondence should be addressed.

However, to increase the anode performance, different chemical and physical strategies have been followed. Park et al. [22] incorporated Mn(IV) and Fe(III) and used covalently linked neutral red to mediate the electron transfer to the anode. Electro-catalytic materials such as polyanilins/Pt composites have also been shown to improve the current generation through assisting the direct oxidation of microbial metabolites [23-25]. Each of these compounds (mediators) has a different potential that possesses in the oxidation-reduction reaction, and their differences result in generate electricity. The difference in the chance of contact between electron shuttles' and anode electrode can also cause the variation in electricity generation. In other views, the higher the concentrations of electron shuttles in the anode phase, the more the electricity generation [26]. SMFC design must be made if SMFC systems are to be used for wastewater treatment. In a two chamber SMFC system, anode and cathode separated by proton exchange membrane. Proton exchange membranes (e.g. Nafion) are not suitable if SMFC systems are applied for wastewater treatment processing because of cost. Finally, a general use of the SMFC might be the removal of organic matter from sediments or other polluted sites. As both the bio-electrochemical activity and microbial metal reducing activity appear to share the same or similar electron transport chains, electrochemical activity of microorganisms may play an important role in the biogeochemical cycling of carbon, nitrogen, metal other organic contaminants. Indeed, the concept of mediator-less MFC studies provide new insights into the function of electrochemical activity bacteria directly associated with practical needs of environmental protection. In past years, environmental pollution control has mainly relied on how fast and feasible process could operate to treat environmental pollutants. The challenges facing the community are many. Every aspect of the SMFC operation can be regarded as sub-optimal at this point, including the anode, the cathode, the membrane, and SMFC design. While these challenges loom as great, the opportunities are equally great. The specific objective in this study was to evaluate the possibility to choose a suitable anode material for field applications by using laboratory experiments as well as to increase power generation. Cost effective SMFC systems can offer a potentially promising remediation technology that also earns surplus energy.

2. Materials and methods

2.1. Sediment collection

Estuarine sediment was used for microbial energy generation. The sediment was collected from the bottom (approximately 10 cm from the sediment–water interface) of Lake Sap-Kyo(N 36°52'9.3", E 126°50'29.12") in a southern province in South Korea. The sampling was performed using a Ponar type grab sampler (2.4-L). All samples including surface water were placed into clean polycarbonate jars (Nalgene, Fisher Scientific) with no headspace gas (i.e. no air) and transported to the laboratory in a cooler box with ice packs. All sediments were passed through a 2-mm sieve to remove plant debris, macro fauna, and other large terrestrial leaves and then homogenized by mixing with a stainless steel spatula prior to use. All samples were kept at 4°C before use.

2.2. Establishing sediment microbial fuel cell (SMFC)

A one-chambered sediment cell was established to analyze the performance of different electrodes. The body of the SMFC was a 500-mL Pyrex beaker. Aliquots of 50-g wet sediment and 120-mL of sea-water were loaded into the lower and upper part of sediment MFC, respectively. The SMFC was monitored for 72 hours. During the runs, water loss due to evaporation was compensated daily by adding distilled water.

2.3. Establishing sediment microbial fuel cell (SMFC)

SMFC, equipping different metallic anodes, was used in respect of graphite felt as common cathode. The electrodes used in this experiment were Zinc (Zn), Aluminium (Al), Copper (Cu), Iron (Fe), and Graphite felt (GF). The dimension of the anode was 150×45×3 mm (length×width×thickness). The cathode (40×35×3 mm) was bare graphite felt and placed parallel to the anode and 4 cm above the sediment–water interface. Both anode and cathode were connected by a platinum wire (internal resistance 20Ω) and an insulated copper wire to an external load. All electrodes had a projected surface area of 0.002 m^2.

2.4. Electricity measurement

Redox potentials were measured using an Ag/AgCl reference electrode (9678BNWP, Thermo Orion, Beverly, MA, USA). The pH was measured using an automatic calibrated pH meter (Model pH 20, Hanna Instruments, USA). Cell voltage was recorded using a multi-meter and a data acquisition system (Model 2700, Keithley Instruments, Cleveland, OH, USA). Current density, i, was calculated as i=I/A=V/RA, where V (mV) is the voltage, I (mA) the current in electrochemical tests, R(Ω) the external resistance, and A(m^2) the projected surface area of the studied electrode. Power density was calculated according to P (mWm^{-2}) =10×iV (10 needed for the given units). Columbic efficiency was calculated as CE $=C_p/C_{th}\bullet 100\%$, where C_p is the total coulombs calculated by integrating the current over time, and C_{th} is the theoretical amount of coulombs available based on the Chemical Oxidant Demand (COD) removed in the SMFC. Current density was also calculated by dividing the current by the apparent surface area of anode.

3. Results and discussion

3.1. Redox potentials of SMFC

For each metallic anode has recorded oxidation reduction potential (ORP) of SMFC for both chambers (anode and cathode) with time. Four types (Zn, Al, Cu, Fe) of metals and graphite felt as non-metallic electrode were examined here. Anodic part was not modified any mediator but covered naturally microbial environment condition. As recent studies show, a number of bacteria have been found to possess the ability to transfer electrons from oxidized fuel (substrate) to a working electrode without a mediator [27, 28], making it possible to establish mediator-less MFCs. To date, many metal reducing bacteria, which are capable of the reduction of solid metal oxides ability including *Clostridium Butyricum* [29], *Geobacter sulfurreducens* [30], *Rhodofoferax ferrireducens* [1] and *Shewanella puterfaciens* [31]. Furthermore, in the SMFC system, growth and metabolism of *S. puterfaciens* were dependent on the presence of active anode-in essence, it served as the electron acceptor for growth and metabolism. Anodic potentials in our cell ranged 184 to -155 mV for Zn, 155 to -150 mV for Al; 148 to -126 mV for Cu; 161 to -42 mV for Fe and 178 to -5 mV for Graphite felt followed for 72 h, whereas the cathodic potentials showed a slow drift. Metallic anode has shown larger ORP change than that in non-metallic graphite electrode. The results in figure 1 show that as time progresses the potential of the anode tends to decrease. It might be due to corrosion decreases over time for changes in the properties of the electrodes' surfaces resulting from biofilm buildup and polarization effect. Generally biofilm deposit promotes corrosion in aerated water not in anaerobic water. Oxidation reduction potential changes were higher for metallic anode than for non-metals because of variation in corrosion rate. Sufficient microbial growth guarantees formation of a certain thickness of biofilm. According to electromotive series Al is more reactive than Zn, and Cu metal is much lower than Fe in potential value. The potential of Al is higher than that of Zn, which is attributed to its microstructure as well as slight changes with type, concentration, and temperature around the metal. In case of Fe and Cu, it might be due to overvoltage (producing hydrogen from H_2S) phenomena in Fe which is not absence in Cu. The microbial fuel cell potential reported in the literature varies widely depending on the type of anodic and cathodic efficiencies of the cell used. The anode reaction potentials reported in the literature vary between -300 and -500 mV_{SCE} [19, 30, and 32]. The open circuit potential (OCP) of the plain graphite anode, iron containing anodes, and manganese containing electrodes are different from each other, indicating that different electron transfer reactions occur at each of these anode type[24].

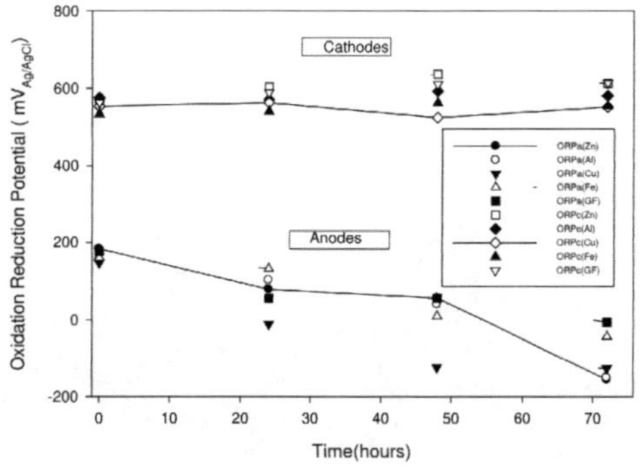

Figure 1. Oxidation reduction potential (ORP) variations in the cathode and anode (against Ag/AgCl) in our sediment microbial fuel cell.

3.2. Power density of SMFC

From the measurements repeated several times, power density of SMFC using different metallic and non-metallic anode was depicted in figure 2. However, SMFCs were started with variety of anode distinctly and was run till 72 hours. We used Zn, Al, Cu, Fe, and graphite felt (GF) as anodic material. Once operated with metallic and non-metallic anode, the deployed fuel cell has achieved its maximum power density of Zn (913 mWm^{-2}), Al (266 mWm^{-2}), Cu (387.8 mWm^{-2}), Fe (646 mWm^{-2}), GF (127 mWm^{-2}) varying with time. At beginning time of SMFC, the power density decreases over time significantly (10 h) for all anodes but interestingly Zn anode was found to increase after that time. Increasing trends of power density of Zn anode have sustained till 45 hours after that drop. Power densities were shown to be different for different anodes probably because of some various catalytic interactions on the electrode surface. Indeed, power densities of metallic anodes were higher than non-metal graphite felt, because corrosion and microbial activity provide electrons to metal whereas only microbial activity provides electrons to non-metal surface. Bacterial adhesion capacity to electrode might cause variation in power density. And the total area of the electrodes including the internal surface area [22], apparent surface area [32, 33], or projected area [19, 34] has been adopted computation of power density in MFCs. Schroder et al. [10] obtained a power density of 6.0 Wm^{-2} for polyaniline–modified platinum as an anode. From literature, it is reported that mediated anode produces higher power than non-mediated one. A similar fuel cell using a plain graphite anode could sustain maximum power of ~20 mW/m^2 (~66 mA/m^2) at 0.30 V while AQDS modified graphite electrodes initially produced five-time greater power, and then the performance gradually decayed over time [14].

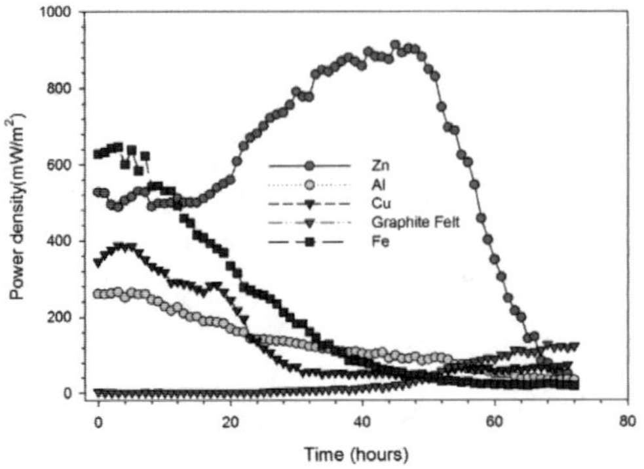

Figure 2. Power density of SMFC according to different anode with respect to time.

3.3. Cell voltage and current density of SMFC

Figure 3 presents the cell voltage and figure 4 represents the current density of SMFC with adding neither external substrates nor electron transport mediators over 72 hours of operation. Immediately after circuit connection (external circuit resistance fixed at 20Ω) was established, the current density sharply increased. Maximum current density of Zn (1175 mA/m^2), Al (360 mA/m^2), Cu(610 mA/m^2), Fe (890 mA/m^2), and Graphite felt (255 mA/m^2) were observed at 45 h, 3 h, 5 h, 3h and 68 h, respectively. Current density and cell voltage are varying within a certain range over the next 72 hours. Figure 4 indicate that the majority of electrical charge was generated in the initial stage of the sediment MFC runs: approximately more than 50% of the total amount current density was delivered within the first 45 hours. Corrosion and microbial activity vary over time and condition of solution. The overall pattern of electricity generation was analogous to that observed by Holms et al. [35, 36], who demonstrated microbially-mediated current production using sediment MFC systems: a rapid increase in current production within the first few days of reactor operation with no lag period, followed by a gradual decrease. Wei and Zhang [37] and Mathis et al. [38] isolated the electricigenic (also referred to as anodophilic and electrochemically-active) bacteria from natural marine sediments and proved that the instant electricity generation in their MFC systems augmented with isolated cultures. Current density among the metallic anodes of Al, Cu and Fe showed a continuous decrease whereas Zn metal showed somewhat different phase. Zn anode showed rising to falling incidents and non-metallic anode graphite felt showed an elevated trend till end time. These results suggest that catalytic or microbial activity on the surface of Zn or graphite felt was higher than Al, Cu and Fe anode. It is observed that voltage losses as like current density over time (Fig.3). According to anode, the observed maximum cell voltages were Zn (777 mV), Al (751 mV), Cu (645 mV), Fe (734 mV) and GF (499 mV) over different time. A number of different bacteria are able to reduce corrosion rates of different materials in several corrosive media. One type of bacteria can shift corrosion potential (E_{corr}) of one metal in the positive direction while another type can shift corrosion potential (E_{corr}) of some other metal in the negative direction [39, 40].

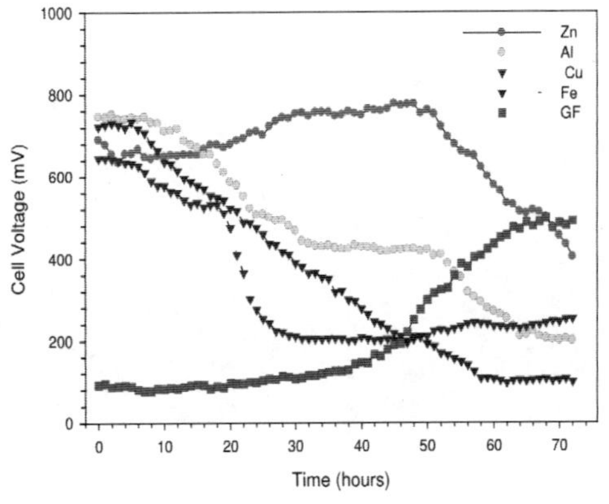

Figure 3. Cell voltages of SMFC with respect to time.

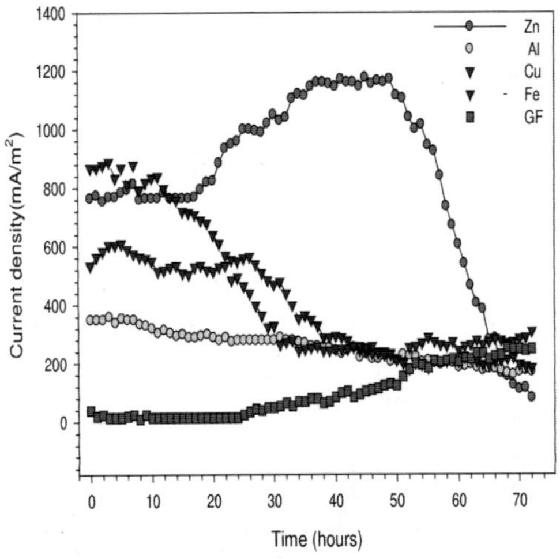

Figure 4. Current density vs time of SMFC operation (surface area 0.002 m^2 and internal resistance 20Ω).

3.4. pH and chemical oxygen demand (COD) of SMFC

pH of SMFC in this work has changed over time at both in anode part and cathode part in reverse way (Figure 5). When pH of anodic part were decreased, at the same time cathode pH increased. The pH variation ranges were 7.5 to 6.4 in anode Zn, 7.3 to 6.8 in Al, 6.6 to 6.4 in Cu, 6.7 to 6.2 in Fe, 7.5 to 6.7 in GF over starting time and ending time, respectively. But, in cathode part, pH changed very little (7.6~7.7). pH reduction indicated that anodic environment is acidic and acidic media reduces the corrosion rate of metal. Metal sulfide or metal oxides are likely to reduce pH value. Acidic environment in anode part is an important issue because MFCs that are able to operate at low pH are technologically advantageous. The proton transport rate from anode to cathode is increases and the kinetic barrier for O_2 reduction to H_2O at the cathode decreases, which leads to the higher current and power densities [41, 42]. In acidic medium acidophilic microorganisms that have colonized the anode and cathode surface can oxidize glucose and other organic compounds in the absence of redox mediators [43]. It is well known that *Acidiphilium spp.* is able to use organic compounds such as glucose and glycerol as electron donors [44, 45, and 43]. Chemical oxygen demand of SMFC was decreased over time in case of all anodes. But losses of organic matter, shown in COD decreases might be facilitative for long time electricity generation. Biomass production could account for additional COD removal, but it was not possible to establish a complete mass balance of COD in SMFC system. Not all the organic matter in a wastewater can be biologically degraded, so that COD removal is typically lower than that for biochemical oxygen demand (BOD) [46].

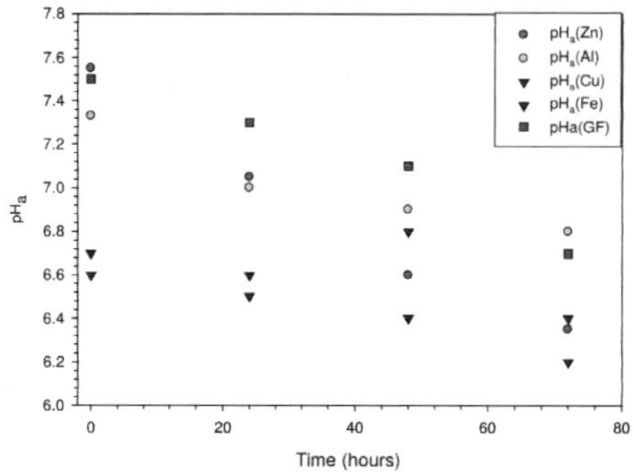

Figure 5. pH changes in anode chamber over time of SMFC (pHa, anodic pH).

4. Conclusions

This research was aimed to evaluate the feasibility of electricity generation of SMFC using different metal and non-metal anodes. From our study we can assume that a mediator less single chamber sediment microbial fuel cell might produce higher power by changing anode material. As for power density Zn was excellent over other metals. Some interfacial reactions or phenomena on metal or non-metal surface need to be investigated further as mentioned in Section 3. Also time duration of the cell and limitation of current generation could be next issues.

5. Acknowledgement

This research was supported by BK21 plus program and authors acknowledge Soonchunhyang University for its partial finance (No. 20130584).

6. Figures

Each figure should have a brief caption describing it and, if necessary, a key to interpret the various lines and symbols on the figure.

7. References

[1] Aderhold J, Davydov V Yu, Fedler F, Klausing H, Mistele D, Rotter T, Semchinova O, Stemmer J and Graul J 2001 *J. Cryst. Growth* **222** 701

[1] SK Chaudhuri, DR Lovley. Electricity generation by direct oxidation of glucose in mediatorless microbial fuel cells[J] 2003 *Nat. Biotechnol.* **21** 1229-1232.

[2] K Rabaey, W Verstraete Microbial fuel cells: novel biotechnology for energy generation [J] 2005 Trends in Biotechnology **23** 291-298, doi:10.1016/j.tibtech.2005.04.008

[3] K Rabaey, KVD Sompel, L Maignien, N Boon, P Aelterman, P Clauwaert, LD Schamphelaire, HT Pham, J Vermeulen, M Verhaege, P Lens, W Verstraete. Microbial fuel cells for sulfide removal[J]. *Environmental Science & Technology* 40 5218-5224, doi:10.1021/es060382u.

[4] S Tsujimura, A Wadano, K Kano, T Ikeda Photosynthetic bioelectrochemical cell utilizing cyanobacteria and watergenerating oxidase 2001 *Enzyme Microb. Technol* 29 225-231.

[5] HJ Kim, HS Park, MS Hyun, IS Chang, M Kim, BH Kim. A mediator-less microbial fuel cell using a metal reducing bacterium, *Shewanella putrefaciense*[J]. Enzyme Microb. Technol., 2002, 30:145-152.

[6] Y Choi, J Song, S Jung, S Kim. Optimization of the performance of microbial fuel cells containing alkalophilic *Bacillus sp.* J. Microbiol. Biotechnol., 2001, 11:863-869.

[7] DH Park, JG Zeikus. Electricity generation in microbial fuel cells using neutral red as an electronophore[J]. Appl. Environ. Microbiol. 2000, 66:1292-1297.

[8] JB McKinlay, JG Zeikus. Extracellular iron reduction is mediated in part by neutral red and hydrogenase in *Escherichia coli*. Appl. Environ. Microbiol. 2004, 70:3467-3474.

[9] E Katz, I Willner, AB Kotlyar. A noncompartmentalized glucose vertical bar O-2 biofuel cell by bioengineered electrode surfaces. J. Electroanal. Chem.,1999, 479:64-68.

[10] U Schroder, J Niessen, F Scholz. A generation of microbial fuel cells with current outputs boosted by more than one order of magnitude. Angew. Chem., Int. Ed. 2003, 42:2880-2883.

[11] KB Gregory, DR Bond, DR Lovley, Graphite electrodes as electron donors for anaerobic respiration Environ. Microbiol 2004, 6:596-604.

[12] SA Lee, Y Choi, SH Jung, S Kim. Effect of initial carbon sources on the electrochemical detection of glucose by *Gluconobacter oxidans* Bioelectrochemistry, 2002, 57:173-178.

[13] BE Logan. Extracting hydrogen electricity from renewable resources Environ. Sci. Technol, 2004, 38:160A-167A.

[14] LM Tender, CE Reimers, HA Stecher, DE Holmes, DR Bond, DA Lowy, K Pilobello, SJ Fertig, DR Lovley. Harnessing microbially generated power on the seafloor Nat. Biotechnol., 2002, 20:821–825.

[15] H Liu, R Ramnarayanan, BE Logan. Production of electricity during wastewater treatment using a single chamber microbial fuel cell Environ. Sci. Technol., 2004, 38:2281-2285.

[16] B Min, BE Logan. Continuous electricity generation from domestic wastewater and organic substrates in a flat plate microbial fuel cell Environ. Sci. Technol., 2004, 38:5809-5812.

[17] J R Kim, B Min, BE Logan. Evaluation of procedures to acclimate a microbial fuel cell for electricity production Appl. Microbial Biotechnol., 2005, 68:23-30.

[18] K Rabaey, N Boon, SD Siciliano, M Verhaege, W Verstraete. Biofuel cells select for microbial consortia that self-mediate electron transfer. Appl. Environ. Microbiol. 2004, 70:5373-5382.

[19] H Liu, BE Logan. Electricity generation using an air-cathode single chamber microbial fuel cell in the presence and absence of a protone exchange membrane Environ. Sci. Technol. 2004, 38:4040-4046.

[20] S Cheng, H Liu, BE Logan. Power densities using different cathode catalysts (Pt and CoTMPP) and polymer binders (nafion and PTFE) in single chamber microbial fuel cells Environ. Sci. Technol., 2006, 40:364-369.

[21] H Liu, SA Cheng, BE Logan. Power generation in fed-batch microbial fuel cells as a function of ionic strength, temperature, and reactor configuration Environ. Sci. Technol. 2005, 39:5488-5793.

[22] DH Park, JG Zeikus. Improved fuel cell and electrode designs for producing electricity from microbial degradation Biotechnol. Bioeng., 2003, 81:348-355.

[23] HA Videla. Manual of Biocorrosion CRC Press, 1996.

[24] PN Froelich, GP Klinkhammer, ML Bender, NA Luedtke, GR Heath, D Cullen, P Dauphis, D Hammond, B Hartman, V Maynard. Early oxidation of organic-matter in pelagic sediments of the eastern equatorial Atlantic-suboxic diagenesis Geochim. Cosmochim. Acta 1979, 43 (7), 1075–1090.

[25] G C Gil, IS Chang, BH Kim, M Kim, JK Jang, HS Park, HJ Kim. Operating parameters affecting the performance of a mediator-less microbial fuel cell Biosens. Bioelectron 2003, 18:327-334.

IOP Conf. Series: Materials Science and Engineering **88** (2015) 012072 doi:10.1088/1757-899X/88/1/012072

[26] B E Logan, B Hamelers, R Rozendal, U Schroder, J Keller, S Freguia, P Aelterman, W Verstraete, K Rabaey. Microbial fuel cells: Methodology and technology Enveronmental Science & Technology, Jul. 2006, 40:5181-5192, doi:10.1021/es0605016.

[27] B H Kim, HJ Kim, MS Hyun, DH Park. Direct electrode reaction of Fe(iii) reducing bacterium, *Shewanella puterfaciens* J. Microbial. Biotechnol., 1999:127-131.

[28] H J Kim, MS Hyun, IS Chang, BH Kim. A fuel cell type lactate biosensor using a metal reducing bacterium, *Shewanella puterfaciens* J. Microbial. Biotechnol., 1999, 9:365-367.

[29] H S Park, BH Kim, HS Kim, HJ Kim, GT Kim, M Kim, IS Chang, YK Park, HI Chang. A novel electrochemically active and Fe(III) reducing bacterium phylogenetically related to *Clostridium butyricum* isolated from a microbial fuel cell 2001, Anaerobe 7:297-306.

[30] D R Bond, DR Lovely. Eelectricity production by *Geobacter sulfurreducens* attached to electrodes Apple. Environ. Microbial. 2003, 69:1546-1555.

[31] H J Kim, HS Park, MS Hyun, IS Chang, M Kim, BH Kim. A mediator-less microbial fuel cell using a metallic bacterium *Shewanella puterfaciens Enzyme* Microb. Technol., 2002, 30:125-152.

[32] H Liu, SA Cheng, BE Logan, Production of electricity from acetate or butyrate using a single-chambered microbial fuel cell Environ. Sci. Technol., 2005, 39:658-662.

[33] H Moon, IS Chang, KH Kang, JK Jang, BH Kim. Residence time distribution in microbial fuel cell and its influence on COD removal with electricity production Biochem. Eng. J. 2005, 27:59-65.

[34] BE Logan, C Murano, K Scott, ND Gray, IM Head. Electricity generation from cysteine in a microbial fuel cell *Water Res*. 2005, 39:942-952.

[35] DE Holmes, DR Bond, DR Lovley. Electron transfer by *Desulfobulbus propionicus* to Fe(III) and graphite electrodes Applied Environmental Microbiology, 2004a, 70:1234–1237.

[36] DE Holmes, DR Bond, RA O'Neill, CE Reimers, LR Tender, DR Lovley. Microbial communities associated with electrodes harvesting electricity from a variety of aquatic sediments Microbial Ecology, 2004b, 48:178–190.

[37] D Wei, X Zhang. Current production by a deep-sea strain *Shewanella sp*. DS1. Current Microbiology 2007, 55:497–500

[38] BJ Mathis, CW Marshall, CE Milliken, RS Makkar, SE Creager, HD May. Electricity generation by thermophilic microorganisms from marine sediment Applied Microbiology & Biotechnology, 2008, 78:147–155.

[39] E Kus, R Abboud, R Popa, KH Nealson, F Mansfeld. The concept of the bacterial battery Corros. Sci., 2005, 47:1063-1069.

[40] F Mansfeld. The interaction of bacteria and metal surfaces; 2007; Electrochimica Acta 52:7670–7680.

[41] JC Biffinger, J Pietron, O Bretschger, LJ Nadeau, GR Johnson, CC Williams, KH Nealson, BR Ringeisen. The influence of acidity on microbial fuel cells containing *Shewanella oneidensis* Biosens. Bioelectron., 2008, 24:900-905.

[42] B Erable, L Etcheverry, A Bergel. Increased power from a two chamber microbial fuel cell with a low pH air-cathode compartment Electrochem. Commun., 2009, 11:619-622

[43] M Malki, AL De Lacey, N Rodríguez, R Amils, VM Fernández. Preferential use of an anode as an electron acceptor by an acidophilic bacterium in the presence of oxygen Appl. Environ. Microbiol. 2008, 74:4472-4476.

[44] DB Johnson, S McGinness. Ferric iron reduction by acidophilic heterotrophic bacteria. Appl. Environ. Microbiol., 1991, 57:207-211.

[45] K Kusel, T Dorsch, G Acker, E Stackebrandt. Microbial reduction of Fe(III) in acidic sediments: Isolation of Acidiphiliun cryptum JF-5 capable of coupling the reduction of Fe(III) to the oxidation of glucose Appl. Environ. Microbiol., 1999, 65:3633-3640.

[46] Grady CPL Jr, Daigger G T, Lim HC. In Biological wastater treatment 2nd ed., New York, Marcek, 1999.

IOP Conf. Series: Materials Science and Engineering **88** (2015) 012073 doi:10.1088/1757-899X/88/1/012073

A computation ANN model for quantifying the global solar radiation: A case study of Al-Aqabah-Jordan

I M Abolgasem[1]*, M A Alghoul*, M H Ruslan, H Y Chan, N G Khrit, K Sopian
Solar Energy Research Institute, Univeristi Kebangsaan Malaysia
Bangi 43600 Selangor DE, Malaysia

*E-mail: : malk_197716@yahoo.com, dr.alghoul@gmail.com

Abstract. In this paper, a computation model is developed to predict the global solar radiation (GSR) in Aqaba city based on the data recorded with association of Artificial Neural Networks (ANN). The data used in this work are global solar radiation (GSR), sunshine duration, maximum & minimum air temperature and relative humidity. These data are available from Jordanian meteorological station over a period of two years. The quality of GSR forecasting is compared by using different Learning Algorithms. The decision of changing the ANN architecture is essentially based on the predicted results to obtain the best ANN model for monthly and seasonal GSR. Different configurations patterns were tested using available observed data. It was found that the model using mainly sunshine duration and air temperature as inputs gives accurate results. The ANN model efficiency and the mean square error values show that the prediction model is accurate. It is found that the effect of the three learning algorithms on the accuracy of the prediction model at the training and testing stages for each time scale is mostly within the same accuracy range.

1. Introduction

Solar energy technologies with appropriate designs can utilize the solar energy effectively to meet human needs in various applications. However, this depends on the availability of information on solar radiation characteristic of the location, where these systems are to be built. In this regard, scientists and engineers developed various approaches and methods to collect information and characterize solar radiation components of a proposed site. There are also approaches that rely on mathematical-based measurements. Nonetheless, the best solar radiation information is obtained through experimental measurements but however this method may be hindered by technical and financial constraints. The assessment of solar energy involves huge challenges and uncertainties, resulting from the complex processes of quantification (prediction), shortage in data measurements, and variability of spatio-temporal parameters. Thus, there is an urgent need to develop smart techniques of assessment that manage to (1) produce reliable and representative information on solar conditions with minimum error and high certainty, and (2) assist in identifying the optimum requirements of estimation that ensure high performances and trusted scheme of assessment.

[1] To whom any correspondence should be addressed.

One of the methods is advanced computation method that diagnosing solar energy and creating better, quicker, and more practical forecasting than any of the traditional methods are the artificial intelligence (AI) techniques. These techniques include [1]: artificial neural network (ANN), fuzzy logic (FL), Adaptive Network based Fuzzy Inference System (ANFIS), and Data Mining (DM). Artificial intelligence methods aim to provide time-and cost-effective framework of automation, recognition, data storage, retrieval and processing, problem solving, communication, system training, and information super-intelligence [2]. Contrary to conventional statistical methods used in scientific investigations, ANNs deal with forecasting and modeling linear and non-linear systems to arbitrary accuracy without the need for implicit assumptions. They are self-training systems where they adaptively construct linkages between a given pattern of input data and particular outputs [3, 4]. Due to the fact that they exhibit robustness, parallel architecture, fault tolerant capabilities, and the ability to work as universal function approximators, ANNs are profoundly used in solving complex science and engineering problems [5], and in comprehending the phenomena of multidimensional informational domains. The aim of this study is to develop an artificial neural network model to estimate monthly and seasonal global solar radiation in Aqaba city, Jordan.

2. Methodology

The methodology of this study includes data collection, data processing, analysis, and interpretation as well. In this work, solar radiation potential in Aqaba city (Jordan) was estimated by two cases of selected neural network model. The modeling results were validated to assess the reliability of the estimation process.

2.1. Database

The surface meteorological parameters for Aqaba were collected from the Jordan meteorological department over a period of two years. The collected information included sunshine duration, maximum and minimum air temperature, relative humidity and global solar-radiation. The global solar radiation component was measured by a pyranometer. The datasets showed that the daily average values of clearness index (H/H_o) were found to vary between approximately (0.0540 - 0.7930), and the sunshine ratio were found to vary between approximately (0.00 - 0.92 hour). During the same period of observation, the daily average values of relative humidity were found to vary between approximately (13-98%).

2.2 ANN-based implementation of solar radiation models

The ANN model selected for forecasting GSR was a simple multilayered feed forward perception. The network architecture consisted of an input and output layers, and two hidden layers between them. Each hidden layer contained a number of neurons. The input layer consisted of the following three data: sunshine ratio, temperature and humidity; while the output layer consisted of the data of clearness index. The neurons in the layers were interconnected with weights characteristic of the information passing through them; the learning algorithm of error back propagation determined the weights. The configuration pattern (model) of training that was presented to the network is in the following form:

Case 1:

$$[\frac{H}{H_0}]_{monthly} = f([\frac{S}{S_0}],[\frac{T_{Min}}{T_{Max}}],[h])_{monthly} \tag{1}$$

Case 2:

$$[\frac{H}{H_0}]_{Seasonal} = f([\frac{S}{S_0}],[\frac{T_{Min}}{T_{Max}}],[h])_{Seasonal} \tag{2}$$

Where Case 1 is monthly GSR and Case 2 is seasonal GSR. (H/H_o) represents the clearness index with H, H_0 are the solar radiation and the maximum solar radiation respectively. (S/S_o) is the sunshine ratio, such that S, S_0 are the sunshine duration and the maximum sunshine duration; (T_{Min} / T_{Max}) is the temperature ratio, with T_{Min}, T_{Max} are the minimum and maximum the temperature respectively, and (h) represent the humidity.(f) is the function model depend on the architecture of the neural network.This procedure will allow predicting the monthly and seasonal GSR of the site.

In the training process, neurons were trained and adjusted using the error back propagation rule (algorithm) to adjust the adaptation of the synaptic weights. The outputs are dependent on variables produced for the corresponding input. This algorithm is a supervised iterative training method commonly applied for multilayer feed forward nets, with nonlinear sigmoidal threshold units. The nonlinear sigmoidal transfer function is:

$$f(x) = \frac{1}{(1+e^{-x})} \tag{3}$$

The iterative process continued until a tolerance level was reached. This procedure was conducted on the training and testing datasets. The following training parameters were set during the training process: Epochs of training was set to 10e-4, Training goal was set to 10e-3, Maximum time of training the network is depend on the learning algorithm, Momentum constant =0.92, number of neurons = (30 15 1). In order to suit the consistency of the model, input and output data were firstly normalized in the (0, 1) range, using the following formula:

$$x' = \frac{x - x_{Min}}{(x_{\max} - x_{Min})} \tag{4}$$

Then input and output data will be returned to the original values after the simulation. Where (x) represent the inputs/outputs of the network, and (x') is normalized inputs or outputs of the network. The value of normalized input or output is 1 when the input or output is $x_{\max}$ and the value of normalized input or output is 0 when the input or output is x_{Min} . The activation function applied in the designed ANN was the sigmoid function. In this study, the "Tansig" transfer function was used in the hidden layer and a linear activation function, and the "Purelin" transfer function in the output layer. The "Tansig" transfer function is defined as:

$$\tan sig(x) = \frac{2}{1+e^{(-2x)}} - 1 \tag{5}$$

2.3 ANN Model Error Analyses

Initially, the collected meteorological data was used for training, validating, and testing the designed feed-forward back propagation neural networks. The estimated value of clearness index was tested for its accuracy. The predicted value resulted was compared with the true (measured) data in order to verify the performance of the prediction model (for training and testing stages). The comparison between measured and estimated value was evaluated by statistical error of mean square error (MSE) to check the stability of the model using by the following formula:

$$MSE = \frac{1}{N}\sum_{i=1}^{N}(X_{Estimated} - X_{Measured})^2 \tag{6}$$

Where (N) is the number of input-output pairs, X is the measured or estimated value of the output. In addition, the efficiency (EFF) of the prediction model for training and testing stage is defined as:

$$Model_{Efficiency}(EFF) = 1 - (\frac{MSE}{Var}) \quad (7)$$

Where variance (Var) equals the square of standard deviation.

The model that gives the lowest errors (MSE) and best EFF values (approach to 1) is considered stable and the most suitable model.

3. Results and Discussions

This section discusses the modeling results of the GSR for Aqaba city when three different types of learning algorithms are applied in the ANN model for monthly and seasonal GSR. The performance of the prediction is evaluated by checking the *(MSE)* and Model efficiency *(EFF)* indicators.

3.1 Modeling results: monthly GSR of Aqaba City

Figure 1 illustrates the results of the monthly GSR modeling when three different types of learning algorithms are applied in the ANN model. The prediction data showed generally excellent accuracy when plotted against the measured data. Table 1 summarizes the **ANN** model efficiency at training and testing stages using different learning algorithms. From the graph and cross-validation report, it can be seen that the average efficiency of the model at testing stage is (0.9103). The learning algorithm of LM had the lowest MSE of 0.0052 and highest efficiency of 0.9363 compared to the other algorithms. GDX has the highest MSE of 0.01 and the lowest efficiency of 0.8775, with SGD being somewhere in between. The model showed high efficiency in predicting monthly GSR under the training and testing stage. Figure 1 shows the testing stage comparison of monthly GSR at different learning algorithms: Train LM, Train SCG and Train GDX respectively.

Table 1: Efficiency of monthly **ANN** GSR model at training and testing stages using different algorithms

[30 15 1]	Training		Testing	
	MSE	EFF	MSE	EFF
Train LM	9.4e-04	0.9878	0.0052	0.9363
Train SCG	7.1e-04	0.9908	0.0068	0.9171
Train GDX	9.6e-04	0.9874	0.0100	0.8775
Average				0.9103

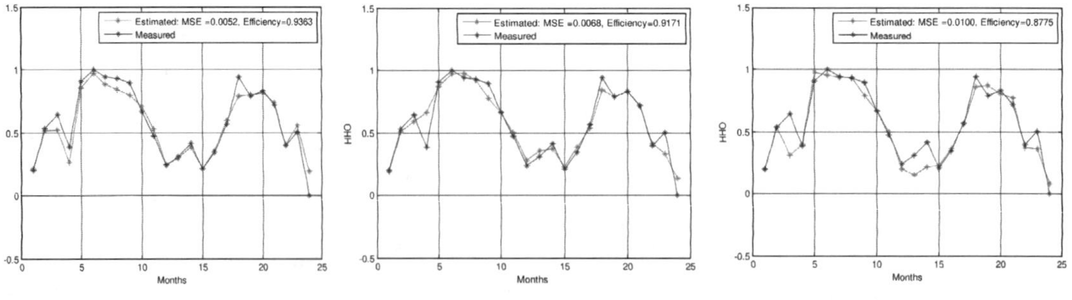

(a) (b) (c)

Figure 1. Testing stage comparison of monthly GSR under Train LM, Train SCG and Train GDX learning algorithms respectively

3.2 Modeling results: Seasonal GSR for Aqaba City

This section discussed the outcomes of modeling the seasonal GSR. Table 2 summarizes the seasonal **ANN** model efficiency at training and testing stages using different learning algorithms.

It is shown that the average testing efficiency of autumn, spring, summer and winter GSR models are 0.876, 0.9275, 0.6450, and 0.9122 respectively. The lowest average testing efficiency is found for summer and the highest for spring. Figures 2-5 show the testing stage comparison of seasonal GSR at different learning algorithms over different time scales.

Table 2: Efficiency of training and testing stages at different learning algorithms of the seasonal ANN GSR model

	Autumn				Spring				Summer				Winter			
[30 15 1]	Training		Testing		Training		Testing		Training		Testing		Training		Testing	
	MSE	EFF	MSE	EFF	MSE	EFF	MSE	EFF	MSE	EFF	MSE	EFF	MSE	EFF	MSE	EFF
Train LM	8.7e-04	0.9548	0.0017	0.8846	6.6e-04	0.9762	0.0019	0.9411	9.7e-04	0.9697	0.0107	0.6213	9.6e-4	0.9755	0.0038	0.9089
TrainSCG	9.8e-04	0.9492	0.0017	0.8832	9.9e-04	0.9645	0.0025	0.9226	9.9e-04	0.9690	0.0078	0.7092	9.9e-4	0.9746	0.0033	0.9192
TrainGDX	0.0012	0.9404	0.0020	0.8610	9.7e-04	0.9654	0.0027	0.9188	0.0087	0.7315	0.0106	0.6047	0.0016	0.9588	0.0038	0.9087
Average				0.8762				0.9275				0.6450				0.9122

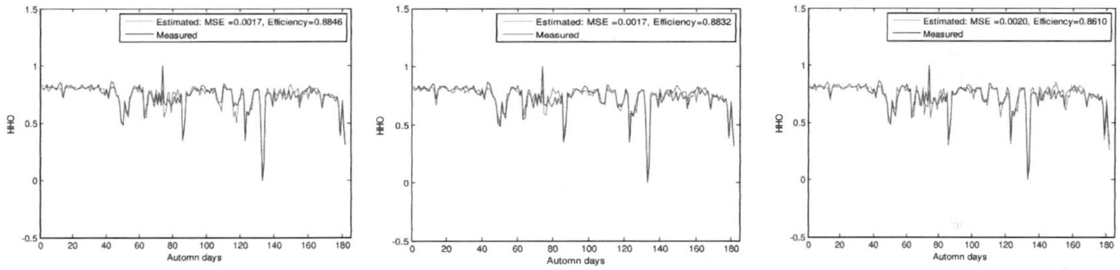

(a) (b) (c)

Figure 2. Testing stage comparison of Autumn GSR model under Train LM, Train SCG and Train GDX learning algorithms respectively

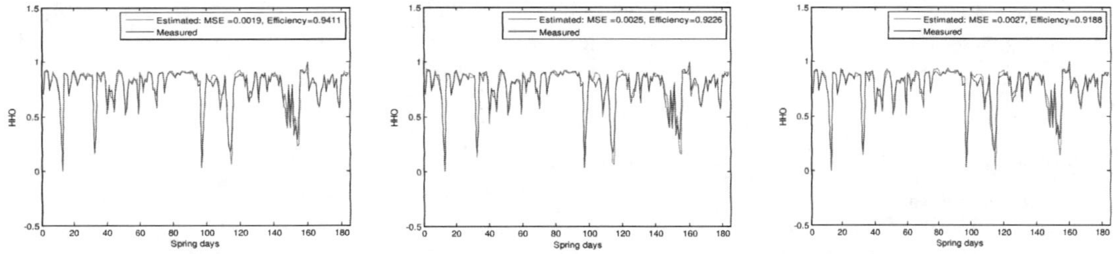

(a) (b) (c)

Figure 3. Testing stage comparison of Spring GSR model under Train LM, Train SCG and Train GDX learning algorithms respectively

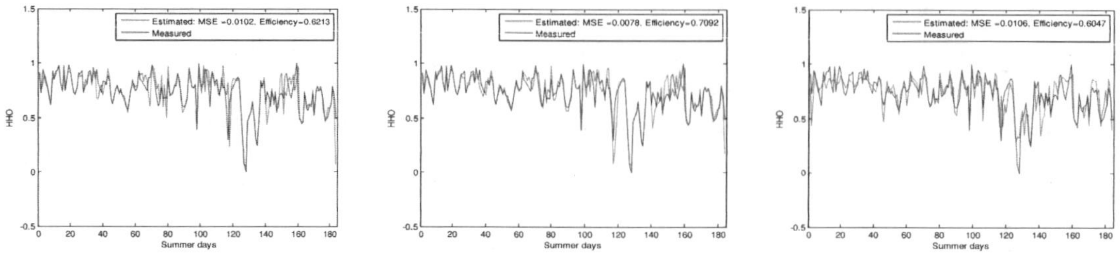

(a) (b) (c)

Figure 4. Testing stage comparison of Summer GSR model under Train LM, Train SCG and Train GDX learning algorithms respectively

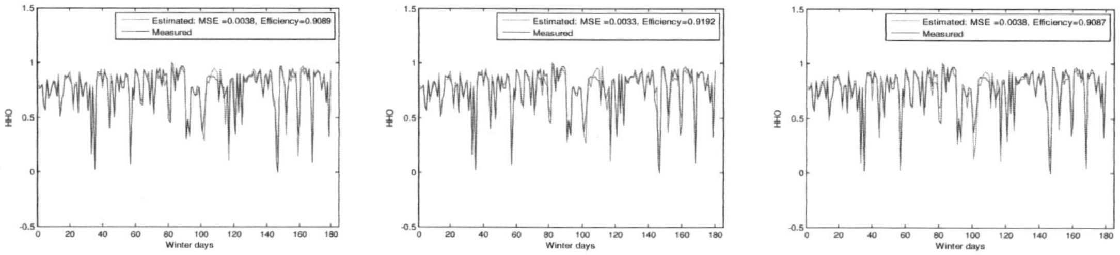

(a) (b) (c)

Figure 5. Testing stage comparison of Winter GSR under Train LM, Train SCG and Train GDX learning algorithms respectively

4. Conclusion

The proposed ANN model performed well in evaluating the monthly and seasonal variations in global solar radiation for Aqaba-Jordan, based on the solar parameters measured during a period of 2 years. The ANN model under different learning algorithms showed high efficiency in predicting GSR over different time scales. It demonstrated a significant role in reducing the error between measured and estimated data to minimum levels. The three learning algorithms show close accuracies of the prediction model at the training and testing stages for each time scale.

5. References

[1] Bourbakis N.G 1992 *Artificial Intelligence Methods and Applications* (Singapore: World Scientific Publishing Co)

[2] Akerkar, R 2005 *Introduction to Artificial Intelligence* (New Delhi: Prentice-Hall)

[3] Mellit A, Drif M, Malek A 2010 EPNN-based prediction of meteorological data for renewable energy systems *Revue des Energies Renouvelables* **13** 25-47

[4] Yang J, Rivard H, Zmeureanu R 2005 Buliding Energy Predication with Adaptive Artificial Neural Networks *Ninth International IBPSA Conference Montréal*

[5] Cybenko, G 1989 Approximation by superposition of a sigmoidal function. *Mathematics of Control Signal and Systems* **2** 303–14.

7th International Conference on Cooling & Heating Technologies (ICCHT 2014) IOP Publishing
IOP Conf. Series: Materials Science and Engineering **88** (2015) 012074 doi:10.1088/1757-899X/88/1/012074

Performance Analysis of Solar-Wind-Diesel-Battery Hybrid Energy System for KLIA Sepang Station of Malaysia

S K A. Shezan[1], R Saidur[2], A Hossain[3], W T Chong[4], M A Kibria[5]
[1,2,3,4,5]Department of Mechanical Engineering
Faculty of Engineering, University of Malaya
50603, Kuala Lumpur, Malaysia.

E-mail: shezan.ict@gmail.com

ABSTRACT: A large number of populations of the world live in rural or remote areas those are geographically isolated. Power supply and uninterrupted fuel transportation to produce electrical power for these remote areas poses a great challenge. Using renewable energy in hybrid energy system might be a pathway to solve this problem. Malaysia is a large hilly land with the gift of renewable energy resources. There is a good chance to utilize these renewable resources to produce electrical power and to limit the dependency on the fossil fuel as well as reduce the carbon emissions. In this perspective, a research is carried out to analyze the performance of a solar-wind-diesel-battery hybrid energy system for a remote area named "KLIA Sepang station" in the state of Selangor, Malaysia. In this study, a 56 kW hybrid energy system has been proposed that is capable to support more than 50 households and 6 shops in that area. Real time field data of solar radiation and wind speed is used for the simulation and optimization of operations using "Homer" renewable energy software. The proposed system can reduce CO_2 emission by about 16 tons per year compared to diesel generator only. In the same time the Cost of energy (COE) of the optimized system is USD 5.126 /kWh.The proposed hybrid energy system might be applicable for other parts of the world where the climate conditions are similar.

1. Introduction

Usage of renewable energy for electricity generation is a priority research topic now a day. Remarkable efforts to expand the energy sources and to intensify the deployment of renewable energy options have been increasing around the world. The first priority in intensifying renewable energy deployment in the 21st century is the combined effects of the depletion of fossil fuels and the awareness of environmental degradation[1]. Therefore, policy makers and researchers are paying more attention to research in this field. For an instance, European Union countries aim to get 20% of its energy from renewable sources by 2020[2]. Renewables include wind, solar, hydro-electric and tidal power as well as geothermal energy and biomass. More renewable energy will enable the EU to cut

[1] To whom any correspondence should be addressed.

greenhouse emissions and make it less dependent on imported energy. Malaysia is blessed with abundant energy resources. It has the world's largest wind sources and the larger Global Solar radiation for its geographical position[3]. During the last two decades, electrical energy consumption in Malaysia increased significantly due to rapid economic development and the absence of energy conservation measures. It is expected that peak loads will reach 60 GW in 2023 which causes total investment may exceed $90 billion. Therefore, there is an urgent need to develop energy conservation policies for sustainable development[4]. Almost all its electricity is produced from the combustion of fossil fuels [5]neglecting the use of its renewable energy resources such as wind and solar to generate electricity. Apart from local conservation efforts, with increasing load demand and global warming, policy makers are looking at environment-friendly type of energy solutions to preserve the earth for the future generations. Wind and photovoltaic energy holds the most potential to meet our energy demands. As a result such power generation systems have been attracted greatly all over the world. However, increasing number of renewable energy sources requires new strategies for their operations in order to maintain or improve the power-supply stability, quality and reliability[6]. Malaysia is a country hilly area. Grid extension to populations living in such areas is costly and unfeasible. In such situations decentralized renewable energy based power generating options can provide feasible alternatives options. These alternatives may include hybrid power systems like wind-pv-diesel, wind-diesel, pv-diesel and others with or without battery backup option. In this research, a complete performance analysis of a solar-wind-diesel-battery hybrid energy system for the remote area of "KLIA Sepang station" in the state "Selangor" of Malaysia is proposed. Focus is given to system engineering, energy production, system stability and reliability using Homer renewable energy software and real time data. Therefore HOMER (Hybrid optimization Model for Electric Renewable), a software developed by National Renewable Energy Laboratory (NREL), USA for micro-power optimization model, has been used to find out the best energy efficient renewable based hybrid system options for the island. It contains a number of energy component models and evaluates suitable technology options based on cost and availability of resources [7]. To design the optimum system HOMER performs thousands of hourly simulations. HOMER also performs sensitivity analysis to see the impact of solar isolation, PV investment cost, and wind speed and diesel fuel price on the COE [8].

2. Data resource and location analysis:

We have got daily solar radiation data for each month of the year 2009 from the Malaysian meteorological department. An estimation of solar insulation on horizontal surface has been done by using well known Angstrom Correlation and the sunshine hour data of Klia Sepang Station, Malaysian Meteorological Department, the nearest meteorological station from Selangor, Malaysia [9]. Also a method has been developed by DLR, Germany which is a combination of DLR / SUNY –model output for Global Horizontal Insulation (GHI) is sampled for 16.3 km spatial resolution, and the calculated data has been collected from the Malaysian Meteorological Department for Malaysia [10]. Figure 1 shows the schematic diagram of hybrid energy system. Seven tidal gauge stations were set up by Malaysian Meteorological department and Malaysian Renewable Energy Committee for the feasibility study of tidal energy [9]. But the result was not in favour. So, only the solar, wind source and average temperature data have been considered to find out the best hybrid options of renewable based efficient system. Figure 1 shows the schematic diagram of hybrid energy system.

Table 1. Yearly wind Power density and wind energy density:

Station	Year	Meteorological		Weibull		V_{mp}	$V_{max,E}$
		P/A	E/A	P/A	E/A		
Klia	2008	20.36	175.91	20.63	178.24	1.43	4.40
Sepang	2009	17.73	153.19	18.16	156.90	0.72	4.55

In the table 1 the parameters have been used as follows P/A as wind Power density, E/A as wind Energy density, V_{mp} as most probable wind speed and $V_{max,E}$ as maximum energy carrying by the wind speed.

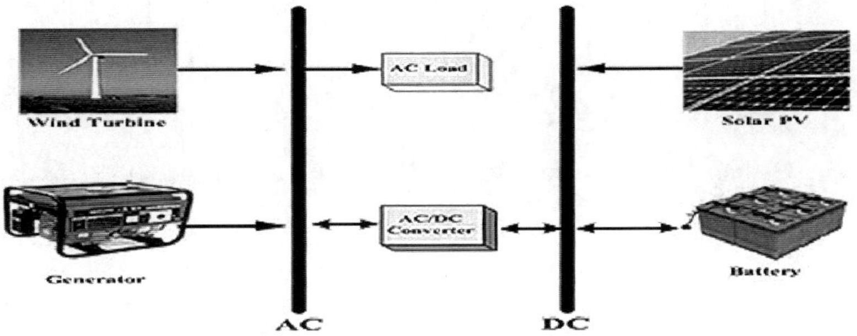

Figure 1. Schematic diagram of hybrid energy system.

Table 1 show the meteorological data of klia sepang station in the year 2008 and 2009 and also defined the Weibull values with power density and energy density.

3. Hybrid Energy System Components

3.1. Solar Energy (Photovoltaic)

As hourly data is available as monthly averaged global radiation data has been taken from Malaysian Meteorological Department [11]. HOMER introduces clearness index from the latitude and longitude information of the selected site. HOMER creates the synthesized 8760 hourly values for a year using the Graham algorithm. Figure 5 illustrates that the solar radiation is high in February, March, April, august, September and October. The cost of PV module including installation has been considered as USD 250 /W for Malaysia. Life time of the modules has been taken as 20 years.5 kW, 18 kW and 30 kW PV modules are considered. The parameters considered for the simulation solar PV are furnished in table 2. Figure 2 shows global horizontal radiation data for KLIA Sepang station of Malaysia.

Table 2. Solar PV array-technical Parameters and Cost assumption

Parameter	Value
Capital Cost	250 $/W
Replacement Cost	200 $/W
Operation and Maintenance Cost	20 $/W
Lifetime	20 Years
Derating factor	80 %
Tracking System	No Tracking System

IOP Conf. Series: Materials Science and Engineering **88** (2015) 012074 doi:10.1088/1757-899X/88/1/012074

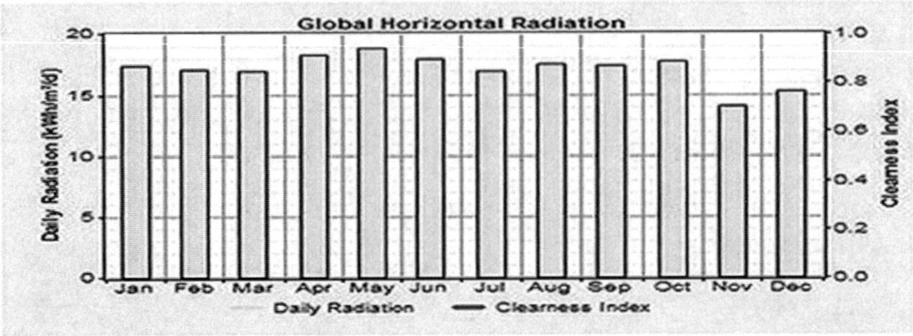

Figure 2. Global Horizontal Radiation for KLIA Sepang Station (2009), Malaysia.

3.2. Wind Energy (Wind Turbines)

When hourly data is not available, hourly data can be generated synthetically from the monthly averages. HOMER's synthetic wind speed data generator is a little more different to use than the solar data because it requires four parameters [12]. Technical and economic parameters for selected wind turbine are furnished in table 3. Figure 3 shows average Wind speed of every month in 2009 of Klia Sepang station of Malaysia. The Weibull value: k value is a measure of distribution of wind speed over the year of 2009 for KLIA Sepang Station. It is important for the average wind speed in a year span. In this study the value of k is taken as 2.

Table 3. Technical and economic parameters of wind turbine.

Parameter	Value
Rated Power	10 kW
Starting Wind Speed	2.860 m/s
Rated Wind Speed	2.04 m/s
Cut-off Wind Speed	2.010 m/s
Capital Cost	150 $/kW
Replacement Cost	120 $/kW
Operation and Maintenance Cost	1 $/kW
Lifetime	15 years

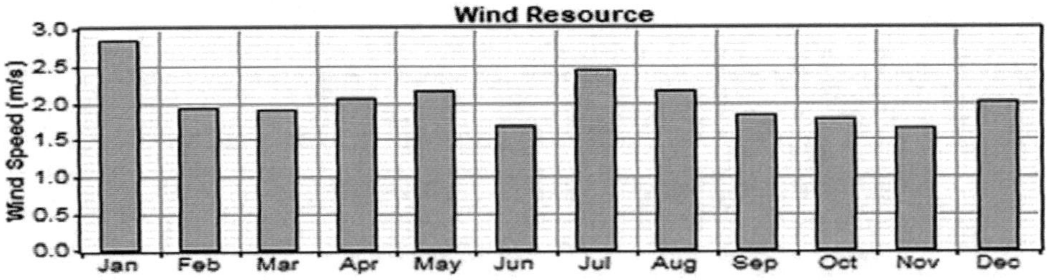

Figure 3. Average Wind speed of every month in 2009 of KLIA Sepang station of Malaysia.

4. Simulation, Optimization results and discussion

Figure 4 shows the energy yield for the feasible hybrid PV-diesel-wind-battery system. Figure 6 shows the details calculation and cost analysis related to energy generated by PV, wind turbine and diesel generator system. Figure 5 shows the optimization and simulation results for PV-diesel-wind turbine-battery system for a solar radiation of 6.09 kWh/m2/d, diesel price of 1$/L, maximum capacity

IOP Conf. Series: Materials Science and Engineering **88** (2015) 012074 doi:10.1088/1757-899X/88/1/012074

shortage of 0 %.(All the currency values were in terms of USD). The optimization results for specific wind speed 2.04 m/s, solar irradiation parameters 6.09 kWh/m2/d and diesel price 1 USD are illustrated in Figure 6. It is seen that a PV, wind turbine, diesel generator and battery hybrid system is economically more feasible with a minimum COE of USD 5.126/kWh and a minimum NPC of 786,822 USD. The hybrid system comprised of 18 kW PV array, two wind turbines (10 kW each), a diesel generator with a rated power of 15 kW (using 3kW) and 25 storage batteries in addition to 3 kW converters is found to be the most feasible system.

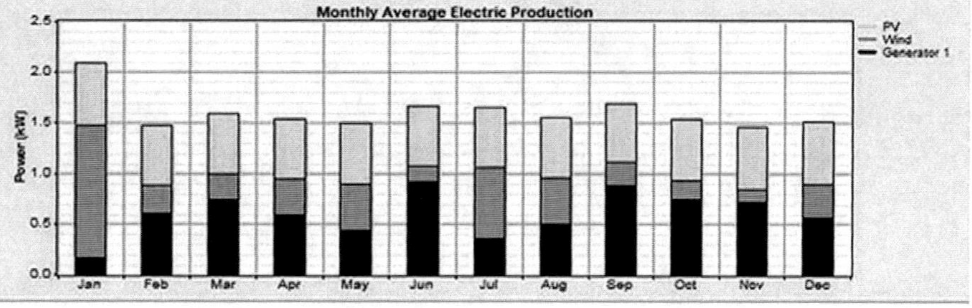

Figure 4. Energy yield for the feasible hybrid PV-diesel-wind-battery system.

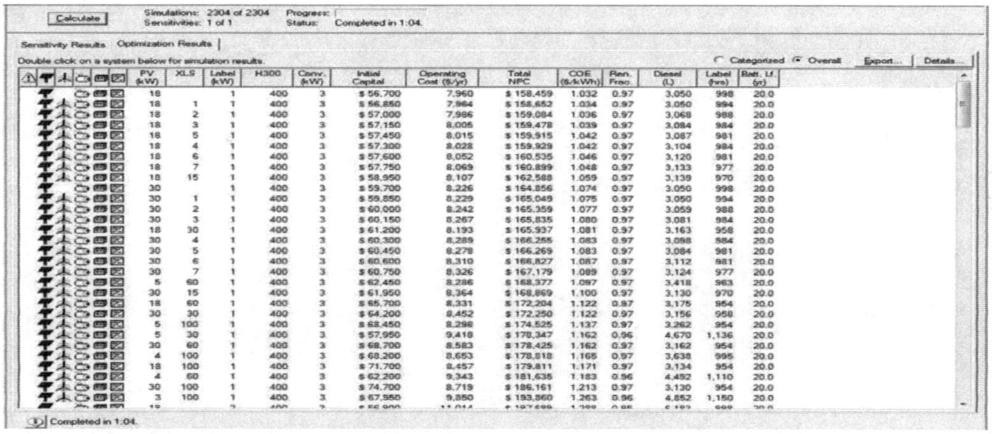

PV (kW)	XLS	Label (kW)	H300	Conv. (kW)	Initial Capital	Operating Cost ($/yr)	Total NPC	COE ($/kWh)	Ren. Frac.	Diesel (L)	Label (hrs)	Batt. Lf. (yr)
18		1	400	3	$ 56,700	7,960	$ 158,459	1.032	0.97	3,050	998	20.0
18	1	1	400	3	$ 56,850	7,964	$ 158,652	1.034	0.97	3,050	994	20.0
18	2	1	400	3	$ 57,000	7,986	$ 159,084	1.036	0.97	3,068	988	20.0
18	3	1	400	3	$ 57,150	8,005	$ 159,478	1.039	0.97	3,084	984	20.0
18	5	1	400	3	$ 57,450	8,015	$ 159,915	1.042	0.97	3,087	981	20.0
18	4	1	400	3	$ 57,300	8,028	$ 159,929	1.042	0.97	3,104	984	20.0
18	6	1	400	3	$ 57,600	8,052	$ 160,535	1.046	0.97	3,120	981	20.0
18	7	1	400	3	$ 57,750	8,069	$ 160,899	1.048	0.97	3,133	977	20.0
18	15	1	400	3	$ 58,950	8,107	$ 162,588	1.059	0.97	3,139	970	20.0
30		1	400	3	$ 59,700	8,226	$ 164,856	1.074	0.97	3,050	998	20.0
30	1	1	400	3	$ 59,850	8,229	$ 165,049	1.075	0.97	3,050	994	20.0
30	2	1	400	3	$ 60,000	8,242	$ 165,359	1.077	0.97	3,059	988	20.0
30	3	1	400	3	$ 60,150	8,267	$ 165,835	1.080	0.97	3,081	984	20.0
18	30	1	400	3	$ 61,200	8,193	$ 165,937	1.081	0.97	3,163	958	20.0
30	4	1	400	3	$ 60,300	8,289	$ 166,255	1.083	0.97	3,098	984	20.0
30	5	1	400	3	$ 60,450	8,278	$ 166,269	1.083	0.97	3,084	981	20.0
30	6	1	400	3	$ 60,600	8,310	$ 166,827	1.087	0.97	3,112	981	20.0
30	7	1	400	3	$ 60,750	8,326	$ 167,179	1.089	0.97	3,124	977	20.0
5	60	1	400	3	$ 62,450	8,286	$ 168,377	1.097	0.97	3,418	963	20.0
30	15	1	400	3	$ 61,950	8,364	$ 168,869	1.100	0.97	3,130	970	20.0
18	60	1	400	3	$ 65,700	8,331	$ 172,204	1.122	0.97	3,175	954	20.0
30	30	1	400	3	$ 64,200	8,452	$ 172,250	1.122	0.97	3,156	958	20.0
5	100	1	400	3	$ 68,450	8,298	$ 174,525	1.137	0.97	3,262	954	20.0
5	30	1	400	3	$ 57,950	9,418	$ 178,347	1.162	0.96	4,670	1,136	20.0
30	60	1	400	3	$ 68,700	8,583	$ 178,425	1.162	0.97	3,162	954	20.0
4	100	1	400	3	$ 68,200	8,653	$ 178,818	1.165	0.97	3,638	995	20.0
18	100	1	400	3	$ 71,700	8,457	$ 179,811	1.171	0.97	3,134	954	20.0
4	60	1	400	3	$ 62,200	9,343	$ 181,635	1.183	0.96	4,492	1,110	20.0
30	100	1	400	3	$ 74,700	8,719	$ 186,161	1.213	0.97	3,130	954	20.0
3	100	1	400	3	$ 67,950	9,850	$ 193,860	1.263	0.96	4,852	1,150	20.0

Figure 5. Optimization results for PV-diesel-wind turbine-battery system.

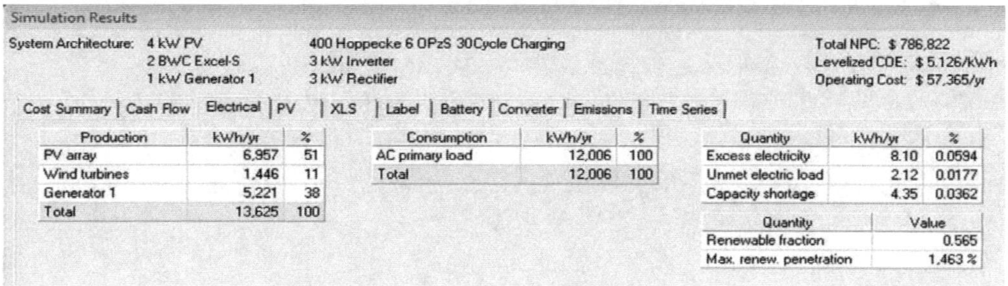

Production	kWh/yr	%
PV array	6,957	51
Wind turbines	1,446	11
Generator 1	5,221	38
Total	13,625	100

Consumption	kWh/yr	%
AC primary load	12,006	100
Total	12,006	100

Quantity	kWh/yr	%
Excess electricity	8.10	0.0594
Unmet electric load	2.12	0.0177
Capacity shortage	4.35	0.0362

Quantity	Value
Renewable fraction	0.565
Max. renew. penetration	1,463 %

Figure 6. Energy generated by PV, wind turbine and diesel generator system, excess electricity, un-metload, capacity shortage and renewable fraction.

5. Conclusion

The study simulates a PV-wind –diesel-battery hybrid energy system in KLIA Sepang Station of Malaysia. The optimized hybrid energy system was developed considering manufacturing cost and efficiency. The result shows that the COE of the optimized system is USD 5.126 /kWh and the NPC of the optimized system is USD 786,822. The total sensitivity analysis, optimization and simulation process has been conducted through HOMER renewable energy software. This system gives better performance than the other system because if any fault occur in PV panel or in wind turbines then the generator can minimize this problem and also we tried to reduce the cost of power generation from the in general hybrid energy systems. In the near future we will try to introduce some more convenient renewable energy model and proper control system for the hybrid energy system for the different area of the world.

6. Acknowledgments

This paper was funded by the UMRG grant of University of Malaya. The project title is "Performance Investigation of Solar-Wind Hybrid Energy System for Malaysian Climate". The project no is RG-158, AET/12.

7. References

[1] Chen, H.H., H.-Y. Kang, and A.H. Lee, *Strategic selection of suitable projects for hybrid solar-wind power generation systems.* Renewable and Sustainable Energy Reviews, 2010. **14**(1): p. 413-421.

[2] Ringel, M., *Fostering the use of renewable energies in the European Union: the race between feed-in tariffs and green certificates.* Renewable energy, 2006. **31**(1): p. 1-17.

[3] Solangi, K., et al., *A review on global solar energy policy.* Renewable and Sustainable Energy Reviews, 2011. **15**(4): p. 2149-2163.

[4] Al-Ajlan, S., et al., *Developing sustainable energy policies for electrical energy conservation in Saudi Arabia.* Energy Policy, 2006. **34**(13): p. 1556-1565.

[5] Alyousef, Y. and P. Stevens, *The cost of domestic energy prices to Saudi Arabia.* Energy Policy, 2011. **39**(11): p. 6900-6905.

[6] Marisarla, C. and K.R. Kumar, *A Hybrid Wind and Solar Energy System with Battery Energy Storage for an Isolated System.* IJEIT, 2013. **3**(3): p. 99-104

[7] Rehman, S., et al., *Feasibility study of hybrid retrofits to an isolated off-grid diesel power plant.* Renewable and Sustainable Energy Reviews, 2007. **11**(4): p. 635-653.

[8] Shaahid, S. and I. El-Amin, *Techno-economic evaluation of off-grid hybrid photovoltaic–diesel–battery power systems for rural electrification in Saudi Arabia—a way forward for sustainable development.* Renewable and Sustainable Energy Reviews, 2009. **13**(3): p. 625-633.

[9] Lee, S. and B. Pradhan, *Landslide hazard mapping at Selangor, Malaysia using frequency ratio and logistic regression models.* Landslides, 2007. **4**(1): p. 33-41.

[10] Shen, W., *Optimally sizing of solar array and battery in a standalone photovoltaic system in Malaysia.* Renewable energy, 2009. **34**(1): p. 348-352.

[11] Zulkifli, N., et al. *Probabilistic analysis of solar photovoltaic output based on historical data.* in *Power Engineering and Optimization Conference (PEOCO), 2014 IEEE 8th International.* 2014. IEEE.

[12] Kaygusuz, K. and A. Sarı, *Renewable energy potential and utilization in Turkey.* Energy Conversion and Management, 2003. **44**(3): p. 459-478.

Institute of Physics
Dirac House, Temple Back
Bristol BS1 6BE UK

ISSN: 1757-8981
ISBN 978-1-5108-1311-3

9 781510 813113

CURRAN ASSOCIATES INC.
proceedings.com